PROTEINS
OF
IRON
METABOLISM

PROTEINS OF IRON METABOLISM

Ugo Testa, M.D.

Instituto Superiore di Sanità
Rome, Italy

CRC Press
Taylor & Francis Group
Boca Raton London New York

CRC Press is an imprint of the
Taylor & Francis Group, an **informa** business

CRC Press
Taylor & Francis Group
6000 Broken Sound Parkway NW, Suite 300
Boca Raton, FL 33487-2742

First issued in paperback 2019

ISBN-13: 978-0-8493-8676-3 (hbk)
ISBN-13: 978-0-367-39697-8(pbk)

Library of Congress Cataloging-in-Publication Data

Testa, Ugo.
    Proteins of iron metabolism / by Ugo Testa.
        p. cm.
    Includes bibliographical references and index.
    ISBN 0-8493-8676-4 (perm. paper)
    1. Iron proteins.   2. Iron--Metabolism. I. Title.
[DNLM: 1. Hepatitis B virus. QW 710 G289h]
QP552.I67 T47 2001
572'.6--dc21

                                                             2001002267

Library of Congress Card Number 2001002267

Visit the Taylor & Francis Web site at
http://www.taylorandfrancis.com

and the CRC Press Web site at
http://www.crcpress.com

Introduction

All living organisms, from the most primitive to the most complex, require iron for metabolic functions. Excess iron produces toxic effects; its capacity to react with oxygen to form hydroxyl radicals may cause oxidative damage to proteins and nucleic acids that will lead to cell death. The maintenance of iron homeostasis is necessary to achieve an optimal balance of iron import and export that ensures release of the required amounts of iron to sustain biochemical activities while preventing the harmful effects of excessive iron uptake. Iron homeostasis is maintained through the coordinated regulation of uptake, storage, and secretion. All three processes are regulated via transcriptional and post-transcriptional mechanisms by proteins that respond to changes in iron availability.

Organisms adopted different mechanisms of control of iron metabolism as they evolved. The limited bioavailability of iron in soil triggered plants to develop two unique iron uptake strategies: one based on the chelation of iron in the microenvironment through the secretion of phytosiderophores (low molecular mass secondary amino acids) and the other based on iron reduction. Fe^{3+} phytosiderophores are taken up by a membrane receptor, yellow stripe *1*, present at the root surface. The expression of this receptor is stimulated under conditions of reduced iron availability.[1]

In bacteria, the synthesis of iron transporters and siderophores increases in low iron environments. The expression of genes that defend against oxidative stress is induced when bacteria are exposed to iron-rich media. Siderophores are low molecular weight compounds secreted to scavenge Fe^{3+} from the environment and facilitate its uptake into cells. Siderophores are secreted by terrestrial and marine bacteria.[2] The synthesis of siderophores and other iron uptake genes is controlled by the Fur repressor protein which acts on the promoters of genes possessing the consensus *Fur*- and Fe^{2+}-binding sequence, GATAATGATAATCATTATC, known also as the *Fur* box.[3]

Mammals have developed elaborated iron-withholding defense systems against microbial infections. Biochemical steps create an environment that contains virtually no free iron. In conditions of scarce iron availability, the development of virulence of some bacterial pathogens, such as *Neisseria, Haemophilus,* and *Yersinia,* is related to their specialized bacterial iron uptake systems. Virulence in fungi is associated with unique iron uptake systems based on the presence of membrane iron permease, whose synthesis is induced by iron deprivation.[4] In enteric bacteria, two pathways serve to control iron homeostasis: one is controlled by the *Fur* gene and responds to intracellular Fe^{2+} and the other is controlled by the PmrA/PmrB genes and responds to extracellular Fe^{3+}.[5]

The situation is more complex in pluricellular animal species that require several genes to accomplish iron metabolism, from absorption to transport, and then to excretion. Another element is the specialization of tissues involved in iron absorption (duodenal enterocytes), iron storage (hepatocytes), iron recycling (monocytes–macrophages), and utilization of high levels of iron required for certain metabolic needs (erythroid cells that require large amounts of iron to sustain high rates of heme and hemoglobin synthesis).

Iron is essential for many biological processes including oxygen transport, DNA synthesis, and electron transport. Iron is a constituent of the heme moiety of hemoglobin; it provides the capacity to bind and carry oxygen. This binding is dependent on oxygen tension in that oxygen is bound by hemoglobin at high oxygen tension levels, and it is delivered to tissues that have low oxygen tension levels. Iron also serves as a constituent of ribonucleotide reductase, the only enzyme involved in the synthesis of deoxyribonucleotides from ribonucleotides — a process that requires iron. Its third main function relates to the activity of the cytochromes of the respiratory chain and the capacity of iron to accept and lose electrons.

Iron is a major constituent of the Earth's crust. It exists mostly in the ferric (Fe^{3+}) form which is almost completely insoluble in water at neutral pH. To overcome these limitations and allow absorption and metabolism, complex iron oxidative–reductive and transport systems evolved in virtually all living forms. The discovery of many molecules involved in these processes allows us to understand the different steps of iron metabolism and the biochemical activities of the different cell types, e.g., the functions of intestinal enterocytes involved in iron metabolism and the roles of monocytes–macrophages in iron recycling.

The different proteins involved in iron metabolism are listed in Table I-1. In addition to structural, biochemical, and functional studies, data on the biologic effects caused by the loss of function of these proteins are also available.

The first step in iron metabolism by mammals is absorption by intestinal enterocytes. Dietary iron is primarily in the form of ferric iron; however, the ferrous form is absorbed much more efficiently by intestinal duodenal cells. Therefore, intestinal cells must possess a system localized at the brush border to reduce ferric iron to ferrous iron.

Ferric reductase activity at duodenal cells was described more than 10 years ago, but the first membrane molecule involved in ferric iron reduction was identified only recently. This protein, called Dcytb (duodenal cytochrome b), is part of the family of cytochrome b561 plasma membrane reductases and is highly expressed on brush border membranes of duodenal enterocytes.[6] The expression of Dcytb is up-modulated by all the known conditions associated with an increase of intestinal iron absorption, such as iron deficiency, chronic anemia, and hypoxia. The structure of Dcytb seems identical to that of cytochrome b 558, a protein isolated from plasma rabbit neutrophil membranes.

The reduction of ferric iron to ferrous iron is required for uptake by animal and plant cells. Low iron availability limits plants growth because iron forms insoluble ferric oxides, leaving as the only source of iron suitable for uptake a small, organically complex fraction present in soil solutions. Iron absorption requires the action of a ferric chelate reductase, which releases iron from organic compounds and reduces the ferric form to ferrous iron. The plant ferric chelate reductase was recently cloned and relates to a family of flavocytochromes that transport electrons across membranes.[7] Ferric reductases also have been cloned in yeast and bacteria.

Ferric iron reduced to ferrous iron on the brush border membranes of duodenal cells can be absorbed by cells through the action of a divalent metal transporter designated DCT1 (divalent cation transporter 1; also called Nramp2). One approach to identifying the gene encoding this protein was characterization of microcytic anemia (mk) mice that exhibit autosomal recessive defects affecting the absorption of dietary iron from the gut lumen into intestinal epithelial cells due to a missense mutation at the DCT1 gene. The mutated DCT1 does not retain iron transport capacity and is inappropriately localized in the cytoplasmic compartments, rather than at the apical brush border compartments of duodenal enterocytes.[8]

The second approach was an expression–cloning assay developed with the specific aim of isolating intestinal mRNAs that would confer iron uptake activity in *Xenopus laevis* oocytes.[9] This iron transporter responsible for the uptake of non-transferrin-bound iron is localized at the cell surfaces and endosomal compartments, thus indicating that it is involved in both the uptake of iron from extracellular sources and iron transport within the cells. It is characterized by two isoforms: one possessing an iron regulatory element in its 3′ untranslated region, predominantly located at the cell membranes, and the other lacking an iron regulatory element and located only at the intracellular compartments.

Nramp2 expression is restricted to the villus compartments of the duodenum; it is absent in the crypts. In iron-loaded animals, Nramp2 expression was predominant at intracellular sites of enterocytes, whereas in iron-depleted animals, Nramp2 expression predominated at the cell membranes.[10] All these studies strongly suggest that Nramp2 is involved in transferrin-independent iron uptake in the duodenum.

In addition to participating in iron absorption at the intestinal brush borders, Nramp2 seems to play a role in mediating absorption of non-transferrin-bound iron at the erythroid cells. These cells, in fact, possess very active systems of iron uptake of both transferrin-bound iron (mediated by transferrin receptors 1 and 2 hyperexpressed in erythroid cells) and non-transferrin-bound iron (mainly mediated by Nramp2).

In addition to the apical surfaces involved in iron uptake from digested foods, the polarized intestinal surface epithelium also exhibits a basolateral membrane involved in iron export from the enterocytes to body compartments. Recent studies revealed possible roles in iron export of a series of proteins located at the basolateral membranes of enterocytes. One of these molecules is ferroportin 1 (also called IREG1 and MTP1). This protein is a membrane spanning transporter that functions as an iron exporter.[11-13] It is absent in the crypt cells and expressed in villus cells. Its expression is controlled by body iron levels and by hypoxia, in that both iron deficiency and hypoxia produce an increase in the expression of ferroportin 1 at the basolateral membranes of enterocytes.

The mechanism responsible for this modulation is unclear even though ferroportin 1 possesses an iron regulatory element (IRE) at the 5′ untranslated region of ferritin mRNA. Preliminary evidence suggests that ferroportin 1 also may play a role in iron recycling at the monocytes–macrophages. It seems to be localized at the intracellular compartments and not at the cell membranes.

The analysis of the molecular defect responsible for sex-linked anemia in mice led to the identification of another molecule involved in iron export at the enterocytes. The gene mutated in sex-linked anemia of mice encodes for a protein whose sequence resembles that of ceruloplasmin. This protein is called hephaestin and it seems to act as an oxidase involved in iron oxidation and transport within enterocytes.[14] This role of hephaestin in iron metabolism of enterocytes is not surprising in that its ceruloplasmin homologue acts as a multicopper ferroxidase whose activity is required for iron recycling in several cellular compartments, including the liver, the blood, and the reticulolendothelial system.[15] Hephaestin is found in the perinuclear areas of vesicular compartments. It is expressed in enterocytes of the villi, but not in the crypts. The role of hephaestin in iron export is relatively clear; however, it is not known whether the oxidase activity of hephaestin is necessary to convert iron to a molecular form required for a selective basolateral iron exporter or for an intracellular iron transporter.

In addition to molecules involved in iron import and export at the villus enterocytes, other molecules present in enterocyte precursors in the crypts create a mechanism that senses body iron stores and helps program the production of mature enterocytes. The exact sensing mechanism is largely unknown, but evidence suggests a role of the HFE membrane protein whose mutation is involved in the genesis of hereditary hemochromatosis.

Hereditary hemochromatosis is a relatively common autosomal-recessive disease (1 of 300 Caucasians has the disease) characterized by increased absorption of dietary iron that results in progressive accumulation of iron in several parenchymal structures. The mutated gene in hereditary hemochromatosis encodes a 348-residue type I transmembrane glycoprotein, HFE, which is homologous to class I MHC molecules and associates with class I light chain β2-microglobulin. Unlike classical class I MHC molecules that function in the immune system by presenting peptide antigens to T cells, HFE does not bind peptides or perform any known immune function.

Evidence indicates that HFE acts as a negative regulator of iron uptake. Mutations of HFE, such as those observed in hereditary hemochromatosis, lead to iron accumulation, while HFE overexpression produces an iron-deficient phenotype. The regulatory role of HFE on iron metabolism is mediated through its interaction with transferrin receptor 1; mutant HFE proteins are unable to interact with this receptor. Crystallographic studies have shown that the interaction of HFE with transferrin receptor 1 is independent of the binding of transferrin to its receptor; as a consequence of the binding of HFE to the receptor, a conformational change of transferrin receptor 1 is induced.[16] As a consequence of the change, the affinity of transferrin receptor 1 for transferrin is decreased and the rate of receptor internalization is significantly reduced, resulting in reduced iron uptake. These findings indicate that HFE can be regarded as a regulator of transferrin receptor function. The absence of function of this regulator in hereditary hemochromatosis results in increased iron uptake and, consequently, iron loading.

The highest levels of expression of HFE are observed in the liver and small intestine, thus suggesting that the main role of this protein is the control of iron absorption. The intracellular localization of HFE shows several interesting qualities. In the epithelial cells of the stomach and colon, the protein is localized to the basolateral surfaces. In the crypt cells, where HFE is highly expressed, its localization is intracellular and perinuclear.

In polarized duodenal epithelial cells, mutated HFE protein was unable to localize at the basolateral and endosomal compartments and remained confined to the endoplasmic reticulum and Golgi complex. The overexpression of the wild-type, but not of the mutated HFE protein, elicited a reduction of ferritin content associated with an increase of transferrin receptor levels.[17] The mechanisms through which HFE controls the rate of intestinal iron absorption remain to be determined.

A model of control of intestinal iron absorption under physiological conditions and in the presence of mutated HFE (hereditary hemochromatosis) was proposed. This model is based on the idea that the HFE/transferrin receptor 1 complex in the basolateral membranes of enterocyte progenitors present in the crypt cells acts as a body iron sensor and plays a key part in programming the iron absorption mechanisms of mature enterocytes. This process is deregulated in hereditary hemochromatosis.[18-20] As a consequence of this deregulation, intestinal villus cells express an iron absorption mechanism characterized by increased Nramp2 expression, reduced ferritin content, and increased iron absorption capacity.[21]

Recent studies based on the analysis of intestinal villi provided evidence of increased expression of Nramp2 in hereditary hemochromatosis. Furthermore, inactivation of both HFE alleles in mice elicited increased intestinal expression of Nramp2. Other iron transporters, such as the transferrin receptor (which stimulates iron transport) and ferroportin exhibit increased expression in HFE hemochromatosis. Based on these observations, the expression of all the intestinal iron absorption machinery is deregulated in hereditary hemochromatosis.

Different molecules are involved in iron uptake at the apical sides of enterocytes and in the transfer of iron from these cells to the circulation through iron transporters located at the basolateral membranes of enterocytes. The modulation of the expression and function of these apical and basolateral transporters may account for the adjustment of the rate of intestinal iron absorption based on physiologic demand.

The so-called mucosal block (the resistance of absorptive enterocytes to acquiring additional iron for several days after a dietary iron bolus) may be related to down-modulation of the Nramp2 apical iron transporter. The control of the rate of intestinal iron absorption excreted by body iron stores may be sensed by crypt cells through the HFE/transferrin receptor complex. In the presence of high iron body stores, this mechanism produces a down-modulation of Nramp2 expression at the villus cells. The mechanism by which the rate of erythropoiesis controls the rate of intestinal iron absorption remains to be determined. It seems logical that a mediator released by bone marrow and transported by plasma is responsible for the transmission of a signal to the intestine.

After iron enters the circulation, it is complexed to and transported to the tissues by transferrin. Transferrin may be involved in the transport of other metals such as aluminium, manganese, copper and cadmium. At physiologic pH levels, such as that observed in plasma, iron is insoluble; its free state can generate dangerous free radicals. Thus, the binding of iron with transferrin is vital. The affinity of transferrin for iron is pH-dependent. Iron is released from transferrin as the physiologic pH level drops below 6.5.

Transferrin is a single-chain glycoprotein produced primarily by the liver and composed of two homologous lobes. Each lobe has two subdomains that delimit a cleft that forms an iron-binding site. The two iron-binding sites exhibit some differences, but both are able to bind iron in association with a carbonate ion.

Crystallographic studies of wild-type transferrin and mutants of transferrin receptors at critical amino acid residues provided considerable insight into the molecular mechanisms involved in the formation of the iron-binding cleft and in the binding and release of iron. Transferrin displays three main functions : (1) it solubilizes ferric iron that is largely insoluble at neutral pH; (2) it binds iron with high affinity, thus avoiding the free state that can generate free radicals; and (3) it functions as an iron supplier for the cells via its interaction with two membrane receptors: transferrin receptors 1 and 2.

Many cells require iron uptake to sustain their metabolism. Iron may be supplied through transferrin or through alternative iron uptake pathways, but the pathway for the uptake of iron-bound transferrin is essential for sustaining erythropoiesis. This conclusion is supported by the analysis of several types of mutants. A first model is represented by hypotransferrinemic mice which have severe serum transferrin deficiencies as a result of a mutation linked to the transferrin locus. These mice are born alive, but die of severe microcytic anemia before weaning.[22,23] This observation indicates that transferrin is essential for supplying iron to maturing erythroid cells. Surprisingly, this defect in erythropoiesis was associated with markedly increased iron uptake by non-hematopoietic tissues as a result of increased intestinal iron absorption and markedly augmented cellular uptake of non-transferrin-bound iron. This suggests that transferrin does not play a fundamental role in supplying iron to non-hematopoietic tissues.

A second model resulted from observation of patients with a rare disease called hereditary atransferrinemia. It is characterized by a deficiency of transferrin synthesis due to mutation of the transferrin gene; clinical findings include hypochromic anemia and iron overload of several tissues, including liver, pancreas, thyroid, myocardium, and kidneys.[24] Clinical manifestations of human atransferrinemia closely resemble the hypotransferrinemic mouse phenotype.

A third model is the study of mice in which the gene encoding murine transferrin receptor 1 was disrupted by gene targeting. Homozygous transferrin receptor-null mice die early in embryogenesis due to severe anemia and hydrops related to apoptosis of neuroepithelial cells.[25] This observation confirms the key role of transferrin-mediated iron uptake for erythropoiesis and suggests a role for this iron uptake pathway in neural development.

Most biological effects of transferrin relate to its capacity to interact with two types of membrane receptors: transferrin receptor 1 (TfR1) and transferrin receptor 2 (TfR2). TfR1 is a typical receptor expressed in virtually all cell types except mature erythrocytes. It is a homodimer consisting of two identical subunits of about 90 kDa joined by two disulfide bonds at cysteine residues 89 and 98 and composed of three domains (a 61-residue amino terminal domain, a 28-residue transmembrane region, and a large extracellular carboxy terminus of 671 residues).

Crystallographic studies indicate that TfR1 monomers are arranged in three distinct globular domains organized in a butterfly-like shape: a protease-like domain near the cell membrane, an intermediate helical domain, and an apical domain.[26] These three domains form a lateral cleft involved in transferrin binding. The apical domain of TfR1 exhibits similarity with the so-called protease-associated domain and the amino acid sequences of all three globular ectodomains exhibit similarity to the sequence of membrane glutamate carboxypeptidase.[27] These observations led to the speculation that TfR1 evolved from a peptidase related to membrane glutamate carboxypeptidase.

The roles of the regions of the TfR1 monomer were analyzed by studying selected mutants. Particularly relevant was an analysis of the amino acid sequences required for ligand-induced receptor internalization. After interaction with TfR1, the transferrin/TfR complex is internalized. TfR is one of the more studied models investigating receptor-mediated endocytosis. Structural studies show that a conserved internalization signal (YTRF) within the 61 amino acid residues of the cytoplasmic portion of TfR1 is critical for internalization.[28]

Phosphorylation and dephosphorylation of TfR1 at Ser 24 and at the other amino acid residues of the cytoplasmic tail play a role in the control of TfR1 internalization. TfR1 is expressed in virtually all tissues except mature erythrocytes. Despite this ubiquity, some tissues, such as erythroid cells, hepatocytes, placental tissues, and rapidly dividing cells, express particularly high levels of TfR1.[29] The highest levels are observed in developing erythroid cells, where high rates of iron uptake are required to sustain high levels of heme and hemoglobin synthesis.[30]

The elevated expression of TfR1 observed in erythroid cells and rapidly proliferating cells depends mainly on transcriptional mechanisms. The proliferation-dependent pathway seems to be related to a transcriptional mechanism involving the binding of transcriptional factors to an AP-1 element in the promoter of the TfR1 gene. The transcriptional mechanisms of TfR1 observed in erythroid cells seem to depend on a balance between the inhibitory activity mediated by the binding of Ets-1 and the stimulatory activity related to the binding of AP-1 and HIF to specific sequences present in the promoter of the TfR1 gene.

The role of post-transcriptional mechanisms play in the modulation of TfR1 expression is dependent on cellular iron levels. Increases in intracellular iron levels elicited a marked down-modulation of TfR1 expression, while the opposite effect was induced by reduction of free iron levels.[31,32] The iron-dependent modulation of TfR1 levels depends on the presence of five stem–loop structures of iron-responsive elements (IREs) present in the unusually large 3′ untranslated region of about 2500 nucleotides of TfR1 mRNA. Iron modulates the stability of TfR1 mRNA via the binding of iron-regulatory proteins (IRPs) to the IREs in TfR1 mRNA. Depending on the intracellular iron concentrations, the IREs can be bound (under low iron conditions) or not bound (under high iron conditions) by IRPs, resulting in the stability or degradation, respectively, of TfR1 mRNA.[33] The roles of IRPs and IREs in the control of iron metabolism were the subjects of intensive studies.

The main function of TfR1 is allowing the entry of transferrin. The internalization of the TfR1/transferrin complex was investigated and produced a model for the study of receptor internalization. The internalization process is known as the transferrin cycle; it is responsible for the uptake of iron by cells and the recycling of transferrin and its receptor after internalization. After endocytosis, iron is released from the TfR1/ferrotransferrin complex in the acidic receptor. They recycle together to the cell surface where the neutral pH of the exterior medium causes the release of free-iron apotransferrin from TfR1, which becomes available for a new internalization cycle after binding with ferrotransferrin.[34]

The transferrin/TfR1 cycle has a half-life of 3 to 5 minutes. During the cycle, most of the transferrin/TfR1 complex returns to the surface; however, a minor amount accumulates in membranes of the tubulovesicular structures in the vicinity of the microtubule organizing center and is defined as a distinct organelle, the pericentriolar recycling endosome. The studies of TfR1 internalization contributed greatly to our understanding of the cellular and molecular steps involved in the receptor-mediated endocytosis that occurs in certain regions of the plasma membrane coated pits.

Ultrastructural and biochemical analyses indicate that the formation of clathrin-coated vesicles containing the internalized transferrin/TfR1 complex has several distinct steps. Clathrin coat assembly is initiated by the binding of coat components to a docking site at the plasma membrane. Through the addition and rearrangement of coat constituents, the initially flat pit gains curvature and becomes progressively invaginated until the neck is greatly restricted. Membrane fission releases the coated vesicle and its cargo into the cell interior.

Several molecules required for receptor-mediated internalization have been identified. Their roles in this process were determined through functional and biochemical analysis, with a focus on various regulatory factors, such as Rab 11, Rab 15, cellubrevin, and syntaxin 13, which control the transport of the endocytosed transferrin/TfR1 complex.

Recent studies suggest an important role of a new protein family, RME-1/EHD, in the transport through the endocytic recycling compartment. The RME-1 proteins participate in the control of protein exit from the endocytic recycling compartment. The loss of function of these proteins causes considerable slowing of transferrin recycling .[35]

TfR1 plays an important role in the placenta by mediating iron uptake from maternal blood to fetal tissues. TfR1 is expressed at the level of the apical (maternal) side of the syncytiotrophoblastic membrane where it mediates the binding of maternal transferrin. After binding to the brush border membrane, the transferrin/TfR1 complex is internalized into endosomes. Iron is subsequently released and transferred to the cytoplasm where it is bound and moved by divalent metal transporter 1 (DMT1 or Nramp2). The subsequent process involves ferrireductase activity provided by a ceruloplasmin-like molecule and the export of iron from basolateral membranes of the syncytiotrophoblasts to the fetus through the action of the iron exporter, IREG1/ferroportin 1.

The distribution of TfR1 in tissues showed a preferential expression in rapidly dividing tissues. The levels of TfR1 are comparable in these tissues. However, the levels of TfR1 are markedly higher in maturing erythroid cells than in all other tissues. This phenomenon is related to the great need of these cells to acquire the high levels of iron required to sustain high rates of heme synthesis necessary for hemoglobin synthesis. The molecular mechanisms responsible for the high levels of expression of TfR1 in erythroid cells involve transcriptional and post-transcriptional events. The molecular mechanisms of the high rate of TfR1 gene transcription in erythroid cells remain largely unknown, although some studies point to a stimulatory role played by HIF-1α and AP-1-like factors; an Ets-1 element seems to play an inhibitory role. The post-transcriptional mechanisms involve high levels of spontaneous activity of iron regulatory proteins 1 and 2 (IRP1 and IRP2) in erythroid cells, with consequent stabilization of TfR1 mRNA.[30]

The high level of expression of TfR1 on some tumor cells has been explored in an attempt to develop new therapeutic approaches for some neoplasias. One interesting approach is local–regional therapy of brain tumors. The injection of transferrin conjugated with bacterial toxins produced significant toxicity in brain tumor cells and limited toxicity in normal brain cells. Other studies used the transferrin/TfR complex to achieve transferrin-mediated gene transfer. This approach may be of value in developing gene therapy protocols.

A second transferrin receptor (TfR2) was recently identified. Its patterns of expression and regulation are different from those reported for TfR1. The gene encoding TfR2 is on chromosome 7q22 and encodes a typical type II transmembrane glycoprotein whose extracellular domain exhibits 66% homology with TfR1; its cytoplasmic tail is highly divergent from that of TfR1.[36] The non-coding 3' sequence of TfR2 showed the absence of IREs, thus suggesting that TfR2 expression is not controlled by the IRP system. A unique quality of TfR2 is its pattern of tissue distribution, which is restricted to hepatocytes[36] and erythroid cells.[30]

Important differences were noted in regulation of the expression of TfR2 and TfR1. TfR2 expression is modulated by cell growth rate and not by iron levels.[37] The lack of regulation of TfR2 by iron was considered as a possible mechanism involved in hepatic iron accumulation in hereditary hemochromatosis. Despite the markedly increased hepatic iron stores noted in hemochromatosis, TfR2 continued to be expressed on the surfaces of hepatocytes and mediated high rates of transferrin and iron uptake. Another difference between the two receptors is that TfR2 is unable to interact with HFE, while TfR1 forms a complex with HFE. The affinity of transferrin for TfR2 is 20-fold lower than its affinity for TfR1.[38]

Interest in TfR2 increased considerably after the observation that mutations of the gene encoding this receptor are responsible for a rare form of hereditary hemochromatosis called HFE3.[39,40] The pathogenetic mechanism by which the mutation of the TfR2 gene causes iron overload remains to be determined.

In addition to the membrane-bound forms, a soluble transferrin receptor (sTfR) was identified in the sera of both humans and animals. sTfR is a truncated monomer of the membrane-bound dimeric TfR; it lacks the first 100 amino acids and circulates bound to its receptor. Recent studies indicate that the evaluation of the serum concentration of sTfR is a valuable non-invasive tool for evaluating the extent of erythropoietic activity *in vivo*.[41] sTfR levels increase in pathologic conditions characterized by increased erythropoietic activity, and decrease when erythropoietic activity declines. Body iron status determines sTfR levels in that iron deficiency is usually associated with considerably elevated sTfR levels.

A molecule designated lactoferrin can bind iron with high affinity. At variance with transferrin, its main function relates to the sequestration of iron in certain tissue compartments and not to iron transport. Lactoferrin may participate in iron absorption and possibly serve as an anti-oxidant, an anti-microbial agent, an anti-inflammatory agent, and an anti-cancer agent. It may also play a role in anti-viral defense.[42]

The three-dimensional structure of lactoferrin is similar to that of transferrin. Lactoferrin has two homologous lobes that represent the N and C terminal halves.

The iron- and carbonate-binding sites are located in a deep cleft between the domains of each lobe. The binding and release of iron by lactoferrin are accompanied by marked changes between the open conformation of the apolactoferrin form and the closed conformation of the iron-bound form. When iron is lost from the N lobes of lactoferrin, the N1 and N2 domains move apart through a rigid-body domain rotation of 50 to 60° that opens the binding cleft.

Despite high levels of structure similarity, lactoferrin differs from transferrin in its binding properties. Particularly significant is the much greater acid stability of iron binding by lactoferrin, as compared to that of transferrin. Transferrin loses its iron in the 6.0 to 4.0 pH range, while iron release from lactoferrin occurs in the range of 4.0 to 2.5. This unique property of lactoferrin is consistent with its ability to retain iron even at low pH, thus exerting significant bacteriostatic activity. The differences in stability of iron binding do not depend on the structures of the iron-binding pockets. The same ligands with the same geometry are found for the iron- and carbonate-binding sites of both proteins, but other differences exist.

Recent studies suggest that the ability of lactoferrin to retain iron at lower pH levels than transferrin is due to interlobe interactions, to the absence in lactoferrin of an attraction analogous to the dilysine pair in transferrin, and to the presence of a basic residue (Arg 210) only in lactoferrin.[43] The breaking of the interdomain H bond upon protonation may trigger the opening of the iron cleft and the subsequent iron release by transferrin. The situation is markedly different with lactoferrin. Iron loss requires a prior change in protein conformation.[44]

Camel lactoferrin displays structures and properties of both lactoferrin and transferrin. The N lobe of camel lactoferrin is similar to that of lactoferrin. They both have three-dimensional structures and iron-releasing properties, The C lobe exhibits similarity with transferrin with respect to its three-dimensional structure and its iron-releasing properties. As a consequence of this hybrid structure, camel lactoferrin shows peculiar iron release activity. It releases 50% of its iron at pH 6.5, and the remaining 50% is released only at pH levels below 4.0.[45]

Lactoferrin is present in high concentrations in human milk (1 to 2 mg/ml) and at particularly high levels in colostrum (>7mg/ml), and seems to influence the nutritional properties of human milk. Human milk provides an infant's nutrient requirements during the first year of life. It is composed of nutritive compounds that promote growth. Human milk proteins are sources of peptides, amino acids, and nitrogen and also stimulate immunologic (immunoglobulin) and non-immunologic (lactoferrin) defenses. Breast-fed infants have fewer and less severe infections than formula-fed infants, possibly because lactoferrin is present at high concentrations in human milk (1 to 2 mg/ml), in low concentrations in cow's milk (0.1 mg/ml), and is absent in soya-based formulas. Evidence suggests that many anti-bacterial, anti-viral, and anti-fungistatic properties of human milk are related to lactoferrin.

Lactoferrin is found in virtually all human mucosal secretions, including milk, tears, semen, and plasma, and in specific granules of polymorphonuclear leukocytes. It rarely provides iron to nourish cells except in neonates. Its function in iron transport may be negligible in a normal healthy organism.

Like transferrin, lactoferrin is bound by specific membrane receptors; lactoferrin-binding sites are different from those involved in transferrin binding and the two ligands do not cross-react with their respective ligands. Different surface molecules are able to bind lactoferrin. Thus, lactoferrin can bind to multiple receptors on hepatocytes, including the asialoglycoprotein receptor and low-density lipoprotein receptor-related protein (LRP), a non-LPR-binding site that functions as a chylomicron remnant receptor. Lactoferrin is also targeted to specific cell surface receptors; these 105-kDa lactoferrin receptors are expressed on activated T cells, platelets, megakaryocytes, dopaminergic neurons, mesencephalon microvessels, and brain endothelial cells.[46]

One important function of lactoferrin is its anti-microbial activity. It can chelate iron, resulting in inhibition of bacterial growth. A second mechanism is a peptide released by peptic digestion of lactoferrin and known as lactoferricin. The anti-microbial action of lactoferricin may be related to its capacity to interact with negatively charged divalent cation-binding sites such as lipopolysaccharides (LPSs) on bacterial cell surfaces. A third mechanism is the ability to inactivate colonization factors of some bacteria, such as *Haemophilus influenzae*. Fourth, lactoferrin interacts with porins on the outer membranes of some Gram-negative bacteria; as a consequence of this binding, the bacterial membranes are destabilized, the permeability of the outer membranes increases, and the bacteria are lysed.[47]

The interaction of lactoferrin and bacteria is complex because several pathogenic species have developed unique mechanisms for acquiring iron directly from lactoferrin through the interaction of lactoferrin and cell surface receptors. These bacteria include *Bordetella pertussis*, *Helicobacter pylori*, *Moraxella catarrhalis*, *Neisseria gonorrhaeae*, *Neisseria meningitides*, *Prevotella nigrescens*, and *Treponema*.

Bacterial lactoferrin receptors are composed of two chains designated lactoferrin-binding proteins A and B (LbpA and LbpB). LbpA exhibits some homology with siderophores, while LbpB shows significant homology with lactoferrin receptors. The activities of these bacterial membrane proteins and the preferential reactivity of convalescent antisera with LbpB suggest possibilities for vaccines. The interaction of lactoferrin and bacteria may represent a tool for developing new vaccines and diagnostic procedures.

In experimental models of bacterial and fungal infections, the *in vivo* injection of radioactively-labelled lactoferrin allowed the imaging of tissues infected by bacteria; little labelling was observed in sterile inflamed tissues.[48]

In *Porphyromonas gengivalis*, an anaerobic Gram-negative bacterium implicated in the genesis of adult periodontitis, another type of membrane receptor is involved in the binding of lactoferrin. Lactoferrin binds to the hemoglobin receptors on the bacterial membranes and causes the release of the receptors from the bacterial surfaces, thus abrogating the capacity of the bacteria to bind the hemoglobin that is its only iron source. The receptor-mediated interaction of lactoferrin with the bacterial surface triggers its bacteriostatic activity.[49]

Lactoferrin is a constituent of neutrophils, particularly of their secondary granules that are histochemically classified into two categories based on peroxidase content: primary azurophilic peroxidase-positive granules and secondary peroxidase-negative granules. These granules are acquired progressively during maturation. Azurophilic granules appear earlier than secondary granules. Lactoferrin is a secondary granule protein and represents a prototype of the genes expressed late in neutrophil maturation. The expression of lactoferrin in granulocyte cells is coordinated at the transcriptional level during neutrophil maturation.

Recent studies defined the regulatory sequences in the promoter of the lactoferrin gene that exert negative and positive controls on the transcription of this gene in granulocytic cells. Sp1 and c/EPB are the positive elements required to activate the lactoferrin gene promoter during myeloid differentiation.[50] The comparative analysis of the molecular mechanisms involved in regulating neutrophil granule protein expression showed that the transcription factor AML-1 is critical for primary granule protein expression, while C/EBPε is essential for secondary granule expression.[51]

Molecular studies partially revealed the molecular mechanisms underlying lactoferrin gene expression in mammary glands. They also clarified the molecular mechanisms involved in the induction of lactoferrin gene expression in reproductive organs, such as endometrium and uterus, in which estrogens and epidermal growth factor induce the expression of this glycoprotein. Studies of the lactoferrin gene promoter identified two gene elements: a steroidogenic factor-binding element and an estrogen response element involved in transactivation of the lactoferrin gene in the mammary gland and uterus.[52]

Ferritin is another protein involved in iron storage in virtually all tissues. Its structure is highly conserved in plants, animals, and bacteria. The ferritin molecule has 24 subunits assembled into a spherical shell characterized by four-fold, three-fold, and two-fold symmetries. Up to 4500 atoms of Fe^{3+} can be stored as an inorganic complex in the cavity of the ferritin molecule. Iron sequestration is an important process. It renders iron atoms non-toxic and biologically available when required. The analysis of the structures of vertebrate and bacterial ferritins showed that each subunit consists of a four-helix bundle and a fifth short (E) helix. The E helix bundle is placed around the four-fold intersubunit symmetry axes of the protein shell and contributes to the formation of a hydrophobic pore.

Mammalians possess H and L ferritin subunits that exhibit similar three-dimensional structures and 50% sequence similarity and seem to display different functions. The H subunit exhibits ferroxidase activity and catalyzes the first step in iron storage, i.e., the oxidation of Fe^{2+} to Fe^{3+}. The L subunit does not possess ferroxidase activity and may be involved in core iron formation in the cavity of the ferritin molecule.

A new type of ferritin subunit in human cells was recently discovered and designated mitochondrial ferritin. It is expressed as a precursor targeted to mitochondria where it is processed into a functional protein exhibiting structural and functional properties similar to those of cytoplasmic ferritins.[53] Unlike cytoplasmic ferritins, mitochondrial ferritin is formed by a homopolymer, similar to ferritins found in bacteria and plants.[53] The evolutionary origin of this ferritin remains unclear. It may have evolved from an ancient ferritin H-like sequence whose 5′ untranslated region mutated to become a mitochondrial leader sequence.

Mammalian ferritins are 24-subunit heteropolymers assembled from two types of subunits that provide different functions. The heavy subunit chain (Fer-H) catalyzes the first step in iron uptake by ferritin by oxidizing ferrous iron. The light subunit chain (Fer-L) promotes the biochemical reactions involved in the accumulation of iron, i.e., the nucleation of ferrihydrite.

The three-dimensional structures of ferritin homopolymers composed of Fer-H and Fer-L chains showed many similarities. However, differences were noted in the hydrophilic residues in the centers of the four-helix bundles associated with ferroxidase activity in Fer-L chains and mineralization of iron within the iron cavity of the molecule.[54]

Escherichia coli ferritin, exhibits only 22% amino acid sequence identity with mammalian Fer-H, but shows considerable similarity to Fer-H chains in its three-dimensional structure. The 24 subunits of *E. coli* ferritin A are folded as four-helix bundles that assemble into hollow shells and possess iron-binding sites at dinuclear centers in the middles of the bundles.[55]

The structure of the ferritin of *Listeria innocua* differs from other known ferritins in that the protein shell is assembled is assembled from 12 identical subunits (each of 18 kDa) to form a hollow structure exhibiting a molecular weight of about 240 kDa and 3:2 tetrahedral symmetry.[56] This dodecameric ferritin possesses iron-binding sites at the interfaces of subunits related by two-fold symmetry.[56] Its iron-binding capacity is much lower than those of mammalian ferritins (500 iron atoms per ferritin molecule).

Ferritins are present also in vegetals. Plant ferritins were thought to be composed of homopolymers. Recent studies show that plant ferritin has two subunits of 28 and 26.5 kDa that interact to form a stable molecule.[57] The 26.5-kDa subunit exhibits 40% homology with mammalian Fer-H and is synthesized as a 32-kDa precursor that contains a unique two-domain N terminal sequence involved in transporting the ferritin precursor to plastids.

The functions of mammalian ferritin chains were investigated using gene targeting or overexpression. Gene targeting showed that mice deficient in Fer-H expression die very early (between days 3.5 and 9 of embryonic development). This indicates that Fer-H expression is essential for cell survival and cannot be replaced by Fer-L.[58] Since Fer-H is present in the heart and central nervous system at day 9 of embryonic development, its loss may result in toxicity in these tissues.

The function of the Fer-H chain was also explored by overexpression of the Fer-H gene in mouse erythroleukemia cells[59] and HeLa cells.[60] The overexpression of Fer-H leads to a phenotype related to iron starvation and characterized by reduction of the free iron pool and reduced heme and hemoglobin synthesis.[59] The cells were resistant to oxidative damage and had multidrug resistance properties.[59] The overexpression of Fer-H in HeLa cells induced an iron-deficient phenotype characterized by reduced growth and a decreased iron pool associated with increased transferrin receptor expression and iron uptake.[60] Cells overexpressing Fer-H showed increased resistance to oxidative stress.[60] The overexpression of Fer-H mutants lacking ferroxidase activity did not induce these changes, thus indicating that ferroxidase activity is essential. A transient decrease of Fer-H in erythroleukemia cells achieved through transfection with an antisense construct elicited an increase of the free iron pool associated with a decrease of transferrin receptor expression and iron uptake.

The expression of ferritin is controlled through translational and transcriptional mechanisms. Translational mechanisms are dependent on the level of cellular iron loading and play a role in the control of ferritin synthesis. In conditions of iron deprivation, the binding of the IRP repressor protein to the 5' IRE elements in the untranslated region of the ferritin mRNA is induced, with consequent repression of mRNA translation. In contrast, in conditions of iron loading, the IRP protein is inactivated and does not bind to the ferritin mRNA.

In addition to the translational mechanisms that mediate the response of ferritin to iron changes, transcriptional mechanisms underly the stimulatory effects of heme on ferritin synthesis[61] and the induction of ferritin gene expression observed in some cell types during differentiation.[62] The induction of ferritin gene transcription observed during monocyte-to-macrophage differentiation may be related to the activation of the NFY transcription factor that binds to the Fer-H gene promoter.[62] The induction of ferritin gene expression observed during cell differentiation may contribute to the loss of c-*myc* which acts as a repressor of ferritin gene transcription.[63]

Ferritin synthesis is greatly induced by inflammatory stimuli, mainly by inflammatory cytokines, including interleukin-1α and β (IL-1α and IL-1β), and tumor necrosis factor-α (TNF-α). The stimulatory action of these cytokines on ferritin synthesis is mediated through complex mechanisms. TNF-α mainly acts by stimulation of ferritin gene transcription. IL-1 acts through post-translational mechanisms dependent on inactivation of IRP through an increase of the free intracellular iron pool and the induction of binding of a cytoplasmic factor to an acute-box element in the 5′ untranslated region of L and H ferritin mRNAs that is different from the IRE element.

The measurement of ferritin serum concentrations is a simple and non-invasive clinical index of body iron stores. Serum ferritin levels are increased in several pathological conditions, particularly those associated with acute or chronic inflammation, that result from defects in iron mobilization from storage sites such as the reticuloendothelial system and consequent accumulation of iron and ferritin.

In addition to its role in iron storage and detoxification, ferritin exhibits other activities that may be physiologically relevant. Studies of the purification and characterization of a folate-catabolizing enzyme indicated that the enzyme is similar, and probably identical, to heavy ferritin.[64] The overexpression of ferritin H (but not of ferritin L) causes marked enhancement of a folate-dependent cytoplasmic enzyme called serine hydroxymethyltransferase.[64] This effect depends on the capacity of ferritin L to reduce the size of the free iron pool.

Ferritin is formed by variable proportions of heavy and light chains (isoferritins). Isoferritins can be acidic (H-rich) or basic (L-rich), depending on the relative proportions of H and L chains. Liver and spleen ferritins are basic because they are composed mainly of light chains. Heart, kidney, and placental ferritins are acidic due to their high content of ferritin H chains. Ferritins are frequently increased in tumor cells and are mainly composed of ferritin H chains. Increased expression of H chains may be a factor in some neoplasias, such as breast carcinoma. Ferritin is one of the molecules released by tumor cells and it exerts an immunosuppressive effect. In fact, ferritin H inhibits T lymphocyte activation through a mechanism that implies the induction of interleukin-10 release.[65]

The brain requires iron for certain processes related to neural development, e.g., myelination. Iron is also potentially toxic and its accumulation in certain types of brain cells is implicated in the genesis of some neurodegenerative diseases. Oligodendrocytes, the cells responsible for the synthesis of myelin in the central nervous system (CNS), have the highest iron and ferritin contents of all the types of cells in the CNS. They can synthesize the transferrin required for oligodendrocyte maturation and myelination and can bind ferritin.[66,67]

Transferrin receptors in the CNS are distributed selectively to the grey matter. It is important to understand iron homeostasis in the brain because disruption of iron metabolism in the brain has been implicated in diseases such as multiple sclerosis, Alzheimer's disease, Parkinson's disease, supraventricular progressive palsy, and stroke. In some diseases, such as in stroke, much of the ischemia-induced damage related to oxygen activated species was dependent upon the intracellular iron cation. In experimental models of focal cerebral ischemia, the ischemic insult produced pronounced increases of both ferritin H and ferritin L mRNAs. The highest ferritin H mRNA levels were observed in cerebral areas that survived, while ferritin L mRNA was observed in areas of tissue damage following the ischemic insult.[68] Ferritin may protect ischemic tissue by holding reactive iron molecules.

Marked iron accumulation occurs at the basal ganglia of the hippocampus, amygdala, or cerebral cortex in several neurodegenerative disorders. However, the rise of iron content in these areas was not accompanied by a concomitant rise of ferritin in neurons. That usually occurs in microglia cells, thus resulting in cellular damage.

The expression of the different genes involved in iron metabolism must progress along a coordinated pathway and be finely tuned to iron-loading status. Regulation was thought to be performed by mechanisms specific for each gene or through common mechanisms and common regulatory elements. The discovery of the common mechanism scheme led to the understanding of one of most complex systems of regulation of gene expression in mammalian cells. This system of regulation operates through post-transcriptional mechanisms. It was discovered as a result of studies aimed at elucidating the molecular mechanisms involved in the regulation of ferritin gene expression by iron.

This system operates through the interaction of two cytoplasmic proteins, iron regulatory proteins (IRP1 and IRP2) with specific sequences, called iron responsive elements (IREs) present either in the 5′ or 3′ untranslated regions of the mRNAs encoding for most genes involved in iron metabolism. The IREs present in the genes involved in iron metabolism consist of a stem–loop structure. The loop consists of a terminal-conserved hexanucleotide region (CAGUGX), while the stem is composed of a base-paired stem-structure interrupted by a conserved unpaired cytosine nucleotide (bulged C).[69] The hairpin loop and the bulged C are critical for IRP binding and the C bulge seems to orient the hairpin for assuming optimal conformation for binding to IRP.[70] Comparison of the structures of IREs for the same proteins in different animal species showed that conservation of sequence identity is more protein specific than species specific.[69] A novel type of IRE was found in the 5′ untranslated region of the mRNA encoding a 75-kDa protein included in Complex I of the mitochondrial respiratory chain. This IRE is composed of a 5-base loop (CAGAG) and an A bulge (rather than C bulge) in the stem.[71] This IRE is bound by a novel type of IRP different from IRP1 and IRP2; its characterization is not final.

IREs are present in the 5′ untranslated regions of many genes involved in iron metabolism: ferritin heavy and light chains, ferroportin 1, erythroid-specific 5-aminolevulinate synthase, and mitochondrial aconitase. IREs are also present in the 3′ untranslated regions of transferrin receptor 1 and Nramp2 mRNAs.[72]

Iron regulatory proteins 1 and 2 (IRP1 and IRP2) bind to IRE-containing RNAs. The structure of IRP1 is identical to that of mitochondrial aconitase, an enzyme of the citric acid cycle involved in the conversion of citrate to isocitrate. Under conditions of iron deprivation IRP1 loses its enzymatic activity as aconitase and acquires optimal IRE-binding activity. The conversion of IRP1 from an enzymatically active protein to an RNA-binding protein is accompanied by the loss of an iron–sulfur cluster required for enzymatic activity.

IRP2 exhibits 50% amino acid homology with IRP1 and binds IREs with an affinity similar to that of IRP1. IRP2 is expressed most tissues, but usually to a lesser degree than IRP1. The highest levels of IRP2 are observed in the CNS and in the intestinal epithelium. IRP2 does not possess aconitase activity. It is proteolytically downgraded in the presence of iron loading.[72]

The role of IRPs in the control of iron metabolism is supported by studies of knockout mice with targeted disruptions of IRP genes. Analysis of the phenotype of IRP2$^{-/-}$ mice showed accumulations of iron in the white matter and the brain nuclei and neurodegeneration.[73] Increased iron, ferritin and ferroportin 1 levels were seen in the intestinal epithelium; increased and unexplained expression of Nramp2 was also observed. Iron metabolism was normal in other tissues. These observations indicate that IRP2 activity is required for iron homeostasis in the brain and intestinal epithelium. Targeting of the IRP1 gene was not reported.

IRP activity may be modulated by different stimuli. The most relevant stimulus for modulation is probably iron level. This type of regulation works only on intact cells; it does not work on cell extracts or purified IRP. This finding implies that iron does not act directly on IRP, but causes biochemical changes that are in turn responsible for changes in IRP activity.

Iron is a key regulator of IRP activity. When iron level is low. the ferritin and TfR mRNAs are recognized by activated IRPs with consequent stabilization of TfR mRNA and inhibition of ferritin mRNA translation. In contrast, when iron is abundant, IRPs are inactivated and thus deprived of their RNA-binding activity, resulting in degradation of TfR mRNA and increased ferritin synthesis. When IRP is bound to TfR mRNA, the half-life of the mRNA is long (>6 hours). The half-life of IRP that is not bound to TfR RNA is considerably shortened (<1 hour). The binding of IRP to TfR mRNA prevents its degradation by nucleases. When IRP is bound to ferritin IRE mRNA, the cap-binding complex (eIF4F) binds to the mRNA, but it is unable to make contact with the 43S preinitiation translational control.[72] The effects of iron on IRP activity are indirect in that the addition of iron to purifiied IRPs was unable to modulate their activity. The treatment of the whole cells with iron salts or iron chelators (agents that modulate cellular iron content) elicited a modulation of IRP activity. IRP activity is modulated by stimuli other than iron, including nitric oxide, oxidative stress, and hypoxia.

Nitric oxide (NO) is generated by nitric oxide synthases, a group of isoenzymes that convert L-arginine to citrulline and NO. NO production is particularly relevant in monocytes–macrophages. The effects of NO production on IRP1/IRP2 activity in these cells were explored. Induction of NO release elicited IRP1 activation associated with the loss of IRP2 activity.[74] The mechanisms underlying these effects are complex and have not been fully explored. Evidence suggests a mixed mechanism dependent on both direct effects of NO on IRP and on indirect effects mediated via a decrease of the free iron pool.[75]

Studies that explored the effect of oxidative stress on IRP activity are particularly relevant because reactive oxygen intermediates are implicated in iron-mediated toxicity. Exposure of different cell types to hydrogen peroxide (H_2O_2) elicited rapid activation of IRP1 from a cytoplasmic aconitase to an IRE-binding protein. The molecular mechanisms of these effects are complex and not fully known. IRP1 activation by H_2O_2 is not a simple consequence of a direct attack on the 4Fe–4S cluster of IRP1; it appears to involve stress response signalling. A recent study described IRP1 activation in an intact liver perfused with a buffer generating low but constant H_2O_2 levels.[76] The levels of H_2O_2 generated in this system were similar to those produced by activated neutrophils/macrophages, thus suggesting that IRP1 activation through oxidative stress may be relevant in inflammatory conditions.[76] The stimulation of IRP1 activity produced a decrease of ferritin synthesis associated with a stimulation of TfR expression. H_2O_2 also affects cellular metabolism of iron through IRP-dependent mechanisms.[77]

Since transferrin and transferrin receptor are among the genes modulated by hypoxia, it seemed logical to evaluate possible effects of a reduction of oxygen tension on IRP activity. The studies showed that IRP activity is modulated by oxygen levels. Tests of the effects of low oxygen tension on IRP activity generated conflicting results. Some studies reported a decrease in IRP activity during hypoxia, while other studies noted an increase. Hanson et al. observed a differential effect of hypoxia on IRP1 and IRP2 in that IRP1 activity was decreased, while IRP2 activity was increased.[78] The increased IRP2 activity was due to a stabilization effect on the IRP2 protein.

Some studies suggest that IRP activity is controlled by cell growth, as suggested by: (1) relatively high levels of IRP activity observed in mitogenically activated lymphocytes and in regenerating hepatocytes; and (2) c-myc stimulation of IRP2 expression associated with a repression of ferritin H expression. These findings suggest that active cell growth requires a high level of free iron obtained through the repression of ferritin H synthesis and a high level of IRP2 expression. The increased free iron pool in turn stimulates IRP activity which allows a high level of TfR1 expression required for sustaining cell growth.

Additional molecules, whose physiological roles have not been fully demonstrated, participate in iron metabolism in certain specialized tissues or subcellular compartments. The study of Friedreich's ataxia, a rare neurodegenerative disease, led to the discovery of frataxin, a protein possibly involved in controlling iron metabolism in the mitochondria. Frataxin is the defective gene in Friedreich's ataxia. It encodes a protein of 210 amino acids associated with the mitochondria[79] and may participate in metabolism of iron in the mitochondria. Deletion of the yeast gene homologue of the mammalian frataxin gene caused marked iron accumulation in the mitochondria. Tissues such as the myocardium that exhibit high frataxin levels under normal conditions exhibit mitochondrial iron overload in Friedreich's ataxia. The exact role of the human frataxin gene in iron metabolism remains to be determined.

It seems logical to assume that it participates in the import and/or export of mitochondrial iron.

A recently identified iron transport protein enhances uptake of non-transferrin-bound iron. This transporter is an additional and independent method of non-transferrin-bound iron uptake. Regulation by this iron transporter is similar to that exerted by TfR1 in that transporter expression is stimulated by iron loading. However, the physiological role of this transport protein in iron metabolism remains to be evaluated.

Another molecule involved in iron metabolism is ceruloplasmin, a multicopper oxidase that contains more than 95% of the plasma copper. Despite its copper-binding capacity, it does not seem to be involved in copper transport. It acts as a ferroxidase that catalyzes the conversion of ferrous to ferric iron. Ceruloplasmin is synthesized in the liver and monocytic cells. Patients with aceruloplasminemia showed marked iron accumulations in most tissues. This indicates that ceruloplasmin is required for normal iron metabolism by virtually all tissues. Ceruloplasmin's role in iron metabolism requires further study.

Recent studies suggest a potential role for hypoxia inducible factor 1 (HIF), a transcriptional regulator of cellular response to hypoxia. A link between HIF and iron metabolism is suggested. HIF acts as a transcriptional regulator of transferrin receptor 1 and transferrin genes. The ceruloplasmin gene contains three hypoxia-responsive elements to which HIF binds, thus stimulating gene transcription. The link between HIF and iron is also related to the mechanism of HIF activation induced by hypoxia. HIF is composed of α and β subunits. The α subunit is rapidly degraded under normoxia conditions and stabilized during hypoxia. The stabilizing effect of hypoxia on HIF-1α is also induced by iron chelators.[80] Iron is required for oxygen sensing by cells and tissues through a prolyl hydroxylase enzyme. In the presence of oxygen and iron, this enzyme targets proline 564, a highly conserved residue in human HIF-1α, and hydroxylates it, allowing interaction with a protein called the von Hippel–Lindau (VHL) tumor suppressor gene product and its subsequent degradation.[81]

Heme oxygenases may be involved in iron metabolism, particularly the potential cytotoxic effects of iron. Heme oxygenase 1 (HO1) is involved in the conversion of heme into bilirubin, carbon monoxide, and iron. HO1 is also known as a heat-shock protein (Hsp32) since various types of stress rapidly induce its synthesis. Studies of cells from mice with targeted deletions of the HO1 gene showed increased sensitivity to stress-induced apoptosis; this phenomenon was completely inhibited by iron chelators, thus suggesting the involvement of iron.[82] HO1$^{-/-}$ cells showed increased iron accumulation due to reduced iron efflux. HO1 participates in iron metabolism. HO1-mediated iron efflux controls cell survival following stress.

A second isoform known as HO2 is constitutively active in cells. It exhibits enzymatic activity almost identical to that of HO1, but has the capacity to bind heme. Studies on mice lacking HO2 showed increased sensitivity to oxidative stress and iron accumulation in the lungs.[83]

Our understanding of the cellular and molecular bases of the proteins involved in iron metabolism led to progress in the study of inherited disorders involving iron metabolism. The most frequent inherited disorder of iron metabolism is hereditary hemochromatosis. It is characterized by the progressive iron overload in various parenchymal tissues and is caused by homozygosity at the point mutation C282Y in the HFE gene. The role of a second mutation observed in a minority of patients remains unclear. However, factors other than the C282Y mutation may contribute to the development of tissue iron overload. About 50% of the patients are homozygous for the C282Y mutation.

The disease is characterized by two defects in iron metabolism: increased intestinal iron uptake and reduced storage capacity in the reticuloendothelial system. Both mechanisms contribute to the development of iron overload in hereditary hemochromatosis. The low iron level in the reticuloendothelial system despite considerable body iron overload represents a hallmark of the disease. The molecular mechanisms responsible for this defect in iron accumulation in the monocytes–macrophages, are unclear, but seem to be related to genetic and non-genetic mechanisms: (1) the reduced iron accumulation at the HH monocytes–macrophages is normalized by the introduction of a normal HFE gene[84]; and (2) the defect in iron accumulation in HH monocytes–macrophages is associated with reduced capacity of these cells to release TNF-α.[85] C282Y homozygotes with polymorphisms of the TNF-α gene associated with high capacity to release TNF-α by monocytes exhibit only a mild phenotype of hereditary hemochromatosis and tissue iron accumulation.

This conclusion is supported by the analysis of animal models of hereditary hemochromatosis. These models clearly showed that mouse strain differences are major determinants of the severity of iron accumulation in HFE knockout mice.[86] This strain-dependent variability is dependent on genetic traits.[87]

A rare form of hereditary hemochromatosis is caused by a genetic defect different from the one responsible for the development of the classic form. The rare form was found to be dependent on the Y250X mutation in the transferrin receptor 2 gene.[39,40]

A unique form of hereditary hemochromatosis is the so-called African iron overload found in sub-Saharan Africa. It is distinct from HLA-linked hemochromatosis and seems to arise from the interaction of increased dietary iron uptake and a genetic defect not associated with the HLA locus.[88] Iron deposition is prevalent in macrophages and hepatic parenchymal cells.

A new type of hereditary hemochromatosis caused by a dominant mutation A495 in the IRE sequence of the H ferritin gene was described recently.[89] This mutation causes reductions in the level of ferritin H gene[87] and in the level of ferritin H which leads to decreased capacity to sequester iron and subsequent tissue iron accumulation.[89] This type of mutation may be responsible for other non-HFE iron overload conditions described in the literature, but not yet characterized from a molecular view.[90] A recent study showed that a mutation in the gene encoding ferroportin is associated with other types of autosomal-dominant hemochromatosis.[91]

Juvenile hemochromatosis is a rare and more severe form. It is an autosomal-recessive disorder characterized by the early onset of severe iron overload in parenchyma and reticuloendothelial cells. It was mapped at chromosome 1. The gene affected by this disease is still unknown.

Another autosomal-dominant disease of iron metabolism due to a mutation of the IRE of the L ferritin gene was recently discovered. It is called hereditary hyperferritinemia/cataract syndrome and is characterized by a combination of elevated serum ferritin concentration and nuclear cataract.[92] The elevation in serum ferritin (due to an increase of L ferritin concentration) is not accompanied by augmentation of iron stores. The disease is caused by different types of point mutations of the IRE at the 5′ untranslated region of the L ferritin gene.

A form of sideroblastic anemia associated with ataxia results from the mutation of the ABC7 mitochondrial gene, an orthologue of the yeast ATM1 gene, whose product gene is involved in iron homeostasis.[93] The genetic defect causes accumulation of iron at the mitochondria of erythroid cells in the absence of increased parenchymal iron stores.

Recent studies implicate the L Ferritin gene mutation in the genesis of a rare neurodegenerative process. It is a dominantly inherited disease characterized by abnormal accumulation of iron in the basal ganglia and symptoms of extrapyramidal disease. This disorder was mapped to chromosome 19 and results from a point mutation of the L ferritin gene (an adenine insertion at position 460 to 461).[94] Preliminary studies suggest that Hallervorden–Spatz syndrome, a genetic defect mapping to chromosome 20, may be caused by a processed pseudogene that may have acquired an active promoter.[94] Whether this dominant form of adult basal ganglial disease is due to a loss or to a gain of function of the mutant LFerritin remains to be determined.

Iron may also play an important role in the control of cell proliferation. The effects of iron in the control of DNA synthesis mainly relate to the absolute need for iron to sustain the activity of ribonucleotide reductase (the only enzyme involved in reducing ribonucleotides to deoxyribonucleotides), particularly of the R2 subunit whose expression is cell cycle dependent. A reduction in the rate of cellular iron uptake is accompanied by a parallel decrease of ribonucleotide reductase activity. Iron chelators exert marked inhibitory effects on ribonucleotide reductase activity and, though this mechanism, inhibit cell proliferation.[95] Iron chelators also inhibit the replication of DNA viruses through a similar mechanism that involves the chelation of cellular iron required for sustaining the activity of viral ribonucleotide reductase.[96] Iron chelators may also have utility treating acquired immunodeficiency syndrome since they inhibit two steps required for virus replication: NF-κB activation and induction of ribonucleotide reductase.[97]

The role of iron in the cell cycle is more complex than simply sustaining the activity of ribonucleotide reductase. The c-*myc* proto-oncogene helps control cell proliferation, differentiation, and apoptosis and regulates the expression of H ferritin and IRP2.[63] Iron availability may influence the levels and the activities of genes involved in the cell cycle, such as cyclin-associated kinases and the retinoblastoma susceptibility gene product.

These observations suggest potential applications of iron chelators in treating certain neoplasias. Iron chelators can markedly decrease N-*myc* expression and induce growth and apoptosis in neuroblastoma cells.[98] This observation prompted clinical trials of deferoxamine, an iron chelator, in conjunction with standard anti-tumor chemotherapy measures to treat neuroblastoma. Other studies propose the use of iron chelators with doxorubicin and growth factor supply.[99] Iron was shown to lessen the incidence and severity of anthracycline-associated cardiac toxicity. The modulation of iron availability for tumor cells may become a unique tool for potentiating the anti-tumor effects of standard chemotherapy.

TABLE I.1
Proteins Involved in Iron Metabolism

Protein	Molecular Weight	Gene Location	Structure	Function	Mutation	Phenotype due to loss of function	Human disease
Ceruloplasmin	132,000	3q21–24	Single chain glycoprotein containing six copper atoms; contains three type 1 copper sites symmetrically arranged in three domains arising from triplication of an ancestral gene	Belongs to family of multicopper oxidases; ferroxidase activity; involved in control of iron efflux	Different types of mutations in humans; gene knockout in mice	Iron loading in central nervous system and liver; diabetes; mild anemia	Aceruloplasminemia
Divalent metal transporter (DMT1 / Nramp2)	90,000	12q13	Single-chain glycoprotein with 12 transmembrane domains; the gene generates two transcripts: one IRE$^+$ encoding a DMT1 isoform localized at the cell membranes and intracellular vesicles; the IRE$^-$ isoform is localized at the cell nucleus	Localized at the apical compartments of duodenal enterocytes; involved in iron transport from the gastrointestinal lumen into enterocytes; involved in iron transport from endosomes to cytoplasm of erythroid cells	Microcytic anemia (mk) mice; *Malvolio* mutation in flies; *Chardonnay* zebrafish mutant; *Belgrade* rat	In mutant mammals, severe hypochromic anemia, reduced reticulocyte iron uptake, and defective intestinal iron absorption	None known
Duodenal cytochrome b	30,000	Unknown	Di-heme protein with six predicted transmembrane domains; four conserved His residues act as heme-binding sites; lacks conventional NADH, NADPH, and flavin-binding motifs	Localized and highly expressed at brush border membranes of duodenal enterocytes; acts as a ferric reductase at cell membranes; uses intracellular reducing cofactors to reduce Fe^{3+} to Fe^{2+}	Unknown	Unknown	None known

continued

TABLE I.1 (continued)
Proteins Involved in Iron Metabolism

Protein	Molecular Weight	Gene Location	Structure	Function	Mutation	Phenotype due to loss of function	Human disease
Fe-ATPase iron transporter	Unknown	Unknown	Unknown	Microsomal iron transporter; co-expressed in tissues with HO1	Unknown	Unknown	None known
Ferritin	440,000		Hollow sphere containing solid iron and constructed from 24 polypeptide helix bundles (subunits) that assemble through non-covalent interactions with 4, 3, and 2 symmetry; structure is conserved in animals, plants, and microorganisms	Iron storage and recycling			
Ferritin H subunit	18,000	11q12		Ferroxidase activity	Fer-H knockout in mice; mutations of IRE sequence in humans	Embryonic death; iron accumulation in hepatocytes and macrophages	Autosomal-dominant iron overload
Ferritin L subunit	18,000	19q13		Iron core formation	Mutations of IRE; mutation in coding sequence (adenine insertion at 460–461)	Hyperferritinemia with cataract; adult-onset basal ganglia disease	Congenital hyperferritinemia with cataract; adult-onset basal ganglia disease

Protein	Molecular weight	Chromosome location	Structure	Function	Mutation/model	Phenotype	Disease
Ferroportin 1 (IREG1)	62,000	2	Single-chain glycoprotein of 570 amino acids with 10 transmembrane domains; localized at the basolateral membranes of polarized cells	May be involved in iron export from cells; expressed in duodenum (export of iron into the circulation), placenta (transport from mother to embryo), and Kupffer cells in the liver (recycling)	Zebrafish mutant weissherbst (weh); mutation in coding sequence in humans	Severe hypochromic anemia associated with reduction of erythroid cells and blockade of erythroid maturation; phenotype may depend on ferroportin 1 in transporting iron from the yolk sac to the embryonic circulation; autosomal-dominant hemochromatosis	Autosomal-dominant hemochromatosis
Frataxin	18,000	9q13	Mitochondrial protein of unknown function formed by five-stranded, antiparallel sheet forming a flat platform that supports two helices; the ensemble forms a compact sandwich	Mitochondrial respiratory function and iron homeostasis; seems to be involved in regulation of mitochondrial iron export	Triplet repeat expansion in humans; gene knockout in mice	Ataxia; cardiomiopathy; iron accumulation in central nervous system and in mitochondria	Friedreich's ataxia
Heme oxygenase 1 (HO1)	30,000	22q12	Membrane-bound protein with a heme binding site between α-helical folds; the heme pocket is formed by one helix on the proximal side and the other on the distal side of the heme	A heat-shock protein induced by stressful stimuli; degradation of heme into bilirubin, iron, and carbon monoxide	HO1 knockout in mice	Iron accumulation in liver and kidney; increased resistance to apoptosis; cytoprotective effect attributable to its augmentation of iron efflux	None known
Hemochromatosis (HFE)	37,200	12q13	Major histocompatibility complex type 1-like glycoprotein; associates in a heterodimer with β2-microglobulin	HFE/β2 microglobulin heterodimer interacts with transferrin receptor 1, reducing its affinity for iron; interaction required for regulation of intracellular iron homeostasis	C282Y, S65C, and H63D mutations in humans; HFE knockout in mice	Increased iron accumulation in several parenchymas associated with decreased iron in macrophages	Hereditary hemochromatosis

continued

TABLE I.1 (continued)
Proteins Involved in Iron Metabolism

Protein	Molecular Weight	Gene Location	Structure	Function	Mutation	Phenotype due to loss of function	Human disease
Hephaestin	155,000	Xq11–12	Homologue of ceruloplasmin; bound to cell membrane	Intracellular ferroxidase involved in iron export from cells; localized at the basolateral membranes of crypt duodenal cells	Sex linked anemia in mice due to gene deletion	Decreased intestinal iron absorption and decreased placental iron transfer	None known
Hepcidin	25,000	19	Cysteine-rich antimicrobial peptide named for its origin in the liver and antimicrobial properties	Prevents iron overload by reducing iron transport in enterocytes and programming macrophages to retain iron	USF2 knockout mice	Multivisceral iron overload	None known
Iron regulatory protein 1	90,000	9	Four-domain protein with a 4Fe–S cluster	Regulates synthesis of ferritin, transferrin receptor 1, DMT1, and ferroportin	Unknown	Unknown	None known
Iron regulatory protein 2	90,000	15	Four-domain protein lacking a 4Fe–4S cluster	Regulates synthesis of ferritin, transferrin receptor 1, DMT1, and ferroportin	IRP-2 knockout mice	Misregulation of iron metabolism; neurodegenerative disease due to iron accumulation in neurons and oligodendrocytes	None known

Protein	Size	Location	Structure	Function			
Lactoferrin	80,000	3p21–23	Glycoprotein with a single chain and two iron-binding sites; has two homologous lobes, N and C terminal halves; each lobe is divided into two domains; iron- and carbonate-binding sites located in a deep cleft between domains	Has several biological functions: (1) bacteriostatic activity largely related to iron chelation; (2) regulation of myelopoiesis; (3) anti-tumor activity	Unknown	Unknown	None known
Melanotransferrin	97,000	3q28	Glycosylphosphatidylinositol-anchored membrane glycoprotein; 37% sequence homology with serum transferrin; a single iron-binding site	Uptake of non-transferrin-bound iron; possible role in protection against membrane–lipid peroxidation	Unknown	Unknown	None known
Natural resistance-associated macrophage protein 1 (Nramp1)	100,000	2q35	Integral membrane protein expressed in lysosomal compartments of macrophages	Recruited to membranes of phagosomes soon after completion of phagocytosis; contributes to defense against infection by extrusion of divalent cations from phagosomes; facilitates iron release from macrophages	Mutations in mice	Mutations at its locus cause susceptibility to infection by antigenically unrelated intracellular pathogens	None known
Stimulator of iron transport (SFT)	338 amino acids	10q21	Single chain glycoprotein with six predicted transmembrane domains; contains REXXE motifs	Transferrin-independent iron uptake; expression controlled by iron levels (increased levels in iron-deprived cells)	Unknown	Unknown	Unknown

continued

TABLE I.1 (continued)
Proteins Involved in Iron Metabolism

Protein	Molecular Weight	Gene Location	Structure	Function	Mutation	Phenotype due to loss of function	Human disease
Transferrin	79,500	3q21	Glycoprotein with a single chain and two homologous iron-binding sites	Binding and transport of iron in plasma and extracellular fluids	Several base-pair deletion in humans; hypotransferrinemic mice (hpx) due to aberrant gene splicing	Severe anemia and iron overload in non-hemopoietic tissues; hpx mice exhibit severe anemia associated with marked iron overload in all non-hematopoietic tissues	Atransferrinemia
Transferrin receptor 1	185,000	3q29	Transmembrane glycoprotein composed of two identical subunits; two transferrin-binding sites/dimers; butterfly-like three-dimensional structure	Receptor-mediated uptake of diferric transferrin; expression in all proliferating tissues; hyperexpressed in erythroid cells; expression down-modulated by iron	Transferrin receptor 1 knockout mice	Microcytic anemia and decreased iron stores in heterozygotes; embryonic death in homozygotes due to severe anemia and apoptosis of nervous tissue	None known
Transferrin receptor 2	215,000	7q22	Transmembrane glycoprotein; dimer; homologue of transferrin receptor 1	Receptor-mediated endocytosis of ferric transferrin; preferentially expressed in erythroid cells and liver; no modulation by iron	Stop codon in humans	Parenchymal iron loading dependent on increased intestinal iron absorption	Type 3 hemochromatosis

Hepcidin, a cysteine-rich antimicrobial peptide synthesized in the human liver,[100] may participate in iron metabolism. Suppressive subtractive hybridization of livers of iron-loaded and control mice revealed that hepcidin mRNA was significantly overexpressed in the livers of iron-loaded animals.[101] Expression was deficient in USF2 knockout mice.[102] Since they showed patterns of iron overload similar to those observed in HFE-deficient animals, hepcidin may function in the same regulatory pathway as HFE. Hepcidin may prevent iron overload by reducing iron transport in enterocytes and programming macrophages to retain iron.

REFERENCES

1. Curie, C. et al. Maize yellow stripe 1 encodes a membrane protein directly involved in Fe(III) uptake, *Nature,* 409: 346–349, 2001.
2. Martinez, J.S. et al. Self-assembling amphiphilic siderophores from marine bacteria, *Science,* 287: 1245–1247, 2000.
3. Hantke, K. Iron and metal regulation in bacteria, *Curr. Opin. Microbiol.,* 4: 172–177, 2001.
4. Ramanan, N. and Wang, Y. A high-affinity iron permease essential for *Candida albicans* virulence, *Science,* 288: 1062–1064, 2000.
5. Wosten, M. et al. A signal transduction system that responds to extracellular iron, *Cell,* 103: 113–125, 2000.
6. McKie, A.T. et al. An iron-regulated ferric reductase associated with absorption of dietary iron, *Science,* 291: 1755–1759, 2001.
7. Robinson, N.J. et al. A ferric-chelate reductase for iron uptake from soils, *Nature,* 397: 694–697, 1999.
8. Gunshin, H., MacKenzie, B., and Berger, U.V. Cloning and characterization of a mammalian proton-coupled metal ion transporter, *Nature,* 388: 482–488, 1997.
9. Fleming, M.D., Trenor, C.C.I., and Su, M.A. Microcytic anemia mice have a mutation in Nramp2, a candidate iron transporter gene, *Nature Genet.,* 16: 383–386, 1997.
10. Canonne–Hergaux, F. et al. Cellular and subcellular localisation of the Nramp2 iron transporter in the intestinal brush border and regulation by dietary iron, *Blood,* 93: 4406–4417, 1999.
11. Donovan, A., Brownlie, A., and Zhou, Y. Positional cloning of zebrafish ferroportin 1 identifies a conserved vertebrate iron exporter, *Nature,* 403: 776–781, 2000.
12. McKie, A.T., Marciani, P., and Rolfs, A. A novel duodenal iron-regulated trasporter, IREG1, implicated in the basolateral transfer of iron to the circulation, *Mol. Cell.,* 5: 299–309, 2000.
13. Abbou, S. and Haile, D.J. A novel mammalian iron-regulated protein involved in intracellular iron metabolism, *J. Biol. Chem.,* 275: 19906–19912, 2000.
14. Vulpe, C.D., Kuo, Y.M., and Murphy, T.L. Hephaestin, a ceruloplasmin homologue implicated in intestinal iron transport, is defective in the *sla* mouse, *Nature Genet.,* 21: 195–199, 1999.
15. Mukhopashyay. C.K., Attleh, Z.K., and Fox, P.L. Role of ceruloplasmin in cellular iron uptake, *Science,* 279: 714–717, 1998.
16. Bennet, M.J., Lebròn, J.A., and Bjorkman, P.J. Crystal structure of the hereditary hemochromatosis protein HFE complexed with transferrin receptor, *Nature,* 403: 46–53, 2000.

17. Ramalingman, T.S. et al. Binding to the transferrin receptor is required for endocytosis of HFE and regulation of iron homeostasis, *Nature Cell Biol.,* 2: 953–957, 2000.

18. Ray, C.N. and Enns, C.A. Iron homeostasis: new tales from the crypt, *Blood,* 96: 4020–4027, 2000.

19. Enns, C.A. Pumping iron: the strange partnership of the hemochromatosis protein, a class I MHC homolog, with the transferrin receptor, *Traffic,* 2: 167–174, 2001.

20. Rolfs, A., and Hediger, M.A. Intestinal metal ion absorption: an update, *Curr. Opin. Gastroenterol.,* 17: 177–183, 2001.

21. Zoller, H., Pietrangelo, A., and Vogel, W. Duodenal metal-transporter (DMT-1, NRAMP-2) expression in patients with hereditary hemochromatosis, *Lancet,* 353: 2120–2123, 1999.

22. Trenor, I.I. et al. The molecular defect in hypotransferrinemic mice, *Blood,* 96: 1113–1118, 2000.

23. Andrews, N.C. Iron homeostasis: insights from genetics and animal models, *Nature Genet. Rev.,* 1: 208–217, 2000.

24. Beutlerm, E. et al. Molecular characterization of a case of atranferrinemia, *Blood,* 96: 4071–4074, 2000.

25. Levy, J.E., Jin, O., and Fujiwara, Y. Transferrin receptor is necessary for development of erythrocytes and the nervous system, *Nature Genet.,* 21: 396–399, 1999.

26. Lawrence, C.M. et al. Crystal structure of the ectodomain of human transferrin receptor, *Science,* 286: 779–782, 1999.

27. Luo, X. and Hofmann, K. The protease-associated domain: a homology domain associated with multiple classes of proteins, *Trends Biochem. Sci.,* 28: 147–148, 2001.

28. Collowan, J.F. et al. Transferrin receptor internalisation sequence YXRF implicates a tight turn as the structural recognition motif for endocytosis, *Cell,* 63: 1061–1072, 1993.

29. Testa, U., Pelosi, E., and Peschle, C. The transferrin receptor, *Crit. Rev. Oncogenesis,* 4: 241–276, 1993.

30. Sposi, N.M. et al. Mechanisms of differential transferrin receptor expression in normal hematopoiesis, *Eur. J. Biochem.,* 267: 6762–6774, 2000.

31. Louache, F. et al. Regulation of transferrin receptors in human hematopoietic cell lines, *J. Biol. Chem.,* 259: 11576–11582, 1984.

32. Ward, J.H., Kushner, J.P., and Kaplan, J. Regulation of HeLa cell transferrin receptors, *J. Biol. Chem.,* 257: 10317–10323, 1982.

33. Mullner, E.W. and Kuhn, L.C. A stem-loop in the 3′ untranslated region mediates iron-dependent regulation of transferrin receptor mRNA stability in the cytoplasm, *Cell,* 53; 815–825, 1988.

34. Dautry–Varsat, A., Ciechanover, A., and Lodish, H.F. pH and the recycling of transferrin during receptor-mediated endocytosis, *Proc. Natl. Acad. Sci. U.S.A.,* 80: 2258–2262, 1983.

35. Lin, S.X. et al. Rme-1 regulates the distribution and function of the endocytic recycling compartment in mammalian cells, *Nature Cell Biol.,* 3: 567–572, 2001.

36. Kawabata, H. et al. Molecular cloning of transferrin receptor 2. A new member of the transferrin receptor-like family, *J. Biol. Chem.,* 274: 20826–20832, 1999.

37. Kawabata, H. et al. Transferrin receptor 2-alpha supports cell growth both in iron-chelated cultured cells and *in vivo, J. Biol. Chem.,* 275: 16618–16625, 2000.

38. West, A.P. et al. Comparison of the interactions of transferrin receptor and transferrin receptor 2 with transferrin and hereditary hemochromatosis protein HFE, *J. Biol. Chem.,* 275: 38135–38138, 2000.

39. Camaschella, C., Roetto, A., and Calì, A. The gene encoding transferrin receptor 2 is mutated in a new type of hemochromatosis mapping to 7q22, *Nature Genet.*, 25: 14–15, 2000.

40. Roetto. A. et al. New mutations inactivating transferrin receptor 2 in hemochromatosis type 3, *Blood,* 97: 2555–2560, 2001.

41. R'zik, S. and Beguin, Y. Serum soluble transferrin receptor concentration is an accurate estimate of the mass of tissue receptors, *Exp. Hematol.*, 29: 677–685, 2001.

42. Steijns, J.M. and Hoijdenk, A.C.M. Occurrence, structure, biochemical properties and technological characteristics of lactoferrin, *Br. J. Nutr.,* 84 (Suppl. 1): S11–S17, 2000.

43. Peterson, N.A. et al. Crystal structure and iron-binding properties of the R210K mutant of the N-lobe of human lactoferrin: implications for iron release from transferrins, *Biochemistry,* 39: 6625–6633, 2000.

44. Abdallah, F.B. and Chahine, J.M. Transferrins: iron release from lactoferrin, *J. Mol. Biol.,* 303: 255–266, 2000.

45. Khan, A.J.U. et al. Camel lactoferrin, a transferrin-cum-lactoferrin: crystal structure of camel apolactoferrin at 2.6 Å resolution and structural basis of its dual role, *J. Mol. Biol.,* 305: 751–761, 2001.

46. Fillebeen, C. et al. Receptor-mediated transcytosis of lactoferrin through the blood–brain barrier, *J. Biol. Chem.,* 274: 7011–7017, 1999.

47. Sallmann, F.R. et al. *Escherichia coli* as specific cell-surface targets of human lactoferrin, *J. Biol. Chem.,* 274: 16017–16114, 1999.

48. Welling, M.M. et al. ^{99m}Tc-labeled antimicrobial peptides for detection of bacterial and *Candida albicans* infections, *J. Nuclear Med.,* 42: 788–794, 2001.

49. Shi, Y., Kong, W., and Nakayama, K. Human lactoferrin binds and removes the hemoglobin receptor protein of the periodontopathogen *Porphyromonas gengivalis*, *J. Biol. Chem.,* 275: 30002–30008, 2000.

50. Gupta–Khanna, A. et al. Sp1 and c/EBP are necessary to activate the lactoferrin gene promoter during myeloid differentiation, *Blood,* 95: 3734–3741, 2000.

51. Borregaard, N. et al. Regulation of human neutrophil granule protein expression, *Curr. Opin. Hematol.,* 8: 23–27, 2001.

52. Zhang, Z. and Teng, C.T. Estrogen receptor-related receptor alpha-1interacts with coactivator and constitutively activates the estrogen response elements of the human lactoferrin gene, *J. Biol. Chem.,* 275: 20837–20846, 2000.

53. Levi, S. et al. A human mitochondrial ferritin encoded by an intronless gene, *J. Biol. Chem.,* in press 2001.

54. Hempstead, P.D. et al. Comparison of the three-dimensional structures of recombinant human H and horse L ferritins at high resolution, *J. Mol. Biol.,* 268: 424–448, 1997.

55. Stillman, T.J. et al. The high-resolution x-ray crystallographic structure of the ferritin (EcFtuA) of *Escherichia coli*: comparison with human H ferritin (HuHF) and the structures of the $Fe^{(3+)}$ and $Zn^{(2+)}$ derivatives, *J. Mol. Biol.,* 307: 587–603, 2001.

56. Ilari, A. et al. The dodecameric ferritin from *Listeria innocua* contains a novel intersubunit iron-binding site, *Nature Struct. Biol.,* 7: 38–43, 2000.

57. Masuda, T., Goto, F., and Yoshihara, T. A novel plant ferritin subunit from soybean that is related to a mechanism in iron release, *J. Biol. Chem.,* in press 2001.

58. Ferreira, C. et al. Early embryonic lethality of H ferritin gene deletion in mice, *J. Biol. Chem.,* 275: 3021–3024, 2000.

59. Epstein, S. et al. H-ferritin subunit overexpression in erythroid cells reduces the oxidative stress response and induces multidrug resistance properties, *Blood,* 94: 3593-3603, 1999.

60. Cozzi, A. et al. Overexpression of wild-type and mutated human ferritin H-chain in HeLa cells, *J. Biol. Chem.*, 275: 25122–25129, 2000.

61. Coccia, E.M. et al. Modulation of ferritin H-chain expression in Friend erythroleukemia cells: transcriptional and translational regulation by hemin, *Mol. Cell Biol.*, 12: 3015–3022, 1992.

62. Marziali, G. et al. Transcriptional regulation of the ferritin heavy-chain gene: the activity of the CCAAT binding factor NF-Y is modulated in heme-treated Friend leukaemia cells and during monocyte-to-macrophage differentiation, *Mol. Cell Biol.*, 17: 1387–1395, 1997.

63. Wu, K.J., Polack, A., and Dalla Favera, R. Coordinated regulation of iron-controlling genes, H-ferritin and IRP-2, by c-*myc*, *Science*, 283: 676–679, 1999.

64. Oppenheim, E.W. et al. Heavy chain ferritin enhances serine hydroxymethyltransferase expression and *de novo* thymidine biosynthesis, *J. Biol. Chem.*, in press 2001.

65. Gray, C.P. et al. Immunosuppressive effects of melanoma-derived heavy-chain ferritin are dependent on stimulation of IL-10 production, *Int. J. Cancer*, 92: 843–850, 2001.

66. Zerpa, G.A. et al. Alternative splicing prevents transferrin secretion during differentiation of a human oligodendrocyte cell line, *J. Neurosci. Res.*, 61: 388–395, 2000.

67. Hulet, S.W. et al. Oligodendrocyte progenitor cells internalise ferritin via clathrin-dependent receptor mediated endocytosis, *J. Neurosci. Res.*, 61: 52–60, 2000.

68. Chi, S.I. et al. Differential regulation of H- and L-ferritin messenger RNA subunits, ferritin protein and iron following focal cerebral ischemia–reperfusion, *Neuroscience*, 100: 475–484, 2000.

69. Theil, E.C. and Eisenstein, R.S. Combinatorial mRNA regulation: iron regulatory proteins and iso-iron-responsive elements (iso-IREs), *J. Biol. Chem.*, 275: 40659–40662, 2000.

70. Meehan, H.A. and Corell, G.J. The hairpin loop but not the bulged C of the iron responsive element is essential for high affinity binding to iron regulatory protein-1, *J. Biol. Chem.*, 276: 14791–14796, 2001.

71. Lin, E., Graziano, J.H., and Freyer, G.A. Regulation of the 75 kDa subunit of mitochondrial complex I by iron, *J. Biol. Chem.*, in press 2001.

72. Eisenstein, R.S. Iron regulatory proteins and the molecular control of mammalian iron metabolism, *Annu. Rev. Nutr.*, 20: 627–662, 2000.

73. La Vaute, T. et al. Targeted deletion of the gene encoding iron regulatory protein-2 causes misregulation of iron metabolism and neurodegenerative disease in mice, *Nature Genet.*, 27: 208–214, 2001.

74. Bogdan, C. Nitric oxide and the regulation of gene expression, *Trends Cell Biol.*, 11: 66–75, 2001.

75. Cairo, G. and Pietrangelo, A. Iron regulatory proteins in pathobiology, *Biochem. J.*, 352: 241–250, 2000.

76. Mueller, S. et al. IRP1 activation by extracellular oxidative stress in the perfused rat liver, *J. Biol. Chem.*, in press 2001.

77. Caltagirone, A., Weiss, G., and Pantopoulos, K. Modulation of cellular iron metabolism by hydrogen peroxide, *J. Biol. Chem.*, in press 2001.

78. Hanson, E.S., Foot, L.M., and Leibold, E.A. Hypoxia post-translationally activated iron regulatory protein, *J. Biol. Chem.*, 19: 5047–5052, 1999.

79. Rotig, A. et al. Aconitase and mitochondrial iron–sulfur deficiency in Friedreich ataxia, *Nature Genet.*, 17: 215–217, 1997.

80. Maxwell, P.H. et al. The tumor suppressor protein VHL targets hypoxia-inducible factors for oxygen-dependent proteolysis, *Nature*, 399: 271–275, 1999.

81. Jaakkola, P., Mole, D.R., and Tian, Y.M. Targeting of HIF-1α to the von Hippel–Lindau ubiquitylation complex by O$_2$-regulated prolyl hydroxylation, *Science*, 292: 468–472, 2001.
82. Montosi, G. et al. Wild-type HFE protein normalizes transferrin iron accumulation in macrophages from subjects with hereditary hemochromatosis, *Blood*, 96: 1125–1129, 2000.
83. Fargion, S. et al. Tumor necrosis factor-α promoter polymorphism influences the phenotypic expression of hereditary hemochromatosis, *Blood*, 97: 3707–3712, 2001.
84. Ferris, C.D., Jaffrey, S.R., and Sawa, A. Haem oxygenase-1 prevents cell death by regulating cellular iron, *Nature Cell Biol.*, 1: 152–157, 1999.
85. Dennery, P.A., Spitz, D.R., and Yang, G. Oxygen toxicity and iron accumulation in the lungs of mice lacking heme oxygenase 2, *J. Clin. Invest.*, 101: 1001–1011, 1998.
86. Fleming, R.E. et al. Mouse strain differences determine severity of iron accumulation in HFE knockout model of hereditary hemochromatosis, *Proc. Natl. Acad. Sci. U.S.A.*, 98: 2707–2711, 2001.
87. Sproule, T.J. et al. Naturally variant autosomal and sex-linked loci ndetermine the severity of iron overload in β2-microglobulin-deficient mice, *Proc. Natl. Acad. Sci. U.S.A.*, 98: 5170–5174, 2001.
88. Kasvosve, I. et al. African iron overload, *Acta Clin. Belg.*, 55: 88–93, 2000.
89. Kato, J., Fuijikawa, K., and Kanda, M. A mutation in the iron-responsive element of H ferritin mRNA, causing autosomal dominant iron overload, *Am. J. Hum. Genet.*, 69: 191–197, 2001.
90. Pietrangelo, A., Montosi, G., and Totaro, A. Hereditary hemochromatosis in adults without pathogenic mutations in the hemochromatosis gene, *New Engl. J. Med.*, 341. 725–732, 1999.
91. Njajou, O.T., Vaessen, N., amd Joosse, M. A mutation in SLC11A3 is associated with autosomal dominant hemochromatosis, *Nature Genet.*, 28: 213–214, 2001.
92. Cazzola, M. and Skoda, R.C. Translational pathophysiology: a novel molecular mechanism of human disease, *Blood*, 95: 3280–3288, 2000.
93. Bakai, S. et al. Human ABC7 transporter gene structure and mutation causing X-linked sideroblastic anemia with ataxia with disruption of cytosolic iron–sulfur protein maturation, *Blood*, 96: 3256–3264, 2000.
94. Curtis, A.R.J. et al. Mutation in the gene encoding ferritin ligtht polypeptide causes dominant adult-onset basal ganglia disease, *Nature Genet.*, in press 2001.
95. Chitambar, C.R. et al. Cellular adaption to down-regulated iron transport into lymphoid leukemic cells: effects on the expression of the gene for ribonucleotide reductase, *J. Cell Physiol.*, 345: 681–685, 2000.
96. Romeo, A.M. et al. Intracellular chelation of iron by bipyridyl inhibits DNA virus replication: ribonucleotide reductase maturation as a probe for intracellular iron pools, *J. Biol. Chem.*, in press 2001.
97. Asbek, B.S. et al. Anti-HIV effect of iron chelators: different mechanisms involved, *J. Clin. Virol.*, 20: 141–147, 2001.
98. Fan, L. et al. Inhibition of N-*myc* expression and induction of apoptosis by iron chelation in human neuroblastoma cells, *Cancer Res.*, 61: 1073–1079, 2001.
99. Tetef, M.L. et al. Phase I trial of 96-hour continuous infusion of dexrazoxane in patients with advanced malignancies, *Clin. Cancer Res.*, 7: 1569–1576, 2001.
100. Park, C.H. et al. Hepcidin, a urinary antimicrobial peptide synthesized in the liver, *J. Biol. Chem.*, 276: 7806–7810, 2001.

101. Pigeon, C. et al. A new mouse liver-specific gene encoding a protein homologous to human antimicrobial peptide hepcidin is overexpressed during iron overload, *J. Biol. Chem.,* 276: 7811–7819, 2001.
102. Nicolas, G. et al. Lack of hepcidin gene expression and severe tissue iron overload in upstream stimulatory factor 2 (USF2) knockout mice, *Proc. Natl. Acad. Sci. U.S.A.,* 98: 8780–8785, 2001.

The Author

Ugo Testa was born in Naples, Italy. He studied medicine at the University of Naples and earned his degree in 1976. He spent 7 years as a research scientist at the Unité pour la Recherche sur les Anémies, an INSERM facility at the Hôpital Henri Mondor in Créteil, France. He joined the Department of Hematology and Oncology of the Instituto Superiore di Sanità in Rome as a research director in 1985. His major research since then has focused on the mechanisms that control iron metabolism and hematopoietic cell differentiation. In addition to this book, Dr. Testa has published more than 250 papers in major scientific journals.

Dr. Testa and his wife, Elvira, live in Rome and are the parents of two children, Federico and Vittoria.

Table of Contents

Chapter 1   Iron Absorption 1

1.1   Introduction 1
1.2   Iron Requirements 2
1.3   Iron Digestion 3
1.4   Iron Absorption 6
1.5   Physiologic Regulation of Iron Absorption 14
1.6   Iron Bioavailability 19
      1.6.1   Enhancers of Nonheme Iron Absorption 20
      1.6.2   Inhibitors of Nonheme Iron Absorption 22
1.7   Absorption of Heme Iron 26
References 28

Chapter 2   Iron and Cell Proliferation: Mechanisms and Applications in
               Cancer Therapy 39

2.1   Introduction 39
2.2   Ribonucleotide Reductase: A Key Enzyme in the Control of DNA
      Synthesis 40
2.3   Iron and Cell Cycle Control 45
      2.3.1   An Overview of the Cell Cycle 45
      2.3.2   Role of Iron in the Control of Cyclin Activation 47
2.4   Inhibition of Cell Proliferation by Agents that Reduce Cellular Iron
      Incorporation 48
      2.4.1   Inhibition of Cell Growth Using Monoclonal Antibodies to
               Transferrin Receptor 48
      2.4.2   Inhibition of Cell Growth by Agents that Interfere with
               Intracellular Iron Incorporation 51
      2.4.3   Inhibition of Cell Proliferation Mediated by Iron Chelators 53
2.5   Iron and Cancer 59
References 61

Chapter 3   Lactoferrin 71

3.1   Introduction 71
3.2   Biochemistry and Molecular Biology of Lactoferrin 74
3.3   Lactoferrin Receptors 88
      3.3.1   Lactoferrin Receptors on Leukemic Lines 88
      3.3.2   Lactoferrin Receptors on Activated T Lymphocytes 89
      3.3.3   Lactoferrin Receptors on Monocytes–Macrophages 91
      3.3.4   Lactoferrin Receptors on the Small Intestinal Brush Border
               Membrane 93

3.3.5   Lactoferrin Receptors on Parenchymal Liver Cells 94
3.3.6   Lactoferrin Receptors on Breast Epithelial Cells..........................96
3.3.7   Lactoferrin Receptors on Platelets...97
3.3.8   Lactoferrin Receptors on Neuronal Cells.....................................97
3.4   Functions of Lactoferrin 98
       3.4.1   Bacteriostatic Activity of Lactoferrin 98
             3.4.1.1   Lactoferricin...102
       3.4.2   Bacterial Lactoferrin Receptors 103
             3.4.2.1   Serum Lactoferrin Concentration and Infectious
                           Diseases..107
       3.4.3   Lactoferrin and the Regulation of Myelopoiesis.........................109
       3.4.4   Lactoferrin as a Gene Regulator...113
       3.4.5   Anti-Tumor Activity of Lactoferrin 114
3.5   Lactoferrin Expression in Neutrophils and Mammary Glands................114
       3.5.1   Lactoferrin Synthesis by Neutrophils..115
       3.5.2   Lactoferrin Synthesis in Mammary Glands..................................119
3.6   Lactoferrin Expression during Development...122
References..123

Chapter 4   Transferrin 143

4.1   Introduction 143
4.2   Comparative Analysis of the Structures of Transferrin Family Proteins....151
4.3   Transferrin Genetic Variants 155
       4.3.1   Transferrin C Alleles 156
       4.3.2   Transferrin B and D Alleles...157
       4.3.3   Association between Transferrin Alleles and Disease 157
4.4   Biochemical Properties of Serum Transferrin..158
       4.4.1   General Properties and Basic Structure.......................................158
       4.4.2   Structural Basis of Transferrin/Transferrin Receptor Interaction....174
       4.4.3   Glycosylation of Transferrin 176
4.5   Transferrin Gene Expression 179
       4.5.1   Regulation of Transferrin Synthesis in Liver, Testis, and
             Brain: Cellular Studies..182
       4.5.2   Regulatory Sequences Involved in the Regulation of Transferrin
             Gene Expression Observed in Different Tissues..........................187
4.6   Transferrin and Microbiology...195
4.7   Transferrin as a Growth Factor...208
4.8   Congenital Atransferrinemia...218
4.9   Transferrin as a Transporter of Drugs 219
References..222

Chapter 5   Transferrin Receptor 249

5.1   Introduction 249
5.2   General Features 251
5.3   Molecular Biology 254

5.4   Transferrin Receptor Glycosylation...258
5.5   Transferrin Receptor Endocytosis..262
       5.5.1   Basic Features 262
       5.5.2   Structural Requirement for Efficient Transferrin Receptor
               Internalization...272
5.6   Transferrin Receptor Phosphorylation..275
5.7   Intracellular Pools of Transferrin Receptors 277
5.8   Effect of CA^{2+} and Oxidation on Transferrin Receptor Internalization 279
5.9   Tissue Distribution and Expression in Selected Cell Types 281
5.10   Regulation of Transferrin Receptors...294
       5.10.1 The Expression of Transferrin Receptors is Correlated with
               Cell Proliferation..294
       5.10.2 Iron-Dependent Regulation of Transferrin Receptors.................297
       5.10.3 Gene Elements Required for Receptor Expression and
               Regulation..301
5.11   Expression and Regulation in Different Cell Types............................305
       5.11.1 Expression in Activated T Lymphocytes 305
       5.11.2 Expression in Monocytes–Macrophages 317
       5.11.3 Expression of Transferrin Receptor during Erythropoietic
               Differentiation and Maturation 324
5.12   Interaction between Transferrin Receptors and HFE: The Class I
       MHC-Related Protein Mutated in Hereditary Hemochromatosis..............332
5.13   A General Overview of Iron Uptake Mechanisms in Various Cell
       Types 334
References...339

Chapter 6   Soluble Transferrin Receptor...371
6.1   Introduction 371
6.2   Generation of Soluble Transferrin Receptor 371
6.3   Circulation in Human Serum...374
6.4   Clinical Relevance..374
References...379

Chapter 7   Alternative Iron Uptake Systems..383
7.1   Transferrin-Independent Iron Uptake 383
7.2   Proton-Coupled Metal Ion Transporter 386
7.3   Mechanisms of Iron Transport in and out of Cells: Role of
       Ceruloplasmin 390
7.4   HFE 395
7.5   Melanotransferrin 399
References...401

Chapter 8   Iron-Responsive Elements and Iron Regulatory Proteins 407

8.1   Introduction: The Discovery of Iron-Responsive Elements (IREs) and
        Iron Regulatory Proteins (IRPs) 407
8.2   Secondary and Tertiary Structures of Iron Response Elements (IREs) 410
8.3   Iron Regulatory Proteins 415
8.4   Role of IRP in Mediating Transferrin Receptor mRNA Stability 422
8.5   Role of IRP in the Control of Ferritin Translation 425
8.6   Evolution of IRPs 426
8.7   Regulation of IRP Activity by Nitric Oxide 429
8.8   Regulation of IRP by Oxidative Stress 432
8.9   Role of Phosphorylation in the Control of IRP Activity 435
8.10 Role of Heme in the Regulation of IRP Activity 436
8.11 IRPs Coordinate *In Vivo* Iron Metabolism 436
8.12 Tissue-Dependent Regulation of IRP Activity 437
8.13 Conclusions 439
References 440

Chapter 9   Ferritin 449

9.1   Introduction 449
9.2   Evolution of Ferritins 450
9.3   Three-Dimensional Structure of Ferritin 457
9.4   Ferroxidase Activity of Ferritin 459
9.5   Role of Phosphate in Iron Oxidation and Iron-Core Formation 461
9.6   Iron Release from Ferritin 462
         9.6.1   Ferritin as an Iron Donor in Erythroid Cells 463
         9.6.2   Ferritin as an Iron Donor in Hepatic Cells 464
9.7   Hemosiderin 466
9.8   Serum Ferritin 467
9.9   Ferritin Genes 474
9.10 Transcriptional Control of Ferritin Gene Expression 477
9.11 Ferritin and Endothelium 484
9.12 Ferritin in Tumor Cells 487
9.13 Ferritin Expression and Function in Selected Tissues 492
         9.13.1 Ferritin Expression and Function in the Eye 492
         9.13.2 Expression and Function of Ferritin in the Skin 494
         9.13.3 Ferritin Expression in the Brain 495
         9.13.4 Ferritin Expression and Function in Macrophages 504
9.14 Unexpected Functions of Ferritin 513
References 515

Index 541

1 Iron Absorption

CONTENTS

1.1 Introduction 1
1.2 Iron Requirements 2
1.3 Iron Digestion 3
1.4 Iron Absorption 6
1.5 Physiologic Regulation of Iron Absorption 14
1.6 Iron Bioavailability 19
     1.6.1 Enhancers of Nonheme Iron Absorption 20
     1.6.2 Inhibitors of Nonheme Iron Absorption 22
1.7 Absorption of Heme Iron 26
References 28

1.1 INTRODUCTION

In adults, the amount of iron lost from the body is relatively small. In fact, iron losses from the gastrointestinal tract, skin, and urinary tract in an adult 70-kg male amount to about 1 mg daily[1] and in a 55-kg menstruating female to about 2 mg daily. Iron losses in post-menopausal women are assumed to equal those observed in men. The losses are generally unregulated, and total body iron stores are regulated by changes in the rate of intestinal iron absorption. The average iron intake in the U.S. and Europe is about 15 to 25 mg/day; iron absorption averages 0. 5 to 1 mg/day in men and 1 to 2 mg/day in women during their reproductive years. Thus, the amount of iron absorbed normally ranges from 3 to 6% of the total amount ingested.

Iron is more readily absorbed in the ferrous state (Fe^{2+}), but the majority of dietary iron is in the ferric form (Fe^{3+}). Iron is minimally absorbed in the stomach, but the acidic gastric secretions play an important role in iron absorption in that they facilitate the solubilization of nonheme iron from food. The important role of the stomach in iron absorption is indicated by the observation that iron deficiency is a troublesome and frequent complication of gastrectomy. Absorption of iron, defined as its movement from the intestinal lumen to the circulation through epithelial cells of the digestive tract, occurs largely from the proximal small intestine. Iron absorption involves two steps: (1) uptake of iron from the intestinal lumen into the mucosa; (2) transfer across the mucosal cells and serosal membrane into the circulation. Much of the iron entering the mucosal cells is not transferred to the plasma, but remains trapped within the cells and is excreted into the lumen when cells are shed.

The capacity for iron absorption is markedly affected by body iron content. If the content is diminished, a high proportion of the available iron is absorbed and as the body iron content rises the rate of iron absorption falls. However, compared to iron absorption during conditions of normal iron homeostasis, the relative enhancement of absorption during iron deficiency is greater than inhibition observed during iron overload.

Furthermore, iron uptake is also controlled by the rate of erythropoiesis. Iron absorption is elevated when the rate of erythropoiesis is increased as extra iron is required to sustain new hemoglobin synthesis. Signals to alter iron absorption are received from the body via the cells of the intestinal crypt and are translated into a change in the rate of iron absorption after the cells have migrated to the villi and differentiated into mature absorptive enterocytes.

1.2 IRON REQUIREMENTS

About 30 to 40 mg of iron is required daily for internal utilization by the body, largely to resynthesize the hemoglobin destroyed along with old red blood cells. A large part (90%) of this iron requirement derives from the recycling of existing internal iron stores. Because this recycling is highly efficient, only 1 to 1. 5 mg of the internal iron is lost from the body and needs to be supplied by iron absorption. Iron is not actively excreted from the body in the urine or through the intestines, but is lost only with cells from the exterior or interior surfaces of the body. Thus, an adult man has obligatory losses corresponding to about 1 mg daily.[1] Two thirds of obligatory iron losses derive from the exfoliation of gastroinestinal mucosal cells; exfoliation of cells from the skin and urinary tract accounts for further daily losses of about 0. 3 mg and 0. 1 mg, respectively. Obligatory iron losses for non-menstruating women amount to about 0. 8 mg/day. On the basis of these data, it is evident that a normal adult needs to absorb about 1 mg iron from the diet on a daily basis to maintain balanced iron stores.

Adult women experience additional losses due to menstruation. Menstrual losses vary markedly among women, but losses for an individual woman usually remain constant from menarche and throughout her fertile life.[2] The median blood loss in menstrual material corresponds to about 30 ml; assuming that 1 g of hemoglobin contains 3.34 mg of iron, these menstrual blood losses correspond to an iron loss of about 16.5 mg. When expressed in terms of median daily loss, this iron loss due to menstruation corresponds for the majority of women to 0.6 mg/day. However, 25%, 10%, and 5% of women have menstrual blood losses corresponding to 52, 84, and 118 ml, respectively[3]; this corresponds to iron losses of 1.05, 1.68, and 2.36 mg/day, respectively. Thus, the median daily iron requirement for menstruating women is higher than for males and ranges from 1.4 mg to 3.2 mg/day. To ensure that these amounts of iron are absorbed from an average Western-type diet without any reduction in the supply of iron to the tissues, a diet must contain about 19 mg available dietary iron (estimating that, at most, 15% of the available iron can be expected to be absorbed from this type of diet).[3]

The amount of iron that must be absorbed to meet the iron requirements of menstruating teenage girls is higher than the amount required for adult women because teenagers require additional iron to sustain body growth.[3] Certain methods of contraception markedly modify menstrual iron losses. Oral contraceptives reduce iron blood losses about 50%, whereas intrauterine devices virtually double menstrual blood losses.

During pregnancy, iron requirements increase because additional iron is needed to replace basal losses, to allow for expansion of maternal red cell mass, and to provide for the needs of the fetus and placenta.[4] It has been suggested that the iron requirements are 1 mg/day in the first trimester, 4 mg/day in the second trimester, and about 6 mg/day in the third trimester.[5] It was calculated that a woman weighing 55 kg has a total iron cost of about 1000 mg throughout pregnancy[6]: about 230 mg to replace basal losses; 450 mg to sustain the increased red cell mass of the mother; 270 to 300 mg in the fetus body; and 50 to 90 mg in products of conception.

A series of studies using different isotopic techniques to measure iron absorption have provided clear evidence that iron absorption increases during pregnancy.[7-9] Thus, in a study carried out by Whittaker et al. it was found that non-pregnant normal women exhibited values of iron absorption corresponding to 10%; at 12 weeks' gestation, iron absorption values corresponded to 8%, a value similar to that observed in non-pregnant women. Later gestation values of iron absorption of 21 and 37% were observed[9] at 24 and 36 weeks, respectively. It is of interest that all the studies carried out on iron absorption during pregnancy provided evidence that a major increase in iron absorption occurs before 24 weeks of gestation.

A normal full-term infant does not need exogenous iron for the first 4 months of life, in that initial iron stores are sufficient to sustain hemoglobin, myoglobin, and iron-containing enzymes. After 4 months of life, 0.9 mg of iron are required daily, of which 0.7 mg are required to sustain body growth and 0.2 mg are required to replace iron losses.[10] During the second year of life, the daily iron requirement is about 0.4 mg/day.

1.3 IRON DIGESTION

The gastrointestinal system is the portal through which nutritive substances, vitamins, minerals, and fluids enter the body. Proteins, fats, and complex carbohydrates are broken down (digested) into absorbable units, principally in the stomach and small intestine. The products of digestion and the vitamins, minerals, and water cross the mucosa and enter the lymph system or the blood (absorption).

Dietary iron does not exist in a free form, but is bound to other components that must be digested before iron can be absorbed. Thus, the process of iron absorption can be subdivided according to Carpenter and Mahoney[11] into three phases: (1) iron digestion; (2) iron intestinal absorption; (3) iron utilization (Figure 1.1). In considering the various processes involved in iron digestion and absorption, it must be taken into account that dietary iron is present in the forms of nonheme and heme iron. Nonheme iron is not present in readily soluble form, and the digestion process must first release it in a soluble form that can be absorbed at the level of the small intestine. The importance of gastric digestion in increasing the bioavability of nonheme iron is a well known phenomenon, as based on a series of observations:

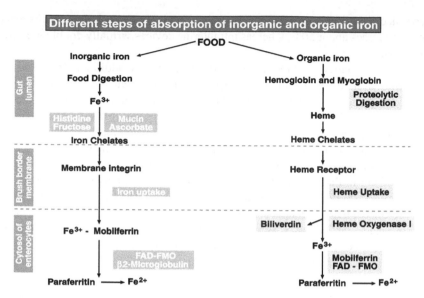

FIGURE 1.1 Schematic representation of the main steps underlying the process of absorption of inorganic and organic iron.

1. Nonheme iron is only scarcely absorbed when injected directly into the duodenum.[12]
2. Anemia is a well recognized complication of partial gastrectomy, the incidence increasing with postoperative interval.[13]
3. Gastrectomized rats made anemic by induced hemolytic anemia absorb only a very low proportion of nonheme dietary iron.[14]

Most inorganic iron in foodstuffs is in a ferric form which precipitates above pH 3 if it is not chelated (Table 1.1). The low pH present in the gastric juice facilitates iron solubility as suggested by the striking correlation observed between the pH of gastric juice and food iron absorption.[15] The role of the acidic pH present in the stomach in nonheme iron absorption is also supported by experiments carried out with H_2 receptor blockers. High doses of these drugs elicited a marked inhibition of gastic acid output, associated with a 42 to 65% reduction in absorption of nonheme food iron.[16]

**TABLE 1.1
Solubility of Inorganic and Organic Iron at Different pH Values**

Iron	pH Values
Inorganic (Fe^{2+})	3–7
Inorganic (Fe^{3+})	2–3
Organic (Heme)	7–11

As a consequence of the gastric digestion process, nonheme iron is solubilized from its originary dietary forms into a common pool that reaches an equilibrium with all dietary constituents to which iron can bind. The existence of a common pool for all dietary nonheme iron was validated by a large number of studies.[17,18] It was also shown that nonheme iron remaining out of this pool is not available for iron absorption.[17,18] Nonheme iron entering this pool is usually present in the form of chelates formed with different types of food constituents including ascorbic acid, citrate, sucrose, and amino acids; is linked to components of gastric secretions such as gastrin or mucin or is bound to other dietary components including phytates, polypeptides, and fiber.[11]

In this context, the role played by endogenous gastric ligands seems particularly relevant in the solubilization of nonheme iron. Early studies were focused to characterize high molecular weight complexes in the stomach and upper intestine that bound iron at acidic pH and prevented its precipitation in the more alkaline milieu of the small intestine.[19]

Recent studies suggest that gastric mucins could play an important role in nonheme iron solubilization. Gastric and duodenal mucin was able to bind iron at acidic pH and maintain the iron soluble at neutral pH.[20] Multiple iron molecules bind to a single molecule of macromolecular mucin. Furthermore, mucin is capable of accepting iron from dietary chelators (iron chelates of ascorbate, fructose, and histidine) at neutral pH, and maintains the iron in a form acceptable for absorption by the intestinal mucosa[20] (Figure 1.1). The competitive binding of iron and other metals to mucin may represent the biochemical basis that explains why dietary iron diminishes the absorption of other metal cations.[21] Finally, the pH-dependent binding of iron to mucin may explain the role of hydrochloric acid in the absorption of iron and explains why gastrectomy and achlorhydria predispose an individual to the development of iron deficiency.[22] Thus, clear experimental evidence indicates that gastric and duodenal mucin delivers inorganic iron to intestinal absorptive cells in an acceptable form for absorption.

Some studies suggest that an iron-binding component present in gastric juice, called gastroferrin, may act as an inhibitor of iron absorption.[23,24] This contention was based on initial observations indicating that the concentration of this protein in the gastric juice was decreased in hemochromatosis and could be, to some extent, responsible for the increased iron absorption observed in this condition.[25] However, this hypothesis was repudiated in several studies showing that gastroferrin was present in the gastric secretions of hemochromatotics.[26]

Heme iron is derived from the hemoglobin, myoglobin, and other heme proteins present in food of animal origin. During the process of digestion, heme iron forms an iron pool in the gastrointestinal tract independent of nonheme iron. This is proven by the observation that heme iron cannot be labeled with inorganic radioiron, but can be labeled with iron-labeled heme (i.e., biosynthetically radioiron-labeled hemoglobin).

Before it can be absorbed, heme must be detached from globin and then be absorbed as free heme (Figure 1.1). Exposure to the acidic pH and proteases of the gastric juice induces the release of heme from globin; heme is then converted to hemin through oxidation of iron contained in heme.[27] The release of heme from globin continues in the duodenum as a consequence of the activity of proteolytic duodenal enzymes such as trypsin.[28] At neutral pH, as observed in the duodenum, iron is scarcely soluble and forms aggregates of high molecular weight that are poorly absorbable.[28] The presence of dietary compounds, peptides, and amino acids derived from the digestion of food proteins, particularly meat proteins, increases heme solubility and then iron absorption.

It was postulated that the majority of heme iron is directly absorbed by intestinal cells under form of heme or hemin. However, the analysis of the products of digestion provided evidence that about 80% of heme present in meat digests is present in the form of low molecular weight nonheme complexes, whereas heme present in hemoglobin remains largely (about 95%) intact at the end of digestion. These studies suggest that dietary iron, largely represented by meat iron, is degraded during digestion in the form of inorganic iron and is then absorbed in this form at the level of the duodenum.[11]

This interpretation is fully supported by studies with the iron chelator desferrioxamine, indicating that this chelator does not inhibit *in vivo* absorption of purified heme iron, but markedly decreases absorption of heme iron from meat.[29,30] In fact, when given orally, desferrioxamine is not absorbed by the gastrointestinal tract, but prevents iron absorption by binding inorganic (nonhemoglobin) iron within the gut lumen.[29,30] When given parenterally, some desferrioxamine binds iron and is excreted in the urine. Another portion of the free chelating agent enters the hepatocytes, binds iron, and is excreted in the bile as ferrioxamine complex.[31-33] The iron chelate cannot return to the plasma unless there is biliary obstruction. Furthermore, a recent study[34] provided evidence that parenterally administered desferrioxamine also enters the small intestinal mucosa, where it binds intracellular iron and reduces iron absorption.

1.4 IRON ABSORPTION

At the level of intestinal mucosa occur a series of finely regulated processes consisting of cell proliferation, commitment, differentiation, digestion, and absorption. These processes occur at specific locations along the crypt to the villus axis. Crypt cells represent the proliferative compartment composed of immature enterocytes, while the brush border represents the non-proliferative compartment composed of functionally mature enterocytes.

Iron absorption occurs mostly in the proximal small intestine where the mucosa remains attuned to body requirements for iron. Studies carried out in rats provided evidence that absorption of iron was uniform throughout the length of the small intestine, while absorption showed a steeply decreasing gradient from the duodenum to the ileum.[35] Manis and Schacter proposed that iron, in physiological amounts, is absorbed by an active transport mechanism consisting of an initial mucosal uptake step, followed by intracellular transfer to the basolateral plasma membrane and release in the portal venous blood.[36] The transfer from the intestine to the blood may be divided into several steps including binding of iron to the brush border membrane, penetration into the mucosal cells, intracellular processing and transport, and release across the basolateral membrane into the blood.

A large number of studies then focused on the initial brush border permeation step, involving uptake experiments with isolated microvillous duodenal membrane vesicles of various animal species.[37-45] In the majority of these studies, a specific interaction of iron with microvillous membranes involving a facilitated transmembrane translocation mechanism was postulated.

The uptake of iron at the intestinal brush border membrane is the rate-limiting step for iron absorption in humans. The majority of nonheme iron absorbed through the brush border surface must be present in a ferrous state in order to cross the enterocyte membrane. This phenomenon is in part due to the solubility of ferrous (not ferric) iron at a pH above 3. Particularly relevant was a study by Han et al.[46] performed in Caco-2 cells, showing that iron must first be reduced by reductases present in the brush border membrane before being absorbed. The biochemical characterization of this ferric reductase activity showed that it was dependent upon NADH or NADPH: the Fe^{3+} reduction to ferric iron occurs directly by catalyzed electron transfer from NADH through enzymes localized in the brush border membrane.[47] In line with these observations, other studies performed *in vivo*[48] and *in vitro*[49] in Caco-2 cells have shown that cysteine and cysteinyl glycine enhance iron uptake primarily by reduction and not by increasing the solubility of iron.

The kinetics of iron absorption was carefully investigated using *in situ* ligated loops of small intestinal mucosa and upper small intestine microvillous membrane vesicles. All these studies provided evidence that the process of intestinal iron absorption exhibits biphasic kinetics.[35,50-53] The initial phase of absorption is very rapid and begins within seconds after iron reaches the mucosal surface; absorption during the second phase occurs at a much slower rate, but continues for a longer time. The kinetics of iron absorption was carefully investigated in human upper small intestine microvillous membrane vesicles.[54] The kinetic activity was very rapid, with an initial phase of maximal and linear uptake over the initial 45-second incubation; thereafter, the uptake curve sharply declined.[50] These authors showed also that ^{59}Fe uptake was optimal at 37°C, whereas no transport was detectable at 0°C. Analysis of saturation kinetics of iron uptake showed a K_m of 315 ± 33 nM and a V_{max} of 361 ± 68 pmol iron × min^{-1} × mg protein.[54] Pretreatment of intestine microvillous membrane vesicles with pronase led to a marked decrease of iron uptake velocity, suggesting that one or more proteins seem to be involved in the iron uptake process.[54]

Studies on the mechanism of intestinal iron uptake are made complex by the difficulty of obtaining the biological material required for these studies. It was recently proposed that the IEC-6 rat intestinal cell line may represent a valuable tool for investigating iron absorption.[55] These IEC-6 rat cells grow and form a monolayer of closely apposed cells, thus mimicking the intestinal mucosa with the apical or lumenal surfaces of the cells exposed to culture media. These cells exhibited biphasic kinetics of iron absorption, similar to that observed in studies with isolated intestinal loops, with a K_m of about 300 nM.

In addition to IEC-6 cells, the Caco-2 cell line represents a useful model for studies of iron absorption (Figure 1.2). Although colonic in origin, Caco-2 cells differentiate *in vitro*, developing brush border membranes and acquiring transport properties similar to those exhibited by enterocytes.[56] A series of studies show that Caco-2 cells represent important models for studying the mechanisms of iron absorption as well as iron bioavailability: (1) exposure of Caco-2 cells to high concentrations of iron leads to a down-modulation of their iron uptake capacity; (2) the expression of transferrin receptors in Caco-2 cells and normal enterocytes is limited to the basolateral membrane; (3) a simulated digestion model showed that meat enhances iron absorption in Caco-2 cells, in line with numerous studies performed in human subjects.

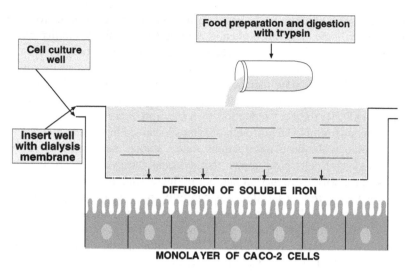

FIGURE 1.2 Caco-2 cell line represents a useful model for the study of the mechanisms of intestinal iron absorption. This shows the use of this cell line to study iron bioavailability in foods: foods are first digested with pepsin and pancreatin bile, and then added to the upper chamber of a two-chamber system. A cell monolayer of Caco-2 cells is attached to the bottom surface of the lower chamber and the upper chamber is separated from the lower chamber by a dialysis membrane. (Reprinted with modifications from Glahn, R.P. et al., *J. Nutr.,* 128, 257-264, 1998. With permission.)

Because the mechanism and regulation of iron absorption have remained unclear, it is not surprising that studies of virtually all the proteins known to be involved in human iron metabolism have advocated possible roles for all of them in intestinal iron uptake. However, recent studies discussed below provided evidence about the presence of a peculiar iron-binding protein in duodenal mucosa.

In the small intestine, the transferrin receptor is found predominantly in the crypts and shows a highly polarized distribution across the intestinal epithelial cells. Essentially all the receptor is found on the basal, lateral, and intracellular membranes, with little or no receptor found on brush border membrane.[57-61] Data from adult rats show that the intestinal receptor level varies inversely with the body iron stores and directly with changes in iron absorption, and that changes in receptor expression may be separated from changes in iron absorption in animals in which hemolysis has been produced experimentally.[60] These data suggest that the transferrin receptor is unlikely to play a direct role in iron absorption, but may play an indirect role by determining the iron status of the epithelial cells when they are formed in crypts. This conclusion is directly supported by studies evaluating in parallel iron absorption capacity and intestinal transferrin receptor expression during murine ontogenesis.

Immunohistology studies showed that in the late fetal intestine, transferrin receptors were expressed at high levels throughout the lengths of the villi, thus correlating with the cell proliferation in this tissue.[61] After birth and the establishment of distinct crypt and villous zones, cell surface transferrin receptors were associated mainly with the actively dividing cells of the crypts, and the differentiated cells of the villi showed few receptors.[61] At later stages of ontogenesis, two major changes in transferrin receptor expression in the developing small intestine were observed: a decrease in receptor expression associated with birth, and an increase at the time of weaning.[61]

In parallel other studies have shown that iron absorption is very high in neonates, and decreases to adult levels upon weaning.[62-65] These observations suggest that the likely role of the transferrin receptor in the small intestine is in supplying iron to the developing epithelial cells in the crypts, and that the receptor does not play a direct role in iron transit across the intestinal epithelium.

Recent studies have indicated that transferrin receptors may exert a peculiar regulation at the level of enterocytes. These studies were focused on evaluating the regulation of transferrin receptor mRNA in response to dietary iron changes. A first study involving *in situ* hybridization showed transferrin receptor mRNA was detected in crypt epithelial cells of the small and large intestine in iron-deficient and normal rats.[66]

Particularly, the study of transferrin receptor expression along the crypt–villus axis showed a progressive decrease from the crypt to the apical region of the villus.[67] In contrast, in iron-loaded rats, transferrin receptor mRNA was detected also in the superficial epithelial cells of the small intestine and colon.[66] In a second study, the localization of ferritin at the level of intestinal microvilli was determined and showed that ferritin mRNA was expressed at the highest levels in the epithelial cells of the crypt. Macrophages localized at the level of the lamina propria and at lower levels in villus epithelial cells, while ferritin protein was undetectable at the level of the crypt region, but present at increasing levels in the apical region of the villus.[68] In conclusion, these observations indicate that in the iron-deficient and normal rat intestine, transferrin receptor was expressed only by proliferating crypt cells, while in iron-loaded animals, surface enterocytes, which contained increased iron stores, expressed transferrin receptor.

Recent studies focused on characterizing the protein whose genetic alteration is involved in hereditary hemochromatosis have shed some light on the possible role of transferrin receptors in iron absorption. These studies show that hereditary hemochromatosis is due to the mutation of a gene encoding for a 343-amino acid protein that exhibits significant homology with the HLA class I molecules and is widely expressed.[69] This gene was designed as HLA-H or HFE. Northern blot analysis of HFE mRNA expression showed that this mRNA species was expressed in virtually all tissues, but particularly in liver and intestine.

The expression of HFE in the gastrointestinal tract was investigated in detail by immunohistochemistry.[70] In the small intestine, the HFE staining was most intense in the epithelial cells in crypts, while the surface epithelial cells in upper portions and tips of the villi did not exhibit reactivity. At the level of single enterocytes, the HFE distribution was peculiar, being primarily intracellular and perinuclear, while neither apical nor basolateral surfaces showed significant reactivity with anti-HFE antibody. It was hypothesized that HFE may act as a barrier to iron transport and in these tissues its loss of function may contribute to the development of hemochromatosis.

A recent study offers some evidence about the mechanisms through which HFE may modulate iron metabolism. In fact, it was shown that wild-type HFE protein forms stable complexes with the transferrin receptor and induces an inhibition of transferrin binding to its receptor by lowering the affinity of the ligand for the receptor. The mutant H63D HFE protein also exhibits the capacity to complex with transferrin receptor, but does not modify the affinity of this receptor for its ligand. Finally, the C282Y mutant protein completely failed to interact with the transferrin receptor.[71]

This observation was further supported by the analysis of the crystal structure of the hemochromatosis protein HFE. This analysis showed that hemochromatosis mutations concern a region involved in pH-dependent interaction with transferrin receptor.[72] Furthermore, it was shown that soluble HFE and transferrin receptor can interact at the basic pH of the cell surface, but not at the acidic pH levels of intracellular vesicles.[72] Finally, it was shown that the stoichiometry of transferrin receptor–HFE (2:1) differed from transferrin receptor–transferrin stoichiometry (2:2). Based on these findings, it was proposed that HFE, transferrin, and transferrin receptor form a ternary complex. These observations clearly support the existence of a molecular link between HFE and a key regulator of iron metabolism, the transferrin receptor.

Many investigators have identified transferrin or "transferrin-like" protein in homogenates of mucosal cells,[73] but the suggestion that it plays a role in iron transport across the mucosal cells has not been proven.[74] Recent studies have shown that duodenal transferrin is not synthesized by mucosal cells, but derives from plasma.[75,76]

Huebers et al. proposed that iron uptake into the enterocyte from the lumen may be mediated, at least in part, by transferrin.[77] These authors envisage a model of intestinal iron absorption whereby an iron-poor mucosal transferrin is secreted into the intestinal lumen, binds iron, and then reenters the enterocyte by receptor-mediated endocytosis. This is based on studies indicating that plasma transferrin instilled into the lumen of an isolated intestinal loop may be taken up into the mucosa. This hypothesis is now not tenable in that:

1. Recent studies carried out in human gastroduodenal biopsies[78] and earlier studies carried out on rat duodenum[75] failed to find any detectable transferrin mRNA, thus suggesting that duodenal cells are unable to synthesize transferrin.
2. As previously mentioned, a large number of studies have failed to detect transferrin receptors on the brush border.
3. Administration of diferric transferrin to achlorhydric humans failed to promote iron absorption.[79]
4. Iron absorption is not affected in children with congenital atransferrinemia[80] or in hypotransferric mice.[81,82]

A further possible role for serum transferrin in iron absorption was suggested by the early experiments of Levine et al.[83] and Evans and Grace,[84] who showed that transferrin enhanced iron release from intestinal cells and membranes that had been pre-loaded with radioactive iron. Levine et al.[83] also found specific binding of transferrin to intestinal cells and provided evidence that transferrin with a low iron saturation bound more avidly. These data suggest that iron-poor transferrin acquires iron from the intestinal cells after bonding to the cell membranes.[84] This hypothesis now appears untenable in the light of our current view of the transferrin/transferrin receptor cycle.

The role of transferrin in intestinal iron absorption was, however, supported by recent studies performed on the Caco-2 cell line. This cell line, when grown as a polarized cell layer, represents a useful model to investigate various phases of iron uptake and transfer. Studies of these cells provided evidence indicating that apo-transferrin and diferric transferrin undergo different endocytic cycles: after internalization, both apotransferrin and diferric transferrin recycled to the cell surface, but apotransferrin exhibited a much slower recycling time than diferric transferrin.[85] As a consequence of this delayed recycling rate, a significant proportion of cellular transferrin receptors are sequestered in the cell interior.[85] The transfer of iron from the apical surfaces to the basal surfaces of Caco-2 cells was facilitated in the presence of apotransferrin.[86] Furthermore, it was shown that apotransferrin is internalized by Caco-2 cells from the basal side and then localizes within the cells at the levels of vesicles localized above the nuclei.[86] On the basis of these observations, it was suggested that apotransferrin could bind newly absorbed iron at the level of an intracellular compartment and accelerate its transfer out of the cell.[86]

A possible role for mucosal ferritin in the mechanism and control of iron absorption has been envisaged. In fact, as originally proposed by Hahn et al.[87] and Granick,[88] duodenal ferritin might function as an iron acceptor and block absorption of unwanted iron. Although this model remained unproven,[89,90] the hypothesis that the amount of mucosal ferritin dictates the extent of iron absorbed by the enterocyte is still intriguing.[91]

Ferritin H and L mRNAs are detected in mucosal cells of the human gastrointestinal tract, with considerably higher levels in the duodenum than in the gastric mucosa.[78] Ferritin levels are similar throughout the duodenum and ileum despite large variations in iron absorption; furthermore, both duodenum and ileum ferritin levels are significantly decreased in iron-deficient animals.[92] These observations are in agreement with the suggestion that ferritin is simply a storage protein and does not regulate iron absorption.[93-95]

Several proteins capable of binding iron have been identified on the membranes of duodenal microvilli.[96] In this context, purification of iron-binding proteins from the rat duodenum microvillous membrane resulted in a dimer composed of two peptides, one of about 90 kDa and the other of about 150 kDa.[97] These two proteins exhibited reactivity with monoclonal antibodies that interact with integrins ($\alpha3$, $\alpha5$, $\beta3$). It is unknown whether the two proteins are able to directly bind and transport iron (Figure 1.3).

Experimental evidence suggests that the two proteins play a role in intestinal iron absorption through binding of mobilferrin. This interaction allows localization of the cytoplasmic domain of mobilferrin in a manner compatible with membrane iron transport (Figure 1.3).

Recent studies led to the identification of other proteins present on membrane microvilli and possibly involved in iron absorption. Among them, the functional role played by Nramp2 seems particularly important. This protein and its role in iron absorption were identified through two different approaches. First, a series of studies

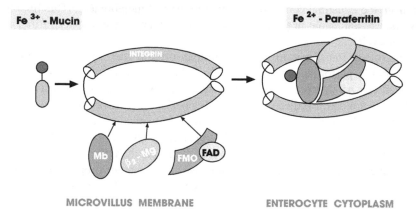

FIGURE 1.3 Model of mucosal iron uptake based first on the interaction between iron and membrane integrins, and then formation of a cytoplasmic complex (paraferritin) which involves integrins, mobilferrin, flavin mono-oxygenase (FMO), and β-2-microglobulin. The association of FMO to this complex induces the reduction of Fe^{3+} to Fe^{2+}. Fe^{2+} is then made available for absorption into the systemic circulation or for cellular metabolism. (Redrawn with modifications from Umbreit, J.N. et al., *Sem. Hematol.*, 35: 13–26, 1998.)

were carried out to identify the gene responsible for microcytic anemia in the mk/mk mouse. Microcytic anemia is a genetic disorder characterized by reduced intestinal iron absorption. Map analysis of the mk/mk mouse genome led to the identification of the Nramp2 gene as the one responsible for microcytic anemia.[98] Mutation of this gene results in the substitution of a charged amino acid for glycine.

Other studies focused on the identification and cloning of nonheme iron transporters also led to the identification of Nramp2.[99] Analysis of the pattern of expression of Nramp2 showed an ubiquitous pattern of expression. Interestingly, at the level of the intestine, Nramp2 expression was highest in the duodenum and decreased toward the colon. After diet-induced iron deficiency, Nramp2 mRNA levels markedly increased in the duodenum, but at a lesser extent in other tissues.[99] Functional studies carried out in oocytes transfected with Nramp2 cDNA showed that this membrane transporter is involved in the transport of Fe^{2+}, Zn^{2+}, Cd^{2+}, Mn^{2+}, Cu^{2+}, and Co^{2+}. It was suggested that Nramp2 may represent a key mediator of intestinal iron absorption.

Studies on mice with microcytic anemia have shown that Nramp2, although expressed in all tissues, fulfills its more relevant function in the intestine, where it contributes to dietary iron absorption.

Recent studies[100,101] suggest that duodenal cells may possess a specific iron-binding protein located at the level of apical cytoplasm and possibly involved in the mechanism of iron absorption. An initial study allowed the identification of an iron-binding protein with an approximate molecular weight of 56,000 Da purified from homogenates of rat duodenal mucosa.[100] This protein was called mobilferrin after the city in which it was isolated and also to distinguish it from other iron-binding proteins.

Its molecular size, electrophoretical mobility, monomeric structure, and amino acid composition distinguished mobilferrin from other known iron-binding proteins such as transferrin, lactoferrin, and ferritin. In fact, mobilferrin had a molecular size of 56 kDa as evaluated on SDS-PAGE electrophoresis and an isoelectric point of 4.7. Mobilferrin bound iron at an optimum pH between 5.6 and 7.6 and reached a saturation level of 1 mol/mol of protein (KD = 1×10^{-6} M). The iron binding was reversible; the dissociation of iron from the protein was favored by low pH values.

In addition to iron, other metals could bind mobilferrin, competing for the same site as iron. The relative binding levels were Fe > Cu > Zn > Co > Pb. This last observation may also explain why iron inhibits the absorption of these metals, and iron-deficient animals absorb increased quantities of certain of these metals from the gut lumen. Using affinity-purified anti-mobilferrin antibodies, it was shown that: (1) the protein was found localized in the apical cytoplasm of the duodenal cells of the rat intestine; (2) the protein was observed only at low levels in specimens from the small distal intestine; and (3) the protein was selectively present in the small intestine and not detected in other tissues. In a recent study, the same authors reported the presence of mobilferrin in human duodenal mucosa[101]; the properties of human mobilferrin are identical to those previously reported for rat mobilferrin.

However, localization of mobilferrin within the cytosol does not explain membrane transport of iron and seems to play an important role in the intracellular mechanism metabolism of iron within duodenal cells. A recent study by Conrad and coworkers[97] provides preliminary evidence that integrins are expressed on mucosal duodenal cells and are able to bind iron. This conclusion is based on the observation that biochemical isolates from microvillous preparations of duodenum from rats dosed with radioiron showed radioactivity concentrated in integrins.[97] The molecular mechanism through which integrins bind iron remains to be determined and may be at least in part mediated by binding to calcium-binding sites present on the alpha chains of integrins.

It is of interest that alpha chains of integrins are known to contain on the cytoplasmic side a receptor for calreticulin, a cytoplasmic protein. Recent studies provided evidence that calreticulin and mobilferrin are homologues, thus offering a biochemical basis to explain the capacity of mobilferrin to bind the alpha chain of the integrin receptor.[102] Pulse chase studies show that iron passages from mucin to integrin and mobilferrin and lastly to transferrin.

Mobilferrin is highly homologous to calreticulin, a soluble endoplasmic reticulum protein comprising the major storage reservoir for inositol triphosphate-releasable calcium.[103] Calreticulin fulfills several biological functions and some of them may be relevant for iron absorption. In fact, calreticulin has the ability to bind cations, particularly iron.[104] It is also of interest that under physiological conditions, repeated cycles of association with and dissociation from calreticulin play an important role in the maturation of transferrin during protein biogenesis.[105]

A series of recent studies partially clarified the possible role of mobilferrin in the intracytoplasmic transport of iron within enterocytes and other cell types. During iron transport, mobilferrin is found in the form of two molecular species in the cytoplasm. One is represented by monomeric mobilferrin, while the other is formed by a multimeric complex of about 520 KDa, where mobilferrin is complexed with integrins, β_2 microglobulin, flavin monooxygenase, and other not yet identified polypeptides[106] (Figure 1.3).

Biochemical studies showed the presence of a NADPH-dependent flavin monooxygenase ferrireductase activity that reduces Fe^{3+} to Fe^{2+}. Interestingly, antibodies against mobilferrin or α-integrin inhibit the ferrireductase activity.[106] Since the inhibition of monooxygenase activity leads to a decreased intestinal iron uptake, it was suggested that paraferritin may play a role in the cytoplasmic transport of iron within enterocytes and in iron delivery to cellular constituents in an appropriate redox state[106] (Figure 1.3). Studies on human erythroleukemia K562 cells[107] and rat reticulocytes[108] have shown the existence of mobilferrin which acts in these cells to mediate cytoplasmic iron transport in the context of an alternate iron uptake pathway involving membrane integrins, mobilferrin, and ferritin.

On the basis of all these observations, a model of protein regulation of iron absorption was postulated.[109] Dietary nonheme iron is solubilized at the acidic pH of the stomach. At acidic pH, solubilized iron combines with mucins and is thus made available for absorption within a tissue milieu characterized by neutral pH (small bowel). Integrins present in the membranes of brush border cells facilitate the binding and then the transfer of iron through the cell membrane; iron is then bound by a cytosolic duodenal protein, mobilferrin. This last protein has the function of an iron shuttle protein within the duodenal cells. Iron is then transferred from mobilferrin to ferritin or to plasma transferrin through an unknown process.

Basolateral transferrin receptors may play an important role in regulating the transfer of iron from plasma into the enterocytes to keep the cells informed of body requirements for iron. Following this theory, the absorptive process appears to be driven by a cascade of differences in the binding constants of these various proteins so that iron moves from plasma mucin to mucosal mobilferrin to plasma transferrin.

Iron that arrives in the duodenum bound to heme (mostly hemoglobin heme) is absorbed directly by the mucosal cells in the form of the intact iron–protoporphyrin complex. Heme that enters the mucosal duodenal cells is then enzymatically split by heme oxygenase and releases iron; the splitting of iron from heme represents the rate-limiting step in the process of heme iron absorption.[110] Since most dietary heme (primarily meat heme) iron is absorbed as low molecular weight nonheme complexes formed during the digestion of heme, it is not surprising that the absorption of dietary heme iron competes with that of dietary nonheme iron. In fact, animals preloaded with inorganic iron scarcely absorb heme and nonheme iron.[111,112] Furthermore, the parenteral administration of the iron chelator desferrioxamine blocks the absorption of both heme and nonheme iron.[34]

1.5 PHYSIOLOGIC REGULATION OF IRON ABSORPTION

The adaption of the body to iron deficiency or iron overload occurs primarily through modulation of the extent of iron absorption by the intestinal mucosa. The condition of iron deficiency is associated with an adaptive response consisting of an increased iron absorption; whereas the opposite response occurs under iron overload conditions. Compared to iron absorption observed under normal conditions, the relative enhancement of absorption during iron deficiency is greater than inhibition during iron overload.

Two factors seem of primary importance in regulating the rate of intestinal iron absorption. The first factor is represented by the level of iron stores. When these stores are depleted, iron absorption is significantly increased; when iron stores are at supranormal levels, iron absorption is decreased. The second factor is the level of erythroid cell production in bone marrow; in fact, intestinal iron absorption is significantly increased when the rate of erythropoiesis is increased, and absorption is decreased when red cell production is depressed.

The roles played by these two factors in regulating iron absorption are directly supported by several experimental evidences. However, the cellular and molecular mechanisms through which these two factors determine the control of intestinal iron absorption are poorly defined.

Regulation of iron absorption may be accomplished by alterations in the number of mucosal cells adapted to absorb iron or changes in the individual mucosal cell's capability to absorb iron or both. The second hypothesis is more popular. Several theories were proposed to explain the biochemical mechanisms undergoing the control of iron absorption by mucosal intestinal cells.[113,114] The most popular was that initially proposed by Hahn,[115] subsequently revised by Granick,[116] and now known as the "mucosal block theory." According to this model, the number of iron acceptor molecules present in the intestinal mucosa is limited; as a consequence, if these acceptors are saturated either by exogenous or endogenous iron, a block of dietary iron absorption takes place.[115]

An extended version of this theory proposed by Granick suggested that the ferritin–apoferritin system in mucosal intestinal cells directly controls the amount of iron absorbed.[116] According to this theory, apoferritin should represent the intestinal iron acceptor molecule, and ferritin was considered an obligatory intermediate during the iron absorption process.[116] This theory lost favor since no apoferritin or ferritin was observed on brush border intestinal membranes. Furthermore, the presence of little apoferritin in the intestinal mucosae of iron-deficient animals and large quantities in specimens from iron-overloaded animals is inconsistent with the hypothesis that this protein acts as a receptor to regulate iron absorption.

In contrast to the mucosal block theory, several studies have shown that iron continues to be absorbed even when tissues are laden with deposited iron. Although ferritin cannot be regarded as an iron-binding acceptor, its concentration in the mucosal intestinal cells significantly decreases in conditions of iron deficiency and markedly increases in conditions of chronic iron overload.[117-119]

The observation that iron localized at the level of epithelial cells of intestinal mucosa was derived from dietary iron and body stores[120,121] led to the hypothesis that the quantity within these absorptive cells regulated iron absorption in the intestinal mucosa. This theory was supported by studies showing a decreased concentration of nonheme iron in small intestinal mucosa obtained from animals with increased iron absorption in association with either iron deficiency, enhanced erythropoiesis, or hypoxia, and increased nonheme iron concentration in the small intestines of iron-overloaded animals with decreased iron absorption.[120-122]

Experiments using isolated duodenal epithelial mucosal cells challenged this hypothesis, and showed less marked differences between the iron concentration in epithelial cell isolates and measurements of nonheme iron concentrations of whole intestinal segments. It was postulated that the differences originally reported were seemingly caused by changes in the quantity of iron in the laminae propria and submucosae of the intestinal specimens rather than in the mucosal absorptive cells.[123]

Ultrastructural methods using Prussian blue staining provided evidence about low levels of stainable iron within the small intestinal mucosal cells of iron-deficient animals in comparison with specimens from normal fasting rats; in contrast, the mucosal cells of iron-overloaded animals were intensely stained.[124] These observations showed that the quantity of iron within small intestinal mucosal cells varies directly with the degree of iron repletion and suggests that this phenomenon may play a role in the regulation of iron absorption.

However, since free ionic iron is a toxic oxidant, it is difficult to believe that ionic iron directly inhibited iron uptake by the intestinal mucosal cells. Thus, it was postulated[119] that this iron might saturate iron-binding receptors of some molecules within the mucosal cells that acted as regulators of iron absorption.

To test this hypothesis, Conrad et al.[119] evaluated the total iron-binding capacity and unsaturated iron-binding capacity of aliquots of homogenates of small intestinal mucosae derived from normal, iron-deficient, and iron-loaded animals. There were no significant differences in the total iron-binding capacities of normal, iron-deficient, or iron-loaded animals; whereas the unsaturated iron-binding capacity varied inversely with the degree of iron repletion of the animals. This is consistent with the hypothesis that the mucosal uptake of iron from the lumen of the gut is regulated by the number of unbound receptors for iron either within the cell or in close relationship to the cell surface.

The existence of an iron-dependent mechanism of control of intestinal iron uptake was also recently demonstrated in human enterocytes isolated from duodenal biopsies.[125] Normal enterocytes were isolated from subjects with normal iron stores, as evaluated according to plasma ferritin levels. The rate of iron uptake was significantly greater in enterocytes isolated from individuals with low serum ferritin levels as compared with those with normal ferritin levels.[125] Interestingly, zinc, lead, cobalt, and manganese added to the incubation medium significantly lowered iron uptake into cells (Fe > Zn > Pb > Co > Mn). This observation suggests that the recently identified metal cation transporters such as Nramp2 may play relevant roles in intestinal iron uptake.

Another theory suggests that iron absorption is regulated by a balance mechanism based on the probability of exchanges occurring between plasma and the various cellular iron pools.[126] The probability that an iron atom will be picked up by a plasma transferrin molecule is considered proportional to the plasma iron turnover rate, and in any given tissue, the probability is related to the ratio of the exchangeable iron in that tissue to the total body exchangeable iron. Thus, iron absorption would be proportional to the ratio of intestinal exchangeable iron to total exchangeable iron multiplied by the plasma iron turnover. Because total exchangeable iron would be directly related to iron stores, the hypothesis would account for the effects of iron deficiency or iron overload on absorptive rate.

The effect of erythropoietic rate on iron absorption is accounted for on the basis of its effect on plasma iron turnover. More recently, an additional theory of iron absorption control was proposed.[11] This theory, while largely based on speculation, could account for iron absorption modulation. The theory is based on the idea that iron absorption is controlled through mechanisms similar to those responsible for iron-dependent control of the synthesis of ferritin, transferrin, and transferrin receptors. Thus, in line with this theory, it can be hypothesized that mucosal cells developing during iron deficiency would be regulated in such a manner that they mature with increased iron-absorbing capabilities (i.e., increased expression and/or activity of iron transporters localized in the apical membrane) (Figure 1.4).

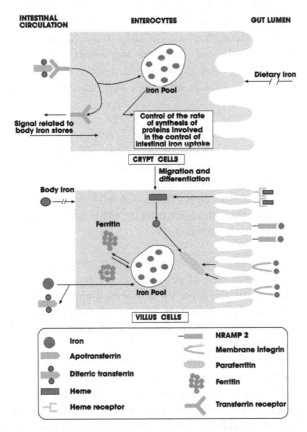

FIGURE 1.4 A current model for the control of intestinal iron absorption. This model envisages that signals coming from the body are capable of modulating the capacity of enterocytes to absorb iron. These signals are received by the enterocytes present at the level of the intestinal crypts and modify the capacity of these cells to synthesize a series of proteins (Nramp2, HFE, transferrin receptors) involved in the control of iron absorption. This process takes place during differentiation and migration of crypt intestinal cells from the crypt to the tops of intestinal villi. In line with experimental evidence, this model implies that the proteins playing a key role in the absorption process are synthesized only after the crypt cells complete the differentiation process and become mature enterocytes localized at the level of the brush border. An important aspect of this model is that immature enterocytes (crypt cells) can respond to body signals and do not acquire iron from the gut lumen, while mature enterocytes acquire iron from the gut lumen but appear virtually unable to receive iron from the body.

Under this condition of iron deficiency, mucosal cells will develop with increased capacity to bind iron required for their growth (i.e., increased expression of transferrin receptors present on the basal membranes of mucosal cells). Mucosal cells developing during periods of iron overload will not fully develop their absorption capabilities[11] (Figure 1.4). These cells will develop contemporaneously with increased ferritin levels; iron removed from the circulation is stored in ferritin within mucosal cells until its excretion with exfoliated cells.

Although the molecular mechanisms regulating iron absorption remain to be determined, it is clear that the entire process of iron absorption is a highly regulated process. This phenomenon is part of a more general mechanism of regulation of intestinal transporters[127] through which the body tunes the absorption of these substances to metabolic needs and faces varying conditions (dietary variations) (Figure 1.4). It is of interest that intestinal transporters provide numerous systems for studying adaptive regulation in eukaryotic cells and afford evidence for some general regulatory patterns:

1. Specific transporters of water-soluble vitamins are down-regulated as dietary levels of substrates increase or their body stores increase.
2. Specific transporters of essential trace minerals are modulated as are the above-mentioned transporters.
3. Specific transporters of sugars and amino acids are usually up-modulated as dietary levels of substrates increase (whereas hepatic receptors for these compounds are down-modulated).

As previously mentioned, the three main factors that modulate the rate of intestinal iron absorption are the level of body iron stores, the rate of erythropoiesis, and hypoxia. Concerning the first of these three factors, it must be pointed out that body iron stores are known to be important because of the close inverse relationship demonstrated between serum ferritin and iron absorption in normal subjects.[128-130] In this context, a recent study carried out by Cook et al. is particularly relevant.[131] These authors examined the relationship of serum ferritin, soluble transferrin receptor, and iron absorption (inorganic iron and dietary nonheme iron) in 234 healthy subjects. With both forms of iron, the correlation with absorption was far lower for transferrin serum receptor (r = 0.369) than for serum ferritin (r = 0. 803) and was no longer significant when subjects with depleted iron stores were excluded.[131] These observations indicate that in normal subjects the level of iron stored into the body represents the main physiological determinant of iron absorption and that, in absence of iron deficiency, tissue transferrin receptor mass, reflected by serum transferrin receptor levels, has no significant influence.[131]

During pregnancy an increase in iron absorption is observed — a phenomenon associated with a decrease of maternal iron stores. A major exception to the rule that an increase in body iron stores leads to a decrease in iron absorption is found in hereditary hemochromatosis, where the pathological defect results in disproportionately high iron absorption relative to iron stores.

Concerning the second factor (rate of erythropoiesis) it must be pointed out that a close parallel exists between intestinal iron absorption and the rate of erythropoiesis.[132]

Iron absorption is enhanced when erythropoiesis is stimulated as a compensatory response to either bleeding or acute hemolysis.[133,134] However, intestinal iron absorption significantly decreases when the rate of erythropoiesis is decreased by hypertransfusion[133] or radiation treatment.[135] In chronic hemolytic states in humans, iron absorption is usually normal.[136] Only in chronic anemias, such as β-thalassemia intermedia and major and sideroblastic anemia, where erythropoiesis is highly ineffective, iron absorption remains persistently high. Iron overload in patients with thalassemia intermedia who received few or no blood transfusions is clearly due to continued enhanced intestinal iron absorption, a phenomenon causally related to the enhanced erythroid mass in these patients.[137]

The third factor is hypoxia. In fact, hypoxia is associated with an increase in iron absorption independent of changes in the rate of erythropoiesis.[138-141] The effect of hypoxia on iron absorption occurs before a response from the erythropoietic bone marrow.[142] Chronic hypoxia elicits a rapid and reversible increase in intestinal iron absorption. Kinetic analysis of Fe^{3+} uptake provides evidence of a two- to three-fold increase in V_{max}, suggesting an increased number of brush border iron carriers.[143,144] In order to rule out the role of erythropoiesis, further studies were carried out in mice in which the erythroid responses were reduced either by splenectomy and marrow ablation or by partial nephrectomy.[117]

In spite of the marked inhibition of erythropoiesis observed in these animals, iron absorption was enhanced and further augmented if the animals were exposed to hypoxic conditions.[143] The effects of hypoxia and changes in erythropoiesis on the absorption of $^{59}Fe^{3+}$ from in situ tied-off duodenal segments were studied in mice.[138] Hypoxia elicited an increase in mucosal uptake within 6 h, while mucosal transfer was unaffected for the first 20 h, thus suggesting independent regulation of the two processes. Mice exposed for 3 days to hypoxia exhibited stimulated erythropoiesis and a two- to three-fold increase in the mucosal uptake of ^{59}Fe.[138]

Another study showed that 3 days of hypoxia elicited a six-fold increase of the rate of iron accumulation during transit along the lower villus and a three-fold maximal accumulation of iron.[144] In partially nephrectomized mice (where no significant reticulocyte response to hypoxia was seen) and in splenectomized and ^{89}Sr-treated mice (with marrow ablation), hypoxia produced a two- to three-fold increase in total mucosal uptake with respect to the controls.[138] It is of interest that in ^{89}Sr-treated mice, the absolute amounts of mucosal iron uptake and the proportions transferred to the carcasses of the animals were significantly lower than in comparable splenectomized controls.[138] These results suggest that hypoxia enhances intestinal iron absorption by a mechanism that is independent of stimulated erythropoiesis, and that the rate of erythropoiesis has an additional effect, particularly on the transfer phase of iron absorption.

1.6 IRON BIOAVAILABILITY

In the context of iron nutrition, variations in food iron bioavailability, largely determined by the dietary source of iron, are of greater importance than the absolute amount of iron ingested. While absorption of the heme iron pool is virtually unaffected by dietary compounds, the iron contained in the common nonheme pool is subject to a balance of promotive and inhibitory luminal factors. Food is generally inhibitory to nonheme iron bioavailability. A paradigm of this general assumption is given by a series of classic observations indicating that bioavailability of nonheme iron is much higher when iron is administered alone as compared with iron administered with food.[145] Thus, individuals with low iron stores usually absorb 40% of a reference dose of radioiron in the form of ferrous ascorbate, while absorption of radioiron is usually within a range of 10 to 20% when the reference dose of ferrous ascorbate is administered together with a standard meal dose.[146]

Biovailability of nonheme iron is largely determined by dietary components that can interact with iron-forming complexes with a large spectrum of solubility and capacity to be absorbed by the intestine. More specifically, the bioavailability of nonheme iron from the diet is dependent on the interaction of the promoters and inhibitors of iron absorption contained within that diet.[147] According to Clydesdale[148] and Fairweather–Tait,[149] dietary ligands that promote iron absorption form soluble and stable complexes able to release iron easily and make it available for uptake by intestine. On the contrary, dietary ligands that inhibit iron absorption form insoluble or soluble complexes that bind iron so tightly that it cannot be taken into the intestinal mucosal cells.

1.6.1 ENHANCERS OF NONHEME IRON ABSORPTION

The major enhancers of nonheme iron absorption from the common pool are meat and organic acids, including ascorbic acid, citric acid, and lactic acid.

Ascorbic acid. Two potent promoters of nonheme iron absorption are meat and ascorbic acid,[150] and a number of studies document the relative quantities of these two substances to be the most important determinants of iron bioavailability from Western-type meals.[151] Ascorbic acid enhances iron absorption from the diet through several independent mechanisms:[152,153]

1. Ascorbic acid converts ferric iron to the ferrous state, which maintains its solubility in the alkaline environments of the upper gut.
2. Ascorbic acid favors acid conditions within the stomach so that dietary iron can be sufficiently solubilized.
3. In the acid environment of the stomach, ascorbic acid forms a chelate with ferric iron and this complex maintains solubility when food enters the alkaline environment of the small intestine.
4. Ascorbic acid prevents the dose-dependent inhibitory effects of polyphenols and phytates on nonheme iron absorption.

To achieve an increased iron absorption, it is important that at least two mechanisms cooperate in a synergistic fashion. Enhancement of iron absorption due to chelation or reduction is only possible after dietary iron has been solubilized in the acid milieu found in the stomach. Second, like other dietary promoters or inhibitors of iron absorption, it is important to point out that final effect of ascorbic acid in promoting iron absorption depends on the types and concentrations of other dietary compounds inhibiting or stimulating iron absorption.

Particularly relevant in this context is a recent study by Siegenberg et al.[153] focused on evaluating the possible effect of ascorbic acid in preventing the inhibitory effects of polyphenols and phytates on nonheme iron absorption. These authors showed that iron absorption decreased progressively in 199 subjects when maize bran containing increasing amounts of phytate phosphorus was given; this inhibitory effect was overcome by 30 mg ascorbic acid. Similar observations were made with tannic acid, with the exception that a higher dose of ascorbic acid was required to prevent this inhibitory effect.[153]

Meat. It is well known that iron present in meat is relatively well absorbed. This phenomenon is largely related to the fact that iron in meat is bound to heme. In fact, Kalpalathika et al.[154] reported 62% heme iron in roast beef. Bioavailable heme plus nonheme iron from 120 g beef totaled 0.67 to 1.13 mg Fe, assuming the beef contained 55% heme iron and an individual with iron stores of 300 mg absorbed 28% heme iron.

However, it must be noted that heme in meat is well absorbed in that is bound to muscle proteins. This conclusion is directly supported by a series of observations showing that iron from heme alone is poorly absorbed, but when supplied as hemoglobin,[155] or with meat,[156] or with soy protein,[157] it is well absorbed. The most likely interpretation of these observations is that proteins maintain heme in a monomeric state, thus preventing the formation of poorly absorbed macromolecular heme polymers.[158]

The peculiar role of meat in promoting iron absorption is particularly well supported by a recent study carried out in sedentary women undergoing a 3-month moderate aerobic excercise program and randomly assigned to four dietary groups.[159] Meat supplements were significantly more effective in protecting hemoglobin and ferritin status than were other types of iron supplements.[159]

The beneficial effect of meat on iron bioavailability is not only related to the fact that meat alone represents a good source of bioavailable food iron. Meat also improves the level of nonheme iron absorption from the common iron pool. This beneficial effect on nonheme iron absorption seems to be specifically related to muscle proteins, since other animal proteins do not display this effect.[158] The final result of this action is that meat facilitates iron absorption from a variety of dietary compounds exhibiting low availability.

The effect of meat on nonheme iron absorption in infants was also investigated.[160] Nonheme iron absorption was significantly increased from a vegetable puree supplemented with 25 g of meat compared with results from pureed vegetables alone. This observation indicates that meat is also an enhancer of nonheme iron absorption in infants and that nonheme iron absorption from weaning foods can be augmented with additional meat.

A large number of studies, reviewed in a series of articles by Cook et al.[161] indicate that meat enhancement of iron absorption operates via multiple mechanisms involving an increase in gastric acidity and iron chelation/solubility. Thus, following this hypothesis, during the early phases of digestion, meat may improve iron absorption by enhancing the production of gastric acid, promoting iron solubilization within the stomach through this mechanism.[162] In subsequent stages of digestion, meat stimulates iron absorption by chelating ferric iron, as directly supported by studies indicating that meat contains one or more factors able to bind ferric iron and maintain it in solution at neutral pH.[163]

1.6.2 INHIBITORS OF NONHEME IRON ABSORPTION

Several dietary compounds that limit nonheme iron absorption from the common iron pool have been identified. These factors include phytates, polyphenols, dietary fiber complexes, calcium, phosphorus-containing compounds, and other minerals.

Phytates. Phytates constitute 1 to 2% of many cereals, nuts, and legumes. Their biological role is preventing accumulation of excessively large amounts of inorganic phosphorus during seed maturation by acting as phosporus stores. Furthermore, phytic acid functions in plants as a vehicle to store cations such as iron, zinc, calcium, copper, and magnesium. The observation that consumption of wheat bran reduces iron absorption led Widdowson and McCance[164] to suspect that phytate may be an important inhibitor of nonheme food iron absorption.

Subsequent investigations with bran confirmed the inhibitory effect, but other studies yelded contradictory results as to the inhibitory nature of phytate. In one study,[165] the reduction of phytate in wheat bran was reported to have no effect on nonheme iron absorption. Monoferric phytate, which represents half of the iron in wheat bran,[166] was reported to be well absorbed. In other studies, the reduction of phytates in wheat bran improved iron absorption.[167] Adding phytic acid to wheat bran rolls inhibited iron absorption dose dependently.[168]

The role of phytate in modifying nonheme iron absorption from soy products was even more unclear as neither the removal of phytate from soy flour by acid washing[169] nor a two-fold variation in the phytate content of soybeans produced under different growing conditions influenced nonheme iron absorption.

A recent study provided definitive evidence that phytates exhibited a marked inhibitory effect on iron absorption.[170] The effect of reducing the phytate in soy protein isolates on nonheme iron absorption was examined in 32 human subjects. Iron absorption increased four- to five-fold when phytic acid was reduced from its native quantity of 5 to 8 mg to <0.01 mg/g of isolate.[170] Also of considerable interest was the observation that even relatively small amounts of residual phytate were strongly inhibitory for iron absorption.[170] Even after removal of virtually all the phytic acid, iron absorption from the soy protein meal was still low compared to absorption observed with other meals.[170]

The mechanism responsible for the inhibitory effect of phytic acid on iron absorption is dependent upon the formation of scarcely soluble and digestible complexes of iron, phytic acid, and proteins or fibers.[171] Torre et al.[172] reviewed the different factors that may affect the bioavailability of phytic acid iron complexes and noted three: (1) dietary content of iron and phytates; (2) existence of different dietary compounds able to form complexes with iron and phytates such as proteins, fibers, and minerals; and (3) the pH value.

The negative effects of phytates on iron absorption can be counterbalanced by other dietary compounds, including meat and ascorbic acid. Interestingly, a recent study showed that vitamin A and β-carotene may form a complex with iron, keeping it soluble in the intestinal lumen and preventing the inhibitory effects of phytates and polyphenols on iron absorption.[173] This conclusion was based on the observation that vitamin A and β-carotene can improve nonheme iron absorption from rice, wheat, and corn in normal subjects.

Polyphenols. Another group of compounds capable of exerting inhibitory effects on iron absorption are the polyphenols or tannins present in certain spices, fruits, vegetables, and tea. A large number of phenol compounds, members of a family of secondary plant metabolites, have been described. These compounds can be separated into two categories: hydrolizable and non-hydrolizable polyphenols. Both types of polyphenols have been shown to produce marked inhibitory effects on nonheme iron absorption.

This was first suggested in 1975, when tea was shown to inhibit iron absorption in humans.[174] Subsequent *in vitro* and *in vivo* work provided evidence that the polyphenol content of tea is the inhibitor.[175] Further studies provided evidence of a negative correlation between the bioavailability of iron from different foods. Polyphenol content was shown *in vitro* for a variety of legumes and spices[176] and *in vivo* for a number of vegetables.[177]

The inhibitory effect of polyphenols on iron absorption is dose dependent.[178] There was a marked decrease in iron absorption over the polyphenol dose range of 10 to 50 mg; at higher polyphenol doses, only a slight additional increase in the inhibition of iron absorption was observed. The inhibitory effect is probably due to the polymerization of polyphenols with iron, and to the consequent formation of insoluble complexes.[175] From a chemical point of view, the presence of galloyl groups, each containing three phenolic hydroxyls, has been hypothesized to be responsible for the binding of iron by polyphenols.[179] As seen with phytates, the inhibitory effects of polyphenols on iron absorption can be counterbalanced by iron absorption enhancers such as meat and ascorbic acid.

Dietary fiber complexes. Dietary fiber can be defined as the dietary component that is resistant to digestion by the endogenous secretions of the upper digestive tract. More particularly, dietary fiber is that portion of a food carbohydrate (usually materials derived from the cell walls of plants such as cellulose, hemicellulose, pectin, lignin, and gums) which cannot be digested by enzymes secreted by the host and normally present in the gastrointestinal tract.

Some evidence indicates that complex carbohydrates reduce the absorption of certain minerals, but recent reviews[180,181] show that the results of human studies are conflicting. Problems arise from the different levels of complex carbohydrates fed — some of the intakes bear little relation to the human dietary situation.[182] The use of different sources (bran, pectin, cellulose, and hemicellulose) and different foods (cereals, fruits, and vegetables) also makes overall interpretation difficult. Concurrent changes in mineral intake associated with consumption of foods high in complex carbohydrates add to the confusion. Furthermore, many foods also contain phytate; thus, it is difficult to separate the effects of these two components. However, all the recent experimental evidence points to phytate rather than complex carbohydrates as a major determinant of iron bioavailability. The situation is further complicated because different methods of analysis were used to measure the intake of complex carbohydrates, and different techniques were employed to assess mineral absorption.

However, in spite of these limitations, results of most human studies have shown that complex carbohydrates do not have significant effects on iron balance,[183] although bran (and not pectin or cellulose) was shown to decrease iron absorption.[184]

Using the intrinsic tag technique and two different radiotracers, Rossander et al. compared iron absorption from meals with and without additional fiber from citrus pectin in the same subjects.[185] Fiber supplementation did not alter the levels of iron absorption in these subjects. In line with these observations, a recent study by Mason et al.[186] provided further evidence that complex carbohydrates do not seem to impair iron absorption. Fifteen adult women were given diets in which the intake of complex carbohydrates (derived from cereals, vegetables, and fruit) was increased from 20 to 30 g over a 12-week period. The changed diet had no influence on the intake of both iron and zinc.

Complex carbohydrates appeared to have no significant effects on iron absorption in animal models. The effects of guar gum supplements on the uptake of iron by isolated loops of rat duodenum isolated *in situ* were studied.[187] In a 10-week experiment, no effect of feeding guar gum on iron retention was observed.[187]

Finally, studies in subjects ingesting diets rich in fiber for long periods of time provided further evidence that complex carohydrates do not significantly impair iron absorption. For example, vegetarians showed no evidence of depressed iron status despite high intakes of complex carbohydrates,[188] and epidemiological studies indicate that populations habitually consuming diets high in complex carbohydrates do not have poor mineral status.[189]

Calcium. An interaction between dietary iron and calcium has been investigated with variable results. Elevated calcium intake has been reported to reduce iron utilization by 0 to 73% in humans.[190-197] In a controlled study in post-menopausal women, whole-body retention of ^{59}Fe from a test meal was significantly lowered by ingestion of calcium carbonate, calcium phosphate, milk, and calcium citrate malate tablets,[190,191] but not by calcium citrate malate tablets in orange juice.[191] However, a recent study provided evidence that over a 12-week period, use of 1 g of calcium carbonate daily with meals did not appear to be detrimental to iron stores in healthy pre-menopausal women.[198]

In a recent study, a rat hemoglobin repletion assay and an *in vitro* digestion procedure were used to assess the effects of four salts (calcium carbonate, calcium sulfate, sodium carbonate, and sodium sulfate) on iron bioavailability.[199] Calcium carbonate had the greatest depressive effect on iron bioavailability, depressing hemoglobin iron gain in a dose-dependent manner. Calcium sulfate and sodium carbonate also depressed hemoglobin iron gain but to a lesser extent, whereas sodium sulfate did not affect hemoglobin iron gain. On the basis of these results, it was concluded that significant interactions of cations, anions, and salt concentration were found, suggesting that both the cations and the anions in calcium carbonate contribute to the iron-depressing effect of this salt.[199]

Differences in iron utilization from meals containing cow milk and human milk may be partly due to their dissimilar calcium contents. Iron absorption doubled in weanling rats when iron-supplemented cow's milk was replaced with supplemented human milk.[200] When calcium chloride was added to human milk to equalize the calcium content, iron absorption from the supplemented human milk was similar to that of cow's milk.[200]

The mechanism responsible for the inhibition of iron absorption mediated by calcium is not clear. One mechanism may be the fact that calcium salts may change the physico-chemical state of iron salts, rendering them less available for iron absorption.

A second mechanism may be that iron and calcium compete for common mucosal receptors or that calcium inhibits the movement of iron from mucosa into circulation.

Phosphorus-containing compounds. Foods with relatively high phosphate content, such as milk and eggs, represent less effective sources of available iron. Particularly relevant to this issue is the study of Greger et al.[201] who reviewed a large number of phosporus-containing compounds and their effects on iron availability. Their study showed that inconsistent results reported in the literature were apparently related to differences in the types of phosphorus compunds, interactions with other nutrients, and methods used to assess iron availability.[201]

Other studies investigated the effect of phosphorus content on iron milk bioavailability. The higher content of phosphate in cow milk, compared with human milk, is one of the major determinants of lower iron availability from cow milk. Cow milk contains 60 to 80 mg of phosphorus/10 g, most of which is phosphate and is associated with casein micelles. Studies in animal models showed that orthophosphates, pyrophosphates, and polyphosphates reduce iron availability.[202,203] Studies in humans provided contradictory results.

Monsen and Cook[194] investigated the effects of calcium and phosphate on absorption of radiolabeled iron from breakfast meals. Iron absorption was not significantly affected by calcium chloride or potassium phosphate alone. However, when combined, these two salts elicited significant inhibition of iron absorption by reducing iron bioavailability.[194] It was suggested that iron may form an insoluble complex with calcium phosphate.

In contrast to this study, Snedeker and coworkers[193] evaluated iron balance in nine subjects fed diets containing different levels of calcium and phosporus, and observed that fecal excretion and retention of iron were not significantly affected by either phosphorus or calcium levels. Here again, as noted in other studies on iron absorption in humans, the discrepancies between the two studies may be largely ascribed to differences in methodologies used to measure availability, types of calcium and phosphorus compounds used, or differing levels of ascorbic acid and other enhancers of iron absorption in the diet.

Other minerals. Several studies indicate that excesses of other minerals in the diet may exert inhibitory effects on iron absorption, and this phenomenon may be in part mediated through a competitive mechanism. In this context, particularly well investigated are the interactions between dietary manganese and iron.

Many investigators observed interactions between iron and manganese. Intestinal absorption of iron and manganese is increased during iron deficiency.[204,205] Mena found that anemic subjects absorbed 7.5% of ingested manganese compared to 3% by normal subjects. The enhanced manganese absorption was inhibited by infusion of iron.[204]

However, the effects of iron overload on manganese absorption are less clear. Iron supplementation did not affect tissue manganese concentrations in calves[206] and chicks,[207] but decreased tissue manganese concentrations in mice,[208] sheep,[209] and rats.[210] Iron supplementation in humans resulted in increased fecal manganese losses and decreased manganese retention.[211]

The effect of manganese supplementation on iron utilization is also debatable. Hartman et al.[212] and Davis et al.[210] observed that animals given supplemental manganese had depressed hematocrit values. Black et al.[213] observed that the ingestion of supplemental manganese depressed liver iron concentrations in chicks. However, Davis et al.[210] and Baker and Halpin[207] observed no changes in liver iron concentrations in rats and chicks, respectively, in response to manganese supplements.

Particularly relevant is a recent study carried out by Davis et al.[214] These authors investigated the relationship of dietary intakes of heme iron, nonheme iron, and manganese on indices of hematological and nutritional status of manganese in 47 women. Increasing dietary iron intake by consuming more nonheme iron had questionable effects on hematological status and negative effects on nutritional status in regard to manganese. In contrast, heme iron intake was positively correlated with hematological status and had no consistent effects on indices of manganese or iron status, possibly because foods that contain significant amounts of manganese (green vegetables and cereals) often contain significant amounts of nonheme iron.

1.7 ABSORPTION OF HEME IRON

As previously mentioned, heme iron is absorbed from food more efficiently than nonheme iron. These differences are mainly related to the difficulty of maintaining nonheme iron in solution and to the presence in foods of inhibitors of inorganic iron absorption, while iron is more soluble and more readily absorbed.

A series of observations in humans and animals show that hemoglobin and other heme-containing proteins are digested to heme and to protein fragments in the lumen of the proximal small intestine; the protein fragments are required to maintain heme solubility.[215] Heme is insoluble at acidic pH and soluble at alkaline pH. Thus, in the acidic pH environment of the proximal small intestine, it is particularly important that heme is complexed with protein fragments that ensure its solubility at acidic pH.[215] In line with this observation, heme is scarcely absorbed from purified preparations, while it is consistently absorbed from hemoglobin and purified preparations of heme administered with peptides and amino acids.

Heme enters enterocytes as an intact molecule (Figure 1.5). This conclusion was supported by experiments in which an iron chelator was administered orally and its effect on the absorption of food nonheme and heme iron was investigated. The iron chelator markedly inhibited the absorption of inorganic iron, but did not modify the extent of absorption of heme iron.[216]

The mechanism through which heme binds to and enters the enterocytes is poorly understood. The molecular mechanisms by which heme traverses the cell membranes are largely unknown. A series of studies carried out on several cell types including enterocytes have shown that the cells internalize heme through a process involving a membrane receptor.[217-221] This conclusion was based on a series of observations indicating that:

1. Heme binding to cells is a saturable phenomenon mediated through high-affinity interactions.
2. Binding to cells is relatively specific for heme as compared to other porphyrins.
3. Heme binding is greatly inhibited or abolished by trypsin pre-treatment of the cells.

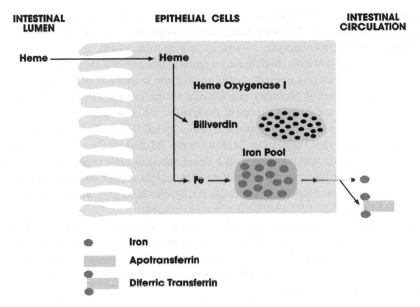

FIGURE 1.5 Schematic mechanism of heme iron absorption by enterocytes. Heme is released from food by different proteolytic enzymes in the stomach and duodenum and then is absorbed by enterocytes present at the level of brush border, probably through interaction with a specific membrane receptor. Heme is degraded within the enterocytes by heme oxygenase 1, giving rise to the release of biliverdin and inorganic iron. Iron is then transported through the basal membrane into the circulation where it is transported by transferrin.

These observations, however, while interesting, did not lead to the identification and characterization of a membrane receptor capable of binding heme.

After internalization into enterocytes, heme is degraded through the action of heme oxygenase and gives rise to iron, biliverdin, and carbon monoxide (Figure 1.5). Several lines of evidence suggest that degradation of heme by heme oxygenase represents an important event in the mechanism of iron absorption by intestinal microvilli:

1. Enterocytes possess elevated levels of heme oxygenase, whose synthesis is increased following exposure of the cells to heme or metal ions.[222-224]
2. Effects of long-term treatment with heme oxygenase inhibitors, such as tin mesoporphyrins, analogous to the anemic condition observed in patients with Crigler–Najjar type I syndrome undergoing treatment in an attempt to ameliorate severe jaundice. This treatment resulted in diminished uptake of ^{59}Fe from radiolabeled heme in the gut.[225]
3. Iron deficiency induced a significant potentiation of the capacity of intestinal microvilli to split heme through heme oxygenase activity.[222]
4. Mice lacking heme oxygenase 1 develop marked serum iron deficiencies.[226]

Mammalian heme oxygenase is represented by two isoforms, heme oxygenase 1 and heme oxygenase 2, encoded by separate genes. Growing evidence indicates that heme oxygenase 2 may have more a regulatory function at the level of the nervous system through the generation of carbon monoxide which acts as a physiological signaling molecule. On the other hand, heme oxygenase 1 is more involved in the antioxidant defense mechanisms as well as in iron metabolism. It is of interest that, at the level of intestinal cells, both heme oxygenase isoenzymes are observed.[227] Heme oxygenase 2 is selectively present at the level of the interstitial cell networks of the small intestine (where it probably plays a role in the control of intestinal motility), while heme oxygenase 1 is distributed at the level of the cells of the intestinal mucosa.

Experiments carried out on intestinal cells labeled with radiolabeled heme have provided evidence that three main proteins were labeled.[215] One was a membrane protein, seemingly corresponding to the heme receptor. The other two were cytoplasmic proteins of lower and higher molecular weights. The low molecular weight protein seemingly corresponds to mobilferrin, while the higher weight protein may correspond to the 520-kDa paraferritin protein complex.

It was clearly shown in fasting subjects that the absorption of nonheme iron and heme iron occurs with comparable efficiency. However, when test doses of nonheme or heme iron were supplied with food, the amount of iron absorption from nonheme iron markedly decreased, while the amount of iron absorbed from hemoglobin remained virtually unmodified.[215]

This phenomenon was largely dependent upon the presence of inhibitors of nonheme iron absorption in food; these compounds, however, did not affect the absorption of iron from heme iron sources. It is well known that meat induces enhancement of iron absorption of both inorganic and organic (heme) iron, a phenomenon related to the capacity of peptides or amino acids released from meat digestion to bind to inorganic and organic iron and maintain the iron in a molecular form suitable for absorption.

The problem of measuring iron absorption from the whole diet was investigated more recently using appropriate methodology.[228,229] The methodology was based on a separate uniform labeling of nonheme iron and heme iron in all meals. Results indicated that the absorption of both heme and nonheme iron is influenced by iron status. In subjects who had iron depletion (i.e., serum ferritin levels <10 ng/ml), the fractional absorption of the two forms of iron was identical, while in subjects with higher serum ferritin concentrations, a decrease of iron absorption was more pronounced for nonheme iron than for heme iron.[230]

This observation indicates that in iron-replete subjects, heme iron absorption represents a greater part of the total iron absorption. Finally, the plotting of absorption values from nonheme and heme iron showed a relatively rapid decrease in heme iron absorption in correspondence with a decrease in nonheme iron absorption, thus indicating effective control of the rates of absorption from these two different iron sources.[230]

REFERENCES

1. Green, R., Charlton, R., and Seftel, H. Body iron excretion in man. A collaborative study, *Am. J. Med.*, 45: 336–353, 1968.
2. Hallberg, L. et al. Menstrual blood loss: a population study. Variation at different ages and attempts to define normality, *Acta Obstet. Gynecol. Scand.*, 45: 320–351, 1966.

3. Hallberg, L. and Rossander-Hulten, L. Iron requirements in menstruating women, *Am. J. Clin. Nutr.,* 54: 1047–1058, 1991.
4. International Nutritional Anemia Consultative Group. *Iron Deficiency in Women,* Washington, Nutrition Foundation, 1–68, 1981.
5. Bothwell, T.H. et al. Nutritional requirement and food iron absorption, *J. Int. Med.,* 226: 357–365, 1989.
6. Hallberg L. Iron balance in pregnancy, in *Vitamins and Minerals in Pregnancy and Lactation,* New York, Raven Press, 115–127, 1988.
7. Hahn, P.E. et al. Iron metabolism in early pregnancy as studied with radioactive isotope ^{59}Fe, *Am. J. Obstet. Gynecol.,* 61: 477–486, 1951.
8. Svenberg, B. Iron absorption in early pregnancy. A study of the absorption of non-haem iron and ferric iron in early pregnancy, *Acta Obstet. Gynecol. Scand.,* 48, Suppl. 69–86, 1975.
9. Whittaker, P.G., Lind, T., and Williams, J.C. Iron absorption during normal human pregnancy: a study with stable isotopes, *Brit. J. Nutr.,* 65: 457–463, 1991.
10. Dollman, P.R. Iron deficiency in infancy and childhood, *Am. J. Clin. Nutr.,* 33: 86–95, 1980.
11. Carpenter, C.E. and Mahoney, A.W. Contributions of heme and nonheme iron to human nutrition, *Crit. Rev. Food Sci. Nutr.,* 31: 333–367, 1992.
12. Rhodes, J., Benton, D., and Brown, D.A. Absorption of iron instilled into the stomach, duodenum and jejunum, *Gut,* 9: 323–330, 1968.
13. Tovey. F.I. and Clark, C.G. Anemia after partial gastrectomy: a neglected curable condition, *Lancet 1,* 956–958, 1980.
14. Murray, M.J. and Stein, N. The integrity of stomach as a requirement for maximal iron absorption, *J. Lab. Clin. Med.,* 70: 673–683, 1967.
15. Bezwoda, W et al. The importance of gastric hydrochloric acid in the absorption of nonheme food iron, *J. Lab. Clin. Med.,* 92: 108–117, 1978.
16. Skikne, B.S., Lynch, S.R., and Cook, J.D. Role of gastric acid in food iron absorption, *Gastroenterology,* 81: 1068–1071, 1981.
17. Hallberg, L., Bioavailability of dietary iron in man, *Annu. Rev. Nutr.,* 1: 123–161, 1981.
18. Charlton, R.W. and Bothwell, T.H. Iron absorption, *Annu. Rev. Med.,* 34: 55–75, 1983.
19. Murray, M.J. and Stein, N. A gastric factor promoting iron absorption, *Lancet 1,* 614–616, 1968.
20. Conrad, M.E., Umbreit, J.N., and Moore, E.G. A role for mucin in the absorption of inorganic iron and other metal cations, *Gastroenterology,* 100: 129–136, 1991.
21. Quatersman, J. Metal absorption and the intestinal mucus layer, *Digestion,* 37: 1–9, 1987.
22. Conrad, M.E. and Schade, S.G. Ascorbic acid chelate in iron absorption: a role for hydrochloric acid and bile, *Gastroenterology,* 55: 35–45, 1968.
23. Davis, P.S., Luke, C.G., and Deller, D.J. Gastric iron-binding protein in iron chelation by gastric juice, *Nature,* 214: 1126–1128, 1967.
24. Multani, J.S. et al. Biochemical characterization of gastroferrin, *Biochemistry,* 9: 3970–3976, 1970.
25. Davis, P.S., Luke, C.G., and Deller, D.J. Reduction of gastric iron-binding protein in haemochromatosis, *Lancet 2,* 1431–1433, 1966.
26. Winter, C.V.A. and Williams, R. Iron-binding properties of gastric juice in idiopathic haemochromatosis, *Lancet 2,* 534–537, 1968.
27. Hazell, T., Ledwaed, D.A., and Neale, R.J. Iron binding from meat, *Brit. J. Nutr.,* 39: 631–640, 1978.

28. Wheby, M.S., Suttle, G.E., and Ford, K.T. Intestinal absorption of hemoglobin iron, *Gastroenterology,* 58: 647–658, 1970.
29. Hwang, Y.F. and Brown, E.B. Effect of desferrioxamine on iron absorption, *Lancet 1,* 137–139, 1965.
30. Martinez-Torres, C. and Larysse, M. Iron absorption from veal muscle, *Am. J. Clin. Nutr.,* 24: 531–541, 1971.
31. Cumming, R.L.C. et al. Clinical and laboratory studies on the action of desferriox-amine, *Brit. J. Haematol.,* 17: 257–263, 1969.
32. Pippard, M.J., Johnson, D.K., and Finch, C.A. A rapid assay for evaluation of iron-chelating agents in rats, *Blood,* 58: 685–692, 1981.
33. Pippard, M.J., Callender, S.T., and Finch, C.A. Ferrioxamine excretion in iron-loaded man, *Blood,* 60: 288–294, 1982.
34. Levine, D.S. et al. Blocking action of parenteral desferrioxamine on iron absorption in rodents and man, *Gastroenterology,* 95: 1242–1248, 1988.
35. Johnson, G., Jacobs, P., and Purves, L.R. Iron-binding proteins of iron-absorbing rat intestinal mucosa, *J. Clin. Invest.,* 71: 1467–1476, 1983.
36. Manis, J.G. and Schachter, D. Active transport of iron by intestine: features of two-step mechanism, *Am. J. Physiol.,* 203: 73–80, 1962.
37. Thomson, A.B.R. and Valberg, L.S. Intestinal uptake of iron, cobalt, and manganese in the iron deficient rat, *Am. J. Physiol.,* 223: 1327–1329, 1972.
38. Easthman, E.J., Bell, J.I., and Douglas, A.P. Iron transport characteristics of vesicles of brush-border and basolateral plasma membrane from the rat enterocyte, *Biochem. J.,* 164: 289–294, 1977.
39. Savin, M.A. and Cook, J.D. Iron transport by isolated rat intestinal mucosal cells, *Gastroenterology,* 75: 688–694, 1978.
40. Raja, K.B. et al. *In vitro* measurement and adaptive response of Fe^{3+} uptake by mouse intestine, *Cell Biochem. Funct.,* 5: 69–76, 1987.
41. Cox, T.M. and O'Donnel, M.W. Studies on the binding of iron by rabbit intestinal microvillous membranes, *Biochem. J.,* 194: 753–759, 1981.
42. Marx, J.J.M. and Aisen, P. Iron uptake by rabbit intestinal mucosal membrane vesi-cles, *Biochim. Biophys. Acta,* 649: 297–304, 1981.
43. Kimber, C.L., Mukherjee, S., and Deller, D.J. *In vitro* iron attachement to the intestinal brush border. Effect of iron stores and other environmental factors, *Dig. Dis.,* 18: 781–791, 1973.
44. Nathanson, M.H., Muir, A., McLaren, G.D. Iron absorption in normal and iron–defi-cient beagle dogs: mucosal iron kinetics, *Am. J. Physiol.,* 249: G439–G448, 1985.
45. Cox, T.M. et al. Iron binding proteins and influx of iron across the duodenal brush border. Evidence for specific lactotransferrin receptors in the human intestine, *Bio-chim. Biophys. Acta,* 588: 120–128, 1979.
46. Han, O. et al. Reduction of Fe (III) is required for uptake of nonheme iron by Caco-2 cells, *J. Nutr.,* 125: 1291–1299, 1995.
47. Ekmeciagly, C., Fetrtag, J., and Marktl, W. A ferric reductase activity is found in brush border membrane vesicles isolated from Caco-2 cells, *J. Nutr.,* 126: 2209–2217, 1996.
48. Taylor, P.G. et al. The effect of cysteine-containing peptides released during meat digestion on iron absorption in humans, *Am. J. Clin. Nutr.,* 43: 68–71, 1986.
49. Glahn, R.P. and Van Campen, D.R. Iron uptake is enhanced in Caco-2 cell monolayers by cysteine and reduced cysteinyl glycine, *J. Nutr.,* 127: 642–647, 1997.
50. Wheby, M.S., Jones, L.G., and Crosby, W.H. Studies on iron absorption. Intestinal regulatory mechanisms, *J. Clin. Invest.,* 43: 1433–1443, 1963.

51. Charlton, R.W. et al. The role of the intestinal mucosa in iron absorption, *J. Clin. Invest.,* 44: 543–553, 1965.
52. Conrad, M.E. and Crosby, W.H. Intestinal mucosal mechanisms controlling iron absorption, *Blood,* 22: 415–424, 1963.
53. Wheby, M.S. and Crosby, W.H. The gastrointestinal tract and iron absorption, *Blood,* 22: 414–424, 1963.
54. Teichmann, R. and Stremmel, W. Iron uptake by human upper small intestine microvillous membrane vesicles, *J. Clin. Invest.,* 86: 2145–2153, 1990.
55. Nichols, G.M. et al. The mechanism of nonheme iron uptake determined in IEC-6 rat intestinal cells, *J. Nutr.,* 122: 945–952, 1992.
56. Garcia, M.N., Flowers, C., and Cook, J.D. The Caco-2 cell culture system can be used as a model to study food iron bioavailability, *J. Nutr.,* 126: 251–258, 1996.
57. Levine, J.S. and Seligman, P.A. The ultrastructural immunocytochemical localization of transferrin receptor and transferrin in the gastrointestinal tract of man, *Gastroenterology,* 86: 1161–1170, 1984.
58. Levine, D.S. and Woods, J.W. Provision of an iron-containing diet does not induce expression of transferrin and transferrin receptors in mouse enterocyte brush border membranes, *Gastroenterology,* 92: 1504–1514, 1987.
59. Banerjee, D. et al. Transferrin receptors in the human gastrointestinal tract. Relationship to body iron stores, *Gastroenterology,* 91: 861–869, 1986.
60. Anderson, G.J., Powell, L.W., and Halliday, J.W. Transferrin receptor distribution and regulation in the rat small intestine. Effect of iron stores and erythropoiesis, *Gastroenterology,* 98: 576–585, 1990.
61. Anderson, G.J. et al. Intestinal transferrin receptors and iron absorption in the neonatal rat, *Brit. J. Haematol.,* 77: 229–236, 1991.
62. Loh, T.T. and Kaldor, I. Intestinal iron absorption in suckling rats, *Biol. Neonate,* 17: 173–186, 1971.
63. Ezekiel, E. Intestinal iron absorption by neonates and some factors affecting it, *J. Lab. Clin. Med.,* 70: 138–149, 1967.
64. Forbes, G.B. and Reina, J.C. Effect of age on gastrointestinal absorption in the rat, *J. Nutr.,* 102: 647–652, 1972.
65. Gallagher, N.D., Mason, R., and Folly, K.E. Mechanisms of iron absorption and transport in the neonatal rat intestine, *Gastroenterology,* 64: 438–444, 1973.
66. Jeffrey, G.P., Basclain, K.A., and Allen, T.L. Molecular regulation of transferrin receptor and ferritin expression in the rat gastrointestinal tract, *Gastroenterology,* 110: 790–800, 1996.
67. Oates, P.S., Thomas, C., and Morgan, E.H. Characterization of isolated duodenal epithelial cells along a crypt-villus axis in rats fed diets with different iron content, *J. Gastroenterol. Hepatol.,* 12: 829–838, 1997.
68. Oates, P.S. and Morgan, E.H. Ferritin gene expression and transferrin receptor activity in intestine of rats with varying iron stores, *Am. J. Physiol.,* 273: G636–G646, 1997.
69. Feder, J.N. et al. A novel MHC class I-like gene is mutated in patients with hereditary hemochromatosis, *Nature Genet.,* 13: 399–408, 1996.
70. Parkkila, S. et al. Immunohistochemistry of HLA-H: the protein deficit in patients with hereditary hemochromatosis reveals unique pattern of expression in gastrointestinal tract, *Proc. Natl. Acad. Sci. U.S.A.,* 94: 2543–2539, 1997.
71. Feder, J.N. et al. The hemochromatosis gene product complexes with the transferrin receptor and lowers its affinity for ligand binding, *Proc. Natl. Acad. Sci. U.S.A.,* 95: 1472–1477, 1998.

72. Lebròn, J.A. et al. Crystal structure of the hemochromatosis protein HFE and characterization of its interaction with transferrin, *Cell,* 93: 111–123, 1998.
73. Bothwell, T.H. et al. *Iron Metabolism in Man.* Oxford, Blackwell Scientific, 1979.
74. Osterloh, K.R.S., Simpson, R.J., and Peters, T.J. The role of mucosal transferrin in intestinal iron absorption, *Brit. J. Haematol.,* 65: 1–3, 1987.
75. Idzerza, R.L. et al. Rat transferrin gene expression: tissue-specific regulation by iron-deficiency, *Proc. Natl. Acad. Sci. U.S.A.,* 83: 3723–3727, 1986.
76. Purves, L.R. et al. Properties of the transferrin associated with rat intestinal mucosa, *Biochim. Biophys. Acta,* 966: 318–327, 1988.
77. Huebers, H.A. et al. The significance of transferrin for intestinal iron absorption, *Blood,* 61: 283–290, 1983.
78. Pietrangelo, A. et al. Regulation of transferrin, transferrin receptor, and ferritin genes in human duodenum, *Gastroenterology,* 102: 802–809, 1992.
79. Bezwoda, W.R. et al. Failure of transferrin to enhance iron absorption in achlorhydric human subjects, *Brit. J. Haematol.,* 63: 749–756, 1986.
80. Heilmeyer L. et al. Kongenitale atransferrinamie bie einem seiben jahre alten kind, *Dtsch. Med. Wochenschr.,* 86: 1475–1481, 1961.
81. Craven, C.M. et al. Tissue distribution and clearance kinetics of non-transferrin bound iron in hypotransferrinemic mouse: a rodent model for hemochromatosis, *Proc. Natl. Acad. Sci. U.S.A.,* 84: 3457–3462, 1987.
82. Simpson, R.J. et al. Iron absorption by hypotransferrinaemic mice, *Brit. J. Haematol.,* 78: 565–570, 1991.
83. Levine, P.H., Levine, A.J., and Weintraub, L.R. The role of transferrin in the control of iron absorption: studies on a cellular level, *J. Lab. Clin. Med.,* 80: 333–341, 1972.
84. Evans, G.W. and Grace, C.I. Interaction of transferrin with iron-binding sites on rat intestinal epithelial cell plasma membranes, *Proc. Soc. Exp. Biol. Med.,* 147: 687–689, 1974.
85. Nunez, M.T. et al. Apotransferrin and holotransferrin undergo different endocytic cycles in intestinal epithelial (Caco-2) cell, *J. Biol. Chem.,* 272: 19425–19431, 1997.
86. Alvarez-Hernandez, X., Smith, M., and Glass, J. The effect of apotransferrin on iron release from Caco-2 cells, an intestinal epithelial cell line, *Blood,* 91: 3974–3979, 1998.
87. Hahn, P.F. et al. Radioactive iron absorption by gastrointestinal tract: influence of anemia, anoxia, and antecedent feeding distribution in growing dogs, *J. Exp. Med.,* 78: 169–188, 1943.
88. Granick, S. Ferritin. IX. Increase of the protein apoferritin in the gastrointestinal mucosa as a direct response to iron feeding. The function of ferritin in the regulation of iron absorption, *J. Biol. Chem.,* 164: 737–746, 1946.
89. Brown, E.B., Dubach, R., and Moore, C.V. Studies on iron transportation and metabolism. XI. Critical analysis of mucosal block by large doses of inorganic iron in human subjects, *J. Lab. Clin. Med.,* 52: 335–355, 1958.
90. Brittin, G.M. and Raval, D. Duodenal ferritin synthesis in iron-replete and iron-deficient rats: response to small doses of iron, *J. Lab. Clin. Med.,* 77: 54–58, 1971.
91. Bjorn-Rasmussen, E. Iron absorption: present knowledge and controversies, *Lancet 1,* 914–916, 1983.
92. Schuman, K. et al. Rat intestinal iron transfer capacity and the longitudinal distribution of its adaptation to iron deficiency, *Digestion,* 46: 35–54, 1990.
93. Bergamaschi, G. et al. The effect of transferrin saturation on internal iron exchange, *Proc. Soc. Exp. Med.,* 183: 66–76, 1986.
94. Brittin, G.M. and Raval, D. Duodenal ferritin synthesis in iron-replete and iron-deficient rats: response to small doses of iron, *J. Lab. Clin. Med.,* 77: 54–63, 1971.

95. Whittaker, P. et al. Relationship of duodenal mucosal ferritin and transferrin to iron absorption in humans, *Blood,* 66: 51a, 1985.
96. Teischman, R. and Stremmel, W. Iron uptake by human upper intestinal microvillus membrane vesicles. Indication for a facilitated transport mechanism mediated by a membrane iron-binding protein, *J. Clin. Invest.,* 86: 2145–2153, 1990.
97. Conrad, M.E. et al. Function of integrin in duodenal mucosal uptake of iron, *Blood,* 81: 517–521, 1993.
98. Fleming, M.D. et al. Microcytic anemia mice have a mutation in Nramp2, a candidate iron transporter, *Nat. Genet.,* 16: 383–386, 1997.
99. Gunshiln, H. et al. Cloning and characterization of a mammalian proton-coupled metal-ion transporter, *Nature,* 388: 482–488, 1997.
100. Conrad, M.E. et al. A newly identified iron binding protein in duodenal mucosa of rats, *J. Biol. Chem.,* 265: 5273–5279, 1990.
101. Conrad, M.E. et al. Newly identified iron-binding protein in human duodenal mucosa, *Blood,* 79: 244–247, 1992.
102. Conrad, M.E. et al. Mobilferrin: a homologue of Ro/SS-A autoantigens and calreticulin, *Blood,* 78: 89a, 1991.
103. Conrad, M.E., Umbreit, J.N., and Moore, E.G. Rat duodenal iron-binding protein mobilferrin is a homologue of calreticulin, *Gastroenterology,* 104: 1700–1704, 1993.
104. Umbreit, J.N. et al. Iron absorption and cellular transport: the mobilferrin/paraferritin paradigm, *Sem. Hematol.,* 35: 12–26, 1998.
105. Wada, I. et al. Promotion of transferrin folding by cyclic interaction with calnexin and calreticulin, *EMBO J.,* 16: 5420–5432, 1997.
106. Umbreit, J.N. et al. Paraferritin: a protein complex with ferrireductase activity is associated with iron absorption in rats, *Biochemistry,* 33: 0400–0409, 1996.
107. Conrad, M.E. et al. Alternate iron transport pathway. Mobilferrin and integrin in K562 cells, *J. Biol. Chem.,* 269: 7169–7173, 1994.
108. Umbreit, J.N. et al. The alternate iron transport pathway: mobilferrin and integrin in reticulocytes, *Brit. J. Haematol.,* 96: 521–529, 1997.
109. Conrad, M.E. and Umbreit, J.N. Iron absorption: the mucin-mobilferrin-integrin pathway. A competitive pathway for metal absorption, *Am. J. Hematol.,* 42: 67–73, 1993.
110. Wheby, M.S. and Spyker, D.A. Hemoglobin iron absorption kinetics in the iron-deficient dog, *Am. J. Clin. Nutr.,* 34: 1686–1695, 1981.
111. Conrad, M.E. et al. Human absorption of hemoglobin iron, *Gastroenterology,* 53: 5–14, 1967.
112. Hallberg, L. and Sovell, L. Absorption of hemoglobin iron in man, *Acta Med. Scand.,* 181: 335–346, 1967.
113. Cook, J.D. Adaptation in iron metabolism, *Am. J. Nutr.,* 51: 301–319, 1990.
114. Flanaghan, P.R. Mechanisms and regulation of intestinal uptake and transfer of iron, *Acta Pediatr. Scand.,* suppl. 361: 21–31, 1989.
115. Hahn, P.F. Radioactive iron and its metabolism in anemia, *J. Exp. Med.,* 74: 197–209, 1941.
116. Granick, S. Iron metabolism, *Bull. N.Y. Acad. Med.,* 30: 81–101, 1954.
117. Ehtechami, C., Elsenhans, B., and Forth, W. Incorporation of iron from an oral dose into the ferritin of the duodenal mucosa and liver of normal and iron deficient rats, *J. Nutr.,* 119: 202–212, 1989.
118. Schuman, K. et al. Increased intestinal iron absorption in rats with normal hepatic iron stores. Kinetic aspects of the adoptive response to parenteral iron repletion, *Biochim. Biophys. Acta,* 1033: 277–285, 1990.

119. Conrad, M.E., Parmley, R.T., and Osterloh, K. Small intestinal regulation of iron absorption in the rat, *J. Lab. Clin. Med.,* 110: 418–426, 1987.
120. Conrad, M.E., Weintraub, L.R., and Crosby, W.H. The role of intestine in iron kinetics, *J. Clin. Invest.,* 43: 963–973, 1964.
121. Chisari, L. and Izak, G. The effect of acute hemorrhage and acute haemolysis on intestinal iron absorption in the rat, *Brit. J. Haematol.,* 12: 611–622, 1966.
122. Weintraub, L.R., Conrad, M.E., and Crosby, W.H. Regulation of the intestinal absorption of iron by the rate of erythropoiesis, *Brit. J. Haematol.,* 11: 432–438, 1965.
123. Balacezzak, S.P. and Greenberg, N.J. Iron content of isolated intestinal epithelial cells in relation to iron absorption, *Nature,* 220: 270–271, 1968.
124. Parmley, L.T. et al. Ferrocyanide staining of transferrin and ferritin-conjugated antibody to transferrin, *J. Histochem. Cytochem.,* 27: 681–685, 1979.
125. Goddard, W.P. et al. Iron uptake by isolated human enterocyte suspensions *in vitro* is dependent on body iron stores and inhibited by other metal cations, *J. Nutr.,* 127: 177–183, 1997.
126. Cavill, I. Internal regulation of iron absorption, *Nature,* 256: 328–330, 1975.
127. Diamond, J.M. and Karasov, W.H. Adaptive regulation of intestinal nutrient transporters, *Proc. Natl. Acad. Sci. U.S.A.,* 84: 2242–2245, 1987.
128. Cook, J.D. et al. Serum ferritin as a measure of iron stores in normal subjects, *Am. J. Clin. Invest.,* 27: 691–687, 1974.
129. Walters, G.O. et al. Iron absorption in normal subjects and patients with idiopathic haemochromatosis: relationship with serum ferritin concentration, *Gut,* 16: 188–192, 1975.
130. Baynes, R.D. et al. Relationship between absorption of inorganic and food iron in field studies, *Ann. Nutr. Metabol.,* 31: 109–116, 1987.
131. Cook, J.D., Dassenko, S., and Skinke, B.S. Serum transferrin receptor as an index of iron absorption, *Brit. J. Haematol.,* 75: 603–609, 1990.
132. Finch, C.A. Erythropoiesis, erythropoietin, and iron, *Blood,* 60: 1241–1246, 1982.
133. Bothwell, T.H., Pirzio-Biroli, G., and Finch, C.A. Iron absorption. I. Factors influencing absorption in the normal subject, *J. Lab. Clin. Med.,* 51: 24–35, 1958.
134. Erlandson, M.E. et al. Studies on congenital hemolytic syndromes. IV. Gastrointestinal absorption of iron, *Blood,* 19: 399–410, 1962.
135. Mendel, G.A. Studies of iron absorption. I. The relationship between the rate of erythropoiesis, hypoxia and iron absorption, *Blood,* 18: 727–735, 1961.
136. Bothwell, T.H. et al. *Iron Metabolism in Man.* Oxford, Blackwell Scientific, 1979.
137. Pippard, M.J. et al. Iron absorption and loading in beta-thalassemia intermedia, *Lancet 2,* 819–821, 1979.
138. Raja, K.B. et al. *In vivo* studies on the relationship between intestinal iron absorption, hypoxia and erythropoiesis in the mouse, *Brit. J. Haematol.,* 68: 373–378, 1988.
139. Reynafarje, C. and Ramos, J. Influence of altitude changes on intestinal iron absorption, *J. Lab. Clin. Med.,* 57: 848–857, 1961.
140. Peschle, C. et al. Independence of iron absorption from the rate of erythropoiesis, *Blood,* 44: 353–362, 1974.
141. Peters, T.J. et al. Mechanisms and regulation of iron absorption, *Ann. N.Y. Acad. Sci.,* 525: 141–148, 1988.
142. Hathorn, M.K.S. The influence of hypoxia on iron absorption in the rat, *Gastroenterology,* 60: 76–86, 1971.
143. Raja, K.B. et al. Relationship between erythropoiesis and the enhanced intestinal uptake of ferric iron in hypoxia in the mouse, *Brit. J. Haematol.,* 64: 587–593, 1986.

144. O'Riordan, D.K. et al. Mechanisms involved in increased iron uptake across rat duodenal brush-border membrane during hypoxia, *J. Physiol.*, 500: 379–384, 1997.

145. Laryisse, M. et al. Food iron absorption: a comparison of vegetable and animal protein, *Blood*, 33: 430–438, 1969.

146. Hallberg, L. Bioavailable nutrient density: a new concept applied to the interpretation of food iron absorption data, *Am. J. Clin. Nutr.*, 34: 2242–2249, 1981.

147. Hallberg, L. Bioavailability of dietary iron in man, *Annu. Rev. Nutr.*, 1: 123–147, 1981.

148. Clydesdale, F.M. Physicochemical determinants of iron bioavailability, *Food Technol.*, 133–161, October, 1983.

149. Fairweather-Tait, S.J. Iron in food and its bioavailability, *Acta Pediatr. Scand.*, Suppl. 361: 12–26, 1989.

150. Monsen, E.R., Hallberg, L., and Larysse, M. Estimation of available dietary iron, *Am. J. Clin. Nutr.*, 31: 134–141, 1978.

151. Hallberg, L. and Rossander, L. Bioavailability of iron from Western-type whole meals, *Scand. J. Gastroenterol.*, 17: 151–160, 1982.

152. Lynch, S.R. and Cook, J.D. Interaction of vitamin C and iron, *Ann. N.Y. Acad. Sci.*, 355: 32–40, 1980.

153. Siegenberg, D. et al. Ascorbic acid prevents the dose-dependent inhibitory effect of polyphenols and phytates on nonheme-iron absorption, *Am. J. Clin. Nutr.*, 53: 537–541, 1991.

154. Kalpalathika, P.V.M., Clark, E.M., and Mahoney, A.W. Heme iron content in selected ready-to-serve beef products, *J. Agric. Food Chem.*, 39: 1091–1093.

155. Turnball, A., Cleton, F., and Finch, C.A. Iron absorption. IV. The absorption of hemoglobin iron, *J. Clin. Invest.*, 41: 1897–1907, 1962.

156. Hallberg, L. et al. Dietary heme iron absorption. A discussion of possible mechanisms for the absorption-promoting effect of meat and for the regulation of iron absorption, *Scand. J. Gastroenterol.*, 14: 769–779, 1979.

157. Lynch, S.R. et al. Soy protein products and heme iron absorption in humans, *Am. J. Clin. Nutr.*, 41: 13–20, 1985.

158. Conrad, M.E. et al. Polymerization and intraluminal factors in the absorption of hemoglobin iron, *J. Lab. Clin. Med.*, 68: 659–668, 1966.

159. Lyle, R.M. et al. Iron status in exercising women: the effect of oral iron therapy vs. increased consumption of muscle foods, *Am. J. Clin. Nutr.*, 56: 1049–1055, 1992.

160. Engelmann, M.D. et al. The influence of meat in nonheme iron absorption in infants, *Pediatr. Res.*, 43: 768–773, 1998.

161. Cook, J.D. and Monsen, E.R. Food iron absorption in human subjects. III. Comparison of the effect of animal proteins on non-heme iron absoprion, *Am. J. Clin. Nutr.*, 29: 859–867, 1976.

162. Zhang, D., Carpenter, C.E., and Mahoney, A.M. A mechanistic hypothesis for meat enhancement of iron absorption: stimulation of gastric secretions and iron chelation, *Nutr. Res.*, 10: 929–939, 1990.

163. Kim, Y., Carpenter, C.E., and Mahoney, A.M. Iron solubilizing capacity of meat and bioavailability of beef–iron complex, *FASEB J.*, 5: A589, 1991.

164. Widdowson, E.M. and McCance, R.A. Iron exchange of adults on white and brown bread diets, *Lancet 1*, 588–591, 1942.

165. Simpson, K.M., Morris, E.R., and Cook, J.D. The inhibitory effect of bran on iron absorption in man, *Am. J. Clin. Nutr.*, 34: 1469–1478, 1981.

166. Morris, E.R. and Ellis, R. Isolation of monoferric phytate from wheat bran and its biological value as an iron source in rats, *J. Nutr.*, 106: 753–760, 1976.

167. Hallberg, L., Rossander, L., and Skanberg, A.B. Phytates and the inhibitory effect of bran on iron absorption in man, *Am. J. Clin. Nutr.,* 45: 988–996, 1987.
168. Hallberg, L., Brune, M., and Rossander, L. Iron absorption in man: ascorbic acid and dose-dependent inhibition by phytate, *Am. J. Clin. Nutr.,* 49: 140–144, 1989.
169. Hallberg, L. and Rossander, L. Effect of soy protein on nonheme iron absorption in man, *Am. J. Clin. Nutr.,* 36: 514–520, 1982.
170. Hurrel, R.F. et al. Soy protein, phytate, and iron absorption in humans, *Am. J. Clin. Nutr.,* 56: 573–578, 1992.
171. Champagne, E.T. Effects of pH on mineral-phytate, protein-mineral-phytate, and mineral-fiber interaction. Possible consequences of atrophic gastritis on mineral bioavailability, *J. Am. Coll. Nutr.,* 6: 499–509, 1988.
172. Torre, M., Rodriguiz, A.R., and Saura-Calixto, F. Effects of dietary fiber and phytic acid on mineral bioavailability, *Crit. Rev. Food Sci. Nutr.,* 1: 1–31, 1991.
173. Garcia-Casal, M.N. et al. Vitamin A and β-carotene can improve nonheme iron absorption from rice, wheat and corn by humans, *J. Nutr.,* 128: 646–650, 1998.
174. Disler, P.B., Lynch, S.R., and Charlton, R.W. The effect of tea on iron absorption, *Gut,* 16: 193–200, 1975.
175. Disler, P.B. et al. The mechanism of the inhibition of iron absorption by tea, *S. Afr. J. Med. Sci.,* 40: 109–116, 1975.
176. Rao, B.S. and Prabhavathi, T. Tannin content of foods commonly consumed in India and its influence on ionisable iron, *J. Sci. Food Agri.,* 33: 1–8, 1982.
177. Gillooly, M., Bothwell, T.H., and Torrance, J.D. The effects of organic acids, phytates and polyphenols on the absorption of iron from vegetables, *Brit. J. Nutr.,* 49: 331–342, 1983.
178. Tuntawiroon, M. et al. Dose-dependent inhibitory effect of phenolic compounds in food on nonheme iron absorption in men, *Am. J. Clin. Nutr.,* 53: 554–562, 1991.
179. Brune, M., Rossander, L., and Hallberg, L. Iron absorption and phenolic compounds: the importance of different phenolic structures, *Eur. J. Clin. Nutr.,* 43: 547–559, 1989.
180. Rossander, L. Effect of dietary fiber on iron absorption in man, *Scand. J. Gastroenterol.,* 22, suppl. 129, 68–72, 1987.
181. Harlan, B.F. Dietary fiber and mineral biovailability, *Nutr. Res. Rev.,* 2: 133–147, 1989.
182. Rattan, J. et al. A high-fiber diet does not cause nutrient deficiencies, *J. Clin. Gastroenterol.,* 3: 389–393, 1981.
183. Kelsay, J.L. Effect of dietary fiber level on bowel function and trace mineral balances in humans, *Cereal Chem.,* 58: 2–5, 1981.
184. Cook, J.D. et al. Effect of fiber on nonheme iron absorption, *Gastroenterology,* 85: 1354–1358, 1990.
185. Rossander, L., Hallberg, L., and Bjorn-Rasmussen, E. Absorption of iron from breakfast meals, *J. Clin. Nutr.,* 32: 2484–2489, 1971.
186. Mason, P.M. et al. The effect of moderately increased intakes of complex carbohydrates (meals, vegetables and fruit) for 12 weeks on iron and zinc metabolism, *Brit. J. Nutr.,* 63: 597–611, 1990.
187. Swindell, T.E. and Johnson, I.T. The effect of prolonged dietary supplementation with guar gum on subsequent iron absorption and retention in rats, *Brit. J. Nutr.,* 57: 245–253, 1987.
188. Anderson, B.M., Gibson, R., and Sabry, J.H. The iron and zinc status of long-term vegetarian women, *Am. J. Clin. Nutr.,* 34: 1042–1048, 1981.
189. Walker, A.R.P. Dietary fiber and mineral metabolism, *Molecular Aspects Med.,* 9: 6–26, 1987.

190. Dawson-Hughes, B., Seligson, F.H., and Hughes, V.A. Effects of calcium carbonate and hydroxypatite on zinc and iron retention in post-menopausal women, *Am. J. Clin. Nutr.,* 44: 83–88, 1986.

191. Deehr, M.S. et al. Effects of different calcium sources on absorption in post-menopausal women, *Am. J. Clin. Nutr.,* 51: 95–99, 1990.

192. Milne, D.B., Gallagher, S.K., and Nielson, F.H. Response of various indices of iron status to acute iron depletion in menstruating women by low iron intake and phlebotomy, *Clin. Chem.,* 36: 487–491, 1990.

193. Snedeker, S.M., Smith, S.A., and Greger, J.L. Effect of dietary calcium and phosphorus levels on the utilization of iron, copper, and zinc by adult males, *J. Nutr.,* 112: 136–143, 1982.

194. Monsen, E.R. and Cook, J.D. Food iron absorption in human subjects. IV. The effect of calcium and phosphate salts on the absorption of nonheme iron, *Am. J. Clin. Nutr.,* 29: 1142–1148, 1976.

195. Seligman, P.A. et al. Measurements of iron absorption from prenatal multivitamin–mineral supplements, *Obstet. Gynecol.,* 61: 356–362, 1983.

196. Hallberg, L. et al. Calcium: effect of different amounts on nonheme- and heme-iron absorption in humans, *Am. J. Clin. Nutr.,* 53: 112–119, 1991.

197. Cook, J.D., Dassenko, S.A., and Whittaker, P. Calcium supplementation: effect on iron absorption, *Am. J. Clin. Nutr.,* 53: 106–111, 1991.

198. Sokol, L.J. and Dawson-Hughes, B. Calcium supplementation and plasma ferritin concentrations in premenopausal women, *Am. J. Clin. Nutr.,* 56: 1045–1048, 1992.

199. Prather, T.A. and Miller, D.D. Calcium carbonate depresses iron bioavailability in rats more than calcium sulfate or sodium carbonate, *J. Nutr.,* 122: 327–332, 1992.

200. Barton, J.C., Conrad, M.E., and Parmley, R.T. Calcium inhibition of inorganic iron absorption in rats, *Gastroenterology,* 84: 90–98, 1983.

201. Greger, J.L. Effects of phosphorus-containing compounds on iron and zinc utilization, in *Nutritional Bioavailability of Iron,* Kies, C., Ed., Washington, American Chemical Society, 107–127, 1982.

202. Mahoney, A.W. and Hendrick, D.G. Some effects of different compounds on iron and calcium absorption, *J. Food Sci.,* 43: 1473–1481, 1978.

203. Zamel, M.B. and Bidari, M.T. Zinc, iron and copper bioavailability as affected by orthophosphates, polyphosphates and calcium, *J. Food Sci.,* 48: 567–579, 1983.

204. Mena, I. Manganese, in *Disorders of Mineral Metabolism,* Bronner, F.L. and Caburn, J.W., Eds., New York, Academic Press, 233–270, 1981.

205. Thomson, A.B.R., Olostunbosum, D., and Valberg, L.S. Interrelationship of intestinal transport system of manganese and iron, *J. Lab. Clin. Med.,* 78: 643–655, 1971.

206. Ho, S.Y. et al. Effects of high but nontoxic dietary manganese and iron on their metabolism by calves, *J. Dairy Sci.,* 67: 1489–1495, 1984.

207. Baker, D.H. and Halpin, K.M. Manganese and iron interrelationship in the chick, *Poultry Sci.,* 70: 146–152, 1991.

208. Hurley, L.S., Keen, C.L., and Lonnerdal, B. Aspects of trace elements interactions during development, *Fed. Proc.,* 42: 1735–1739, 1983.

209. Ivan, M. and Hidroglou, M. Effect of dietary manganese on growth and manganese metabolism in sheep, *J. Dairy Sci.,* 63: 385–390, 1980.

210. Davis, C.D., Ney, D.M., and Greger, J.L. Manganese, iron and lipid interactions in rats, *J. Nutr.,* 20: 507–519, 1990.

211. Kies, C. et al. Manganese availability for humans: effect of selected dietary factors, in Kies C., ed., *Nutritional Bioavailability of Manganese,* ACS Symposium Series 35, Washington, American Chemical Society, 112–122, 1987.

212. Hartman, R.H., Matrone, G., and Wise, G.H. Effect of high dietary manganese on hemoglobin formation, *J. Nutr.*, 57: 429–439, 1955.
213. Black, J.R. et al. Effect of dietary manganese and age on tissue trace mineral composition of broiler-type chicks as bioassay of manganese souces, *Poultry Sci.*, 64: 688–693, 1984.
214. Davis, C.D., Malecki, E.A., and Greger, J.L. Interactions among dietary manganese, heme iron, and nonheme iron in women, *Am. J. Clin. Nutr.*, 56: 926–932, 1992.
215. Urel, C. and Conrad, M.E. Absorption of heme iron, *Sem. Hematol.*, 35: 27–34, 1998.
216. Wheby, M.S., Suttle, G.E., and Ford, K.T. Intestinal absorption of hemoglobin iron, *Gastroenterology,* 58: 647–654, 1970.
217. Tenhunen, R. et al. An intestinal receptor for heme: its partial characterization, *Int. J. Biochem.,* 12: 713–716, 1980.
218. Grasbeck, R. et al. Spectral and other studies on the intestinal haem receptor of the pig, *Biochim. Biophys. Acta,* 700: 137–142, 1982.
219. Galbraith, R.A., Sassa, S., and Kappas, A. Heme binding to murine erythroleukemia cells. Evidence for a heme receptor, *J. Biol. Chem.,* 260: 12198–12202, 1985.
220. Majuri, R. Heme-binding plasma membrane proteins of K562 erythroleukemia cells: adsorption to heme-microbeads, isolation with affinity chromatography, *Eur. J. Haematol.,* 43: 220–225, 1989.
221. Galbraith, R.A. Heme binding to Hep G2 human hepatoma cells, *J. Hepatol.,* 10: 305–310, 1990.
222. Raffin, S.B. et al. Intestinal absorption of hemoglobin iron-heme cleavage by mucosal heme oxygenase, *J. Clin. Invest.,* 54: 1344–1352, 1974.
223. Rosenberg, D.W. and Kappas, A. Characterization of heme oxygenase in the small intestine epithelium, *Arch. Biochem. Biophys.,* 274: 471–480, 1989.
224. Cable, J.W., Cable, E.E., and Bonkovsky, H.L. Induction of heme oxygenase in intestinal epithelial cells, *Mol. Cell. Biochem.,* 129: 93–98, 1993.
225. Boni, R.E. et al. Tin-mesoporphyrin inhibits heme oxygenase activity and heme absorption in the intestine, *Pharmacology,* 47: 318–329, 1993.
226. Poss, K.D. and Tonegawa, S. Heme oxygenase 1 is required for mammalian iron reutilization, *Proc. Natl. Acad. Sci. U.S.A.,* 94: 10919–10924, 1997.
227. Miller, S.M. et al. Heme oxygenase 2 is present in interstitial cell networks of the mouse small intestine, *Gastroenterology,* 114: 239–244, 1998.
228. Gleerup, A. et al. Iron absorption from the whole diet: comparison of the effect of two different distributions of daily calcium intake, *Am. J. Clin. Nutr.,* 61: 97–104, 1995.
229. Hultén, L. et al. Iron absorption from the whole diet. Relation to meal composition, iron requirements and iron stores, *Eur. J. Clin. Nutr.,* 49: 794–808, 1995.
230. Hallberg, L., Hulten, L., and Gramatkovski, E. Iron absorption from the whole diet in men: how effective is the regulation of iron absorption, *Am. J. Clin Nutr.,* 66: 347–356, 1997.

2 Iron and Cell Proliferation: Mechanisms and Applications in Cancer Therapy

CONTENTS

2.1 Introduction 39
2.2 Ribonucleotide Reductase: A Key Enzyme in the Control of DNA
Synthesis.. 40
2.3 Iron and Cell Cycle Control 45
    2.3.1 An Overview of the Cell Cycle 45
    2.3.2 Role of Iron in the Control of Cyclin Activation............................... 47
2.4 Inhibition of Cell Proliferation by Agents that Reduce Cellular Iron
Incorporation 48
    2.4.1 Inhibition of Cell Growth Using Monoclonal Antibodies to
    Transferrin Receptor 48
    2.4.2 Inhibition of Cell Growth by Agents that Interfere with
    Intracellular Iron Incorporation.. 51
    2.4.3 Inhibition of Cell Proliferation Mediated by Iron Chelators 53
2.5 Iron and Cancer... 59
References... 61

2.1 INTRODUCTION

Iron is an essential element for all living organisms and plays a key role in multiple biological reactions. Among these reactions the most important are: (1) the mitocondrial respiratory chain mediating electron transport, a process involved in the generation of the energy required for cellular metabolism; and (2) the control of cell growth, largely mediated by actively maintaining ribonucleotide reductase, the only enzyme catalyzing the transformation of ribonucleotides into deoxyribonucleotides.

However, iron is also a potentially toxic element in that it can be involved in the generation of toxic free radicals. The dangers of severe iron overload are well known. However, even moderate iron overload may have harmful consequences. Increased available body iron stores may increase the risk of cancer through two different mechanisms:[1]

1. Iron can catalyze the production of oxygen radicals, and may generate carcinogenic molecular species.
2. Iron may be a limiting nutrient for the proliferation of cancer cells in the living organism; moderate or high iron stores may enhance the chances that a cancer cell will become a clinically apparent neoplasm.

The fundamental role of iron in cell metabolism and its potential toxicity imply that cellular iron metabolism must be a carefully regulated process. Perturbations of iron availability have important consequences for cellular proliferation. In this chapter, the role played by iron in the control of cellular proliferation is reviewed, with emphasis on the manipulations of iron availability that may have potential applications in cancer therapy.

2.2 RIBONUCLEOTIDE REDUCTASE: A KEY ENZYME IN THE CONTROL OF DNA SYNTHESIS

The production of deoxyribonucleotides directly from ribonucleotides was suggested initially by the observation that in animals injected with cytidine labeled both in cytosine and ribose, the ratio of labels was conserved in the deoxycytidine of DNA.[2] An enzyme that reduces ribonucleosides directly to the corresponding deoxyribonucleosides was subsequently discovered in extracts of *Escherichia coli* and identified in the cells of virtually all mammalians.[3] This enzyme, ribonucleoside diphosphate reductase(ribonucleotide reductase), catalyzes the following reaction:

$$riboNDP + reductant - SH_2 \rightarrow deoxyriboNDP + reductant - (S\text{-}S)$$

The consequence of this reaction is that the hydroxyl group at the $2'$ position of the ribose moiety of a ribonucleoside diphosphate is replaced by hydrogen(see Figure 2.1). The electrons required for the reduction of carbon 2 of the ribose moiety

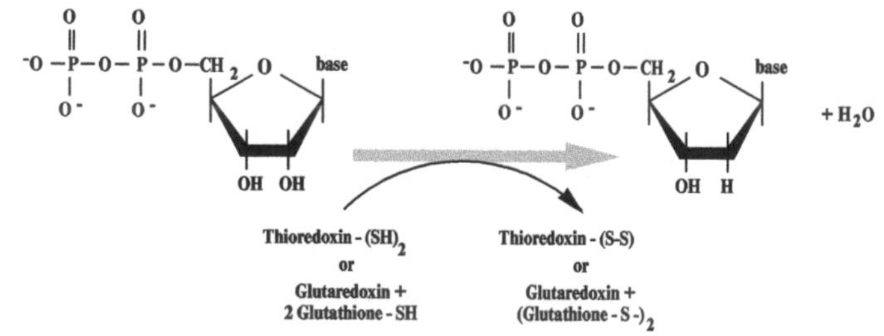

FIGURE 2.1 Ribonucleotide reduction to $2'$-deoxyribonucleotide by ribonucleoside diphospate reductase.

are provided by redox active dithiols present on the large subunit of the enzyme.[4] At the end of the reaction, the two sulfhydryl groups are oxidized and must be restored in their reduced form by a thiol coenzyme. This last reaction occurs with NADPH via a thioredoxin or glutaredoxin system or via lipoic acid, as recently suggested.

Hydroxyurea and related compounds scavenge the tyrosyl radical of ribonucleotide reductase and thereby inactivate the enzyme. As a consequence of ribonucleotide reductase inactivation, hydroxyurea inhibits DNA synthesis in mammalian cells in culture.[5] Prolonged cultivation of mammalian cells in culture in the presence of increasing concentrations of hydroxyurea may result in the development of a marked resistance to the drug associated with a very pronounced increase in the activity of ribonucleotide reductase, largely due to overproduction of the R1 and R2 subunits of the enzyme.[6]

Interestingly, resistance to hydroxyurea leads to a marked increase in the expression of the genes encoding for ferritin chains.[7]

Three classes of ribonuleotide reductases have been distinguished according to structural, functional and evolutionary criteria. Class I and Class III enzymes consist of two homodimeric subunits. The large homodimer carries catalytic and allosteric sites, while the small homodimer is responsible for radical generation. Class II enzymes consist of a single monomeric or homodimeric protein and need the presence of a separate cofactor, adenosylcobalamin, for the generation of the radical.[8]

Mammalian cells, DNA viruses of the herpes group, and some prokaryotes, notably *Escherichia coli*, contain Class I heterodimeric iron containing ribonucleotide reductase of the $\alpha_2\beta_2$ type (see Figure 2.2). Ribonucleotide reductase of *Escherichia coli*

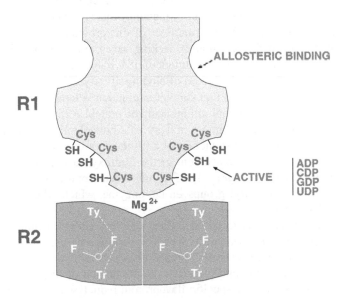

FIGURE 2.2 Schematic representation of the structure of ribonucleotide reductase of *Escherichia coli*.

was characterized in detail and its structure is similar to that of the mammalian enzyme. This enzyme consists of a heterodimer complex composed by two R1 subunits and two R2 subunits.

Protein R1 is a dimer of 171 kDa constituted by two identical α subunits of 85, 7 kDa with two identical substrate binding sites and two different kinds of effector binding sites. One effector binding site confers substrate specificity, and the other regulates overall catalytic efficiency. R1 also contains thiol groups (redox-active cysteines) essential for enzyme activity.

R2 protein is smaller than R1 and is composed of two identical β subunits (each of 43.4 kDa). R2 contains an essential tyrosyl radical which is stabilized by an adjacent dinuclear iron center and is thought to initiate the radical-based reaction.

Electron spin resonance spectroscopy has been used to characterize the radical as a protein-derived tyrosyl radical in the vicinity of the iron center.[9] Site-directed mutagenesis has shown that the radical is located at Tyr 122.[10] The structure of the R2 protein was considerably elucidated through determination of the crystal structure of this protein at 2.2 Å resolution.[11] This analysis provided evidence that the R2 subunits of ribonucleotide reductase had an unusual α-helical structure — a novel arrangement having an eight-helix bundle as its main motif.[11]

Two main additional features of the R2 subunit are: (1) two iron centers that are about 25 Å apart in the dimeric molecule; and (2) Tyr 122, which harbors the stable free radical necessary for the activity of ribonucleotide reductase, is buried inside the protein 10 Å from the surface. It is thus protected from reducing molecules and is located 5 Å from the closest iron atom.[11]

The development of a system for production of recombinant ribonucleotide reductase subunits greatly contributed to the understanding of the molecular structure of ribonucleotide reductase.[12] In fact, this system, coupled with site-directed mutagenesis, may allow the production of mutant forms and identification of amino acid sequences involved in different functions of this protein.

A truncated form of the R2 subunit, lacking seven amino acid residues in the carboxyl terminus, generated similar amounts of tyrosyl-free radical as the wild-type protein, but was completely devoid of binding affinity to the R1 subunit.[13] This observation clearly demonstrates that subunit interaction is totally dependent on the seven outermost carboxyl-terminal amino acids of protein R2.

The mechanism of the enzymatic activity of ribonucleotide reductase was in part elucidated. The enzymatic reduction of a ribonucleotide occurs with the retention of configuration by the replacement of the hydroxyl at C-2′ with a hydrogen derived from the solvent (Figure 2.1). The immediate source of reducing power comes from pairs of sulfhydryl groups on the enzyme, which, after oxidation, are restored by a thiol coenzyme.

The first such coenzyme to be identified was thioredoxin, a ubiquitous, heat stable, 12-kDa protein with two sulfhydryl groups in nearby cysteine residues that are oxidized to cystine in the enzymatic reaction. A second hydrogen-donor system is represented by glutathione and glutaredoxin. Reduction of oxidized thioredoxin and glutathione is achieved by specific flavoprotein reductases employing NADPH as the electron donor.

The most attractive model of ribonucleotide reductase activity was proposed by Stubbe et al.[14] This scheme proposes the formation of a very short-lived substrate radical, leading to the abstraction of a hydrogen atom from the 3' carbon, possibly by the tyrosyl radical of the enzyme, and restoration of the enzyme radical following the loss of the 2' hydroxyl. In favor of this mechanism is the observation that interaction between the reductase and the substrate analog, 2' azido-CDP, results in the transient appearance of an electron paramagnetic response signal localized to the nucleotide.

The iron center of the R2 subunit plays a key role in enzyme activity, notably at the level of the generation and maintenance of the tyrosyl radical. Two different mechanisms may lead to the inactivation of the R2 subunit by loss of the tyrosyl radical. Treatment of R2 with a radical scavenger (e.g., hydroxyurea) results in loss of reductase activity due to the formation of met-R2 which lacks the tyrosyl radical. Reduction of a single electron leads to destruction of the tyrosyl radical and maintains the Fe^{3+} center intact.[15]

Treatment of R2 with iron chelators results in reductase inhibition due to the formation of apoR2 which lacks both iron and radical.[16] Active R2 can be reconstituted from both apoR2 and met-R2. Reconstitution from apoR2 requires addition of Fe^{2+} in the presence of oxygen.[15] The simultaneous oxidation of Fe^{2+} and Tyr 122 reforms both the Fe^{3+} center and the tyrosyl radical. The reconstitution of an active R2 from met-R2 involves a complex enzymatic system.

The radical does not seem to be directly involved in the abstraction of the hydrogen at C-3' since it is deeply buried in the protein globulus about 10 Å distant from the protein surface and thus cannot interact with the catalytic site on R1. Five cysteine residues present in the R1 subunit represent the active thiols. Two cysteines near the COOH-terminus, Cys 754 and Cys 759, appear to accept electrons from reduced thioredoxin and are thought to relay them to a second pair, Cys 225 and Cys 462, that are directly involved in the reduction of C-2'.[16] The fifth cysteine, Cys439, may form a thiyl radical and abstract the hydrogen from C-3'.[17]

Recent views of the mechanism of action of the enzyme indicate that a thiyl radical located at Cys 439 of the enzyme abstracts the hydrogen from C-3' and generates an activated substrate radical that is reduced by two redox-active cysteines (Cys 225 and Cys 462).[18] The three-dimensional structure of the large R1 subunit largely supports this view on the enzyme activity.[19] The thiyl radical of the large R1 subunit is generated from the tyrosyl radical of the small R2 subunit through a process of long-range electron transfer.[20]

Ribonucleotide reductase reduces all four common ribonucleotides. Coordinated synthesis of deoxyribonucleotides is based on a complex allosteric control of their substrate specificity which enables the reductase to attain a fine adjustment to the needs for DNA synthesis. Binding of ATP to the overall activity sites (see Figure 2.2) makes the enzyme active, whereas binding of dATP to these sites renders the enzyme inactive. Binding of ATP, dATP, dTTP, or dGTP to the substrate specificity sites determines the rate of the reaction and the affinity for nucleosides diphosphates formed by the interaction of R1 and R2 subunits in the substrate binding sites (Figure 2.2). When ATP is bound in the substrate specificity site, reduction of UDP and CDP are favored. Binding of dTTP stimulates GDP reduction, and binding of dGTP leads to ADP reduction.

This pattern of regulation of ribonucleotide reductase is based on *in vitro* experiments on purified *Escherichia coli* reductase. However, the pattern appears to apply also to animal cells. Thus, it is well kown that thymidine added to animal cells in culture at concentrations above 1 mM blocks DNA synthesis. Pools of dTTP, dGTP, and dATP are increased 25-, 10-, and two-fold, respectively, but dCTP is decreased 10-fold. Thus, it appears evident that raising the dTTP level by thymidine administration stimulates the reduction of GDP and ADP and inhibits the reduction of CDP. The reduced dCTP level may block DNA synthesis; indeed, the addition of deoxycytidine overcomes the blocking action.

The levels of ribonucleotide synthesis markedly change during the cell cycle. Thus, quiescent mammalian cells lack the enzyme. Induction of DNA synthesis leads to a steady synthesis of the R1 subunit throughout the growth cycle and a striking increase of R2 occurring at the beginning of the S phase. A half-life of 3 hours for R2 protein contrasts with 15 hours for R1 protein. Quantitation of the level of mRNAs encoding for R1 and R2 subunits provided evidence that both were very low in G0/G1-phase cells, showed a pronounced increase as cells progressed into S phase, and then declined when cells moved into G2+M phase.[21-23] Thus, the relative lack of cell-cycle dependent variation in R1 protein level may simply be a consequence of the long half-life of the R1 protein.

Transcription studies[23] showed that R2 gene transcription is induced at early time points (i.e., 1 hour) after mitogenic activation of the cells, but results in the synthesis of a short, premature R2 transcript as long as cells are in the G1 phase. S-phase specific synthesis of mature full-length R2 mRNA then occurs as a result of a cell cycle phase release from the transcriptional block as cells proceed into the S phase.

Recent studies indicate that ribonucleotide reductase, in addition to providing deoxyribonucleotides for DNA synthesis, also displays other biological functions that can be summarized as follows. Ribonucleotide reductase activity may be induced outside the S phase by DNA damage either from UV light or DNA cross-linking agents.[24] Ribonuleotide reductase activity is higher in malignant tissues than in their normal counterparts; tumor promoters significantly increase R2 subunit levels. Furthermore, studies based on forced expression of recombinant R2 or R1 ribonucleotide reductase subunits in mouse fibroblasts have provided some interesting findings showing that:

1. R2 overexpression leads to increased *in vitro* and *in vivo* tumorigenic potential.[25]
2. R1 overexpression induces a tumor suppressor activity which results in reduced tumorigenicity of *ras*-transformed fibroblasts.[26]

Several drugs capable of inhibiting ribonucleotide reductase have been developed. The majority of these drugs are hydroxybenzhydroxamic acid derivatives which inhibit ribonucleotide reductase interference with the free tyrosyl radicals (radical scavengers) required for enzyme activity.

2.3 IRON AND CELL CYCLE CONTROL

Iron plays an important role in the control of various phases of the cell cycle. As outlined above, iron is required for the activity of ribonucleotide reductase and, through this mechanism, is required for progression through the S phase of the cell cycle. However, recent studies indicate that iron also plays an important role in the progression through the G1 phase of the cell cycle.

2.3.1 AN OVERVIEW OF THE CELL CYCLE

To better understand the requirements for iron during the cell cycle, a brief outline of the cell cycle is presented.

According to the timing of DNA synthesis, it is possible to divide the cycles of eukaryotic cells into four distinct phases (Figure 2.3). The M phase corresponds to mitosis which is usually followed by cell division. The M phase is followed by the G1 (gap1) phase, which can be defined as the interval between mitosis and start of DNA synthesis. During this phase, the cells are metabolically active, have an active protein synthesis, but do not replicate their DNA.

The G1 phase is followed by the S (synthesis) phase, characterized by DNA synthesis which progressively leads to DNA replication. The S phase ends with replication of DNA content and is followed by the G2 (gap2) phase, which is characterized by the synthesis of proteins required for mitosis.

The proliferation of animal cells is controlled by external stimuli and mediated by growth factors. The cell cycle progression, however, requires both the stimulus of the growth factor and the presence of nutrients. A key control in the cell cycle is operated during the late G1 phase at the level of a decision point called the *restriction point*. The cells have to pass through this restriction point to be definitely committed to proliferation. If competent growth factors are not available when the cell is in the

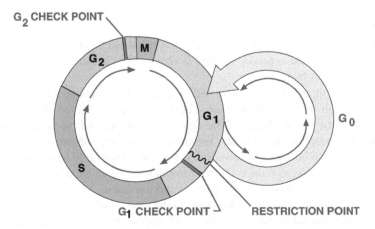

FIGURE 2.3 Regulation of the cell cycle. The cell cycle can be subdivided in four phases: G1, S, G2, and M. Cycle progression is controlled at the level of the restriction point and at the level of the check points.

G1 phase, progression through the cell cycle is blocked and the cell stops at the level of the restriction point. These cell cycle-arrested cells enter a quiescent state of the cycle called G0. Cells in the G0 phase have a reduced rate of protein synthesis and remain in this state until an appropriate growth factor stimulus is originated.

In addition to the restriction point, two important cell cycle checkpoints control the regularity of the cell cycle. A first G2 checkpoint acts as a sensor of unreplicated DNA, which generates a cell cycle arrest signal when unreplicated DNA is found. Another important cell cycle checkpoint operates at the level of the late G1 phase; the cell cycle arrests when damaged DNA is detected. This function is mediated by the p53 protein, whose synthesis is rapidly induced following DNA damage.

Studies carried out in recent years led to the discoveries of several proteins that act as regulators of cell cycle progression. The cell cycles of mammals are controlled through complex regulatory networks of cell-cycle related proteins called cyclins and cdc (cell division cycle) kinase proteins. These proteins are activated during different phases of the cell cycle and form complexes of one cyclin with one cdk that mediate phosphorylation events required for cell cycle progression. Progression from the G1 to the S phase is regulated principally by Cdk2 and Cdk4 in association with cyclins D and E. Thus, complexes of Cdk2, Cdk4, and Cdk6 with D-type cyclins (D1, D2, and D3) drive the cell through the G1 restriction point; cyclin E is expressed later in G1 and the cyclin E/Cdk2 complexes induce biochemical events required for the G1 to S transition and the start of DNA synthesis.

At later phases of the cell cycle, complexes of cyclin A with Cdk2 and cyclin B with Cdc2 function in progression of cells through the S phase and the transition from G2 to M, respectively. The activity of Cdk depends on the formation of a complex with the appropriate cyclin and the activity of the cyclin/Cdk complexes requires a phosphorylation event occurring at the level of a Cdk threonine residue catalyzed by a Cdk-activating kinase. More importantly, the activity of cyclin/Cdk complexes is regulated by Cdk inhibitors that bind to and inactivate these complexes. Various types of Cdk inhibitors are known. They are classified according to their molecular weights (p15, p16, p21, p27) and their capacity to bind to different cyclin/Cdk complexes.

A key event in the control of cell cycling is related to the activation of D cyclins. The synthesis of these types of cyclins is absent in quiescent cells and is rapidly induced by growth factors.

Cyclin D compounds associated with a specific Cdk mediate the progression of the cells through the restriction point. The key role of cyclin D/Cdk complexes in the control of the initial phases of the cell cycle is supported by another important function: the phosphorylation of the tumor suppressor retinoblastoma protein (Rb). Phosphorylated Rb becomes inactive and is not able to bind E2F transcription factors that become available to induce the transcription of genes required for cell proliferation. The cyclic changes in the phoshorylation status of Rb protein occurring during the cell cycle represent a key step in the control of cell proliferation. These changes in the phosphorylation status of Rb imply also that this protein is hypophosphorylated in quiescent cells and thus capable of interacting with and inactivating E2F transcription factors.

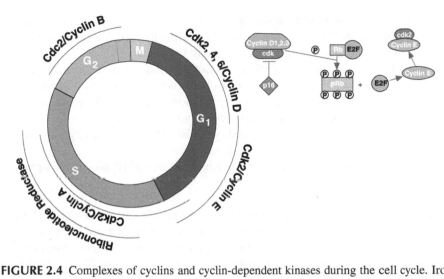

FIGURE 2.4 Complexes of cyclins and cyclin-dependent kinases during the cell cycle. Iron modulates the transcription of the Cdc2 gene of Cdck2, and particularly of cyclinA/Cdk2 kinase activity. Furthermore, iron affects the phosphorylation status of p110 retinoblastoma (Rb) protein.

2.3.2 ROLE OF IRON IN THE CONTROL OF CYCLIN ACTIVATION

Recent studies suggest that iron may play a role in the control of the activities of cyclin/Cdk complexes (Figure 2.4). These studies arose from the observation that iron chelators arrested cells at a point in the G1 phase, notably earlier than the G1/S boundary.

Studies performed on normal human T lymphocytes stimulated with phorbol dibutyrate and ionomycin have shown that these cells are blocked in the G1 phase by deferoxamine.[27] Analysis of cell-cycle related genes in deferoxamine-treated T lymphocytes showed: (1) reduced accumulation of cellular RNA content; (2) reduced synthesis of the cell cycle regulatory kinase p34 Cdc2; and (3) normal production of p53 proteins, whose synthesis is induced 3 to 6 hours after mitogen addition.[27-29]

Similar results were observed in neuroblastoma cells where deferoxamine induces cell cycle arrest at the level of the G1 phase of the cycle. Furthermore, deferoxamine-treated neuroblastoma cells exhibited decreases in both p34 Cdc2 immunoreactive proteins as well as kinase activity.[30] It is of interest that the block of the cell cycle by iron chelators before it reaches the G1/S boundary is further supported by experiments involving sequential addition of an iron chelator and aphidicolin, an agent that specifically inhibits cell cycle progression at the G1/S boundary. Cells treated with deferoxamine and released into aphidicolin exhibit arrest at the G1/S interface, while a shift from aphidicolin into deferoxamine resulted in entry into the S phase.[30]

A subsequent study carried out in human activated T lymhocytes provided a detailed analysis of the effect of deferoxamine on cyclin expression.[31] Results showed that:

1. Iron chelation inhibits transcription of the Cdc2 gene (which normally occurs just before S-phase entry), but has virtually no effect on expression of the Cdk2 gene (which occurs in early G1 phase).
2. In spite of the absence of effects on Cdk2 transcription, the iron chelator significantly reduced the level of Cdk2 protein and markedly decreased Cdk2 kinase activity. This phenomenon mostly concerns the second wave of Cdk2 activity that occurs after entry into the S phase.
3. Iron chelation moderately reduced cyclin E/Cdk2 kinase activity and completely abolished cyclin A/Cdk2 kinase activity
4. Deferoxamine markedly reduced retinoblastoma p110 protein phosphorylation.

Experiments performed on MDA-MB-453 human breast cancer cells showed that iron chelators (either deferoxamine or mimosine) induced a cell cycle block for most of the cells (87%) at the level of the G1 phase of the cycle. This phenomenon was associated with a marked or complete inhibition of cyclin E- and cyclin A-associated kinase activity.[32]

Other studies performed on human myelocytic leukemia ML-1 cell lines confirmed the G1 cell cycle block following iron chelator treatment, but failed to demonstrate the association of this phenomenon with reduced retinoblastoma phosphorylation.[33]

2.4 INHIBITION OF CELL PROLIFERATION BY AGENTS THAT REDUCE CELLULAR IRON INCORPORATION

The iron requirement for cellular proliferation is related in large part to a requirement for DNA synthesis. As mentioned above, activity of one Fe-requiring enzyme, ribonucleotide reductase, has been proposed as the critical step that makes Fe necessary for DNA synthesis. Agents that interfere with iron metabolism inhibit cellular proliferation by three different mechanisms:

1. Monoclonal antibodies to the transferrin receptor inhibit transferrin-mediated iron uptake by interfering with transferrin binding to the receptor at the cell surface.
2. Certain metals that bind to transferrin, such as gallium (Ga) and indium (In) appear to mainly interfere with intracellular Fe incorporation.
3. Iron chelators, such as desferrioxamine, deplete intracellular iron with a preference for removal of iron present in the "chelatable" pool, with the subsequent release of ferrioxamine to the extracellular milieu.

2.4.1 INHIBITION OF CELL GROWTH USING MONOCLONAL ANTIBODIES TO TRANSFERRIN RECEPTOR

Monoclonal antibodies (mAbs) against transferrin receptors can be subdivided in two categories according to their capacity to interfere or not interfere with transferrin binding and internalization.[34]

Classic examples of antibodies that react with the transferrin receptor, but are unable to prevent the binding of transferrin to its receptor are antibodies OKT9 and B3/25. In contrast, others, such as antibody 42/6, are able to prevent the binding of transferrin to its receptor.

Initial studies carried out by Trowbridge indicate that the 42/6 mAb was able to inhibit the growth of the human T-leukemic cell line CCRF-CEM *in vitro*.[35] These cells, grown for several days in the presence of this antibody, accumulated in the S phase of the cell cycle, and the addition of iron salts or transferrin was unable to overcome this inhibitory effect. Subsequent studies have shown that this antibody can also inhibit cell growth of mitogen-activated T lymphocytes[36] as well as the KG-1 and K562[37] leukemic lines. However, iron salts in these cells were partially able to overcome the inhibitory effect of the 42/6 mAb.

It is of interest that all the antibodies that exhibited antiproliferative activity were polymeric. Thus, 42/6, a mAb against human transferrin receptor, is an IgA,[35] whereas mAbs R17208 and REM 17.2 against murine transferrin receptors are IgM.[38,39] However, the mechanisms responsible for the inhibition of transferrin internalization displayed by these antibodies are different in that: (1) 42/6 mAb prevents the binding of transferrin to its receptor, and through this mechanism inhibits iron uptake; and (2) neither of the mAbs to murine transferrin receptors inhibits transferrin binding, but they induce an extensive cross-linking of the receptors to the cell surface, thus preventing iron uptake and inhibiting cell growth.[40-42]

A large group of mAb anti-human transferrin receptors were recently obtained by Trowbridge and his coworkers and used for immunization of purified recombinant human transferrin receptor produced in a baculovirus expression system.[43] Studies carried out with these mAbs provided evidence that some combinations of two or more mAbs inhibit the *in vitro* and *in vivo* growth of tumor cells.[44]

Using the recombinant human transferrin receptor as an immunogen, monoclonal antibodies against defined epitopes of the human transferrin receptor cytoplasmic tail were obtained. One of these three monoclonal antibodies (H68.4), which binds proximally to the carboxy-terminal side of the II68.4 epitope, blocks receptor internalization and does not label the receptor in clathrin-coated pit proteins.[45]

Bispecific monoclonal antibodies to human transferrin receptor and tumor associated antigens were obtained by chemical cross-linking; some of these antibodies, particularly when supplied with iron chelators, markedly inhibited tumor growth.[46]

Sauvage et al.[47] conducted a series of studies to investigate the feasibility of *in vivo* administration of mAb anti-transferrin receptor to inhibit cell growth.[47] Among these antibodies, mAb R17208 has been shown to inhibit the *in vivo* growth of a transplantable murine AKR leukemia, SL-2, and increase the life spans of tumor-bearing mice. *In vivo* administration of this anti-murine transferrin receptor mAb resulted in only limited toxicity manifested by reduction of erythroid and myeloid colony-forming units in the bone marrow and splenomegaly associated with increased erythropoiesis in the spleen.[47]

Other studies of murine tumor growth have shown that anti-transferrin mAbs and iron chelators synergize in eliciting inhibition of tumor growth.[48] Interestingly, this synergism was observed even when anti-transferrin mAbs unable to inhibit tumor cell growth (i.e., IgG antibodies) were administered together with an iron chelator.[48]

Another experimental therapeutic approach based on the use of anti-transferrin mAbs involves the preparation of immunoliposomes containing a drug in their interior and coated with an mAb anti-transferrin receptor. This approach was initially used to deliver daunomycin to the rat brain.[49] The basis of this approach is the observation that brain microvascular endothelium possesses transferrin receptors. Thus, the immunoliposome injected intravenously is bound by brain endothelium, endocytosed, and accumulated in the brain. Studies of *in vivo* pharmacokinetics and tissue distribution of daunomycin anti-transferrin receptor immunoliposomes, compared to distribution of free daunomycin, showed that it allowed a higher accumulation of the drug at the level of the spleen and liver, associated with a lower accumulation at the level of the heart, lung, and kidney.[50]

A conjugate of doxorubicin and transferrin may also be used to overcome the resistance of tumor cells to this drug. An L929 cell line selected for doxorubicin resistance was completely resistant to the free drug, while it was sensitive to the cytotoxic effect of the conjugated transferrin–doxorubicin.[51] Analysis of the intracellular distribution in membranes, cytoplasm and nucleus indicated the drug supplied through the conjugate selectively accumulated at the level of the cytoplasm.[51]

Similar conclusions were derived from a study in which an immunoliposome containing doxorubicin and coated with an anti-transferrin receptor mAb was used to overcome doxorubicin resistance in K562 cells.[52] This immunoliposome accumulated in the cell at the level of the juxtanuclear region. Interestingly, while free doxorubicin was rapidly excreted from the cells, the doxorubicin administered via the immunoliposome remained within the cell.

Another set of experiments focused on evaluating the effect of mAb anti-transferrin receptors coupled with toxins as possible anti-tumor agents. Recombinant toxins are hybrid cytotoxic proteins made by recombinant DNA technology and designed to selectively kill cancer cells.[53] The cell-targeting moiety can be a growth factor or a single-chain, antigen-binding protein. The toxic moiety is a protein of bacterial or plant toxin. Immunotoxins are similar in concept, but are composed of antibodies chemically linked to toxins.

In this context, a number of studies were carried out using human ovarian carcinoma cell lines as targets for transferrin receptor antibodies linked to *Pseudomonas* exotoxin.[54-56] *In vivo* experiments provided evidence that intraperitoneal injection of the immunotoxin (anti-transferrin receptor antibody conjugated with *Pseudomonas* toxin or ricin A chain) protected nude mice injected with ovarian tumor cells from the development of malignant tumoral ascites and markedly prolonged animal survival. Other studies carried out by intraperitoneal administration of an immunotoxin made with anti-transferrin receptor and a recombinant form of *Pseudomonas* exotoxin caused regression of solid carcinomas growing as subcutaneous xenografts.[57]

Alternatively, immunotoxins were constructed by conjugating the anti-transferrin receptor mAb B3/25 to a ribosome-inactivating protein, saporin-6. The B3/25-S06 toxin displayed a marked cytotoxicity against leukemic cell lines and fresh leukemia blasts.[58]

The distribution of transferrin receptors on normal tissues limits the clinical use of immunotoxins made with mAb anti-transferrin receptors and a toxin. In fact, it is expected that the intravenous administration of such an immunotoxin would result in significant toxicity for tissues expressing elevated levels of transferrin receptors (i.e., red cell precursors, placenta, liver, alveolar macrophages, actively proliferating tissues).

On the basis of these considerations, experimental *in vivo* studies have been limited to regional intracavitary therapy. Additional studies provided further evidence that the intraperitoneal administration of an immunotoxin composed of anti-trans-ferrin receptor mAb linked to a recombinant toxin to nude mice bearing peritoneal human tumors[59,60] or to mice bearing syngeneic peritoneal tumors[61] resulted in significant anti-tumor activity. Furthermore, these studies showed two important points: (1) a significant anti-tumor effect was observed only when the immunotoxins were administered intraperitoneally and not intravenously; and (2) successful treat-ment required a relatively small tumor burden.

These preclinical studies were largely predictive of the possible clinical appli-cations based on the use of immunotoxins with an antibody part represented by an mAb anti-transferrin receptor. Two recent studies showed that this approach may have important applications in the treatment of brain tumors.

A pilot study of intraventricular therapy with immunotoxin 454A12-rRA (a conjugate of an mAb against the human transferrin receptor and recombinant ricin A chain) was performed in eight patients with leptomeningeal spread of systemic neoplasia. The administration of this toxin was well tolerated up to the dose of 38 µg. At higher doses corresponding to 120 µg of immunotoxin a transient CNS inflammatory response was observed; it was responsive to steroid treatment, and no systemic toxicity was observed.[62]

In four of the eight patients undergoing this treatment, a greater than 50% reduction of tumor cell counts was observed; however, none of these patients achieved a complete response. A second trial was based on intralesional injection (high-flow interstitial microinfusion) of an immunotoxin represented by a conjugate of human transferrin and a genetic mutant of diphtheria toxin (transferrin–CRM107).[63] Nine of the 15 treated patients exhibited at least 50% reduction of tumor size, including two complete responses.[63] Interestingly, three patients exhibited clinical responses lasting more than 120 weeks. This treatment too was associated with low toxicity.

Finally, *in vitro* studies suggest that blocking anti-transferrin receptor mAb could be used in combination with other anti-receptor antibodies to inhibit the growth of tumor cells. This approach was assayed *in vitro* in myeloma cells and showed that anti-transferrin receptor and anti-interleukin-6 receptor mAbs have additive anti-proliferative effects.[64] Interestingly, the combined addition of the two antibodies inhibited the growth of myeloma cell lines scarcely sensitive to anti-transferrin receptor or anti-interleukin-6 receptor mAb alone.[64]

2.4.2 INHIBITION OF CELL GROWTH BY AGENTS THAT INTERFERE WITH INTRACELLULAR IRON INCORPORATION

Gallium, a Group IIIA metal, has gained clinical importance recently as a pharma-cologic agent for treatment of hypercalcemia and certain malignancies, such as lymphoma and bladder cancer.[65,66] Although gallium is not a transition element, many aspects of its chemistry resemble those of iron. In fact, the ionization potentials, ionic radii, and the coordination numbers of the aqueous +3 cations are similar for both metals.

The remarkable similarity between gallium and iron allows for the interaction of gallium with various iron-containing proteins. Gallium avidly binds to the iron transport protein transferrin, resulting in stable transferrin–gallium complexes,[67] that can be incorporated into cells via transferrin receptor-mediated endocytosis.[68,69]

The cytotoxicity of gallium can be enhanced significantly when cells are exposed to gallium as transferrin–gallium rather than gallium nitrate.[70] However, studies performed in a series of leukemic cell lines showed that while gallium inhibited iron uptake in all these cells, it displayed a variable effect on transferrin receptor expression (i.e., an increase in the myeloid HL-60 cell line, compared with a decrease in CCRF-CEM lymphoid cells).[71]

Exposure of cells to transferrin–gallium results in a decrease in iron uptake, an increase in cellular transferrin receptors, and a decrease in ferritin content — findings consistent with cellular iron deprivation.[70] In addition to the inhibition of transferrin-dependent iron uptake, gallium is also able to inhibit transferrin-independent iron uptake by tumor cells.[70]

The uptake of iron by cells can be blocked by gallium, resulting in cellular iron deprivation and an arrest in cellular proliferation *in vitro*.[70] The block in DNA synthesis is due, in part, to inhibition of ribonucleotide reductase,[72,73] the enzyme responsible for the synthesis of deoxyribonucleotides. In fact, the ribonucleotide reductase M2 subunit electron spin resonance spectroscopy signal in cell cytoplasmic extracts was markedly inhibited in Ga-treated cells. The signal was restored to normal within 10 minutes of exposure of these cytoplasmic extracts to ferrous ammonium sulfate. Thus, gallium appears to inhibit ribonucleotide reductase by interfering with the incorporation of iron into its non-heme iron-containing R2 subunit.[74]

In certain cells, the gallium-induced block in cell growth can be overcome by transferrin-iron, hemin, or iron salts.[70,72] Tumor cells may become gallium resistant through a process that involves a reduction in gallium uptake; this resistance may be overcomed by transferrin addition.[75] Patients treated with gallium nitrate infusion have been shown to develop tissue iron deficiency,[76] suggesting that the interaction of gallium with iron metabolism is not limited to *in vitro* cell culture conditions.

An important strategy in the treatment of tumors is combining drugs that act synergistically to inhibit the growth of tumor cells. In line with this view, some studies were done using gallium in combination with other anti-proliferative drugs. Some studies indicated that gallium acts synergistically with hydroxyurea and fludarabine (inhibitors of ribonucleotide reductase) to inhibit the growth of leukemic HL-60 cells.[73,77]

Similarly, gallium and interferon-α act in synergism by inhibiting the growth of T-lymphoblastic leukemic CCRF-CEM growth through action on cellular iron uptake.[78]

Gallium nitrate has some important applications in oncology. The most important is its use in the radiodiagnosis of tumors, particularly lymphomas.[79-81] It was clearly shown that gallium scans predict outcomes in patients with aggressive non-Hodgkin's lymphoma treated with high-dose chemotherapy[79] and patients undergoing high-dose chemotherapy and autologous stem-cell transplantation for non-Hodgkin's lymphoma.[80] In contrast, gallium scans are not predictive of the outcomes of Hodgkin's disease.[81] Gallium nitrate is also an active agent in the treatment of advanced carcinoma of the urothelium[82] and for the treatment of bone metastases.[83]

2.4.3 Inhibition of Cell Proliferation Mediated by Iron Chelators

Deferoxamine, a trihydroxamate siderophore (Figure 2.5) derived from *Streptomyces pilosus*, is currently the most clinically useful iron chelator available. Each molecule of deferoxamine can complex with one ferric ion to form ferrioxamine. Since its introduction in 1960, deferoxamine has been used extensively in the prevention and

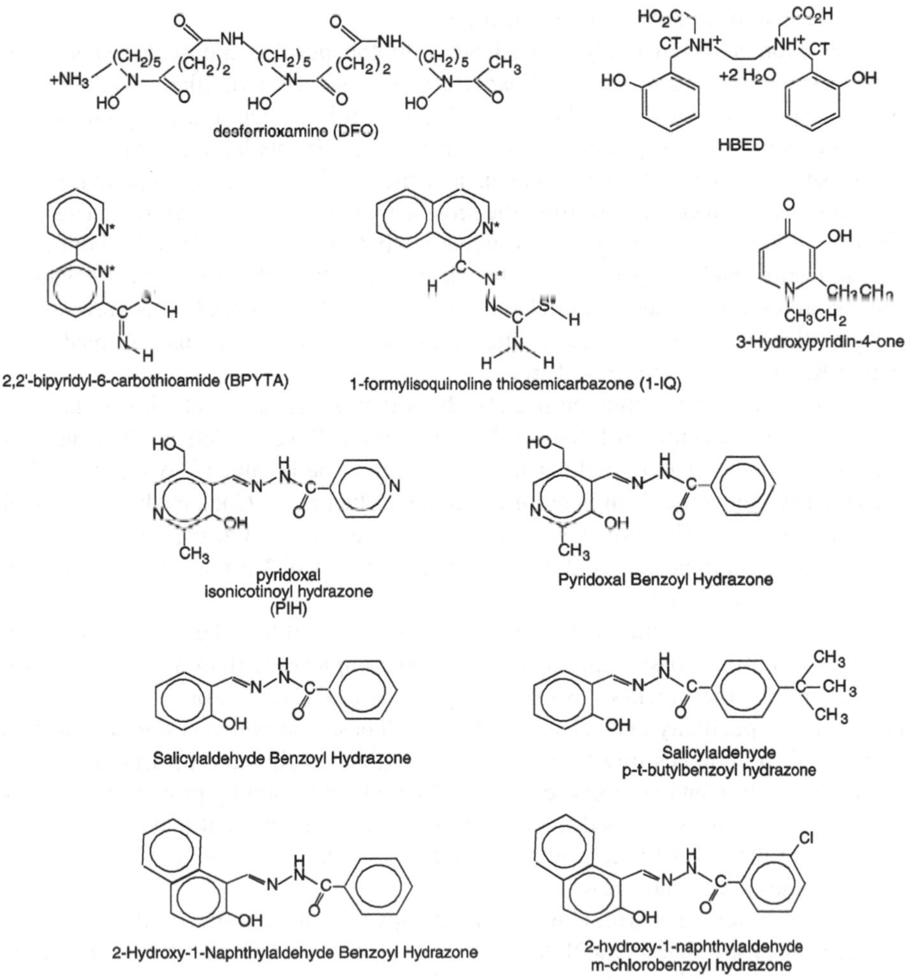

FIGURE 2.5 Structures of the main iron chelators studied both for their iron-chelating and anti-proliferative properties.

treatment of iron overload in chronically transfused children. Deferoxamine does not remove iron from transferrin; it enters the cells and acts on the intracellular iron pool, which can be regarded as the iron in transit between different cellular compartments.

Several studies have shown the iron deprivation due to the addition of deferoxamine had profound effects on DNA synthesis.[84-88] When added to *in vitro* growing normal and neoplastic cells, deferoxamine behaves like an S-phase inhibitor, determining a block of cell proliferation at the beginning of the S phase. This phenomenon is associated with elevated levels of expression of transferrin receptors. The biochemical mechanism of this anti-proliferative effect is the inhibition of ribonucleotide reductase activity, thereby preventing cells from completing the S phase of cell cycle. Although there has been convincing demonstration of the effects of deferoxamine on both ribonucleotide reductase activity *in vitro*[84] and intracellular levels of deoxyribonucleoside triphosphates,[89,90] direct *in vivo* effects on ribonucleotide reductase have not been demonstrated.

However, a recent study carried out with the purified mouse R2 subunit of ribonucleotide reductase provided experimental evidence about the possible mechanism responsible for the inhibition of ribonucleotide reductase activity elicited by iron chelators.[8] In fact, *in vitro* studies using pure, ^{59}Fe-labeled recombinant mouse R2 protein showed that its iron center is labile at physiological temperatures and that iron is spontaneously lost from the protein even in the absence of iron chelators.[8] On the basis of these findings it was suggested that iron chelators inhibit ribonucleotide reduction and cell growth, not by directly attacking the iron-radical center of the R2 protein, but instead by chelating the intracellular iron pool. This prevents the regeneration of the iron-radical center in apo-R2 protein continuously formed from active R2 protein by the loss of iron.

Furthermore, deferoxamine has also been demonstrated to inhibit cellular proliferation independently of DNA synthesis in some cellular models,[88,91,92] suggesting that other iron-dependent cellular mechanisms may be involved.[93] A recent study[28] provided evidence that iron is required for the induction of Cdc2 cyclin during cell cycle progression. This conclusion is based on the observation that the addition of deferoxamine completely inhibited the induction of Cdc2 protein in lymphocytes stimulated with mitogens.[28]

Similar anti-proliferative effects were also displayed by other compounds that, like deferoxamine, possess the capacity to form complexes with iron.[94,95] Particularly interesting are the bidentate 3-hydroxypyridin-4-one (HPO) iron chelators which exhibit high specificity and selectivity for iron. These compounds have rapid access across cell membranes due to their low molecular weights and neutral charges in both the iron-free and complexed forms. When added to rapidly proliferating tumor cells, these chelators arrested the cell cycle in a dose-dependent manner causing a blockade at the G1/S border, inhibited ribonucleotide reductase, and then blocked DNA synthesis and cell growth.[12]

Another interesting iron chelator is 2,2′-bipyridyl-6-carbothioamide. Its chelating properties are due to its N-N-S tridentate ligand system and the compound is therefore comparable to α-(N)-heterocyclic carboxaldehyde thiosemicarbazones which are potent inhibitors of ribonucleotide reductase.[95]

It is of interest that the iron chelators HPO[94] and parabactin[96] (a microbial siderophore) have been shown to be more potent synchronization agents than either deferoxamine or hydroxyurea when tested on leukemic cell lines. Therefore, these efficient cell-synchronizing iron chelators could be of value to enhance the effects of other cytotoxic antitumor drugs.

Recent studies focused on evaluating the iron-chelating and anti-proliferative activity of the pyridoxal isonicotinoyl hydrazone (PIH) family of iron chelators[97] (Figure 2.5). These studies showed that analogues derived from pyridoxal benzoyl hydrazone exhibited elevated iron-chelating activities, but produced relatively low anti-proliferative effects. On the contrary, an analogue derived from salicylaldehyde benzoyl hydrazone exhibited high iron chelation activity and elevated capacity to inhibit cell proliferation.[97] According to these observations, it was proposed that the first group of compounds exhibits characteristics suitable for treatment of iron overload, while the second group is potentially suitable for cancer treatment.[97]

Analysis of the mechanism of action of the PIH analogues that are most active as proliferation inhibitors showed that on a molar basis they are more active than desferrioxamine in preventing iron uptake from transferrin and in inhibiting cell proliferation. Furthermore, these effects blocked different cell types at various phases of the cell cycle, even though the most prevalent block was at the level of G0/G1.[98]

The effects of iron chelators on cell proliferation are largely dependent on their capacity to chelate a labile cytoplasmic iron pool. Iron is mainly acquired from the milieu via transferrin through a receptor-mediated process. Endocytosed transferrin in the acid milieu of endosomes releases iron which is then translocated into the cytoplasm where it enters into a low molecular weight cytoplasmic pool. This iron pool represents the direct target of iron chelators, and for this reason is also known as the cytoplasmic pool of chelatable iron.[99] The addition of slowly permeating chelators, such as desferrioxamine, elicited a progressive decline of this cytoplasmic iron pool, while highly lipophilic iron chelators, such as 2,2'-bipyridyl, induced a rapid decline of this iron pool.[99]

The removal of iron from this pool by chelators is a process requiring energy; however, the removal of iron mediated by some lipophilic chelators, such as 2-hydroxy-11-naphthylaldheyde isonicotinoyl hydrazone is an energy-independent process, probably occurring through simple diffusion or passive transport.[100]

In addition to these molecules, two other iron chelators, deferipone and HBED (N,N-bis2-hydroxybenzyl-ethylenediamide-N,N-diacetic acid), were actively investigated for the chronic treatment of transfusional iron overload[101] (Figure 2.5). Since these chelators are apparently more active than desferrioxamine in removing iron, they also could be potentially used as anti-proliferative agents in cancer therapy.

Other studies have shown that deferoxamine synergizes with anti-tumor drugs in inhibiting tumor cell growth. More particularly, it was shown that deferoxamine enhances cytosine arabinoside, methotrexate, and daunorubicin cytotoxicity.[102] The synergism of each of these drugs with deferoxamine may be explained by considering the biochemical mechanism mediating the action of deferoxamine and of the cytotoxic drug. Deferoxamine-mediated inhibition of ribonucleotide reductase lowers the intracellular pool of deoxyuridylate (dUNP), the substrate of thymidilate synthetase, which is the enzyme inhibited by methotrexate. Thus, the combined depletion of the dUNP pool by deferoxamine and decreased thymidilate synthetase may act synergistically to deplete dTTP pools and result in potentiated inhibition of DNA synthesis.

The cytotoxic effect of methotrexate in the S phase of the cell cycle may be enhanced by deferoxamine, which inhibits cell proliferation in the early S phase. Similar considerations may explain the synergism between cytosine arabinoside or daunorubicin and deferoxamine. Furthermore, it was shown that deferoxamine may also strongly potentiate the anti-proliferative effects of gallium.[103]

Recent studies indicate that iron chelators, in addition to blocking the proliferation of cancer cells, may also induce apoptosis. Studies performed on HL-60,[104] CCCRF-CEM,[105] T lymphocytes,[106] and lymphoma cells[107] showed that iron chelation mediated either by desferrioxamine or deferipone resulted first in inhibition of proliferation and then in apoptosis. Interestingly, this effect is restricted only to proliferating cells in that the addition of desferrioxamine to quiescent T lymphocytes did not induce apoptosis.[106]

The apoptosis induced by iron chelation seems to require p53. The iron chelators increase p53 levels through a post-transcriptional mechanism.[108] The mechanism through which iron chelators trigger p53 synthesis is seemingly represented by DNA damage. In fact, iron deprivation is associated with accumulation of single-strand breaks in DNA. However, recent studies have in large part clarified the mechanisms responsible for p53 induction by iron chelators.[109] They found that iron chelators mediate induction of p53 protein accumulation through a mechanism dependent upon induction of hypoxia-inducible factor 1α (HIF-1α). This protein interacts with and stabilizes p53 (in control cells, p53 protein half-life was <30 min; in desferrioxamine-treated cells, it was >120 min).[109]

Based on these experimental results, initial clinical studies were carried out to assess the potential application of iron chelators to tumor therapy. These studies focused on three types of tumors, including leukemia, neuroblastoma, and hepatoma since clear evidence indicated that iron could play an important role in the control of the proliferation of these cancer cells. An additional consideration that made these studies reliable was the observation that *in vitro* anti-proliferation effects of deferoxamine were observed at concentrations readily attainable *in vivo*. In fact, pharmacokinetics studies have shown that healthy controls develop peak plasma levels of about 15 µmol/L after a single intramuscular injection of 10 mg/kg body weight.

The first clinical assays with deferoxamine were carried out in a few patients with acute leukemia.[110,111] Estrov et al.[104] administered deferoxamine to an infant with neonatal acute leukemia, the form of leukemia related to the B lymphoid lineage. The intravenous infusion of deferoxamine to this patient elicited a block of blast cell growth associated with a partial change of membrane phenotype (induction of the expression of myelomonocytic markers). The combined administration of deferoxamine and cytosine arabinoside to this patient induced a marked leukemic cytoreduction.

A second study was carried out by Dezza et al.[111] to evaluate the effects of deferoxamine on progression of leukemia refractory to conventional antiblastic therapy[105] In a patient with blast crisis of chronic myelogenous leukemia, deferoxamine induced a marked reduction in circulating blasts. In contrast, in a patient with acute nonlymphocytic leukemia and transfusional iron overload, no effect of defer-oxamine was observed. The loss of activity of deferoxamine in the second patient was ascribed to the presence of iron overload, a condition in which deferoxamine is converted to ferrioxamine and thus cannot act on target leukemic cells.

A second series of clinical trials was carried out in neuroblastoma patients. Neuroblastoma is the fourth most common tumor in infants less than one year of age. Seventy percent of patients have widespread disease at diagnosis. Their chances of survival are dismal. Fewer than 25% are free of disease 2 years after diagnosis, and median survival is less than 1 year.

Despite improvements in diagnostic, surgical, and radiotherapeutic techniques, and the development of chemotherapeutic regimens, prognosis for patients with metastatic neuroblastoma has changed little over the past 30 years. These facts highlight the need to better understand the biology of this tumor in order to develop new strategies with which to treat it.

Particularly interesting are the studies concerning ferritin content of neuroblas-toma tumors. Among children with advanced neuroblastoma, serum concentrations of ferritin appear to discriminate among subsets of patients with different prognoses. Patients with supranormal concentrations tend to fare worse than those with high normal concentrations, and both these groups do significantly worse than children with concentrations in the low/normal range.[112]

In neuroblastoma patients at a given stage of the disease, wide variations have been found among ferritin concentrations, which have proven to have greater prog-nostic significance than other tumor markers such as the cathecolamine metabolites (vanilmandelic acid, homovanillic acid) and neuron-specific enolase. Progression-free survival at 24 months of follow-up for patients with Stage III (regionally extensive) disease and normal ferritin levels was 76%, and with elevated ferritin levels (140–600 ng/ml) it was 23%.[112] For patients with Stage IV (metastatic) disease, progression-free survival rates were 27% and 3% with normal and elevated ferritin levels, respectively.[112] In contrast, no direct relationship has been found between prognosis and either serum iron levels or total iron binding capacity.[112]

These observations suggest that ferritin could stimulate the growth of neuroblas-toma cells. This hypothesis was directly confirmed by experiments in which neuro-blastoma cells were grown in vitro in the presence of increasing concentrations of ferritin.[113] Under these experimental conditions, ferritin at concentrations higher than 50 ng/ml elicited a significant enhancement of neuroblastoma cell growth.

Characteristics of neuroblastoma isoferritins appear to be variable. In one report, ferritin extracted from tumor samples consisted of equal parts H and L chains, similar to ferritin in heart tissue.[114] In another report from the same investigators, isoferritin found in serum samples of patients with neuroblastoma, although heterogeneous, consisted predominantly of light chains.[115] In addition, it has been demonstrated that neuroblastoma cells not only contain and secrete ferritin,[116,117] but synthesize it as well.[117]

The peculiar role played by iron in the control of the growth of neuroblastoma cells was directly supported by several studies showing particular sensitivity of neuroblastoma cell lines to the anti-proliferative activity of deferoxamine.[118,119] A recent study carried out on a neuroblastoma cell line clearly confirmed the particular sensitivity of these cells to the anti-proliferative effect of deferoxamine as compared to that exhibited by other types of tumor cells.[30] Furthermore, it was shown that deferoxamine blocked neuroblastoma cells to the G1 phase of the cell cycle, thus suggesting that iron, in addition to being necessary for DNA synthesis, is required for G1 to S phase progression.[30]

This conclusion was supported by the observation that deferoxamine markedly inhibited the expression of the cyclin Cdc2.[30] However, recent studies have questioned whether iron removal from neuroblastoma cells by iron chelators reresents the main mechanism through which these agents inhibit the growth of these tumor cells. In fact, using a series of different iron chelators, it was shown that no relationship between the efficacy of iron chelation and the inhibition of DNA synthesis existed.[120]

Taking into account these findings, clinical studies were performed to evaluate the possible efficacy of deferoxamine in the treatment of neuroblastoma. In an early study, Donfrancesco et al.[121] administered a single 5-day course of deferoxamine to 9 patients with neuroblastomas. Within 2 days of completion of treatment, responses were observed in 7 patients without signs of drug toxicity. The responses consisted of a decrease in bone marrow infiltration and, in one patient, a significant reduction of tumor cell mass.

In a second study, 13 patients with Stage III (3 patients) or Stage IV (10 patients) neuroblastomas were teated with an iron chelation–cytotoxic therapy regimen[122] that consisted of 5 days of administration of deferoxamine followed by 3 days of cyclo-phosphamide, etoposide, carboplatin, and thiotepa therapy. This regimen caused moderate to severe myelotoxicity. Objective responses were observed in 12 of 13 patients. Three of four partial responses were achieved in previously treated, relapsed patients, and seven of eight complete responses were achieved in previously untreated patients.

A recent study has questioned the real impact of iron chelation therapy on neuroblastoma status through changes in iron metabolism. Using a xenograft model of human neuroblastoma, it was shown that treatment with iron chelators (either desferrioxamine or deferipone) at doses capable of inducing iron depletion does not affect tumor engraftment, latency or size.[123]

Other studies focused on evaluating the effects of iron chelators on the growth of human hepatocellular carcinoma cells. These studies were stimulated by clinical observations. The most relevant is that idiopathic hemochromatosis, a genetic dis-order characterized by iron overload, is associated with a markedly increased risk of the development of primary liver cancer and other malignant diseases. Although in idiopathic hemochromatosis, development of hepatocellular carcinoma often is associated with existing cirrhosis, there have been reports of hepatocellular carci-nomas arising in non-cirrhotic livers.

Studies carried out in experimental animals showed that transplanted hepatoma tumors grew faster and larger in mice fed an iron-rich diet than in those that were fed a low-iron diet.[124] In vitro studies have shown that deferoxamine exerted cyto-toxic effects against hepatocellular carcinoma cell lines.[125] When three hepatocellular carcinoma cell lines, PLC/PRF/5, Hep G2, and Hep 3B were incubated in tissue culture medium mixed with graded amounts of deferoxamine for 48 to 96 hours, more than 50% cell deaths in PLC/PRF/5 cells and 30 to 50% cell deaths in Hep

G2 and Hep 3B cells were observed 48 to 72 hours after exposure to deferoxamine.[124] These obsevations were confirmed by *in vivo* studies showing that the administration of deferoxamine to nude mice transplanted with hepatocellular carcinoma cells elicited a significant impairment of tumor growth.[126]

2.5 IRON AND CANCER

It was proposed that one of the crucial events that occurs in a tumor cell is the alteration of cell membrane, resulting in increased internal concentration of nutrients that regulate cell growth.[127] Several studies provided evidence that one of these nutrients is iron.

Several tumors offer different examples of such a phenomenon:

1. Small-cell lung cancer cells synthesize transferrin through an autocrine mechanism and this phenomenon may represent the strategy used by these cells to actively proliferate even in the presence of a reduced vascularization.[128]
2. Melanoma cells synthesize a 97-kDa membrane protein that is similar both from stuctural and functional points of view to transferrin.[129]
3. Several tumor cells are able, unlike normal cells, to acquire iron from iron salts in the absence of transferrin[130-132] or synthesize elevated levels of transferrin receptors, whereas corresponding normal cells do not express these receptors.[133,134]
4. Aggressive lymphomas have been found to display an epitope of transferrin receptor that is not seen in indolent lymphomas or in normal cells.[135]
5. The preferential colonization of bone marrow by prostatic carcinoma has been attributed to the passage of cells from the prostate to the spine via specialized paravertebral blood vessels and proliferation in the marrow of the metastatic cells under the stimulation of the high concentrations of available transferrin in bone marrow.[136,137]

Other studies[138] have shown that increases of body iron stores may enhance the risk of cancer by two different mechanisms

1. Iron can catalyze the synthesis of oxygen radicals that may deprive cells of their reducing activity and then their capacity to neutralize harmful oxidant species released by toxic agents or radiations.
2. Iron may be a limiting nutrient for tumor cells and that increases the probability that a tumor can develop to become a clinically detectable tumor.

Some studies provided direct evidence that high iron stores increase the risk of the development of tumors. This conclusion is based on experimental studies carried out on animals and epidemiological studies in humans. The animal studies were based on a single approach intended to evaluate the effect of iron supplementation or iron deprivation on the growth of tumor cells inoculated into syngeneic mice. These studies showed that: (1) iron supplementation favored tumor development and more rapid death[139]; and (2) tumors grow more slowly in mice on low-iron diets.[140]

Other studies focused on evaluating the effects of iron on the development of tumors in mice inoculated with chemical carcinogens.[141,142] These studies showed that rats receiving dietary iron supplementation exhibited the greatest tumor burdens, whereas the rats receiving iron-deprived diets had fewer tumors than animals on the normal diet.[141,142]

Few epidemiological studies have investigated the association of body iron stores and risk of cancer.[143-146] A particularly relevant study was carried out by Stevens et al.[143] analyzing cancer incidence on a sample of 14,407 subjects. These subjects were screened according to several biomedical parameters including serum iron and total iron-binding capacity.

Among 3355 men who remained alive and were not diagnosed with cancer within 4 years, Stevens noted a significant direct association between serum iron/transferrin saturation and the risk of developing cancer over an approximately 10-year follow-up period. More particularly, cancers of the colon, bladder, esophagus, and lung were most strongly related to transferrin saturation.[143] For cancer of the colon, risk in each quartile of transferrin saturation relative to a baseline of 1.0 for those with transferrin saturation values under 22.8% increased to 1.76 in those at 22.9 to 36.7% levels, and to 4.69 in those with values above 36.7%.

This last observation was supported by a more recent study of Willet et al.[147] showing that red meat intake was significantly related to risk of colon cancer: women consuming red meat every day have a risk of developing colon cancer 2.5 times higher than those who consume red meat once a month.[147] These authors suggested iron as a possible reason for their reported association of red meat with colon cancer risk.[148]

The findings of Stevens et al. were confirmed in a more recent survey carried out on 3287 men and 5269 women participating in the first National Health and Nutrition Examination Survey who had transferrin saturation determinations at enrollment (1971–1975), remained alive and cancer-free for at least 4 years, and were followed through 1988 to determine cancer outcome.[149] This study provided evidence of elevated cancer risk in subjects with moderately elevated iron levels; this pattern was observed equally in men and in women.

Finally, a more recent study carried out on a relatively large sample of a Finnish population[149] (41,276 subjects) screened for total iron-binding capacity (TIBC), transferrin saturation, and serum iron concentration, and followed for a period of 14 years. During this period, 2469 subjects developed primary cancers. Excess risks of colorectal and lung cancers were found in subjects with transferrin saturation levels exceeding 60%; the relative risks were estimated to correspond to 3.04 and 1.51 for colorectal and lung cancer, respectively, as compared to subjects having lower iron levels.[150] It was also observed that the risk of stomach cancer was inversely related to serum iron and transferrin saturation and directly related to TIBC.

Additional studies are required before it is clear whether and to what extent moderate elevations in body iron stores increase cancer risk.

A recent study further confirmed the existence of a relationship betweeen iron intake and the risk of colorectal cancer.[151] A study based on the follow-up of 14,407 persons, of whom 194 developed colorectal cancer over a period of 15 years, showed that iron may confer increased risk of colorectal cancer and the localization of risk may be related to some extent to the mode of epithelial iron exposure. Luminal exposure to iron increases the risk of developing a cancer in the proximal region of colon, while humoral exposure to iron augments the risk of cancer in the distal region of colon.[151]

However, some recent studies have questioned the existence of a relationship between iron intake and the risk of developing colon cancer. Iron intake does not seem to increase the recurrence of colon adenomas.[152] The existence of such a linkage may be in part due to the iron supplementation given to many patients with colorectal cancer due to early symptoms of iron shortage.[153]

It was suggested that increased hepatic iron load may represent a risk factor for the development of hepatocellular carcinoma. Iron overload may lead to liver cirrhosis and to an increased risk of developing primary hepatocellular carcinoma. Whether iron is of pathogenic importance for carcinogenic process, or whether the increased cancer risk results solely from the cirrhotic process is unknown. Some recent studies favor the second hypothesis. In fact, 229 patients' homozygotes for hereditary hemochromatosis were prospectively studied. During follow-up, 28 of 97 patients with cirrhosis, but none of the patients without cirrhosis, developed hepatocellular carcinoma.[154] The effect of dietary iron on hepatocarcinogenesis in animal models, as evaluated, indicates that iron neither acts as an initiator nor as a promoter of hepatic carcinogenesis.[155]

REFERENCES

1. Stevens, R.G. and Neriishi, K. Iron and oxidative damage in human cancer, in *Biological Consequences of Oxidative Stress,* Spatz, L. and Bloom A.D., Eds., Oxford, Oxford University Press, 138–161, 1992.
2. Reichard, P. The biosynthesis of deoxyribose, in *Ciba Lectures in Biochemistry,* New York, Wiley, 1967.
3. Reichard, P. Interactions between deoxyribonucleotide and DNA synthesis, *Ann. Rev. Biochem.,* 57: 349–374, 1988.
4. Mass, S.S. et al. Mechanism-based inhibition of a mutant *Escherichia coli* ribonucleotide reductase (cysteine 225–serine) by its substrate CDP, *Proc. Natl. Acad. Sci. U.S.A.,* 86: 1485–1490, 1989.
5. Bianchi, V., Pontis, E., and Reichard, P. Interrelations between substrate cycles and *de novo* synthesis of pyrimidine deoxyribonucleoside triphosphates in 3T6 cells, *Proc. Natl. Acad. Sci. U.S.A.,* 83: 986–990, 1986.
6. Akerblom, R. et al. Overproduction of free radical of deoxyribonucleotide reductase in hydroxyurea-resistant mouse fibroblast 3T6 cells, *Proc. Natl. Acad. Sci. U.S.A.,* 78: 2159–2163, 1981.
7. McClarty, G.A. et al. Increased ferritin gene expression is assoiated with increased ribonucleotide reductase gene expression and the establishment of hydroxyurea resistance in mammalian cells, *J. Biol. Chem.,* 265: 7539–7547, 1990.
8. Reichard, P. The evolution of ribonucleotide reduction, *TIBS,* 22: 81–85, 1997.
9. Harder, J. and Follmann, H. Identification of a free radical and oxygen dependence of ribonucleotide reductase in yeast, *Free Radic. Res. Commun.,* 10: 281–286, 1990.
10. Larsson, A. and Sjoberg, B.M. Identification of the stable free radical tyrosine residue in ribonucleotide reductase, *EMBO J.,* 2037–2040, 1986.

11. Norlund, P. et al. Three-dimensional structure of the free radical protein of ribonu-cleotide reductase, *Nature,* 345: 593–598, 1990.
12. Mann, G.J. et al. Purification and characterization of mouse and herpes simplex virus ribonucleotide reductase R2 subunit, *Biochemistry,* 30: 1939–1947, 1991.
13. Filatov, D. et al. The role of herpes simplex virus ribonucleotide reductase small subunit carboxyl terminus in subunit interaction and formation of iron-tyrosyl center structure, *J. Biol. Chem.,* 267: 15816–15822, 1992.
14. Stubbe, J., Ator, M., and Krenitsky, T. Mechanism of ribonucleotide diphosphate reductase from *Escherichia coli.* Evidence for 3'-C-H bond cleavage, *J. Biol. Chem.,* 258: 1625–1632, 1983.
15. Lassmann, G., Thelander, L., and Graslund, A. EPR stopped-flow studies of the reaction of the tyrosyl radical of protein R2 from ribonucleotide reductase with hydroxyurea, *Biochim. Biophys. Res. Commun.,* 188: 879–887, 1992.
16. Petersson, L. et al. The iron center in ribonucleotide reductase from *Escherichia coli.* *J. Biol. Chem.,* 288: 6706–6715, 1980.
17. Reichard, P. From RNA to DNA. Why so many ribonucleotide reductases? *Science,* 260: 1773–1777, 1993.
18. Licht S., Gerfen, G.J., and Stubbe, J. Thiyl radicals in ribonucleotide reductase, *Science,* 271: 477–481, 1996.
19. Uhlin, U. and Eklund, H. Structure of ribonucleotide reductase protein R1, *Nature,* 370: 533–539, 1994.
20. Sjoberg, B.M. Ribonucleotide reductase, *Nucleic Acids Mol. Biol.,* 9: 192–221, 1995.
21. Bjorklund, S. et al. S-phase-specific expression of mammalian ribonucleotide reduc-tase R1 and R2 mRNAs, *Biochemistry,* 29: 5452–5458, 1990.
22. Albert, D.A. and Rozengurt, E., Synergistic and coordinate expression of the genes encoding ribonucleotide reductase subunits in Swiss 3T3 cells: effect of multiple signal-transduction pathways, *Proc. Natl. Acad. Sci. U.S.A.,* 89: 1597–1601, 1992.
23. Bjorklund, S., Skogman, E., and Thelander, L. An S-phase specific release from a transcriptional block regulates the expression of mouse ribonucleotide reductase R2 subunit, *EMBO J.,* 11: 4953–4959, 1992.
24. Filattov, D. et al. Induction of the mouse ribonucleotide reductase R1 and R2 genes in response to DNA damage by UV light, *J. Biol. Chem.,* 271: 23698–237704, 1996.
25. Fan, H., Villeegas, G., and Wright, J.A. Ribonucleotide reductase R2 component is a novel malignancy determinant that cooperates with activated oncogenes to determine transformation and malignant potential, *Proc. Natl. Acad. Sci. U.S.A.,* 93: 14036–14040, 1996.
26. Fau, H. et al. The R1 component of mammalian ribonucleotide reductase has malig-nancy-suppressing activity as demonstrated by gene transfer experiments, *Proc. Natl. Acad. Sci. U.S.A.,* 94: 13181–13186, 1997.
27. Terada, N., Lucas, J.J., and Gelfand, E.W. Differential regulation of the tumor sup-pressor molecules. Retinoblastoma susceptibility gene product (Rb) and p53 during cell cycle progression of normal human T cells, *J. Immunol.,* 147: 698–704, 1991.
28. Terada, N. et al. Definition of the roles for iron and essential fatty acids in cell cycle progression of nomal human T lymphocytes, *Exp. Cell. Res.,* 204: 260–267, 1993.
29. Lucas, J.J. et al. Regulation of synthesis of p34 cdc2 and its homologues and their relationship to p110 Rb phosphorylation during cell cycle progression of normal human T cells, *J. Immunol.,* 148: 1804–1810, 1992.
30. Brodie, C. et al. Neuroblastoma sensitivity to growth inhibition by deferrioxamine. Evidence for G1 block in the cell cycle, *Cancer Res.,* 53: 3968–3075, 1993.

31. Lucas, J.J. et al. Effect of iron-depletion on cell cycle progression in normal human T lymphocytes: selective inhibition of the appearance of the cyclin A-associated compartment of the p33 gdk2 kinase, *Blood*, 86: 2286–2280, 1995.
32. Kulp, K.S., Green, S.L., and Vulliet, P.R. Iron deprivation inhibits cyclin-dependent kinase activity and decreases cyclin D/Cdk4 protein levels in synchronous MDA-MB-453 human breast cancer cells, *Exp. Cell Res.*, 229: 60–68, 1996.
33. Fukuchi, K. et al. G1 accumulation by iron deprivation with deferoxamine does not accompany change of pRb status in ML-1 cells, *Biochim. Biophys. Acta*, 1357: 297–303, 1997.
34. Trowbridge, I.S. et al. Transferrin receptors: structure and function, *Biochem. Pharmacol.*, 31: 925–941, 1984.
35. Trowbridge, I.S. and Lopez, D. Monoclonal antibody to transferrin receptor blocks transferrin binding and inhibits human tumor growth *in vitro*, *Proc. Natl. Acad. Sci. U.S.A.*, 79: 1175–1179, 1982.
36. Mendelsohn, J., Trowbridge, I.S., and Castagnola, J. Inhibition of human lymphocyte proliferation by monoclonal antibody to transferrin receptor, *Blood*, 62: 821–829, 1983.
37. Teatle, R. et al. Role of transferrin, Fe, and transferrin receptors in myeloid leukemia cell growth. Studies with an anti-transferrin receptor monoclonal antibody, *J. Clin. Invest.*, 75: 1061–1072, 1985.
38. Trowbridge, I.S. et al. Murine cell surface transferrin receptor: studies with anti-receptor monoclonal antibody, *J. Cell Physiol.*, 112: 403–410, 1982.
39. Lesley, J.F. and Schulte, R.J. Selection of cell lines resistant to anti-transferrin receptor antibody: evidence for a mutation in transferrin receptor, *Mol. Cell. Biol.*, 4: 1675–1681, 1984.
40. Lesley, J.F. and Schulte, R.J. Inhibition of cell growth by monoclonal anti-transferrin receptor antibodies, *Mol. Cell. Biol.*, 5: 1814–1821, 1985.
41. Teatle, R., Castagnola, J., and Mendelsohn, J. Mechanism of growth inhibition by antitransferrin receptor monoclonal antibodies, *Cancer Res.*, 46: 1759–1763, 1986.
42. Lesley, J.F., Schulte, R., and Woodes, J. Modulation of transferrin receptor expression and function by anti-transferrin receptor antibodies and antibody fragment, *Exp. Cell Res.*, 182: 215–233, 1989.
43. Domingo, D.L. and Trowbridge, I.S. Characterization of the human transferrin receptor produced in a baculovirus expression system, *J. Biol. Chem.*, 163: 13386–13392, 1988.
44. White, S. et al. Combinations of anti-transferrin receptor monoclonal antibodies inhibit human tumor cell growth *in vitro* and *in vivo*. Evidence for synergistic anti-proliferative effects, *Cancer Res.*, 50: 6295–6301, 1990.
45. White, S. et al. Monoclonal antibodies against defined epitopes of the human transferrin receptor cytoplasmic tail, *Biochim. Biophys. Acta*, 1136: 28–34, 1992.
46. Hsieh-Ma, T. et al. *In vitro* tumor growth inhibition by bispecific antibodies to human transferrin receptor and tumor-associated antigens is augmented by the iron chelator deferoxamine, *Clin. Immunol. Immunopharmacol.*, 80: 185–193, 1996.
47. Sauvage, C.A. et al. Effects of monoclonal antobodies that block transferrin receptor function on the *in vivo* growth of a syngeneic murine leukemia, *Cancer Res.*, 47: 747–753, 1987.
48. Kemp, J.D. et al. Synergistic inhibition of lymphoid tumor growth *in vitro* by combined treatment with the iron chelator deferoxamine and an immunoglobulin G monoclonal antibody against the transferrin receptor, *Blood*, 76: 991–995, 1990.

49. Huwgler, J., Wu, D., and Pardridge, W.M. Brain drug delivery of small molecules using immunoliposomes, *Proc. Natl. Acad. Sci. U.S.A.,* 93: 14164–14169, 1996.

50. Huwler, J., Yang, J., and Pardridge, W.M. Receptor mediated delivery of daunomycin using immunoliposomes: pharmacokinetics and tissue distribution in the rat, *J. Pharmacol. Exp. Ther.,* 282: 1541–1546, 1997.

51. Lai, B.T., Gao, J.P., and Lanks, K.W. Mechanism of action and spectrum of cell lines sensitive to a doxorubicin–transferrin conjugate, *Cancer Chemother. Pharmacol.,* 41: 155–160, 1998.

52. Suzuki, S. et al. Modulation of doxorubicin resistance in a doxorubicin-resistant human leukemia cell by an immunoliposome targeting transferrin receptor, *Brit. J. Cancer,* 76: 83–89, 1997.

53. Pastan, I. and Fitzgerald, D. Recombinant toxins for cancer treatment, *Science,* 259: 1173–1177, 1991.

54. Pirker, R. et al. Anti-transferrin receptor antibody linked to *Pseudomonas* exotoxin as a model immunotoxin in human ovarian carcinoma cell lines, *Cancer Res.,* 45: 751–759, 1985.

55. Fitzgerald, D., Willingham, M.C., and Pastan, I., Antitumor effects of an immunotoxin made with *Pseudomonas* exotoxin in a nude mouse model of human ovarian cancer, *Proc. Natl. Acad. Sci. U.S.A.,* 83: 6727–6731, 1986.

56. Fitzgerald, D. et al. Antitumor activity of an immunotoxin in a nude mice model of human ovarian cancer, *Cancer Res.,* 47: 1407–1416, 1987.

57. Batra, J.K. et al. Antitumor activity in mice of an immunotoxin made with antitranferrin receptor and a recombinant form of *Pseudomonas* exotoxin, *Proc. Natl. Acad. Sci. U.S.A.,* 86: 8545–8550, 1989.

58. Cazzola, M. et al. Cytotoxic activity of an anti-transferrin receptor immunotoxin on normal and leukemic human hematopoietic progenitors, *Cancer Res.,* 51: 536–541, 1991.

59. Griffin, T.W. et al. Antitumor activity of intraperitoneal immunotoxin in a nude mouse model of human malignant mesothelioma, *Cancer Res.,* 47: 4266–4272, 1987.

60. Scott, F.C. et al. An immunotoxin composed of a monoclonal antitransferrin receptor antibody linked by a disulphide bond to the ribosome-inactivating protein gelonin: potent *in vitro* and *in vivo* effects against human tumors, *J. Natl. Cancer Inst.,* 79: 1163–1170, 1987.

61. Bjorn, M.J. and Groetsema, G., Immunotoxins to the murine transferrin receptor: intracavitary therapy of mice bearing syngeneic peritoneal tumors, *Cancer Res.,* 47: 6639–6647, 1987.

62. Laske, D.W., Murasrkok, M., and Oldfield, E.H. Intraventricular immunotoxin therapy for leptomeningeal neoplasia, *Neurosurgery,* 41: 1039–1049, 1997.

63. Laske, D.W., Youle, R.J., and Oldfield, E.H. Tumor regression with regional distribution of the targeted toxin TF-CRM107 in patients with malignant bone tumors, *Nat. Med.,* 3: 1362–1368, 1997.

64. Taetle, R. et al. Effects of combined anti-growth factor receptor treatment on *in vitro* growth of multiple myeloma, *J. Natl. Cancer Inst.,* 86: 450–455, 1994.

65. Foster, B.J. et al. Gallium nitrate: The second metal with clinical activity, *Cancer Treat. Rep.,* 70: 1311–1319, 1988.

66. Warrel, R.P. Clinical trials of gallium nitrate in patients with cancer-related hypercalcemia, *Sem. Oncol.,* 18: 26–31, 1991.

67. Harris, W.R. and Pecoraro, V.L., Thermodynamic binding constants for gallium transferrin, *Biochemistry,* 22: 292–299, 1983.

68. Larson, S.M. et al. Common pathway for tumor cell uptake of gallium-67 and iron-59 via a transferrin receptor, *J. Natl. Cancer Inst.,* 64: 41–53, 1980.

69. Chitambar, C.R. and Zivkovic, Z. Uptake of gallium-67 by human leukemic cells: demonstration of transferrin receptor-dependent and transferrin-independent mechanisms, *Cancer Res.,* 47: 3929–3934, 1987.

70. Chitambar, C.R. and Seligman, P.A. Effects of different transferrin forms of transferrin receptor expression, iron uptake and cellular proliferation of human leukemic HL60 cells: mechanisms responsible for the specific cytotoxicity of transferrin–gallium, *J. Clin. Invest.,* 78: 1538–1546, 1986.

71. Ul-Haq, R. and Chitambar, C.R. Modulation of transferrin receptor mRNA by transferrin–gallium in human myeloid HL-60 and lymphoid CCRF-CEM leukemic cells, *Biochem. J.,* 294: 873–877, 1993.

72. Chitambar, C. and Sax, D., Regulatory effects of gallium on transferrin-independent iron uptake by human leukemic HL-60 cells, *Blood,* 80: 505–511, 1992.

73. Chitambar, C. et al. Inhibition of leukemic HL-60 cell growth by transferrin–gallium: effect on ribonucleotide reductase and demonstration of drug synergy with hydroxyurea, *Blood,* 72: 1930–1936, 1988.

74. Chitambar, C. et al. Inhibition of ribonucleotide reductase by gallium in murine leukemic L1210 cells, *Cancer Res.,* 51: 6199–6206, 1991.

75. Chitambar, C.R. and Wereley, J.P., Resistance to the anti-tumor agent gallium nitrate in human leukemic cells is associated with decreased gallium/iron uptake, increased activity of iron regulatory protein-1, and decreased ferritin production, *J. Biol. Chem.,* 272: 12151–12157, 1997.

76. Narasimhan, J., Antholine, W.E., and Chitambar, C., Effect of gallium on the tyrosyl radical of the iron-dependent M2 subunit of ribonucleotide reductase, *Biochem. Pharmacol.,* 44: 2403–2412, 1992.

77. Lundberg, J.L. and Chitambar, C., Interaction of gallium nitrate and iron chelators: effects on the proliferation of human leukemic HL-60 cells, *Cancer Res.,* 50: 6466–6472, 1990.

78. Chitambar, C.R., Wereley, J.P., and Ul-Haq, R. Synergistic inhibition of T-lymphoblastic leukemic CCRF-CEM cell growth by gallium and recombinant human alpha-interferon through action on cellular iron uptake, *Cancer Res.,* 54: 3224–3228, 1994.

79. Janicek, M. et al. Early restaging gallium scans predict outcome in poor-prognosis patients with aggressive non-Hodgkin's lymphoma treated with high-dose CHOP chemotherapy, *J. Clin. Oncol.,* 15: 1631–1637, 1997.

80. Vose, J.M. et al. Single-photon emission computed tomography gallium imaging versus computed tomography: predictive value in patients undergoing high-dose chemotherapy and autologous stem-cell transplantation for non-Hodgkin's lymphoma, *J. Clin. Oncol.,* 14: 2473–2479, 1996.

81. Salloum, E. et al. Gallium scans in the management of patients with Hodgkin's disease: a study of 101 patients, *J. Clin. Oncol.,* 15: 518–527, 1997.

82. Dreicer, R. et al. Vinblastine, ifosfamide, and gallium nitrate: an active new regimen in patients with advanced carcinoma of the urothelium. A phase II trial of the Eastern Cooperative Oncology Group (E5892), *Cancer,* 79: 110–114, 1997.

83. Warrel, R.P. Gallium nitrate for the treatment of bone metastases, *Cancer,* 80: 1680–1685, 1997.

84. Hoffbrand, A.V. et al. Effect of iron deficiency and desferrioxamine on DNA synthesis in human cells, *Brit. J. Haematol.,* 33: 517–525, 1976.

85. Robbins, E. and Pederson, T., Iron: its intracellular localization and possible role in cell division, *Proc. Natl. Acad. Sci. U.S.A.,* 66: 1244–1248, 1970.

86. Lederman, H.M. et al. Deferoxamine: a reversible S-phase inhibitor of human lymphocyte proliferation, *Blood,* 64: 748–754, 1984.

87. Foa, P. et al. Inhibition of proliferation of human leukemic cell populations by deferoxamine, *Scand. J. Haematol.*, 36: 107–115, 1986.

88. Bomford, A. et al. The effect of desferrioxamine on transferrin receptors, the cell cycle and growth rates of human leukemic cells, *Biochem. J.*, 236: 243–249, 1986.

89. Ganeshaguru, K. et al. Effect of various iron chelating agents on DNA synthesis on human cells, *Biochem. Pharmacol.*, 29: 1275–1279, 1980.

90. Barankiewicz, J. and Cohen, A. Impairment of nucleotide metabolism by iron-chelating deferoxamine, *Biochem. Pharmacol.*, 36: 2343–2347, 1987.

91. Reddel, R.R., Hedley, D.W., and Sutherland, R.L., Cell cycle effects of iron depletion on T-47D human breast cancer, *Exp. Cell. Res.*, 161: 277–284, 1985.

92. Becton, D.L. and Roberts, B., Antileukemic effects of deferoxamine on human myeloid leukemic cell lines, *Cancer Res.*, 49: 4809–4815, 1989.

93. Sussman, H.H. Iron and tumor cell growth, in *Iron in Immunity, Cancer and Inflammation*, DeSouza, M. and Brock, J.H., Eds., Chichester, Wiley, 261–282, 1989.

94. Hayes, K.P., Hider, R.C., and Porter, J.B. Cell cycle synchronization and growth inhibition by 3-hydroxypyridin-4-one iron chelators in leukemia cell lines, *Cancer Res.*, 52: 4591–4599, 1992.

95. Nocentini, G. et al. 2-2′-Bipyridyl-6-carbothioamide and its ferrous complex: their *in vitro* antitumoral activity related to the inhibition of ribonucleotide reductase R2 subunit, *Cancer Res.*, 53: 19–26, 1993.

96. Bergman, R.J. and Ingeno, M.J., Microbial iron chelator-induced cell cycle synchronization in L1210 cells: potential in combination therapy, *Cancer Res.*, 47: 6010–6016, 1987.

97. Richardson, D.R., Tran, E.H., and Ponka, P., The potential of iron chelators of the pyridoxal isonicotinoyl hydrazone class as effective antiproliferative agents, *Blood*, 86: 4295–4306, 1995.

98. Richardson, D.R. and Milnes, K., The potential of iron chelators of the pyridoxal isonicotinoyl hydrazone class as effective antiproliferative agents,II: the mechanism of action of ligands derived from salicylaldhehyde benzoyl hydrazone and 2-hydroxy-1-naphthylaldehyde benzoyl hydrazone, *Blood*, 89: 3025–3038, 1997.

99. Breuer, W., Epsztejn, S., and Cabantchik, Z.I. Iron acquired from transferrin by K562 cells is delivered into a cytoplasmic pool of chelatable iron (II), *J. Biol. Chem.*, 270: 24209–24215, 1995.

100. Richardson, D.R. Mobilization of iron from neoplastic cells by some iron chelators is an energy-dependent process, *Biochim. Biophys. Acta*, 1320: 45–47, 1997.

101. Bergeron, R.J., Wiegand, J., and Brittenham, G.M. HBED: a potential alternative to deferoxamine for iron chelating therapy, *Blood*, 91: 1446–1452, 1998.

102. Estrov, Z. et al. Synergistic antiproliferative effects on HL-60 cells: deferoxamine enhances cytosine arabinoside, methotrexate, and daunorubicin cytotoxicity, *Am. J. Ped. Hematol. Oncol.*, 10: 288–291, 1988.

103. Seligman, P. et al. Effects of agents that inhibit cellular iron incorporation on bladder cancer cell proliferation, *Blood*, 82: 1608–1617, 1993.

104. Estrov, Z. et al. *In vitro* and *in vivo* effects of deferoxamine in neonatal acute leukemia, *Blood*, 69: 757–603, 1987.

105. Fukuchi, K. et al. Iron deprivation-induced apoptosis in HL-60 cells, *FEBS Lett.*, 350: 139–142, 1994.

106. Haq, R.U., Wereley, J.P., and Chitambar, C.R., Induction of apoptosis by iron deprivation in human leukemic CCRF-CEM cells, *Exp. Hematol.*, 23: 428–432, 1995.

107. Hileti, D., Panayiotidis, P., and Hoffbrand, A.V. Iron chelators induce apoptosis in proliferating cells, *Brit. J. Haematol.*, 89: 181–187, 1995.

108. Fukuchi, K. et al. Iron deprivation results in an increase in p53 expression, *Biol. Chem. Hoppe–Seyler,* 376: 627–630, 1995.

109. An, W.G. et al. Stabilization of wild-type p53 by hypoxia, *Nature,* 392: 405–408, 1998.

110. Dezza, L. et al. Effects of desferrioxamine on normal and leukemic human hematopoietic cell growth: *in vitro* and *in vivo* studies, *Leukemia,* 3: 104–110, 1989.

111. Hann, H.W., Evans, A.E., and Siegel, S.E. Prognostic importance of serum ferritin in patients with stages III and IV neuroblastoma: the Children Cancer Studies Group experience, *Cancer Res.,* 45: 2843–2848, 1985.

112. Blatt, J. and Wharton, V. Stimulation of growth of neuroblastoma cells by ferritin *in vitro, J. Lab. Clin. Med.,* 119: 139–143, 1992.

113. Hann, H., Levy, H.M., and Evans, A.E. Serum ferritin as a guide to therapy in neuroblastoma, *Cancer Res.,* 40: 1411–1413, 1977.

114. Hann, H., Stahlbut, M.W., and Evans, A.E. Basic and acidic isoferritins in the sera of patients with neuroblastoma, *Am. J. Clin. Oncol.,* 3: 86–92, 1984.

115. Watanabe, N., Niitsu, Y., and Koseki, J., Ferritinemia in nude mice bearing various human carcinomas, in *Carcinoembryonic Proteins,* Lehman, F.G., Ed., Amsterdam, Elsevier-North Holland, 273–291, 1984.

116. Hann, H., Stahlbut, M.W., and Millman, I., Human ferritins present in the sera of nude mice transplanted with human neuroblastoma or hepatocellular carcinoma, *Cancer Res.,* 44: 3898–3901, 1984.

117. Hann, H.L., Stahlbut, M.W., and Evans, A.E., Source of increased ferritin in neuroblastoma: studies with concavalin A-Sepharose binding, *J. Natl. Cancer Inst.,* 76: 1031–1033, 1986.

118. Blatt, J. and Stittely, S., Antineuroblastoma activity of deferoxamine in human cell lines, *Cancer Res.,* 47: 1749–1750, 1987.

119. Helson, C. and Helson, L. Deferoxamine and human neuroblastoma and primitive neuroectodermal tumor cell lines, *Anticancer Res.,* 12: 481–484, 1992.

120. Richardson, D.R. and Ponka, P., The iron metabolism of the human neuroblastoma cell: lack of relationship between the efficacy of iron chelation and the inhibition of DNA synthesis, *J. Lab. Clin. Med.,* 124: 660–671, 1994.

121. Donfrancesco, A. et al. Effects of a single course of deferoxamine in neuroblastoma patients, *Cancer Res.,* 50: 4929–4930, 1990.

122. Donfrancesco, A. et al. Deferoxamine, cyclophosphamide, etoposide, carboplatin, and thiotepa (D-CECaT): a new cytoreductive chelation-chemotherapy regimen in patients with advanced neuroblastoma, *Am J. Clin. Oncol.,* 15: 319–322, 1992.

123. Selig, R.A. et al. Failure of iron chelators to reduce tumor growth in human neuroblastoma xenografts, *Cancer Res.,* 58: 473–478, 1998.

124. Hann, H.L., Stahlaut, M.W., and Blumberg, B.S. Iron nutrition and tumor growth: decreased tumor growth in iron-deficient mice, *Cancer Res.,* 48: 4168–4170, 1988.

125. Hann, H.L., Stahlhut, M.W., and Hann, C.L. Effects of iron and deferoxamine on cell growth and *in vitro* ferritin synthesis in human hepatoma cell lines, *Hepatology,* 11: 566–569, 1990.

126. Hann, H. et al. Antitumor effect of deferoxamine on human hepatocellular carcinoma growing in athymic nude mice, *Cancer,* 70: 2051–2056, 1992.

127. Holley, R.W. A unifying hypothesis concerning the nature of malignant growth, *Proc. Natl. Acad. Sci. U.S.A.,* 69: 2840–2845, 1972.

128. Vostreys, M., Moran, P.L., and Seligman, P.A., Transferrin synthesis by small cell lung cancer acts as an autocrine regulator of cellular proliferation, *J. Clin. Invest.,* 82: 231–241, 1988.

129. Brown, J.P., Hewick, R.W., and Hellstrom, I. Human melanoma associated antigen p97 is structurally and functionally related to transferrin, *Nature,* 296: 171–174, 1982.

130. Seligman, P.A. et al. Transferrin-independent iron uptake supports B lymphocyte growth, *Blood,* 78: 1526–1531, 1991.

131. Basset, P., Quesneau, Y., and Zwiller, J. Iron-induced L1210 cell growth: evidence of a transferrin-independent iron transport, *Cancer Res.,* 46: 1644–1650, 1986.

132. Titeux, M. et al. The role of iron in the growth of human leukemic cell lines, *J. Cell Physiol.,* 121: 251–260, 1984.

133. Faulk, W.P., Hsi, B.L., and Stevens, P.J., Transferrin and transferrin receptors in carcinoma of the breast, *Lancet 2,* 390–392, 1980.

134. Wrba, F. et al. Transferrin receptor(TfR) expression in breast carcinoma and its possible relationship to prognosis. An immunohistochemical study, *Virchows Arch. Pathol. Anat. Histopathol.,* 410: 69–73, 1986.

135. Esserman, L., Takahashi, S., and Rojas, V. An epitope of the transferrin receptor is exposed on the cell surface of high-grade but not low-grade human lymphomas, *Blood,* 74: 2718–2726, 1989.

136. Chackal-Roy, M. et al. Stimulation of human prostatic carcinoma cell growth by factors present in human bone marrow, *J. Clin. Invest.,* 84: 43–50, 1989.

137. Chackal-Roy, M. and Zetter, B.R., Selective stimulation of prostatic carcinoma cell proliferation by transferrin, *Proc. Natl. Acad. Sci. U.S.A.,* 89: 6197–6201, 1992.

138. Stevens, R.G. Iron and cancer, in *Iron and Human Disease,* Lauffer, R.B., Ed., Boca Raton, CRC Press, 333–347, 1992.

139. Bergeron, R.J., Streiff, R.R., and Elliott, G.T., Influence of iron on *in vivo* proliferation and lethality of L1210 cells, *J. Nutr.,* 115: 369–379, 1985.

140. Hann, H.L., Stahlut, M.W., and Blumberg, B.S., Iron nutrition and tumor growth: decreased tumor growth in iron-deficient mice, *Cancer Res.,* 48: 4168–4172, 1988.

141. Thompson, H.J. et al. Effect of dietary iron deficiency or excess on the induction of mammary carcinogenesis by 1-methyl-1-nitrosourea, *Carcinogenesis,* 12: 111–118, 1991.

142. Siegers, C.P. et al. Dietary iron enhances the tumor rate in dimethylhydrazine-induced colon carcinogenesis in mice, *Cancer Lett.,* 4: 251–256, 1988.

143. Stevens, R.G. et al. Body iron stores and the risk of cancer, *New Engl. J. Med.,* 319: 1047–1051, 1988.

144. Stevens, R.G., Beasley, R.P., and Blumberg, B.S. Iron-binding proteins and risk of cancer in Taiwan, *J. Natl. Cancer Inst.,* 76: 605–613, 1986.

145. Selby, J.V. and Friedman, G.D. Epidemiological evidence of an association of body iron stores and risk of cancer, *Int. J. Cancer,* 41: 677–683, 1988.

146. Stevens, R.G. et al. Iron-binding proteins, hepatitis B virus, and mortality in the Solomon Islands, *Am. J. Epidemiol.,* 118: 550–555, 1983.

147. Willet, W.C. et al. Relation of meat, fat and fiber intake to the risk of colon cancer in a prospective study among women, *New Engl. J. Med.,* 323: 1664–1670, 1990.

148. Willet, W.C. et al. Correspondence, *New Engl. J. Med.,* 326: 201–202, 1992.

149. Stevens, R.G. et al. Moderate elevation of body iron level and increased risk of cancer occurrence and death, *Int. J. Cancer,* 56: 364–369, 1994.

150. Knekt, P. et al. Body iron stores and risk of cancer, *Int. J. Cancer,* 56: 379–382, 1994.

151. Wurzelmann, J.I. et al. Iron intake and the risk of colorectal cancer, *Cancer Epidemiol. Biomarkers Prev.,* 5: 503–507, 1996.

152. Treng M. et al. Dietary iron and recurrence of colorectal adenomas, *Cancer Epidemiol. Biomarkers Prev.,* 6: 1029–1032, 1997.

153. Ullen, H. et al. Supplementary iron intake and risk of cancer: reversed causality? *Cancer Lett.,* 114: 215–216, 1997.
154. Fargion, S. et al. Prognostic factors for hepatocellular carcinoma in genetic hemochromatosis, *Hepatology,* 20: 1426–1431, 1994.
155. Stal, P. et al. The effects of dietary iron on initiation and promotion in chemical hepatocarcinogenesis, *Hepatology,* 21: 521–528, 1995.

3 Lactoferrin

CONTENTS

3.1   Introduction 71
3.2   Biochemistry and Molecular Biology of Lactoferrin..................................74
3.3   Lactoferrin Receptors..88
    3.3.1   Lactoferrin Receptors on Leukemic Lines 88
    3.3.2   Lactoferrin Receptors on Activated T Lymphocytes.......................89
    3.3.3   Lactoferrin Receptors on Monocytes–Macrophages 91
    3.3.4   Lactoferrin Receptors on the Small Intestinal Brush Border
          Membrane...93
    3.3.5   Lactoferrin Receptors on Parenchymal Liver Cells 94
    3.3.6   Lactoferrin Receptors on Breast Epithelial Cells...........................96
    3.3.7   Lactoferrin Receptors on Platelets...97
    3.3.8   Lactoferrin Receptors on Neuronal Cells......................................97
3.4   Functions of Lactoferrin 98
    3.4.1   Bacteriostatic Activity of Lactoferrin 98
          3.4.1.1   Lactoferricin..102
    3.4.2   Bacterial Lactoferrin Receptors 103
          3.4.2.1   Serum Lactoferrin Concentration and Infectious
                  Diseases...107
    3.4.3   Lactoferrin and the Regulation of Myelopoiesis............................109
    3.4.4   Lactoferrin as a Gene Regulator...113
    3.4.5   Anti-Tumor Activity of Lactoferrin 114
3.5   Lactoferrin Expression in Neutrophils and Mammary Glands..................114
    3.5.1   Lactoferrin Synthesis by Neutrophils 115
    3.5.2   Lactoferrin Synthesis in Mammary Glands...................................119
3.6   Lactoferrin Expression during Development..122
References..123

3.1   INTRODUCTION

Lactoferrin is a 80,000-kDa protein found in several glandular epithelial tissues and human neutrophils where it is localized into secondary granules.[1,2] It is a member of the family of iron-binding proteins that includes transferrin, with which it shares 60% homology at the amino acid level. The exact biological role of lactoferrin remains to be determined. However, in spite of this limitation, lactoferrin has been implicated in several biological processes including: (1) host microbial defense through both direct and indirect actions on the neutrophil function; and (2) regulation of granulopoiesis through a negative feedback mechanism.

Lactoferrin is normally found in serum at microgram levels, but is more abundant in different types of external secretions including seminal fluid, tears, sweat, milk, and nasal and genital secretions. Lactoferrin is also found in bile, seemingly derived from the pancreas, and in the secondary granules of neutrophils.

The protein is particularly abundant in human milk, where it reaches a concentration of about 1 mg/ml; even higher concentrations (about 7 mg/ml) are observed in the colostrum. However, it has to be underlined that the majority of iron found in human milk is not linked to lactoferrin, but to lipid and to low molecular weight compounds. As a consequence, milk lactoferrin is only 6 to 8% saturated with iron, presumably because of the difficulty in gaining access to iron in the lipid fraction or casein micelles.

Although lactoferrin is found in the milk of other species, it is completely absent in certain species such as rats, rabbits, and dogs. The milks of rabbits and rats do, however, contain significant amounts of transferrin. The milk of the dog contains neither protein, even though its iron content is particularly high. Lactoferrin is mainly synthesized in the secretory organs in which the protein is found; is also synthesized in neutrophils, at the myelocyte and metamyelocyte stages. The production of lactoferrin is significantly enhanced in inflammatory processes of the mammary gland, pancreas, and parotid salivary gland.

Lactoferrin, identified originally as the major iron-binding constituent of milk, and serving a major physiological role in iron storage and transport in lactation, has also been implicated in a wide variety of other functions involving a number of organ systems. Lactoferrin is a significant component of the secondary granules of neutrophils and is thought to play a major role in antimicrobial activity of granulocytic cells and milk by chelating iron required for microbial growth.[3,4]

Lactoferrin contributes to the antibacterial activities of neutrophils by two different mechanisms: (1) in the apo form it may perform a withholding function and prevent growth of phagocytosed bacteria; and (2) in iron lactoferrin form, it may provide iron that can catalyze the production of free radicals that lead to microbial killing within phagosomes of phagocytic cells. The affinity of lactoferrin for iron is greater than that of transferrin; thus, lactoferrin removes available free iron and bound iron from plasma or milk. After release from granulocytes and saturation with iron, lactoferrin is removed from the circulation by the reticuloendothelial system through specific receptors for iron-saturated lactoferrin.[5] This is considered one of the mechanisms involved in the generation of hyposideremia in acute inflammatory disease.

Neutrophil degranulation is considered to be the only source of circulating lactoferrin[6]; as a consequence, serum lactoferrin levels are evaluated as a measure of inflammation or the acute phase response, and of the size of the total granulocyte pool.

Lactoferrin is somewhat related to the transferrin protein family. Lactoferrin displays several structural and functional similarities to transferrin. In spite of these similarities, however, lactoferrin differs from transferrin in some important functional aspects, including its relative affinity for iron, propensity to release iron, and

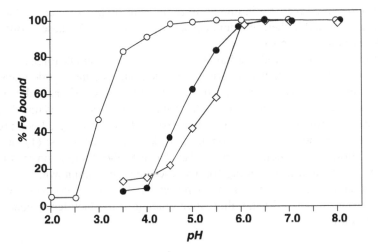

FIGURE 3.1 The pH-dependent release of iron from lactoferrin and transferrin. In the figure the pH-dependent release of iron from wild-type lactoferrin (open circle), N-terminal half of human lactoferrin (square) and native transferrin (closed circle) is shown. The percentage of iron saturation is calculated by comparing the absorbance at 465 nm with that of the fully iron-saturated molecule. (Redrawn from Day, C.L. et al., *J. Biol. Chem.,* 267: 13857–13862, 1992.)

receptor binding properties. Human lactoferrin binds iron much more strongly than does human transferrin, by a factor of about 260. Lactoferrin retains iron at pH values distinctly lower than transferrin: in fact, human transferrin releases iron at pH values below 6, whereas human lactoferrin does not release iron until the value is reduced below pH 4 (Figure 3.1).

Hematopoietic regulatory functions have also been ascribed to lactoferrin. Broxmeyer et al.[7] have shown that Fe-saturated lactoferrin binds to receptors on macrophages and inhibits their release of GM-CSF. Furthermore, *in vivo* lactoferrin decreases hemopoiesis in normal mice and dampens rebound myelopoiesis in animals treated with cyclophosphamide. Other studies showed that lactoferrin was able to promote neutrophil adhesiveness, reduced lysozyme regeneration, stimulated NK activity, suppressed *in vitro* primary immune responses, enhanced growth of certain lymphoid cells in culture, and decreased the proliferation of hemopoietic progenitors through an indirect mechanism.[8]

Many of these effects seem to be mediated through lactoferrin inhibition of IL-1 production. Finally, it was shown that lactoferrin catalyzes hydroxyl radical production from superoxide anion and hydrogen peroxide. This activity was detected both in milk lactoferrin and neutrophil lactoferrin, and required iron saturation. Production of hydroxyl radical may be involved in antimicrobial activity as well as tissue injury occurring at sites of inflammation.[9]

All these varied functions of lactoferrin have been attributed to a single molecule. Lactoferrins isolated from human milk and human neutrophils have been reported to have the same molecular weight, isoelectric point, reactivity with monospecific antibody, and clearance from circulation.[10] However, multiple isoforms of lactoferrin have been identified recently in human milk and neutrophils that vary in their iron-binding capacity and in their biochemical activity, most likely because of differences in post-translational glycosylation.[11]

Furmansky et al. showed that human milk ribonuclease and lactoferrin exhibited identical properties: molecular weight, isoelectric point, partial peptide maps, NH_2-terminal amino acid sequences, and reactivity with a panel of antibodies of anti-milk ribonuclease and anti-lactoferrin. However, milk ribonuclease is unable to bind iron and exhibits ribonuclease activity. On the basis of these findings, it was proposed that human milk and neutrophils contain three lactoferrin isoforms: (1) lactoferrin A, the traditional iron-binding form; (2) lactoferrin B, the RNase, non-iron-binding form; and (3) the non-iron-binding form whose function is unknown.

A further element of heterogeneity related to the recent discovery of a new lactoferrin species derived from alternative splicing of the lactoferrin gene will be discussed below.

3.2 BIOCHEMISTRY AND MOLECULAR BIOLOGY OF LACTOFERRIN

Lactoferrin shares some of its biochemical properties and biological functions with the family of transferrin proteins. The biochemical structure of lactoferrin was carefully elucidated both for its sequence and molecular structure. Early studies focused on evaluating its basic biochemical features. Evaluations of the molecular weight of lactoferrin by gel filtration, ultracentrifugation, and polyacrylamide gel electrophoresis have revealed values in the range of 80 to 92 kDa. The elucidation of the complete amino acid sequence of lactoferrin shows unequivocal values of 82 kDa[1]; approximately 8% is attributed to two heterogeneous sugar chains.

Several methods have been used to purify lactoferrin from plasma, milk, and granulocytes. The methods most commonly used involved standard multi-step purification procedures, immobilized metal-ion affinity chromatography, aminohexyl affinity chromatography, or thin-layer ion exchange chromatography. Lactoferrin has an isoelectric point (pI) much higher than those of transferrins; values in the range of 8.4 to 9.0 were obtained. The high isoelectric point of lactoferrin may explain the property of this protein to complex with other molecules; lactoferrin also polymerizes in the presence of calcium ions.

Particularly relevant are the studies carried out to characterize the molecular complexes formed by the interaction of lactoferrin with other proteins contained in human milk. The formation of multi-molecular complexes of lactoferrin and other compounds is related to the incredible aptitude of this protein to bind to various components, including plastic.

Spik et al.[12] observed that the electrophoretic mobility of purified human lacto-ferrin (slow β_2 globulin behavior) and of human lactoferrin as it exists in human milk (β_1 globulin behavior) is significantly different. This phenomenon is due to the fact that lactoferrin present in human milk gives rise to several molecular complexes: (1) lactoferrin–sIgA secretory component; (2) lactoferrin–lysozyme (two molecules of lysozyme bound to one molecule of lactoferrin); and (3) lactoferrin–glycopeptides.

The formation of molecular complexes between lactoferrin and other proteins may have functional consequences of some importance in that: (1) the association of human lactoferrin with specific sIgA inhibited the growth of a pathological *E. coli* in contrast to sIgA and lactoferrin alone; and (2) the role of the associated lactoferrin–lysozyme is indicated by experiments showing that bacteria submitted to lysozyme action are agglutinated by lactoferrin.

An important feature of lactoferrin is its much greater acid stability when compared with transferrin. Iron release for lactoferrin begins at pH 4.0 and is complete at pH 2.5, whereas iron release for transferrin begins 2 pH units higher, starting at 6.0 and ending at pH 4.0 (Figure 3.1). The reasons for this characteristic difference are not fully known, but they may have very important functional significance. In fact, if the primary mechanism of iron release is a lowering of pH, then the different pH stabilities of the two molecules may have fundamental importance in determining the different roles of lactoferrin and transferrin *in vivo*.

Following interactions with iron, lactoferrin and transferrin, exhibit changes in their absorption spectra within the visible region, resulting in a characteristic increase of absorption that peaks at 465 nm.[13] Interestingly, the N terminal half of human lactoferrin produced from cloned cDNA also showed this increase in the absorption spectrum within the visible region, with a peak, however, at 454 nm (Figure 3.2).

The process of interaction of iron and lactoferrin follows complex kinetics, involving a fast kinetic process characterized by an exponential increase in absorbance at 465 nm lasting 0.2 seconds, followed by a second exponential increase in absorbance lasting about 10 seconds; finally, a very slow process, lasting >2 hours, characterized by a slight increase in absorbance is observed (Figure 3.3).

The determination of the primary amino acid sequence of lactoferrin led to the understanding of the molecular and functional organization of this molecule.[1] The protein consists of a single polypeptide chain of relative molecular mass close to 82,000 kDa to which two glycans are attached through N-glycosidic linkages (Figures 3.4a and 3.4b). The two N-acetylactosaminic-type glycans are heterogeneous and differ from those of other transferrins. Like all transferrins, lactoferrin possesses two metal-binding sites, each of which can bind a ferric ion (Fe^{3+}) with a bicarbonate ion.

Three-dimensional structures for human lactoferrin, rabbit serum transferrin, and chicken and duck ovotransferrins, show that the three proteins exhibit essentially the same folding, as well as very similar iron-binding sites, in line with their extensive homology at the level of the amino acid sequence. Sequence studies, partial proteolytic analysis, and, particularly, crystallographic studies of lactoferrin demonstrated that this protein, like the other transferrins, is composed of two homologous lobes, each containing a single ferric ion (Figure 3.5).

Crystallographic data on human lactoferrin have shown that each lobe consists of two domains, domain I (residues 1 to 90 and 252 to 320 for the N terminal lobe and residues 343 to 444 and 609 to 672 for the C terminal lobe) and domain II (residues 91 to 251 for the N terminal lobe and residues 445 to 608 for the C terminal lobe), and that the iron-binding site is located in the inter-domain cleft (Figure 3.5).

One bicarbonate ion and four amino acids are involved in the iron-binding site. Three of the protein ligands correspond to Tyr 93, Tyr 191, and His 252, and are

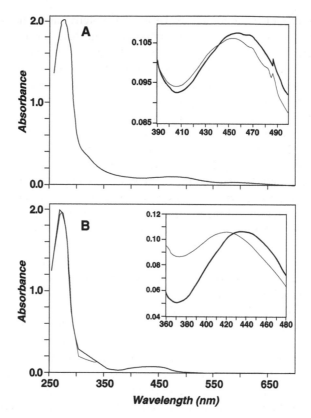

FIGURE 3.2 Absorption spectra of native human lactoferrin (thick line) and N-terminal half of human lactoferrin (thin line). Panel A shows the absorption spectrum for iron-saturated samples of Fe$_2$ lactoferrin (thick line) and Fe lactoferrin N (thin line). Panel B shows the absorption spectra for copper-saturated samples of native Cu$_2$ lactoferrin (thick line) and Cu lactoferrin N (thin line). (Redrawn from Day, C.L. et al., *J. Biol. Chem.*, 267: 13857–13862, 1992.)

located in domain II. The fourth ligand corresponds to Asp 61 and belongs to domain I[1]; the bicarbonate ion is linked to an arginine residue (Figure 3.5).

Analysis of the iron-binding capacities of different proteolytic fragments (corresponding to the N and C terminal iron-binding sites) of lactoferrin provided evidence that:

1. Asp 61 does not take part in the the stability upon protonation of the iron complex of lactoferrin.
2. Full removal of the glycan moiety leads to the loss of iron-binding capacity.
3. The N and C terminal iron-binding sites exhibit lower stability toward protonation (they easily lose iron, thus lowering pH) in isolated lobes than they do in native lactoferrin or in the reassociated N–C tryptic complex, suggesting that the interactions between the lobes stabilize the two iron-binding sites.

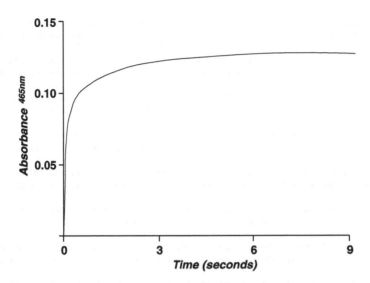

FIGURE 3.3 Kinetics of absorbance at 465 nm following incubation of human lactoferrin with iron. The absorption of iron to lactoferrin was determined by stopped-flow technique and detected by absorption spectroscopy at 400 to 500 nm. (Redrawn from Packdaman, R. et al., *Eur. J. Biochem.*, 254: 144–153, 1998.)

Evidence for differences between the two iron binding sites of lactoferrin has come from a number of physico-chemical studies. The clearest demonstration of the non-identities of the two metal-binding sites was provided by the differing abilities of the sites to release iron by protonation. Thus, it was shown during iron resaturation that between pH levels 6.2 and 4.8, only the acid-labile site of human lactoferrin binds iron.

Occupancy of iron-binding sites by iron requires binding of an anion. The protein binds iron with extremely high equilibrium constants, but readily releases iron as needed by cells. The anion is thought to play a crucial role in iron binding and release. The lactoferrin provides four ligands for the metal, leaving two open sites for coordination of the anion and/or water. A variety of spectroscopic techniques showed that the anion-binding site is in close proximity to the metal-binding site.[14]

Some studies, especially those exploring steric effects, have suggested that the anion binds simultaneously to the metal and the protein. Studies of the interaction of iron with lactoferrin using electron spin echo envelope modulation (ESEEM, a powerful tool for identifying nuclear spins in the vicinity of the paramagnetic center) provided strong and definitive evidence that the carbonate anion binds directly to the metal in iron and copper complexes of lactoferrin.[15,16]

The primary structures of the two glycans linked to the single polypeptide chain of lactoferrin were determined. This analysis showed that the structures of glycans of human milk lactoferrin differ from those of leukocyte lactoferrin. Thus, glycans present in human milk lactoferrin are of the sialyl-N-acetyl-lactosaminic type and are a-1,6-fucosylated at the N-acetylglucosamine residue linked to asparagine.[17] They are highly heterogeneous, owing to a variable number of sialic acid residues

Proteins of Iron Metabolism

```
   1 CTT GTC TTC CTC GTC CTG CTG TTC CTC GGG GCC CTC GGA CTG TGT CTG GCT GGC CGT AGG
     Leu Val Phe Leu Val Leu Leu Phe Leu Gly Ala Leu Gly Leu Gly Leu Ala Gly Arg Arg

  61 AGA AGG AGT GTT CAG TGG TGC GCC GTA TCC CAA CCC GAG GCC ACA AAA TGC TTC CAA TGG
     Arg Arg Ser Val Gln Trp Gys Ala Val Ser Gln Pro Glu Ala Thr Lys Cys Phe Gln Trp

 121 CAA AGG AAT ATG AGA AAA GTG CGT GGC CCT CCT GTC AGC TGC ATA AAR AGA GAC TCC CCC
     Gln Arg Asn Met Arg Lys Val Arg Gly Pro Pro Val Ser Cys Ile Lys Arg Asp Ser Pro

 181 ATC CAG TGT ATC CAG GCC ATT GCG GAA AAC AGG GCC GAT GCT GTG ACC CTT GAT GGT GGT
     Ile Gln Cys Ile Gln Ala Ile Ala Glu Asn Arg Ala Asp Ala Val Thr Leu Asp Gly Gly

 241 TTC ATA TAC GAG GCA GGC CTG GCC CCC TAC AAA CTG CGA CCT GAT GCG GCG GAA GTC TAC
     Phe Ile Tyr Glu Ala Gly Leu Ala Pro Tyr Lys Leu Arg Pro Val Ala Ala Glu Val Tyr

 301 GGG ACC GAA AGA CAG CCA CGA ACT CAC TAT TAT GCC GTG GCT GTG GTG AAG AAG GGC GGC
     Gly Thr Glu Arg Gln Pro Arg Thr His Tyr Tyr Ala Val Ala Val Val Lys Lys Gly Gly

 361 AGC TTT CAG CTG AAC GAA CTG CAA GGT CTG AAG TCC TGC CAC ACA GGC CTT CGC AGG ACC
     Ser Phe Gln Leu Asn Glu Leu Gin Gly Leu Lys Ser Cys His Thr Gly Leu Arg Arg Thr

 421 GCT GGA TGG AAT GTC CCT ATA GGG ACA CTT CGT CCA TTC TTG AAT TGG ACG GGT CCA CCT
     Ala Gly Trp Asn Val Pro Ile Gly Thr Leu Arg Pro Phe Leu Asn Trp Thr Gly Pro Pro

 481 GAG CCC ATT GAG GCA GCT GTG GCC AGG TTC TTC TCA GCC AGC TGT GTT CCC GGT GCA GAT
     Glu Pro Ile Glu Ala Ala Val Ala Arg Phe Phe Ser Ala Ser Cys Val Pro Gly Ala Asp

 541 AAA GGA CAG TTC CCC AAC CTG TGT CGC CTG TGT GCG GGG ACA GGG GAA AAC AAA TGT GCC
     Lys Gly Gln Phe Pro Asn Leu Cys Arg Leu Cys Ala Gly Thr Gly Glu Asn Lys Cys Ala

 601 TTC TCC TCC CAG GAA CCG TAC TTC AGC TAC TCT GGT GCC TTC AAG TGT CTG AGA GAC GGG
     Phe Ser Ser Gln Glu Pro Try Phe Ser Try Ser Gly Ala Phe Lys Cys Leu Arg Asp Gly
                                                                                   Lys

 661 GCT GGA GAC GTG GCT TTT ATC AGA GAG AGC ACA GTG TTT GAG GAC CTG TCA GAC GAG GCT
     Ala Gly Asp Val Ala Phe Ile Arg Glu Ser Thr Val Phe Glu Asp Leu Ser Asp Glu Ala

 721 GAA AGG GAC GAG TAT GAG TTA CTC TGC CCA GAC AAC ACT CGG AAG CCA GTG GAC AAG TTC
     Glu Arg Asp Glu Try Glu Leu Leu Cys Pro Asp Asn Thr Arg Lys Pro Val Asp Lys Phe

 781 AAA GAC TGC CAT CTG GCC CGG GTC CCT TCT CAT GCC GTT GTG GCA CGA AGT GTG AAT GGC
     Lys Asp Cys His Leu Ala Arg Val Pro Ser His Ala Val Val Ala Arg Ser Val Asn Gly

 841 AAG GAG GAT GCC ATC TGG AAT CTT CTC CGC CAG GCA CAG GAA AAG TTT GGA AAG GAC AAG
     Lys Glu Asp Ala Ile Trp Asn Leu Leu Arg Gln Ala Gln Glu Lys Phe Gly Lys Asp Lys

 901 TCA CCG AAA TTC CAG CTC TTT GGC TCC CCT AGT GGG CAG AAA GAT CTG CTG TTC AAG GAC
     Ser Pro Lys Phe Gln Leu Phe Gly Ser Pro ser Gly Gln Lys Asp Leu Leu Phe Lys Asp

 961 TCT GCC ATT GGG TTT TCG AGG GTG CCC CCG AGG ATA GAT TCT GGG CTG TAC CTT GGC TCC
     Ser Ala Ile Gly Phe Ser Arg Val Pro Pro Arg Ile Asp Ser Gly Leu Tyr Leu Gly Ser

1021 GGC TAC TTC ACT GCC ATC CAG AAC TTG AGG AAA AGT GAG GAG GAA GTG GCT GCC CGG CGT
     Gly Tyr Phe Thr Ala Ile Gln Asn Leu Arg Lys Ser Glu Glu Glu Val Ala Ala Arg Arg
```

FIGURE 3.4a Nucleotide sequence of human lactoferrin cDNA.

```
1081   GCG CGG GTC GTG TGG TGT GCG GTG GGC GAG CAG GAG CTG CGC AAG TGT AAC CAG YGG AGT
       Ala Arg Val Val Trp Cys Ala Val Gly Glu Gln Glu Leu Arg Lys Cys Asn Gln Trp Ser

1141   GGC TTG AGC GAA GGC AGC GTG ACC TGC TCC TCG GCC TCC ACC ACA GAG GAC TGC ATC GCC
       Gly Leu Ser Glu Gly Ser Val Thr Cys Ser Ser Ala Ser Thr Thr Glu Asp Cys Ile Ala

1201   CTG GTG CTG AAA GGA GAA GCT GAT GCC ATG AGT TTG GAT GGA GGA TAT GTG TAC ACT GCA
       Leu Val Leu Lys Gly Glu Ala Asp Ala Met Ser Leu Asp Gly Gly Try Val Tyr Thr Ala

1261   GGC AAA TGT GGT TTG GTG CCT GTC CTG GCA GAG AAC TAC AAA TCC CAA CAA AGC AGT GAC
       Gly Lys Cys Gly Leu Val Pro Val Leu Ala Glu Asn Tyr Lys Ser Gln Gln Ser Ser Asp
       Cys

1321   CCT GAT CCT AAC TGT GTG GAT AGA CCT GTG GAA GGA TAT CTT GCT GTG GCG GTG GTT AGG
       Pro Asp Pro Asn Cys Val Asp Arg Pro Val Glu Gly Tyr Leu Ala Val Ala Val Val Arg

1381   AGA TCA GAC ACT AGC CTT ACC TGG AAC TCT GTG AAA GGC AAG AAG TCC TGC CAC ACC GCC
       Arg Ser Asp Thr Ser Leu Thr Trp Asn Ser Val Lys Gly Lys Lys Ser Cys His Thr Ala

1441   GTG GAC AGG ACT GCA GGC TGG AAT ATC CCC ATG GGC CTG CTC TTC AAC CAG ACG GGC TCC
       Aal Asp Arg Thr Ala Gly Trp Asn Ile Pro Met Gly Leu Leu Phe Asn Gln Thr Gly Ser

1501   TGC AAA TTT GAT GAA TAT TTC AGT CAA AGC TGT GCC CCT GGG TCT GAC CCG AGA TCT AAT
       Cys Lys Phe Asp Glu Tyr Phe Ser Gln Ser Cys Ala Pro Gly Ser Asp Pro Arg Ser Asn

1561   CTC TGT GCT CTG TGT ATT GGC GAC GAG CAG GGT GAG AAT AAG TGC GTG CCC AAC AGC AAC
       Leu Cys Ala Leu Cys Ile Gly Asp Glu Gln Fly Glu Asn Lys Cys Val Pro Asn Ser Asn

1621   GAG AGA TAC TAC GGC TAC ACT GGG GCT TTC CGG TGC CTG GCT GAG AAT GCT GGA GAC GTT
       Glu Arg Tyr Tyr Gly Tyr Thr Gly Ala Phe Arg Cys Leu Ala Glu Asn Ala Gly asp Val

1681   GCA TTT GTG AAA GAT GTC ACT GTC TTC CAG AAC ACT GAT GGA AAT AAC AAT GAG GCA TGG
       Ala Phe Val Lys Asp Val Thr Val Leu Gln asn Thr asp Gly Asn Asn Asn Glu ala Trp

1741   GCT AAG GAT TTG AAG CTG GCA GAC TTT GCG CTG CTG TGC CTC GAT GGC AAA CGG AAG CCT
       Ala Lys Asp Leu Lys Leu Ala Asp Phe Ala Leu Leu Cys Leu Asp Gly Lys Arg Lys Pro

1801   GTG ACT GAG GCT AGA AGC TGC CAT CTT GCC ATG GCC CCG AAT CAT GCC GTG GTG TCT CGG
       Val Thr Glu Ala Arg Ser Cys His Leu Ala Met Ala pro Asn His Ala Val Val Ser Arg

1861   ATG GAT AAG GTG GAA CGC CTG AAA CAG GTG TTG CTC CAC CAA CAG GCT AAA TTT GGG AGA
       Met Asp Lys Val Glu Arg Leu Lys Gln Val Leu Leu His Gln Gln Ala Lys Phe Gly Arg

1921   AAT GGA TCT GAC TGC CCG GAC AAG TTT TGC TTA TTC CAG TCT GAA ACC AAA AAC CTT CTG
       Asn Gly Ser Asp Cys Pro Asp Lys Phe Cys Leu phe Gln Ser Glu Thr Lys Asn Leu Leu

1981   TTC AAT GAC AAC ACT GAG TGT CTG GCC AGA CTC CAT GGC AAA ACA ACA TAT GAA AAA TAT
       Phe Asn Asp Asn Thr Glu Cys Leu Ala Arg Leu His Gly Lys Thr Thr Tyr Glu Lus Tyr

2041   TTG GGA CCA CAG TAT GTC GCA GGC ATT ACT AAT CTG AAA AAG TGC TCA ACC TCC CCC TCC
       Leu Gly Pro Gln Tyn Val Ala Gly Ile Thr Asn Leu Lys Lys Cys Ser Thr Ser Pro Ser

2101   TGG AAG CCT GTG AAT TC 2117
       Trp Lys Pro Val Asn
```

FIGURE 3.4b Deduced amino acid sequence of lactoferrin.

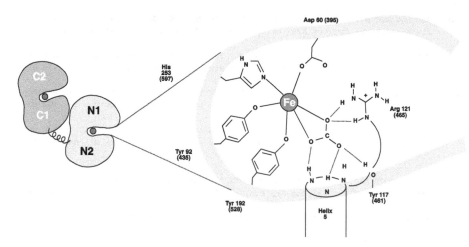

FIGURE 3.5 Left: Schematic representation of diferric human lactoferrin with indication of the N and C lobes, subdivided respectively into N1 and N2 and C1 and C2 domains. The iron-binding sites are also indicated, with the iron atoms shown in small dark circles. Right: Enlarged view of the iron-binding site, with indication of the metal and anion-binding sites for lactoferrin. Amino acid residues involved in binding iron or the synergistic anion are indicated for the N lobe (equivalent C-lobe amino acid residues are shown in parentheses) .

A

NeuAc (α 2-6) - Gal (β 1-4) GlcNAc (β 1-2) - Man (α 1-3)

NeuAc (α 2-6) - Gal (β 1-4) GlcNAc (β 1-2) - Man (α 1-6)

$\rangle$ Man (β 1-4) - GlcAC (β 1-4) GlcNAc β1- Asn

(α 1-6)

Fuc

B

NeuAc (α 2-6) - Gal (β 1-4) GlcNAc (β 1-2) - Man (α 1-6)

NeuAc (α 2-6) - Gal (β 1-4) GlcNAc (β 1-2) - Man (α 1-6)

$\rangle$ Man (β 1-4) - GlcAC (β 1-4) GlcNAc - Asn

FIGURE 3.6 Primary structure of the oligosaccharide chains isolated from milk (A) and leukocyte (B) lactoferrin. (Reprinted from Derisbourg, P. et al., *Biochem. J.,* 269: 821–825, 1990. With permission.)

(one or two) and to the presence or absence of a fucose residue a-1,3-linked to the external N-acetylglucosamine residue.[17]

Leukocyte lactoferrin contains two identical glycans possessing biantennary structures of the disiasylated N-acetyl-lactosaminic type identical to those found in human transferrin.[18] The absence of a-1,3- and a-1,6-linked fucose residues and the structural homogeneity of leukocyte lactoferrin glycans constitute the two major differences from human milk lactoferrin[17,18] (Figure 3.6). These observations indicate that the specificity of the glycan structure of lactoferrin depends on its origin and may be related to the absence or presence in the cells of some specific glycosyl-transferases that link glycan units to the neosynthesized protein.

Comparative studies of the primary structures of lactoferrin glycans provided evidence that lactoferrin glycans are specific to species.[19] Recent studies of molecular dynamics showed preliminary evidence about the conformation of lactoferrin glycans: both glycan antennae of lactoferrin are organized in two coplanar loops rolled in contrary directions and oriented perpendicularly to the plane of the di-N-acetyl-chitobiose residue, leading to a "lobster-like conformation".[20]

Human lactoferrin possesses three potential sites for N-glycosylation: Asn 138, Asn 479, and Asn 624.[21] Studies of mutants exhibiting mutation at the level of each of the Asn residues provided evidence that the majority of lactoferrin molecules use Asn 138 and Asn 479 for glycosylation, while all three Asn residues are glycolysated only in a minority of lactoferrin molecules.[21]

The possible role of lactoferrin glycosylation on anti-bacterial and anti-inflammatory activities was also investigated. The affinities of natural human lactoferrin and glycosylated and non-glycosylated recombinant human lactoferrin for both human lysozyme and bacterial lipopolysaccharide did not differ,[22] thus suggesting that lactoferrin glycosylation had no effect on anti-bacterial and anti-inflammatory activities.[22] However, unglycosylated lactoferrin exhibited a markedly more pronounced susceptibility to trypsin degradation than glycosylated lactoferrin.[22]

The three-dimensional structure of iron-saturated lactoferrin has been determined at 3.2 Å resolution, and shows the protein folded into two equal-sized lobes, connected by a short alpha-helix constituting the N and C terminal halves of the polypeptide chain.[23] The folding pattern of each lobe of lactoferrin reveals the existence of domains I and II. The two domains delimit a cleft containing the iron-binding site. Domain I is based on a b sheet of 4 parallel and 2 anti-parallel strands, and domain II on a b sheet of 4 parallel and 1 anti-parallel strands. This places the C termini of the parallel b strands and the N termini of the a helices close to the interface between the two domains.[23] The two lobes comprise residues 1 to 332 (N terminal lobe) and 344 to 703 (C terminal lobe) with a 3-turn connecting helix, composed of residues 333 to 343.

Each lobe is capable of binding one Fe^{3+} ion very tightly in the presence of a carbonate or bicarbonate anion ($K = 10^{20}$ M^{-1}). Each lobe has the same folding, based on two domains of similar supersecondary structure, with the iron site at the domain interface. The two iron-binding sites are about 4.2 nm apart and are buried at the inner end of the deep interdomain cleft, with iron atoms at least 1.0 nm from the exterior of the molecule. Each iron atom is coordinated by protein ligands: two tyrosines, one histidine, and one aspartate.

More precisely, the coordination sphere of the iron atoms involves four ligands from the protein, namely one carboxylate oxygen, two phenolate oxygens, one imidazole nitrogen, and two bidentate oxygen ligands from the anion (carbonate or bicarbonate). The formation of each high affinity site for Fe^{3+} derives from the fit of Fe^{3+} and CO_3^{2-} ions when enclosed within the lactoferrin protein, and from the types of ligands. Three anion oxygen ligands (one carboxylate from Asp, two phenolates from Tyr) match the 3+ charge of the metal ion, and the 2− charge on the CO_3^{2-} is matched by a positive charge from an arginine side chain and a helix N terminus.

The four protein ligands occupy four octahedral positions about the iron atom: in the N lobe, Asp 60 and Tyr 192 are *trans* to one another, while Tyr 92 and His 253 occupy *cis* positions. In the C lobe, the corresponding residues are Asp 395, Tyr 528, Tyr 435, and His 597. The two remaining *cis*-octahedral positions are left vacant for the two non-protein ligands.

The protein ligands are widely spaced along the polypeptide backbone; Asp 60 is located in the main part of domain N1, at the junction of b sheet c and helix 3. Tyr 192 is in domain N2 at the N terminal extremity of helix 7, pointing toward the middle of the cleft; Tyr 92 and His 253 come from the interconnecting backbone strands that cross between the two domains at the back of the iron site.

At variance with many other metal-binding proteins, the essential metal-binding ligands of lactoferrin are distributed along several exons: for the N lobe, His 61 is on exon 3, Tyr 93 is on exon 4, Tyr 191 is on exon 5, and His 252 is on exon 7. This may reflect the fact that iron is not necessary for the correct folding of the apotransferrin polypeptide chain, whereas the metal-binding site may facilitate protein folding in many other metal-binding proteins.

Mutagenesis of lactoferrin gene by site-directed mutagenesis and production of recombinant mutant lactoferrin molecules with a single amino acid residue allowed researchers to prove definitely the crucial roles of Asp, Tyr, and His in the formation of iron-binding sites. Mutation of both Tyr ligands (which were mutated to Ala) of either the NH_2 or COOH-terminal domain allowed a complete loss of iron-binding capacity.[24] Furthermore, the loss of Tyr residues, particularly at the level of COOH-terminal lobes, produces a reduced iron-binding stability.[24] However, equivalent studies of site-directed mutagenesis have shown inequivalences of the two tyrosine ligands in the N lobe of transferrin, indicating that a mutation of Tyr 188 was sufficient to abolish iron-binding capacity.[25]

Mutagenesis of Asp residue also provided some interesting information. In humans, lactoferrin Asp 60 is involved not only in iron binding, but also provides important stabilizing interactions in the "closed" structure through a carboxylate oxygen atom that forms hydrogen bonds linking two helices in domains I and II. The substitution of serine for the aspartic acid at position 60 results in the loss of this metal-binding ligand. In fact, the substituted serine does not coordinate to iron directly, but a water molecule fills the iron coordination site and takes part in interdomain hydrogen bonding.[26]

Another set of experiments focused on evaluating the contribution of a conserved His ligand to iron (within the iron-binding pocket) in lactoferrin by oligonucleotide-directed mutagenesis. His residue 253 located at the level of the N terminal half molecule of human lactoferrin has been changed to other amino acids, including Gly, Ala, Pro, Thr, Leu, Met, and Tyr. All these substitutions destabilized iron binding irrespective of whether or not these replacements were potential iron ligands. All these mutants exhibited iron loss at pH below 6 or 7, while wild-type lactoferrin showed iron loss only at pH below 5[27] (Figure 3.7).

Some evidence indicates that a significant conformational change occurs as iron is bound; the molecule becomes markedly more compact. Each lobe is further subdivided into two domains, suggesting that the conformational change involves a transition from "open" to "closed" configuration with the two domains closing over the metal as it binds.[23]

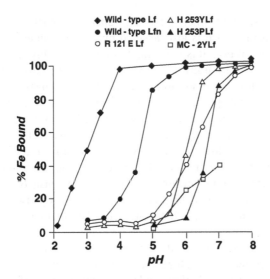

FIGURE 3.7 pH dependence of iron release from wild-type lactoferrin (wild-type Lf), wild-type N-terminal half of human lactoferrin (wild-type Lfn) and lactoferrin mutants in which His 253 is replaced by Tyr (H253Y), His 253 is replaced by Phe (H253P), Arg 121 is replaced by glutamic acid (R121E), and both Tyr residues of the COOH terminal are replaced by Ala (MC-2Y). (Reprinted from References 24 through 27.)

This last point was tested by analyzing the crystal structure of apolactoferrin[28,29] and the X-ray solution scattering of apolactoferrin following iron binding[30]. In the first of these two studies, Anderson et al.[29] analyzed at 2.8 Å resolution the structures of crystals of human apolactoferrin in its deglycosylated form. This analysis showed that two remarkable features are apparent in the apolactoferrin structure: (1) the binding cleft in the N lobe is wide open, following a rotation of 53° of the N2 domain relative to the N1 domain; (2) the binding cleft in the C lobe, although no metal ion is present, is present in a closed conformation. This one open–one closed structure is surprising and may be related to the fact that an equilibrium exists between open and closed forms in solution and that the observed structure is selected by structure packing.[23]

However, when X-ray solution scattering was used for studying the structural changes that take place upon uptake and release of iron from lactoferrin, different conclusions were reached about the conformation of the binding cleft in the C lobe.[30] In fact, this study showed that in the presence of iron, both iron-binding sites are found in a closed configuration, consistent with crystallographic data; however, in the absence of iron, both the N and C terminal iron-binding sites open substantially.

It was also shown that each lobe of lactoferrin exhibits a marked resemblance to a group of bacterial periplasmic proteins. These proteins can exist in three structural forms: liganded closed, liganded open, and unliganded open. An examination of the N lobes of apolactoferrin and iron lactoferrin showed the same hinge-bending, ligand-induced conformational change as with bacterial periplasmic binding proteins.

It is also of considerable interest that the transport of bacterial ferri-siderophores across the periplasmic space is also assured by a periplasmic-binding protein. On the basis of this observation, it is tempting to suggest that this protein may represent the bacterial equivalent of transferrin. Structural studies have also contributed to explanation that lactoferrin binds iron very tightly and releases it only at low pH values. In fact, the *in vitro*-produced N terminal half of human lactoferrin synthesized from cloned cDNA is able to bind iron reversibly, but loses iron in the 4.0 to 6.0 pH range, compared with a range of 4.0 to 2.5 for native lactoferrin[31] (Figure 3.1).

This finding suggests that the easier release of iron from the N terminal half of lactoferrin, compared with native lactoferrin, results from the absence of stabilizing contacts between the N and C terminal halves.[25] On the basis of this observation, it was suggested that the typical difference in pH stability between lactoferrin and transferrin is primarily due to differences in the interactions between N and C terminal halves.[31]

Recently, the three-dimensional structure of the recombinant N terminal lobe (Lf$_N$) of human lactoferrin at high resolution was determined,[32] offering a possible structural explanation for the higher tendency to release iron of the N terminal half molecule, compared to native lactoferrin. The structure of Lf$_N$ resulted essentially the same way as the N lobe of intact lactoferrin, being folded into two similar a/b domains, with the Fe^{3+} and CO$_3^{2-}$ bound in a specific site in the interdomain cleft.

At the C terminus, however, the conformation of residues 321 to 333 is changed. In native lactoferrin, residues 321 to 332 form a helix crossing between the domains at the back of the iron site; in Lf$_N$, residues to 321 to 326 have an extended conformation, forming a third interdomain b strand, and residues 328 to 333 appear disordered.[33] This conformational change is attributed to the loss of stabilizing interactions from the C lobe and seems to be responsible for the more facile iron release properties of Lf$_N$.[33]

As previously mentioned, the metal- and anion-binding sites present in the two halves of the lactoferrin molecule are also very similar, with the same metal ligands (2 Tyr, 1 Asp, and 1 His) in equivalent orientations, and the same anion interactions. This similarity also extends between lactoferrin and transferrin. The binding sites in the N and C terminal halves of each molecule do not have identical properties, and differ in detail in their binding strength, acid stability, and facility of iron release. Differences in the N terminal and C terminal sites also become more pronounced when metals other than iron or anions other than carbonate are substituted, as shown by crystallographic studies.[34,35]

The anion-binding site in the N terminal lobe of lactoferrin is formed by residues 121 to 124 at the N terminus of a helix, the side chain of Arg 121, and the side chain of Thr 117 in a preceding loop[36]; a functionally and structurally equivalent site is present in the C terminal lobe. The arginine residue is conserved in all molecules of the transferrin family, with the exception of human melanotransferrin and *Manduca sexta* transferrin, where it is changed to Ser and Thr, respectively, leading to a loss of iron-binding capacity.

The role of this Arg residue in lactoferrin binding was directly evaluated by site-directed mutagenesis of Arg 121 in the N terminal half molecule of human lacto-ferrin.[36] Mutation of Arg 121 to either Ser or Glu causes a 0.5 Å displacement of the synergistic carbonate ion as compared to its position in the wild-type protein. As a consequence of these changes, the mutants retain the ability to bind iron, but with decreased stability.[36] The affinities of both N and C sites of lactoferrin for bicarbonate are low, thus indicating an extremely weak interaction. However, this weak interaction is immediately followed by a proton loss, which stabilizes the lactoferrin/synergistic anion complex and makes the protein more available for interaction with bicarbonate.[37]

The interaction of bicarbonate and lactoferrin is favored by alkaline pH (>7.5), while the anion binding capacity of transferrin remains constant in the pH range from 7.40 to 9.0.[37] This observation suggests that lactoferrin preferentially inteacts with carbonate in metal media, while transferrin interacts only with bicarbonate.

Interestingly, bovine transferrin exhibits a peculiar behavior in that it binds iron more weakly than human lactoferrin and releases iron more readily. The analysis of its amino acid sequence reveals that it has two Lys residues equivalent to those of human transferrin that form the so-called "di-lysine trigger", but it behaves as a typical lactoferrin with respect to iron-induced iron release. Furthermore, it has four glycan chains, as compared with three for human lactoferrin, with only one glyco-sylation site (Asp 476) in common.

Analysis of the three-dimensional structure of bovine lactoferrin greatly contributed to understanding the peculiar behavior of this protein. The folding of bovine lactoferrin is very similar to that of human lactoferrin, but this protein differs in the extent of closure of the two domains of each lobe, and in the relative orientation of the two lobes.[38] These modified interdomain interactions, as well as the presence of two lysine residues behind the N lobe iron site, contribute to understanding the lower affinity for iron of bovine lactoferrin as compared to human lactoferrin.

The lactoferrin gene has been localized to the human chromosome 3, band 3q21-q23, a region that also contains the human transferrin gene family cluster.[39,40] Current evidence suggests that this family of genes evolved first through intragenic duplication,[41] followed by further duplication events that resulted in a family of genes encoding proteins with divergent tissue expressions and functions.

The gene encoding human lactoferrin was cloned and its sequence determined.[42-46] The amino acid sequence derived from DNA sequence is virtually identical with that obtained by peptide mapping.[1] It is also of interest that the sequence of lactoferrin cDNA derived from neutrophils is identical to that isolated from mammary gland tissue.[43-46] This finding confirms that glandular and neutrophil lactoferrins derive from the same gene, and do not represent different isoforms. The full length cDNA of lactoferrin encodes a protein with a signal peptide of 19 amino acids followed by a mature protein of 692 residues.[42-43] Analysis of genomic libraries showed that the lactoferrin gene spans along about 35 Kb[43]; however, the complete exonic and intronic organization of the whole human lactoferrin gene was not completely determined.

However, the complete structure of the mouse lactoferrin gene was determined recently.[47] Mouse lactoferrin gene is composed of a single gene copy of approximately 30 Kb in size. The gene is organized in 17 exons separated by 16 introns; the exons range in size from 48 base pairs to 190 base pairs, whereas the introns range from 0.2 Kb to 4.3 Kb.[47]

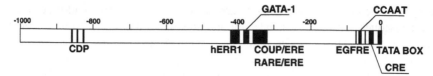

EGFRE: 5'GGGCAACAGGGCGGGCAAAGC 3'
COUP/ERE: 5'TCTCACAGGTCAAGGCGATC 3'
GATA: 5'AAGATAG 3'
hERR1:   5'ACCTGCCCTAACTGGCTCCTAGGCACCTTCAAGGTCATCTG 3'

FIGURE 3.8 Human lactoferrin gene promoter. The relevant sites from a functional view are indicated by boxes. The nucleotide sequences of some of the boxes are shown below.

The structural analysis of the mouse lactoferrin gene revealed that it shares a similar intron–exon distribution pattern with both human transferrin and chicken ovotransferrin.

Recently, the complete gene encoding human lactoferrin was isolated from a cosmid library.[47] The human lactoferrin gene spans along 24.5 Kb. The lactoferrin transcript was assembled from 17 exons separated by introns of variable size. *In situ* fluorescence hybridization showed that the human lactoferrin gene was located in region 3p21,3.[47a]

A novel form of lactoferrin was discovered recently: D lactoferrin, characterized by the absence of the signal peptide (encoded by exon 1) and by a predicted amino acid structure identical to that of canonical lactoferrin, with the exception of the absence of the first 25 amino acids from the N terminus of the mature protein.[48] D lactoferrin originates through an alternative splicing of the lactoferrin gene which uses an alternative promoter located at the level of the junction between exons 1 and 2; the sequence of this promoter is only in part identical to that of the canonical lactoferrin promoter.[48] The biological function of D lactoferrin remains unknown; however, of interest is the finding that D lactoferrin mRNA is expressed by a variety of tissues, including breast, spleen, bone marrow, placenta, testis, kidney, and brain, while it is absent in the tumor counterparts of these tissues.[48] Finally, since many functions have been ascribed to lactoferrin, it is reasonable to assume that some of its functions are selectively mediated by D lactoferrin.

The promoter region of the lactoferrin gene was cloned and sequenced,[45] and produced several interesting findings (Figure 3.8) including:

1. A region of –350 bp with respect to the transcription start site is highly homologous to the consensus sequence (5'-GGTCANNNTGACC-3') of the estrogen response element, from which it differs by only two of 13 bp (GGTCAAGGCGATC). This finding is consistent with the estrogen inducibility of lactoferrin in reproductive epithelium.[49,50]
2. A hexamer, CTGGGA, that has been found in several genes involved in the acute phase response (fibrinogen, haptoglobin, α_1-antitrypsin) is found at –280 bp.

3. A TATA box element is found at −30 bp.
4. A binding site for the zinc finger transcription factor GATA-1, currently felt to be a predominantly erythroid-specific factor, is located at −367 bp; the presence of this binding site is surprising, but its role, if any, in the control of lactoferrin gene expression remains to be determined.

Also, the promoter of the mouse lactoferrin gene at position −329 to −341 contained a sequence highly homologous to the consensus sequence of the estrogen responsive element[51]; this sequence seems to be essential for lactoferrin expression in estrogen-sensitive cells (i.e., uterus and mammary gland). More recently, at sequences between −75 and −53, a mitogen-responsive unit composed of two elements, the EGF response element (CREB) and the AP1 binding element (CRE) was identified.[52] This region of the lactoferrin promoter is responsible for the induction of this gene by epidermal growth factor.[53]

The sequence of the human lactoferrin gene promoter was also reported by another group of investigators.[54] The inspection of this sequence showed two new findings as compared to the previous study:[55] (1) two half-palindromic estrogen response elements (GGTCA) are present at 5′ with respect to the complete estrogen response element (ERE); and (2) a sequence resembling the chicken ovalbumin upstream promoter transcription factor (COUP-TF) binding site (GTCTCACAG-GTCA). This sequence and the ERE share five nucleotides (GGTCA) (Figure 3.8).

These authors also provided evidence that a synthetic nucleotide including the sequence of COUP/ERE acted as an enhancer in response to estrogen stimulation. *In vitro* DNAaseI footprint analysis showed binding of the estrogen receptor on the ERE sequence.[55] Studies of mutants obtained by site-directed mutagenesis showed that COUP-TF does not interact with the estrogen receptor.

A different situation is found in the mouse lactoferrin gene, where the COUP-TF overlapping with ERE acts as a competitive repressor for estrogen receptor-mediated activation of the lactoferrin gene.[56] Therefore, the molecular mechanisms of the estrogen action that govern lactoferrin gene expression differ between mouse and human. A RARE (retinoic acid receptor element) was identified in the human lactoferrin gene promoter located in the sequence region −349 to −337 and composed of two identical AGGTCA-like motifs arranged as a direct repeat with 1-bp spacing; this element is strongly activated by RXR homodimers.[58] Interestingly, the lactoferrin RARE overlaps an ERE in which both RARE and ERE have a common AGGTCA-like motif.[57]

The presence within the lactoferrin promoter of individual features characteristic of both housekeeping and inducible promoters could contribute to the differential regulation of lactoferrin gene in different cell types. The lactoferrin gene sequence seems to be highly conserved during evolution; in fact, murine,[59] porcine,[60] and bovine[61] lactoferrins exhibit high levels of homology with human lactoferrin, corresponding to 70%, 71%, and 69% amino acid identity, respectively.

The recombinant lactoferrin protein was expressed in baby hamster kidney cells transfected with full-length lactoferrin cDNA.[62] The recombinant protein synthesized in this cellular system was highly glycosylated and exhibited physico-chemical properties (i.e., molecular weight, absorption spectrum, and pH-dependent release of iron) identical to those displayed by the native protein.[62]

The production of human lactoferrin in *Aspergillus nidulans* was also described.[63] The human lactoferrin cDNA was expressed under the control of the ethanol-inducible alcohol dehydrogenase promoter. The recombinant lactoferrin produced by this system is indistinguishable from native lactoferrin. Using an identical expression system, recombinant murine lactoferrin was produced.[64] The recombinant protein was immunologically identical to native lactoferrin, was correctly processed at its N terminus, and was appropriately glycosylated.

Transgenic mice harboring human lactoferrin cDNA fused to regulator elements of the bovine aS1 casein gene were produced. Recombinant human lactoferrin expressed in the milk of transgenic mice was identical to natural human milk-derived lactoferrin.[65] A recent study showed that recombinant human lactoferrin produced in the baculovirus expression system, although slightly defective in its glycosylation pattern, exhibited a pattern of binding to its receptor present on T cells that was fully comparable to that displayed by native lactoferrin.[66]

The successful cloning and expression of human lactoferrin provide the appropriate tools for a detailed study of the protein. Analogous to studies carried out with ferritin, site-directed mutagenesis can be used to study specific regions of the protein implicated in iron binding and release, and, in particular, the marked conformational change that occurs in the N terminal half of the molecule during iron uptake and release.

3.3 LACTOFERRIN RECEPTORS

The biological effects of lactoferrin, as well as those of other transferrins, are mediated through the interaction of the protein with specific membrane receptors. The membrane receptors responsible for binding lactoferrin seem to be specific for this protein and do not interact with other transferrins (serum transferrin or ovotransferrin) or unrelated proteins. The main biological function dependent upon the interaction of lactoferrin with its receptor consists of mediating the internalization of the protein into the cells. No evidence suggests that the interaction of lactoferrin with its receptor mediates the release of cellular signals or mediators. At variance with transferrin, lactoferrin receptors equally bound apo and iron-saturated forms of lactoferrin. The tissue distribution of lactoferrin receptors seems to be more restricted as compared to that of transferrin receptors. The molecular structure of lactoferrin receptor remains to be determined, but biochemical data suggest that is not related to the transferrin receptor. The main biological and biochemical features of lactoferrin receptors are analyzed below for the different cell types in which these receptors are expressed.

3.3.1 LACTOFERRIN RECEPTORS ON LEUKEMIC LINES

The binding of lactoferrin to neoplastic lines, particularly to human leukemic lines, was also reported.[67,68] These studies showed that both erythroid[67] and granulo-monocytic cell lines[68] possess elevated numbers of lactoferrin binding sites (i.e., $>10^6$ binding sites per cell) with apparent receptor affinities for the ligand in the range of 10^{-6} to 10^{-7} M. The lactoferrin binding sites expressed on these cells seem to be trypsin-sensitive and do not cross-react with transferrin. Interestingly, these leukemic cells seem to be able to bind lactoferrin much more than apolactoferrin.[67] Lactoferrin binding to these cells was not influenced by iron status,[67] whereas this phenomenon was clearly observed for the transferrin receptor in the same cell lines.[67]

HL-60 promyelocytic cells induced to differentiate to monocytes–macrophages by TPA showed a two-fold increase of lactoferrin binding capacity.[68] This observation was also confirmed in the THP-1 promonocytic cell line, where macrophage induction was associated with a marked increase of the number of high-affinity lactoferrin receptors.[69]

The binding of lactoferrin to human erythroleukemic K-562 cells was characterized in detail,[70] providing evidence that these cells possess around 10^7 binding sites per cell and bind lactoferrin with a constant of affinity of 4×10^{-6} M. Preliminary characterization of the membrane protein that binds lactoferrin showed a band of about 120 kDa.[70] After binding at the cell surface, lactoferrin is internalized in a temperature-dependent fashion and is then immunologically detectable as a DNA-linked protein in nuclear extracts.[70]

3.3.2 LACTOFERRIN RECEPTORS ON ACTIVATED T LYMPHOCYTES

Lactoferrin receptors were particularly studied in T lymphocytes. Initial studies showed that resting T lymphocytes do not possess lactoferrin receptors, but they progressively synthesize these receptors after mitogenic activation with phytohemagglutinin (PHA). During T lymphocyte activation, the kinetics of lactoferrin receptor expression parallels that of the transferrin receptor. Furthermore, lactoferrin and transferrin are equally active in sustaining the proliferation of human T lymphocytes stimulated with PHA and grown under serum-free conditions[71]; the growth-promoting activity of lactoferrin was maximal when the protein was only partially (10 to 30%) iron-saturated.

The binding of lactoferrin to human T lymphocytes was not affected by iron saturation. All major lymphocyte subsets, i.e., TCR α/β^+ T cells, TCR γ/δ^+ T cells, B cells, and NK lymphocytes, express lactoferrin receptors after activation.[72] Interestingly, the proportion of activated γ/δ lymphocytes expressing lactoferrin receptors is higher than that of α/β T lymphocytes; in line with this observation, lactoferrin potentiated the proliferation of γ/δ^+ T lymphocytes triggered by polyclonal activators.[72]

The affinity of lactoferrin for binding sites present on PHA-activated T lymphocytes seems to be higher ($K_d = 0.83 \times 10^{-7}$) than that observed on other cell types. Preliminary characterization of lactoferrin binding sites expressed on activated T lymphocytes showed that they are represented by two proteins of 100 and 110 kDa, respectively.[71]

Subsequent studies were focused on identifying the part of the lactoferrin molecule involved in binding with the T lymphocyte receptor. Using large protein fragments obtained by tryptic hydrolysis of human lactoferrin demonstrated that the N terminal domain I of the protein could interact with the mitogen-stimulated lymphocyte receptor.[73,74] Further proteolysis of the 30-kDa N-tryptic fragment of the protein allowed identification of a 6-kDa peptide (residues 4 to 52) which is able to inhibit the binding of the whole lactoferrin to the PHA-activated lymphocytes.[75]

Molecular modeling experiments performed with agents able to interact with specific amino acids provided evidence that residues 4 to 6, 28 to 34, and 38 to 45 could be involved in mediating the interaction between lactoferrin and its receptor.[74,75] Thus, derivatization of human lactoferrin on Lys 264[74] or Lys 74[75] was able to inhibit the binding of the protein to the membrane receptor.

Experiments of derivatization of human lactoferrin with FITC or heterobifunctional reagents showed that the two loop-containing regions of human lactoferrin (residues 28 to 34 and 38 to 45) are involved in mediating the interaction between the lymphocyte lactoferrin receptor and lactoferrin.[76] Furthermore, a 6-kDa peptide corresponding to residues 4 to 52 inhibited the binding of lactoferrin to lymphocytes.[76] However, the N terminal residues of Arg 2, Arg 3, and Arg 4 do not seem to play a role in mediating the binding of lactoferrin to lymphoid cells; in fact, removal of this basic cluster from the lactoferrin molecule increases the affinity of this ligand for the receptor present on lymphoid cells.[77]

Experiments using chlorate, which inhibits sulfation, reduced the number of binding sites for both native lactoferrin and the lactoferrin Arg 3 mutant, but not for the Arg 5 mutant (the mutant that lacks the first four Arg residues). These results suggested that: (1) 80,000 binding sites per lymphoid cell, mainly sulfated molecules, are dependent upon the presence of Arg 2, Arg 3, and Arg 4 residues, but not upon Arg 5; and (2) interaction with about 20,000 binding sites per lymphoid cell, seemingly representing the true lactoferrin receptors, does not require the presence of the first N terminal cluster of basic amino acids of human lactoferrin.[78]

It is worth mentioning that lactoferrin does not exhibit homology with transferrin at the level of the amino acid residues — a finding in line with the absence of cross-reactivity of transferrin and lactoferrin for binding to their respective membrane receptors. Finally, the comparison of the primary and tertiary structures of human lactoferrin and serotransferrin, which bind to specific cell receptors, shows that these regions, which are likely to be involved in protein–receptor interactions, possess specific structural features.

Careful comparison of the three-dimensional structure of the Nt peptide (4 to 52) of human lactoferrin and rabbit serotransferrin showed that the two peptides exhibit similar, but not identical structures. The peptide of rabbit serotransferrin is three amino acids longer and that of the Ct part of human lactoferrin is three amino acids shorter. Since rabbit (or human) serum transferrin and the Ct lobe of human lactoferrin are not recognized by the lactoferrin receptor, it may be suggested that the human lactoferrin loop present in the N terminal lobe and absent in the C terminal lobe is involved in the receptor binding site (Figure 3.9).

The comparative analysis of the primary structure of the polypeptide chain covering residues 4 to 52 from bovine and murine lactoferrin indicates a high percentage of homology with the human lactoferrin polypeptide chain; as expected, murine and bovine lactoferrin inhibited binding of human lactoferrin to human lactoferrin receptor.

A - K A T T V R W C A V S N S E E E K C L R W Q N E M R K V G G

¹G R R R R S V Q W C A V S Q P E A T K C F Q W Q R N M R K V R G

A - P R K N V R W C T I S Q P E W F K C R R W Q W R M K K L G A

P P L S C V K K S S T R Q C I Q A I V T

P P V S C I K R D S P I Q C I Q A I A E⁵²

P S I T C V R R A F A L E C I R A I A E

FIGURE 3.9 Amino acid sequence analysis of human, murine, and bovine lactotransferrins covering regions 1 to 52. The human sequence is shown in the middle, the murine sequence at the top, and the bovine sequence at the bottom. Amino acid residues are numbered according to the human lactoferrin sequence.

3.3.3 Lactoferrin Receptors on Monocytes–Macrophages

Van Snick et al.[79] showed that human lactoferrin caused hypoferremia in rats during acute inflammation, possibly by capturing the iron from transferrin. These authors also showed that iron-saturated lactoferrin might deliver its iron into intracellular ferritin in mouse macrophages by a receptor-mediated mechanism.[80,81] Subsequent studies confirmed that human monocytes–macrophages possess lactoferrin receptors.[82-85] These studies provide evidence that:

1. Human monocytes–macrophages possess a high number of lactoferrin receptors (2 to 6 × 10⁶ binding sites per cell) with a calculated K_d of 2 to $4 \times 10^{-6}\ M$.
2. High-affinity and low-affinity components are clearly demonstrated by Scatchard plot analysis of the binding data (Figure 3.10).
3. Monocyte lactoferrin receptor exhibits similar affinity for diferric lactoferrin and apolactoferrin.
4. The binding of lactoferrin on these cells is specific for this protein in that several other proteins, including transferrin, are unable to compete with lactoferrin for binding to its receptor.
5. The binding of lactoferrin to monocytes is a calcium-dependent process.
6. The fucose polymer fucoidan decreases lactoferrin binding by 80 to 90%. The binding of lactoferrin to monocytes–macrophages considerably increases following the activation of these cells.

The binding of lactoferrin to monocytes–macrophages has been reported to elicit several biologic effects including inhibition of colony-stimulating factor release,

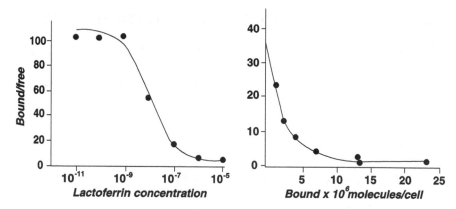

FIGURE 3.10 Binding of lactoferrin to human monocytes. Cells were incubated with fixed amounts of (^{125}I)-lactoferrin and increasing amounts of cold lactoferrin. Data are shown as a competition curve (left) or a Scatchard plot of binding data (right).

enhancement of phagocytosis and killing of intracellular parasites, inhibition of phorbol ester-stimulated PGE2 production, inhibition of hydroxyl radical formation, and promotion of intracellular iron storage. At least some of these biological effects imply that lactoferrin has been internalized following its binding to monocyte membrane receptor.

In vitro binding studies carried out at 37°C indicate that after the binding to the membrane receptors, the majority of detectable lactoferrin was present in the cytosol; the iron initially bound to lactoferrin was found after lactoferrin uptake in the cytosol bound to ferritin.[86,87] Lactoferrin in the cytoplasm seems to enter a lysosomal compartment of relatively low pH (<5.0). After internalization into monocytes–macrophages, lactoferrin is recycled, remains normal in terms of molecular weight and iron-binding capacity, and is slightly modified (as shown by a slight, but significant decrease in the isoelectric point from 8.9 to 8.8), thus becoming unable to rebind to macrophage receptors.[88]

This observation strongly suggests that, at variance from transferrin, lactoferrin cannot operate in a cyclic manner to deposit iron into monocytes–macrophages. The internalization of lactoferrin by monocytes–macrophages may be particularly relevant in modulating the multiplication of intracellular bacterial pathogens. In fact, it was shown that the binding and internalization of apolactoferrin markedly inhibited intracellular multiplication in mononuclear phagocytes of *Legionella pneumophila*, whereas an opposite effect was elicited by iron-saturated lactoferrin.[89]

Lactoferrin is felt to play an important role in host defense by acting to withhold iron from pathogenic bacteria during infection. By binding iron and then being internalized by mononuclear phagocytes, lactoferrin may contribute to the decrease in serum iron, the so-called hypoferremic response seen in infection and inflammatory states. Furthermore, lactoferrin functionally activates macrophages and stimulates the secretion of tumor necrosis factor-α and interleukin-8 as well as the production of nitric oxide. This action of lactoferrin on macrophage cytokine release helps mediate the antibacterial effects of this molecule.[90]

3.3.4 LACTOFERRIN RECEPTORS ON THE SMALL INTESTINAL BRUSH BORDER MEMBRANE

Uptake of lactoferrin-bound iron by human duodenal biopsies was first demonstrated by an *in vitro* assay,[91] and the survival of lactoferrin in the digestive guts of human newborns was highlighted.[92] A series of studies provided evidence that:

1. Lactoferrin is quite resistant to proteolysis.
2. A significant amount of immunologically intact lactoferrin is able to survive passage through the intestinal tract of the exclusively breast-fed infant and can be detected in the feces.[92,93]
3. Lactoferrin has been shown to resist proteolysis in *in vitro* studies using proteolytic enzymes, such as pepsin, trypsin, and chymotrypsin, giving rise to large fragments.
4. Intact lactoferrin and two large fragments (one corresponding to the 3 to 283 amino acid sequence and the other to the 284-COOH terminus sequence; the two fragments are bound together giving rise to a molecule with a weight identical to that of the entire lactoferrin molecule) have been detected in the urine of pre-term infants receiving human milk; this urinary lactoferrin arises from maternal lactoferrin absorbed across the gut.[94,95]

These studies indicate that maternal lactoferrin is only partially digested during passage in the gut and a significant proportion of it is absorbed, particularly in pre-term infants.

In addition to these physiological observations, a specific receptor has been evidenced on the small intestinal brush borders of several animal species including rabbits, monkeys, mice, and rats.[96-100] It was also found that the intestinal mucosal surface is covered with lactoferrin derived from goblet cells. The lactoferrin receptor present on the intestinal brush border was to some extent characterized for its main biological and biochemical properties. The most careful characterization was carried out for mouse small-intestinal brush border lactoferrin receptor,[99,100] and showed that: (1) the receptor is strictly specific for lactoferrin with an affinity constant (K_a) of about $3 \times 10^{-6}\,M$ (similar for both apolactoferrin and iron-saturated lactoferrin); (2) the binding of lactoferrin was not inhibited by fucose; (3) optimal lactoferrin binding was observed at pH 5.5 to 6.0; (4) bivalent cations, particularly Ca^{2+}, are required for optimal lactoferrin binding to brush border membrane vesicles; (5) the receptor is highly glycosylated, composed of a single-chain polypeptide of 130 kDa and with pI of 5.8; and (6) the receptor is recognized by ConA and PHA-L and susceptible to treatment with N-glycanase and endo-N-acetyl-b-D-glycosimanidase B, suggesting the presence of glycans of the bi- and/or tri-antennary n-acetyllactosaminic type.

Using the HT-29 enterocyte-like differentiable cell line, it was demonstrated that the level of lactoferrin receptor expression on enterocytes is modulated by iron concentration; addition of an iron chelator to these cells elicited a two-fold increase in the number of lactoferrin receptors.[101]

In spite of intensive efforts involved in the identification and characterization of intestinal lactoferrin receptors, their role and the role of lactoferrin in mediating iron absorption in newborns remain to be fully demonstrated. Human milk lactoferrin is saturated with iron to a low degree corresponding to 3 to 5% of its capacity.[102] However, during passage through the gastrointestinal tract, the pH changes and release of iron from other binding ligands during digestion may alter the iron saturation of lactoferrin when it reaches the small intestine.

In this context, initial studies showed that breast-fed infants maintain adequate iron status through 6 to 9 months of age, whereas formula-fed infants consuming similar amounts of iron require iron supplementation after about 4 months.[103,104] In agreement with these initial observations, Simes et al.[105] found that none of 36 exclusively breast-fed infants exhibited iron deficiency anemia at 9 months. Iron status of these infants was similar to a control group receiving iron-supplemented formula. In contrast, low-birth-weight infants, born with significantly lower iron stores, require iron supplementation at an earlier age.

In a study of 15 exclusively breast-fed preterm infants, Iwai et al.[106] found that about 60% showed signs of iron deficiency at 4 months, and 86% were iron deficient at 6 months. A large number of studies have attempted to determine the bioavaibility of iron from various milks and formulas.[107-120] These studies conclude that iron absorption from cow milk-based formulas is low, 5 to 20%, whereas iron uptake from human milk is considerably higher, 40 to 70%.[84-89]

Closer examination of these studies reveals that values for iron absorption are typically highly variable. Despite the wide variation in iron absorption, it is commonly accepted that iron absorption from human milk is 2.5 times higher than from cow milk formulas. In an attempt to explain the high bioavability of human milk iron, lactoferrin has been suggested as a possible enhancer of iron absorption.[117] However, recent clinical trials have failed to demonstrate any improvement in iron absorption in infants fed formula milk supplemented with bovine lactoferrin.[118,119] Studies in infant rhesus monkeys showed that the bioavailability of iron bound to lactoferrin was no better than that of inorganic iron (ferrous sulfate).[120]

3.3.5 LACTOFERRIN RECEPTORS ON PARENCHYMAL LIVER CELLS

Intravenously injected lactoferrin is rapidly removed from circulation of rats, mice, and rabbits.[121-124] The rapid clearance was found to be mainly a result of the association of lactoferrin to the liver. The cell type involved in the hepatic association, however, remained controversial until recently. In fact, initially Preels et al.[121] reported that parenchymal cells were responsible for the liver association of lactoferrin, whereas others claimed a major role for sinusoidal liver cells.[122,123] A recent study gives a clear demonstration that liver parenchymal cells are responsible for more than 90% of the lactoferrin binding *in vivo* observed at level of the liver.[124] Scatchard analysis of binding data indicates that isolated parenchymal liver cells bind lactoferrin with low affinity ($K_a = 10 \times 10^{-6}\ M$) and high capacity ($20 \times 10^6$ binding sites per cell).[125]

Lactoferrin binding could not be inhibited by competitors for asialoglycoprotein and fucose receptors.[125] This finding is particularly relevant in that it indicates that:

(1) parenchymal liver cells possess specific receptors for lactoferrin; (2) oligosaccharide lactoferrin chains are not responsible for lactoferrin binding to liver parenchymal cells; and (3) clearance of lactoferrin is not mediated through binding with liver glucose receptors. This last item is worthy of mention since a number of hepatic receptors (i.e., asialoglycoprotein receptor and fucose receptor) mediate glycoprotein clearance through recognition of their terminal carbohydrate units.

Four lines of evidences suggest that lactoferrin may interact with liver parenchymal cells through interaction with the receptor for chylomicron remnants (the interaction of chylomicrons with endothelial cell-bound lipoprotein lipase and high-density lipoproteins results in the formation of chylomicron remnants):

1. Lactoferrin inhibits the uptake of ^{125}I-labelled chylomicron remnants by whole liver by about 50%.[126]
2. Lactoferrin inhibits the association of apoE-bearing lipoproteins to the chylomicron remnant receptors on parenchymal liver cells.[127]
3. Both apoE lipoproteins and lactoferrin possess very positively charged arginine/lysine-rich sequences: residues 134 to 150 in human apoE and four clustered arginine residues at the N terminus of the lactoferrin polypeptide chain.[125]
4. Derivatization of arginine residues on lactoferrin resulted in a significantly lower liver association, which was the result of a large reduction of the association of lactoferrin to the parenchymal cells.[125]

Subsequent studies with a lactoferrin molecule lacking 14 N-terminal amino acids showed that the four-arginine cluster of lactoferrin at positions 2 to 5 is involved in the massive, low-affinity association of lactoferrin with the liver, possibly to proteoglycans, but it is not essential for the inhibition of lipoprotein remnant uptake.[128] Conversely, the Arg–Lys sequence at positions 25 to 31 may mediate the high-affinity binding of lactoferrin and block the binding of b-VDL to the remnant receptor with high efficiency.[128]

However, since the number of binding sites for lactoferrin is much higher than the number of binding sites for chylomicron remnants, one has to assume that the remnants are bound to multiple receptors or that the remnant receptor may not be the only receptor system involved in the hepatic association of lactoferrin. Additional studies indicate that lactoferrin is able to interact also with the third member of the LDL receptor gene family, called gp330, whose function and biological ligands are unknown.[129] It was also shown that lactoferrin blocked the lipoprotein receptor-related, protein-dependent stimulation of cholesterol synthesis in cultured human fibroblasts elicited by apoprotein E-b VLDL or lipoprotein lipase-b VLDL complexes.

Recent studies led to a characterization of the interaction of lactoferrin with this LDL-related receptor. In fact, it was determined that lactoferrin binds to the LDL receptor-related protein (LRP), which is a large 600-kDa scavenger receptor present on the surfaces of many cell types, including hepatocytes, fibroblasts, macrophages, and neurons. In addition to lactoferrin, this receptor binds a number of protein ligands, such as α2 macroglobulin, plasminogen activator, and lipoprotein lipase, with relatively high affinity. Interestingly, all these ligands, including lactoferrin, bind to the same region of the receptor and their binding is completed by a 39-kDa receptor-associated protein,[130] which acts as a chaperone for LPR.

More recent studies based on the analysis of the interaction of bovine lactoferrin with rat hepatocytes led to the identification of two classes of lactoferrin binding sites: one is high-affinity and Ca^{2+}-dependent, while other is low-affinity and Ca^{2+}-independent.[131] Only the high-affinity binding sites are competent to mediate lactoferrin endocytosis.[131]

The biochemical characterization of these high-affinity lactoferrin binding sites showed that they correspond to a 45-kDa membrane protein. This 45-kDa purified protein possesses the capacity to specifically bind to lactoferrin in solution, and this binding is inhibited by an excess of cold ligand or by monospecific anti-p45 antibodies.[132] This p45 protein shares amino acid sequence homology and immunoreactivity with the RHL-1 subunit of the Gal/GalNAc asialoglycoprotein receptor.[133]

Recent studies have shown that glycosylated and deglycosylated lactoferrins equally bind to the RHL-1 receptor.[134] Moreover, β-lactose, but not sucrose, competed vigorously for ^{125}I-lactoferrin endocytosis by hepatocytes, thus indicating that lactoferrin binds at or near the carbohydrate-recognition domain of RHL-1.[134]

Interestingly, the expression of this Ca^{2+}-dependent lactoferrin receptor on hepatocytes was considerably increased following iron loading with ferric ammonium citrate. This phenomenon was observed 16 to 24 hours after ferric ammonium citrate addition and consisted of a two- to six-fold increase of both lactoferrin binding and endocytosis.[135]

The upregulation of hepatic lactoferrin receptor expression following iron loading is due to a post-transcriptional event, as shown by experiments using transcription (actinomycin D) and protein synthesis (cycloheximide) inhibitors.[134] Experiments performed with lactoferrin proteolytic fragments containing either the N or C lobes showed that the lactoferrin determinants responsible for binding to the Ca^{2+}-dependent receptor on hepatocytes is present within the C lobe of lactoferrin.[136]

Finally it must be emphasized that the clearance of circulating lactoferrin was carefully investigated in several animal species. These studies clearly showed that radiolabelled lactoferrin is rapidly removed from circulation after injection in mice and rabbits. In mice, lactoferrin clears rapidly in the absence of competing ligands, with a half-life of less than one minute.[124] Competing levels of ligands specific for the hepatic galactose receptor, the hepatic fucose receptor, and the mononuclear phagocyte system pathway recognizing mannose, N-acetylglucosamine, and fucose did not block radiolabelled lactoferrin clearance in vivo. The lactoferrin cleared from circulation resulted in more than 90% fixed to the liver, mainly by parenchymal cells.[125]

3.3.6 LACTOFERRIN RECEPTORS ON BREAST EPITHELIAL CELLS

Lactoferrin receptors on epithelial SV-40 immortalized cell lines derived from non-malignant human breast, benign mastopathies, hst oncogene-transformed HBL-100 cells, and breast carcinoma cells were detected.[137] The magnitude and the affinity of binding were similar in normal and breast cancer cells. The level of binding was similar to that detected in activated T lymphocytes; furthermore, the lactoferrin receptor expressed on epithelial breast cells is recognized by an antibody raised against the lymphocyte lactoferrin receptor,[137] suggesting that the lactoferrin breast cell receptor possesses common antigenic determinants to those found in activated lymphocyte receptor.

It is of interest that one breast carcinoma cell line (MDA-MB231) exhibited only low-affinity receptors, not recognized by antibodies specific for lymphocyte lactoferrin receptor. These observations suggest that the canonical receptor has been modified or replaced by another receptor. The peculiar lactoferrin receptor present on these cells could be related to particular properties of these cells. In fact, these cells, at variance with the other breast carcinoma cell lines, do not express estradiol and progesterone receptor genes. Since it has been demonstrated in breast tissue that the amount of lactoferrin mRNA is inversely related to that of estrogen receptor protein, it may be suggested that the hormones that regulate biosynthesis of lactoferrin can also influence the expression of lactoferrin receptors of normal and malignant breast cells.[137]

3.3.7 Lactoferrin Receptors on Platelets

It has been shown recently that the KRDS tetrapeptide, which corresponds to loop 39 to 42 of the lactoferrin molecule and is thus an integral part of the lactoferrin binding site, displays an inhibitory action on platelet aggregation.[138-140] It is of interest that the KRDS signal sequence has been demonstrated to be structurally homologous to the RGDS signal sequence[139] by which the glycoprotein IIb-IIIa complex, expressed at the platelet surface, recognizes the fibrinogen molecule.

These observations prompted studies focused on evaluating the possible presence of specific membrane receptors for lactoferrin on platelets. Using a lactoferrin fluorescent probe (by coupling fuorescein to the glycan moiety of the lactoferrin molecule), a specific, saturable, and reversible binding of lactoferrin to non-activated platelets was shown.[141] The putative human platelet receptor was purified and exhibited homology with the lymphocyte lactoferrin receptor.[141] The entire lactoferrin molecule and two synthetic peptides, KRDS and CFQWQRNMRKVRGPPVSC, covering the two lactoferrin loops (28 to 34 and 39 to 42) that belong to the receptor binding site exhibited inhibitory activity on ADP-induced platelet aggregation. The action of the KRDS tetrapeptide was 25 and 16,000 times lower than the activities of the octodecapeptide and the lactoferrin molecules, respectively. Finally, it was shown that the inhibitory effect of lactoferrin on platelet aggregation is mediated through a mechanism that requires the direct binding of the molecule to its own membrane receptor and not to the platelet glycoproteins IIb-IIIa.[141]

3.3.8 Lactoferrin Receptors on Neuronal Cells

Few studies have evaluated lactoferrin receptor expression on nervous system cells. Lactoferrin expression was investigated in the mouse brain by immunochemistry and reverse transcription. Both techniques showed that lactoferrin was expressed in the pituitary gland, the hippocampus, and the cerebral cortex.[142]

Particular attention was focused on lactoferrin receptors in the mesencephalon. These studies were stimulated by the observation that iron accumulates in the substantia nigra in Parkinson's disease, and this phenomenon cannot be related to an increased expression of transferrin receptors.

Immunohistological analysis showed that lactoferrin receptors are preferentially expressed at the level of cerebral nuclei, while the expressioin was much lower in regions containing bundles of nervous fibers. At the microscopic level, pronounced lactoferrin receptor immunoreactivity was observed in the apical dendrites, axons, and perikarya of neurons in the substantia nigra.[143] In some regions, lactoferrin receptors are also observed on cerebral microvasculature and on glial cells. In Parkinsonian patients, lactoferrin receptor immunoreactivity on neurons was increased, particularly in regions of the mesencephalon where the disease leads to a severe loss of dopaminergic neurons.[143]

3.4 FUNCTIONS OF LACTOFERRIN

3.4.1 BACTERIOSTATIC ACTIVITY OF LACTOFERRIN

Most pathogens face the problem of extremely low availability of free iron in the infected host as a result of sequestering of iron by specific proteins, mainly transferrin and lactoferrin in the extracellular spaces and ferritin within the cell cytoplasm. Iron scavenging caused by these proteins reduces free-iron concentration below $10^{-12}\ M$, which is much lower than the iron concentration required for bacterial growth (0.05 to 0.5 M).

To overcome this physiologic iron deficiency, host pathogens have developed iron uptake mechanisms. The genes involved in these mechanisms are usually regulated by the availability of iron in that they are repressed in the presence of iron. In many bacterial species, these mechanisms are based on the synthesis and secretion of small compounds, usually of <1,000 Da, called siderophores, which display high affinity for ferric iron (Fe^{3+}). The majority of them are either hydroxamates or catecholates–phenolates. They are capable of removing transferrin- or lactoferrin-bound iron to form ferrisiderophore complexes, which in turn are recognized by specific iron-repressible membrane receptors and internalized into the bacterium where iron is released.

This bacterial mechanism has been described in many bacterial species, such as *Vibrio anguillarum*, *Pseudomonas aeruginosa*, *Klebsiella pneumoniae*, and *Escherichia coli*. However, other pathogens, such as *Neisseria gonorrhoeae*, *Neisseria meningitidis*, *Haemophilus influenzae*, *Actinobacillus pleuropneumoniae*, and *Pasteurella haemolytica* do not secrete detectable siderophores when grown in an iron-deficient medium, but produce outer membrane proteins that bind directly and specifically to lactoferrin or transferrin, thereby allowing iron transport into the bacterial cell. The molecular mechanisms that govern this type of iron uptake are not yet elucidated.

Lactoferrin has broad spectrum antimicrobial properties and is widely considered an important component of host defense mechanisms against microbial infections. The bacteriostatic and, in some cases, bactericidal activity of lactoferrin is exerted against a large number of Gram-positive and Gram-negative bacteria and yeasts. Lactoferrin is present in high concentration in the specific granules of neutrophils and is released into the phagosomes after microbial ingestion.

The antimicrobial activity of lactoferrin is dependent upon its iron-free state and is commonly attributed to its ability to bind and sequester iron. By depriving microorganisms of the iron needed for growth, lactoferrin can be bacteriostatic or bactericidal.[144-153]

The antimicrobial properties of apolactoferrin seem to be favored by a low pH and the use of organisms in the early exponential growth phase.[149,153] Bullen and Armstrong have shown that when neutrophil lactoferrin is saturated, bacteria that survive the killing process are able to grow within the cells.[154] Particularly relevant in this regard are the studies of Byrd and Horwitz[155-157] on the effect of lactoferrin on *Legionella pneumophila* intracellular multiplication in non-activated and interferon γ-activated human monocytes. These authors clearly showed that apolactoferrin completely inhibited *L. pneumophila* multiplication in non-activated monocytes, and enhanced the capacity of interferon γ-activated monocytes to inhibit *L. pneumophila* intracellular multiplication.[155-157]

In contrast, iron-saturated lactoferrin had no effect on the rapid rate of *L. pneumophila* multiplication in non-activated monocytes and reversed the capacity of activated monocytes to inhibit *L. pneumophila* intracellular multiplication.[157] The explanation of these phenomena depends on the capacity of monocytes to bind and internalize lactoferrin. The internalization of iron-saturated lactoferrin provides monocytes with an iron source, which is required for the growth of the bacteria and can inhibit the anti-microbial activity of interferon γ by limiting iron availability for monocytes. The internalization of apolactoferrin into monocytes leads to intracellular iron chelation, thus preventing the growth of *L. pneumophila.*

Other studies provided evidence that iron is released from phagocytosed bacteria (*E. coli*) and can be taken up by apolactoferrin.[158] Most of the iron in *E. coli* is present in iron–sulfur proteins, with a small fraction in heme proteins. When ^{59}Fe-labeled *E. coli* organisms were phagocytosed by neutrophils, bacterial iron was soon released and detected both inside the neutrophils and in the surrounding medium; if apolactoferrin is present, it binds most of the iron released by phagocytosed bacteria.[158] These observation further suggest that neutrophil apolactoferrin may function to trap iron from ingested microorganisms, enabling its removal from the sites of inflammation. This may prevent iron from catalyzing undesirable oxidative reactions and makes it unavailable for growth of microorganisms that survive the killing process.

Other studies focused on the effects of lactoferrin on microbicidal systems that are oxygen-dependent. A typical feature of phagocytic cells (neutrophils and monocytes) is their response to stimulation with a marked increase in oxygen consumption; this respratory burst is required for optimal microbicidal activity.[159] The majority of the extra oxygen consumed is converted to highly reactive oxygen species that contribute to the destruction of ingested microorganisms and extracellular targets.[159] The superoxide anion (O_2^-) and H_2O_2 are formed by the respiratory burst and interact through a series of iron-catalyzed reactions (Haber–Weiss reactions) to form a powerful oxidant, the hydroxyl radical (OH·).[159] The overall reaction is:

$$H_2O_2 + O_2 \rightarrow O_2 + OH^- + OH·$$

The hydroxyl radical is one of the most powerful oxidants known and may contribute to the toxic activities of phagocytes. The formation of OH· by the Haber–Weiss reaction is limited by the very low solubility of ferric iron at neutral pH.[159] This reaction is favored by some iron chelators and inhibited by others.[159] Ambruso and Johnston[160] suggest that lactoferrin can aid oxidation killing by neutrophils by catalyzing a Fenton-type reaction ($H_2O_2 + Fe^{2+} \rightarrow >Fe^{3+} + OH^- + OH·$) between superoxide and H_2O_2.

Subsequent studies have demonstrated that iron bound to lactoferrin does not catalyze this reaction.[161,162] Lactoferrin may in fact limit hydroxyl radical production by neutrophils provided with an exogenous iron catalyst,[163] and an anti-oxidant role for lactoferrin has been proposed.[164] Other studies suggest that apolactoferrin or partially-saturated lactoferrin may inhibit the iron-catalyzed Haber–Weiss reaction, presumably by the chelation of free iron in an unreactive form.

These apparently contrasting findings on the effect of lactoferrin on hydroxyl radical formation have also been confirmed in recent studies using improved spectrometric technologies to detect radical formation. Thus, a set of studies indicate that lactoferrin may contribute to limiting hydroxyl radical formation in phagocytic cells. A study by Britigan and coworkers showed that both lactoferrin and myeloperoxidase are able to inhibit hydroxyl radical formation by neutrophils activated by either zymosan or phorbol ester myristate acetate.[165] Since lactoferrin is predominantly secreted extraphagosomally, and myeloperoxidase is deposited mostly within phagosomes, it is conceivable that the two compounds might have relatively different effects on intra- and extra-phagosomal hydroxyl radical production.

The same authors[85] showed that apolactoferrin and, to a lower extent, iron-saturated lactoferrin inhibit monocyte radical formation. The authors also showed that the reduced hydroxyl radical formation elicited by lactoferrin may have a protective effect for neutrophils in that it reduces iron-dependent peroxidation of phagocyte membranes. These observations suggest that the binding of apolactoferrin to neutrophils and monocytes could complement the roles of cellular anti-oxidant enzymes and other factors in protecting phagocytes from auto-oxidative injury.

Other studies suggest that lactoferrin may also stimulate hydroxyl radical formation by phagocytic cells. Thus, Klebanoff and coworkers[166] showed that apolactoferrin and apotransferrin favor the auto-oxidation of Fe^{2+} with the formation of oxidant species such as H_2O_2 and OH· which resulted in toxicity for *E. coli* bacteria. However, this phenomenon required rather precise chemical conditions which can be difficult to observe under physiological conditions:

1. An acidic pH (4.5 to 5.5)
2. Fe^{2+} at a concentration of $10^{-5}\ M$
3. Apolactoferrin or apotransferrin over a narrow concentration range above which activity is lost[166]

In another study, Britigan and coworkers have shown that lactoferrin and transferrin may be structurally modified by bacterial proteases, and then generate proteolytic peptides (catalytic iron complexes) able to generate hydroxyl radicals.[167] However, this property seems to be more pertinent to transferrin than to lactoferrin.

Another series of studies focused on evaluating the bactericidal effects of lacto-ferrin on Gram-negative bacteria. Several lines of evidence suggest that lactoferrin may have effects on Gram-negative bacteria in addition to those resulting from nutritional deprivation of iron. Some investigators found that the bactericidal activity of lactoferrin is enhanced with concurrent exposure of the microrganisms to immu-noglobulins. However, a precise explanation for this phenomenon was not offered.

A series of studies by Arnold and coworkers indicate that lactoferrin may have direct bactericidal activity against a panel of Gram-negative bacteria.[147-149] As in bacteriostasis elicited by lactoferrin, this effect is inhibited by iron saturation of lactoferrin. Since lactoferrin is able to interact with the membranes of bacteria, it was postulated that membrane destabilization may contribute to the bactericidal effect of lactoferrin.[151] This hypothesis was confirmed by a series of studies carried out by Ellison and his associates.[168-170]

In a first study, these authors showed that lactoferrin causes LPS release from enteric Gram-negative bacteria and alters the permeability of the Gram-negative outer membrane.[168] This effect is blocked by iron saturation of lactoferrin. It is of interest that lactoferrin affects the Gram-negative outer membrane in a manner similar to synthetic metal chelators. Bacterial lipopolysaccharide release induced by lactoferrin is inhibited by the addition of Ca^{2+} and Mg^{2+}.[169]

Another set of experiments indicated that lactoferrin makes Gram-negative bac-teria (usually resistant to the killing by lysozyme) sensitive to the bactericidal activity of lysozyme.[170] Recent studies have shown synergic anti-staphylococcal properties of lactoferrin and lysozyme. On some bacterial strains, the combined action of the two molecules resulted in bacteriostasis; in other strains, bacteriolysis was not observed.[171]

These last observations may have important physiological implications, since both lactoferrin and lysozyme are present together in high levels in mucosal secre-tions and neutrophil granules and it is probable that their interaction contributes to host defense.

Recent studies indicate also a functional synergism between lactoferrin and defensins, small, cysteine-rich, cationic peptides expressed by a number of epithelia. This observation suggests an important role of these two molecules in the mecha-nisms of natural immunity-protecting secretory epithelia.[172]

Finally, a recent study provided the first direct evidence of an anti-bacterial effect of lactoferrin different from iron deprivation. In fact, lactoferrin prevented *Haemo-philus influenzae* colonization through an effect on IgA1 protease and Hap adhesins of this bacterium. Both of these molecules facilitate *Haemophilus influenzae* colo-nization and disease. Human lactoferrin efficiently extracted the IgA1 protease preparation from the bacterial outer membrane, specifically degraded the Hap adhes-ins, and abolished Hap-mediated adherence.[173] These activities were inhibited by serine protease inhibitors, suggesting that the N lobe of lactoferrin may contain serine protease activity.[173]

The antibacterial activity of lactoferrin may be significantly impaired by the binding of molecules whose concentrations are abnormally increased in some patho-logical conditions. This is the case of glucose-modified proteins bearing advanced glycation end (AGE) products, whose concentrations are markedly increased in diabetes mellitus; these proteins bind to both lactoferrin and lysozyme, inhibiting the bactericidal activities of both molecules.[174,175] The binding of proteins with AGEs and lactoferrin is mediated by two binding domains, each containing a 17- to 18-amino acid cysteine-bounded loop motif that is highly hydrophilic.[175] This observation also contributes to the understanding of the biochemical mechanisms responsible for the increased susceptibility to bacterial infections observed in diabetes mellitus.

3.4.1.1 Lactoferricin

The iron-sequestering capacity of lactoferrin seems to play a relevant role in mediating the bacteriostatic activity of this molecule. Recent studies have shown that a peptide fragment of lactoferrin near the N terminus of the molecule exhibits more bactericidal activity than the intact lactoferrin itself.[176] This peptide was named lactoferricin, and was shown to possess bactericidal activity against a large spectrum of bacteria. Interestingly, bovine lactoferricin showed more pronounced bactericidal activity than human lactoferricin. Bovine and human lactoferricin are released from whole lactoferrin by pepsin-mediated proteolytic cleavage.

Recent studies have shown that lactoferricin is generated in the human stomach from ingested lactoferrin.[176a] This observation suggests that physiologically functional quantities of human lactoferricin may be generated in the stomachs of breast-fed infants and, in adults, from lactoferrin secreted into saliva.

From a structural view, lactoferricin is composed of 25 amino acids: two cysteine residues are linked by a disulfide bridge, whose integrity, however, is not essential for bactericidal activity.[176] This amino acid sequence corresponds in part to a surface helix on the lactoferrin molecule which is different from the area of iron binding and corresponds to the region involved in LPS binding.[177] Furthermore, a synthetic 11-amino acid peptide corresponding to residues 20 to 30 of the α-helical region (NH₂-FQWQRNMRKVR-COOH) possesses the capacity to both bind LPS and kill bacteria[178]; the RKVR region within this peptide forms the glycosaminoglycan binding site of lactoferrin.

Some studies have partially clarified the mechanism responsible for the bactericidal activity displayed by lactoferricin. From a three-dimensional structural point of view, lactoferricin shows an extended hydrophobic surface, composed of hydrophobic amino acids, and surrounded by many hydrophilic and positively-charged residues, that give the lactoferricin molecule an amphipathic character.[179] This structure is similar to those of a vast number of cationic peptides that exert their antimicrobial activities through membrane disruption.

In line with this observation, flow cytometry studies have shown that lactoferricin, as well as its derived synthetic 11-amino acid peptide, acts at the level of bacterial membrane. After incubation with lactoferricin, a loss of membrane potential was observed (after 10 minutes), followed by loss of membrane integrity (after 30 minutes) with irreversible damage to bacteria as shown by rapid loss of viability.[180] According to these observations, it was proposed that lactoferricin acts through disruption of the outer bacterial membrane, resulting in lysis.[180]

Since lactoferricin originates from pepsin-mediated proteolytic degradation of lactoferrin, it was proposed that this peptide may play an antimicrobial role within the infant gastrointestinal tract.[181] This hypothesis is supported by recent studies showing that lactoferrin administered orally to mice is partially degraded to molecular forms, including lactoferricin, which contain receptor-binding regions and antimicrobial active centers. These lactoferrin degradation products are found in the feces of mice fed milk supplemented with lactoferrin.[182] The bactericidal activity of lactoferricin is mediated through direct and indirect mechanisms. In fact, in addition to its direct bacteriolytic action, lactoferricin also enhances the phagocytic activities of neutrophils.[183]

Lactoferricin also was shown to cooperate with anti-microbial agents in mediating inhibitory activity on bacterial or fungal growth. Thus, lactoferricin as well as lactoferrin combined with azole antifungal agents shows synergistic anti-fungal activity against *Candida albicans*.[184] Interestingly, lactoferricin was able to inhibit the growth of azole-resistant strains of *Candida albicans*.[185] These observations imply a potentail role for lactoferricin in anti-microbial therapy.

3.4.2 BACTERIAL LACTOFERRIN RECEPTORS

In recent years, particular attention focused on the study of the interaction of lactoferrin with bacterial pathogens. The studies briefly reviewed below indicate that a wide spectrum of bacteria are able to bind lactoferrin. The outer membrane proteins interacting with lactoferrin are different in the various types of bacteria. However, the observation that the type of membrane protein that binds lactoferrin is identical in groups of bacteria exhibiting similar biological properties is remarkable.

The interaction between lactoferrin and *Shigella flexneri*, the most common etiological agent of bacillary dysentery, was carefully investigated. The ability of *S. flexneri* to compete for iron within the host is one of the factors that influence pathogenicity. Under iron stress conditions, *S. flexneri* may produce one or more siderophores, for acquiring Fe^{3+} iron; *S. flexneri* may also utilize various host iron compounds, including iron-saturated lactoferrin, transferrin, hemin, and hematin *in vitro*. Furthermore, a 101-kDa heme-binding protein in *S. flexneri* has been identified. Certain peptides derived from lactoferrin are able to inhibit the adherence of *S. flexneri* to enterocytes.[186]

It was also shown that lactoferrin is able to interact with *S. flexneri*:[187] 5×10^3 lactoferrin binding sites per bacterium were detected, with a K_a of $0.7 \times 10^{-6} M$. Lactoferrin interacts with a membrane protein of *S. flexneri* of 39 kDa which seems to correspond to a membrane porin.[187] The binding of lactoferrin to the *S. flexneri* outer membrane porin may alter the permeability of the outer bacterial membrane, facilitating the killing of the bacteria by antibiotics.

Haemophilus influenzae has been shown to utilize transferrin-bound iron for growth, and the demonstration of transferrin iron utilization in almost all invasive strains, but not in most non-typable avirulent isolates, is consistent with an important role of iron in pathogenesis.[188] Furthermore, *H. influenzae* can utilize iron or heme from serum protein hemopexin or from hemoglobin complexed with haptoglobin. Subsequent studies provided evidence that *H. influenzae* is able to bind lactoferrin as well as transferrin[189]; the binding between lactoferrin and *H. influenzae* is mediated through interaction with a bacterial membrane protein of about 105 kDa, distinct from the transferrin-binding protein.[189]

It is of interest that a protein of identical molecular weight was shown to mediate the interaction between lactoferrin and *Neisseria meningitidis*.[190] The similarity in size and binding specificity of the lactoferrin receptor in *H. influenzae* to the lactoferrin in meningococcal strains suggests a structural and functional conservation. Particularly interesting is the observation that the expression of both *H. influenzae* and *N. meningitidis* lactoferrin and transferrin receptors is modulated by bacterial iron availability, i.e., iron starvation increases the synthesis of both receptors, while the opposite effect is elicited by iron loading.[168,169]

Bordetella pertussis and *B. bronchiseptica* are Gram-negative bacilli that adhere to and colonize the epithelium of the upper respiratory tract. The capacity of these organisms to colonize the mucosal surface of the upper respiratory tract implies that these bacteria have developed iron uptake mechanisms allowing growth in the infected host. It seems that these bacteria derive their iron from host lactoferrin present on the respiratory mucosa. This mechanism is made possible by the presence of specific lactoferrin receptors present on the outer membranes of these bacteria.[191]

The *Bordetella* membrane proteins responsible for lactoferrin binding exhibit a molecular weight corresponding to 27 and 32 kDa. These two proteins seem to relate to the family of IROMPs (iron-repressed outer membrane proteins) in that their synthesis is up-modulated by iron deprivation. These two proteins are able to bind also transferrin, even if to a lower degree than lactoferrin.[191]

The *Bordetella* lactoferrin-binding proteins may be the functional analog of the *Neisseria gonorrhoeae* major iron-binding protein of 37 kDa.[192] It remains to be determined whether this protein is able to bind lactoferrin. However, recent studies questioned the roles of lactoferrin and transferrin receptors of *Bordetella* in mediating iron uptake by these bacteria. It was shown that both *B. pertussis* and *B. bronchiseptica* secrete a hydroxamate siderophore under conditions of iron stress.[193,194] A *B. bronchiseptica* mutant lacking siderophore production was incapable of using lactoferrin as an iron source, although it was still able to bind lactoferrin.[195] This observation suggests that *Bordetella* uses a siderophore for removal of iron from lactoferrin and transferrin rather than relying upon a receptor for these host iron-binding proteins.

Escherichia coli is a predominant member of the normal intestinal flora and also a frequent causative agent of human intestinal infections. Lactoferrin inhibits the growth of *E. coli* and is more effective with immunoglobulin G1 and immunoglobulin A.[196] Lactoferrin may damage the outer membrane of *E. coli* by releasing lipopolysaccharides and facilitating the entry of certain antibiotics.[168,197]

Lactoferrin absorption to *E. coli* seems to enhance antibacterial effects. Lactoferrin-binding studies on *E. coli* strains isolated from human intestinal infection provided evidence that these bacteria possess membrane receptors (about 5.4×10^3 per bacterium with a K_a of $1.7 \times 10^{-7} M$) able to interact with this protein.[198] Apolactoferrin and iron-saturated lactoferrin are equally bound by *E. coli*. Subsequent studies showed that the capacity of *E. coli* strains to bind lactoferrin correlates with their colicin susceptibility.[199] Finally, the *E. coli* proteins able to interact with lactoferrin were identified as porins OmpC and OmpF. It is of interest that lactoferrin also binds to porins on *Aeromonas hydrophila* and *Salmonella*.[200]

The ability to bind and acquire iron from either human transferrin or lactoferrin (and hemoglobin/hemoglobin–haptoglobin) is shared by three pathogenic bacterial species in the family Neisseriaceae: *Neisseria meningitidis*, *Neisseria gonorrhoeae*, and *Moraxella catarrhalis*. Competitive binding assays indicated that these three bacterial species have separate receptors for transferrin and lactoferrin; this has been verified by isolation of mutants deficient in transferrin binding or lactoferrin binding.[201-203] In addition, the analysis of the two receptors isolated by affinity chromatography with one or the other ligand showed that they have different compositions.[204,205] The lactoferrin receptor present on these bacteria was identified as a single, high molecular weight, iron-repressible, outer membrane protein.

Studies intended to determine the regions of the lactoferrin molecule interacting with the bacterial receptor demonstrated that the interaction between lactoferrin and the receptors in these three bacterial species does not involve the N-linked oligosaccharides,[206] thus indicating that surface amino acid residues of this glycoprotein mediate the binding interaction.

In a second study, it was shown that preparations of isolated C and N lobes of human lactoferrin are capable of binding to the bacterial lactoferrin receptors in competitive and direct binding assays.[207] This contrasts with bacterial transferrin receptors, which bind only to C lobe fragments of human transferrin, indicating that the bacterial lactoferrin and transferrin receptors differ in their interactions with their respective glycoprotein ligands and may differ in the mechanisms of iron removal.

Other studies led to the identification of the genes encoding for lactoferrin receptors of the three pathogenic bacterial species in the family Neisseriaceae.[208-210] The most recent studies showed that the lactoferrin receptors of the bacterial species are heterodimers composed of two subunits of identical molecular weight: LbpA and LbpB. LbpA protein exhibits a great homology at the level of sequence between the different Neissieriaceae and migrates as a single band of about 105 kDa; LbpB also showed a high degree of homology and encodes a lipoprotein.

Interestingly, LbpB showed the presence of a conserved RGD motif used as a site for attachment of bacteria to eukaryotic cells, and this finding suggests that LbpB may act as an adhesin.[210] LbpA and LbpB genes are transcribed as a polycistronic message, beginning at the Fe-regulated promoter identified 5′ to the gene LbpB.[211] Both these genes are regulated by iron in that iron deprivation leads to an induction of their expression, while the opposite effect is elicited by iron loading. Mutants of *Neisseria meningitidis* with deficient LbpA expression showed markedly reduced lactoferrin-binding capacity, thus showing that LbpB alone was unable to confer to bacteria the capacity of binding iron. In contrast, mutants lacking LbpB continued to bind lactoferrin. These mutants, however, grow less efficiently than wild-type bacteria in the presence of lactoferrin, thus suggesting that LbpB, while not involved in lactoferrin binding, plays a facilitative role for *in vitro* iron acquisition.[212]

Similar observations were also made in *Neisseria gonorrhoeae*: LbpB⁻/LbpA⁺ mutants of these bacteria exhibited normal lactoferrin binding and grew normally under conditions where lactoferrin was the only iron source, while LbpB⁻/LbpA⁻ and LbpB⁺/LbpA⁻ mutants showed reduced lactoferrin-binding capacity and were unable to grow when lactoferrin was the only iron source.[212] However, in spite of these observations, it was shown that recombinant LbpB was able to bind lactoferrin in an affinity binding assay.[212]

The exact role of lactoferrin in the pathogenesis of some bacterial infections remains to be established. In fact, many pathogens are able to survive in the mucosal environment without being able to bind lactoferrin or use it as an iron donor. A significant number of gonococcal strains are LbpA⁻, and thus are unable to bind lactoferrin, but are able to cause urethritis.[211]

Recent studies suggest that lactoferrin may be active also against fungal infections, particularly those caused by *Candida albicans*. It is known that neutrophils play a key role in the defense against *C. albicans*, a major opportunistic pathogen that causes local or disseminated disease in immunocompromised and debilitated subjects. Compelling experimental and clinical evidence has shown neutropenia to be a critical factor in the outbreak of these infections. Several studies have shown that lactoferrin was able to inhibit *in vitro* growth of *C. albicans*.[213-216] It is of interest that mannoproteins extracted from *C. albicans* are able to activate neutrophils that release components of primary and secondary granules, resulting in inhibition of *C. albicans* growth; the addition of neutralizing anti-lactoferrin antibodies abrogates this inhibitory activity on the growth of *C. albicans*.[216]

Particularly interesting are the observations related to lactoferrin binding to a bacterium apparently involved in the genesis of gastritis and peptic ulcer. *Helicobacter pylori* is a curved or spiral Gram-negative microaerophilic bacterium which was first isolated from human gastric biopsies, and represents an etiologic agent of gastritis and peptic ulcer. This bacterium does not produce siderophores and requires human lactoferrin for growth in media lacking other iron sources[217]; however, neither human transferrin, bovine lactoferrin, nor hen ovotransferrin served as a source of iron for these bacteria.[217] This observation may be of relevance in the pathogenesis of peptic ulcer and gastritis. One study[218] has shown that lactoferrin is well expressed in stomach resections with superficial or atrophic gastritis, detected with high frequency (80 to 85%) in gastric tumors, and virtually undetected in normal specimens.

Immunocytochemical studies showed localization of lactoferrin in cells of both antral and body glands; on the other hand, histological studies showed that *Helicobacter pylori* was consistently seen in the intercellular junctions of gastric epithelial and gastric mucus-secreting glands. All these observations suggest that iron acquisition by the lactoferrin receptor system of *H. pylori* might explain stomach colonization by this bacterium and play a major role in the virulence of infection.

The region of the lactoferrin molecule responsible for its bactericidal properties has been recently identified.[219,220] Tomita and coworkers discovered that enzymatic hydrolysates of both human and bovine lactoferrin prepared with porcine pepsin exhibited bactericidal properties more potent than those of undigested lactoferrin.[219] The antimicrobial sequence was found to consist mainly of a loop of 18 amino acid residues formed by a disulfide bond between cysteine residues 20 and 37 of human lactoferrin or residues 19 and 36 of bovine lactoferrin (Figure 3.9).[220] This peptide does not contain amino acids involved in iron binding or in glycosylation. This observation suggests that the identified bactericidal domain of lactoferrin may contribute to the anti-microbial activity of lactoferrin by a mechanism distinct from metal chelation.

Distinctive structural features of this anti-microbial domain of lactoferrin are the relatively high proportions and asymmetric clusterings of basic amino acid residues. This structure is reminiscent of cationic peptides such as magainins, cecropins, and defensins that kill sensitive microorganisms by inducing an increase in cell membrane permeability.[221] It is of interest that the modulation of the bactericidal activity of the isolated peptide is similar, but not completely identical, to that of the entire lactoferrin molecule. Saturation of lactoferrin with iron abolishes its bactericidal activity, whereas the activity of the isolated bactericidal domain is diminished about 2.5-fold, but not abolished, in the presence of iron salts.[214] Ca^{2+} ions inhibit the bactericidal activity of lactoferrin at concentrations (10^{-4} M) lower than those required (>10^{-3} M) to inhibit the activity of the antimicrobial peptide of lactoferrin.[220]

3.4.2.1 Serum Lactoferrin Concentration and Infectious Diseases

During an infectious process, the serum concentration of lactoferrin sharply increases. In fact, the infectious bacterial agents activate polymorph neutrophils to release antibacterial factors such as elastase and cathepsin G from primary or azurophilic granules and lactoferrin from secondary or specific granules. In parallel, monocytes are activated to produce cytokines like tumor necrosis factor (TNF), interleukin-6 (IL-6), and interleukin-1 (IL-1), and all three cytokines seem to be important mediators of septic shock.

Bacterial lipopolysaccharide (LPS) is one of the major stimuli for TNF and IL-1, and the elevation of these cytokines is caused by the release of LPS during Gramnegative bacterial infection. TNF and IL-1 may act on the neutrophils to release lactoferrin.[222] *In vitro* studies showed that incubation of total blood in the presence of bacteria, such as *E. coli*, resulted in a rapid release of both TNF and lactoferrin.[223,224] It is of interest that the addition of exogenous TNF may replace the bacteria in inducing lactoferrin release by neutrophils.[224] *In vivo* studies showed that:

1. Administration of chemotactic peptides to rabbits elicited a rapid increase in the serum concentration of lactoferrin.[219]
2. Plasma lactoferrin levels reflect granulocyte count and activation *in vivo*.[225]
3. Plasma levels of lactoferrin increase in healthy persons who receive low doses of endotoxin intravenously.[226,227]

The increase in lactoferrin starts 90 minutes after endotoxin administration, when TNF and IL-6 reach their highest plasma levels (Figure 3.11).[227] Plasma levels of lactoferrin markedly increase in persons who receive bolus injections of TNF; the rise of the protein is more rapid as compared to that observed after LPS administration.[228] The rapid release of lactoferrin occurring after endotoxin administration seems to be one of the factors responsible for the initial and transient neutropenia observed in these patients. In fact, lactoferrin stimulates the adherence of neutrophils to vascular endothelium.[229]

This interpretation is supported by the observation that lactoferrin infusion determines neutropenia.[229] Administration of granulocytic colony-stimulating factor

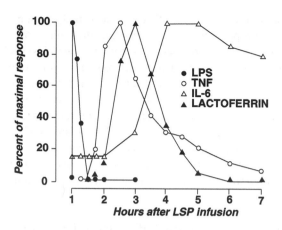

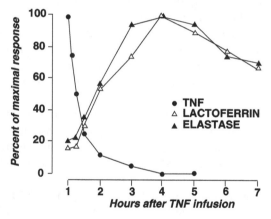

FIGURE 3.11 Plasma levels of lactoferrin, TNF-α, and IL-6 in healthy persons who received low doses of endotoxin intravenously. (Reprinted from Van Deventer, S.J.H. et al., *Blood,* 76: 2520–2526, 1990. With permission.)

(G-CSF) 24 hours before LPS challenge produces a marked decline of the release of inflammatory cytokines elicited by endotoxin; since G-CSF administration leads to a marked increase in the concentration of plasmatic lactoferrin, it is conceivable that this rise may represent one of the mechanisms through which G-CSF attenuates the inflammatory response.[230]

G-CSF administration to healthy donors causes, in addition to a marked rise of the number of circulating neutrophils and monocytes, a moderate increase of lymphocytes; these lymphocytes exhibit a reduced blastogenic response to mitogenic lectins, a phenomenon at least in part caused by the lactoferrin release from neutrophils induced by G-CSF.[230a] In line with this observation, it was found that G-CSF administration causes a decrease of neutrophil lactoferrin content.

Several observations provided evidence that plasma lactoferrin levels are increased in patients with severe infections.[231] However, in patients with sepsis, the level of plasma lactoferrin does not correlate with the clinical outcome, whereas the plasma levels of other molecules released by neutrophils, such as elastase, seem to be related to prognosis.[232]

It is of interest that viral illness is usually associated with reduced plasma lactoferrin levels. Since patients with viral diseases usually have normal neutrophil numbers, the reduced level of lactoferrin observed in these conditions has been ascribed to a defective release.[232]

Recent studies provide clear evidence that lactoferrin exerts a protective effect against endotoxin-mediated shock, a phenomenon related to the capacity of lactoferrin to interact with LPS and to inhibit the binding of LPS to its membrane receptor present on monocytes–macrophages (the CD14 antigen). LPS stimulates monocytes–macrophages to produce several cytokines: in the presence of large amounts of LPS, an excess release of these mediators leads to septic shock.

The interaction between bacterial LPS and monocytes–macrophages is mediated by a membrane receptor, the CD14 antigen (a glycosylphosphatidylinositol-anchored membrane protein present on monocytic lineage). Some serum proteins interacting with LPS modulate the biologic effects of lipopolysaccharides; in this context, particularly relevant is a 60-kDa acute-phase protein known as LPS-binding protein (LBP), which binds the lipid portion of LPS and facilitates its binding to CD14 receptor, thus inducing the activation of monocytes. In contrast to LBP, lactoferrin inhibits the interaction of LPS with CD14. In fact, lactoferrin is able to interact with the lipid A moiety of LPS[233,234] and thus inhibit the LPS-induced release of cytokines by monocytes–macrophages.[235,236]

Flow cytometry studies showed that lactoferrin markedly inhibited the binding of LBP/LPS to CD14 in a concentration-dependent manner; this inhibitory action may be explained by a competition between LBP and lactoferrin to bind LPS.[237] The capacity of lactoferrin to inhibit *in vitro* the binding of LPS to CD14 offers the opportunity to modulate *in vivo* the inflammatory process through infusion of lactoferrin. This hypothesis is directly supported by recent studies showing that lactoferrin may significantly reduce the mortality induced by endotoxin challenge in different animal models.[238,239]

Recent studies also indicate that the evaluation of urinary lactoferrin levels is of diagnostic interest. In fact, values of urinary lactoferrin ranging from 30 to 60 ng/ml are observed in healthy subjects or in patients without urinary tract infections, while mean values corresponding to 3300 pg/ml are observed in specimens from patients with urinary tract infections.[239a] The predictivity of a positive urinary lactoferrin test was >96% for a diagnosis of urinary tract infection.

3.4.3 LACTOFERRIN AND THE REGULATION OF MYELOPOIESIS

Mammalian blood cells have definite life spans that are considerably shorter than the life span of the entire organism. The maintenance of constant numbers of cells in the peripheral blood is achieved by the proliferation and differentiation of progenitor and precursor cells located primarily in the bone marrow. These progenitors and precursors are all derived from a common self-maintaining population of stem cells established early in embryogenesis.

Progenitor cells are the immediate progeny of multipotent stem cell differentiation and possess limited further differentiation potential; they are also known as lineage-restricted or committed progenitors in recognition of this characteristic. They give rise to colonies of morphologically recognizable differentiation progeny in semisolid clonal cultures.

The clonogenic cells were operationally defined as colony-forming units and post-fixed by letters denoting the cell types to which they gave rise, e.g., CFU-GM for cells generating colonies of granulocytes and macrophages. These progenitors are committed to specific lineages and are functionally defined as colony- or burst-forming units (CFUs or BFUs), i.e., progenitors of the erythroid series (BFU-E and CFU-E), the granulocyte–monocyte lineage (CFU-GM), and multipotent CFU for the GM, erythroid, and megakaryocytic lineages (CFU-GEMM). The progenitors in turn differentiate into morphologically recognizable precursors that mature to terminal elements circulating in peripheral blood.

The survival, proliferation, and differentiation of progenitor cells *in vitro* is absolutely dependent on the continous presence of a family of growth factors called colony-stimulating factors (CSFs). CSFs exert either a multilineage or unilineage stimulus. Interleukin-3 (IL-3) acts on the early progenitor pool, i.e., multipotent (CFU-GEMM), early erythroid (BFU-E), GM (CFU-GM), and megakaryocytic progenitors. Granulo-monocytic colony stimulating factor (GM-CSF) exerts similar effects, but possibly stimulates progenitors at a more distal differentiation stage.

Erythropoietin (Ep), G-CSF, IL-5, and M-CSF are largely specific for end-stage progenitors of the erythroid (CFU-E), granulocytic (CFU-G), eosinophilic (CFU-Eo), and monocytic (CFU-M) lineages, respectively.

Recent studies indicate also an important role for a group of "permissive" hemopopietic growth factors, including *kit* ligand, IL-6, IL-11, IL-9, and basic fibroblast growth factor (b-FGF). Under stringent culture conditions, these growth factors exert little activity, but potentiate the stimulatory activity of CSFs, particularly those exhibiting multilineage activity.

Several cell types in culture are able to produce one or more of the CSFs either constitutively or, more commonly, following a variety of different stimuli. G-CSF and M-CSF are produced by fibroblasts and endothelial cells following stimulation with interleukin-1, endotoxin, or tumor necrosis factor (TNF); G-CSF and M-CSF by monocytes following phorbol ester and γ-interferon treatments; G-CSF by monocytes exposed to IL-3 and M-CSF; and GM-CSF and IL-3 by activated lymphocytes.

The investigations of the effects of lactoferrin in the control of myelopoiesis have their origins in the postulate that mature cells of a given lineage (e.g., neutrophils) contain factors that limit the proliferative activity of their progenitors and precursors, and that this feedback regulatory mechanism might be perturbed in leukemias.[240] The investigation of such a feedback mechanism has been the object of many studies focused on examining neutrophils and neutrophil extracts for granulopoietic inhibitory factors.

In this context, lactoferrin represents one of the candidates for a feedback regulator of neutrophil production. In 1975, Baker, Broxmeyer, and Galbraith reported a series of experiments in which neutrophils inhibited granulocyte macrophage colony growth *in vitro*.[241] Subsequent studies showed that neutrophil extracts were active only in assays that included no exogenous sources of colony-stimulating factors.[242-244] These observations suggested that the inhibitory factor present in neutrophils functions to inhibit the production of CSFs rather than to inhibit CFU-GMs. In further studies, Broxmeyer and his coworkers identified this inhibitory activity as lactoferrin.[245] *In vitro* lactoferrin produces its myelopoietic inhibitory effect by decreasing the production/release of colony stimulating factors from monocytes–macrophages and by decreasing the release of IL-1β from monocytes.[246-249] Lactoferrin inhibits, in addition to IL-1β, the release of TNF-α by mononuclear cells.[250]

These observations suggest that the inhibition of granulopoiesis mediated by lactoferrin is secondary to suppression of IL-1 and TNF required for production of both GM-CSF and G-CSF from fibroblasts and endothelial cells. The suppressive activity of lactoferrin was related to the iron saturation state of the molecule; fully iron-saturated lactoferrin was most active, native lactoferrin (8 to 20% iron-saturated) was less active, and apo (iron-depleted) lactoferrin had little or no activity.

Recent studies have in part clarified the molecular mechanisms responsible for the inhibition of GM-CSF production elicited by lactoferrin. Experiments performed on embryonic fibroblasts transfected with the lactoferrin gene showed that lactoferrin inhibits GM-CSF production induced by IL-1β through a decrease of GM-CSF gene transcription.[251] It was proposed that this inhibitory effect of lactoferrin on GM-CSF transcription could be related to the binding of lactoferrin at the level of DNA sequences present in the promoter of the GM-CSF gene.

In vivo, lactoferrin decreases the cycling rates and absolute numbers of progenitor cells in bone marrow and spleen; both these effects are dose-dependent and time-related.[252-254] This activity is similar to that expected from its action *in vitro*. The administration of colony-stimulating factors to mice pretreated with lactoferrin overcomes the suppressive effects of this molecule.[254-255] Other molecules capable of abrogating the myeloinhibitory activity of lactoferrin include interleukin-6 and bacterial lipolysaccharide. Interleukin-6 seems to abrogate the effect of lactoferrin through stimulation of IL-1 production.[256] The inhibitory effect of lipopolysaccharide is mediated by direct interaction of this molecule with lactoferrin: the lactoferrin-lipopolysaccharide complex binds to LPS receptors and not to lactoferrin receptors.

However, other stuides have shown that LPS binding to monocytes is inhibited by lactoferrin: this phenomenon is due to the competition of lactoferrin with LPS for the binding to the serum LPS-binding protein.

In addition to *in vitro* observations, studies of cancer patients undergoing high-dose chemotherapy or bone marrow transplantation provide further evidence in favor of the role of lactoferrin in regulating myelopoiesis. Several studies provided evidence that lactoferrin may represent a fairly accurate indicator of neutrophil kinetics.[257] Elevated levels were found in subjects with neutrophilia, while spontaneous or chemotherapy-induced neutropenia[260-268] is associated with extremely low plasma lactoferrin levels.

It has been suggested that an increase of plasma lactoferrin concentration is a predictor of neutrophil regeneration, preceding a detectable elevation in peripheral neutrophil count by a number of days.[259-260] Other studies have clearly indicated that, during chemotherapy, plasma lactoferrin first decreases down to levels virtually undetectable, and then increases strictly in parallel with neutrophil recovery.[261-263] One study provided evidence that cancer patients undergoing autologous stem cell transplantation following high-dose chemotherapy exhibited pronounced rises in their G-CSF plasma levels preceding neutrophil recovery by about one week.[264] Furthermore, the peak level of G-CSF strictly correlated with the peak level of neutrophils.[264]

In a subsequent study,[265] these authors showed that in patients undergoing autologous stem cell transplantation:

1. Plasma lactoferrin levels sharply decline in the the first days following transplantation.
2. Plasma lactoferrin levels become virtually detectable 7 to 10 days after transplantation at the moment when plasma G-CSF levels sharply increase.
3. About 12 to 14 days following transplantation, the plasma lactoferrin and circulating neutrophils start to increase and reach a plateau at 13 to 16 days after transplantation.

See Figure 3.12. These observations strongly suggest that the rapid and marked decline in lactoferrin concentration occurring after transplantation may represent a stimulus for inducing G-CSF production and subsequent granulocyte recovery.

Patients with chronic myeloid leukemia possess lactoferrin that is apparently less active in inhibiting myelopoiesis than that extracted from normal neutrophils. This conclusion is based on a series of observations indicating that crude extracts of neutrophils from patients with chronic myelogenous leukemia were less active than crude extracts of neutrophils from normal donors in suppressing the release of CSFs from normal human blood mononuclear cells.[266-268] However, from the assessment of crude extracts of chronic myeloid leukemia neutrophils, it was not possible to determine whether decreased suppressor activity was due to a quantitative deficiency of lactoferrin, the presence of an inactivator of lactoferrin or of its action, and/or a qualitative deficiency of lactoferrin.

Subsequent studies have shown that chronic myeloid leukemia is associated with triple defects in lactoferrin–cell interactions consisting of: (1) abnormal quantitative effects, such as decreased amounts of lactoferrin; (2) abnormal qualitative effects, such as functional inactivity of the lactoferrin molecule; and (3) decreased sensitivity of patient cells to the actions of normal lactoferrin molecules.[269] As a consequence of these observations, it was suggested that chronic myeloid leukemia may be the first disease in which it is possible to evaluate the clinical efficacy of lactoferrin, administered alone or in combination with myelosuppressive cytokines, such as interferon-γ.[270]

Patients with lactoferrin deficiency (a condition discussed elsewhere) exhibit increased or normal numbers of circulating neutrophils. The first clinical condition (lactoferrin deficiency associated with neutrophilia) is consistent with the proposed role of lactoferrin as a negative regulator of myelopoiesis; however, the lack of increased leukocytes in other patients with lactoferrin deficiency has led some authors to the suggestion that lactoferrin might not be a physiological regulator of blood cell production.[271]

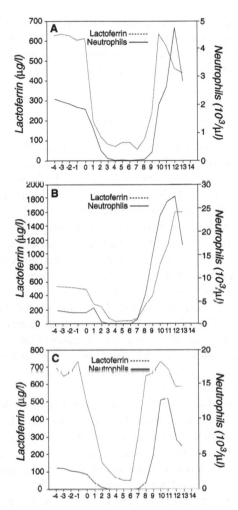

FIGURE 3.12 Plasma lactoferrin levels in patients undergoing autologous hematopoietic stem cell transplantation (HSCT). Patients in panel A have undergone HSCT without exogenous cytokine administration, while patients in panels B and C have undergone HSCT in combination with G-CSF (B) or GM-CSF (C) administration. (Reprinted from Testa, U. et al., *Blood*, 89: 2615–2617, 1997. With permission.)

3.4.4 LACTOFERRIN AS A GENE REGULATOR

As mentioned in the section on lactoferrin receptors on leukemic lines, lactoferrin is first bound at the level of cell surface by K-562 cells,[70] then internalized, and finally transported into the nucleus where it binds DNA. Some studies suggest that the interaction between lactoferrin and DNA may play a relevant role in the transcriptional control of gene expression.[272] In fact, it was shown that lactoferrin interacts with high affinity with DNA and, using the CASTing method, it was shown that it preferentially binds to the nucleotide sequence GGCACTTGC.[272] Transfection experiments in K-562 cells showed that lactoferrin enhanced the expression of reporter genes containing in their promoter regions the above-mentioned sequence.[272] It was proposed that lactoferrin may affect the expression of several target genes via this mechanism and that such events may underly many of the biological activities of lactoferrin.

3.4.5 ANTI-TUMOR ACTIVITY OF LACTOFERRIN

Studies have shown that lactoferrin, in addition to its antimicrobial activities, also exhibited anti-tumor activity. In this context, the anti-tumor effects of lactoferrin in established tumor models in mice were evaluated.[273] The *in vivo* effects of lactoferrin were evaluated in mice injected with either methylcolanthrene-induced fibrosarcoma or v-*ras*-transformed fibroblasts; in both the tumor models, lactoferrin injections elicited marked inhibition of tumor growth.[273] Interestingly, this effect was displayed both by apolactoferrin and iron-saturated lactoferrin, but was specific for lactoferrin in that other iron-containing proteins, including transferrin, had no effect on tumor growth. It was suggested also that the anti-tumor effect of lactoferrin could be mediated via natural killer (NK) lymphocytes.[273]

In line with this hypothesis, a recent study has shown that NK lymphocytes, and particularly those exhibiting high expression of the CD-56 antigen, bind lactoferrin and, in consequence, are induced to moderate increases of their anti-tumor cytotoxic activity.[274] These observations were also confirmed by *in vivo* studies showing that lactoferrin administration potentiated both NK and LAK (lymphokine-activated killer) anti-tumor cytotoxic activity.[275]

Anti-tumor activity of lactoferrin was also observed in models of tumor chemo-prevention; in fact, lactoferrin reduced the occurrence of colon tumors (adenoma and carcinoma) in rats treated with the carcinogen azoxymethane.[276] More recent studies have shown that the peptide-derived lactoferricin, in addition to its anti-microbial activities, also possesses anti-tumor activity. Lactoferricin B inhibited experimental tumor metastasis (lung and liver colonization). Interestingly, this anti-tumor activity was also displayed by apo-lactoferricin, but not by iron-saturated lactoferricin.[277]

In other tumor models lactoferricin, but not lactoferrin, exhibited anti-tumor activity. Thus, it was shown that the addition of bovine lactoferricin to monocytic leukemia cells THP-1 elicited a rapid and marked apoptosis, while bovine lactoferrin, even at high doses, had no effect.[278] The apoptotic effect of lactoferricin was inhibited by reduced glutathione or by N-acetylcysteine, thus suggesting a role of reactive oxygen species in this phenomenon.

3.5 LACTOFERRIN EXPRESSION IN NEUTROPHILS AND MAMMARY GLANDS

Immunoperoxidase technique was used for detecting lactoferrin in routine histolog-ical sections of human tissues. Lactoferrin was found in lactating breast tissue, neutrophils, bronchial glands, and duodenal epithelial cells.[279] Major producers of lactoferrin are neutrophils and lactating mammary glands and this topic will be described in detail below. Concerning the presence of lactoferrin in the duodenum, it must be emphasized that this protein is detected in some intestinal epithelial cells of the absorptive type, usually localized to the tips of villi.[279]

3.5.1 Lactoferrin Synthesis by Neutrophils

Human peripheral blood neutrophils contain two main populations of cytoplasmic granules that are formed in the bone marrow at different stages of maturation. The azurophil granules are packaged in promyelocytes and are identified via light and electron microscopy by their peroxidase activity; these granules contain myeloperoxidase and several proteinases, including elastase and cathepsin G.

Specific granules are formed later, during the myelocyte stage, and have been categorized by electron microscopy by an absence of peroxidase activity and by the presence of lactoferrin. In contrast to peroxidase-positive granules, peroxidase-negative granules fuse readily with the plasma membrane and empty their content to the exterior. Biochemical analysis of purified granules separated by density centrifugation also distinguishes two main granule classes: (1) a fast-moving population of large granules that contains myeloperoxidase and lysosomal enzymes and (2) a slower-sedimenting population of granules that lacks myeloperoxidase and lysosomal enzymes, but contains lactoferrin, lysozyme, collagenase, and vitamin B_{12}-binding protein.

These granules have distinct functions, and their release appears to be under differential control, with proteins contained in primary granules (i.e., elastase) requiring stronger stimuli for release than those contained in secondary granules (i.e., lactoferrin). Lactoferrin is considered one of the major and specific constituents of secondary granules, and the localization of lactoferrin has been studied by light and electron microscopic immunocytochemistry. Light microscopic demonstration of lactoferrin in cells after the myelocyte stage of differentiation[280,281] and electron microscopic observation of the lack of secondary granules in lactoferrin-deficient neutrophils[271,282] support the concept that lactoferrin is localized in secondary granules. Using antibodies against myeloperoxidase and lactoferrin, it has been clearly shown that myeloperoxidase is present starting from or even before the myeloblast and promyelocytic stages, whereas lactoferrin appears at the myelocyte stage, when secondary granules are formed.[283,284]

It is of interest that after exposure of neutrophils to phorbol esters, a marked decline in the number of cytoplasmic granules reacting with anti-lactoferrin mAb was observed,[284] whereas those reacting with anti-myeloperoxidase are virtually unmodified. This observation may be explained by the finding that phorbol esters induce selective exocytosis of specific granules in neutrophils.[285] A morphometrical study showed that the mean size of lactoferrin-positive granules was significantly greater in myelocytes and metamyelocytes than in mature circulating neutrophils.[284] This observation indicates that lactoferrin granules decrease in size as the cells mature.

Peroxidase-negative, specific granules may be subdivided into three different subsets according to their contents of lactoferrin and gelatinase:

1. Granules that contain lactoferrin but not gelatinase (about 15% of the total)
2. Granules that contain both lactoferrin and gelatinase (about 60% of the total)
3. Granules that contain gelatinase but not lactoferrin (25% of the total)[286]

These subsets of secondary granules also differ in the extent to which they release their contents following a stimulus by a neutrophil activator.[287] This heterogeneity in secondary granules can be largely explained by differences in timing of biosynthesis of the proteins localized in granules.[288,289]

The kinetics of lactoferrin expression during neutrophil maturation was also investigated by biochemical and molecular techniques. Immunoprecipitation studies on metabolically-labeled human bone marrow cells provided evidence that lactoferrin is actively synthesized in myelocytes and metamyelocytes, whereas myeloblasts and promyelocytes are unable to synthesize this protein[290]; the synthesis of lactoferrin is stopped at terminal stages of neutrophil maturation (bands and polymorphonuclear neutrophils).[290] These findings were confirmed by studies carried out at mRNA level.[31]

In another set of experiments, the biosynthesis of lactoferrin and its processing were carefully investigated in immature myeloid cells.[291] This study showed that: (1) lactoferrin was synthesized in the form of the final mature protein (i.e., it was not synthesized as a larger precursor) and it was found not to be phosphorylated; and (2) lactoferrin seems to be processed like proteins destined for secretion, i.e., the neosynthesized protein is transported through the medial and transcisternae of the Golgi apparatus, where the protein is highly glycosylated.

Recent studies have in part clarified the molecular mechanisms responsible for lactoferrin expression in myeloid cells. The nucleotide sequence of the lactoferrin promoter and its 5′ flanking region has been shown to contain a number of transcription factor binding motifs which include a TATA-like sequence, two CCAAT boxes, an AP-2 site, PU.1 boxes, GATA-1 sites, and overlapping COUP-TF and RAR elements. The latter two elements play an important functional role in the expression of lactoferrin gene in mammary cells, but no functional data are available indicating the significance of RARE in myeloid-specific lactoferrin gene expression.

However, a recent study showed the identification of a negative regulatory element within the lactoferrin gene promoter that is recognized by the CCAAT displacement protein (CDP/cut).[292] The mechanism through which CDP/cut represses the expression of the lactoferrin gene remains to be clarified; however, it is particularly interesting that CDP/cut protein synthesis is lost during the initial stages of myeloid precursor maturation: at the moment CDP/cut protein synthesis is blocked, the synthesis of lactoferrin starts. Interestingly, CDP/cut also regulates expression of the phagocyte-specific cytochrome heavy chain gene (gp91-phox) .

Recent studies have clarified the molecular mechanisms responsible for induction of lactoferrin synthesis in leukemic cells by retinoic acid. It was shown that retinoic acid induces terminal maturation of some types of myeloid leukemic cells, particularly of promyelocytic leukemia cells. Some leukemic cell lines offer a cellular model to study the induction of cell maturation by retinoic acid; following induction of cell diffrentiation of 32D cells, a marked increase in lactoferrin expression is observed.[293] The induction of lactoferrin gene by retinoic acid is mediated by activation of the transcriptional activity of the lactoferrin gene dependent upon activation of the RARE box of the promoter of lactoferrin gene.[58] The stimulatory effect of retinoic acid on lactoferrin expression may be inhibited by estrogens: this phenomenon is dependent upon the composite nature (RARE/ERE) of the RARE box of the lactoferrin promoter which implies a functional interaction among retinoid receptors, the estrogen receptor, and their ligands.

The synthesis of all proteins localized within secondary granules (lactoferrin, transcobalamin I, gelatinase, and collagenase) is orchestrated through similar molecular mechanisms.[294] These mechanisms mainly operate to restrict the synthesis of these proteins to limited stages of myeloid maturation. Particularly, the synthesis of the repressor CDP protein during early stages of myeloid maturation prevents the transcription of these genes in immature myeloid cells; at later stages of myeloid development (myelocytes), CDP protein expression is stopped and lactoferrin, gelatinase, and collagenase genes are expressed.[295,296]

In a rare hereditary syndrome, neutrophil-specific granule deficiency, a marked deficit of neutrophil lactoferrin was observed. Neutrophil-specific granule deficiency is a rare congenital disorder characterized by recurrent infections of skin and deep tissues by a variety of bacterial and fungal pathogens, without any increased propensity for viral infection.[297,298]

Morphological examination of neutrophils from these patients revealed absent specific or secondary granules on Wright's stain and multiple nuclear abnormalities including blebs, clefts, and bilobed nuclei.[139,297,298] In vitro functional disturbances of chemotaxis, bactericidal activity, and disaggregation have also been described in neutrophils of patients affected by this syndrome.[299,300] After subcellular fractionation of the granule components of specific granule deficiency neutrophils on a sucrose gradient, the primary granule fraction is seen as a single broad band that is less dense than normal, and the band of the expected density for specific granules is absent.[301-303] These abnormal binding patterns are associated with: (1) the absence or deficiency of a subset of neutrophil secretory proteins usually found in specific granules such lactoferrin and vitamin B_{12}-binding protein; (2) deficiency of some proteins such as defensin, associated with primary azurophilic granules; (3) normal amounts of other primary granule proteins such as myeloperoxidase.

It is believed that the marked lactoferrin deficiency greatly contributes to the defective bactericidal activity observed in this syndrome. Molecular studies[304] provided evidence that: (1) metabolically-labeled specific granule deficiency nucleated marrow cells showed no detectable synthesis of lactoferrin; and (2) transcripts of the expected size for lactoferrin were detectable in the nucleated marrow cells of specific granule deficiency patients, but were markedly diminished in abundance when compared with normal nucleated cells and marrow cell RNA. It is of interest that this deficiency of lactoferrin mRNA was specifically observed in neutrophils of these patients, but not in other tissues which synthesize lactoferrin.[304]

These findings were also confirmed in a more recent study, in which the steady-state level of mRNA encoding granule proteins was measured in a patient with neutrophil granule deficiency. This patient exhibited a marked decrease of the bone marrow mRNA levels for lactoferrin, transcobalamin I, neutrophil collagenase, neutrophil gelatinase, and defensins.[305]

The molecular mechanisms responsible for absent synthesis of lactoferrin and deficiency in the synthesis of several other granule proteins have not been defined. Since the deficiency seen in specific granule deficiency neutrophils involves several secretory proteins from different granule compartments, it is likely that the primary defect affects a common regulatory mechanism controlling the production of these proteins, rather than a mutation of each gene coding for a deficient protein. More precisely, specific granule deficiency may reflect a primary defect in gene expression at the level of transcription, possibly in a shared transcription factor common to the affected subset of protein.

The presence of lactoferrin in leukemic cells was also investigated. Screening studies using specific antibodies provided evidence that lactoferrin is a useful marker of M2 and M4 myeloid acute leukemias.[306,307] As expected, M1 acute myeloid leukemias, when blocked at an early myeloblastic differentiation stage, are always negative for lactoferrin.[306,307] Thus, the existence of lactoferrin in a significant number of acute myeloid leukemic cells indicates some level of myeloid maturation. In another study, the presence of lactoferrin granules was studied in a group of patients at diagnosis, after achievement of complete remission, and, in some patients, at the time of relapse.[308] At the time of the initial diagnosis, leukemic cells were negative or scarcely positive for lactoferrin; after remission, neutrophils showed strong reactivity with anti-lactoferrin antibody. Interestingly, neutrophils and bone marrow myelocytes become lactoferrin-negative before the onset of the relapse (blast cell infiltration) and this may serve to predict relapse.[308]

A great number of studies focused on the study of acute promyelocytic leukemia (APL, or M3 following the French–American–British FAB classification). APL harbors the unique reciprocal chromosomal translocation t (15; 17), which involves the retinoic receptor a (RARa) gene on chromosome 17 and PML, the gene encoding a putative transcription factor on chromosome 15; this translocation determines the formation of the fusion gene PML/RARa which encodes the fusion protein PML/RARa, a pathognomic abnormality associated with APL. This fusion protein determines the block of APL cells at the level of promyelocytes and also determines the capacity of APL cells to differentiate *in vitro* and *in vivo* in the presence of high concentrations of retinoic acid, representing the first successful model of differentiation-induction therapy for human cancer. Following induction with retinoic acid, APL blasts differentiate up to the stage of band neutrophils; although these cells are morphologically identical to mature neutrophils, they exhibit a marked secondary-granule deficiency (i.e., these cells show very low reactivity with anti-lactoferrin antibodies and at ultrastructural level show a cytoplasm with numerous myeloperoxidase-positive primary granules, but no secondary granules).[309] Similar results were obtained in the NB-4 continuous cell line derived from an APL patient.[310]

Finally, another study showed that the induction of elevated levels of v-*myb* expression in progenitor cells abrogates the capacity of these cells to differentiate in the presence of G-CSF and to synthesize high levels of lactoferrin mRNA.[311]

3.5.2 LACTOFERRIN SYNTHESIS IN MAMMARY GLANDS

As reported in the introduction to this chapter, large quantities of lactoferrin (1 to 2 mg/ml) are found in human breast milk.[312] Lactoferrin is one of the defense factors present in human milk. Interestingly, the inclusion in human milk of lactoferrin and other antimicrobial agents does not show geographic/ethnic variation.[313]

The synthesis of lactoferrin by mammary gland is limited to the lactation period, as indicated by immunoperoxidase analysis of sections of human non-lactating and lactating mammary glands.[312,313] Biochemical analysis of the various protein constituents of fluid aspirated from the nipples of nonpregnant women revealed the absence of lactoferrin in 98% of the cases.[314] Experimental studies in mammary explants from mice in mid-pregnancy, cultured in a synthetic medium containing insulin and hydrocortisone, responded to the addition of prolactin by the synthesis of milk proteins.[315] One of these proteins is lactoferrin, whose synthesis increases in response to prolactin in a dose-response manner.

The synthesis of lactoferrin was recently studied in detail in cultures of murine mammary lines. Lactoferrin was not expressed in adherent monolayer cultures under standard subconfluent conditions on plastic dishes, but its synthesis was induced when the cells were grown on basement membrane gel.[316] Additional experiments showed that lactoferrin synthesis was observed when mammary cells exhibited a rounded morphology, but was inhibited when these cells showed a flat, spread morphology.[316] The regulation of lactoferrin expression under these conditions was dependent upon both transcriptional and post-transcriptional mechanisms. A similar pattern of lactoferrin expression was also observed in early passages of normal human mammary epithelial cells.

The role of lactoferrin in iron absorption from human milk is not fully understood. The discovery that infants absorb iron from human milk (containing high amounts of lactoferrin) more efficiently than from cow milk (containing low quantities of lactoferrin) led to the hypothesis that lactoferrin promotes iron absorption. In favor of this interpretation, Kawakami et al.[317] found that iron from bovine lactoferrin was more effective than ferrous sulfate in improving hematocrit in anemic rats. Absorption of iron from iron-saturated lactoferrin was similar to that from ferrous sulfate in weanling rats fed with a purified diet.[318]

However, studies by Fransson and Lonnerdal[319] shed doubt on the importance of lactoferrin as a major promoter of iron absorption. In fact, these authors found that only a small proportion of iron in human milk was bound to lactoferrin and only 1 to 4% of the protein was saturated with iron. This observation is incompatible with a role for lactoferrin in iron absorption from human milk. Furthermore, an *in vitro* study using everted duodenal sacks from rats provided evidence that apolactoferrin reduced mucosal transfer as well as serosal uptake of iron.[320] Finally, adult subjects absorbed less iron from simulated human milk when iron-saturated lactoferrin was added.[321]

In line with this observation, De Vet and Van Gool[322] reported a negative relationship between concentration of duodenal lactoferrin and absorption of iron in human adults. On the basis of these findings, it was suggested that lactoferrin may not play a major role in mediating iron absorption from milk. The efficient absorption of human milk iron may be due to low concentrations of inhibitory compounds (casein phosphate and calcium) or the presence of compounds (amino acids) that enhance iron availability.[323]

In mammary gland and in uterus, the production of lactoferrin is regulated by hormones. As previously mentioned, prolactin stimulates lactoferrin synthesis in mammary gland.[315] Lactoferrin was induced by estrogen in a time- and dose-dependent fashion in murine[51] and human[324] uterus, but was not affected by estrogen in the mammary gland.

Differences were found in the expression of lactoferrin in mammary glands and uteri of adult murine females during lactation.[51] A high level of lactoferrin was detected in uterine epithelial cells one day after parturition, but it disappeared quickly thereafter. In contrast, the levels of lactoferrin in the mammary gland sharply increased after parturition and remained at high levels during the entire lactation period.[51] Lactoferrin levels were also studied in the first period of murine pregnancy and during the preimplantation period.[325] The levels of the protein were particularly high during the first two days of gestation, but sharply disappeared thereafter. This finding suggests that uterine lactoferrin may play an important role in early pregnancy.[325]

There is also evidence that estrogens play an important role *in vivo* in the control of lactoferrin synthesis in human endometrium. Several lines of evidence suggest such a mechanism:[326]

1. In normal ovulatory cycles, the mean serum lactoferrin concentration during the proliferative phase was significantly higher than in the secretory phase.
2. In gonatropin-induced cycles, lactoferrin levels were distinctly higher than the midcycle levels in normal ovulatory cycles.
3. Immunohistochemical analysis of the endometrium revealed greater expression of lactoferrin in proliferative endometrium than in secretory endometrium.

These observations are compatible with the hypothesis that estradiol plays a role in the regulation of lactoferrin expression in human endometrium.

The modulation of lactoferrin expression by estrogen in the uterus is mediated through a transcriptional mechanism.[49-51] Structural and functional analyses of the promoter regions of various estrogen-regulated genes reveal a common *cis*-acting DNA sequence that is responsible for estrogen regulation.[327] The estrogen receptor, upon binding to the consensus palindromic estrogen-responsive motif (5′-GGT-CANNNTGACC-3′ "perfect" estrogen-responsive-element (ERE)) is able to confer estrogen-stimulated transcription.

It has been shown, however, that in a majority of the estrogen-responsive genes, the ERE is an imperfect palindromic structure. Sequence analysis of the 5′ flanking region of the mouse lactoferrin gene reveals an imperfect palindromic ERE sequence located at −341 to −329 (mouse lactoferrin ERE) overlapping with a sequence resembling the COUP (chicken ovalbumin upstream promoter) element, located at −349 to −337 (mouse lactoferrin COUP) from the transcription initiation site of the lactoferrin gene.[50,328] The mouse lactoferrin COUP/ERE sequence (5'-AAGTGTCA-CAGGTCAAGGTAACCCACAAAT-3') conferred estrogen-stimulated transcription to both a homologous or heterologous promoter.[328]

A great contribution to the understanding of the mechanisms through which estrogens control lactoferrin expression derived from the analysis of mice rendered deficient in estrogen receptor expression (mice ER$^{-/-}$) through genetic techniques of homologous recombination. Wild-type mice treated with a single estradiol injection showed a 350-fold induction of lactoferrin mRNA levels, while ER$^{-/-}$ females showed no detectable responses.[329] Interestingly, catecholestrogen and an environmental estrogen, chlordecone, are able to stimulate lactoferrin mRNA levels in ER$^{-/-}$ mice.[330] This observation indicates that lactoferrin expression in murine uterus is modulated also through a distinct estrogen-signaling pathway that mediates the effects of physiological and environmental estrogens and does not involve ERa.[330] This conclusion was further supported by the observation that the stimulatory effect of catecholestrogen and xenoestrogen is not inhibited by selective anti-estrogens.[331]

While estrogen stimulates uterine lactoferrin synthesis, an opposite effect is displayed by progesterone, which exerts an inhibitory effect on lactoferrin synthesis.[332] The stimulatory effect of estrogen on lactoferrin synthesis is lost in estrogen-receptor knock-out mice, while the inhibitory effect of progesterone on lactoferrin synthesis is lost in mice lacking progesterone receptors.[332]

Lactoferrin synthesis was also investigated in benign and malignant breast diseases. Analysis of the breast secretions in women affected by benign breast diseases showed that the secretions usually do not contain lactoferrin and strictly resemble the protein patterns observed in secretions derived from normal mammary glands.[314] By contrast, a large percentage (57%) of secretions from women with breast carcinoma contained lactoferrin.[314] These findings were confirmed by immunohistological studies of breast carcinoma sections.[333] Recent studies suggest that lactoferrin may represent a useful marker of low-grade breast cancer.[334]

Of some interest is the observation of Furmanski that a lactoferrin isoform possesses ribonuclease activity.[11] RNase activity was noted in human milk in 1976 and subsequently ascribed to lactoferrin.[335] This observation may be relevant in the context of the breast carcinoma biology.

Das et al.[335] evaluated breast milk for inhibitors of reverse transcriptase in ethnic groups with different risks for developing breast cancer. They observed that RNase activity was lowest in Pausi women, an inbred group living in North America, that has a high risk of breast cancer. Forty percent of the milk in Pausi women did not inhibit avian myeloblastosis virus reverse transcriptase. Since viral synthesis may start at the level of cell membrane,[336] this function may be relevant in terms of antiviral defense. A lack of RNase may lead to increased viral invasion and then to malignant transformation.

Lactoferrin expression was significantly increased in malignant diseases of the uterus. Immunohistochemical studies showed that lactoferrin, expressed in normal cycling endometrium by a restricted number of glandular epithelial cells located deep in the zona basalis, is overexpressed in the majority (>75%) of endometrial adenocarcinomas, while only a minority (10%) of endometrial hyperplasia cases showed increased levels of lactoferrin.[337]

The level of lactoferrin expression in endometrial adenocarcinomas did not correlate with tumor grade or stage; however, a striking inverse correlation between the presence of progesterone receptors and degree of lactoferrin expression was observed.[337] A different situation is observed for cervical carcinoma. Lactoferrin was clearly expressed by the normal cervical epithelium, but is strikingly down-regulated upon neoplastic transformation of the endocervix, as early as in adenocarcinoma *in situ*.[338] Interestingly, neoplastic transformation of the endocervix was also accompanied by a marked reduction in estrogen receptor expression. According to these observations, it was proposed that lactoferrin may represent a cancer-specific marker of endocervical adenocarcinomas.[338]

Lactoferrin is also observed in male reproductive systems. In fact, lactoferrin and transferrin are present in human seminal plasma.[339] Some lactoferrin molecules in seminal plasma are free and others are associated with not-yet-identified lactoferrin binding molecules.[340] Lactoferrin is secreted by prostate and seminal vesicles, and is absent from testes and epididymes.[341] Interestingly, immunohistochemical studies showed a peculiar distribution of lactoferrin within the human prostate; some of the glands contained exclusively positive cells and others were completely lactoferrin-negative.[341] A more detailed immunohistochemical study showed that within the normal human prostate, lactoferrin primarily localized at the level of the epithelium of the central zone. More particularly, lactoferrin staining occurred in numerous individual cells scattered throughout the epithelium as well as within multiple intraepithelial lumens.[342]

A different situation was observed in other animal species. Studies on murine and porcine male sexual organs showed that epididymis was able to synthesize both lactoferrin mRNA and protein.[343] It was also shown that lactoferrin is bound to the sperm surface during the transit through the epididymis.[343] In mice epididymal cells, lactoferrin was unglycosylated, while porcine epididymal lactoferrin was glycosylated. Lactoferrin synthesis in epididymis is stimulated by 17-β-estradiol, but not by testosterone.

The exact role of lactoferrin in seminal plasma is not known. During ejaculation, lactoferrin binds to sperm and appears to represent a major component of the sperma-coating antigens.[344-347] Lactoferrin binding to sperm is reversible. Lactoferrin is gradually released from spermatozoa after they penetrate the cervical mucus.[348] More recent studies have shown the existence of an innate natural antibody present in the fertilization milieu that is reactive with a cryptic sequence of lactoferrin exposed on the sperm head surface.[349] The biochemical mechanisms responsible for the binding of lactoferrin to sperm during ejaculation, as well as those involved in its release from sperma within the female reproductive tract, remain unknown.

Although the precise role of lactoferrin in semen is unclear, lactoferrin concentrations are similar in normal and in oligospermic or azoospermic semen samples, thus excluding a role for a disturbed synthesis of this protein in these processes.[350]

3.6 LACTOFERRIN EXPRESSION DURING DEVELOPMENT

Few studies have been performed to characterize the pattern of lactoferrin expression during embryogenesis. A recent study allowed us to define the spatio-temporal pattern of expression of lactoferrin during normal murine embryonic development.[351] Three stages of lactoferrin expression during embryogenesis have been defined:

1. In the preimplantation embryo, lactoferrin is already expressed at the two-cell stage stage and continues to be expressed up to the blastocyst stage when expression markedly declines.
2. Late in gestation, corresponding to half of the gestational period, there is a progressive induction of lactoferrin synthesis first in fetal liver, then in spleen and bone marrow, in concert with the onset of myelopoiesis in these organs.
3. In the third stage, lactoferrin synthesis is detected in a variety of glandular epithelial cells, in line with the pattern observed in adults.

These observations suggest that lactoferrin, in addition to its well known role in anti-microbial defense systems, may play also a role in the preimplantation stage.[351]

The mechanisms involved in induction of lactoferrin synthesis during ontogenesis remain largely unknown. However, a recent study suggests that estrogens and retinoids could be involved in such a phenomenon.[352] In fact, it was shown that untreated murine embryonic cells do not express lactoferrin mRNA and protein, while these cells are induced to high levels of lactoferrin expression after treatment with retinoids or estrogens.[352] A role for estrogen in lactoferrin induction during ontogenesis is also suggested by studies performed in the developing mouse uterus.[353] These studies showed that lactoferrin synthesis is induced in epithelial uterine cells only at very late periods of gestation, and a significant increase in lactoferrin synthesis in these cells is observed at parturition in correspondence with the marked rise of circulating estrogens.[353] Treatment of pregnant mice with estrogens accelerated the induction of lactoferrin synthesis in fetal uterine cells.[354]

REFERENCES

1. Metz-Boutique, M. et al. Human lactotransferrin: amino acid sequence and structural comparison with the other transferrins, *Eur. J. Biochem.*, 145: 659, 1984.
2. Sanchez, L., Calvo, M., and Brock, J.H. Biological role of lactoferrin, *Arch. Dis. Childhood*, 67: 657, 1992.
3. Weinberg, E.D. Iron withholding: a defense against infection and neoplasia, *Physiol. Rev.*, 64: 65, 1984.
4. Bullen, J.J., Rogers, H.G., and Leigh, L. Iron-binding proteins in milk and resistance to *Escherichia coli* infection in infants, *Brit. Med. J.*, 1: 69, 1972.
5. Van Snick, J.L., Masson, P.L., and Hermans, J.F. The involvement of lactoferrin in the hyposideremia of the acute inflammation, *J. Exper. Med.*, 140: 1068, 1974.
6. Bennet, R.M. and Kokorinski, T. Lactoferrin turnover in man, *Clin. Sci.*, 57: 453, 1979.
7. Broxmeyer, H.E. et al. Identification of the lactoferrin as the granulocyte-derived inhibitor of colony-stimulating activity production, *J. Exp. Med.*, 148: 1052, 1978.
8. Broxmeyer, H.E. et al. Functional activities of acidic isoferritins and lactoferrin *in vitro* and *in vivo*, *Blood Cells*, 10: 397, 1984.

9. Ambruso, D.R. and Johnston, R.B. Lactoferrin enhances hydroxyl radical production by human neutrophils, neutrophil particulate fractions, and an enzymatic generation system, *J. Clin. Invest.,* 67: 352, 1981.

10. Moguilevsky, N., Retegui, L.A., and Masson, P.L. Comparison of human lactoferrins from milk and neutrophilic leukocytes. Relative molecular mass, isoelectric point, iron binding properties and uptake by the liver, *Biochem. J.,* 229: 353, 1985.

11. Furmanski, P. et al. Multiple molecular forms of human lactoferrin, *J. Exp. Med.,* 170: 415, 1989.

12. Spik, J. et al. Characterization and biological role of human lactoferrin complexes, in *Human Milk Banking.* Williams, A.F. and Baum J.D., Eds., New York, Raven Press, 133–143, 1984.

13. Legrand, D. et al. Properties of the iron-binding site of the N-terminal lobe of human and bovine lactotransferrins, *Biochem. J.,* 266: 575–581, 1990.

14. Bertini, I. et al. *J. Biol. Chem.,* 261: 1139–1146, 1986.

15. Eaton, S. et al. Comparison of the electron spin echo envelope modulation (ESEEM) for human lactoferrin and transferrin complexes of copper (II) and vanadyl ion, *J. Biol. Chem.,* 264: 4776–4881, 1989.

16. Eaton, S. et al. Electron spin echo envelope modulation evidence for carbonate binding to iron (III) and copper (II) transferrin and lactoferrin, *J. Biol. Chem.,* 265: 7138–7141, 1990.

17. Spik, J. et al. Primary structure of the glycans from human transferrin, *Eur. J. Biochem.,* 121: 413–419, 1982.

18. Derisbourg, P. et al. Primary structure of glycans isolated from human leukocyte lactoferrin, *Biochem. J.,* 269: 821–825, 1990.

19. Spik, G., Coddeville, B., and Montreuil, J. Comparative study of the primary structures of sero-, lacto- and ovotransferrin glycans from different species, *Biochimie,* 70: 1459–1469, 1988.

20. Daucher, M. et al. Molecular dynamics simulations of a monofucosylated biantennary glycan of the N-acetyllactosamine type: the human lactotransferrin glycan, *Biochimie,* 74: 63–74, 1992.

21. Van Berkel, P.H. et al. Heterogeneity in utilization of N-glycosylation sites Asn 624 and Asn 138 in human lactoferrin: a study with glycosylation-site mutants, *Biochem. J.,* 319: 117–122, 1996.

22. Van Berkel, P.H. et al. Glycosylated and unglycosylated human lactoferrins both bind iron and show identical affinities toward human lysozyme and bacterial lipolysaccharide, but differ in their susceptibilities toward tryptic proteolysis, *Biochem. J.,* 312: 107–114, 1995.

23. Anderson, B.F. et al. Structure of human lactoferrin at 3.2 Å resolution, *Proc. Natl. Acad. Sci. U.S.A.,* 84: 1769–1775, 1987.

24. Ward, P.P., Zhou, X., and Conneely, O.M. Cooperative interactions between the amino- and carboxyl-terminal lobes contribute to the unique iron-binding stability of lactoferrin, *J. Biol. Chem.,* 271: 12790–12794, 1996.

25. He, Q.Y. et al. Inequivalence of the two tyrosine ligands in the N-lobe of human serum transferrin, *Biochemistry,* 36: 14853–4860, 1997.

26. Faber, H.R. et al. Altered domain closure and iron binding in transferrins: the crystal structure of the Asp 60–Ser mutant of the amino-terminal half-molecule of human lactoferrin, *J. Mol. Biol.* 256: 352–363, 1996.

27. Nicholson, H. et al. Mutagenesis of the histidine ligand in human lactoferrin: iron binding properties and crystal structure of the histidine 253–methionine mutant, *Biochemistry,* 36: 341–346, 1997.

28. Norris, G., Baker, H., and Baker, E. Preliminary crystallographic studies on human apolactoferrin in its native and deglycosylated forms, *J. Mol. Biol.,* 209: 329–339, 1989.
29. Anderson, B. et al. Apolactoferrin structure demonstrates ligand-induced conformational change in transferrins, *Nature,* 344: 784–787, 1990.
30. Grossmann, G. et al. X-ray solution scattering reveals conformational changes upon iron uptake in lactoferrin, serum and ovo-transferrins, *J. Mol. Biol.,* 225: 811–825, 1992.
31. Day, C.L. et al. Studies of the N-terminal half of human lactoferrin produced from the cloned cDNA demonstrate that interlobe interactions modulate iron release, *J. Biol. Chem.,* 267: 13857–13866, 1992.
32. Lee, D.A. and Goodfellow, J.M. The pH-induced release of iron from transferrin investigated with a continuum electrostatic model, *Biophys. J.,* 74: 2747–2759, 1998.
33. Day, C.L. et al. Structure of the recombinant human lactoferrin at 2.0 Å resolution, *J. Mol. Biol.,* 232: 1084–1110, 1993.
34. Shownge, M.S. et al. Anion binding by human lactoferrin: results from crystallographic and physicochemical studies, *Biochemistry,* 31: 4451–4458, 1992.
35. Smith, C.A. et al. Metal substitution in transferrins: the crystal structure of human copper-lactoferrin at 2.1 Å resolution, *Biochemistry,* 31: 4527–4533, 1992.
36. Faber, H.R. et al. Mutation of arginine 121 in lactoferrin destabilizes iron binding by disruption of anion binding: crystal structures of R121S and R121E mutants, *Biochemistry,* 35: 14473–14479, 1996.
37. Packdaman, R. and Chahine, M.E.H. Transferrin. Interaction of lactoferrin with hydrogen carbonate, *Eur. J. Biochem.,* 249: 149–155, 1997.
38. Moore, S.A. et al. Three-dimensional structure of diferric bovine lactoferrin at 2.8 Å resolution, *J. Mol. Biol.,* 274: 222–236, 1997.
39. McCombs, J.L. et al. Chromosomal localization of human lactotransferrin gene by *in situ* hybridization, *Cytogenet. Cell Genet.,* 47: 16–25, 1988.
40. Teng, C.T. et al. Assignment of the lactotransferrin gene to human chromosome 3 and to mouse chromosome 9, *Somat. Cell. Mol. Genet.,* 13: 689–695, 1987.
41. Park, I. et al. Organization of the human transferrin gene: direct evidence that it is originated by gene duplication, *Proc. Natl. Acad. Sci. U.S.A.,* 82: 3149–3155, 1985.
42. Rado, T.A., Wei, X., and Benz, E.J. Isolation of lactoferrin cDNA from a human myeloid library and expression of mRNA during normal and leukemic myelopoiesis, *Blood,* 70: 989–993, 1987.
43. Rey, M. et al. Complete nucleotide sequence of human mammary gland lactoferrin, *Nucleic Acid Res.,* 18: 5288, 1991.
44. Powell, M.J. and Ogden, J.E. Nucleotide sequence of human lactoferrin cDNA, *Nucleic Acid Res.,* 18: 4013, 1991.
45. Johnston, J. et al. Lactoferrin gene promoter: structural integrity and nonexpression in HL-60 cells, *Blood,* 79: 2988–3006, 1992.
46. Panella, T. et al. Polymorphism and altered methylation of the lactoferrin gene in normal leukocytes, leukemic cells, and breast cancer, *Cancer Res.,* 51: 3037–3043, 1991.
47. Cunningham, G.A., Headon, D.R., and Conneely, O.M. Structural organization of the mouse lactoferrin gene, *Biochem. Biophys. Res. Comm.,* 189: 1725–1731, 1992.
47a. Kim, S.J. et al. Structure of the human lactoferrin gene and its chromosomal localization, *Mol. Cells,* 8: 663–668, 1998.
48. Siebert, P.D. and Huang, B.C.B. Identification of an alternative form of human lactoferrin mRNA that is expressed differentially in normal tissues and tumor-derived cell lines, *Proc. Natl. Acad. Sci. U.S.A.,* 94: 2198–2203, 1997.

49. Cohen, M.S. et al. Preliminary observations on lactoferrin secretion in human vaginal mucus: variation during the menstrual cycle, evidence of hormonal regulation and implication for infection with *Neisseria gonorrhoeae, Am. J. Obstet. Gynecol.,* 157: 1122–1129, 1987.

50. Pentecost, C.T. and Teng, C.T. Lactotransferrin is the major estrogen inducible protein of mouse uterine secretions, *J. Biol. Chem.,* 262: 10134–10140, 1986.

51. Liu, Y. and Teng, C. Characterization of estrogen-responsive mouse lactoferrin promoter, *J. Biol. Chem.,* 266: 21880–21885, 1991.

52. Shi, H.P. and Teng, C.T. Characterization of a mitogen-response unit in the mouse lactoferrin gene promoter, *J. Biol. Chem.,* 269: 12973–12980, 1994.

53. Shi, H. and Teng, C. Promoter-specific activation of mouse lactoferrin gene by epidermal growth factor involves two adjacent regulatory elements, *Mol. Endocrinol.,* 10: 732–741, 1996.

54. Teng, C. et al. Mouse lactoferrin gene. Promoter-specific regulation by EGF and cDNA cloning of the EGF-response-element binding protein, *Adv. Exp. Med. Biol.,* 443: 65–78, 1998.

55. Teng, C.T. et al. Differential molecular mechanism of the estrogen action that regulates lactoferrin gene in human and mouse, *Mol. Endocrinol.,* 6: 1969–1981, 1992.

56. Liu, Y., Yang, N., and Teng, C.T. COUP-TF acts as a competitive repressor for estrogen receptor-mediated activation of the mouse lactoferrin gene, *Mol. Cell. Biol.,* 13: 1836–1846, 1993.

57. Yang, N., Shigeta, H., and Teng, C.T. Estrogen-related receptor, hERR1, modulates estrogen receptor-mediated response of human lactoferrin gene promoter, *J. Biol. Chem.,* 271: 5795–5804, 1996.

58. Lee, M.O., Liu, Y., and Zhang, X.K. A retinoic acid response element that overlaps an estrogen response element mediates multihormonal sensitivity in transcriptional activation of the lactoferrin gene, *Mol. Cell. Biol.,* 15: 4194–4207, 1995.

59. Teng, C.T. et al. Lactoferrin gene expression in the mouse uterus and mammary gland, *Endocrinology,* 124: 992–999, 1989.

60. Pierce, A. et al. Molecular cloning and sequence analysis of bovine lactotransferrin, *Eur. J. Biochem.,* 196: 177–184, 1991.

61. Lydon, J. et al. Nucleotide and primary amino acid sequence of porcine lactoferrin, *Biochim. Biophys. Acta,* 1132: 97–99, 1992.

62. Stowell, K.M. et al. Expression of cloned human lactoferrin in baby hamster kidney cells, *Biochem. J.,* 276: 349–355, 1991.

63. Ward, P.P. et al. An inducible expression system for the production of human lactoferrin in *Aspergillus nidulans, Gene,* 122: 219–223, 1992.

64. Ward, P.P. et al. Expression and characterization of recombinant murine lactoferrin, *Gene,* 204: 171–176, 1997.

65. Nuijens, J.H. et al. Characterization of recombinant human lactoferrin secreted in milk of transgenic mice, *J. Biol. Chem.,* 272: 8802–8807, 1997.

66. Salmon, V. et al. Characterization of human lactoferrin produced in the baculovirus expression system, *Protein Expr. Purif.,* 9: 203–210, 1997.

67. Yamada, Y. et al. Lactoferrin binding by leukemia cell lines, *Blood,* 70: 264–270, 1987.

68. Miyazawa, K. et al. Lactoferrin–lipolysaccharide interactions. Effect of lactoferrin binding on monocyte/macrophage-differentiated HL-60 cells, *J. Immunol.,* 146: 723–729, 1991.

69. Eda, S., Kikugawa, K., and Beppu, M. Binding characteristics of human lactoferrin to the human monocytic leukemia cell line THP-1 differentiated into macrophages, *Biol. Pharm. Bull.,* 19: 167–175, 1997.

70. Garré, C. et al. Lactoferrin binding sites and nuclear localization in K562 cells, *J. Cell Physiol.*, 153: 477–482, 1992.
71. Mazurier, J. et al. Expression of human lactotransferrin receptors in phytohemagglutinin-stimulated human peripheral blood lymphocytes, *Eur. J. Biochem.*, 179: 481–487, 1989.
72. Mancheva-Nilsson, L., Hammerstrom, S., and Hammerstrom, M.L. Activated human gamma delta T lymphocytes express functional lactoferrin receptors, *Scand. J. Immunol.*, 46: 609–618, 1997.
73. Rochard, E. et al. The N-terminal domain I of human lactotransferrin binds specifically to phytohemagglutinin-stimulated peripheral blood human lymphocyte receptors, *FEBS Lett.*, 255: 201–204, 1989.
74. Legrand, D. et al. Inhibition of the specific binding of human lactotransferrin to human peripheral blood phytohemagglutinin-stimulated lymphocytes by fluorescein labelling and location of the binding site, *Biochem. J.*, 276: 733–738, 1991.
75. Legrand, D. et al. Molecular interactions between human lactotransferrin and the phytohemagglutinin-activated human lymphocyte lactotransferrin receptor lie in the two loop-containing regions of the N-terminal domain I of human lactotransferrin, *Biochemistry*, 31: 9243–9251, 1992.
76. Mazurier, J. et al. Study on the binding of lactotransferrin (lactoferrin) to human PHA-activated lymphocytes and non-activated platelets. Localisation and description of the receptor-binding site, *Adv. Exp. Med. Biol.*, 317: 111–119, 1994.
77. Legrand, D. et al. The N-terminal Arg[2], Arg[3], Arg[4] on human lactoferrin interact with sulphated molecules, but not with the receptor present on Jurkat human lymphoblastic cells, *Biochem. J.*, 327: 841–846, 1997.
78. Legrand, D. et al. Role of the first N-terminal basic cluster of human lactoferrin (R2R3R4R5) in the interactions with the Jurkat human lymphoblastic T-cells, *Adv. Exp. Med. Biol.*, 443: 49–55, 1998.
79. Van Snick, J.L., Masson, P.L., and Heremans, J.F. The involvement of lactoferrin in the hyposideremia of acute inflammation, *J. Exp. Med.*, 140: 1068–1084, 1974.
80. Van Snick, J.L. and Masson P.L. The binding of human lactoferrin to mouse peritoneal cells, *J. Exp. Med.*, 144: 1568–1580, 1976.
81. Van Snick, J.L., Markowetz, B., and Masson, P.L. Ingestion and digestion of human lactoferrin by mouse peritoneal macrophages and the transfer of its iron into ferritin, *J. Exp. Med.*, 146: 817–827, 1977.
82. Bennet, R.M. and Davis, J. Lactoferrin binding to human peripheral blood cells: an interaction with a B-enriched population of lymphocytes and a subpopulation of adherent mononuclear cells, *J. Immunol.*, 127: 1211–1216, 1981.
83. Burgens, H.S. et al. Receptor binding of lactoferrin by human monocytes, *Brit. J. Haematol.*, 54: 383–391, 1983.
84. Campbell, E.J. Human leukocyte elastase, cathepsin G, and lactoferrin: family of neutrophil granule glycoproteins that bind to an alveolar macrophage receptor, *Proc. Natl. Acad. Sci. U.S.A.*, 79: 6941–6945, 1984.
85. Britigan, B.E. et al. Uptake of lactoferrin by mononuclear phagocytes inhibits their ability to form hydroxyl radical and protects them from membrane autoperoxidation, *J. Immunol.*, 147: 4271–4277, 1991.
86. Moguilevsky, N., Masson, P.L., and Courtoy, P.J. Lactoferrin uptake and iron processing into macrophages: a study in familial haemochromatosis, *Brit. J. Haematol.*, 66: 129–136, 1987.
87. Birgens, H. et al. Lactoferrin-mediated transfer of iron to intracellular ferritin in human monocytes, *Eur. J. Haematol.*, 41: 52–57, 1988.

88. Birgens, H. and Kristensen, L. Impaired receptor binding and decrease in isoelectric point of lactoferrin after interaction with human monocytes, *Eur. J. Haematol.*, 45: 31–35, 1990.

89. Bird, T. and Horwitz, M. Lactoferrin inhibits or promotes *Legionella pneumophila* intracellular multiplication in nonactivated and interferon gamma-activated human monocytes depending upon its degree of saturation, *J. Clin. Invest.*, 88: 1103–1112, 1991.

90. Sorimachi, K. et al. Activation of macrophages by lactoferrin: secretion of TNF-alpha, IL-8 and NO, *Biochem. Mol. Biol. Int.*, 43: 79–83, 1997.

91. Cox, T.M. et al. Iron-binding proteins and influx of iron across the duodenal brush border. Evidence for specific lactoferrin receptors in human intestine, *Biochim. Biophys. Acta*, 588: 120–128, 1979.

92. Spik, G. et al. Characterization and properties of the human and bovine lactotransferrins extracted from faeces of newborn infants, *Acta Pediatr. Scand.*, 71: 979–985, 1982.

93. Davidson, L.A. and Lonnerdal, B. Persistence of human milk proteins in the breast-fed infant, *Acta Pediatr. Scand.*, 76: 733–740, 1987.

94. Goldman, A.S. et al. Molecular forms of lactoferrin in stool and urine from infants fed human milk, *Pediatr. Res.*, 27: 252–255, 1990.

95. Hutchens, T.M., Henry, J.F., and Yip, T.T. Structurally intact (78-kDa) forms of maternal lactoferrin purified from urine of preterm infants fed human milk: identification of a trypsin-like proteolytic cleaveage event *in vivo* that does not result in fragment dissociation, *Proc. Natl. Acad. Sci. U.S.A.*, 88: 2994–2998, 1991.

96. Mazurier, J., Montreuil, J., and Spik, G. Visualization of lactoferrin brush-border receptors by ligand blotting, *Biochim. Biophys. Acta*, 821: 453–460, 1985.

97. Davidson, L.A. and Lonnerdal, B. Specific binding of lactoferrin to brush border membrane: ontogeny and effect of glycan chain, *Am. J. Physiol.*, 254: 4580–4585, 1988.

98. Kawakami, H., Dosako, S., and Lonnerdal, B., Iron uptake from transferrin and lactoferrin by rat intestinal brush-border membrane vesicles, *Am. J. Physiol.*, 258: 4535–4539, 1990.

99. Hu, W.L. et al. Lactotransferrin receptor of mouse small-intestinal brush border, *Biochem. J.*, 249: 435–441, 1988.

100. Hu, W.L. et al. Isolation and partial characterization of a lactotransferrin receptor from mouse intestinal brush border, *Biochemistry*, 29: 535–541, 1990.

101. Mikogami, T., Marianne, T., and Spik, G. Effect of intracellular iron depletion by picolinic acid on expression of the lactoferrin receptor in human colon carcinoma subclone HT9-18-C1, *Biochem. J.*, 208: 391–397.

102. Fransson, G.B. and Lonnerdal, B. Iron in human milk, *J. Pediatr.*, 96: 380–384, 1980.

103. Saarinen U.M. Need for iron supplementation in infants on prolonged breast feeding, *J. Pediatr.*, 93: 177–180, 1978.

104. Duncan, B. et al. Iron and the exclusively breast-fed infant from birth to 6 months, *J. Pediatr., Gastroenterol Nutr* 4: 421–425, 1985.

105. Simes, M.A., Salmenpera, L., and Perheentupa, J. Exclusive breast feeding for 9 months: risk of iron deficiency, *J. Pediatr.*, 104: 196–199, 1984.

106. Iwai, Y., Nakao, Y., and Mikawa, H. Iron status in low birth weight infants on breast and formula feeding, *Eur J. Pediatr.*, 145: 63–65, 1986.

107. Garry, P.J. et al. Iron absorption from human milk and formula with and without iron supplementation, *Pediatr. Res.*, 15: 822–826, 1981.

108. Heinrich, H.C. et al. Ferrous and hemoglobin-^{59}Fe absorption from supplemented cow's milk in infants with normal and iron depleted stores, *Z. Kinderheilkd.*, 120: 251–258, 1975.

109. McMillan J., Landaw, S., and Oski, F. Iron sufficiency in breast-fed infants and the availability of iron from human milk, *Pediatrics,* 58: 686–691, 1976.
110. McMillan, J.A. et al. Iron absorption from human milk, simulated human milk, and proprietary formulas, *Pediatrics,* 60: 896–900, 1977.
111. Oski, F.A. and Landaw, S.A. Inhibition of iron absorption from human milk by baby food, *Am. J. Dis. Child.,* 134: 459–460, 1980.
112. Rios, E. et al. The absorption of iron as supplements in infant cereal and infant formulas, *Pediatrics,* 55: 686–693, 1975.
113. Saarinen, U.M. and Siimes, M.A. Iron absorption from infant milk formula and the optimal level of iron supplementation, *Acta Pediatr. Scand.,* 66: 719–722, 1977.
114. Saarinen, U.M. and Siimes, M.A. Iron absorption from breast milk, cow's milk, and iron-supplemented formula: an opportunistic use of changes in total body iron determined by hemoglobin, ferritin, and body weight in 132 infants, *Pediatr. Res.,* 13: 143–147, 1979.
115. Saarinen, U.M., Siimes, M.A. and Dallman, P. Iron absorption in infants: high bioavailability of breast milk as indicated by the extrinsic tag method of iron absorption and by the concentration of serum ferritin, *J. Pediatr.,* 91: 36–39, 1977.
116. Stekel, A. et al. Absorptoion of fortification iron from milk formulas in infants, *Am. J. Clin. Nutr.,* 43: 917–922, 1986.
117. Lonnerdal, B. Iron in breast milk, in *Iron Nutrition in Infancy and Childhood.* Vol. 4, Nestle Nutrition Workshop Series, Stekel, A., Ed., New York, Raven Press, 95–118, 1984.
118. Fairweather-Tait, S.J. et al. Lactoferrin and iron absorption in newborn infants, *Pediatr. Res.,* 22: 651–654, 1987.
119. Schulz-Lell, G. et al. Iron availability from an infant formula supplemented with bovine lactoferrin, *Acta Pediatr. Scand.,* 80: 155–158, 1991.
120. Davidson, L.A., Litov, R.E., and Lonnerdal, B. Iron retention from lactoferrin-supplemented formulas in infant rhesus monkeys, *Pediatr. Res.,* 27: 176–180, 1990.
121. Preels, J.P. et al. Hepatic receptor that specifically binds oligosaccharides containing fucosyl a-1-3-N-acetylglucosamine linkages, *Proc. Natl. Acad. Sci. U.S.A.,* 75: 2215–2219, 1978.
122. Retegui, L.A. et al. Uptake of lactoferrin by the liver. I. Role of the reticuloendothelial system as indicated by blockade experiments, *Lab. Invest.,* 50: 323–328, 1984.
123. Courtoy, P.J. et al. Uptake of lactoferrin by the liver. II. Endocytosis by sinusoidal cells, *Lab. Invest.,* 50: 329–334, 1984.
124. Imber, N.J. and Pizzo, S.V. Clearance and binding of native and defucosylated lactoferrin, *Biochem. J.,* 212: 249–257, 1988.
125. Ziere, G.J. et al. Lactoferrin uptake by rat liver cells, *J. Biol. Chem.,* 267: 11229–11235, 1992.
126. Huettinger, M. et al. Characteristics of chylomicron remnant uptake into rat liver, *Clin. Biochem.,* 21: 87–92, 1988.
127. VanDijk, M. et al. Recognition of chylomicrons remnants and beta-migrating very-low-density lipoproteins by the remnant receptor of parenchymal liver cells is distinct from the liver a_2-macroglobulin, *Biochem. J.,* 279: 863–870, 1991.
128. Ziere, G.J., Bijsterbosch, H., and Van Berkel, J. Removal of 14 N-terminal amino acids of lactoferrin enhances its affinity for parenchymal liver cells and potentiates the inhibition of beta-very-low-density lipoprotein binding. *J. Biol. Chem.,* 268: 27069–27075, 1993.
129. Willnow, T.E. et al. Low density lipoprotein-receptor-related protein and gp330 bind similar ligands, including plasinogen activator-inhibitor complexes and lactoferrin, an inhibitor of chylomicron remnant clearance, *J. Biol. Chem.,* 267: 26172–26181, 1992.

130. Vash, B. et al. Three complement-type repeats of the low-density lipoprotein receptor-related protein define a common binding site for RAP, PAI-1 and lactoferrin, *Blood,* 42: 3277–3285, 1998.

131. McAbee, D. et al. Endocytosis and degradation of bovine apo- and holo-lactoferrin by isolated rat hepatocytes are mediated by recycling calcium-dependent binding sites, *Biochemistry,* 32: 13749–13760, 1993.

132. Bennat, D.J. and McAbee, D.D. Identification and isolation of a 45-kDa calcium-dependent lactoferrin receptor from rat hepatocytes, *Biochemistry,* 36: 8359–8366, 1997.

133. Bennat, D.J., Ling, Y.Y., and McAbee, D.D. Isolated rat hepatocytes bind lactoferrins by the RHL-1 subunit of the asialoglycoprotein receptor in a galactose-independent manner, *Biochemistry,* 36: 8367–8376, 1997.

134. McAbee, D.D., Bennet, D.J., and Ling, Y.Y. Identification and analysis of a Ca-dependent lactoferrin receptor in rat liver. Lactoferrin binds to the asialoglycoprotein receptor in a galactose-independent manner, *Adv. Exp. Med. Biol.,* 443: 113–121, 1998.

135. McAbee, D.D. and Ling, Y.Y. Iron-loading of cultured adult rat hepatocytes reversibly enhances lactoferrin binding and endocytosis, *J. Cell Physiol.,* 171: 75–86, 1997.

136. Sitaram, M.P. and McAbee, D.D. Isolated rat hepatocytes differentially bind and internalize bovine lactoferrin N- and C-lobes, *Biochem. J.,* 323: 815–822, 1997.

137. Rochard, E. et al. Characterization of lactotransferrin receptor in epithelial cell lines from non-malignant human breast, benign mastopathies and breast carcinomas, *Anti-cancer Res.,* 12: 2047–2052, 1992.

138. Raha, S. et al. KRDS, a tetrapeptide derived from lactotransferrin, inhibits binding of monoclonal antibody against glycoprotein IIb-IIIa on ADP-stimulated platelets and megakaryocytes, *Blood,* 72: 172–178, 1988.

139. Drouet, L. et al. The antithrombotic effect of KRDS, a lactotransferrin peptide, compared with RGDS, *Nouv. Rev. Fr. Hématol.,* 32: 59–62, 1990

140. Mazoyer, E. et al. KRDS, a new peptide derived from human lactotransferrin inhibits platelet aggregation and release reaction, *Eur. J. Biochem.,* 194: 43–49, 1992.

141. Lavengle, B. et al. Binding of lactotransferrin to platelet-receptor induces inhibition of platelet aggregation, *Eur. J. Biochem.,* 1993.

142. Fillebeen, C. et al. Lactoferrin is synthesized by mouse brain tissue and its expression is enhanced after MPTP treatment, *Adv. Exp. Med. Biol.,* 443: 293–300, 1998.

143. Faucheux, B.A. et al. Expression of lactoferrin receptors is increased in the mesen-cephalon of patients with Parkinson's disease, *Proc. Natl. Acad. Sci. U.S.A.,* 92: 9603–9607, 1995.

144. Masson, P.L. et al. Immunohistochemical localization and bacteriostatic properties of an iron-binding protein from bronchial mucus, *Thorax,* 21: 538–547, 1966.

145. Oram, J.D. and Reiter, B. Inhibition of bacteria by lactoferrin and other iron-chelating agents, *Biochim. Biophys. Acta,* 170: 351–359, 1968.

146. Bullen, J.J., Gogers, H.J., and Leigh, L. Iron-binding proteins in milk and resistance to *Escherichia coli* infection in infants, *Brit. Med. J.,* 1: 69–75, 1972.

147. Arnold, R.R., Cole, M.F., and McGhee, J.R. A bactericidal effect for human lacto-ferrin, *Science,* 197: 263–265, 1977.

148. Arnold, R.R., Brewer, M., and Gauthier, J.J. Bactericidal activity of human lactoferrin: sensitivity of a variety of microorganisms, *Infect. Immun.,* 28: 893–890, 1980.

149. Arnold, R.R. et al. Bactericidal activity of human lactoferrin: inlfuence of physical conditions and metabolic state of the target microorganism, *Infect. Immun.,* 32: 655–663, 1982.

150. Arnold, R.R. et al. Bactericidal activity of human lactoferrin: differentiation from the stasis of iron deprivation, *Infect. Immun.*, 35: 792–800, 1982.

151. Bortner, C.A., Miller, R.D., and Arnold, R.R. Bactericidal effect of lactoferrin on *Legionella pneumophila*, *Infect. Immun.*, 51: 373–381, 1986.

152. Kalmar, J.R. and Arnold, R.R. Killing of *Actinobacillus actinomycetemcomitans* by human lactoferrin, *Infect. Immun.*, 56: 2552–2560, 1988.

153. Bortner, C.A., Anold, R.R., and Miller, R.D., Bactericidal effect of lactoferrin on *Legionella pneumophila*: effect of the physiological state of the organism, *Can. J. Microbiol.*, 45: 1048–1059, 1989.

154. Bullen, J.J. and Armstrong, J.A. The role of lactoferrin in the bactericidal function of polymorphonuclear leukocytes, *Immunology*, 36: 781–788, 1989.

155. Byrd, T.F. and Horwitz, M.A. Interferon gamma-activated human monocytes down regulate transferrin receptors and inhibit the intracellular multiplication of *Legionella pneumophila* by limiting the avaibility of iron, *J. Clin. Invest.*, 83: 1457–1465, 1989.

156. Byrd, T.F. and Horwitz, M.A. Chloroquine inhibits the intracellular multiplication of *Legionella pneumophila* by limiting the avaibility of iron, *J. Clin. Invest.*, 88: 351–357, 1991.

157. Byrd, T.F. and Horwitz, M.A. Lactoferrin inhibits or promotes *Legionella pneumophila* intracellular multiplication in nonactivated and interferon gamma-activated human monocytes depending upon its degree of iron saturation, *J. Clin. Invest.*, 88: 1103–1112, 1991.

158. Molloy, A. and Winterbourn, C. Release of iron from phagocytosed *Escherichia coli* and uptake by neutrophil lactoferrin, *Blood*, 75: 984–989, 1990.

159. Klebanoff, S.J. Oxygen metabolites from phagocytes, in *Inflammation: Basic Principles and Clinical Correlates*, Gallin, J.I. et al., Eds., New York, Raven Press, 541–587, 1992.

160. Ambruso, D.R. and Johnston, R.B. Lactoferrin enhances hydroxyl radical production by human neutrophils, neutrophil particulate fractions, and an enzymatic generating system, *J. Clin. Invest.*, 67: 352–361, 1981.

161. Winterbourn, C.C. Lactoferrin-catalyzed hydroxyl radical production: additional requirement for a chelating agent, *Biochem. J.*, 210: 15–24, 1983.

162. Baldwin, D.A., Jenny, E.R., and Aisen, P. The effect of human serum transferrin and milk lactoferrin on hyroxyl radical formation from superoxide and hydrogen peroxide, *J. Biol. Chem.*, 259: 13391–13398, 1984.

163. Britigan, B.E. et al. Stimulated human neutrophils limit iron-catalyzed hydroxyl radical formation as detected by spin trapping techniques, *J. Biol. Chem.*, 261: 17026–17033, 1986.

164. Gutteridge, J.M.C., et al. Inhibition of lipid peroxidation by the iron-binding protein lactoferrin, *Biochem. J.*, 199: 259–269, 1981.

165. Britigan, B.E. et al. Neutrophil degranulation inhibits potential hydroxyl-radical formation, *Biochem. J.*, 264: 447–455, 1992.

166. Klebanoff, S.J. and Waltersdorph, A.M., Peroxidant activity of transferrin and lactoferrin, *J. Exp. Med.*, 172: 1293–1303, 1990.

167. Britigan, B.E. and Edeker, B.L. *Pseudomonas* and neutrophil products modify transferrin and lactoferrin to create conditions that favor hydroxyl radical formation, *J. Clin. Invest.*, 88: 1092–1102, 1991.

168. Ellison, R., III, Giehl, T., and LaForce, M. Damage of the outer membrane of enteric Gram-negative bacteria by lactoferrin and transferrin, *Infect. Immun.*, 56: 2774–2781, 1988.

169. Ellison, R., III et al. Lactoferrin and transferrin damage of the Gram-negative outer membrane is modulated by Ca^{2+} and Mg^{2+}, *J. Gen. Microbiol.*, 136: 1437–1446, 1990.

170. Ellison, R., III and Giehl, T. Killing of Gram-negative bacteria by lactoferrin and lysozyme, *J. Clin. Invest.*, 88: 1080–1091, 1991.

171. Leitch, E.C. and Willcox, M.D. Synergic antistaphylococcal properties of lactoferrin and lysozyme, *J. Med. Microbiol.*, 47: 837–842, 1998.

172. Bals, R. et al. Human beta-defensin 2 is a salt-sensitive peptide antibiotic expressed in human lung, *J. Clin. Invest.*, 102: 874–880, 1998.

173. Qin, J. et al. Human milk lactoferrin inactivates two putative colonization factors expressed by *Haemophilus influenzae*, *Proc. Natl. Acad. Sci. U.S.A.*, 95: 12641–12646, 1998.

174. Schmidt, A.M. et al. Regulation of human mononuclear phagocyte migration by cell surface binding proteins for advanced glycation end products, *J. Clin. Invest.*, 91: 2155–2168, 1993.

175. Li, Y.M., Tan, A.X., and Vlanara, H. Antibacterial activity of lysozyme and lactoferrin is inhibited by binding of advanced glycation modified proteins to a conserved motif, *Nat. Med.*, 1: 1057–1061, 1995.

176. Bellamy, W. et al. Identification of the bactericidal domain of lactoferrin, *Biochim. Biophys. Acta*, 1121: 130–136, 1992.

176a. Kuwata, H. et al. Direct evidence of the generation in human stomach of an antimicrobial peptide domain (lactoferricin) from ingested lactoferrin, *Biochim. Biophys. Acta*, 1429: 129–141, 1998.

177. Ellas-Rochard, E. et al. Lactoferrin–lipopolysaccharide interaction: involvement of the 28–34 loop region of human lactoferrin in the high affinity binding to *Escherichia coli* O55B5 lipopysaccharide, *Biochem. J.*, 312: 839–845, 1995.

178. Odell, E.W. et al. Antibacterial activity of peptides homologous to a loop region in human lactoferrin, *FEBS Lett.*, 382: 175–178, 1996.

179. Hwang, P.M. et al. Three-dimensional structure of lactoferricin B, an antimicrobial peptide derived from bovine lactoferrin, *Biochemistry*, 37: 4288–4298, 1998.

180. Chapples, D.S. et al. Structure-function relationship of antibacterial synthetic peptides homologous to a helical surface region on human lactoferrin against *Escherichia coli* serotype O111, *Infect. Immun.*, 66: 2434–2440, 1998.

181. Jones, E.M. et al. Lactoferricin, a new antimicrobial peptide, *J. Appl. Bacteriol.*, 77: 208–214, 1994.

182. Kuwata, H. et al. The survival of ingested lactoferrin in the gastrointestinal tract of adult mice, *Biochem. J.*, 334: 321–323, 1998.

183. Miyauchi, H. et al. Bovine lactoferrin stimulates the phagocytic activity of human neutrophils: identification of its active domain, *Cell Immunol.*, 187: 34–37, 1998.

184. Wakabayashi, H. et al. Cooperative anti-*Candida* effects of lactoferrin or its peptides in combination with azole antifungal agents, *Microbiol. Immunol.*, 40: 821–825, 1996.

185. Wakabayashi, H. et al. Inhibition of hyphal growth of azole-resistant strains of *Candida albicans* by triazole antifungal agents in the presence of lactoferrin-related compunds, *Antimicrobiol. Agents Chemother.*, 42: 1587–1591, 1998.

186. Izhar, M., Nuchamowitz, Y., and Mirelman, D., Adherence of *Shigella flexneri* to guinea pig intestinal cells is mediated by musal adhesin, *Infect. Immunol.*, 35: 1110–1118, 1982.

187. Tigyi, Z. et al. Lactoferrin-binding proteins in *Shigella flexneri*, *Infect. Immun.*, 60: 2619–2626, 1992.

188. Herrington, D.A. and Sperling, P.F. *Haemophilus influenzae* can use transferrin as a sole source for required iron, *Infect. Immunol.*, 48: 248–251, 1985.

189. Schryvers, A.B. Identification of the transferrin- and lactoferrin-binding proteins in *Haemophilus infuenzae*, *J. Med. Microbiol.*, 29: 121–130, 1989.
190. Schryvers, A.B. and Morris, L.J. Identification and characterization of the lactoferrin-binding protein from *Neisseria meningitidis*, *Infect. Immun.*, 56: 1144–1149, 1988.
191. Menozzi, F., Gautier, C., and Locht, C. Identification and purification of transferrin- and lactoferrin-binding proteins of *Bordetella pertussis* and *Bordetella bronchiseptica*, *Infect. Immun.*, 59: 3982–3988, 1991.
192. Berish, S.A. et al. Molecular cloning and characterization of the structural gene for the major iron-regulated protein expressed by *Neisseria gonorrhoeae*, *J. Cell Biol.*, 171: 1535–1546, 1990.
193. Agiato, L.A. and Dyer, D.W. Siderophore production and membrane alterations by *Bordetella pertussis* in response to iron starvation, *Infect. Immun.*, 60: 117–123, 1992.
194. Gorringe, A.R., Woods, G., and Robinson, A. Growth and siderophore production by *Bordetella pertussis* under iron-restricted conditions, *FEMS Microbiol. Lett.*, 66: 101–106, 1990.
195. Agiato-Foster, L.A. and Dyer, D.W. A siderophore production mutant of *Bordetella bronchiseptica* cannot use lactoferrin as an iron source, *Infect. Immun.*, 61: 2698–2702, 1993.
196. Spik, G. et al. Bacteriostasis of milk-sensitive strains of *Escherichia coli* by immunoglobulins and iron-binding proteins in association, *Immunology*, 35: 663–671, 1978.
197. Nikaido, H. Outer membrane barrier as a mechanism of antimicrobial resistance, *Antimicrob. Agents Chemother.*, 33: 1831–1836, 1989.
198. Naidu, S.S. et al. Specific binding of lactoferrin to *Escherichia coli* isolated from human intestinal infections, *APMIS*, 99: 1142–1150, 1991.
199. Gado, I. et al. Correlation between human lactoferrin binding and colicin susceptibility in *Escherichia coli*, *Antimicrob. Agents Chemother.*, 35: 2538–2543, 1991.
200. Kishore, A.R. et al. Specific binding of lactoferrin to *Aeromonas hydrophila*, *FEMS Microbiol. Lett.*, 83: 115–120, 1991.
201. McKenna, W.R. et al. Iron uptake from lactoferrin and transferrin by *Neisseria gonorrhoeae*, *Infect. Immun.*, 56: 785–791, 1988.
202. Schyvers, A.B. and Lee, B.C. Comparative analysis of the transferrin and lactoferrin binding proteins in the family Neisseriaceae, *Can. J. Microbiol.*, 35: 409–415, 1989.
203. Blanton, K.J. et al. Genetic evidence that *Neisseria gonorrhoeae* produces specific receptors for transferrin and lactoferrin, *J. Bacteriol.*, 172: 5225–5235, 1990.
204. Lee, B.C. and Schyvers, A.B. Specificity of the lactoferrin and transferrin receptors in *Neisseria gonorrhoeae*, *Mol. Microbiol.*, 2: 827–829, 1988.
205. Lee, B.C. and Bryan, L.E. Identification and comparative analysis of the lactoferrin and transferrin receptors among clinical isolates of gonococci, *J. Med. Microbiol.*, 28: 199–204, 1989.
206. Alcantara, J., Padda, J.S., and Schryvers, A.B. The N-linked oligosaccharides of human lactoferrin are not required for binding to bacterial lactoferrin receptors, *Can. J. Microbiol.*, 38: 222–230, 1992.
207. Yu, R. and Schryvers, A.B. Regions located in both the N-lobe and C-lobe of human lactoferrin participate in the binding interaction with bacterial lactoferrin receptors, *Microb. Pathog.*, 14: 343–353, 1993.
208. Biswas, G.D. and Sparling, P.F. Characterization of LbpA, the structural gene for a lactoferrin receptor in *Neisseria gonorrhoeae*, *Infect. Immun.*, 63: 2958–2967, 1995.
209. Petterson, A. et al. Molecular characterization of the 98-kilodalton iron-regulated outer membrane protein of *Neisseria meningitidis*, *Infect. Immun.*, 61: 4724–4733, 1993.

210. Du, R.P. et al. Cloning and expression of the *Moraxella catarrhalis* lactoferrin receptor genes, *Infection,* 66: 3656–3665, 1998.

211. Lewis, L.A. et al. Identification and molecular analysis of LbpBA, which encodes the two-component meningococcal lactoferrin receptor, *Infect. Immun.,* 66: 3017–3023, 1998.

212. Biswas, G.D. et al. Identification and functional characterization of the *Neisseria gonorrhoeae* lbpB gene product, *Infect. Immun.,* 67: 455–459, 1999.

213. Ormon, J.D. and Reiter, B. Inhibition of bacteria by lactoferrin with lysozyme in granules of human polymorphonuclear leukocytes, *Infect. Immun.,* 6: 761–766, 1968.

214. Kirpatrick, C.H. et al. Inhibition of growth of *Candida albicans* by iron-unsaturated lactoferrin: relation to host-defense mechanisms in chronic mucocutaneous candidiasis, *J. Infect. Dis.,* 124: 539–543, 1971.

215. Palma, C. et al. Identification of a mannoprotein fraction from *Candida albicans* that enhances human polymorphonuclear leukocyte functions and stimulates lactoferrin in PMNL inhibition of candidal growth, *J. Infect. Dis.,* 166: 1103–1112, 1992.

216. Palma, C. et al. Lactoferrin release and interleukin-1, interleukin-6, and tumor necrosis factor production by human polymorphonuclear cells stimulated by various lipopolysaccharides: relationship to growth inhibition of *Candida albicans, Infect. Immun.,* 60: 4604–4611, 1992.

217. Husson, M.O. et al. Iron acquisition by *Helicobacter pylori*: importance of human lactoferrin, *Infect. Immun.,* 61: 2694–2697, 1993.

218. Luquani, Y.A. et al. Expression of lactoferrin in human stomach, *Int. J. Cancer,* 49: 684–687, 1991.

219. Tomita, M. et al. Enzymatic hydrolysates of bovine lactoferrin have bactericidal properties more potent than undigested lactoferrin, *J. Dairy Sci.,* 74: 4137–4142, 1991.

220. Bellamy, W. et al. Identification of the bactericidal domain of lactoferrin, *Biochim. Biophys. Acta,* 1121: 130–136, 1992.

221. Hill, C.P. et al. Crystal structure of defensin HNP-3, an amphiphilic dimer: mechanisms of membrane permeabilization, *Science,* 251: 1481–1485, 1991.

222. Koivuranta-Vara, P., Bauda, D., and Goldstein, I.M. Bacterial-lipopolysaccharide-induced release of lactoferrin from human polymorph leukocytes: role of monocyte-derived tumor necrosis factor alpha, *Infect. Immun.,* 55: 2956–2961, 1987.

223. Gutteberg, T., Dalaker, K., and Vorland, L. Early response in neonatal septicemia. The effect of *Escherichia coli, Streptococcus agalactiae* and tumor necrosis factor on the generation of lactoferrin, *APMIS,* 98: 1027–1032, 1990.

224. Gutteberg, T. et al. Early response in septicemia in newborns and their mothers, *APMIS,* 99: 602–608, 1991.

225. Lash, J. et al. Plasma lactoferrin reflects granulocyte activation *in vivo, Blood,* 61: 885–888, 1983.

226. Van Deventer, S.J.H. et al. Experimental endotoxaemia in humans: analysis of cytokine release and coagulation, fibrinolytic and complement pathways, *Blood,* 76: 2520–2526, 1990.

227. Van Deventer, S.J.H., Hack, C.E., and Wolbink, G.J., Endotoxin-induced neutrophil activation. The role of complement revisited, *Prog. Clin. Biol. Res.,* 367: 101–107, 1991.

228. VanderPoll, P. et al. Effects on leukocytes after injection of tumor necrosis factor into healthy humans, *Blood,* 79: 693–698, 1992.

229. Oseas, R. et al. Lactoferrin: a promoter of polymorphonuclear leukocyte adhesiveness, *Blood,* 57: 939–946, 1981.

230. Paykrt, D., et al. Modulation of cytokine release and neutrophil function by granulocyte colony-stimulating factor during endotoxemia in humans, *Blood,* 90: 1415–1424, 1997.

230a. Rutella, S. et al. Inhibition of lymphocyte blastogenic response in healthy donors treated with recombinant human granulocyte colony-stimulating factor (G-CSF): possible role of lactoferrin and interleukin-1 receptor antagonist, *Bone Marrow Transplant,* 20: 355–364, 1997.

231. Boxer, L.A. et al. Neutropenia induced by systemic infusion of lactoferrin, *J. Lab. Clin. Med.,* 99: 866–872, 1982.

232. Baynes, R.D. et al. Plasma lactoferrin content: differential effect of steroid administration and infective illness: lack of effect of ambient temperature at which specimens are collected, *Scand. J. Haematol.,* 37: 353–359, 1986.

233. Appelmelk, B. et al. Lactoferrin is a lipid A binding protein, *Infect. Immun.,* 62: 2628–2632, 1994.

234. Elass-Rochard, E. et al. Lactoferrin–LPS interactions: involvement of the 28–34 loop region of human lactoferrin in the high-affinity binding to *Escherichia coli* 055: B5 lipopolysaccharides, *Biochem. J.,* 312: 839–845, 1995.

235. Crouch, S.M., Slater, K.J., and Flechter, J. Regulation of cytokine release from mononuclear cells by the iron-binding protein lactoferrin, *Blood,* 80: 235–240, 1992.

236. Mattsby-Baltzer, I. et al. Lactoferrin or a fragment thereof inhibits the endotoxin-induced interleukin-6 response in human monocytic cells, *Pediatr. Res.,* 40: 257–261, 1996.

237. Elass-Rochard, E. et al. Lactoferrin inhibits the endotoxin interaction with CD-14 by competition with the lipopolysaccharide-binding protein, *Infect. Immun.,* 66: 486–491, 1998.

238. Zagulski, T. et al. Lactoferrin can protect mice against a lethal dose of *Escherichia coli* in experimental infection *in vivo, Brit. J. Exp. Pathol.,* 70: 697–704, 1989.

239. Lee, W.J. et al. The protective effects of lactoferrin feeding against endotoxin lethal schock in germ-free piglets, *Infect. Immun.,* 66: 1421–1426, 1998.

239a. Arao, S. et al. Measurement of urinary lactoferrin as a marker of urinary tract infection, *J. Clin. Microbiol.,* 37: 553–557, 1999.

240. Osgood, E.E. A unifying concept of the etiology of the leukemias, lymphomas, and cancers, *J. Natl. Cancer Inst.,* 18: 155–166, 1957.

241. Baker, F.L., Broxmeyer, H.E., and Galbraith, P.R. Control of granulopoiesis in man. III. Inhibition of colony formation by dense leukocytes, *J. Cell Physiol.,* 86: 337–342, 1975.

242. Broxmeyer, H.E., Moore, M.A., and Ralph, P. Cell-free granulocyte colony inhibiting activity derived from human polymorphonuclear neutrophils, *Exp. Hematol.,* 5: 87–102, 1977.

243. Broxmeyer, H.E. Inhibition *in vivo* of mouse granulopoiesis by cell free activity derived from human polymorphonuclear neutrophils, *Blood,* 51: 889–901, 1978.

244. Mendelsohn, N. et al. Isolation of a granulocyte colony inhibitory factor derived from human polymorphonuclear neutrophils, *Biochim. Biophys. Acta,* 533: 238–247, 1978.

245. Broxmeyer, H.E., Smithyman, A., and Eger, R.R. Identification of lactoferrin as granulocyte-derived inhibitor of colony-stimulating activity production, *J. Exp. Med.,* 148: 1052–1067, 1978.

246. Broxmeyer, H.E. Lactoferrin acts on Ia-like antigen-positive subpopulations of human monocytes to inhibit production of colony stimulatory activity *in vitro, J. Clin. Invest.,* 64: 1717–1720, 1979.

247. Zucali, J.R., Broxmeyer, H.E., and Ulatowski, J.A., Specificity of lactoferrin as an inhibitor of granulocyte–macrophage colony-stimulating activity production from fetal mouse liver cells, *Blood,* 54: 951–954, 1979.
248. Broxmeyer, H.E. and Platzer, E. Lactoferrin acts on IA and I-E/c antigen positive subpopulations of mouse peritoneal macrophages in the absence of T-lymphocytes and other cell types to inhibit production of granulocyte-macrophage colony stimulatory factors *in vitro, J. Immunol.,* 133: 306–314, 1984.
249. Zucali, J.R. et al. Lactoferrin decreases monocyte induced fibroblast production of myeloid colony stimulating activity by suppressing monocyte release of interleukin-1, *Blood,* 74: 1531–1536, 1989.
250. Crouch, S.P.M., Slater, K.J., and Fletcher, J. Regulation of cytokine release from mononuclear cells by the iron-binding protein lactoferrin, *Blood,* 80: 235–240, 1992.
251. Penco, S. et al. Lactoferrin down-modulates the activity of the granulocyte macrophage colony-stimulating factor promoter in interleukin-1b-stimulated cells, *J. Biol. Chem.,* 270: 12263–12268, 1995.
252. Gentile, P. and Broxmeyer, H.E. Suppression of mouse granulopoiesis by administration of human lactoferrin *in vivo* and the comparative action of human transferrin, *Blood,* 61: 982–993, 1983.
253. Broxmeyer, H.E. et al. The opposing action *in vivo* on murine myelopoiesis of purified preparations of lactoferrin and the colony stimulating factors, *Blood Cells,* 13: 31–48, 1987.
254. Broxmeyer, H.E. et al. The comparative effects *in vivo* of recombinant murine IL-3, natural murine colony stimulating factor-1 and recombinant murine granulocyte-macrophage colony stimulating factor on myelopoiesis in mice, *J. Clin. Invest.,* 79: 721–730, 1987.
255. Broxmeyer, H.E., Williams, D.E., and Hangoc, G. Synergistic myelopoietic actions *in vivo* of combinations of purified natural murine colony stimulating factor-1, recombinant murine interleukin-3, and recombinant murine granulocyte–macrophage colony stimulating factor administered to mice, *Proc. Natl. Acad. Sci. U.S.A.,* 84: 3871–3875, 1987.
256. Gentile, P. and Broxmeyer, H.E. Interleukin-6 ablates the accessory cell-mediated suppressive effects of lactoferrin on human hematopoietic progenitor cell proliferation *in vitro, Ann. N.Y. Acad. Sci.,* 628: 74–86, 1991.
257. Hansen, N.E., Malquist, J., and Thorell, J. Plasma myeloperoxidase and lactoferrin measured by radioimmunoassay: relationship to neutrophil kinetics, *Acta Med. Scand.,* 198: 437–443, 1975.
258. Olofsson, T. et al. Serum myeloperoxidase and lactoferrin in neutropenia, *Scand. J. Haematol.,* 18: 73–80, 1977.
259. Brown, R.D., Rickard, K.A., and Kronenberg, H. Immunoradiometric assay of plasma lactoferrin, *Pathology,* 15: 27–31, 1983.
260. Brown, R.D., Rickard, K.A., and Kronenberg, H. Early detection of granulocyte regeneration after marrow transplantation by plasma lactoferrin, *Transplantation,* 37: 423–434, 1984.
261. Baynes, R.D. et al. Relationship of plasma lactoferrin content to neutrophil regeneration and bone marrow infusion, *Scand. J. Haematol.,* 36: 79–84, 1986.
262. Oberg, G. and Venge, P. Bone marrow regeneration after therapy-induced hypoplasia monitored by serum measurements of lactoferrin, lysozyme and myeloperoxidase, *Scand. J. Haematol.,* 37: 130–136, 1986.
263. Suzuki, T. et al. Plasma lactoferrin levels after bone marrow transplantation monitored by a two-site enzyme immunoassay, *Clin. Chim. Acta,* 202: 111–118, 1992.

264. Baiocchi, G. et al. Autologous stem cell transplantation: sequential production of hematopoietic cytokines underlying granulocyte recovery, *Cancer Res.*, 55: 1294–1303, 1993.
265. Testa, U. et al. Autologous stem cell transplantation: exogenous granulocyte colony-stimulating factor or granulocyte–macrophage colony-stimulating factor modulate the endogenous cytokine levels, *Blood*, 89: 2615–2617, 1997.
266. Broxmeyer, H.E., Mendelson, N., and Moore, M.A.S. Abnormal granulocyte feedback regulation of colony stimulating activity-producing cells from patients with chronic myelogenous leukemia, *Leuk. Res.*, 1: 3–12, 1977.
267. Broxmeyer, H.E. et al. A subpopulation of human polymorphonuclear neutrophils contains an active form of lactoferrin capable of inhibiting production of granulo-cyte–macrophage colony stimulating activities by human monocytes, *J. Immunol.*, 125: 903–909, 1980.
268. Broxmeyer, H.E. et al. Lactoferrin, transferrin and acidic isoferritins. Regulatory molecules with potential therapeutic value in leukemia, *Blood Cells*, 9: 83–105, 1983.
269. Broxmeyer, H.E. et al. Qualitative functional deficiency of affinity-purified lactoferrin from neutrophils of patients with chronic myelogenous leukemia, and lactoferrin/H-ferritin-cell interactions in a patient with lactoferrin-deficiency with normal numbers of circulating leukocytes, *Pathobiology*, 59: 26–35, 1991.
270. Broxmeyer, H.E. Suppressor cytokines and regulation of myelopoiesis, *Am. J. Ped. Hematol. Oncol.*, 14: 22–30, 1992.
271. Breton-Gorius, J. et al. Lactoferrin deficiency as a consequence of a lack of specific granules in neutrophils from a patient with a recurrent infections, *Am. J. Pathol.*, 99: 413–428, 1980.
272. He, J. and Furmanski, P. Sequence specificity and transcriptional activation in the binding of lactoferrin to DNA, *Nature*, 373: 721–724, 1995.
273. Bezault, J. et al. Human lactoferrin inhibits growth of solid tumors and development of experimental metastases in mice, *Cancer Res.*, 54: 2310–2312, 1994.
274. Damiens, E. et al. Effects of human lactoferrin on NK cell cytotoxicity against hae-matopoietic and epithelial tumour cells, *Biochim. Biophys. Acta*, 1402: 277–287, 1998.
275. Shou, H., Kim, A., and Golub, S.H. Modulation of natural killer and lymphokine-activated killer cell cytotoxicity by lactoferrin, *J. Leukocyte Biol.*, 51: 343–349, 1992.
276. Sekine, K., Watanabe, E., and Tsuda, H. Inhibition of azoxymethane-initiated colon tumor by bovine lactoferrin administration in F344 rats, *Jpn. J. Cancer Res.*, 88: 523–526, 1997.
277. Yoo, Y.C. et al. Bovine lactoferrin and lactoferricin, a peptide derived from bovine lactoferrin, inhibit tumor metastasis in mice, *Jpn. J. Cancer Res.*, 88: 184–190, 1997.
278. Yoo, Y.C. et al. Apoptosis in human leukemic cells induced by lactoferricin, a bovine milk protein derived peptide: involvment of reactive oxygen species, *Biochem. Biophys. Res. Commun.*, 237: 624–628, 1997.
279. Mason, D.Y. and Taylor, C.R. Distribution of transferrin, ferritin and lactoferrin in human tissues, *J. Clin. Pathol.*, 31: 316–327, 1983.
280. Mason, D.Y., Farrel, C., and Taylor, C.R. The detection of intracellular antigens in human leukocytes by immunoperoxidase staining, *Brit. J. Haematol.*, 31: 361–370, 1975.
281. Pryzwansky, K.B. et al. Immunocytochemical distinction between primary and secondary granule formation in developing human neutrophils: correlations with Romanovsky stains, *Blood*, 53: 179–185, 1979.
282. Miyauchi, J. et al. Lactoferrin-deficient neutrophil polymorphonuclear leukocytes in leukemias: a semiquantitative and ultrastructural cytochemical method, *J. Clin. Pathol.*, 36: 1397–1405, 1983.

283. Cramer, E. et al. Ultrastructural localization of lactoferrin and myeloperoxidase in human neutrophils by immunogold, *Blood,* 65: 423–432, 1985.
284. Myiauchi, J. and Watanabe, Y. Immunocytochemical localization of lactoferrin in human neutrophils, *Cell Tissue Res.,* 247: 249–258, 1987.
285. Estensen, R.D., White, J.G., and Holmes, B. Specific degranulation of human polymorphonuclear leukocytes, *Nature,* 248: 347–349, 1974.
286. Kjeldsen, L. et al. Structural and functional heterogeneity among peroxidase-negative granules in human neutrophils: identification of a distinct gelatinase containing granule subsets by combined immunocytochemistry and subcellular fractionation, *Blood,* 82: 3183–3191, 1993.
287. Sengelov, H., Kjeldsen, L., and Borregaard, N. Control of exocytosis in early neutrophil activation. *J. Immunol.,* 150: 1535–1543, 1993.
288. Borregaard, N. et al. Biosynthesis of granule proteins in normal human bone marrow cells. Gelatinase is a marker of terminal neutrophil maturation, *Blood,* 85: 812–817, 1995.
289. Arnlijots, K. et al. Timing, targeting and sorting of azurophil granule proteins in human myeloid cells, *Leukemia,* 12: 1789–1795, 1998.
290. Rado, T.A. et al. Lactoferrin biosynthesis during granulocytopoiesis, *Blood,* 64: 1103–1109, 1984.
291. Olsson, I. et al. Biosynthesis and processing of lactoferrin in bone marrow cells, a comparison with processing of myeloperoxidase, *Blood,* 71: 441–447, 1988.
292. Khanna-Gupta, A. et al. CCAAT displacement protein (CDP/cut) recognizes a silencer element within the lactoferrin gene promoter, *Blood,* 90: 2784–2795, 1997.
293. Valtieri, M. et al. Cytokine-dependent granulocyte differentiation. Regulation of proliferative and differentiative responses in a murine progenitor cell line, *J. Immunol.,* 138: 3829–3837, 1987.
294. Berliner, N. Moelcular biology of neutrophil differentiation, *Curr. Opin. Hematol.,* 5: 49–53, 1998.
295. Greuber, T., Johnston, J., and Berliner, N. Cloning and expression of cDNA encoding mouse neutrophil gelatinase: demonstration of coordinate secondary granule protein gene expression during terminal neutrophil maturation, *Blood,* 82: 3192–3197, 1993.
296. Lawson, N.D., Khanna-Gupta, A., and Berliner, N. Isolation and characterization of the cDNA for mouse neutrophil collagenase: demonstration of shared negative regulatory pathways for neutrophil secondary granule protein gene expression, *Blood,* 91: 2517–2524, 1998.
297. Strauss, R.G. et al. An anomaly of neutrophil morphology with impaired function, *New Engl. J. Med.,* 290: 478–484, 1974.
298. Gallin, J.I. Neutrophil specific granule deficiency, *Annu. Rev. Med.,* 36: 263–274, 1985.
299. Boxer, L.A. et al. Lactoferrin deficiency associated with altered granulocyte function, *New Engl. J. Med.,* 307: 404–410, 1982.
300. Falloon, J. and Gallin, J.I. Neutrophil granules in health and disease, *J. Allergy Clin. Immunol.,* 77: 653–662, 1986.
301. Gallin, J.I. et al. Human neutrophil-specific granule deficiency: a model to assess the role of neutrophil-specific granules in the evolution of inflammatory response, *Blood,* 59: 1317–1329, 1982.
302. Borregaard, N.L. et al. Anomalous neutrophil granule distribution in a patient with lactoferrin deficiency, *Am. J. Hematol.,* 18: 225–230, 1985.
303. Ohno, Y., Seligman, B.E., and Galin, J.I. Cytochrome b translocation to human neutrophil plasma membranes and superoxide release, *J. Biol. Chem.,* 260: 2409–2414, 1985.

304. Lomax, K.J. et al. Selective defect in myeloid cell lactoferrin gene expression in neutrophil specific granule deficiency, *J. Clin. Invest.*, 83: 514–519, 1989.
305. Johnston, J.J., Boxer, L.A., and Berliner, N. Correlation of messenger RNA levels with protein defects in specific granule deficiency, *Blood*, 80: 2088–2091, 1992.
306. Miyauchi, J. et al. Lactoferrin-deficient neutrophil polymorphonuclear leucocytes in leukemias: a semiquantitative and ultrastructural cytochemical study, *J. Clin. Pathol.*, 36: 1397–1405, 1983.
307. Davey, F.R. et al. Immunophenotyping of acute myeloid leukemia by immuno-alkaline phosphatase(APAAP) labeling with a panel of antibodies, *Am. J. Haematol.*, 26: 157–166, 1987.
308. Schofield, K.P., Stone, C.W., and Stuart, J. Quantitative cytochemistry of blood neutrophils in acute myeloid leukemia, *Brit. J. Haematol.*, 54: 261–268, 1983.
309. Miyauchi, J. et al. Neutrophil secondary granule deficiency as a hallmark of all trans-retinoic acid-induced differentiation of acute promyelocytic leukemia cells, *Blood*, 90: 803–813, 1997.
310. Khanna-Gupta, A. et al. NB-4 cells show bilineage potential and an aberrant pattern of neutrophil secondary granule protein gene expression, *Blood*, 84: 294–302, 1994.
311. Patel, G. et al. V-myb blocks granulocyte colony-stimulating factor-induced myeloid cell differentiation but not proliferation, *Mol. Cell. Biol.*, 13: 2269–2276, 1993.
312. Lonnerdal, B. Iron and breast milk, in *Iron Nutrition in Infancy and Childhood*, Stekel, A., Ed., New York, Raven Press, 95, 1984.
313. Goldman, A.S., Chedda, S., and Garofalo, R. Evolution of immunologic functions of the mammary gland and the postnatal development of immunity, *Pediatr. Res.*, 43: 155–162, 1998.
314. Sanchez, L.M. et al. Identification of the major protein components in breast secretions from women with benign and malignant breast diseases, *Cancer Res.*, 52: 95–100, 1992.
315. Green, M.R. and Pastewka, J.V. Lactoferrin is a marker for prolactin response in mouse mammary explants, *Endocrinology*, 103: 1510–1513, 1978.
316. Close, M.J. et al. Lactoferrin expression in mammary epithelial cells is mediated by changes in cell shape and actin cytoskeleton, *J. Cell Sci.*, 110: 2861–2871, 1997.
317. Kawakami, H., Hiratsuka, M., and Dosako, S. Effects of iron-saturated lactoferrin on iron absorption, *Agric. Biol. Chem.*, 52: 903–912, 1988.
318. Frausson, G.B., Keen, C.L., and Lonnerdal, B. Supplementation of milk with iron bound to lactoferrin using weanling mice. I. Effects of hematology and tissue iron, *J. Pediatr., Gastroenterol Nutr* 2: 693–701, 1983.
319. Fransson, G.B. and Lonnerdal, B. Iron in human milk, *J. Pediatr.*, 96: 380–388, 1980.
320. De Laey, P., Masson, P.L., and Heremans, J.F. The role of lactoferrin in iron absorption, *Protides Biol. Fluids*, 16: 627–635, 1968.
321. McMillan, J.A. et al. Iron absorption from human milk, simulated human milk, and proprietary formulas, *Pediatrics*, 60: 896–901, 1977.
322. De Vet, B.J. and Van Gool, J. Lactoferrin and iron absorption in the small intestine, *Acta Med. Scand.*, 196: 393–402, 1974.
323. Jackson, L.S. and Lee, K. The effect of dairy products on iron availability, *Crit. Rev. Food Sci. Nutrit.*, 31: 259–270, 1992.
324. Tourville, D.R. et al. The human reproductive tract: immunohistological localization of gA, gG, gM, secretory "piece" and lactoferrin, *Am. J. Obstet. Gynecol.*, 108: 1102–1108, 1970.
325. McMaster, M.T. et al. Lactoferrin in the mouse uterus: analyses of preimplantation period and regulation by ovarian steroids, *Mol. Endocrinol.*, 6: 101–111, 1992.

326. Kelver, M.E. et al. Estrogen regulation of lactoferrin expression in human endometrium, *Am. J. Reprod. Immunol.,* 36: 243–247, 1996.

327. Kumar, V. and Chambon, P. The estrogen receptor binds tightly to its responsive element as ligand-induced homodimer, *Cell,* 55: 145–156, 1988.

328. Liu, Y. and Teng, C.T. Estrogen response module of the mouse lactoferrin gene contains overlapping chicken ovalbumin upstream promoter transcription factor and estrogen receptor-binding elements, *Mol. Endocrinol.,* 6: 355–364, 1992.

329. Korach, K.S. et al. Estrogen receptor gene disruption: molecular characterization and experimental and clinical phenotypes, *Recent Progr. Horm. Res.,* 51: 159–186, 1996.

330. Das, S.K. et al. Estrogenic responses in estrogen receptor-alpha deficient mice reveal a distinct estrogen signaling pathway, *Proc. Natl. Acad. Sci. U.S.A.,* 94: 12786–12791, 1997.

331. Das, S.K. et al. Differential spatiotemporal regulation of lactoferrin and progesterone receptor genes in the mouse uterus by primary estrogen, catechol estrogen, and xenoestrogen, *Endocrinology,* 139: 2905–2915, 1998.

332. Tibbetts, T.A. et al. Mutual and intercompartmental regulation of estrogen receptor and progesterone receptor in the mouse uterus, *Biol. Reprod.,* 59: 1143–1152, 1998.

333. Charpin, C. et al. Localization of lactoferrin and nonspecific cross-reacting antigen in human breast carcinomas, *Cancer,* 55: 2612–2617, 1985.

334. Cambell, T. et al. Expression of lactoferrin in normal and malignant human breast, *Proc. Am. Assoc. Cancer Res.,* 31: 209–229, 1990.

335. Das, M.R. et al. Human milk samples from different ethnic samples from different ethnic groups contain RNase that inhibits, and plasma membrane that stimulates, reverse transcription, *Nature,* 262: 802–805, 1976.

336. Padhy, L.C. et al. Role of plasma membranes in stimulation of RNA-directed DNA synthesis, *Nature,* 262: 805–807, 1976.

337. Walmer, D.K. et al. Malignant transformation of the human endometrium is associated with overexpression of lactoferrin messenger RNA and protein, *Cancer Res.,* 55: 1168–1175, 1995.

338. Farley, J. et al. Neoplastic transformation of the endocervix associated with down-regulation of lactoferrin expression, *Mol. Carcinog.,* 20: 240–250, 1997.

339. Tauber, P.F. et al. Components of human split ejaculates, *J. Reprod. Fertil.,* 43: 249–267, 1975.

340. Thaler, C.J. et al. Lactoferrin binding molecules in human seminal plasma, *Biol. Reprod.,* 43: 712–717, 1990.

341. Wichmann, L., Vaalasti, T., and Tuchimaa, P. Localization of lactoferrin in the male reproductive tract, *Int. J. Androl.,* 12: 179–186, 1989.

342. Reese, J.H. et al. Distribution of lactoferrin in the normal and inflamed human prostate: an immunohistochemical study, *Prostate,* 20: 73–85, 1992.

343. Yu, L.C. and Chen, Y.H. The developmental profile of lactoferrin in mouse epididymis, *Biochem. J.,* 296: 107–111, 1993.

344. Jin, Y.Z. et al. Direct evidence for the secretion of lactoferrin and its binding to sperm in the porcine epididymis, *Mol. Reprod. Develop.,* 47: 490–496, 1997.

345. Hekman, A. and Rumpke, P. The antigens of human seminal plasma, with special reference to lactoferrin as a spermatozoa-coating antigen, *Fertil. Steril.,* 20: 312–323, 1969.

346. Robert, T.K. and Boettcher, B.L. Identification of sperma-coating antigen, *J. Reprod. Fertil.,* 18: 347–350, 1969.

347. Li, T.S. and Shulman, S. Immunoelectrophoretic analysis of human seminal plasma fractions after fractionations by various methods, *Int. J. Fertil.,* 16: 87–100, 1971.

348. Goodman, S.A. and Young, L.G. Immunological identification of lactoferrin as shared antigen on radioiodinated human sperm surface and in radioiodinated human seminal plasma, *J. Reprod. Immunol.,* 3: 99–108, 1981.
349. Broer, K.H., Dauber, U., and Hirschlauser, C. The presence of lactoferrin (SCA) on human sperm head during *in vitro* penetration through cervical mucus, *IRCS Med. Sci.,* 5: 362–363, 1977.
350. Rodman, T.C. et al. An innate natural antibody is reactive with a cryptic sequence of lactoferrin exposed on sperm head surface, *Proc. Soc. Exp. Biol. Med.,* 216: 404–409, 1997.
351. Buckett, W.M. et al. Seminal plasma lactoferrin concentrations in normal and abnormal semen samples, *J.SAndrol.,* 18: 302–304, 1997.
352. Waed, P.P. et al. Restricted spatiotemporal expression of lactoferrin during murine embryogenesis, *Adv. Exp. Med. Biol.,* 443: 91–100, 1998.
353. Geng, K. et al. Induction of lactoferrin expression in murine ES cells by retinoic acid and estrogen, *Exp. Cell Res.,* 245: 214–220, 1998.
354. Newbold, R.R., Hanson, R.B., and Jefferson, W.N. Ontogeny of lactoferrin in the developing mouse uterus: a marker of early hormone response, *Biol. Reprod.,* 56: 1147–1157, 1997.

4 Transferrin

CONTENTS

4.1 Introduction 143
4.2 Comparative Analysis of the Structures of Transferrin Family Proteins....151
4.3 Transferrin Genetic Variants 155
    4.3.1 Transferrin C Alleles...156
    4.3.2 Transferrin B and D Alleles...157
    4.3.3 Association between Transferrin Alleles and Disease 157
4.4 Biochemical Properties of Serum Transferrin...158
    4.4.1 General Properties and Basic Structure...158
    4.4.2 Structural Basis of Transferrin/Transferrin Receptor Interaction....174
    4.4.3 Glycosylation of Transferrin..176
4.5 Transferrin Gene Expression 179
    4.5.1 Regulation of Transferrin Synthesis in Liver, Testis, and
           Brain: Cellular Studies...182
    4.5.2 Regulatory Sequences Involved in the Regulation of Transferrin
           Gene Expression Observed in Different Tissues...........................187
4.6 Transferrin and Microbiology...195
4.7 Transferrin as a Growth Factor...208
4.8 Congenital Atransferrinemia 218
4.9 Transferrin as a Transporter of Drugs 219
References..222

4.1 INTRODUCTION

Since Fe^{3+} is the molecular iron form observed in oxygenized fluids at neutral pH under physiological conditions, the iron bioavailability to both animal and plant cells is greatly complicated. This complication is essentially related to two main phenomena: (1) extremely low solubility of ferric ions in oxygenized fluids at neutral pH; and (2) the very low permeability of cell membranes to iron in its trivalent state.

In aqueous solution, the two principal oxidation states of iron, ferrous (Fe^{2+}) and ferric (Fe^{3+}) exist, respectively, as the hydrated ions, $Fe(H_2O)_6^{2+}$ and $Fe(H_2O)_6^{3+}$. In the absence of oxygen or other oxidants, ferrous salts give solutions of ferrous aqua ion $Fe(H_2O)_6^{2+}$. In contrast, between pH 5 and 9, ferric salts lose protons (hydrolyze) and polymerize, precipitating hydrated ferric oxides.

In biological media as in neutral waters, hydrated ferrous ions are real species, whereas hydrated ferric ions are rare except at low pH.

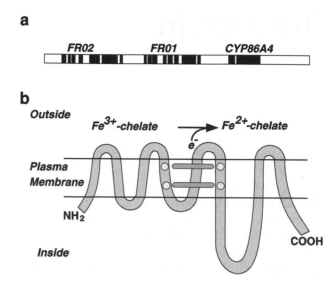

FIGURE 4.1 (a) Genomic organization of plant ferri-chelate reductase genes (FR01 and FR02). (b) Hypothetical structure of plasma membrane-associated FR02 protein. Four histidine residues (dots) coordinate two intramembrane heme groups (bars). (Reprinted with modifications from Robinson, N.J. et al., *Nature,* 397: 694–697, 1999. With permission.)

To overcome this problem, biological systems have developed the capacity to synthesize proteins or other compounds able to bind iron and transport it to the cells. Thus, in primitive biological systems, transport is carried out through organic chelates such as ferric citrate, or ferric galate for plants, and through siderophore chelates for bacteria. Transferrin supplies iron to animal cells and a few protozoa and bacteria. For certain bacteria and protozoa, transferrin and, to some extent, ferritin supply iron to the cells.

Iron uptake mechanisms used by plants to acquire iron from soils have been recently clarified. Low availability of iron limits plant growth because iron forms insoluble ferroxides, leaving only a small, organically-complexed fraction in soil solutions. The enzyme ferric-chelate reductase is required for the majority of plants to acquire soluble iron.

Ferric chelate reductase enzymes belong to a superfamily of flavocytochromes that transport electrons across membranes. They possess intramembranous binding sites for heme and cytoplasmic binding sites for nucleotide cofactors that donate and transfer electrons. Two closely located ferric-chelate reductase genes, FR01 and FR02, are observed at the level of the plant genome[1] (Figure 4.1). The expression of ferric-chelate reductase is greatly stimulated when iron is deficient.

Plants are usually exposed to limiting concentrations of iron in soils. They at times must face stress conditions due to heavy metal contamination. This condition, as well as high or low temperatures, water restriction, wounding, and infection by viruses, bacteria or fungi, determines the rapid accumulation of oxygen-reactive species that play a major role in the damage of crop plants. Production of transgenic plants that ectopically express iron-binding protein resulted in the development of plants tolerant to oxidative damage and pathogens.[2]

Similar mechanisms to acquire iron from ocean water have been developed by phytoplankton. Dissolved iron is present in very low concentrations in sea water (10^{-10} moles/kg), largely bound to organic compounds. To face this very low iron availability, phytoplankton developed a system to obtain iron to sustain energetic metabolism. In fact, fresh water diatoms, members of a eukaryotic, photosynthetic, unicellular group of organisms, acquire iron through the use of non-specific cell-surface enzyme (ferrireductase) to liberate inorganic Fe^{2+} from the porphyrin complex present in sea water.[2a] However, prokaryotic phytoplankton (i.e., cyanobacteria) acquire iron through a different mechanism involving cell-surface siderophore receptors, probably comparable to those observed in other bacteria.[2a]

Transferrins are members of a family of glycoproteins present in the biological fluids of vertebrates and invertebrates that exhibit the important property of reversibility in binding iron.[3,4] Transferrins are formed by a group of evolutionary related proteins:[5]

1. Transferrin (siderophilin), a metal-binding beta-globulin present in vertebrate blood plasma and in other extracellular fluids (including cerebrospinal fluid, milk, semen, and amniotic fluid).
2. Ovotransferrin (conalbumin) in bird and reptile oviduct secretions and egg whites.
3. Lactoferrin (or lacto-transferrin) of mammalian extracellular secretions (milk, pancreatic juice, and tears) and present intracellularly in leukocytes.
4. An integral plasma membrane protein of human malignant melanoma cells and some fetal tissues known as p97.
5. B-lym, a transforming protein of 65 amino acid residues which shares a high degree of sequence identity with the amino-terminal domains of other transferrin sequences including ovotransferrin, lactoferrin, and p97; the region of homology is encoded entirely by exon 2 in the human transferrin gene.
6. Invertebrate iron-binding proteins in the blood of some crustaceae and urochordates.

Transferrin plays a key role in mediating and regulating the transport of iron from the site of absorption (intestinal mucosa) to the sites of utilization by virtually all tissues, particularly those dividing actively, and bone marrow (where transferrin is highly bound by developing erythroid cells that need large amounts of iron to sustain high rates of heme and hemoglobin synthesis).[5] The main physiological role of transferrin is providing iron required to sustain erythropoiesis and, particularly, for hemoglobin production. Mice with hereditary hypotransferrinemia are markedly anemic[6]; mice made deficient in transferrin receptor gene by gene targeting are markedly anemic.[7] These observations clearly indicate that transferrin is mainly required for normal erythroid development and transferrin cannot be replaced by other iron transporters for that function.

The particular need of all dividing tissues for transferrin is largely related to the absolute need for iron to sustain the activity of ribonucleotide reductase, the only enzyme converting ribonucleotides in deoxyribonucleotides. These *in vivo* observations are also confirmed by *in vitro* studies showing that transferrin is a fundamental constituent of serum-free culture media.

Absolute need for a plasma iron transporter extends to vertebrates and chordates, since all these animals carry transferrin in their blood, hemolymph, or comparable biological fluids. A series of studies have shown that primitive vertebrates, such as the lamprey,[8] hagfish,[9,10] and the tench, *Tinca tinca* Linnaeus[11] have transferrin-like molecules with molecular masses of approximately 78 kDa. The lamprey *Geotria australis* Grey possesses a peculiar tetrameric form built up by four 78-kDa subunits with two iron-binding sites per subunit.[12] *Xenopus* species have also been shown to possess the capacity to synthesize transferrin-like molecules with a molecular weight of about 85 kDa.[13]

Iron-binding proteins similar to transferrin were also observed in invertebrates. Iron-binding proteins from a crab (150 kDa) and a tarantula (80 to 100 kDa) have been described[14,15] and a transferrin-like protein was isolated from a tunicate, *Pyura stolonifera*.[16]

Although all vertebrate transferrins are 80 kDa and bind two ferric ions, the *Pyura* protein is 40 kDa and binds a single ferric ion.[16] An iron-binding protein of 77 kDa has been isolated and carefully characterized for the adult sphinx moth *Manduca sexta*.[17] Molecular studies have shown that:[17]

1. This protein binds ferric ions *in vivo* and *in vitro* and has a secondary structure similar to those of human serum transferrin and human lactoferrin.
2. Sequence comparison showed that this protein possesses significant similarity to the vertebrate transferrins, with the greatest areas of similarity around the two iron-binding sites, although the insect protein seems to contain only one such functional site.

A transferrin protein was also identified and sequenced in another insect, the cockroach *Blaberus discoidalis*.[18] The study of optical and electron paramagnetic resonance spectroscopic properties showed that transferrin from the *Blaberus discoidalis* displayed high similarity to vertebrate transferrin: (1) iron binding dependent on bicarbonate anion binding; (2) tyrosyl residues involved in iron binding; and (3) pH-dependent release of iron.[19] These observations suggest that cockroach transferrin behaves as an authentic member of the transferrin superfamily.

The comparative analysis of the amino acid structure of *Blaberus discoidalis* transferrin as compared to that of *Manduca* and of humans revealed several interesting findings:[18]

1. One-hundred twenty-one residues are conserved in the three proteins and, particularly, cysteine residues at 19 positions.
2. In the N terminal lobe, three of the four amino acid (human proteins as references: Asp 63, Tyr 95, Tyr 188, and His 249) involved in the iron binding are conserved in both the insect proteins.

3. In the C terminal lobe, all four ligands to iron (human proteins as references: Asp 387, Tyr 426, Tyr 517, and His 585) are maintained in the cockroach sequence but, notably, Tyr 517 and His 585 are not represented in the monoferric *Manduca sexta* transferrin.

4. The anion-binding amino acids present in vertebrate transferrins (Arg 124 and Arg 456 in human serum transferrin) are mirrored by Arg 25 and Arg 471 in the cockroach protein, whereas the second arginine residue has been replaced by a neutral residue in *Manduca sexta*.

The possible physiological significance of one- versus two-sited iron-binding proteins in insects remains to be determined. These observations suggest that the insect and vertebrate transferrins descend from a sequence that was present in the last common ancestor of vertebrates and arthropods. These observations also suggest that transferrins are maintained in many insects, since the progenitor of moths and cockroaches gave rise to most of the present-day insect orders. Thus, the ensemble of the observations on transferrins of different animal species indicates that this protein was relatively well conserved during evolution (Figure 4.2).

Tranferrin is one of the most studied human plasma proteins. Based on early studies, the iron in plasma is associated with a protein fraction.[20] Transferrin was later isolated by Schade and Caroline[21] and shown to be the major iron-carrying protein in blood plasma.

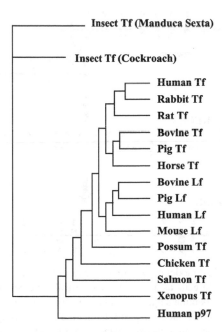

FIGURE 4.2 Phylogenetic analysis of transferrins and lactoferrins. The phylogenetic tree was constructed according to the sequences of the different proteins of the transferrin family. (Redrawn with modifications from Demmer, J. et al., *Biochim. Biophys. Acta,* 1445: 65–74, 1999.)

Transferrins belong to a family of evolutionarily related proteins. Analysis of the amino acid sequences of several members of this protein family indicates that these proteins are composed of two homologous regions corresponding to the NH_2-terminal and COOH-terminal moieties.[22] X-ray crystallographic studies of rabbit serum transferrin and human lactoferrin have shown that the two proteins are folded in two globular domains corresponding to the sequence homology regions.[23,24] These two domains show essentially the same polypeptide chain folding, with an iron-binding site present in each lobe.

Analysis of the organization of human transferrin and of the ovotransferrin genes indicates that the segments encoding the two protein domains are composed of similar numbers of exons. Moreover, introns interrupt the coding sequence, giving rise to homologous exons of similar size in the 5' and 3' regions of the genes. These observations strongly suggest that the present-day transferrins arose by gene duplication, and it has been proposed that this gene family evolved by a series of independent gene duplications from a common ancestor derived from a primordial gene by internal duplication[25] (Figures 4.3 and 4.4).

Subsequent studies of characterization of additional insect transferrins led to additional hypotheses about the function of these molecules. A protein of 65 kDa was purified from the hemolymph of adult female *Sarcophaga peregrina* flies.[26] The analysis of the cDNA encoding this protein, as well as the inspection of its deduced amino acid sequence, showed that this protein is a transferrin; this transferrin binds

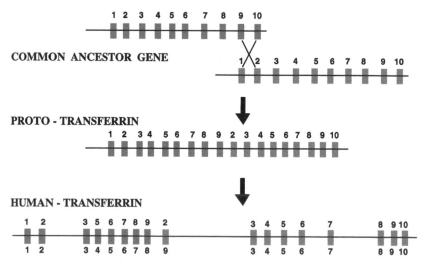

FIGURE 4.3 A hypothetical model to explain the origin of the present-day transferrin genes. Using the human transferrin gene as a model, an ancestral gene composed of 10 exons, with exon 1 coding for a signal peptide sequence and exon 10 coding the for 3' region of the gene is hypothesized. This common ancestor gene duplicated as a consequence of an intragenic cross-over, generating a duplicated ancestor gene (proto-transferrin). During evolution, exon 4 was deleted on the 5' side of the gene and homologous exons evolved independently, leading to the present-day human transferrin gene with its 17 exons. (Redrawn with modifications from Park, I. et al., *Proc. Natl. Acad. Sci. U.S.A.*, 82: 3149–3153, 1985.)

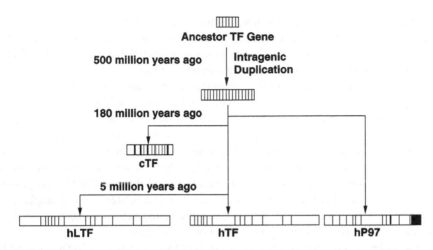

FIGURE 4.4 Evolution of the transferrin family proteins from a common primordial ancestor gene.

only one atom of iron/molecule in its N terminal lobe, while the C terminal lobe was unable to bind iron.[26] *Sarcophaga* transferrin was found to transport iron into eggs and it was thus suggested that its main function is equivalent to an intercellular iron transporter during embryogenesis.

In another study, a 66-kDa protein secreted from cells derived from the mosquito *Aedes aegypti*, was found to have similarity to insect transferrin.[27] The analysis of the amino acid sequence of mosquito transferrin showed that it conserves iron-binding residues in the N terminal lobe, but not in C terminal lobe, and is thus able to bind only one atom/molecule.[27] The synthesis of mosquito transferrin is up-regulated when mosquito cells are incubated with killed bacterial cells or when *Aedes aegypti* encapsulates filarial worms.[27]

These observations suggest that mosquito transferrin plays a role similar to vertebrate lactoferrin in sequestering iron from invading organisms; furthermore, the peculiar structure of the C terminal lobe (which does not possess an iron-binding site and is unable to interact with transferrin receptors) might represent the structural basis for a mechanism for evading pathogens that signals transferrin receptors to tap sequestered iron.[27]

Recently, the gene encoding *Drosophila melanogaster* transferrin was cloned.[28] The deduced amino acid sequence showed that this transferrin is composed of 641 amino acids containing a signal peptide of 29 amino acids. Like other insect trans-ferrins, *Drosophila* transferrin appears to have only a single iron-containing site located in the N terminal lobe, while the C terminal lobe lacks the iron binding pocket.[28] As observed for mosquito transferrin, *Drosophila* transferrin synthesis is greatly stimulated by bacterial infection.

Particularly interesting are the observations on the transferrin-like molecules found in algae. In fact, a transferrin-like molecule was found in the unicellular green alga *Dunaliella salina*.[29,30] This 150-kDa transferrin-like molecule is associated with the plasma membrane of the unicellular alga. The analysis of the structure of this molecule showed a unique structure, distinct from other transferrins in containing three rather than two internal repeats.[29] This protein was able to bind iron and its expression increased when algae were grown in iron-deficient media.[30] The factors required for binding of iron to alga transferrin were similar to those identified for animal transferrins: requirement for carbonate–bicarbonate ions, high-specificity and affinity for Fe^{3+}, and pH dependency with low binding at acid pH.[30]

This observation is particularly important in that it shows that transferrins are also present and function in iron transport in plant systems. In fact, other observations indicate that transferrin may exist and function in iron uptake in other algae. Thus, similar iron uptake systems may exist in microalgae as well as in phytoplankton. According to these observations it may be concluded that transferrin-like proteins may contribute to ubiquituous iron uptake mechanisms in algae and phytoplankton. This mechanism is particularly relevant in view of the fact that iron availability is a limiting factor for marine phytoplankton photosynthesis.

Examples of three-lobe transferrin were found among animals. In fact, a 155-kDa proteinase inhibitor, pacifastin, from plasma of the fresh water crayfish, *Pacifasacus leniusculus*, was found to be composed of two covalently-linked subunits: (1) the heavy chain of about 105 kDa is structurally related to transferrins, containing three transferrin lobes, two of which are able to bind iron; and (2) the light chain of about 44 kDa is structurally related to low molecular weight proteinase inhibitors.[31] Two of the three lobes of the heavy chain, 1 and 3, are similar to the N terminal lobe of transferrin, and 2 is similar to the C terminal lobe.

Four of the five amino acid residues required for iron binding are conserved in lobes 2 and 3. These two lobes, but not lobe 1, are in fact able to bind iron.[32] This molecule represents the first known example of a protein that has the combined properties of a transferrin-like molecule and a proteinase inhibitor. Since transferrin under physiological conditions is only 30% saturated with iron in normal serum, there is substantial iron-binding capacity for other metal ions that enter the blood.

In addition to iron, transferrin is also capable of binding many other metal ions, such as Al^{3+}, Cr^{3+}, Mn^{2+}, Mn^{3+}, Cu^{2+}, Co^{2+}, Ga^{3+}, Zn^{2+}, and Cd^{2+}. Among these metal ions, Fe^{3+} is bound most strongly and certainly represents the only metal bound by transferrin under physiological conditions. The capacity of transferrin to bind metal ions such as Ga^{3+} and In^{3+} has been exploited in nuclear medicine for tumor radio-imaging or tumor radiotherapy using radioactive isotopes of these elements.

The strength of metal binding to transferrin seems to be dependent on the size of the metal ion, being optimal for Fe^{3+} (radius of 0.65 Å), weaker for slightly smaller, such as Ga^{3+} (0.62 Å), or slightly larger, such as In^{3+} (0.80 Å) ions and weaker for very large, e.g., lanthanide (0.86 to 1 Å) ions.

Recent studies indicate metal properties more than metal size are responsible for the binding to transferrin. In fact, it was shown that bismuth (Bi^{3+}), a large metal ion (ionic radius of 1.03 Å), binds tightly to both N and C lobes of human transferrin, with an affinity comparable to those observed for Ga^{3+} and In^{3+}. However, although Bi^{3+} binds relatively strongly to transferrin, it can be displaced by Fe^{3+}.[33]

More recent studies showed that Bi^{3+} binds preferentially at the level of the C lobe. In fact, with Fe^{3+} in the N lobe of transferrin, Bi^{3+} binds to the C lobe as strongly as it does in apotransferrin, whereas the binding of Bi^{3+} to the N lobe of transferrin with Fe^{3+} bound to the C lobe is much weaker than binding to the isolated N lobe.

Transferrin is essentially synthesized in the liver and the protein is secreted into the plasma, reaching different organs via the bloodstream. Transferrin is also synthesized in the testes (by Sertoli cells) and in the brain.

Storage and distribution of iron are essential for cell metabolism. Transferrin serves at least two major roles:

1. Transferrin in body fluids, such as serum, has an important role in bacteriostasis; because of its strong affinity for iron ($k_D = 10^{-22}$ mol^{-1}) and its low iron saturation (25 to 35%) under physiological conditions, circulating transferrin limits the supply of free iron to many bacteria that require iron for proliferation.
2. Transferrin transports iron across cell membranes for storage and for use in the synthesis of iron-containing proteins.

The roles of transferrin, transferrin receptors, and ferritin are interwined for the purpose of iron management. The transferrin–iron complex is bound by transferrin receptors in cell membranes and is endocytosed. Under acidic conditions, iron is released from endosomes to ferritin for storage or, alternatively, is released for the synthesis of other iron-containing proteins.

4.2 COMPARATIVE ANALYSIS OF THE STRUCTURES OF TRANSFERRIN FAMILY PROTEINS

As previously mentioned, four major classes of transferrins are known:

1. Serotransferrin, found in serum and extracellular fluids, which is the carrier protein for metabolic iron.
2. Lactoferrin or lactotransferrin, found in abundance in milk and in a variety of extracellular fluids and in specific granules of polymorphonuclear leukocytes; plays an important role in defense against infection.
3. Ovotransferrin, present in avian egg white, and specified by the same gene as serotransferrin, but differing from the serotransferrin of the same species only in its gylcan part; exhibits anti-bacterial properties.
4. The p97 cell-surface protein present in the majority of human melanomas, but only present in trace amounts in normal tissues; exhibits a considerable sequence homology with human serotransferrin and is also known for this sequence homology as melanotransferrin.

The amino acid sequences of human serum transferrin,[34-36] hen ovotransferrin,[37-38] human lactoferrin,[39] and melanotransferrin p97[40] have been determined. All these proteins consist of 680 to 700 amino acids that can be subdivided in two different halves exhibiting 35 to 40% homology. Each lobe is further organized into two dissimilar sub-domains of about 160 amino acids, separated by a cleft that contains the metal-binding site.

The sequences of the four transferrins show significant homologies. Two types of homologies have been identified: internal homology and homology between transferrins. Concerning the internal homology, a comparison of the two halves of the lactoferrin sequence shows 125 identical amino acid residues in corresponding positions (37%). A similar comparison of human serum transferrin and hen ovotransferrin showed 41% and 33% internal homology, respectively.

Concerning the homology between transferrins, the sequence homologies between human lactoferrin and human serum transferrin or hen ovotransferrin are 59% and 49%, respectively. Between human serum and hen ovotransferrin, the homology is 51%; between human serum transferrin and p97 melanotransferrin, the homology is 37%.

Among these homologies, particularly relevant are those concerning the localization of disulfide bridges.[41] In fact in the four transferrins, the half-cysteine residues are not randomly distributed; the majority of them are located in quite homologous positions. Thus, there are six conserved disulfide bridges in the N terminal half, all essentially in the same positions as six of the nine present in the C terminal half. The remaining three cysteine residues are also in similar positions in the four transferrins, but they are characterized because they link positions in the sequences distant from the regions containing the six highly conserved bridges.

A second major feature of the comparison of the structures of transferrins is linked to their sequence homology in the amino acid region involved in mediating iron binding. In fact, the major biologic function of trasferrins is to bind iron and to transport it into the cells. This function is mediated at a molecular level by the presence of a peculiar amino acid sequence involved in the binding of iron. Since all four transferrins are able to bind iron, the conservation of the amino acid residues involved in these binding sites is expected.

Detailed crystallographic structures are available only for human lactoferrin and rabbit serum transferrin, but the marked similarities of these structures suggest that the ligand structures of the binding sites are probably common to all transferrins.

It is believed that each transferrin lobe is able to bind one atom of iron and each of the two iron-binding sites contains two or three tyrosine residues, one or two histidine residues, and one arginyl residue (required to fix the concomitantly bound bicarbonate anion). More particularly, each site offers one aspartyl carboxylate donor oxygen, the phenolate oxygens of two tyrosines, and one histidyl nitrogen to its bound metal; the two remaining ligand atoms required to coordinate iron are provided by oxygens of the carbonate co-anion. Charge transfer from tyrosyl oxygen donors to bound iron is responsible for the typical salmon-pink color of iron-saturated transferrins, whereas the corresponding apotransferrins are colorless. The comparison of the structures of the four transferrins provided evidence in favor of the existence of one putative iron-binding site present in each lobe.

Finally, the comparison of the glycosylation sites of the transferrins provided evidence about the existence of consistent divergencies. Thus, in human lactoferrin, asparagine residues 137 and 490 were glycosylated. Human serum transferrin contained two code sequences Asn–Lys–Ser (residues 428 to 430) and Asn–Val–Thr (residues 635 to 637); both were glycosylated and located in the C terminal moiety. In hen ovotransferrin, three code sequences were located in the C terminal part, but only one asparagine, residue 490, was glycosylated. Thus, important differences in the localization of glycosylation sites were observed in these three transferrins.

All these observations can be easily explained by assuming that the transferrin proteins evolved from a common ancestor protein and served an important function during evolution. Several observations support this hypothesis, as does the observation that the family of transferrin-like genes is located on the same chromosome. Thus, human chromosome 3 encodes serum transferrin, lactoferrin, and p97 melanotransferrin. Each transferrin gene contains 17 exons separated by 16 introns. Of the 17 exons, 14 constitute seven homologous pairs that code for the corresponding regions in the N and C lobes of the protein. The first exon codes for a signal peptide, necessary for the secretion of transferrin molecule, while the last two exons encode a sequence unique to the C terminal lobe.

The evolution of the transferrin gene family can be described as follows, according to Bowman et al.[41] About 200 to 500 million years ago, a gene encoding a polypeptide corresponding to a single-lobe primordial transferrin was amplified through a phenomenon of intragenic duplication, giving rise to a two-fold larger gene with two homologous domains[42] (Figures 4.3 and 4.4). Bi-lobed transferrin offers an important biological advantage over a primordial mono-lobed transferrin. In fact, the enlargement of the transferrin molecule prevents the loss of transferrin through glomerular ultrafiltration and subsequent loss of both iron and protein through urinary excretion. Indeed, isolated domains of hen transferrin injected into circulatory systems of mice are rapidly excreted in the urine.[43]

At a molecular level, the initial intragenic duplication of the transferrin gene was a result of an unequal cross-over between exons 9 and 2 of the ancestor gene.[44] After a series of additional mutations of the amplified gene, transferrin, lactoferrin, p97 melanotransferrin, and a transferrin pseudogene[45] evolved. Insertion of an identical exon (exon 11) occurred in human and chicken transferrin genes, but not in the gene encoding human p97 antigen, thus indicating that p97 is an older gene in the evolution of the transferrin gene family.[40]

Other arguments suggest that p97 melanotransferrin may represent a primordial transferrin. Williams[42] postulated that the ancestral transferrin gene may have arisen in urochordates as a cell membrane-anchored 40-kDa protein. As explained above, the existence in circulation of a low molecular weight transferrin would be impossible. Therefore, the anchorage may have enabled it to serve as a membrane-associated iron receptor; p97 can be viewed as the rudimentary form at the evolutionary stage after duplication of the gene occurred, but before the protein gained entrance to the blood stream.

During evolution of the transferrin family, elongation of the genes occurred, not only by intragenic duplication, but also by increases in the lengths of their introns. Thus, while the coding regions of ovotransferrin and serum human transferrin are nearly the same length, 2.3 kb, the mammalian gene is 33.5 kb in length compared to only 10.5 kb for the chicken gene. This is due to the expansion of several introns in the mammalian transferrin gene.

Evidence in favor of the evolution of the proteins of the transferrin family was also derived from the analysis of the structures of transferrins present in primitive animal species. The oldest existing species in which transferrin-like molecules have been found are the tobacco hornworm,[17] the cockroach *Blaberus discoidalis*,[18] the tarantula *Dugesiella hentzi*,[15] the Dungeness crab *Cancer register*,[14] and the tunicate *Pyura stolonifera*.[16] Among these transferrins of primitive animals, the best example of primitive transferrin is present in the ascidian sea squirts *Pyura stolonifera*[16] and *Pyura haustor*.[41] The characterization of transferrin present in these prevertebrates showed several interesting findings:[12] (1) the transferrin molecule exhibits a molecular weight of about 41 kDa and fixes one iron atom per molecule; (2) upon iron binding, the protein develops a peak of light absorption at 450 nm, near the peak absorption at 465 to 470 nm observed for vertebrate transferrins; (3) iron binding is associated with concomitant bicarbonate binding; and (4) the *Pyura* transferrin is able to act as an iron-donating species for rat reticulocytes, a phenomenon indicating that the protein is able to interact with membrane transferrin receptors.

Molecular studies showed that the DNA encoding the *Pyura* transferrin exhibits a significant homology with the DNA encoding human transferrin[41]; this finding indicates that the transferrin gene sequences have been largely conserved during the 400 to 500 million years separating prochordates and human beings. In confirmation of this view is the observation that human transferrin cDNA was successfully used to detect *Pyura* transferrin mRNA transcripts selectively localized in digestive cells.

These findings suggest that *Pyura* transferrin is closely related to the putative transferrin precursor gene encoding a single iron-site molecule. Finally, the ability of *Pyura* transferrin to interact with mammalian transferrin receptors has been interpreted as suggesting that the transferrin receptor may be of origin as ancient as transferrin.[16]

A recent study carried out in *Rana catasbesiana* led to the identification of a protein related to the transferrin family that could have derived from a transferrin ancestor through an evolutionary process different from the one that gave rise to transferrin.[46] This protein was called saxiphilin for its property of binding the neurotoxin saxitoxin. Saxiphilin has a molecular weight of 91 kDa and its higher molecular weight as compared to that of transferrin is due to the insertion of 144 residues, corresponding to repetitive elements of thyroglobulin. Pairwise sequence alignment of saxiphilin with various transferrins reveals amino acid identity as high as 51%. Sequence analysis indicates that saxiphilin is an evolutionary relative of the transferrin family, but differs in two major respects: (1) saxiphilin has substitutions of nine of the 10 highly conserved residues that form the two Fe^{3+}/HCO_3-binding sites of transferrin; and (2) saxiphilin has an insertion of 144 amino acids.

These findings led to the hypothesis that saxiphilin originated from an ancestor of the transferrin family, but diverged to perform a different function. The unique ability displayed by this protein to bind saxitoxin and the similarity of its tissue distribution to that of transferrin indicates that its physiological role may consist of the transport of an endogenous organic molecule different from Fe^{3+}.

The ensemble of these observations concerning the structures of transferrins isolated from different animal species, as well as the structures of the different proteins of the transferrin family, led to a hypothetical evolutionary scenario in which at least three gene duplications occurred during evolution.[46a] An initial duplication event occurred before the separation of athropods and chordates, because insect transferrins have duplicate N and C termini. This duplication led to the generation of a proto-transferrin gene.

A second duplication event occurred in the branch leading to vertebrates, before the emergence of land animals. As a consequence of this duplication event, the formation of human melanotransferrin occurred. This even preceded any other type of vertebrate transferrin.

The third duplication event occurred before the appearance of mammals, and lactoferrin and transferrin are the resulting products (see Figures 4.2 and 4.4). Lactoferrin arose by duplication of the transferrin gene at relatively late stages of evolution.

The evolutions of transferrin and lactoferrin were different, as supported by the observation that lactoferrins are not found in non-mammalian species. Analyses of the structures of different mammalian transferrins showed high degrees of sequence homology, with amino acid residues particularly conserved at the levels of the iron-binding regions and of the Cys residues.

Particularly interesting is the phylogenetic analysis of the structure of possum transferrin which suggests that possum transferrin is placed uniquely on a phylogenetic branch independent of eutherian transferrins and lactoferrins.[46b] This finding suggests that the gene duplications involved in the generation of transferrin and lactoferrin genes occurred after the marsupial–eutherian split.[46b]

Further compexity was added to the phylogenetic development of transferrin family genes by a recent study reporting the identification of a murine membrane-bound transferrin-like protein.[46c] Phylogenetic analysis suggested that this membrane-bound transferrin-like gene diverged from the common ancestor gene earlier than the genes encoding the other transferrins such as serum transferrin, lactoferrin, and ovotransferrin. Interestingly, this membrane-bound transferrin possesses an iron-binding site only at the level of its N terminal domain. This peculiar membrane transferrin is expressed at high levels in cartilage, where it may play a relevant role in the embryonic development of this tissue.

4.3 TRANSFERRIN GENETIC VARIANTS

Inherited polymorphisms of transferrin have been observed in all the human populations studied.

Genetically-determined, electrophoretically-detectable variation in the transferrin present in human serum was first reported by Smithies in 1957[47] using starch gel electrophoresis for serum protein separation. Since then, large numbers of data about the world distribution of transferrin alleles was accumulated from different human populations.

The most common transferrin phenotype in all human populations was designated Tf C. Genetic variants that migrate faster than C at alkaline pH are known as Tf B and those that migrate more slowly are termed Tf D.

Using starch or agarose gel electrophoresis, more than 20 transferrin variants have been discovered.[48-50] The synthesis of these variants is controlled by multiple alleles at a single genetic locus. The amino acid substitutions present in some of these variants have been determined (see Figure 4.5):

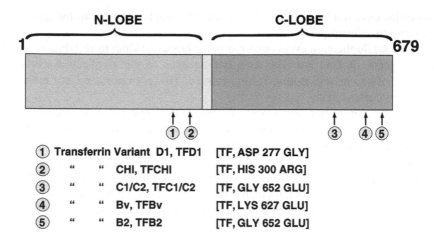

FIGURE 4.5 Nucleotide substitutions of the main transferrin allelic variants. The amino acid mutation for each of these variants is indicated.

1. In Tf D1 at position 277, Asp is replaced by Gly.
2. In TfB at position 627, Lys is replaced by glutamic acid.
3. In TfC2 at position 570 Pro, is replaced by Ser.
4. In Tf DChi at position 652, His is replaed by Arg.
5. In Tf B2 at position 652, Glu is replaced by Gly.
6. Tf D_{Evans} contains a Gly-to-Arg change at amino acid 394.
7. Tf B_{Show} contains an Ile-to-Arg change at amino acid 378 or 381.

The majority of these variants exhibit iron-binding capacity, as evaluated in terms of maximal iron-binding capacity and affinity for iron, virtually identical to characteristics displayed by the common type Tf C. Similarly, most transferrin variants exhibit normal capacities to interact with membrane transferrin receptors. Of the two variants described, one exhibited decreased affinity for iron binding,[51] and the other displayed a reduced capacity to bind membrane transferrin receptor.[52]

The first of these two variants is able to bind two atoms of iron, but the iron in the C terminal binding site is bound abnormally. The iron-free C terminal domain of the variant transferrin is found to be less stable than the normal domain to thermal and urea denaturation. Moreover, even the common C alleles have been shown to differ in their iron-binding capacities: the iron-binding capacity of transferrin from subjects with the 1-1 subtypes of Tf C is significantly higher than the capacities of the 2-2 and 2-1 subtypes.[53]

Transferrin polymorphism in different human populations was intensively investigated using a high-resolution electrophoretic separation method, i.e., isoelectric focusing. The alleles more carefully characterized are transferrins C, B, and D.

4.3.1 TRANSFERRIN C ALLELES

Transferrin C was initially considered to be a single variant. However, using isoelectric focusing, two groups of investigators independently provided evidence about the heterogeneity of transferrin C, thus identifying two common subtypes, Tf C1 and Tf C2.[54,55] A third common subtype of Tf C was later identified, and called Tf C3.[56] The distribution of these three Tf C alleles in different human populations was studied and reviewed by Kamboh and Ferrel.[50]

The two common Tf C1 and Tf C2 alleles were observed in all populations studied. Tf C1 frequency is usually higher in blacks, Micronesians, Polynesians, Melanesians and Australian aborigines (ranging from 80 to 95%) than in white populations (ranging from 78 to 85%). In Asia, the Tf C1 allele frequency is highly variable, reaching minimum values in some Indian populations (65%), some Indonesian groups (50%), and in Malaysia, Singapore, and Taiwan populations (70%).

Interestingly, the Tf C2 allele frequency is usually complementary to that observed for the Tf C1 allele. Thus, the lowest values of Tf C2 allele were observed in Australian aborigines (3%), Melanesians (4%), Pygmies of Africa (3%), and in some Amerindian groups (1%). In contrast, the Tf C2 frequency is 13 to 19% in European populations, ranges in Asians from 13 to 34%, and in some Indian tribal populations, reaches its maximum (34%).

The Tf C3 allele also exhibits a characteristic distribution, appearing more frequently among European populations and American whites (4 to 8%), less frequently in Asiatic Indians (1 to 4%), and only sporadically among U.S. blacks, Amerindians, and East Asians. The peculiar world distribution of Tf C3 may be explained by hypothesizing that it arose because of mutation in European populations and, as a consequence of population growth and movement, was introduced to other populations.

Particularly interesting is the world distribution of Tf C4 allele, which is restricted to New World populations (Amerindians) and South Asia Indians. Among Amerindians, the frequency of Tf C4 ranges from 8 to 18%; this suggests that it may represent a marker of these indigenous populations. On the basis of available data, it is not certain whether this mutation arose in these groups after separation from their ancestral Mongoloid populations or whether it already existed in the ancestral population at low frequency, then increased in frequency in Indians due to genetic drift or other factors.[50]

The other Tf C alleles are very rare and exhibit restricted ethnic and geographic distributions.

4.3.2 Transferrin B and D Alleles

Some transferrin variants exhibit gene frequencies above 1%. Those reaching polymorphic frequencies are Tf D1, Tf DChi, Tf B2, and Tf BO-1. The Tf D1 variant is distributed at relatively high frequencies among Australian Aborigines and Melanesians, at lower frequency in black Americans and Africans, and sporadically in East and Southeast Asians. The Tf DChi allele is a marker of Mongoloid populations and is widely distributed among East Asians and their related Amerindian populations in the New World.

4.3.3 Association between Transferrin Alleles and Disease

Transferrin polymorphism and other polymorphisms seem to be maintained through natural selection. Thus, in humans, both the Tf C2[57] and Tf C3[58] alleles were found to be associated with increased risk for spontaneous abortion, and Tf C2 also showed a highly significant association with premature birth.[59]

In addition to the apparent relation to reproductive hazards, Tf C2 was reportedly associated with phototoxic eczema.[60] These findings led to the hypothesis that transferrin variants C2 and C3 may act as enhancers of cytotoxic and genotoxic damage, possibly by influencing the concentration or distribution of iron, and thereby the formation of oxygen-free radicals.

Because these radicals are believed to be involved in the pathogenesis of rheumatoid arthritis, it was predicted that the transferrin C2 variant would be associated with rheumatoid arthritis; this turned out to be the case.[61] It was also predicted that individuals with the transferrin C2 variant would be more susceptible to chromosome damage when exposed to ionizing radiation. A pilot study of chromosome breaks in irradiated short-term lymphocyte cultures provided some suggestive evidence to support this hypothesis.[62]

In addition to these observations, it was also proposed that transferrin polymorphism may also be related to possible protective effects of certain transferrin alleles against microbial infections[63] or differences in fetal growth-promoting effects of transferrin,[64] but direct evidence in favor of these hypotheses is lacking.

A recent study suggests a protective role for the transferrin C3 variant against smoking-associated lung cancer.[65] This was based on the observation that the frequency of Tf C3 was very low (0.03) among individuals who developed smoking-related lung cancer, while the distribution in the corresponding healthy population was 17 to 25%. This conclusion was questioned by a study showing significantly decreased frequency of the transferrin C3 variant among patients with small lung cancers in Northern Sweden, but not in Southwestern Sweden.[66]

Based on these observations, it was suggested that the association between transferrin C alleles and lung cancer may be secondary and dependent on linkage disequilibrium with allelic variants of tumor-associated genes known to map to the same position (3q21) as the transferrin gene.[66]

Finally, the association between the different transferrin C alleles and some diseases does not seem to be dependent upon differential iron-binding capacity. In fact, individuals with transferrin C2 and C3 types exhibit similar levels of transferrinemia, iron saturation, sideremia, and total iron-binding capacity.

4.4 BIOCHEMICAL PROPERTIES OF SERUM TRANSFERRIN

4.4.1 GENERAL PROPERTIES AND BASIC STRUCTURE

Human serum transferrin consists of a single polypeptide chain containing 679 amino acid residues and two N-linked complex-type glycan chains, which results in a calculated molecular weight mass of 79,570 kDa. The transferrin molecule can be subdivided into two homologous domains, the N terminal domain (residues 1 to 336) and the C terminal domain (residues 337 to 679) with the carbohydrate moieties in the C terminal domain at positions 413 and 611 (Figure 4.6).

The presence of two domains with high degrees of internal homology in the molecule has been demonstrated by partial proteolysis of a number of transferrin

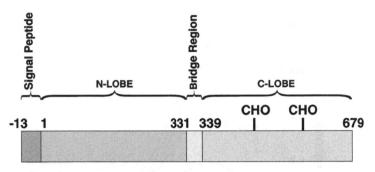

FIGURE 4.6 Schematic representation of transferrin structure. Four regions are outlined, including the signal peptide, the N lobe, the bridge region, and the C lobe.

species to produce fragments containing about half of the molecule and a single metal-binding site and by amino acid sequence studies[67].

Pure diferric transferrin displays an a 465- to 480-Å absorption ratio of 0.046 with a ratio 1.4 at 412 to 465 Å. Sodium dodecyl sulfate electrophoresis in polyacrylamide in the absence or presence of reducing agents shows the existence of one protein band with a molecular weight of about 80 kDa. Crossed immunoelectrophoresis against an anti-serum anti-transferrin displays a single precipitation line in the position of transferrin. One milligram of the iron-saturated protein contains 1.4 μg iron.

Transferrin has the ability to bind two ions of Fe^{3+} per molecule with the incorporation of a bicarbonate anion in each of the two specific Fe^{3+}-binding sites. Transferrin is also capable of binding other transition metal ions such as Co^{2+}, Mn^{3+}, Cr^{3+}, and Cu^{2+}.

The binding of iron by apotransferrin is a complex process that involves the displacement of an anion from or in the vicinity of the metal-binding site of apotransferrin[68] (the apo-anion); the concomitant binding of an anion in close proximity to the iron (the synergistic anion)[69]; the binding of other anions at a site more distant from the iron (the non-synergistic anion)[70]; and finally a change in conformation of the protein such that after iron is bound, it becomes more compact[71] and more resistant to urea denaturation,[72] proteolytic attack and heat denaturation,[73] and proton exchange.[74]

Spectroscopic analysis of transferrin showed that iron binding to apotransferrin is also a slow process characterized by a typical increase of the absorbance in the visible range, with a peak at 461 to 465 nm (Figure 4.7). This process requires the presence of apotransferrin, of iron compound, and an anion (usually bicarbonate). The entire process is slow, requiring about 30 minutes, during which an initial product with an absorbance different from that of ferric transferrin is formed. The absorbance changes to that of ferric transferrin over time, suggesting that the initial product is a mixed ligand complex and that the spectrum changes when the donating ligand is displaced (Figure 4.7).

It is of interest that, on addition of iron to apotransferrin, the absorbance spectrum of the initial product and its subsequent shift are independent of the accompanying ligand:[75]

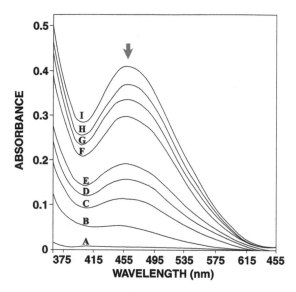

FIGURE 4.7 Visible spectra generated by binding of iron from ATP–Fe by human apotransferrin. The absorption scans were taken before iron addition (A), immediately after iron addition (B), at 1-minute intervals (C to E), at 5-minute intervals (F, G), after 10 minutes (H), and after 60 minutes (I). (Redrawn with modifications from Weaver, J. et al., *Acta Haematol.,* 84: 68–71, 1990.)

1. The initial binding of iron to apotransferrin in its open conformation is followed by a change in conformation, with a concomitant change in the absorbance spectrum.
2. The initial binding of iron to apotransferrin involves a complex of the iron and protein via the apo-anion. The same initial complex is formed regardless of the ligand delivering the iron, and the subsequent change in the absorbance spectrum is consequent to the breaking of protein–anion bonds and the evolution of protein–metal bonds.

Metal binding to transferrin induces the formation of two ultraviolet absorption peaks at 242 and 295 nm, which can be deduced from the spectrum difference between iron-saturated transferrin and apotransferrin. These absorptions are attributed to the complexing of metal with the tyrosine ligands that perturb the π to π^* transitions of the aromatic rings. When the metal ions impair the d electrons, the binding with transferrin determines the intense visible absorption in the 410 to 500 nm range, which is attributed to the transfer of charge between phenolate (π) and the metal ions ($d\pi^*$). These absorption bands in the visible regions are considered indicators for metal binding and, particularly, for specific metal–tyrosine binding in transferrin.

The ability of transferrin to bind other transition metals in addition to iron provided the opportunity to investigate the comparative stabilization and unfolding properties of the different metal–ion complexes. Two studies[76,77] showed greater stabilization and increased resistance to protein unfolding for Fe^{3+}–transferrin complexes compared to

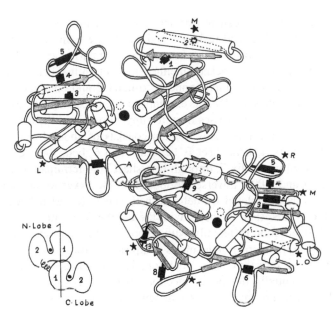

FIGURE 4.8 Schematic representation of the three-dimensional structure of a transferrin molecule. The diagram is based on data for human lactoferrin, but a similar structure was observed for human transferrin. Helices are shown as cylinders, β-strands as arrows, iron atoms as (bullets), probable anion sites as (open circles), disulfide bridges as (rectangles), and carbohydrate branching sites as letters: L, lactoferrin; T, serum transferrin; O, ovotransferrin; and M, melanotransferrin. The N terminal half (N lobe) is at the top; the C terminal half (C lobe) at the bottom; their relative spatial orientation is shown in the inset on the left. The two domains of each lobe are labeled 1 and 2. A hydrophobic region joins the two lobes. (Redrawn with modifications from Baker, E.N. et al., *TIBS,* 12: 350–353, 1987.)

Cu^{2+}–transferrin complexes as determined by isothermal denaturation in the presence of urea.

X-ray crystallographic studies of rabbit serum transferrin,[78] the diferric form of human lactoferrin,[79] and the apo form of human lactoferrin[80] led to the determination of the structure of transferrin at near-atomic resolution and in part to elucidate the molecular mechanisms involved in iron binding. As indicated by low-resolution x-ray crystallographic studies,[81] the transferrin molecule is bilobal, and both lobes exhibit similar conformations, in line with the observation that they have similar amino acid sequences.

Rabbit serum transferrin has a general three-dimensional structure very similar to lactoferrin. The structure of lactoferrin was determined in detail, and we assume the structure can apply also to serum transferrin (Figure 4.8).

The N lobe is connected to the C lobe through a three-turn alpha helix. The two lobes exhibit basically the same structures, built by a network of parallel and rare anti-parallel beta sheets connected by alpha helices to non-alpha-helical, non-beta sheet structures joining the two structural elements (see Figure 4.8). It is believed that disulfide bridges, six in the N lobe and ten in the C lobe, play an important role in the three-dimensional structure of the molecule.

Each of the two lobes can be further subdivided into two domains, each of approximately 160 amino acid residues, and exhibiting a similar super-secondary structure based on a central core of five or six irregularly twisted beta sheets of similar topology with helices packed on either side. N terminal beta sheet A is linked by helix 1 to beta sheet B, which connects to helix 2 and continues through beta sheet C, helix 3, and beta sheet D. This ensemble constitutes the first part of domain 1.

A long beta sheet E then crosses from the upper domain of the molecule to enter the lower domain of the lobe. This domain consists of about 160 residues in the following order: helix 4, beta sheet F, helices 5 and 6, beta sheet G, which leads after a variable turn into helix 7, through beta sheet H to terminate in helices 8 and 8a. Beta sheets I and J then traverse the lobe from the lower to the upper domains where domain 1 is completed by helix 9, beta sheet K, and helices 10 and 11. Helix 12 of the N lobe then connects with the C lobe. In contrast, helix 12 in the C lobe is positioned quite differently with regard to helices 10 and 11.

In summary, a transferrin molecule is composed of four nearly similar domain structures organized into two lobes. Each lobe is constituted by two domains connected by two potentially flexible connections, beta sheets E, I, and J, and helices 10 and 11. This structure may allow movements of the domains with respect to one another.

One of the main biological properties of transferrin is its capacity to fix iron. As previously stated, each transferrin molecule can be subdivided into two globular domains, each containing one iron-binding site with a K_m of approximately 10^{22} M^{-1}. The concomitant binding of an anion (under physiological carbonate or bicarbonate conditions) is essential for iron binding at each site. The function of the anion is to lock the iron into place in the molecules by serving as a bridging ligand between iron and protein, thus making the coordination site on the iron atom unavailable to water. In the absence of the anion, this coordination site is hydrated, and the iron rusts in the form of polynuclear complexes.

Although iron is the most important metal bound by transferrin *in vivo*, transferrins are capable of binding many other metal ions such as Zn^{2+}, Ga^{3+}, Co^{2+}, Cd^{2+}, Ni^{2+}, Al^{3+}, In^{3+}, Gd^{3+}, Cr^{3+}, Cu^{2+}, and Mn^{2+}. The two iron-binding sites are virtually independent since the affinity of one site changes little with the occupancy of the other site. At physiological pH,[82,83] the binding site in the N terminal domain has less than one-twentieth the affinity for iron of the binding site located in the C terminal domain. Thus, the binding site in the C terminal domain has been termed the acid-stable site. Under physiological conditions, the effective affinity constants of the N and C sites amount to 1 and 6×10^{22} M^{-1}, respectively.[82,83]

A large amount of experimental evidence has shown that the two metal-binding sites of transferrin differ in their thermodynamic, kinetic, spectroscopic, and chemical properties. Evidence in favor of major stability of the C lobe iron-binding site was directly supported by small-angle x-ray solution scattering.[84] This technique revealed the conformational changes that occur in transferrin as the pH is progressively lowered from 7.4 to 5.5 and iron is released. The data derived from these studies indicate that N lobe iron release precedes release by the C lobe and is nearly complete at pH 7.0.[84]

This phenomenon corresponds to a conformational change involving opening of the N lobe iron-binding cleft. Most studies suggest that the N terminal, acid-labile site of human serum transferrin is preferentially occupied in the circulation, binds iron less strongly, and releases iron with greater facility. Despite the evidence in favor of the differences in physical and chemical properties of the two iron-binding sites of transferrin, it is not clear whether the two sites differ in their physiological properties, particularly their capacity to donate iron to developing erythroid cells.

This issue was recently investigated by an appropriate technical approach[85] (urea gel electrophoresis combined with Western immunoblotting): more particularly, the relative ability of each iron-binding site of rabbit transferrin in delivering iron to reticulocytes was investigated.

Interestingly, the two sites can be made to release iron at the same or differing rates, depending on the experimental conditions. An atmosphere of 5% CO_2 increased the effectiveness of iron donation by the acid-labile site to reticulocytes, while the presence of 20% serum enhanced the iron-donating ability of the acid-stable C terminal site.[85]

Considerable efforts have focused on understanding the molecular mechanisms responsible for the binding of iron by transferrin. Spectroscopic and chemical modification studies intended to establish the nature of the iron-binding ligands of transferrin revealed two histidine residues,[86] two or three tyrosine residues,[87,88] a water molecule or hydroxyde ion,[89] and a carbonate or bicarbonate ion[90-92] (Figure 4.9). The carbonate or bicarbonate is bound to the protein by a cationic group implied to be an imidazole of histidine by proton NMR spectroscopy of the guanidium group of an arginine by chemical modification.

However, only the analysis of the three-dimensional structures of transferrins allowed the unequivocal identification of the iron-binding sites. Again, the most

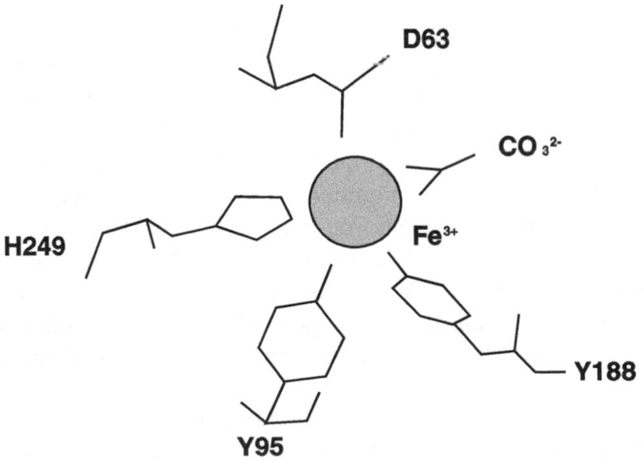

FIGURE 4.9 Schematic representation of the metal-binding site in the N lobe of human serum transferrin. The iron ligands are Tyr 95, Tyr 188, Asp 63, and His 249. The binding site of the synergistic anion carbonate is also shown.

accurate data are available for lactoferrin and rabbit serum transferrin. The iron-binding sites for human lactoferrin are located 4.2 nm apart, in the interdomain cleft of each lobe.[91-94] The iron-atom binding sites are highly anionic, with a preponderance of "hard" oxygen ligands, thus explaining the much stronger binding of Fe^{3+} than Fe^{2+}.

The ligands, which are identical in both iron-binding sites, are three anionic oxygen atoms furnished by two tyrosyl phenolates: aspartyl carboxylate and a neutral histidyl imidazole nitrogen. In the N lobe, these are Asp 60, Tyr 92, Tyr 192, and His 253; and in the C lobe, Asp 395, Tyr 528, Tyr 435, and His 597. The four protein ligands occupy four octahedral positions about the iron atom; the two remaining cis-octahedral positions are left vacant for two non-protein ligands.

The identical distribution of protein ligands was observed in serum rabbit transferrin. Unlike other metal-binding proteins, the protein ligands of transferrins are distributed among several exons. This finding may underline the fact that iron is not necesssary for the correct folding of the apotransferrin molecule, whereas in other metal-binding proteins the interaction between the metal and the protein is required for the correct folding of the protein.

The anion-binding site of transferrin seems to involve a carbonate anion, which binds as a bidentate ligand to the iron atom, the side chain Arg 121, and the N terminus of helix 5, residues 121 to 136. One oxygen atom is bound to iron and forms a hydrogen bond with Arg 121; the second also binds to iron and forms a hydrogen bond with the NH of residue 123, while the third oxygen makes two hydrogen bonds with NH (124) and the side chain Thr 117. The major role of helix 5 in defining this anion-binding site in the pocket created by the side chain of Arg 121 and the main-chain atoms of residues 122 to 125 (and of the corresponding residues 465 to 469 in the C lobe) is underlined by its invariance in all transferrins that have been sequenced. Residues 121, 123, 124, and 125 are totally invariant in both halves of each transferrin molecule, and residue 122 is always Thr or Ser.

The structure of human apolactoferrin determined by x-ray diffraction[9] reveals interesting differences when it is compared to the diferric protein. Whereas the C lobe is still present in a closed configuration, as it is in iron-bound lactoferrin, the N lobe has an open configuration, with the two domains largely separated. The N2 domain has undergone a rotation of about 53° relative to the N1 domain. The open configuration of the binding cleft that is formed exposes three basic amino acid residues that were buried in the closed configuration: i.e., Arg 121, Arg 210, and Lys 301. The sequence of binding to the open iron-binding site may be described as follows: (1) the carbonate anion is first attracted into the bottom of the open interdomain cleft by the positive charges on Arg 121 and Arg 210; and (2) the carbonate anion then binds to the N2 domain. It must be pointed out that the binding of the anion as the initial step in binding iron to transferrin is also suggested by chemical and kinetic arguments.[93,94]

As a consequence of this initial carbonate binding, four iron-binding ligands are now in place (Tyr 92, Tyr 192, and the two carbonate oxygens), and the iron can then bind to the N2 domain. The iron-binding process is completed by rotation of the N2 domain, closing the cleft with Asp 60; His 253 completes the iron coordination, and Asp 60 further links the two domains by hydrogen bonding.[79,80]

The initial stage of iron binding by a transferrin molecule is Fe^{3+} binding in the open iron-binding pocket. Crystallographic evidence about this state of transferrin was obtained with a technique in which a transferrin N-lobe crystal grown as the apo form was soaked with Fe^{3+} nitriloacetate and its structure was solved at 2.1 Å resolution.[95] The analysis of this Fe^{3+}-soaked form of transferrin showed that it possesses the same overall open structure as apotransferrin. However, electron density mapping clearly showed the presence in Fe^{3+}-soaked transferrin of an iron atom with the coordination by two protein ligands provided by two Tyr residues.[95] These observations are consistent with the view that the two Tyr residues represent the protein ligands for the initial Fe^{3+} entry in the intact transferrin with the iron-binding pocket in an open conformation.

The detailed nature of the conformational changes in the transferrin protein family has been carefully defined only in the crystal structure of apolactoferrin, where a domain rotation of 54° determines the formation of an open N lobe binding cleft. Several sets of data suggest that a similar conformational change should occur in serum transferrin. However, the presence of several distinct structural and functional differences between lactoferrin and transferrin, and particularly the tendency of transferrin to release iron from its N lobe at a significantly higher pH value (5.5) than lactoferrin (3.0), may underly the occurrence of different conformational changes. Furthermore, structural differences of the two proteins, such as amino acid differences involving the presence of an additional disulfide bond between Cys 137 and Cys 331 in human transferrin that is not present in lactoferrin, may influence the conformational change.

Recent studies of the conformations of both iron-free and iron-loaded forms of the recombinant N terminal half molecule of human transferrin provided definitive evidence that iron binding to transferrin is accompanied by conformational changes similar to those described for lactoferrin.[96,97] More particularly, these studies showed that comparison of apotransferrin with the holo form of transferrin showed that a large "rigid-body" domain movement of 63° takes place in apotransferrin, giving rise to an open binding cleft.[95,96] The extent of domain opening is highly comparable to that observed for lactoferrin and is independent of the presence of the C lobe.

In addition to this general change in transferrin conformation, some local adjustments occur, particularly concerning the Asp 63 and His 249 residues which act as iron ligands, and whose conformations are changed to form salt bridges with Lys 296 and Glu 283, respectively, at the level of the binding cleft of apotransferrin.[96,97] These salt bridges must break for iron binding to occur. The change in the conformations of Asp 63 and His 249 seems to be specific for transferrin, since corresponding residues of apolactoferrin remain essentially unmodified in the iron-loaded form of the molecule.

In concluson, crystallographic studies did not reveal any changes in the C lobe; its cleft remained closed even in the absence of iron. However, a recent study based on the crystallographic analysis of bilobal hen apotransferrin showed that both the N and C lobes of this protein assume an open conformation upon iron release.[97a] Upon opening, the domains of the N lobe rotate away from one another by 53°, and a corresponding 35° rotation occurs in the C lobe. As a consequence of these conformational changes, both lobes of apo-ovotransferrin expose all four iron-coordinating amino acid residues to solvent, thus indicating that either lobe is accessible for iron entry.

In contrast, x-ray solution scattering of different transferrin species as well as N and C terminal fragments with and without iron,[98] showed that both lobes undergo large-scale conformational changes in solution, i.e., both interdomain clefts are open before iron uptake and close when iron is incorporated into the binding site. These structural studies were extended further to the uptake of non-physiological metals[99] that also bind to transferrin although with lower binding affinities than iron. Some of the metals, i.e., Cu, induce conformational changes in a manner similar to iron. These experiments have led to the suggestion that the "correct" coordination of metal ion is crucial for inducing the closed conformation, and that the closed conformation is likely to be of functional importance for recognition of transferrin by the receptor.

X-ray scattering has also provided evidence that Asp 63 plays an important role in inducing the conformational change upon metal binding.[100] This was made possible through the analysis of mutants of the amino terminal half molecule of human serum transferrin, a fragment of 337 residues incorporating a single iron-binding site. Thus, two mutants, Asp 63 to Ser 63 and Asp 63 to Cys 63, both of which show weaker iron binding than the wild-type, have been examined by x-ray solution scattering in both the apo and holo forms. In the case of the Asp 63 to Ser 63 mutant, the cysteine residue is blocked by a disulfide bond formation that is not available to bind iron. As a consequence, the lobe remains open, even in the presence of iron.

In the case of Asp 63 to Ser 63, the mutation of an uncharged residue weakens the strength of the hydrogen bonds between the two domains and reduces the closure of the interdomain cleft.[100]

The four iron-binding ligands are also conserved in both lobes of ovotransferrin and in the N lobe of melanotransferrin. However, while the N lobe of melanotransferrin is projected outside from the cell surface and binds iron as well as other transferrins, in the C lobe, Asp 407 is changed to Ser, Arg 477 to Ser, and Thr 473 and Thr 478, to Ala and Pro, respectively. On the basis of these observations, it is expected that the C domain does not bind iron and may exhibit other biological properties. The C terminal disulfide 9 is absent in melanotransferrin; this may allow the C terminal segment to project from the molecule and thus anchor the protein to the cell membrane.

Domain 2 contributes four of the five protein ligands involved in iron binding, and it can be isolated from the N lobe of human lactoferrin and duck ovotransferrin as a single-chain iron-binding fragment of 18 kDa[101]; the duck ovotransferrin fragment has been crystallized.[102]

Site-directed mutagenesis coupled with the analysis of the resulting mutants allowed us to define the functional roles of the amino acid residues most critical for transferrin function. Particular attention was focused on tyrosine residues, those involved in iron binding and those located outside the iron-binding pocket.

Initial studies of lactoferrin have shown that when the two liganding tyrosines in a lobe were mutated to alanine, the resulting mutant failed to bind iron. Similar studies were carried out at the level of the N lobe of transferrin: when the liganding tyrosines were mutated individually to phenylalanine, the resulting mutant Y95F

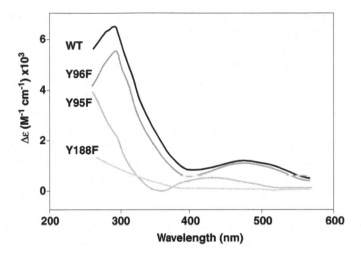

FIGURE 4.10 Visible spectra for the iron saturated complexes of human transferrin 2N (wild-type) and Y96F, Y95F, and Y188F mutants. (Reprinted with modifications from He, Q.Y. et al., *Biochemistry,* 36: 14853–14860, 1997. With permission.)

showed a weak affinity for iron and no affinity for copper. Mutant Y188F completely lost the capacity to bind iron, but formed a stable complex with copper.[103] This observation clearly indicates that Tyr at position 188 is strictly required for iron binding by transferrin (Figure 4.10).

The explanation for these findings may depend on the different positions of the two Tyr residues in the iron-binding pocket. Thus, the loss of Tyr 95 ligand does not destroy coordination because the binding to iron is still able to bring the two domains together to close the cleft by linking the Asp 63 ligand on domain 1 and the Tyr 188 ligand on domain 2. In contrast, the loss of the only ligand on domain 2 determines failure to close the iron-binding cleft.[103]

The role of Tyr 95 in iron binding is also supported by a recent study based on the x-ray scattering analysis of wild-type transferrin and Y95H mutant. In the YtyrH mutant, the mutated residue maintains the iron too far away for Asp 63 to coordinate, thus blocking the large-scale hinge-bending movement and leaving the transferrin molecule in an intermediate conformation of cleft closure.[104]

The mutation studies on tyrosine residues that located out-of-the pocket iron binding also provided other interesting information. Tyrosine at position 85 is a non-liganding residue located within a hydrogen-bound network, near the transferrin iron-binding site. When Tyr 85 was mutated to phenylalanine, iron release from the Y85F mutant was much more pronounced than release from the parent protein.[105] Mutation of the Tyr 85 leads to the elimination of the hydrogen bond between Tyr 85 and Lys 296, thus interfering with the so-called di-lysine (Lys 296–Lys 206) trigger, that affects the iron-binding stability of the protein.

Mutations of other amino acid residues present at the level of this region of the transferrin molecule (the "second shell") have pronounced effects on the iron-binding capacity of the molecule; this is the case with the Glu 83-to-Ala mutant,[105] which lost the capacity to bind iron. Glu 83 must form a strong H-bond linkage with His 249 in order to hold His 249 in the proper position for iron binding. This ability is lost in the Glu 83-to-Ala mutant.

Several studies focused on the role of Asp 63 residue in transferrin iron binding. Studies on Asp 63 mutants showed that this residue is an essential ligand for strong iron binding to transferrin. In fact, mutations of this aspartic acid residue to either Ser, Glu, or Ala elicited pronounced weakness of the ferric iron-binding capacity.[106]

Subsequent studies clarified that mutations of Asp 63 facilitated iron release from transferrin, as shown by experiments where wild-type and Asp 63 mutants of transferrin were incubated in the presence of an iron chelator or at decreasing pH values.[107] Low angle x-ray scattering studies of the N lobe mutant transferrin Asp 63 to Ser showed that the iron-binding pocket is open, thus suggesting that Asp 63 acts as a "trigger" for pocket closure.[108]

The role of Asp residues present within the iron-binding pocket was also investigated; Asp 63 at the level of the N lobe and the corresponding Asp 342 residue at the level of C lobe were mutated. To probe the conformations of the transferrin mutants, the interactions of wild-type transferrin, and single and double transferrin mutants were investigated. Previous studies clearly show that transferrin receptors discriminate among iron-saturated, monoferric, and iron-free transferrin, and exhibit the highest affinity for iron-saturated transferrin and the lowest affinity for apotransferrin.

Iron-saturated Asp double mutants exhibited an affinity for transferrin receptors markedly lower than iron-loaded wild-type transferrin and comparable to that displayed by wild-type apotransferrin.[109]

Another set of experiments focused on evaluating the roles of Lys 206 and 296 in iron release from transferrin. It was hypothesized that in iron-loaded transferrin, Lys 206 and Lys 296 interact with each other and form the di-lysine motif that helps hold together the N lobe transferrin subdomains (Figure 4.11). Both lysines are protonated at acid pH, and the repulsion of the two positively charged lysines, located at the levels of opposite domains, may represent a driving force that helps open the iron pocket and release iron. Transferrin mutants with the two lysine residues mutated to alanine displayed less pH sensitivity than the wild-type and remained in the iron-loaded form at pH 5.6.[110] This study clearly indicates that Lys 206 and Lys 296 are involved in the pH-dependent release of iron from transferrin.

This observation was confirmed in a more recent and detailed study. Thus, when Lys 206 or Lys 296 is mutated to glutamate or glutamine, release of iron is much slower than in wild-type transferrin[110a] (Figure 4.12). This phenomenon is due to the loss in the mutants of the driving force mediating the cleft opening (di-lysine trigger). The di-lysine pair contributes to the formation of the anion-binding site; more particularly, a Lys 296–Tyr 188 interaction is required for the formation of the anion-binding site.

Analysis of the anion binding, measured through the study of the UV-vis difference spectra after addition of the sulfate anion, showed that: (1) wild-type human transferrin 2N exhibited the typical spectral absorption, with two negative absorption bands at 245 and 298 nm and a small positive band at about 276 nm; (2) the K206Q mutant exhibited a pattern of difference absorption spectra similar to that observed for wild-type transferrin, but the absorptivity was much lower; and (3) the K296E

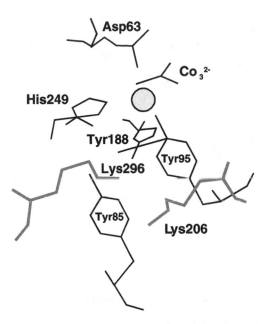

FIGURE 4.11 A schematic diagram showing the iron-binding site and the lysine trigger in the N terminal lobe of rabbit serum transferrin as derived from the analysis of the crystal structure.

mutant showed no detectable absorbance (Figure 4.12). The lack of anion binding in Lys 296 mutants implies that this Lys residue is a major constituent of the anion-binding site, while the reduced anion binding in Lys 206 mutants implies that this second Lys site is a minor contributor to the anion-binding site.[110a]

Periplasmic proteins are prokaryotic proteins involved in periplasmic iron transport; they are structurally and functionally homologous to transferrins. Each lobe of a transferrin molecule exhibits a strong similarity with a group of periplasmic proteins previously characterized.[111] These proteins can exist in three different structural-conformational forms: liganded closed, liganded open, and unliganded open.

A comparison of the N lobes of apolactoferrin and iron lactoferrin shows the same hinge-bending, ligand-induced conformational change as in bacterial periplasmic-binding proteins. Particularly relevant is the observation that the transport of ferrisiderophores across the periplasmic space in bacteria is assured by periplasmic-binding protein which could represent the bacterial equivalent of transferrin.

X-ray studies have shown that periplasmic proteins undergo conformational changes following iron binding, similar to changes observed for transferrin.[112] In addition to the analysis of the three-dimensional structure, biochemical studies also support a conformational study of periplasmic proteins following iron binding. Previous studies have shown that binding of iron to either lobe of human transferrin is accompanied by a decrease in the sensitivity of the protein to trypsinization.[113] A similar phenomenon was observed for periplasmic FbpA protein. In fact, binding of Fe^{3+} to the periplasmic FbpA protein significantly decreases the ability of trypsin to digest wild-type protein.[114]

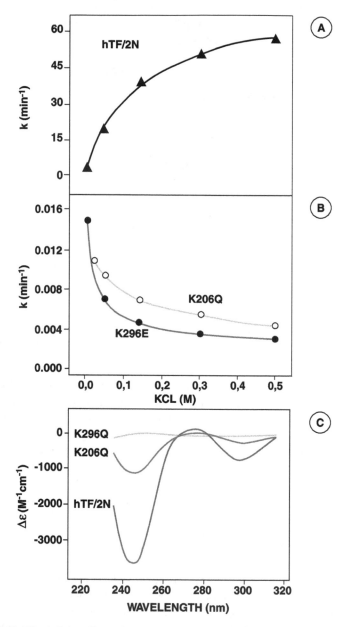

FIGURE 4.12 Plots of the effect of anion addition (chloride) on iron release from wild-type hTF/2N (A); lysine mutants K206Q and K296E with EDTA 4 m*M* at pH 5.60 and 25°C (B); UV-vis spectra obtained after addition of a 2 m*M* sulfate to a solution of wild-type human transferrin 2N and lysine mutants K2O6Q and K296E (C). The reaction was performed at pH 7.40, 25°C. (Redrawn with modifications from He, Q.Y. et al., *Biochemistry*, 38: 9704–9711, 1999.)

It is of interest that the N lobe iron-binding site is thermodynamically less stable and more acid-labile than the C lobe iron-binding site; furthermore, it releases iron faster[115] and it can accept oxalate as a synergistic anion when Cu^{2+} is bound. The C lobe iron-binding site cannot do that.[116] The N lobe preferentially binds Al^{3+}, as shown by experiments using recombinant N lobe human apotransferrin.[117] In summary, evidence for a role of the transferrin receptor in modulating release of iron from transferrin has been shown. The receptor retards the rate of iron release from diferric transferrin at pH 7.4 and accelerates release at pH 5.6,[118] thereby impeding iron release at cell surfaces and promoting iron release within acidified endocytic vesicles. This phenomenon primarily involves the C terminal iron-binding site,[119] and may help explain the prevalence of circulating N terminal monoferric transferrin in human circulation.

Isothermal titration calorimetry represents an important technique to investigate the binding of iron to the N and C sites of transferrins.[120-123] Initial studies carried out on ovotransferrin provided evidence that the binding of ferric ion occurs in two kinetic steps at both iron-binding sites of ovotransferrin.[54] Soon after injection, ferric ion (attached to the chelator) binds weakly; later, in a much slower process, bicarbonate binds at the level of the binding site, replacing the iron chelator. Once bicarbonate insertion is complete, binding to both iron-binding sites becomes much stronger. During this process, the energy and kinetics for binding ferric ion to the N site are significantly different from those for binding to the C site. The first ferric ion binds preferentially to the N site,[120] a phenomenon probably related to the site's easier availability for contact with iron and to faster binding of bicarbonate at the level of this site.

Subsequent studies have shown that the situation is drastically different for human serum transferrin. As discussed earlier, chelated ferric ion binds preferentially to the C site of human transferrin. The presence of chelated iron at the C site is so strong that monoferric transferrin (prepared by mixing Fe nitriloacetate with apotransferrin in a 1:1 *molar* ratio in the presence of bicarbonate at neutral pH and then dialyzing out the iron chelator) shows virtually no detectable binding at the N site.

Studies using calorimetric techniques provided a possible explanation for the preferential binding of iron to the C site of human serum transferrin. Based on these studies, the overwhelming preference of iron to bind first to the C site of apotransferrin arises in large part because of the strong tendency for bicarbonate anion to preinsert at the C site of apotransferrin before the addition of chelated iron, kinetically aiding the C site in achieving its final state once the chelated iron is added.[121]

In contrast, the insertion of bicarbonate at the N site occurs predominantly after iron binding and the insertion process is significantly slower.

In a later group of experiments, the ultrasensitive differential scanning calorimetry technique was employed to study domain interaction[122] and the effects of single amino acid substitutions on the structure and iron-binding affinity of the half transferrin molecule.[123] The first study showed that human apotransferrin exhibited separate differential scanning calorimetry (DSC) transitions for the two domains; two major transitions with T_m values of 57.6 and 68.3°C were seen. The transition of lower T_m occurs in the C terminal domain and the transition of higher T_m in the

N terminal domain.[122] Chelated ferric ions bind strongly to each site of transferrin and produce changes in T_m by as much as $30°C$.[122]

The second study was made on transferrin mutants originated by site-directed mutations located on the surface of the binding cleft. The mutants are D63S, D63C, G65R, H207E and K206Q. The analysis of these mutants by calorimetric studies provided evidence that:

1. Asp 63 is critical for tight binding of iron; in fact, the replacement of Asp 63 with either serine or cysteine causes a decrease in iron binding affinity of five to six orders of magnitude.
2. In contrast, the H207E and K206Q mutations elicited enhancements of iron-binding affinity.

It is worth noting that, in spite of the large differences in binding affinity, the apparent stabilities of the apo forms of the mutant proteins are very similar. UV spectroscopy studies have provided evidence that transferrin, in the absence of a metal ion, is able to bind different types of inorganic anions including HCO_3^-, Cl^-, NO_3^-, HPO_4^{2-}, ClO_4^-, and SO_4^{2-}. Only HCO_3^- functions as a synergistic anion promoting the formation of a metal–anion–transferrin ternary complex.

According to these observations, two different types of anions have been defined: synergistic and non-synergistic. Synergistic anions, such as carbonate, bind to the transferrin molecule and promote iron binding; non-synergistic anions, such as phosphate, interact with apotransferrin but do not promote iron binding.

Since bicarbonate binds to transferrin 2N, it was suggested that anion–transferrin binding is restricted to anions with charges of at least –2. However, a recent study based on difference UV data showed that monovalent anions bind weakly to apotransferrin.[124] It was also shown that binding of non-synergistic anions to apotransferrin determines the interference with metal binding by competing directly with the binding of the synergistic anion.[124] These observations imply that inorganic anions bind to apotransferrin in the vicinity of the iron-binding site.

Harris et al.[125] proposed that the primary anion-binding site is approximately the same as the site occupied by the synergistic carbonate anion in the Fe–CO_3–transferrin ternary complex and thus is mainly represented by Arg 124. They noted also that diphosphate anions bind to apotransferrin much more strongly than phosphate, and this finding suggests that the second phosphate group may bind to an additional site, that in the case of the N lobe of human transferrin may be Lys 206 and Lys 296. In line with this hypothesis it was observed that diphosphate anions bind more strongly with the N terminal site and that mutants K206A and K296A weakly bind diphosphates.[126]

The release of iron from transferrin can be achieved by different methods. The simplest and most widely used consists in dialyzing transferrin at pH 5.5 in the presence of an iron chelator. However, it must be noted that iron release from transferrin at acid pH in the absence of an iron chelator is a slow process. This observation may have important implications in understanding the process by which iron supply to the cells is mediated by transferrin through binding to membrane transferrin receptors.

It is commonly accepted that transferrin releases its iron into the cell within an endosomal compartment exhibiting a low pH (5.5). However, the spontaneous release of iron from transferrin at pH 5.5 is a slow process and transferrin remains within the cell only a few minutes — and only a fraction of the time within the endosome. Furthermore, the acid-stable site of transferrin does not release its iron at pH 5. Thus, it is obligatory to assume the existence of other factors in addition to proton attack on the ligands of the iron-binding site to explain how iron is so rapidly removed from transferrin within the cell.

These factors may include association of trivalent iron by naturally occurring iron chelators, such as amino acids, nucleotides, phosphoglycerates, and low molecular weight iron-binding proteins.[127-132] Among these molecules, ATP (adenosine triphosphate) may play a relevant role.

There is mounting evidence that ATP may be involved, particularly in the reticulocytes. ATP induces the release of iron from transferrin at concentrations that prevail in reticulocyte cytosol[133]; an ammonium sulfate fraction of reticulocyte cytosol enhances the releasing activity of ATP[131]; and most of the low molecular weight cytosolic iron in the reticulocytes is an ATP–Fe complex.[134]

An alternative possibility could be a mechanism recently described that involved the acceleration of iron removal from transferrin at pH 5.50, but not at pH 7.40, the level at which this protein is bound with the transferrin receptor.[91-92]

Other studies have shown that one important factor modulating the kinetics of iron release from transferrin involves the presence of simple non-chelating agents. Thus, at pH 7.40, iron release from transferrin is dependent upon ionic strength: release from either site to a synthetic tricatecholate-sequestering agent extrapolates to zero as the ionic strength of the supporting buffer nears zero.[135-136]

A further study showed that the effects of ionic strength may be rationalized by postulating a kinetically active anion-binding site on the transferrin molecule.[137] For release of iron to occur, the kinetic site must be occupied by a simple non-chelating, non-synergistic anion, such as chloride or perchlorate. Such occupancy is enhanced when ionic strength (or anion concentration) is increased, accounting for the dependence of observed release rates on ionic strength.

Another study shows that the kinetic effects of anion binding and complex formation with the receptor are each preserved in the presence of each other and are therefore dependent.

The release of iron from transferrin has been studied with a variety of chelating agents, including pyrophosphate, aminophosphonic acids, amino carboxylic acids, catecholates, and hydroxamates. Since the stability constants of ferric transferrin are around 10^{20}, it is virtually certain that simple dissociation of ferric ion is too slow to account for the observed rates of iron exchange between transferrin and low molecular weight chelating agents. Thus, it is reasonable to assume that the exchange of ferric ion must proceed via some type of mixed-ligand intermediate.

One study based on ultraviolet spectroscopy provided evidence that chelating agents may bind directly to transferrin.[125] On the basis of these observations, it is proposed that the primary binding site for the iron chelators and phosphonic acids includes the protein groups that bind the synergistic bicarbonate anion required for formation of a stable ferric transferrin complex.[125]

Furthermore, chelators with two phosphonate groups can simultaneously bind to cationic amino acid side chains that extend into the cleft between the two domains of each lobe of transferrin; from an inspection of the ferric transferrin crystal structure, the most likely anion binding residues in the cleft are Arg 632 and Lys 534 in the C terminal lobe and Lys 206 and Lys 296 in the N terminal lobe.

It is of interest that the same amino acid residues have been recently involved in the mechanism of iron release from transferrin triggered by lowering pH. Inspection of the crystallographic structures of human serum transferrin, rabbit serum transferrin and hen ovotransferrin showed the existence of a Lys–Lys arrangement that represents a common feature in the N lobe of all serum transferrins. The short 2.3 Å interaction between the N2 atoms of Lys 206 and Lys 296 appears to be an example of a low-barrier hydrogen bond where the proton is bound equally to both nitrogen atoms.[137]

In the closed structure of iron-saturated transferrin, both these Lys residues are probably neutral; however, at acid pH, like that found in the endosome (pH 5.0), both lysine residues are protonated and then positively charged. Similarly, in the C lobe, a Lys 534, Asp 634, and Arg 632 triple forms an interaction and gives rise to a second type of pH-sensitive triggering mechanism for opening the transferrin domains, thus allowing iron release.

The two hydrogen bonds involving the aspartic residue that bridges the lysine and arginine residues would be broken or weakened at pH values near the pK_a (pH 4.5) that results in neutral aspartic acid; this could, in turn, lead to an opening of the C lobe domains due to the opposing positive charges of the lysine and arginine residues.

Particularly interesting is the observation that neither lobe of human lactoferrin possesses the Lys–Lys couple, or the Lys–Asp–Arg triple; this observation is in line with the general assumption that the main function of lactoferrin is to serve as an iron-chelating anti-microbial species with no requirement for a pH-dependent iron-release mechanism.[138]

In spite of these improvements in the understanding of the process of iron release from transferrin, the mechanism of iron removal from transferrin remains poorly understood. In the case of some iron chelators, such as 1,2-dimethyl-3-hydroxypy-ridine-4-one (L1), 1-hydroxy-2-pyridinone (1H2P), and acetohydroxamic acid (AHA), a simple kinetics of iron removal from transferrin was observed.[138a] Iron removal from transferrin resulted in the disappearance of the absorption peak at 465 nm and the appearance of an absorbance peak at 375 to 400 nm (Figure 4.13).

Using N terminal monoferric transferrin and C terminal monoferric transferrin preparations, iron removal from the N terminal site was faster than from the C terminal site (Figure 4.13). The rate and the kinetics of iron removal from both N terminal and C terminal sites were modified by the addition of anions; this effect was markedly different for different types of anions.[138a]

4.4.2 STRUCTURAL BASIS OF TRANSFERRIN/TRANSFERRIN RECEPTOR INTERACTION

To deliver iron to the cells, diferric transferrin binds to its membrane receptor. Iron uptake is a process that involves the binding of transferrin to its receptor, then endocytosis of the transferrin–receptor complex, followed by the release of iron from

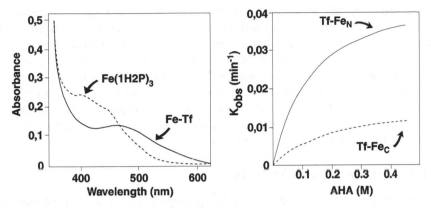

FIGURE 4.13 Left: Visible spectrum of iron-loaded transferrin incubated in physiologic buffer or in the presence of an iron chelator (1-hydroxypyridin-2-one, 1H2P). Right: Plots of observed rate constants versus the concentration of the iron chelator acetohydroxamic acid (AHA). (Reprinted with modifications from Li, Y. et al., *Biochim. Biophys. Acta,* 1387: 89–102, 1998. With permission.)

transferrin at the acidic pH of the endosome. Iron is then utilized for cellular metabolism, while transferrin and its receptor are recycled back to the cell surface.

Little is known about the regions of transferrin and transferrin receptor involved in their interaction. Recent studies have shown that both N and C lobes of transferrin are involved in the interaction with the receptor. This conclusion derives from studies of the binding of either proteolytic fragments[139] containing the N or C lobe of transferrin or the recombinant N or C lobe domain of transferrin.[140] These studies have shown that:

1. Monoclonal antibodies interacting either with the N lobe or C lobe of transferrin produce a marked decline of the level of transferrin binding.
2. The N lobe, in the absence of the C lobe, did not bind to the transferrin receptor and was unable to mediate iron uptake.
3. The C lobe shows a measurable level of binding, although the amount is considerably less than that seen when equimolar N lobe is also present.
4. When equimolar amounts of N lobe and C lobe are added together, levels of binding to receptor and iron uptake only slightly inferior to those mediated by wild-type transferrin are observed.

These studies clearly suggest that both lobes of transferrin contact the receptor. The elucidation of the three-dimensional structure of the transferrin receptor allows us to propose a model to explain the interaction between transferrin and its receptor. In fact, this analysis showed that the transferrin receptor monomer contains three distinct domains, so that the transferrin receptor dimer possesses a butterfly-like shape.[140a] The transferrin receptor molecule forms a bowl-like depression at the top of the molecule and two lateral-facing clefts. The central bowl is too small to allow the interaction with two molecules of transferrin, while the lateral clefts seem to be more adequate sites for mediating interaction with transferrin.

According to this model, several other features of the transferrin–transferrin receptor complex have been outlined: (1) the surface interaction of transferrin with the transferrin receptor mostly implies the C1 domain of the transferrin molecule, with a contribution by part of the N1 domain, at the level of its COOH-terminus; (2) glycosylation sites, which are not involved in the interaction of transferrin and transferrin receptor, are located in a region of the transferrin molecule away from transferrin receptor.[140] This three-dimensional model of transferrin may also explain the mechanism responsible for the facilitated iron release from transferrin following its interaction with the transferrin receptor.

4.4.3 GLYCOSYLATION OF TRANSFERRIN

Human serum transferrin is an iron-binding glycoprotein containing two asparagine glycosylation sites (A and B) in the C terminal part of its single polypeptide chain.[138,141] The primary structures of glycans of transferrin have been studied in detail and it has been established that:

1. They are of the N-acetyllactosaminic type.
2. They constitute a mixture of the biantennary and triantennary types.
3. They are fully sialylated and not fucosylated.[142]

Since one biantennary glycan or one of the tri- or tri′-antennary glycans may be located at either site A or site B, three main types of carbohydrate molecular variants have been characterized: (1) transferrin I, possessing two triantennary glycans; (2) transferrin II, with one biantennary glycan and one triantennary glycan; and (3) transferrin III, with two biantennary glycans (Figure 4.14).

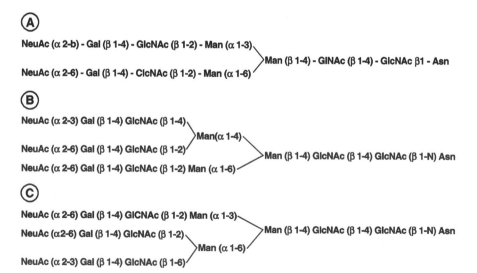

FIGURE 4.14 Primary structures of the biantennary (A) and triantennary (B and C) glycans isolated from human transferrin.

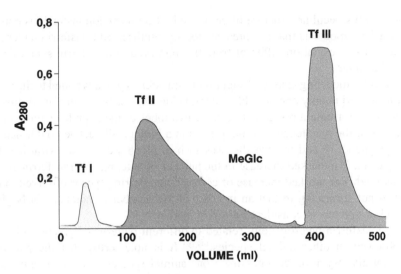

FIGURE 4.15 Affinity chromatography of human serum transferrin on a Con A–Sepharose column. This procedure allows the separation of three transferrin carbohydrate molecular variants called TfI, TfII, and TfIII. (Redrawn with modifications from Leger, D. et al., *Biochem. J.,*. 257: 231–238, 1989.)

The three variants, fully resolved by concavalin A (Con A)-crossed immunoaffinity electrophoresis, are present in normal serum, at relative percentages of 0 to 1%, 15 to 19%, and 80 to 84%, respectively[143] (Figure 4.15). There is no evidence that each of these transferrins variants plays a peculiar biological role. The binding of the three transferrin carbohydrate variants to the receptor of the syncytiotrophoblast plasma membranes was determined. The number of binding sites remained constant, and an increase in the number of triantennary glycans and a decrease up to six-fold in the affinity constant were observed.

Immunoblotting studies provided evidence that each of the three serotransferrin variants binds to a similar placental membrane receptor. Furthermore, similar values of K_a and numbers of receptors were found when the experiments were performed with the completely desialylated carbohydrate variants of serotransferrin, thus indicating that the sialic acid residues and the sugar moieties are probably not involved in the interaction of the serotransferrin variants with the placental membrane receptor.[143]

A more recent study showed that partially deglycosylated transferrin is capable of binding to the transferrin receptor as well as fully glycosylated transferrin; however, aglycosylated transferrin exhibited a markedly reduced capacity to bind to its receptor and donate iron.[144]

Low amounts of asialotransferrin were detected in normal serum. This asialo glycoform of transferrin cannot persist in blood due to the existence of specific hepatic clearance mechanisms. However, in cerebrospinal fluid, 30% of transferrin is present in an unsialylated form. The asialotransferrin is also known as β_2-transferrin and can be distinguished from sialylated transferrin by agarose gel electrophoresis.

Analysis of the glycan structure of this asialotransferrin showed a complex-type agalodiantennary oligosaccharide with bisecting N-acetyglucosamine and proximal fucose.[144a] This peculiar structure of cerebrospinal fluid asialotransferrin is compatible with the hypothesis that it represents locally synthesized transferrin undergoing a process of glycosylation different from that observed in transferrin synthesized at the level of liver.

Molecular modeling studies[145] suggest that transferrin glycans bound to the protein are immobilized in only one possible conformation, the "broken-wing" conformation; this structure, forming a bridge between the two lobes of the peptide chain, probably contributes to the maintenance of the protein in a biologically active conformation.

In pregnancy[143] and in liver diseases such as alcoholic cirrosis, viral hepatitis, and hepatoma[146] a marked increase in the number of antennae of transferrin glycans was observed. The marked increase of complex transferrin glycans observed in liver tumors is not surprising in that an increase of the number of antennae in N-glycosidic-linked glycans is considered a tumor marker.[147,148]

The transferrin heterogeneity related to different patterns of glycosylation was also observed in other animal species.[149,150] It is noteworthy that the pattern of transferrin glycosylation changes from one animal species to another: thus, horse serotransferrin glycans are the N-acetyllactosaminic biantennary type,[149] whereas rat serotransferrin glycans are diantennary, consisting of a trimannosyl-N, N'-diacetyl-chitiobiose core that is frequently fucosylated.[150]

Few pathological conditions are associated with abnormalities of transferrin glycosylation. Particularly interesting are the studies of alcoholism. In recent years, a number of studies indicated a new diagnostic marker of alcohol abuse unrelated to any of the conventional markers. This marker is called carbohydrate-deficient transferrin, and consists of one or two isoforms of transferrins deficient in their terminal trisaccharides.[151] This abnormality seems to represent an ideal marker of alcoholism in that it fulfills all the criteria for such a marker:

1. It detects a biochemical defect specifically related to the presence or metabolism of ethanol.
2. It depends on the amount of ethanol consumed.
3. It is sensitive enough to detect consumption levels associated with somatic and psychiatric risk.
4. It returns to normal levels during abstinence (half-life of the marker is about two weeks).

This transferrin abnormality measures the effects of accumulated alcohol consumption, appearing after regular intake of 50 to 80 g of ethanol per day for at least 1 week.[151] About 2500 individuals were examined; the clinical sensitivity and the specificity (97%) were very high. The biochemical mechanism behind this abnormality of transferrin glycosylation is unknown, but an acetaldhyde-mediated inhibition of glycosyl transfer has been suggested.[151]

Another pathological condition associated with a marked increase of the concentration of abnormally glycosylated transferrin is represented by carbohydrate-deficient glycoprotein syndrome — a new inborn multisystemic syndrome with major nervous system involvement. Clinically, the disease is characterized by psychomotor retardation, olivoponto–cerebellary hypoplasia, peripheral neuropathy, and retinal pigmental degeneration, accompanied by early hepatopathy and several organ dysfunctions.

Biochemically, this disorder is characterized by a complex carbohydrate deficiency in glycoproteins, which is most pronounced in transferrin,[152,153] resulting in very high levels of carbohydrate-deficient transferrin. Carbohydrate-deficient transferrin concentration is also increased in 25% of healthy carriers (heterozygotes) of carbohydrate-deficient glycoprotein syndrome. Biochemical studies provided evidence that the transferrin present in carbohydrate-deficient glycoprotein syndrome exhibits some abnormalities:[154] (1) it is usually of the disialotransferrin type, whereas transferrin is mostly tetrasialylated in normal subjects; and (2) the disialotransferrin possesses sugar chains of reduced size in that it is missing one of two N-linked sugar chains.

Recent studies led to the identification of various types of carbohydrate-deficient glycoprotein syndromes (CDGS). Depending on the type of CDGS, the carbohydrate side chains of glycoproteins are either truncated or completely missing from the protein core. The hypoglycosylation is determined by isoelectric focusing of serum transferrin. The most common form, CDGS type Ia, is caused by phosphomannose-matase (PMM) deficiency[155]; serum transferrin isoelectric focusing showed increased asialo- and disialotransferrin isoforms in three patients.

A variant form of CDGS type I is represented by CDGS type Ib, characterized by a deficiency of phosphomannose isomerase.[156] Interestingly, the isoelectric focusing pattern in CDGS type Ib is identical to that observed in CDGS type Ia, but the clinical presentation is different in that neurological manifestations are absent, while a protein-losing enteropathy represents the main clinical symptom.

CDGS type II is due to a deficiency in the N-acetyl-glucosaminyltransferase II activity which leads to the formation of hypoglycosylated transferrin, particularly to desialylated forms of the molecule.[157]

Type V CDGS is due to a deficit of dolichyl-P-Glc:Man9GlcNac2-PP-dolichyl glucosyltransferase.[158,159] The limited availability of glucosylated lipid-linked oligosaccharides observed in this syndrome leads to the incomplete usage of N-glycosylation sites in glycoproteins. Serum transferrin in these patients is characterized by lack of one or both of the two N-linked oligosaccharides.

Defective glycosylation of serum transferrin was also observed in galactosemia. Serum transferrin isolated from patients with deficiencies of galactose-1-phosphate uridyl transferase exhibited a major defect in glycosylation: the presence of truncated glycan chains which were deficient in galactose and sialic acid.[160]

Abnormally glycosylated transferrin is consistently observed in hepatocellular carcinoma.[161] The transferrin synthesized by hepatoma cells contains two complex-type asparagine-linked sugar chains, as does normal transferrin, but exhibits several abnormalities of the sugar chains: (1) increase of highly branched sugar chains, particularly those with the Galb1>4(Fuca1>3) GlcNAcb1> and Neu5Aca2>3Galb1>GlcNAcb1> groups; and (2) presence of a bisected biantennary sugar chain not detected in normal transferrin.[161]

4.5 TRANSFERRIN GENE EXPRESSION

The complete amino acid sequence of transferrin was determined by Mac Gillivray et al.[35] and the cDNA was cloned and sequenced by Yang et al.[34] Subsequently, the organization of the human transferrin gene was determined by Park et al.[25]

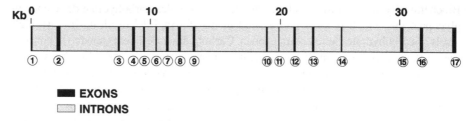

EXONS
INTRONS

FIGURE 4.16 Structural organization of human transferrin gene. The entire locus spans more than 30 Kb. The positions of exons and introns are indicated. Exon numbers are indicated below the bars corresponding to the transferrin gene.

Analysis of the organization of the human transferrin gene indicated that the segments encoding the two protein domains are composed of a similar number of exons; moreover, introns interrupt the coding sequence, creating homologous exons of similar size in the 5′ and 3′ regions of the genes. In fact, human transferrin gene contains 17 exons separated by 16 introns (Figure 4.16). Of the 17 exons, 14 constitute 7 homologous pairs that code for the corresponding regions in the N and C lobes of the protein (Figure 4.16). The first exon encodes for a signal peptide, necessary for the secretion of transferrin molecule, while the last two exons encode a sequence unique to the C terminal. The gene encoding human transferrin was mapped to human chromosome 3, within the region of 3q21–25[162,163] (Figure 4.17).

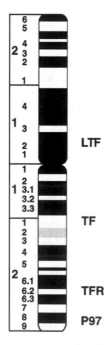

FIGURE 4.17 Mapping of human chromosome 3 showing genes involved in iron metabolism: lactoferrin (LTF), transferrin (TF), transferrin receptor (TFr), and melanotransferrin (p97).

The long arm of chromosome 3 also contains the genes for the transferrin receptor and p97. The exact assignments of these three loci have been refined by analysis of a tumor cell line in which chromosome 3 has been rearranged.[164] These studies showed that the transferrin gene is located at 3q21, the transferrin receptor gene is at 3q26, and the p97 gene is near 3q28.[164] A non-processed pseudogene for transferrin has also been identified and localized to chromosome 3.[163] The transferrin pseudogene possesses the general sequence organization of the parental gene across the region spanning exons 7 to 12 and displays high nucleotide identity, although it does not contain a region homologous to exon 11.

The regulation of transferrin synthesis was carefully investigated at the cellular and molecular levels. Since the transferrin gene is expressed in different tissues, its regulation was investigated in different tissues. These studies provided evidence that the transferrin gene is differentially regulated in tissues.

The liver is the main source of transferrin synthesis,[165-167] although the transferrin gene was reported to be expressed by other tissues deprived of a circulating source of transferrin, such as Sertoli cells in testes,[168] and behind the blood–brain barrier in oligodendrocytes,[169] choroid plexus cells,[170,171] and cerebellum.[171]

Transferrin synthesis has also been reported in placenta, stomach, and spleen; it is produced in low levels in heart and kidney[171] and in high levels in lactating mammary glands.[172] The transferrin mRNA contents of various tissues of normal rats included liver (6500 molecules per cell), testis (114 per cell), brain (83 per cell), spleen (11 per cell), and kidney (5 per cell).[165] Transferrin synthesis in these tissues was also investigated during development. Mouse transferrin could be detected as early as the 2- to 4-cell stage and in early blastocysts of embryogenesis.

Transferrin synthesis during liver development was investigated in great detail. It starts with endodermal cells of the visceral yolk sac.[173] The levels of transferrin synthesis in this tissue are markedly higher than those observed in adult liver, suggesting that the transferrin required for embryonal development is synthesized here before the full development of fetal liver.

This transferrin may represent the immediate source of iron needed for hemoglobin synthesis in developing cells in the blood islands of the visceral yolk sac. In mice, fetal liver transferrin mRNA is detectable as early as 11 days after gestation.[174] During fetal life, transferrin mRNA expression slightly increases, reaching in the immediate post-natal period the adult levels of synthesis.[175] Transferrin mRNA levels in mouse spleen, heart, muscle, kidney, and lung increase during fetal life, reaching maximum levels in the period just preceding birth;[22] these levels drop very rapidly after birth and remain at very low levels during adult life.

Finally, the kinetics of transferrin production in testis and brain during development differ as compared to kinetics described in other tissues; in fact, the level of transferrin expression in these tissues is very low during fetal life and starts to increase only after birth, reaching plateau levels about 1 log lower than those observed in liver.[176]

Transferrin synthesis was also detected in mouse placenta.[177] In fact, transferrin mRNA in placenta at the 12th day of pregnancy corresponded to 1 to 2% that of the adult liver, and at 16 days of pregnancy, was approximately 25% of the concentration in adult liver. Immediately before birth (19th or 20th day of pregnancy), the transferrin mRNA in placenta reached almost the same level as that of adult liver.[177]

Recent studies indicate that transferrin is also synthesized at the level of the lung.[178] High levels of transferrin mRNA were detected in the airway epithelia lining the entire respiratory trees in adult mice, while transferrin gene was expressed at high levels in submucosal glands of normal human lung tissues.[178] Transferrin synthesis by lung tissue is inhibited following ventilatory oxygen support.

4.5.1 REGULATION OF TRANSFERRIN SYNTHESIS IN LIVER, TESTIS, AND BRAIN: CELLULAR STUDIES

The principal site of plasma transferrin production is the liver. This was documented by *in vivo* and *in vitro* studies. *In vitro* studies have shown that liver cells are able to synthesize large amounts of transferrin both at mRNA and protein levels.[164-166] The overall rate of transferrin synthesis in hepatocytes corresponds to about one fifth the rate of albumin synthesis.

In rat liver *in vitro* studies, rates of synthesis of transferrin calculated from the incorporation of (^{14}C)-leucine into precursor pools were comparable to the values of total synthesis observed by *in vivo* incorporation studies.[179,180] Ultrastructural studies carried out on *in vitro*-cultivated hepatocytes provided evidence about the intracellular localization of transferrin in the endoplasmic reticulum and Golgi apparatus.[181]

The transferrin protein initially formed in hepatocytes is a protein transferrin precursor 19 to 20 amino acids longer than the mature protein, but is later split by proteolysis, generating the final transferrin then secreted into the plasma.[182,183] The molecule also undergoes glycosylation before entering circulation. Inhibition of proteolysis prevents secretion, but inhibition of glycosylation has little effect.[182]

Transferrin synthesis starts very early in embryogenesis at the level of the hepatic primordium. *In situ* studies of transferrin mRNA localization in tissues of murine embryos of different ages showed that a clear positivity was observed in the hepatic primordia, thus indicating that transferrin is an early marker of hepatic differentiation prior to the formation of the liver.[184]

Knight et al. demonstrated that steroid hormones and iron status modulate liver transferrin synthesis. These authors showed that treatment with either iron chelators or estrogens caused an increase in the rate of transferrin mRNA of about two-fold magnitude. Treatment with both agents resulted in a synergistic stimulation of transferrin synthesis, whereas iron loading did not block the stimulatory effect of estrogen on transferrin synthesis.[165-167] These results suggest that iron deficiency and estrogen interact with the liver transferrin gene through separate regulatory mechanisms. This was confirmed by a series of molecular studies discussed below.

Estrogen can also regulate the expression of transferrin gene in a similar fashion in transgenic mice. Additional mechanisms controlling liver transferrin synthesis are represented by various factors, including cell density,[185] dexamethasone,[185] human growth hormone,[186] temperature,[187] and retinoids.[188]

Transferrin synthesis in hepatic cells is also modulated by cytokines and hypoxic stimuli. Transferrin is down-regulated during the acute phase response in human serum and hepatoma HepG2 cells.[189] Transferrin transcription significantly decreased in response to pro-inflammatory cytokines, such as interleukin-6.[189] Other studies suggested a possible physiological role of oxygen in the control of transferrin gene expression.

Initial studies reported an up-regulation of transferrin serum protein concentrations in mice and rats[191] exposed to hypobaric hypoxia for 1 to 3 days. Subsequent studies arose from observations showing that oxygen level represents a key mechanism of control of the rate of erythropoiesis (through control of the level of erythropoietin synthesis) and intestinal absorption.

Since hypoxia was shown to increase erythropoiesis, it seemed conceivable that an increase in plasma iron transport capacity was required to sustain the increased level of erythropoiesis induced by hypoxia via stimulation of erythropoietin synthesis. In line with this observation, hypoxia was found to induce a marked stimulation of transferrin gene transcription. The molecular basis of this oxygen-dependent regulation of transferrin gene expression was investigated using a cell culture mode capable of expressing transferrin mRNA in an oxygen dependent manner.[192] Within the liver-specific transferrin gene promoter, a 32-base pair hypoxia-responsive element was identified. This element contained two hypoxia-inducible factor-1 (HIF-1) binding sites.[192] Both binding sites function as oxygen-regulated enhancers of transferrin gene expression.

Transferrin synthesis in testis was actively investigated. Transferrin is an important marker for Sertoli cell function since its synthesis and secretion may directly affect the germ cells. Transferrin was demonstrated to be an important Sertoli cell secretion product by Skinner and Griswold.[193]

In situ hybridization has confirmed that transferrin is synthesized specifically by Sertoli cells in the testes, and this synthesis varies with the stages of the cycle of the seminiferous epithelium.[194] An extensive network of tight junctions between adjacent Sertoli cells constitutes the diffusion-tight blood–testis barrier that divides the seminiferous tubule into anterior (apical or adlumenal) and exterior (basal) compartments and prevents the access of serum transferrin to the apical compartment. It has been proposed that testicular transferrin synthesized and secreted by Sertoli cells has the function of delivering iron to germ cells.[195]

Huggenvik et al. proposed a model in which iron delivered to the basal sides of Sertoli cells by serum transferrin is bound by testicular transferrin synthesized into the cytoplasm of Sertoli cells for subsequent delivery to the apical compartment.[195] Recently, techniques described to study the polarity of epithelial cell functions have been applied to the culture of Sertoli cells in a twin-chamber culture system providing clear evidence that transferrin is vectorially secreted, with predominant secretion at the apical side.

It must be emphasized that the delivery of iron is critical to the process of spermatogenesis, and the means by which the expression of transferrin in testes is regulated is of great importance. This phenomenon was carefully investigated and several modulators able to modify the rate of transferrin synthesis in Sertoli cells were identified.

Sertoli cells respond to FSH (follicle-stimulating hormone), insulin, retinol, testosterone, fibroblast growth factor, insulin-like growth factor and epidermal growth factor.[194-202] Maximum stimulation occurs when the Sertoli cells are stimulated with a combination of these agents (i.e., FSH + insulin + retinol).

Other studies have provided evidence that the interaction of germ cells with Sertoli cells stimulates the latter to secrete transferrin: this phenomenon seems to be mediated by low molecular weight soluble proteins secreted by germ cells.[203,204] Cytokines released by immune cells present in testes modulate the capacity of Sertoli cells to secrete transferrin in response to FSH.[205]

It is of interest that all these stimuli favor secretion from the apical sides of Sertoli cells. Some cytokines stimulate transferrin synthesis by Sertoli cells. Thus, interleukin-6 released by immune cells present in the testes[206,207] and tumor necrosis factor-α secreted by immune cells and/or germ cells[208] induce a significant enhancement of transferrin synthesis by Sertoli cells. The stimulatory effect of TNF-α on transferrin synthesis was considerably stimulated by the contemporaneous addition of retinoic acid, thus suggesting that these two molecules stimulate transferrin synthesis in Sertoli cells through different mechanisms.

Sertoli cells are heterogeneous in their capacity to synthesize transferrin *in vivo* and *in vitro* under stimulation with FSH.[209,210] More particularly, transferrin was found to be released minimally from early-stage segments of the Sertoli epithelium, and maximally from later-stage segments.[210] However, only early-stage segments respond to FSH with a markedly increased rate of transferrin secretion.[210]

It is important to understand the role played by factors released by peritubular cells in modulating transferrin synthesis by Sertoli cells.[211,212] Several studies have provided evidence that peritubular cells that surround the seminiferous tubules produce a paracrine factor, termed PModS (peritubular factor that *mod*ulates *S*ertoli cell function), that exhibits dramatic effects on Sertoli cell functions *in vitro*.

Peritubular myoid cells are target cells for androgens, and *in vitro* studies suggest that the production of PmodS by these cells is increased by androgens. The paracrine factor PModS stimulates 20-day-old rat Sertoli cell function *in vitro*, including transferrin production, to a greater extent than any single agent previously identified, including FSH.[211]

Another study carried out during pubertal development provided evidence that PModS and Sertoli cells cooperate in the prepubertal period to promote Sertoli cell differentitation and synthesis of transferrin, and PModS may act in the adult testes to maintain optimal Sertoli cell function and synthesis of transferrin.[212]

Although two factors with PmodS activity have been purified from peritubular cell-conditioned medium, their identities and exact physiological roles remain elusive. A recent study suggests that heregulin differentiation factors (particularly HER-3 and HER-4) may be somewhat responsible for PmodS activity.[213]

Transferrin is also synthesized at the sites of development of female germinal cells. Thus, transferrin[214,215] or a transferrin-like protein[216] is present in the follicular fluids of growing human follicles. The follicular concentration of transferrin exceeds that present in serum, which suggests either local synthesis or a preferential concentration mechanism.[216,217] Transferrin is also required to sustain the *in vitro* growth of granulosa cells[218]; transferrin receptors are present throughout follicle maturation.[217]

The differentiation and proliferation of granulosa cells are controlled by gonadotropins from the hypothalamic–pituitary unit, but may also be influenced by intragonadal peptides, growth factors, and hormone-binding proteins produced by or already present in the follicular fluids of growing follicles. In this complex hormonal network of granulosa cell function and differentiation, transferrin seems to play a role in that it acts as a negative modulator of the stimulatory actions of FSH and IGF-I on granulosa cells.[219]

The brain is separated from the rest of the body by the blood–brain barrier and the blood–cerebrospinal fluid barrier. The blood–brain barrier prevents brain cells from having free access to serum transferrin. This observation raised a question about the origin of transferrin present in the central nervous system. Recent studies provided evidence that transferrin is synthesized *in situ* in the central nervous system; the level of transferrin synthesis in nervous tissue is significantly lower than that observed in hepatic tissue.

Transferrin mRNA synthesis in the brain was first demonstrated by Lee et al.,[220] in the human brain, by Levin et al.,[175] and finally by Idzerda et al.[167] in the rat brain. Several studies were devoted to identifying the cell types involved in transferrin synthesis in the central nervous system. These studies revealed that some differences were observed in different animal species.

In humans, immunostaining experiments using antibodies specific for human transferrin showed that reactivity is confined to areas where oligodendrocytes are present.[221,222] At the level of central cortex in the gray matter, the transferrin immunoreaction product is predominantly located in small, round cells dispersed throughout layers II through VI of the cerebral cortex and corresponding to perineuronal oligodendrocytes. Pyramidal cells in this site are negative or only weakly positive. In rats and mice, the cells that synthesize the majority of the transferrin mRNA made in the adult brain are oligodendrocytes[169] and choroid plexus,[170,171,223,224] based on tissue RNA extracts.

These findings also were confirmed by immunohistochemistry studies showing selective staining of oligodendrocytes and epithelial cells of the choroid plexus in the central nervous system of adult rats.[225-228] Quantitative studies of transferrin in the adult rat central nervous system found that this glycoprotein is more abundant in white matter than in gray,[226] and is dramatically reduced in mutant rats that lack mature populations of oligodendrocytes.[227,228] Particularly relevant are the results obtained in studies of the synthesis of transferrin in rat choroid plexus.

The choroid plexus, an extension of the ependymal layer covering the ventricular surface of the brain, is a single structure composed of a monolayer of cuboidal epithelial cells with tight junctions underlined by mesenchymal fibroblasts and basement membrane-bordering capillaries. The apical membrane of the epithelial cells faces the cerebrospinal fluid, and the basolateral membrane rests on mesenchymal fibroblasts facing the blood. The choroid plexus has two main functions: it forms the blood–central nervous system barrier; and it is the major site of cerebrospinal fluid production, secreting two-thirds of this biologic liquid. Studies of rat choroid plexus have shown that this organ synthesizes transferrin in large quantities, about as much as the liver on a per-gram tissue basis.[228]

Upon incubation of choroid plexus pieces with (^{14}C)-leucine *in vitro*, about 2% of newly synthesized protein in cells and 4% of the radioactive protein secreted into the medium were found to be transferrin.[171] However, studies carried out on different animal species have shown that the capacity of choroid plexus to synthesize transferrin is species-specific.[229] These studies show that:

1. Large proportions of transferrin mRNA in choroid plexus were found only in rats.
2. Small proportions of transferrin mRNA were observed in RNA in choroid plexus from mice, dogs, and rabbits.
3. No transferrin mRNA was detected in choroid plexus from humans, sheep, pigs, cows, and guinea pigs.

This study clearly indicates that strict species specificity is observed for the pattern of cerebral expression of the transferrin gene.[229]

The distribution of transferrin in human brain is not coincident with distributions of transferrin receptors and melanotransferrin. In fact, transferrin receptors and melanotransferrin are localized at the level of capillary endothelium, while transferrin is mainly localized to glial cells.[230]

A series of studies have shown that developing neurons also synthesize transferrin, but at low levels. The secreted protein was immunoprecipitated from chick spinal cord cultures by Stamatos et al.[231]

Immunohistochemical examination of rat myelinated and non-myelinated peripheral nerves provided evidence about cytoplasmic accumulation of transferrin in Schwann cells of the myelinated sciatic nerve, but not in the myelinated sciatic nerve or in the myelinated cervical sympathetic trunk.[232] These observations are supported by other studies showing the absolute requirement for transferrin by *in vitro* grown neuroblasts,[233] the existence of transferrin receptors in developing neurons,[234] and the modification of neuronal physiological properties elicited by transferrin (increases in intracellular calcium and of (^{3}H)-glutamate).[235]

All the above studies were carried out *in vivo* and thus describe the *in vivo* patterns of transferrin gene expression in central nervous systems of different animal species. De Los Monteros and coworkers[236] studied the synthesis of transferrin protein and transferrin mRNA expression in the three main populations of nervous system: astrocytes, neurons, and oligodendrocytes grown *in vitro*.

All three cell types are able to synthesize transferrin *in vitro*, even if oligodendrocytes display this property to a major extent.[236] Thus, all three major brain cell types have the potential of synthesizing transferrin in cell culture. These observations strongly suggest the existence of factors which, *in vivo*, restrict the expression of transferrin solely to oligodendrocytes in the adult brain.

A recent study investigated the synthesis of transferrin in the developing human brain. Immunocytochemical studies showed that transferrin was present in oligodendrocytes, astrocytes, and neurons.[237] Transferrin-positive neurons appeared at 18 weeks and 22 weeks of gestation in Purkinje cells and the pontine reticular formation and pontine nuclei, respectively. In parallel, transferrin-positive glia cells also appeared.[237] It was suggested that this early appearance of transferrin in the brain is required to transport the iron required for the synthesis and functioning of dopamine, serotonin, and γ-aminobutyric acid (GABA).[237]

The possible action of transferrin at the level of the brain is the regulation of oligodendrocyte development. *In vitro* studies have shown that transferrin acts as a stimulator of myelin synthesis. In fact, intracranial injection of apotransferrin in young rats induces increased myelination; this effect is mediated through stimulation of the expression of certain specific myelin protein genes.[238]

Concerning the regulation of transferrin synthesis in nervous system tissue, two main observations were made. First, transferrin mRNA levels in astrocyte cultures seem to be under hormonal control, as hydrocortisone reduces these levels.[236] This suggests that this hormone may also be a physiological regulator *in vivo*. In primary cultures of epithelial cells of rat choroid plexus, treatments with cAMP analogs decreased the levels of synthesized and secreted transferrin in the medium. However, serotonin increases these levels and transferrin mRNA values.[171] These findings suggest that the production and secretion of transferrin in the central nervous system are regulated by hormones and neurotransmitters.

Finally, some studies show that lymphoid cells are able, at least to some extent, to synthesize transferrin. Thus, peripheral blood mononuclear cells have been shown to express transferrin, with the activity residing in the T cell population.[239,240] *In situ* hybridization provided further evidence that activated CD4+ lymphocytes are the producers of transferrin.[241] It was suggested that in the sequence of autocrine events leading to T lymphocyte activation, IL-2 induces the synthesis of transferrin; subsequently, IL-2 receptors and transferrin receptors are induced, thus leading to DNA synthesis and cellular proliferation.[241] However, it must be pointed out that the levels of transferrin synthesized by CD4+ lymphocytes are low and probably not sufficient to sustain T lymphocyte proliferation. Transferrin is produced in low amounts by normal human fibroblasts and its synthesis is significantly decreased following malignant transformation of these cells.[242]

4.5.2 REGULATORY SEQUENCES INVOLVED IN THE REGULATION OF TRANSFERRIN GENE EXPRESSION OBSERVED IN DIFFERENT TISSUES

Several studies have been performed to identify the regulatory transcriptional elements implicated in regulation of transferrin gene expression in the liver, brain, and testes. In early experiments, acting DNA elements required for liver-specific expression of human transferrin gene were identified by transient and stable expression assays in hepatoma and epithelial carcinoma cells. Deletion analysis of the 5′ DNA sequences of the gene led to the definition of four functionally different regions:[243]

1. A cell-type specific promoter localized between positions −145 and -45 that interacts with two nuclear factors and is sufficient for liver-specific expression.
2. A distal promoter region from −620 to −125 base pairs containing positive and negative *cis*-acting elements that regulate the promoter activity.
3. A negative-acting region between −1.0 and −0.6 kilobase pairs that downregulates transcription from the transferrin promoter.
4. An enhancer located between −4.0 and −3.3 kilobase pairs that is more active in hepatoma than in HeLa cells (Figure 4.18).

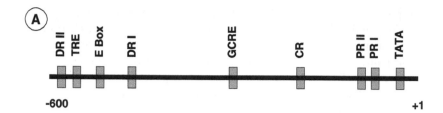

FIGURE 4.18 Schematic representation of the promoter region (A) and 5′ flanking region of the human transferrin gene (B). Regions relevant to the regulation of transferrin gene expression through binding of transcriptional factors are outlined by boxes.

The 5′ DNA sequences required for human transferrin expression in the liver have been determined.[244,245] Transgenic animals containing only –0.2 or –0.3 Kb of the transferrin 5′ flanking region were scarcely expressed. Transgenes containing transferrin 5′ sequences –0.67, –1.2, or –3.5 Kb were expressed at high levels in liver as well as in brain.[245] Furthermore, the DNA sequence between –300 and –670 bp of the 5′ flanking region was required for liver and brain expression of the transferrin gene. Finally, transgenes containing only –139 bp of the transferrin 5′ flanking region are scarcely expressed, but still retain their liver specificity, thus suggesting the presence of a liver-specific transcriptional element between –139 and +50.[244]

In vitro binding assays using the first 620 nucleotides of the 5′ flanking region of the transferrin gene and nuclear extracts from liver cells revealed the existence of five protein-binding sites[246] (Figure 4.18): minimal cell-specific promoter; PRI (from –76 to –57 bp) and PRII (from –103 to –83 bp); and two distal promoters, DRI (from –480 to –454 bp) and DRII. The first two elements, PRI and PRII, comprise part of the tissue-specific promoter, and their integrity is required for full promoter activity.[243] It is of interest that with *in vitro* transcription assays, only the PRI site seems to be required for optimal transcriptional activity,[247] whereas in transfection assays, both PRI and PRII sites are required for optimal transcriptional activity.

Expression of transferrin in Sertoli cells and in liver cells is dependent upon the proximal promoter. However, at variance with liver cells, in cultured Sertoli cells a –34 to –18 bp TATA box binding factor is sufficient to initiate basal levels of transferrin gene transcription.[248] Efficient expression in Sertoli cells is related to the association of two factors binding to the transferrin 5′ flanking region between 82 and −1 or between −153 and −52 bp; these two regions can be identified as the PRI and PRII regions previously mentioned. However, the functional roles of these two regions seem to be different in Sertoli cells and in liver cells in that in liver cells the only mutation of these two regions that substantially modifies the promoter activity is the PRII mutation.[249]

At variance with hepatic and nervous tissues, in Sertoli cells the PRII region of the transferrin promoter binds a cyclic-AMP (cAMP) response element binding protein.[250] In spite of this finding, PRII shows no homology to the consensus cyclic-AMP response element which is important for the induction of transferrin synthesis in Sertoli cells elicited by FSH.

In addition to the proximal promoter, the expression of transferrin in Sertoli cells requires the integrity of an E-box response element, located at different positions in human (−506 to −501) and murine (−327 to −322) transferrin promoter.[251,252] This E-box response element can be activated by bHLH factors, such as E12, present in Sertoli cells. Inhibition of Sertoli bHLH factors by Id factors determines suppression of Sertoli-differentiated functions, such as transferrin synthesis.[251,252] This observation suggests that bHLH transcription factors play a key role in regulating transferrin synthesis in Sertoli cells. The E-box region, in cooperation with the cyclic-AMP response element located at the level of the PRII site, is also important for conferring to the transferrin promoter the capacity to be induced by FSH.[253]

Deletion experiments using diffent constructs of the murine and human 5′ region upstream the transferrin promoter showed that, in addition to the above-mentioned regions, other elements were involved in the control of transferrin gene expression in Sertoli cells. An additional E-box was located upstream of the other E-box at the level of the −1800 nucleotide. One region, not yet characterized, located between nucleotides at positions −2600 to −3000 seemed to exert an inhibitory effect on transferrin transcription. Finally, an enhancer element at position −3300 to −3600 seemed to be present.[252]

The nuclear proteins of liver cells capable of interacting with PRI and PRII sites have been characterized. The protein that binds to the PRII site is a member of the C/EBP family of transcription factors, as shown by DNAaseI footprinting and supershift experiments.[134,254]

The C/EBP transcription factors are members of the leucine zipper family; their role consists of facilitating the coordinated expression of proteins required by a number of different tissues. C/EBP-α, C/EBP-β, and C/EBP-γ transcription factors were described.

C/EBP-α is abundantly expressed in liver cells and functions as a strong positive modulator of transferrin gene transcription. However, targeted disruption of the gene encoding C/EBP-α in adult mouse liver did not result in any significant decrease of the rate of transferrin transcription, thus suggesting that this transcription factor does not play a relevant role *in vivo* in the transcription of the transferrin gene in liver cells.[255]

The protein that binds to the PRI site in mice was named PYBP.[256] It does not seem to be implicated in the cell-type specific expression of the transferrin gene, but may be involved in transcriptional activation. This factor was called PTB in human cells.[257]

Another liver protein able to interact with PRI was recently identified as HNF-4 (hepatocyte nuclear factor 4) a novel member of the steroid hormone receptor super-family.[258] It seems to be involved in the expression of a series of hepatocyte-specific genes. This transcription factor is restricted to few tissues including liver, kidney, and intestine. Additional evidence supports an important role of HNF-4 in the control of liver transferrin transcription. In fact, it was shown that drugs called hypolipidemic peroxisome proliferators suppress transferrin transcription in liver.[258] This inhibitory effect depends on the binding of the complex between the hypolipidemic drug and its receptor at the level of the PRI site of the transferrin promoter, thus inhibiting the binding of HNF-4.[259]

An additional transcriptional factor capable of interacting with the PRI region of the transferrin promoter is COUP-TF, a member of the steroid receptor family that functions either as a positive or negative modulator of transcription.[260] HNF-4 and COUP-TF function separately as transcription activators of the transferrin gene; the major activating species is HNF-4, and the two factors showed antagonistic interactions with the two proteins binding to the same site.

Transfection experiments carried out in hepatoma cells showed antagonistic effects between proteins binding to the PRI sites and those binding to the PRII site. Thus, the HNF-4- and C/EBP-α-mediated activation of transferrin gene transcription is partially repressed by equimolar amounts of the factor binding to the adjacent site.[260]

Interestingly, experiments with nuclear extracts derived from Sertoli cells provided evidence that:[261] (1) distinct combinations of Sertoli proteins SP-A and SP-D and COUP-TF bind to the PRI site, while SP-α and SP-β bind to the PRII site; and (2) the testes protein interacting with PRI could not be identified as C/EBP, HNF-4, or PTB.

As noted above, in the 5' flanking region of the transferrin gene, in addition to the promoter regions, a 300-bp long enhancer, located 3.6 kb upstream of the cap site of the human transferrin gene was identified. Competition footprint analysis, gel retardation assays, and transient expression studies in hepatoma and HeLa cells showed that the enhancer is composed of two structural and functional domains, A (nucleotides 1 through 86) and B (nucleotides 87 through 291).[262]

Domain A seems to act as a proto-enhancer able, when multimerized, to stimulate the transcription of an heterologous promoter in both HeLa and hepatoma cells, and contains an octanucleotide 5'-TGTTTGCT-3' sequence capable of interacting with two different nuclear proteins. One of these two nuclear proteins was identified as HNF-3-α, a member of a liver-enriched transcriptional factor family.[263,264] A direct role of HNF-3-α in sustaining the transcription of transferrin in liver cells is directly supported by experiments showing that over-expression of a truncated HNF-3-α factor in hepatoma cells produces a dramatic decrease of transferrin transcription.[265]

Targeted disruption of the gene encoding HNF-3-γ resulted in a marked decline in the level of liver transferrin transcription.[266] This last observation further supports the role of hepatocyte nuclear factor transcription in the control of transferrin gene transcription, and suggests that the other members of the HNF family (i.e., HNF-β and HNF-γ), in addition to HNF-α, are involved in the control of the transcription of this gene.

Domain B contains four binding sites interacting with several liver nuclear proteins. The binding of each of these proteins requires the contemporaneous binding of all the other proteins. The function of this enhancer consists of blocking the activity of a downstream negative element, but it has no enhancer activity by itself. The full enhancer activity on the transferrin gene promoter requires the comtemporaneous presence of both A and B domains.

Interestingly, in Sertoli cells, the domain B enhancer is completely inactive (i.e., unable to bind testes nuclear protein), whereas domain A is able to bind a protein different from those identified as capable of interacting in the nuclei of liver cells with domain A enhancer.[248] The absence of a functionally active enhancer of transferrin gene in Sertoli cells may in part explain the lower capacity (1/10) of these cells to synthesize transferrin, as compared to liver cells in which the enhancer is fully active.

As mentioned above, at the level of the brain, both glial and neuronal cells have the potential for synthesizing transferrin. DNA sequences have been identified that mediate transcription of the transferrin gene in primary cultured oligodendrocytes, in neuroblastoma cell lines, and in choroid plexus epithelial bases.[267,268] In all three cell types, the transcription of transferrin gene is governed by different types of transcription factors interacting with three adjacent boxes present at the level of the proximal −164 to +1 promoter region (Figure 4.18).

The PRI site, located more proximally to the start codon of transcription, binds the chicken ovalbumin upstream promoter transcription factor. C/EBP-α binds to the PRII site, and the central region I-binding protein (CRI-BP), which belongs to the cyclic-AMP response element binding protein (CREB) family, interacts with the CRI site. In line with these observations, the 0.67-Kb fragment of the regulatory region of the transferrin promoter allows brain-specific expression of transgenes.[269]

It was also shown that a 5′ upstream region stimulates the expression of the transferrin promoter: in neuronal cells, the −1140 to −872 region exerts a stimulatory action on transferrin promoter.[267] This region contains two different sequences called upstream region I (URI) and upstream region II (URII), that interact with a member of the steroid/retinoid receptor family and nuclear factor I (NFI), respectively.[270]

These observations suggest that the regulation of transferrin gene expression is a complex phenomenon that depends on a balance between the effects of different *trans*-acting transcriptional factors whose expression seems to be cell-specific.

In the hepatic system, the transferrin gene may use two liver-enriched transcriptional factors, C/EBP-α and HNF-α in the promoter, and HNF-3-α and other ubiquitous factors in the enhancer to allow a high level of expression of this gene (Figure 4.19).

Furthermore, between these two regions, a positive/silencer region was found. The stimulatory role of the positive region is inhibiting the activity of the silencer. In Sertoli cells, where the enhancer is less relevant in the control of transferrin expression, the regulatory promoter elements PRI and PRII participate differently in the mechanism controlling transferrin gene transcription.

The transcription factors interacting with these sequences are different from C/EBP-α and HNF-4 and, particularly relevant, is the role of CREB-BP interaction with PRII (Figure 4.20).

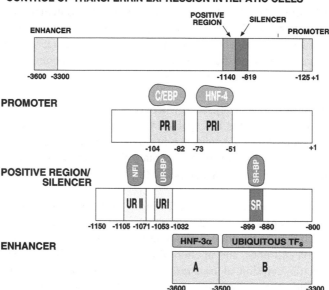

FIGURE 4.19 Main DNA elements involved in the control of transferrin expression in hepatic cells. A promoter region (+1 to –130), a silencer region (–819 to –1000), a positive region (–1000 to –1140), and an enhancer region (–3300 to –3600) were identified. At the level of the promoter region, two nucleotide binding boxes were identified (PRI and PRII). At the level of the silencer region, a single nucleotide binding box involved in the binding of a protein (not well characterized and called silencer region binding protein) was identified. At the level of the positive region and counteracting the inhibitory activity of the silencer region, two binding regions were detected (URI and URII). Finally, two domains were observed at the level of the enhancer region (A and B). The final level of transferrin transcription in liver cells depends on the balance between stimulatory and inhibitory influences mediated by different regions of the promoter and different transcription factors.

In addition to these sites, a region located upstream of the promoter contains E-box sequences involved with the PRII site in the control of hormonal-dependent transferrin expression in Sertoli cells (Figure 4.20). In neuronal cells, where the enhancer is inactive, the rate of transferrin transcription is controlled by a promoter region and a positive region. The promoter region contains three binding sites (PRI, PRII, and CRI) involved in the binding of transcription factors different from those found in liver (Figure 4.21).

Transferrin gene expression is also modulated by iron. The body defends itself against toxic levels of iron by increasing its storage capacity for the metal and decreasing the ability of iron to gain entry into intracellular compartments. As a consequence, in the presence of iron overload, the ferritin storage protein increases in concentration, while the production of transferrin and transferrin receptor decreases; an opposite phenomenon is observed in the presence of iron deficiency.

In rats, diet-induced iron deficiency stimulates liver transferrin synthesis at the level of transcription, whereas chronic iron overload does not affect levels of serum

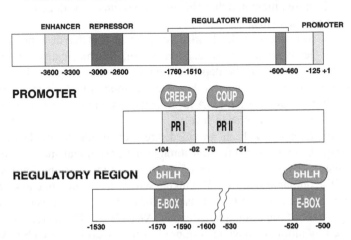

FIGURE 4.20 Main DNA elements involved in the control of transferrin expression in Sertoli cells: a promoter region (+1 to –125), a regulatory region, a repressor region, and an enhancer. The promoter region contains two binding sites called PRI and PRII, interacting with COUP and CREB-BP, respectively. The regulatory region includes two E-box binding sites. The repressor region and enhancer region have been characterized only from a functional point of view and not in terms of binding capacity with nuclear proteins.

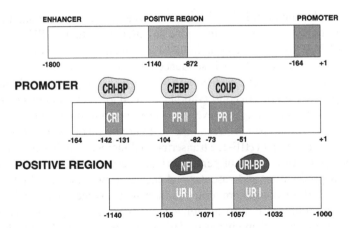

FIGURE 4.21 Main DNA elements involved in the control of transferrin expression in the brain. A promoter region (+1 to –164) and a positive regulatory region (–872 to –1140) have been identified. At the level of the promoter region, three nucleotide-binding boxes were identified (PRI, PRII, and CRI). At the level of the positive regulatory region, two nucleotide-binding boxes corresponding to URI and URII, respectively, have been identified.

transferrin.[167] However, in cultured rat hepatocytes, levels of secreted transferrin decrease with iron treatment and this phenomenon seems to be dependent upon post-transcriptional mechanisms.[271]

In humans, iron overload leads to a decrease in transferrin expression, while iron deficiency leads to an increase. Here again, the mechanism of transferrin modulation by iron seems to be dependent upon post-transcriptional mechanisms. Humans with iron overload related to hereditary hemochromatosis and normal individuals have similar transferrin mRNA levels in the liver; however, hemochromatosis patients have decreased transferrin levels.

A study by Cox and Adrian provided direct experimental evidence that iron modulates transferrin gene expression through a translational mechanism.[272] This conclusion was derived from studies of transgenic mice carrying chimeric genes composed of the human transferrin 5'-flanking sequences (from −620 to 46 bp) fused to the chloramphenicol acetyl transferase (CAT) reporter gene. Liver expression of the protein product of the human transferrin-CAT (hTF-CAT) transgenes was markedly decreased by intraperitoneal injections of iron; however, liver hTF-CAT mRNA levels were unmodified by the iron treatment.[265]

Since the only transferrin sequences present in the hTF-CAT mRNA corresponds to the first 46 bp of the 5' untranslated region, it is assumed that this region must contain an iron regulatory element. Analysis of this sequence revealed a potential stem–loop structure containing 9 of 11 conserved bases of the iron regulatory element, previously reported in the untranslated regions of ferritin and transferrin receptor mRNAs.

The overall conformation of the human transferrin 5' untranslated region stem–loop predicted by computer modeling differs from the ferritin and transferrin receptor iron regulatory element by the presence of a larger loop and different stem pairing.

RNA–cytoplasmic protein binding assays revealed the presence and the specific binding of liver cytoplasmic protein able to interact with the IRE sequence of transferrin mRNA. In iron-treated animals, a significant decrease of the binding of this cytoplasmic protein to the IRE sequence of transferrin was observed.[272]

As outlined above, recent studies have shown that transferrin synthesis is stimulated by hypoxic stimuli. This observation is related to recent findings that virtually all cells are able to sense oxygen levels and respond to hypoxic stimuli.

A universal mechanism of response to hypoxia involves the expression of hypoxia-inducible factor 1 (HIF-1, a transcriptional factor), which acts as an activator of the transcription of a series of genes, including erythropoietin, which stimulates red blood cell production, and vascular endothelial growth factor, which stimulates angiogenesis. Both responses are aimed at improving oxygen delivery.[273]

Transferrin synthesis is also modulated according to oxygen levels through a transcriptional mechanism mediated by HIF-1. At the level of the transferrin gene enhancer, located at position −3600 to −3300 with respect to the transcription start site, a 32-base pair hypoxia-responsive element containing two HIF-1 binding sites (whose sequences are TGCACGTA and CGCACGTA, respectively) was identified.[274] The binding of HIF-1 to these sites is greatly enhanced by hypoxia.[274]

Finally, a recent study showed a peculiar element of regulation of transferrin expression. Transferrin mRNA is one of the few mRNA species that has a short poly (A) tail (<20-nt poly (A) tail).[275] This property is genetically determined by an element homologous to the albumin gene PLEB present within the terminal exon of the transferrin gene.[275]

4.6 TRANSFERRIN AND MICROBIOLOGY

One common factor among the complex and not fully understood interactions between a bacterial pathogen species and its hosts is the capacity of the bacterium to grow in host tissues. In extracellular mammalian body fluids, due to the presence of transferrin and lactoferrin, free ionic iron level is maintained at very low levels (about 10^{-18} M) that cannot sustain the growth of bacteria.

All microorganisms, with the exception of certain lactobacilli that utilize manganese and cobalt as biocatalysts in place of iron, require iron for growth. To overcome this major limitation, bacteria developed mechanisms mediating the uptake of transferrin or lactoferrin.

These mechanisms have been intensively investigated in Gram-negative bacteria that require iron for growth and acquire this iron either through synthesis and secretion of low-molecular-weight iron chelators known as siderophores or through the uptake of transferrin.

Recent studies have shown that Gram-negative bacteria possess the capacity to uptake iron required for growth through three pathways involving different membrane receptors: (1) iron uptake from transferrin or lactoferrin; (2) iron uptake from iron siderophores synthesized and secreted by bacteria; and (3) uptake from iron in the form of heme bound to hemopexin and hemoglobin[276] (Figure 4.22).

Recent investigations have elucidated the mechanisms of bacterial iron uptake through the pathway of siderophores. Since siderophores are present at extremely low concentrations, they must bind to outer membrane-specific receptors. These receptors bind siderophore–iron complexes and promote their active transport into the periplasm. Recent crystallographic studies[277,278,278a] have shown that siderophore–iron first binds to surface-located loops of the siderophore receptor FhuA, then, following a conformational change of the receptor, the ligand is partitioned from the external medium into the external pocket by its affinity for aromatic residues. The subsequent step is transport of ferrichrome iron to the periplasm, producing disruption of the iron-binding site and interaction between FhuA and TonB. TonB is a cytoplasmic membrane involved in providing energy to transport iron from host iron-binding proteins across the outer membrane into the periplasmic space in a number of Gram-negative bacteria.

Pathogenic bacteria that multiply successfully in body fluids to establish extracellular infections have evolved high-affinity uptake systems that allow them to compete effectively with the iron-binding glycoproteins, transferrin or lactoferrin of the host for essential iron.[279-282] The best-understood systems whereby bacterial pathogens assimilate iron from these iron-binding proteins are those that depend on the production of siderophores. However, some pathogens, such as *Haemophilus*

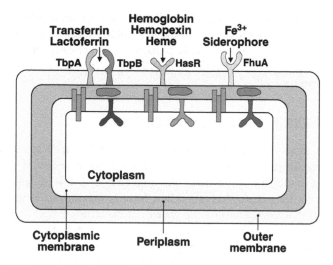

FIGURE 4.22 Iron uptake pathways of Gram-negative bacteria. These bacteria utilize different iron sources: (1) iron bound to transferrin and lactoferrin; (2) heme bound to hemoglobin or hemopexin of the host; and (3) siderophores synthesized by the bacteria. The iron-bound forms are transported into bacteria through interactions with specific receptors present on the outer membrane: (1) receptors for lactoferrin and transferrin composed of two subunits (TBPA and TBPB); (2) HasR, the receptor for hemoglobin and hemopexin; and (3) FhuA, the receptor for ferrichrome-iron (siderophore iron-bound). Bacterial iron uptake is an active process requiring energy derived from the proton gradient across the cytoplasmic membrane, a process mediated by a proteic complex formed by TonB, ExbB, and ExbD. At the level of periplasm, specific proteins, such as FhuD, act as intermediate iron transporters, each specific for each type of iron (transferrin iron, heme iron, siderophore iron). These intermediate iron transporters then deliver the iron compounds to other iron transporters located at the level of the cytoplasmic membrane. Each cytoplasmic transporter is specific for each type of iron compound and translocates iron into the cytoplasm through a transport process requiring ATP. (Redrawn with modifications from Braun, V. et al., *TIBS,* 24: 104–109, 1999.)

and *Neisseria* species, use siderophore-independent receptor-mediated iron uptake systems involving a direct interaction between the bacterial cell surface and the iron-binding protein. These systems are now under investigation, and they are distinguished from the known siderophore-mediated mechanisms by their high degree of specificity for the iron-binding glycoproteins.[283-286]

This has implications for explanations of host specificities of these pathogens and for the development of animal models.[287] Although the molecular mechanisms of iron uptake in such microorganisms are not fully understood, considerable evidence shows that saturable membrane receptors specific for either transferrin or lactoferrin are involved.[288-291]

These transferrin and lactoferrin receptors are distinct entities, and their expression has been shown to be regulated by iron. However, the biochemical identities of the receptors remain unclear, although lactoferrin- and transferrin-binding proteins have been identified among the several iron-regulated proteins found in the envelopes of *Neisseria meningitidis, Neisseria gonorrhoeae,* and *Haemophilus influenzae.*[284-286,291-295]

Transferrin receptor function has been characterized in the pathogenic *Neisseria* species.[285,291,296,297] Transferrin utilization and binding are iron repressible,[285,291,298-300] suggesting the participation of one or more characterized iron-repressible outer membrane proteins (Frps)[300,301] in receptor function. The receptor specifically recognizes human transferrin and not transferrin from other mammalian sources.[285,296] It does not effectively discriminate between iron-loaded transferrin and apotransferrin.[291,296]

Competition experiments indicate the presence of discrete receptors for transferrin and lactoferrin. Both gonococcal and meningococcal transferrin receptors are saturable at about 1 μM protein,[291,296] reflective of an estimated affinity that is approximately 500-fold lower than that of the human transferrin receptor.[302] The copy number of the bacterial receptors is about 2800 to 2900 molecules per CFU.[291,296]

Receptors for transferrin vary among different strains of *Neisseria meningitidis*, both in number (700 to 4700 receptors per bacterium) and in their affinity constants for the protein (K_a ranged from 0.7×10^{-7} to 4×10^{-7} M).[298] Neither receptor numbers nor affinity constants were significantly different in carrier and invasive *Neisseria* strains, although the K_a values seem somewhat higher in the invasive strains.[298]

A set of genetically-linked transferrin receptor function mutants of *Neisseria gonorrhoeae* was recently described.[296] These mutants were incapable of binding transferrin or utilizing transferrin-bound iron, while utilization of lactoferrin was unaffected. Specific loss of lactoferrin receptor function resulted from mutation of another locus, designated lrf.[296] Another unlinked mutation (tlu) prevented transferrin and lactoferrin utilization without affecting ligand binding.

The tlu mutation indicated the presence of non-receptor protein that was common to both transferrin and lactoferrin utilization pathways without affecting ligand binding. The trf, lrf, and tlu gene products were not identified.[296]

Two different meningococcal iron-repressible proteins (Frps) have been identified by their ability to bind human transferrin and thus may be part of a transferrin receptor.

A 95-kDa Frp can be isolated from total membrane preparations by transferrin affinity purification[302] and has been designated TBP1 (transferrin binding protein 1, also known as TBPA).[303,304] A smaller Frp, varying in molecular mass (from 68 to 86 kDa) depending on the strain examined, is also isolated by transferrin affinity purification.[293] This protein, designated TBP2 (transferrin-binding protein 2, also known as TBPB),[304,305] retains its ability to bind transferrin after sodium dodecyl sulfate–polyacrylamide gel electrophoresis (SDS-PAGE) and electroblotting.[285,293]

Recent studies in part clarified the structures of TBP1 and TBP2. Both gonococcal and meningococcal TBP1 bind to transferrin only in affinity purification in which the starting materials are membrane fragments. All efforts to reconstitute transferrin binding by TBP1 after SDS-PAGE have failed, and this finding may indicate that ligand binding to TBP1 requires the maintenance of a peculiar protein conformation in the membrane that SDS-PAGE destroys.

TBP1 also binds transferrin when produced as a recombinant protein in *Escherichia coli*. Using a polyclonal antiserum raised against gonococcal TBP1, it was possible to isolate and purify the TBP1 protein and then the gene encoding gonococcal TBP1.[305] TBP1 and TBP2 are present in a ratio of about 1:1.[306]

The analysis of the deduced amino acid sequence of TBP1 strongly suggests that this protein serves as a transferrin receptor because of:

1. The presence of a signal sequence indicating that TBP1 probably functions beyond the cytoplasmic membrane
2. The presence of seven homologous domains also present in the *E. coli* TonB-dependent outer membrane receptors
3. The inclusion in domain 7 of an amphipathic motif preceding a terminal phenylalanine residue, a structural characteristic considered essential for outer membrane localization.[307]

The second observation is particularly relevant in that it suggests that TBP1 is a member of a class of proteins that serve as specific receptors for necessary nutrients such as iron–siderophore complexes and vitamin B_{12}.

An important additional observation further supports the role of TBP1 as a transferrin receptor. The cloned TBP1 gene was able to repair the defect in transferrin receptor function when transfected into Trf *Neisseria* mutants.[307] Finally, a number of observations suggest that bacterial transferrin receptor is different from the human receptor in that:

1. The neisserial receptor recognizes only human transferrin and the human receptor recognizes the transferrins from several mamalian sources, thus suggesting that the eukaryotic and prokaryotic receptors may recognize different domains of transferrin.
2. The complex transferrin receptor is internalized in eukaryotes, whereas *Neisseria* species do not internalize transferrin.
3. Anti-human transferrin receptor antibodies do not recognize the surfaces of *Neisseria* species.[307]
4. No significant homology was found in the sequence of TBP1 compared with the sequence of the human transferrin receptor.[307,308]

The importance of TBP1 in mediating the binding of human transferrin and in determining the virulence of *Neisseria* bacteria is supported by additional observations. Mutants of *Neisseria gonorrhoeae* that are defective in TBP1 function and then possess a markedly reduced uptake of iron from transferrin are avirulent in mouse subcutaneous chambers.[309] Transfection of the mutants with wild-type DNA resulted in formation of recombinant bacteria that restored their capacity to utilize human transferrin and are virulent in mice.[309]

A recent study[310] using gold-labeled human transferrin to investigate the interaction between transferrin and live meningococci provided evidence about the expression of TBPs on meningococci grown *in vivo* from organisms derived without laboratory culturing from cerebrospinal fluid of patients.

A further interesting observation on TBP1 was provided by Griffiths et al.[311] who showed identical NH_2-terminal sequences of TBP1 in three strains of *Neisseria meningitidis* and a cross-reactivity of anti-peptide antibodies raised against the NH_2-terminal sequence of TBP1 with *Neisseria meningitidis*, *Neisseria gonorrhoeae*, and *Haemophilus influenzae*.

A study by Cornelissen and coworkers provided definite clear evidence that TBP1 mediates transferrin binding.[312] These authors cloned the gene encoding for gonococcal TBP1 behind an inducible promoter in *Escherichia coli*.[312] The resultant recombinant strain was capable of binding human and gonococcal transferrin with the same specificity. However, *E. coli* expressing TBP1 did not internalize transferrin-bound iron or grow on transferrin as a sole iron source.[312]

These observations may be interpreted by assuming that gonococcal TBP1 is a transferrin receptor and that it does not require the presence of TBP2 to specifically bind human transferrin to the cell surface. However, additional genes, perhaps including TBP2, are required to release or transport iron from transferrin.

The other *Neisseria* protein capable of binding transferrin is TBP2. TBP2 or TBPB is considered a peripheral outer membrane lipoprotein anchored to the membrane via N terminal fatty acids. This protein is capable of binding transferrin after SDS-PAGE and electroblotting. It exhibits considerable molecular and antigenic heterogeneity in different clinical isolates of *Neisseria meningitidis*.[313,314] Most strains examined have TBP2 proteins with molecular weight around 78 to 85 kDa; a few contain a binding protein of about 68 kDa. The size of this protein seems unrelated to the serogroup or serotype of the microrganism.

Analysis of the sequences of TBP2 isolated from different gonococcal strains showed that: (1) significant sequence diversity exists among gonococcal and meningococcal TBP2s; and (2) the diversity in sequence is localized particularly at the level of a "hypervariable" domain.[315] Antiserum raised against purified TBP2 from one strain of *Neisseria meningitidis* cross-reacted on immunoblotting with the TBP2 proteins of all meningococcal isolates examined, as well as with the TBP2 of *Neisseria gonorrhoeae* and of *Haemophilus influenzae*.[316]

The analysis of the amino acid sequence of TBPB (TBP2) provided evidence of a bilobed structure.[316a] Alignment of amino acid sequences from the N terminal and C terminal halves of TBPB revealed a bilobed structural arrangement with several regions of identity.[316a] The bilobed nature of this protein is supported by the observation that the N terminal and C terminal halves of TBPB are able to bind human transferrin.

TBP2 seems to differ from human transferrin receptor as shown by two observations: (1) anti-TBP2 antiserum failed to react on immunoblotting with purified human transferrin receptor; and (2) monoclonal antibodies to human transferrin receptor did not react with whole cells or with electroblots of *Neisseria* or *Haemophilus influenzae*.

It must be pointed out that studies suggest the existence of a third protein involved in mediating iron uptake by these bacteria. In fact, a third iron-regulated protein, named Fe-binding protein (Fbp, 37 kDa), may act as a shuttle for the relocation of transferrin iron from the outer membrane to the inner membrane through the periplasmic space.[317]

TBPs were purified by affinity chromatography from *Neisseria meningitidis*. These purified proteins formed a 300-kDa heterodimeric complex composed of two

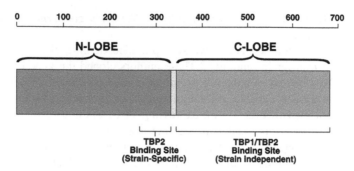

FIGURE 4.23 Regions of human transferrin involved in the interaction with meningococcal transferrin-binding proteins.

TBPA molecules and one TBPB molecule.[318] One molecule of this complex binds one or two molecules of transferrin. This observation suggests that the transferrin receptor present on the surface of *Neisseria* bacteria is formed by a heterodimer composed by two TBP1 molecules and one TBP2 molecule.

Recent studies have achieved the identification of TBP2 domains involved in the binding of human transferrin. The N and C terminal domains bind human transferrin about 10 to 1000 times less, respectively, than full-length TBP2.[319] Host specificity of TBP2 is mediated only by the N terminal domain.[319]

A series of studies were devoted to the analysis of the regions of human transferrin involved in the binding of *Neisseria* transferrin receptors. Analysis of proteolytic fragments of human transferrin provided evidence that the region of human transferrin involved in binding to the TBP of *Neisseria* is localized to the C terminal lobe of transferrin.[320] This conclusion was confirmed via another approach using chimeric transferrins constructed from human transferrin and bovine transferrin, which shares 70% amino acid identity with human transferrin, but is not bound by meningococcal TBPs. The study found that the primary binding site of human transferrin to both TBPA and TBPB corresponded to the transferrin C terminal lobe.[321] However, the TBP of some strains of meningococcus binds also to the N terminal lobe of human transferrin[322] (Figure 4.23).

It would be advantageous for iron acquisition by meningococci and gonococci to differentiate between apotransferrin and iron-saturated transferrin. In line with this statement, it was observed that both *Neisseria meningitidis*[323] and *Neisseria gonorrhoeae*[324] preferentially bind holotransferrin.

Finally, there is considerable interest in incorporating transferrin-binding and other iron-regulated membrane proteins into protein-based vaccines against group B meningococcal disease. If these proteins are to play a useful role in such vaccines, and especially, if the vaccines are to be broadly cross-protective, then their design must be based on a thorough understanding of the antigenic structures and expression of the protein antigens involved.

Another set of studies focused on the analysis of the transferrin-binding proteins of *Haemophilus influenzae*. *Haemophilus* species colonize the mucosa of the upper respiratory tracts of 50 to 80% of adults and children. *Haemophilus influenzae* type B is by far the most important human pathogen. *In vitro* studies have shown an absolute requirement for heme that serves as a source of both iron and porphyrins.[325-327] However, while heme can satisfy all the iron requirements of *Haemophilus influenzae*, it is present in only trace amounts in serum.[328] Furthermore, concentrations of heme that satisfy porphyrin requirements do not supply enough iron for growth *in vitro*. Thus, *Haemophilus influenzae* must possess additional iron-sequestering mechanisms.

Studies have shown that *Haemophilus influenzae* can use human transferrin as its sole source of iron,[329,330] but does not appear to produce any siderophores. The capacity of this bacterium to utilize transferrin as an iron source is dependent upon its capacity to bind transferrin through a receptor-like interaction.[331]

Subsequent studies have shown that *Haemophilus influenzae*, like *Neisseria*, possesses two types of transferrin-binding proteins, similarly designated TBP1 and TBP2. The molecular size of TBP1 was variable in different strains of *Haemophilus influenzae*.[316] Thus, RM7004, NU968 and NU20 strains had TBP2 of 90 kDa, whereas in Eagan strains, transferrin-binding activity was primarily associated with a protein with a molecular weight of 76 kDa. Finally, *Haemophilus influenzae* type B RM926 showed the presence of two transferrin-binding proteins with molecular weights of 72 and 80 kDa.[316]

All these *Haemophilus influenzae* strains expressed a protein with a molecular weight of about 105 kDa that did not bind transferrin on electroblotting, but was always co-extracted with the transferrin-binding proteins by the affinity extraction procedure; this is considered equivalent to TBP1 in *Neisseria* species.[316]

The two genes encoding TBP1 and TBP2 of *Haemophilus influenzae* have been recently cloned, and their structures are similar in different strains.[332,333] TBP1 appeared to be a member of the TonB-dependent family of outer membrane proteins, while TBP2 is lipid modified by signal peptidase II.[332] Comparison of the sequences of TBP1 and TBP2 genes from different bacterial species showed only a limited sequence homology. Mutations of TBP2 protein induced a marked decline of the transferrin-binding capacity by *Haemophilus influenzae*.

The expression of both TBP1 and TBP2 is modulated by the concentration of available iron. Continuous growth of *Haemophilus influenzae* in the presence of an iron source leads to a repression of TBP1 and TBP2 synthesis, while an opposite effect is mediated by iron deprivation.[334] These effects are mediated through transcriptional control of TBP1 and TBP2. This conclusion is supported by the identification of motifs of ferric uptake repressor (Fur) binding in the promoter regions of both TBP1 and TBP2 genes.[335]

Finally, it is of interest that in laboratory-adapted isolates of *Haemophilus influenzae* type B, such as strain Eagan, expression of TBPs occurs only when iron is absent from the growth medium. In contrast, examination of a small number of fresh clinical isolates of *Haemophilus influenzae* type B from the blood and cerebrospinal fluid of patients with meningitis has shown that TBPs are expressed constitutively, i.e., even in the presence of excess free iron.[336]

This finding was validated through the analysis of a large number of clinical isolates of *Haemophilus influenzae*.[337] Seventy-eight commensal isolates and 78 isolates from invasive infections were examined for their transferrin-binding capacity. Of the 78 invasive isolates, 91% were capable of binding transferrin, with 73%

binding transferrin constitutively. In contrast, only 14% of the commensal isolates bound transferrin constitutively.[337] These observations strongly suggest that, while not a universal characteristic, detectable transferrin-binding was clearly associated with *Haemophilus influenzae* isolates from invasive infections. In line with this conclusion, TBP1 and TBP2 proteins are constantly and constitutively expressed in *Haemophilus influenzae* bacteria derived from patients with otitis media.[338]

In addition to the capacity to acquire iron through the binding of transferrin, *Haemophilus influenzae* also possesses the capacity to bind siderophores and to uptake iron through this mechanism. This last iron transport mechanism is mediated by the iron-transporter ferric ion-binding protein (hFBP) that belongs to a protein superfamily that includes human transferrin.[339] The analysis of the primary structure and the quaternary structure of hFBP, showed several interesting findings and suggested an evolutionary process of development for bacterial ferric ion-binding proteins. The function of these proteins is the transport of iron across the bacterial periplasmic space.

Analysis of the three-dimensional structure of hFBP showed a structure similar to those observed for bacterial periplasmic-binding proteins that transport a wide variety of nutrients, including sugars, amino acids, and ions. Furthermore, the comparison of crystal structures of lactoferrin, transferrin, and ovotransferrin with the structures of bacterial periplasmic proteins provided evidence that they share a common structural organization, consisting of two globular domains connected by a pair of antiparallel β-strands forming a "hinge" between the two domains.[339]

According to these findings, it was proposed that both bacterial periplasmic proteins and eukaryotic transferrins arise from a common ancestor gene that existed before the divergence of prokaryotes and eukaryotes (i.e., about 1500 million years ago)[339] (Figure 4.24).

Since the anion-binding site represents the most broadly conserved structure among eukaryotic transferrins and bacterial periplasmic proteins, it was likely derived from the common ancestor gene. Therefore, it was hypothesized that the common ancestor protein was likely an anion-binding protein, rather than an iron-binding protein.[339]

In spite of these structural similarities, hFBP showed only about 10% sequence identity with eukaryotic transferrins. In this context, three main differences were observed at the level of the iron-binding site between hFBP and transferrin:

1. In hFBP, the iron-binding site is more exposed than the correspondent sites of lactoferrin and transferrin.
2. hFBP uses the anion phosphate as a ligand for iron chelation, while the transferrins utilize the anion carbonate.
3. Although similar amino acid residues are used to bind iron in both hFBP and transferrins, three residues are located in different parts of the polypeptide chains.

The function of hFBP is to bind and release Fe^{3+}. It is hypothesized that the *in vivo* release of Fe^{3+} from hFBP is promoted by binding to a permease on the bacterial cytoplasmic membrane. *In vitro* hFBP, like transferrin, releases iron at low

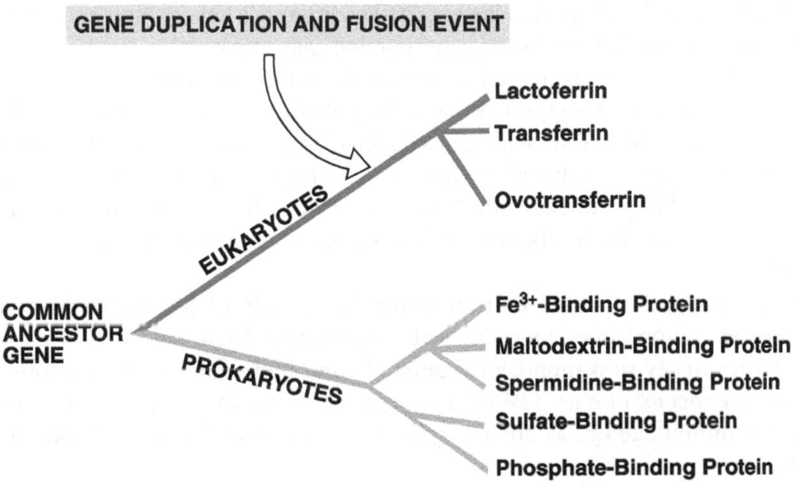

GENE DUPLICATION AND FUSION EVENT

Lactoferrin

Transferrin

Ovotransferrin

EUKARYOTES

COMMON
ANCESTOR
GENE

PROKARYOTES

Fe^{3+}-Binding Protein

Maltodextrin-Binding Protein

Spermidine-Binding Protein

Sulfate-Binding Protein

Phosphate-Binding Protein

FIGURE 4.24 Evolutionary analysis of the development of prokaryotic and eukaryotic members of the transferrin superfamily, based on structural and crystallographic analysis of eukaryotic and prokaryotic transferrins. This analysis suggests that the structural conservation of the anion-binding site throughout the transferrin superfamily strongly indicates that the common ancestor of these genes was likely an anion-binding protein, rather than an iron-binding protein. The structural comparison of all the known crystallographic structures indicates that hFBP shares more structural similarity with sulfate-binding protein, phosphate-binding protein, spermidine-binding protein, and maltodextrin-binding protein, than it does with other structurally-known periplasmic binding proteins or eukaryotic transferrins. (Redrawn with modifications from Bruns, C.M. et al., *Nature Struc. Biol.*, 4: 919–924, 1997.)

pH. At variance with eukaryotic transferrin, pH-mediated release in hFBP seems to involve protonation of one of the Fe^{3+} ligands or one of the phosphate ligands.

Other bacteria species such as *Bordetella pertussis*[340] and *Pasteurella haemolytica*[341] exhibit the capacity to bind transferrin and use this glycoprotein as an iron source. The membrane proteins of these bacteria able to bind transferrin were recently characterized. TBP1 and TBP2 genes cloned from *Pasteurella haemolytica* were sequenced and expressed in *E. coli*. These two genes were organized in a putative operon arrangement of TBP1 and TBP2. The transcription of the two genes is coordinated. The deduced amino acid sequences of TBP1 and TBP2 proteins showed domains of homology with corresponding regions of TBP proteins isolated from *Neisseria* or *Haemophilus*.[342]

Few studies have provided evidence that some Gram-positive bacteria are able to bind and use transferrin as an iron source. Initial studies have shown that *Staphylococcus aureus* is able to grow in human serum, utilizing transferrin as an iron source. Subsequent studies indicated that staphylococci are able to secrete siderophores, such as staphyloferrin A and staphyloferrin B.[343] Another siderophore was recently identified, purified, and termed aureochelin.[344]

Other studies have provided evidence that *Staphylococcus epidermidis* is able to remove iron from transferrin.[345] This is due to the presence on the surface of *S. aureus* and *S. epidermidis* of a 42-kDa transferrin-binding protein that exhibits considerable transferrin species specificity. The protein is able to bind human, rabbit, and rat transferrins, but not bovine and porcine transferrins.[346]

Since this TBP was expressed *in vivo* on the surface of staphylococci during an active infection, it was suggested that it may contribute to the virulence of these bacteria.[347] The 42-kDa TBP determines the binding of transferrin and the acquisition through this receptor-mediated process of iron required to sustain the growth of staphylococci.[348] Experiments with N lobe and C lobe fragments of human transferrin have shown that the N lobe is the primary site of interaction with *Staphylococcus* TBP.[348]

Interestingly, the human transferrin-binding protein of *Staphylococcus aureus* possesses an epitope in common with the human transferrin receptor, as shown by the cross-reactivity of a monoclonal antibody anti-human transferrin receptor with the *Staphylococcus aureus* TBP.[349] This epitope is not immunogenic, but another epitope is immunogenic, as shown by its reactivity with the sera of convalescent patients.[349]

A recent study provided additional evidence about other bacterial species that bind transferrin through a membrane receptor different from TBP1 and TBP2. *Prevotella nigrescens* and *Prevotella intermedia* are able to bind human transferrin through a 37-kDa membrane protein.[350] The expression of this receptor, strictly specific for transferrin, is down-modulated when cells are grown in the presence of hemin.

Other studies have indicated that parasites also possess the capacity to bind transferrin. Studies investigating the role of iron in *Trypanosoma cruzi* infection[351] and intracellular growth in peritoneal macrophages[352] showed that both iron excess and depletion affected growth and pathogenicity of *T. cruzi*. *T. cruzi* amastigotes, but not trypomastigotes, have receptors for human transferrin.[353] These receptors exhibited K_d of $2.8 \times 10^{-6}\,M$ and 200 kDa of apparent molecular weight. Other studies[354] have shown that *Leishmania infantum* also possesses transferrin binding sites and exhibits an affinity of binding for transferrin ($K_d = 2.2 \times 10^{-8}\,M$) comparable to the K_d observed for transferrin receptors on other cell types.

It is of interest that there are at least three significant differences betweeen *Trypanosoma cruzi* and *Leishmania* transferrin receptors:

1. Monoclonal antibody B3/25, specific for human transferrin receptor, cross-reacted with *Trypanosoma cruzi* but not with *Leishmania* transferrin receptor.
2. Unlike the *Trypanosoma* amastigote receptor, *Leishmania* transferrin receptor has a K_d of $2.2 \times 10^{-8}\,M$, a value similar to that reported for mammalian cells.
3. *Leishmania* transferrin receptors appear to be present on both life stages of *Leishmania*, in contrast to the absence of receptors for transferrin on trypomastigotes, an invasive but not dividing form of *Trypanosoma cruzi*.

Other studies have indicated that iron and/or transferrin are required to sustain the growth of *Schistosoma mansoni*.[355] However, the mechanism by which transferrin

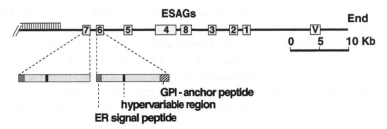

FIGURE 4.25 Schematic representation of a trypanosomal variant surface glycoprotein (VSG) gene expression site. In the upper section, the numbers in boxes indicate expression site-associated genes (ESAGs), V indicates the VSG gene, and End indicates the chromosome terminus. In the lower section the enlargement shows the structures of ESAG6 and ESAG7, which encode the two subunits of trypanosome transferrin receptor. (Reprinted with modifications from Bitter, W. et al., *Nature,* 391: 499–503, 1998. With permission.)

is able to deliver its load of iron to *Schistosoma* cells is still in question. In fact, studies of transferrin binding to schistosomula demonstrated that the binding sites lack specificity and have an essential unsaturable binding capacity.[355]

Particular attention was focused recently on the study of the structure and function of *Trypanosoma* transferrin receptors. A pivotal observation was made by Schell et al. who purified transferrin-binding proteins from *Trypanosoma brucei* and showed that one of the components present in their purified preparation is encoded by ESAG6, one of the genes in the variable surface glycoprotein (VSG) expression site[356] (Figure 4.25).

Subsequent studies led to the identification of a second component of the *Trypanosoma* transferrin receptor, pESAG7, and to the discovery that the purified receptor is constituted by a heterodimer of pESAG6/pESAG7 present in equimolar amounts.[357] Experiments carried out through expression of recombinant ESAG6 and ESAG7 unequivocally showed that the formation of a heterodimer between these two molecules is a prerequisite for transferrin binding.[358,359]

ESAG6 encodes a glycosylated protein of about 52 kDa that is cell surface-associated through a glycosylphosphatidylinositol anchor at its C terminus (Figure 4.25). The ESAG7 protein encodes a 42-kDa glycoprotein located intracellularly. This protein lacks the glycosylphosphatidylinositol modification; however, part of ESAG7 protein remains attached to the trypanosome surface by holding onto the ESAG6 subunit. ESAG6 and pESAG7 amino acid sequences are almost identical and it must be assumed that the interaction between these two proteins at the level of few amino acid residues specific to each subunit is necessary to generate the ligand-binding site. The genes encoding ESAG6 and ESAG7 proteins are located upstream of the variant surface glycoprotein gene in a polycistronic expression site.

Studies of transferrin binding have shown that each *Trypanosoma* organism possesses on its surface about 3000 binding sites.[360] The transferrin receptor present on *Trypanosoma* binds with equal affinity both iron-loaded and iron-free transferrin[361]; furthermore, bloodstream forms of *Trypanosoma brucei* require only small amounts of iron delivered from transferrin.[361] Transferrin receptor present on the surface of *Trypanosoma brucei* is only weakly anchored to the cell surface and does not spread over the entire surface; it is localized selectively at the level of the flagellar pocket.[360] The complex transferrin/transferrin receptor, after binding at the level of cell surface, is internalized and routed to lysosomes where transferrin is proteolytically degraded. Degradation products are then released from the cell, while iron remains cell-associated. The heterodimeric transferrin receptor is recycled to the membrane of the flagellar pocket, ready for another cycle of transferrin binding and uptake.[360]

Since there are about 20 variable-surface glycoprotein genes, each with one copy of ESAG6 and ESAG7, it is evident that *Trypanosoma* can produce up to 20 different transferrin receptors. More particularly, the ESAG6 and ESAG7 genes are expressed in telomeric expression sites along with the VSG gene. There are up to 20 of these expression sites per trypanosome nucleus, but usually only one is active at a given time. The most plausible explanation of the biological significance of the variability of *Trypanosoma* transferrin receptors is that this parasite has evolved multiple alternative versions of its transferrin receptor to extend its capacity to infect a wide range of hosts.

A recent study directly supports this hypothesis[362] and showed that different expression sites encode transferrin receptors that are similar, but not identical. These small differences in the sequences of the different transferrin receptors synthesized by *Trypanosoma brucei* have pronounced effects on the binding affinity of these receptors for transferrins of different mammalian species and on the capacity of *Trypanosoma* to grow in the sera of these animals.[362] It was thus evident that the capacity to switch between different transferrin receptor genes allows *Trypanosoma* to meet the differences in transferrin sequences of its numerous hosts.[362]

Recent studies have partially clarified the structure and function of the *Leishmania* transferrin receptor. Purification of the receptor by affinity chromatography on sepharose–transferrin demonstrated that it is constituted by a single 70-kDa chain; the receptor does not form heterodimers and binds transferrin with high affinity.[363] *Leishmania chagasi* was able to acquire iron from different sources including hemin, ferrilactoferrin, and ferritransferrin. This ability may allow this protozoan parasite to survive in the different microenvironments it encounters in its insect and mammalian hosts.[364]

Comparative analysis of transferrin and lactoferrin receptors of *Leishmania donovani* provided several interesting findings: (1) lactoferrin binding is independent, whether or not the protein contains iron, and is not inhibited by transferrin; and (2) transferrin binding occurs preferentially with iron-loaded transferrin rather than with apotransferrin and is inhibited by the presence of lactoferrin.[365]

Particularly interesting are the studies of the mechanisms of iron utilization of malaria parasites. Several observations suggest that iron is essential for their development. Evidence indicates that the parasites in erythrocytes do not obtain iron from hemoglobin, as suggested by two observations: (1) the total hemin level in parasitized erythrocytes does not vary during parasite development; and (2) no parasite enzyme is able to degrade hemin. Thus, iron apparently cannot be liberated from hemin and instead is sequestered in infected red cells as hemozoin, the characteristic pigment associated with malarial infection.

IRON TRANSPORT IN MYCOBACTERIA

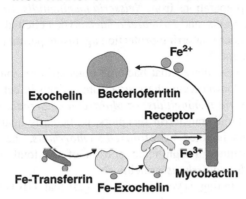

FIGURE 4.26 Mechanism of iron uptake and transport by mycobacteria.

If iron bound to transferrin is the source of ferric ions for malaria parasites within mature erythrocytes, then the parasite must synthesize its own transferrin receptor and localize it on the surfaces of infected cells, because the receptors for transferrin are lost during erythrocyte maturation. Binding studies provided direct evidence that malaria parasites are able to bind transferrin, and this phenomenon is mediated by a 93-kDa membrane protein.[366]

Other microorganisms, however, incorporate iron through a mechanism not involving transferrin binding. One example is *Mycobacterium tuberculosis*. This bacterium releases high-affinity iron-binding siderophores called exochelins; these molecules have structures similar to those of other iron-binding molecules located in the walls of *Mycobacterium tuberculosis* and called mycobactins.[367]

The process of iron uptake by *Mycobacterium tuberculosis* implies two steps: (1) exochelins rapidly remove iron from transferrin or lactoferrin; and (2) ferri-exochelins transfer iron to mycobactins in the cell walls of the bacteria[367] (Figure 4.26).

Histoplasma capsulatum, a fungal pathogen causing a large spectrum of diseases, develops in monocytes, using a peculiar strategy exploiting the physiologic acidification occurring in phagosomes/lysosomes. This acidification is used to acquire iron from transferrin. This observation explains why raising the pH of monocytes containing *Histoplasma capsulatum* with the weak base chloroquine results in killing and digestion of the intracellular fungus.[368] The effect of chloroquine is reversed by iron nitriloacetate, but not by iron-saturated transferrin.

The acidification of the phagosomes by *Histoplasma capsulatum* does not involve accumulation of the enzyme vacuolar ATPase at the level of the endosomal pathway.[369] The receptor proteins involved in iron acquisition pathways are potentially important candidates for the development of a vaccine because of their good accessibility at the surfaces of the bacteria and their role in disease causation.

Based on this rationale, extensive efforts have focused on the development of vaccines based on bacterial transferrin proteins. Several studies were devoted to the analysis of the potential roles of transferrin receptors from the pathogenic families

Pasterellaceae and Neisseriaceae. The studies showed that injection of purified transferrin receptor preparations from *Neisseria meningitidis*, *Haemophilus influenzae*, *Moraxella catarrhalis*, and *Pasteurella haemolytica* elicited in host animals the production of antibodies conferring protection against experimental infections originated by these bacteria.[370-372]

Additional evidence showed that bacterial transferrin receptors are also immunogenic in humans. Convalescent-phase sera from patients infected with *Haemophilus influenzae*, *Neisseria meningitidis*, or *Moraxella catarrhalis* contain antibodies directed against bacterial transferrin receptors, particularly against TBPB.[373-375] In convalescent patients infected by *Moraxella catarrhalis*, the immune response against TBPB constitutes a significant portion of the total detectable immune response to *Moraxella catarrhalis* proteins.[375]

In line with this finding, several studies suggest that TBPB (more than TBPA) is an ideal candidate for the development of anti-bacterial vaccines. TBPB is an outer-membrane lipoprotein component of the bacterial membrane well accessible on the surfaces of bacteria. Recombinant TBPB has been produced from various bacteria and used for the development of anti-bacterial vaccines.

Injection of recombinant TBPB purified from Neisseriaceae, *Haemophilus influenzae*, or *Moraxella catarrhalis* in experimental animals elicited a significant antibody response.[376-378] In all these cases the anti-TBPB antibodies produced significant bactericidal activity and a protective response *in vivo* against bacterial challenge. Despite these encouraging observations, the variability in the TBPB structure may in part limit the efficacy of this approach. In fact, antibodies raised against anti-TBPB purified from *Haemophilus influenzae* were highly protective against homologous isolates, but showed variability in the recognition of TBPB from heterologous isolates.[378]

Based on these observations, initial clinical studies of vaccination with preparations of TBPB purified from *Neisseria meningitidis* have been performed. To date, these studies allow only a preliminary evaluation of the safety and immunogenicity of such a vaccine.

4.7 TRANSFERRIN AS A GROWTH FACTOR

Iron is a fundamental requirement for cellular proliferation.[379] Iron uptake by proliferating cells is mediated by binding of transferrin to specific cell surface receptors. Once internalized, iron transferrin remains bound to the receptor in an intracellular endosomal compartment where the acidic milieu allows for iron release into the cytosol; apotransferrin is then recycled back to the cell surface and released outside the cells, where it becames available for a new cycle of transferrin binding.

Iron released from the endosomal compartment is briefly found in a cytosolic chelatable iron pool in which it may be available for metabolic processes requiring iron or sequestered by ferritin. The iron requirement for cellular proliferation is in large part related to a requirement for DNA synthesis.[380]

Activity of one iron-requiring enzyme, ribonucleotide reductase, has been proposed as the crucial step that makes iron strictly required for sustaining DNA synthesis.[381] Agents that interfere with iron metabolism inhibit cellular proliferation by three different mechanisms:

1. Monoclonal antibodies to the transferrin receptor inhibit transferrin-dependent iron uptake by interfering with transferrin binding to the receptor at cell surface.[382,383]
2. Certain metals that bind to transferrin, such as gallium and indium, appear to mainly interfere with intracellular iron incorporation.[384,385]
3. Iron chelators, such as desferrioxamine, deplete intracellular iron with a preference for removal of iron in the "chelatable" pool, with the subsequent release of ferrioxamine to the extracellular milieu.[386]

The chelatable iron pool, also called the "labile iron pool", is an ill-defined compartment of cellular iron whose existence was supposed to explain several aspects of iron metabolism. This iron pool is localized in the cytosol and contains metabolically active forms of intracellular iron. Several functions have been assigned to the chelatable iron pool, including: cellular iron transport; expression of iron regulatory genes; control of the activity of iron-containing proteins, including ribonucleotide reductase; and catalysis of Fenton reactions.

The evaluation of this labile iron pool remained elusive for a long time. However, it was shown that calcein may have use as a fluorescent probe to measure the labile iron pool and the concentration of cellular free iron.[387] Measurement of the labile iron pool with a fluorescent calcein probe showed that this pool has at least two kinetic components. The first component is the large part of the labile iron pool; the second component is minor and slowly achieves equilibrium with calcein.[388]

Traditionally, the growth-promoting effect of transferrin has been ascribed to its function as an iron vehicle. However, data accumulated during recent years indicates that this view is simplistic in that transferrin sustains proliferation through additional mechanisms.

According to de Jong et al.,[389] the effects of transferrin on cell growth can be distinguished as early and late effects. Classification is related to the timing of the effects of transferrin after mitogenic stimulation of the cells.

Three types of early effects of transferrin on cell growth are observed: (1) rearrangement of transferrin receptors soon after mitogenic stimulation; (2) transfer of reducing equivalents from cytosol to extracellular electron acceptors; and (3) activation of protein kinase C.

Rearrangement of transferrin receptors occurs at early time points after mitogenic stimulation, resulting in redistribution of these receptors from the intracellular pool to the cell membrane. As a consequence of this phenomenon, an increase in the number of surface transferrin receptors (increased transferrin-binding capacity) was observed.[390,391] An example of this phenomenon was described in growth factor-starved HeLa cells. Incubation with insulin- or platelet-derived growth factor led to a rapid ($t_{1/2}$ = 3 minutes) increase in surface transferrin receptor numbers. Subsequent studies have shown that this action requires calcium ions and does not require the presence of transferrin (ligand-independent).[392] This rapid redistribution of transferrin receptors occurs only in cell types in which sizeable intracellular pools of these receptors are present.

The second early effect of transferrin — the modulation of a cell's ability to transfer reducing equivalents from the cytosol to extracellular electron acceptors such as transferrin — received much attention and was experimentally verified recently. This mechanism was proposed by Sun et al.[393] The interaction of transferrin with receptors on the cell surface causes a reduction of transferrin-bound iron by means of a cellular membrane and transferrin receptor-associated electron transfer system: NADH-diferric transferrin reductase. Reduction of transferrin iron is accompanied by proton release with an H^+/e^- ratio over 50. H^+ generated by oxidation of NADH, and simultaneous reduction of ferric ion presented by transferrin exchanges with external Na^+ through tightly coupled Na^+/H^+ antiporters. Stimulation of this transmembrane redox system is postulated to result in a transient alkalinization of the cytoplasm.[393] Alkalinization of the cytoplasm is an early effect induced by several growth factors. The first evidence in favor of this mechanism comes from a study by Ellem and Kay.[394] These authors demonstrated that ferricyanide can sustain the growth of human melanoma cells under serum-free conditions.[394] Ferricyanide is an impermeable, unnatural, substrate for this redox system. This effect was apparently due to the ability of ferricyanide to function as a sink for electrons furnished through the plasma membrane NADH ferricyanide oxidoreductase.

The growth-promoting effect of ferricyanide cannot be ascribed to its iron content, since non-iron extracellular oxidants were later shown to have similar effects, provided their reduction potential was more positive than −125 mV.[395]

Subsequent studies focused on demonstrating the existence of and characterizing a plasma membrane system exhibiting NADH oxidoreductase activity. Transplasma membrane NADH-receptor oxidoreductase activity was found in many cell types from both plants and animals.[396,397] These enzymes transfer electrons from internal NADH to external electron acceptors[396,397] (Figure 4.27). The molecular structure of this putative trans-plasma membrane oxidoreductase has not been established.

Transferrin stimulates the redox system and induces proton efflux via the Na^+/H^+ antiport,[398,399] whereas some antibodies to the transferrin receptor (those blocking the binding of transferrin) inhibit the redox system.[400] Cell growth stimulators also stimulate NADH ferricyanide oxidoreductase,[401,402] whereas growth inhibitors such as adriamycin, bleomycin, and retinoic acid are inhibitors of the redox system.[403]

The ligand-activated NADH oxidase of plasma membrane is different from the NADH oxidase of mitochondria in that:

1. It is activated by specific ligands for plasma membrane receptors such as diferric transferrin and growth factors.
2. It is not stimulated by cytochrome C.
3. It is not inhibited by cyanide, azide, or rotenone.[404]

Recent studies[405] suggest that coenzyme Q may act at level of cellular membrane as an electron transporter. This conclusion is based on three issues:

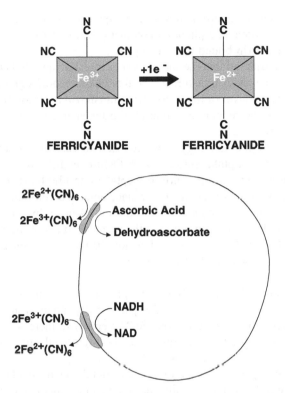

FIGURE 4.27 Membrane oxireductase reactions involving ferricyanide. Top: chemical structures of ferricyanide and ferrocyanide. Bottom: reactions involving reduction of ferricyanide coupled with oxidation of ascorbate to dehydroascorbate or NADH to NAD.

1. Extraction of coenzyme Q from the membrane decreases NADH dehydrogenase and NADH oxygen reductase; the addition of coenzyme Q to the extracted membranes restores the activity.
2. Analogs of coenzyme Q inhibit NADH dehydrogenase and oxidase activity and the inhibition is reversed by added coenzyme Q.
3. Ferricyanide reduction by transmembrane electron transport is inhibited by coenzyme Q analogs and restored with added coenzyme Q.

It has been proposed that in hepatocytes and other cell types, iron uptake is a membrane process dependent on reductive iron release from transferrin at the cell membrane. Evidence in favor of the concept that transferrin reductase mediates iron uptake from transferrin by hepatocytes is that iron uptake is inhibited by ferricyanide and other membrane-impermeable electron acceptors and is stimulated by low oxygen concentrations.[406,407]

This hypothesis was questioned on the basis of the observation that the reduction potential of transferrin-bound iron is considerably lower than that of NADH, so the reduction of iron by NADH is not possible.[408,409] This argument takes into account the properties of transferrin iron at neutral pH. However, the NADH-dependent transplasma membrane electron flow is accompanied by release of protons at the level of the cell membrane, a phenomenon that could lower the pH in the microenvironment of transferrin bound to its receptor.

It has been considered that the interaction of transferrin with its receptor induces conformational changes of transferrin, and this phenomenon together with the lowering of the pH at the level of the transferrin–transferrin receptor microenvironment cooperates to raise the reduction potential of transferrin iron to a level that allows it to be reduced by NADH.

Recently, a new methodology was proposed to carefully assess the extent of redox-dependent iron uptake into cells.[410] Oshiro et al.[410] used $^{55}Fe_2$-transferrin bound to Sepharose beads to measure iron uptake into fibroblasts mediated through a redox process. With this system, iron cannot enter the cells through endocytosis of transferrin, but must be reduced first and then taken up by the cells. Using this approach these authors carefully defined the major features of the redox-dependent iron uptake system that shows several major differences when compared to the transferrin endocytosis-dependent iron uptake system:

1. Ammonium chloride inhibits (through alkalinization of the endosome) Fe-Tf-dependent, but not Fe-Tf-Sepharose dependent iron uptake.
2. Iron uptake and reduction from Fe-Tf-Sepharose increases with temperature hyperbolically, differing from the sigmoidal curve observed for receptor-mediated endocytosis.
3. The optimal pH for iron reduction and uptake using Fe-Tf-Sepharose is acidic pH (about 5.50), while that for receptor-mediated endocytosis is neutral (about 7.40).
4. The subcellular distribution of iron incorporated from Fe-Tf-Sepharose in part differs from that incorporated from Fe-Tf.

The reduction of ferricyanide by plasma membrane reductase can be greatly stimulated by the addition of ascorbate and the oxidized form of ascorbate, dehydroascorbate.[411,412] Although several ideas have been proposed, the exact mechanism of this enhancement by ascorbate remains to be elucidated. It was suggested that the stimulation of ferricyanide reduction by ascorbate was due to a plasma membrane-localized ascorbate free-radical reductase (AFR).[412]

Some observations have suggested a possible involvement of AFR reductase in the mechanism of the transferrin-independent iron uptake system. In fact, ferricyanide has been shown to inhibit uptake of radiolabeled iron from ferric chelates, while ascorbate enhanced radiolabeled iron uptake.[413] However, this effect does not imply the activity of an AFR reductase, as indicated by the finding that the effect was abolished by addition of ascorbate oxidase.[412]

The third early effect of transferrin — the induction of protein kinase C activity — received only limited experimental attention. In fact, in studies carried out in the CCRF-CEM leukemic cell line, Boldt and coworkers found that the addition of Fe_2-transferrin (but not of apotransferrin or of a soluble iron salt such as ferric ammonium citrate) elicited a marked stimulation of protein kinase C activity.[414,415] This phenomenon was accompanied by increased association of the kinase at the cell membrane and seems to be dependent on increased transcription and synthesis of the β-isozyme of protein kinase C.[415] Finally, these authors excluded the possibility that the protein kinase C activation elicited by Fe^2-transferrin could be related to induction of cytosolic Ca^{2+} concentration.

The late and more important effects of transferrin in modulating cellular proliferation relate to the capacity of this glycoprotein to transport iron into the cells through interaction with a specific cell membrane receptor. For this reason, the late effects of transferrin in the control of cellular proliferation are strictly related to the expression and function of transferrin receptors. The importance of iron and of transferrin, the physiological vehicle of this element in the control of cell growth, is in large part related to the activity of ribonucleotide reductase which catalyzes the conversion of ribonucleotides to deoxyribonucleotides.

Mammalian ribonucleotide reductase consists of two non-identical subunits termed M1 and M2.[416] The M1 subunit contains the substrate and the effector-binding sites, while the M2 subunit contains non-heme bridged iron atoms and a tyrosyl-free radical that gives a characteristic signal with electronic spin resonance spectroscopy.[417] The activity of ribonucleotide reductase is strictly related to the tyrosyl radical and the presence of iron. Removal of this iron results in the loss of both the tyrosyl radical ESR signal and enzyme activity.[417,418]

A recent study of normal human T lymphocytes showed that iron may play an additional role in the control of cell proliferation.[419] This study was based on defining the roles of some nutrients in allowing the induction of cell-cycle related events in human T lymphocytes stimulated with phorbol dibutyrate and ionomycin.[419] Treatment of T cells with these agents in serum-free medium is sufficient to induce entry into the cell cycle, as evidenced by an increase in RNA content and immediate induction of early gene products, such as c-*fos*.

Essential fatty acids are required for further progression through the G1 phase of the cell cycle, as evidenced by doubling of RNA content and phosphorylation of the p110Rb protein. However, iron is also required for cell cycle progression in that its presence is strictly required for DNA synthesis and for the induction of kinase p34cdc2, a key cell cycle regulatory protein.

Although the mechanisms by which iron modulates the induction of the synthesis of this cyclin remain to be elucidated, these findings suggest that iron may play an important role in other cell cycle regulatory events.

Recent studies suggest that iron may play a role in c-*myc* induction during cell proliferation.[420,421] C-*myc* is a transcription factor that functions by means of heterodimerization with the product of the *max* gene, exerting either a stimulatory or inhibitory effect on target genes involved in the control of cell proliferation, differentiation, and apoptosis. The evidence that iron is required for c-*myc* activation is based on the observation that treatment with iron chelators leads to a marked decline in c-*myc* levels. It is of interest that c-*myc* acts as a modulator of genes involved in iron metabolism; in fact, it acts as a stimulator of IRP2 expression and an inhibitor of H-ferritin gene expression.[420] The stimulation of IRP2 expression by c-*myc* may stimulate transferrin receptor synthesis; its expression is required for optimal proliferation.

Several studies suggest that a strict relationship exists among iron, transferrin, transferrin receptors, and cellular proliferation. The first line of evidence is the expression of transferrin receptors during the cell cycle. A large number of studies have shown that proliferating tissues actively synthesize and express transferrin receptors, whereas quiescent tissues or confluent cells in culture express low levels of transferrin receptors.[422-424]

Particularly well investigated is the expression of transferrin receptors in T lymphocytes in relation to cell growth.[425-428] In quiescent T lymphocytes, the gene encoding transferrin receptor is transcriptionally inactive. Mitogenic agents such as lectins or specific antigens activate the transcription of the transferrin receptor gene. Interleukin-2 then elicits a series of cellular and molecular events leading to depletion of intracellular iron. The depletion causes an activation of IRF and a stabilization of transferrin receptor mRNA, thus leading to a high rate of synthesis of this receptor.

During T lymphocyte mitogenesis, the induction of transferrin receptor expression occurs in the G1 phase of the cell cycle, as suggested by studies using calcium channel blockers.[429] In fact, Neckers et al.[429] showed that diltiazem, a calcium channel blocker, arrested the growth of normal T cells in the G1 phase through prevention of calcium influx and also prevented transferrin receptor expression.

Additional evidence indicating that transferrin and its receptor are required to sustain cell proliferation derives from studies using a battery of monoclonal antibodies to transferrin receptors. A series of studies indicated that some monoclonal antibodies against human or murine transferrin receptors are able to inhibit iron uptake and block cell growth in tissue culture.[430-432] Interestingly, all the monoclonal antibodies inhibiting cell growth are polymeric in that they pertain either to the IgM or IgA types of immunoglobulins.[431,432]

These antibodies inhibit cell proliferation through different mechanisms that either block the binding of transferrin to its receptor or cross-link transferrin receptors to the cell membrane, thus interfering with their internalization.[433-435]

Additional antibodies to transferrin receptors were obtained by using purified recombinant human transferrin receptor.[436] Studies with these antibodies provided evidence that some of these antibodies exhibited synergistic effects in inhibiting cell proliferation; combinations of the antibodies inhibited growth of tumor cell lines completely insensitive to antibodies added individually.[436-438]

Proliferating cells usually have elevated levels of transferrin receptors, while quiescent cells display little or no transferrin receptor expression. One exception to this rule is represented by mature erythroid cells. In fact, when these cells reach the stage of non-proliferating orthochromatic erythroblasts, they continue to synthesize high levels of transferrin receptors; the high levels of transferrin receptor expression are maintained when orthochromatic erythroblasts mature to reticulocytes through nuclear expulsion. Finally, when reticulocytes terminally mature to erythrocytes, they cease transferrin receptor synthesis and lose their membrane transferrin receptors through a process of extrusion of the receptors and other membrane structures.

The second exception is the villous cytotrophoblast, which is highly proliferative during the initial phases of pregnancy and to a lesser extent in term placenta. Immunohistochemical studies have shown the apparent absence of transferrin receptors on the surfaces of these cells.[439-441] After differentiation to syncytiotrophoblasts, these cells synthesize relatively high levels of transferrin receptors, a phenomenon seemingly related to the transfer function of this metal–fetal barrier compartment.[442]

Muscle cells represent a third notable exception. Mature muscle cells (myotubes) continue to synthesize high levels of transferrin receptors, despite their low levels of DNA synthesis.[443] It was suggested that the elevated levels of transferrin receptors in myotubes may represent a phenomenon largely related to the peculiar iron requirement of these cells for sustaining myoglobin synthesis.[444]

Several studies have indicated that transferrin plays a fundamental role in the control of cell growth. However, the question remains whether transferrin or the iron bound to transferrin is the active agent. Studies performed in the 1950s and 1960s provided preliminary evidence that iron is a necessary constituent of tissue culture media. The first convincing experimental evidence in favor of the role of iron in the induction and maintenance of DNA synthesis was from studies by Robbins and Pedersen in 1970.[445] These authors reported inhibition of DNA synthesis in HeLa cells by the iron chelator deferoxamine, an observation later confirmed in other cell types using the same or other iron chelators. Additional evidence about the role of iron in the control of cell growth derived from studies indicating the requirement for transferrin to sustain the proliferation of mammalian cells in culture.[446-450]

As a consequence of these studies, transferrin has been included among the nutrient factors required for sustaining cell growth in serum-free media.[451] The question of whether proliferating cells need transferrin as a growth factor or require iron released from transferrin was answered unequivicocally by using iron chelates able to supply iron to cells without using transferrin.

Transferrin-independent iron uptake followed by efficient cellular utilization was first demonstrated with ferric complexes of pyridoxal isonicotinoyl hydrazone (PIH) and salicylaldehyde isonictinoyl hydrazone (SIH), using erythroid cells[452,453]; similar observations were made using soluble iron salts, such as ferric ammonium citrate and ferric ammonium sulfate.[454] It was later shown that Fe–PIH or Fe–SIH can replace transferrin in supporting proliferation of mouse embryonic kidney cells,[455,456] and similar results were obtained in numerous other cell types.[457-464] These results clearly indicate that the only function of transferrin in supporting cell proliferation is to supply cells with iron.

Iron compounds substituting for transferrin are often required at much higher concentrations than iron in the form of transferrin, presumably because the mechanism of iron uptake from transferrin is much more efficient than uptake from other iron-containing compounds.

Despite these findings, some studies suggest that the supplying of iron to the cells cannot be accepted as the full explanation of the transferrin growth-stimulating effects. Particularly relevant are the studies carried out in GH_1 rat pituitary tumor cells. Thyroid hormone-dependent GH_1 rat pituitary tumor cell growth in a serum-free chemically defined medium required a serum-derived mediator, initially called thyromedin, and subsequently identified as transferrin.[465] The pituitary growth stimulatory activity was lost if apotransferrin was saturated with iron.[466-468] The only known physiological function of apotransferrin was serving as a carrier/detoxifier of iron. This study indicates a new role of apotransferrin in hormone-dependent pituitary cell growth.

Immunohistochemical studies have shown that specific populations of rat[469] and human[470] anterior pituitary are capable of synthesizing and secreting transferrin.

Pituitary adenomas are constantly positive for transferrin. Interestingly, a positive linear correlation between tumor growth fraction and transferrin content was observed in these tumors, thus suggesting that transferrin may act as a growth-promoting factor for pituitary tumors.

Relevant also is a study with rat thyroid follicular cells (FRTL5 cells).[471] The growth-promoting effect of transferrin in these cells, as measured by (^{3}H)-thymidine incorporation, was mainly confined to cAMP-dependent mitogenic pathways activated by thyroid-stimulating hormone and dibutyryl cAMP.

Transferrin had no apparent effect on cAMP-independent stimulation of DNA synthesis induced by insulin and IGF-1. Desferrioxamine easily abolishes the stimulatory effect of cAMP-dependent mitogens, whereas stimulation by the cAMP-independent mitogens is much more resistant to desferrioxamine treatment.[471]

In addition to its effect on the growth of thyroid cells, transferrin also plays a part in the induction of thyroperoxidase, a glycosylated hemoprotein that has a key role in thyroid hormone synthesis.[472] In fact, iron supply mediated by transferrin and optimal heme synthesis are strictly required for induction of thyroperoxidase activity.[472]

A particularly interesting study focused on the growth-stimulating effect of transferrin and hybridoma cell lines.[473] This study showed that transferrin exerts its growth-promoting activity only in its iron-saturated form. Nevertheless, several results suggest that the growth-stimulating activity of transferrin may involve more than supplying iron to the cells:

1. Incubation of cells in the presence of the optimum growth-stimulating concentration of ferric citrate or transferrin does not result in the same intracellular iron levels.
2. A high intracellular iron level does not represent a sufficient condition for sustaining a high rate of cell growth, unless the medium contains transferrin.[473]

The proliferation of some types of tumor cells is particularly related to transferrin and this partially explains some peculiar aspects of the biology of these tumor cells, such as the preferential diffusion of tumors in selected tissues. The preferential colonization of particular tissues by certain metastatic tumor cells is mediated by a combination of highly specific cellular interactions including cell adhesion, cell migration, and cell proliferation.

Preferential spreading of carcinoma prostatic cells to bone tissue is dependent upon preferential dissemination of cells to spinal bone via paravertebral vessels. Once they migrate to the bone, prostatic carcinoma cells actively proliferate, a phenomenon dependent upon the presence in the bone marrow of factors that stimulate the growth of the tumor cells.

One of these factors has been identified as transferrin[474,475]; in fact, transferrin purified from bone marrow markedly stimulates the proliferation of human prostatic carcinoma cells.[474,475] The proliferation of prostatic carcinoma cells is equally stimulated by bone marrow and plasma transferrin, either in the apo- or iron-saturated form. *In vivo* studies have shown that the main effect of transferrin on prostatic tumor cells is stimulation of tumor metastasis more than enhancement of the tumor.[476]

The requirement for transferrin is not peculiar to prostate cancer cells; it is also displayed by normal prostate cells, as shown by *in vitro* organ cultures.[477] Transferrin also acts as an antagonist of growth-inhibitory effect of the drug suramin on hormone-refractory human prostate cancer cells.[478] The ensemble of these observations suggests that transferrin exerts a stimulatory role on the metastatic growth of prostate cancer cells.

In other types of tumor cells, transferrin seems to act as an autocrine nutrient growth factor sustaining the proliferation of tumor cells. Particularly relevant are the studies carried out on small lung cancer cells[479]. Small cell lung cancer is the prototype of tumors secreting in an autocrine manner a variety of growth factors, such as transferrin, insulin-like growth factor I, and bombesin that can stimulate the growth of these cells.

Small lung cancer cells grown in serum-free medium lacking transferrin synthesize and secrete transferrin. The rate of transferrin synthesis by these cells is inhibited if they are exposed to iron-saturated transferrin or hemin, and synthesis is markedly increased when these cells enter active phases of the cell cycle; this event precedes a rise in transferrin receptor expression.[479] It was suggested that the synthesis of transferrin by small lung cancer cells may offer these cells the capacity to proliferate in poorly vascularized tissular areas, where plasma transferrin is scarcely available.[479]

In addition to serving as a growth factor, transferrin may also be required for the induction of differentiation of some tumor cells. This is the case with ML-1 human myeloblastic leukemia cells which differentiate to monocyte–macrophage-like cells by the sequential actions of competence and progression factors.[480]

Tumor necrosis factor-α, transforming growth factor-β, and phorbol esters were found to induce competence, whereas a 77-kDa glycoprotein isolated from mitogen-stimulated human leukocyte-conditioned medium and identified as a transferrin isoform initiated progression.[480]

Soluble iron salts or iron chelates cannot replace transferrin in mediating the promotion of cellular differentiation, thus suggesting that transferrin may act on these cells as a differentiation/progression factor, its signal transduced via the transferrin receptor.[480] In other leukemic cells, such as promyelocytic HL-60 cells, transferrin is always required for inducing the differentiation either toward the granulocytic (by retinoic acid addition) or monocytic (by vitamin D_3 addition) lineages, but it can be replaced by soluble iron salts.[481]

Among the growth-promoting effects of transferrin, particularly relevant are those on the neural and muscular tissues. The interactions of neurons and their target cells have been intensively investigated and several studies carried out in different experimental systems have provided clear evidence that these neurotrophic interactions exist and are mediated by specific growth factors, a prototype of which is nerve growth factor.

Among these neurotrophic factors studied were the positive interactions exerted by nervous tissue on the development of muscle cells. These studies were originated from the observation that denervation of voluntary muscle leads to atrophy.

Studies reviewed by Oh et al.[482] and Beach et al.[483] have shown that extracts of adult nerve cells, embryonic central or peripheral nervous system cells, and serum of adult chickens contained a factor capable of replacing embryo extract in myogenic development in tissue culture. The active substance in these preparations has been shown to be identical to transferrin.[482,483]

The myotrophic factor was clearly identified as transferrin, but the issue of whether the transferrin required for myogenesis *in vivo* and for maintenance of the differentiated adult muscle by the motor nerve arises from the nerve or is supplied to the muscle from the circulation is controversial. Studies using a battery of anti-chicken transferrin monoclonal antibodies showed that the transferrin present in the nervous system differs from that observed in peripheral blood and is similar to that detected in embryonic tissues.[484]

Some studies suggest that transferrin may play an important role in the control of proliferation of endothelial cells.[485] This effect of transferrin seems to be related to a peculiar mechanism dependent upon an inhibitory effect of iron on the activity of endothelial nitric oxide (NO) synthase. This action of iron leads to a decrease of endothelial NO content, an inhibitor of endothelial cell proliferation.[485]

4.8 CONGENITAL ATRANSFERRINEMIA

Congenital atransferrinemia is a rare genetic disease characterized by very reduced or absent transferrin synthesis which leads to reduced delivery of iron to the marrow and then to the development of hypochromic anemia.[486] Few patients exhibiting this syndrome were reported in the medical literature. The first patient reported was 3-month old girl who had profound hypochromic anemia that was resistant to most treatments. When she was 7 years of age, studies by Heilmeyer et al.[487] demonstrated that her plasma contained only a trace amount of transferrin. Few additional cases of hereditary atransferrinemia have been repored.[488-493]

The major laboratory findings are marked reductions of total iron-binding capacity (20 to 69 mg/dl) and of transferrin (0 to 9 mg/dl).[488-493] Analysis of the low amounts of transferrin observed in these patients by radial immunodiffusion or by isoelectric focusing suggested that the main defect observed in atransferrinemia is a quantitative and not a qualitative defect.

Iron metabolism studies in several of these patients showed normal or enhanced iron absorption from the gastrointestinal tract, normal to accelerated plasma iron clearance, and decreased incorporation of iron into hemoglobin. The administration of purified transferrin or normal plasma to atransferrinemia patients resulted in an increase of hemoglobin preceded by a marked rise of reticulocytosis. One patient who received purified transferrin was alive after 17 years.

In addition to severe anemia, atransferrinemia is characterized by iron overload, which often is exacerbated by treatment with iron or blood transfusions.[494] Symptoms are associated with increased iron deposition in the liver and reticuloendothelial system. Some patients exhibit severe hepatic hemosiderosis, as documented by liver biopsy or autopsy.[487,495,496] In line with these observations, serum ferritin levels are clearly increased in patients with congenital atransferrinemia.[497]

The exact cause of congenital atransferrinemia is not entirely clear. Because parents of affected children have low levels of transferrin, transmission of an autosomal recessive trait is strongly suggested. The putative carriers usually have no symptoms and are not anemic. A recent study[496] involved the family of a patient with previously reported atransferrinemia.[492] The analysis of transferrin levels and the electrophoretic pattern of transferrin provided evidence that the nature of transferrin deficiency in this family was familial hypotransferrinemia, not congenital atransferrinemia. Trace amounts of transferrin were observed in the patient; one sister and one brother exhibited transferrin deficiency; another brother had normal transferrin values; finally, both parents presented transferrin values corresponding to half those of the normal control.

Analysis of the isoelectric focusing of transferrin in the different members of this family allowed us to understand the possible origin of the marked transferrin deficiency observed in this patient. The child's mother was heterozygous for the normal allele and for the "null" allele; the child's father was heterozygous for the normal and a variant transferrin allele (electrophoretically detectable).

In the three transferrin-deficient siblings, the isoelectric focusing patterns revealed only variant transferrin, and all their serum levels were less than 20 mg/dl. These siblings were compound heterozygotes, transmitting a "variant" allele from their father and a "null" allele from their mother.[496] The results of the genetic analysis strongly suggest that this syndrome should be more appropriately called familial hypotransferrinemia and not congenital atransferrinemia.

One animal model for congenital atransferrinemia is a mutant strain of hypotransferrinemic mice (trf[hpx]). The strain exhibits transferrin deficiency associated with cytopathological features similar to the human condition, including a hypochromic, microcytic anemia and marked hepatic parenchymal iron loading[496a]. This mutant strain exhibits a virtually complete absence of circulating transferrin as a consequence of a splicing defect in transferrin mRNA. The excessive liver iron accumulation observed in this mouse strain is due to increased intestinal iron absorption. The ensemble of these symptoms is due to transferrin deficiency since transferrin administration to the mice elicited an increase in iron delivery to the marrow, raised hemoglobin levels, reduced liver iron, and decreased intestinal iron absorption.[496a]

4.9 TRANSFERRIN AS A TRANSPORTER OF DRUGS

A series of studies have shown that transferrin may be used to transport drugs or other compounds to cells. Under this system, a drug conjugated with transferrin is transported inside the cells through interaction with the transferrin receptor and subsequent internalization of the complex formed by transferrin conjugate and transferrin receptor. A number of studies focused on the potential roles of transferrin conjugates as therapeutic agents in the treatment of tumors. This approach takes advantage of the observation that several tumor types express elevated levels of transferrin receptors and may thus represent possible targets for therapy with transferrin conjugates. Based on these observations, it was proposed that the transferrin uptake system may be used to specifically transport a drug to selected cell types.

Initial studies of anti-transferrin receptor monoclonal antibodies conjugated with a bacterial toxin showed an elevated toxicity against several types of tumor cell lines. Interestingly, the antibody conjugates were 10,000 times more effective than the toxin and the uncoupled antibody.

Subsequent studies performed by Raso and coworkers demonstrated that transferrin conjugated with a bacterial toxin exhibited a toxicity comparable to that displayed by the anti-transferrin receptor conjugate.[498] A large number of reports confirmed that other conjugates of transferrin with other anti-neoplastic drugs and toxin proteins, such as doxorubicin, adriamycin, immunotoxins, diphtheria toxin, and cholera toxin exert potent cytotoxic activities. In a few cases, particular models of cancer treatment with transferrin conjugates have been developed up to the stage of clinical trial.

Particularly relevant are the studies of human gliomas that served as the basis for pre-clinical studies. Initial studies showed a marked sensitivity of human glioma cell lines to anti-human transferrin receptor–ricin A-chain immunotoxin; the cytotoxic effect of this conjugate was potentiated by monensin.[499]

Subsequent studies extended these observations to the level of primary tumor cells and showed that glioblastoma multiforme and medulloblastoma are extremely sensitive to the cytotoxic effects elicited by transferrin receptor-targeted immunotoxins.[500] A conjugate of transferrin and diphtheria toxin is effective for the treatment of experimental glioma tumors in mice; treatment with the transferrin conjugate elicited a very pronounced decrease in tumor growth after 30 days of treatment, and induced complete eradication of the tumor in a significant proportion of the animals.[501] The animals were treated with intratumoral injections of the transferrin conjugate to limit the general toxicity related to the administration of the toxin.

Toxicology studies of an immunotoxin composed of transferrin coupled to recombinant ricin A-chain toxin administered into the subarachnoid space showed only minimal inflammation within the cerebrospinal fluid and no signs of systemic toxicity.[502] Furthermore, studies of high-flow interstitial infusion into the brain showed that transferrin conjugates are able to disperse to relatively large brain areas; the spread of the protein ranged from 2 to 3 cm.[503]

On the basis of these pre-clinical observations a clinical trial was conducted in patients with malignant brain tumors refractory to conventional therapy, using a transferrin conjugated with a mutant of diphtheria toxin infused through high-flow interstitial microinfusion. At least 50% tumor reductions occurred in 9 of 15 patients, including two complete responses.[504] The treatment was well tolerated, with minimal signs of toxicity. While these results need to be confirmed on a large number of patients, they are very interesting and suggest that transferrin conjugates may be of value in the treatment of brain tumors.

These observations in oncological models showed that whether transferrin or anti-transferrin receptor monoclonal antibodies are used to target drugs via the transferrin uptake pathway, the technique represents an effective approach for drug delivery.

One interesting potential application of this system of drug delivery to the brain is the possibility of bypassing the blood–brain barrier. This is based on the observation that monoclonal antibodies to rat transferrin receptors administered intravenously bind preferentially to capillary endothelial cells in the brain; anti-transferrin receptor conjugated with methotrexate first binds to brain endothelial cells and then crosses the blood–brain barrier followed by diffusion to the brain parenchyma.[505]

This system may be used also for the delivery to the brain of neurotrophic factors, molecules that cannot traverse the brain capillary wall.[506] In this case, the role of the carrier anti-transferrin receptor antibody is two-fold: it enables the neurotrophic factor to cross the blood–brain barrier and also improves the pharmacokinetics of the neurotrophic factor.[506]

Intravenous injection of anti-transferrin receptor–nerve growth factor chemical conjugate prevented the loss of striatal choline acetyltransferase-immunoreactive neurons in a rat model of Huntington's disease and reversed the age-related cognitive dysfunction.[507]

The construction of chimeric molecules composed of either transferrin or anti-transferrin receptor monoclonal antibody fused with a protein or a drug is a promising approach that has some intrinsic limitations. This approach requires the construction of specific chimeric molecules for each application and is cumbersome. To overcome this important limitation, universal delivery systems that abrogate the need for a specific construct for each application have been developed.

Recent experimental developments arose from these initial observations. It was possible to express and purify a single chain Fv antibody–streptavidin fusion protein encoding the variable regions of the heavy- and light-chain murine OX-2 monoclonal antibody to transferrin receptor.[508] This antibody vector was bivalent, able both *in vitro* and *in vivo* to bind the transferrin receptor and biotin-conjugated drugs.[508] Interestingly, this antibody–avidin fusion protein showed much longer serum half-life and better brain uptake than the chemical conjugate of anti-transferrin receptor monoclonal antibody, and avidin.[509] This antibody vector, after *in vivo* injection, resulted in the labeling of brain capillaries and may be thus used for non-invasive neurotherapeutic delivery to the brain of different types of drugs.[508]

The construction of transferrin-drug conjugates is based on the use of molecules that form a bridge between transferrin and the drug. An alternative approach is incorporating the drug into the structure of transferrin. Two different biochemical techniques may be used:

1. The iron-binding site of transferrin may be modified to bind a drug molecule instead of binding iron.
2. The drug may be incorporated into the structure of transferrin by using recombinant protein engineering techniques.

The second technique was used to construct a transferrin molecule containing a sequence cleavable by the human immunodeficiency virus type 1 protease. Mutants containing this sequence in appropriate regions of the transferrin molecule retain transferrin function and exhibit the protease cleavable sequence in an exposed region of the transferrin mutant.[510] This approach may represent a useful tool for developing a novel class of therapeutic agents for the treatment of a wide spectrum of diseases.

Finally, a unique additional application of transferrin involves its use for the development of new techniques of gene transfer and gene therapy. This application involves a high-efficiency delivery system that uses receptor-mediated endocytosis to carry DNA macromolecules into cells. This was accomplished by conjugating transferrin to polycations (such as polylysine) that bind nucleic acids.[511] The technique was efficaciously used for gene transfer of normal[512] and leukemic[513] hemopoietic cells. These studies showed that with this system, the rate of gene transfer is dependent on the level of transferrin receptor. Thus, agents such as iron chelators that stimulate transferrin receptor density improve the efficacy of the gene transfer procedure.

The transferrin receptor-mediated gene transfer was improved by coupling adenovirus to transferrin–polylysine/DNA complexes[514] or by adding polylysine-conjugated peptides derived from the N terminal sequence of the influenza virus hemagglutinin subunit HA-2.[515]

These observations suggest that transferrin-mediated delivery of DNA may represent an important tool for gene therapy studies.

REFERENCES

1. Robinson, N.J. et al. A ferric-chelate reductase for iron uptake from soils, *Nature*, 397: 694–697, 1999.
2. Deak, M. et al. Plants ectopically expressing the iron-binding protein, ferritin, are tolerant to oxidative damage and pathogens, *Nature Biotechnol.*, 17: 192–196, 1999.
2a. Hutchins, D.A. et al. Competition among marine phytoplankton for different chelated iron species, *Nature*, 400: 858–861, 1999.
3. De Jong, G., Van Dijk, J.P., and Van Eijk, H.G. The biology of transferrin, *Clin. Chim. Acta*, 190: 1–46, 1990.
4. Bowman B. Transferrin, in *Hepatic Plasma Proteins*, New York, Academic Press, 124–143, 1992.
5. Huebers, H.A. and Finch, C.A. The physiology of transferrin and transferrin receptors. *Physiol. Rev.*, 67: 520–582, 1987.
6. Huggenvik, J.I. A splicing defect in the mouse transferrin gene leads to congenital atransferrinemia, *Blood*, 74: 482–486, 1989.
7. Levy, J.E. et al. Transferrin receptor is necessary for development of erythrocytes and the nervous system, *Nature Genet.*, 21: 396–399, 1999.
8. Boffa, G.A. et al. Immunoglobulins and transferrin in marine lamprey sera, *Nature*, 214: 700–702, 1967.
9. Aisen, P. and Leibman, A. *Transport by Proteins*, Berlin, Walter De Gruyter, 277–290, 1978.
10. Webster, R.O. and Pollara, B. Isolation and partial characterization of transferrin in sea lamprey *Petromyzon marinus*, *Comp. Biochem. Physiol.*, 30: 509–527, 1969.
11. Van Eijk, H.G. et al. Isolation and analysis of transferrin from different species, *Scand. J. Haematol.*, 9: 267–277, 1972.
12. Macey, D.J. and Potter, I.C. Iron levels and major iron binding proteins in the plasma of ammocetes and adults of Southern Hemisphere lamprey *Geotria australis* Gray, *Comp. Biochem. Physiol.*, 72A: 307–312, 1982.
13. Schonne, E. et al. The transferrins from *Xenopus laevis*, *Xenopus borealis* and their hybrids, *Comp. Biochem. Physiol.*, 91B: 489–495, 1988.
14. Huebers H.A. et al. Characterization of an invertebrate transferrin from the Crab Cancer Register, *J. Comp. Physiol.*, 148B: 101–109, 1982.
15. Lee, M.Y. et al. Iron metabolism in a spider, *Dugesiella hentzi*, *J. Comp. Physiol.*, 127: 349–354, 1978.

16. Martin, A.W. et al. A mono-sited transferrin from a representative deuterostome: the ascidia *Pyura stolonifera* (subphylum urochordata), *Blood,* 64: 1047–1052, 1984.
17. Bartfeld, N.S. and Law, J.H. Isolation and molecular cloning of transferrin from the tobacco hornworn, *Manduca sexta*. Sequence similarity to the vertebrate transferrins, *J. Biol. Chem.,* 265: 21684–21691, 1990.
18. Januror, R.C. et al. Transferrin in cockroach: molecular cloning, characterization, and suppression by juvenile hormone, *Proc. Natl. Acad. Sci. U.S.A.,* 90: 1320–1324, 1993.
19. Gasdaska, J.R. et al. Cockroach transferrin closely resembles vertebrate transferrins in its metal-ion-binding properties: a spectroscopic study, *J. Inorg. Biochem.,* 64: 247–258, 1998.
20. Bartan, G. and Schales O. Chemischen aufbau und physiologische Bedeutung des leicht abspaltbaren Bluteisens, *Hoppe-Seyler's Z. Physiol. Chem.,* 248: 96–116, 1937.
21. Schade, A.L. and Caroline, L. An iron binding component in human blood plasma, *Science,* 104: 340–341, 1946.
22. Mac Gillivray, R.T.A. et al. The primary structure of human serum transferrin, *J. Biol. Chem.,* 258: 3545–3546, 1983.
23. Gorinsky, B. et al. Evidence of the bilobal nature of diferric rabbit plasma transferrin, *Nature,* 281: 157–158, 1979.
24. Anderson, B.F. et al. Structure of human lactoferrin at 3.2 Å resolution, *Proc. Natl. Acad. Sci. U.S.A.,* 84: 1769–1773, 1987.
25. Park, I. et al. Organization of the human transferrin gene: direct evidence that is originated by gene duplication, *Proc. Natl. Acad. Sci. U.S.A.,* 82: 3149–3153, 1985.
26. Kurama, T., Kurata, S., and Natori, S. Molecular characterization of an insect transferrin and its selective incorporation into eggs during oogenesis, *Eur. J. Biochem.,* 228: 229–235, 1995.
27. Yoshiga, T. et al. Mosquito transferrin, an acute-phase protein that is up-regulated upon infection, *Proc. Natl. Acad. Sci. U.S.A.,* 94: 12337–12342, 1997.
28. Yoshiga, T. et al. *Drosophila melanogaster* transferrin, *Eur. J. Biochem.,* 260: 414–420, 1999.
29. Fisher, M. et al. A structurally novel transferrin-like protein accumulates in the plasma membrane of the unicellular green alga *Dulaniella salina* grown in high salinities, *J. Biol. Chem.,* 272: 1565–1570, 1997.
30. Fisher, M., Zamir, A., and Pick, U. Iron uptake by the halotolerant alga *Dulaniella* is mediated by a plasma membrane transferrin, *J. Biol. Chem.,* 273: 17553–17558, 1998.
31. Liang, Z. et al. Pacifastin, a novel 155-kDa heterodimeric proteinase inhibitor containing a unique transferrin chain, *Proc. Natl. Acad. Sci. U.S.A.,* 97: 6682–6686, 1997.
32. Lee, J.Y. et al. Molecular cloning and evolution of transferrin cDNAs in salmonids, *Mol. Mar. Biol. Biotechnol.,* 7: 287–293, 1998.
33. Li, H., Sadler, D.J., and Sun, H. Unexpectedly strong binding of a large metal ion (Bi^{3+}) to human serum transferrin, *J. Biol. Chem.,* 271: 9483–9489, 1996.
34. Yang, F., Lum, J.B., and McGill, J.R. Human transferrin: cDNA characterization and chromosomal localization, *Proc. Natl. Acad. Sci. U.S.A.,* 8: 2752–2756, 1984.
35. Mac Gillivray, R.T.A. et al. The primary structure of human serum transferrin. The structures of seven cyanogen fragments and the assembly of the complete structure, *J. Biol. Chem.,* 258: 3543–3553, 1983.
36. Mac Gillivray, R.T.A. et al. The complete amino acid sequence of human serum transferrin, *Proc. Natl. Acad. Sci. U.S.A.,* 79: 2504–2508, 1982.
37. Williams, J. Iron-binding fragments from the carboxyl-terminal region of hen ovotransferrin, *Biochem. J.,* 149: 237–244, 1975.

38. Williams, J. et al. The primary structure of hen ovotransferrin, *Eur. J. Biochem.*, 122: 297–303, 1982.
39. Metz-Boutigue, M.H. et al. Human lactotransferrin: amino acid sequence and structural comparisons with other transferrins, *Eur. J. Biochem.*, 145: 297–303, 1982.
40. Rose, T.M. et al. Primary structure of the melanoma-associated antigen p97 (melanotransferrin) deduced from mRNA sequence, *Proc. Natl. Acad. Sci. U.S.A.*, 83: 1261–1266, 1986.
41. Bowman, B.H., Young, F., and Adrian G.S. Transferrin, evolution and genetic regulation of expression, *Adv Genet* 25: 1–36, 1988.
42. Williams, J. The evolution of transferrin, *TIBS*, 7: 394–397, 1982.
43. Williams, J., Grace, S.A., and Williams, J.M. Evolutionary significance of the renal excretion of transferrin half-molecule fragments, *Biochem. J.*, 201: 417–419, 1982.
44. Schaeffer, E. et al. Organization of the human serum transferrin gene, in *Proteins of Iron Storage and Transport*, Montreuil, G.J. et al., Eds., Amsterdam, Elsevier, 361–364, 1995.
45. Schaeffer, E. et al. Complete structure of the human transferrin gene. Comparison with autologous chicken gene and human pseudogene, *Gene*, 56: 109–116, 1987.
46. Morabito, M.A., Leewellyn, L.E., and Moezydlowski, E.G. Expression of saxiphilin in insect cells and localization of the saxitoxin-binding site to the C-terminal domain homologous to the C-lobe of transferrins, *Biochemistry*, 34: 13027–13033, 1995.
46a. Escriva, H. et al. Rat mammary-gland transferrin: nucleotide sequence, phylogenetic analysis and glycan structure, *Biochem. J.*, 307: 47–55, 1995.
46b. Demmer, J. et al. Cloning and expression of the transferrin and ferritin genes in a marsupial, the brushtail possum (*Trichsaurus vulpecula*), *Biochim. Biophys. Acta*, 1445: 65–74, 1999.
46c. Nakamasu, K. et al. Membrane-bound transferrin-like protein (MTf): structure, evolution and selective expression during chondrogenic differentiation of mouse embryonic cells, *Biochim. Biophys. Acta*, 1447: 258–264, 1999.
47. Smithies, O., Variations in human serum beta-globulins, *Nature*, 180: 1482–1483, 1957.
48. Giblett, E.R. *Genetic Markers in Human Beings*, Oxford, Blackwell, 1969.
49. Buettner-Janush, J. Evolution of serum protein polymorphism, *Hum. Hered.*, 37: 47–68, 1970.
50. Kamboh, M.I. and Ferrel, R.E. Human transferrin polymorphism, *Hum. Hered.*, 37: 65–81, 1987.
51. Evans, R.W., Williams, J., and Moreton, K. A variant of human transferrin with abnormal properties, *Biochem. J.*, 201: 19–26, 1982.
52. Young, S.P. et al. Abnormal *in vitro* function of variant human transferrin, *Brit. J. Haematol.*, 56: 131–147, 1984.
53. Wong, C.T. and Saha N. Effects of transferrin genetic phenotype on total iron binding capacity, *Acta Haematol.*, 75: 215–218, 1986.
54. Kuhnl, P. and Spielmann, W. Transferrin: evidence for two common subtypes of the Tf C allele, *Hum. Genet.*, 43: 91–95, 1978.
55. Thymann, M. Identification of a new serum protein polymorphism as transferrin, *Hum. Genet.*, 43: 225–229, 1978.
56. Kuhnl, P. and Spielmann W. A third common allele in the transferrin system, Tf C3, detected by isoelectric focusing, *Hum. Genet.*, 50: 193–198, 1979.
57. Beckman, G., Beckman, L., and Sikstrom, C. Transferrin C subtypes and spontaneous abortion, *Hum. Hered.*, 30: 316–319, 1980.

58. Weitkamp, L.R. and Schcter, B.Z. Transferrin and HLA: spontaneous abortion, neural tube defects, and natural selection, *New Engl. J. Med.,* 313: 925–932, 1985.

59. Anconi, P. et al. Transferrin C subtypes in extremely premature newborn infants, *Pediatr. Res.,* 16: 1022–1024, 1985.

60. Beckman, L. et al. Transferrin C subtypes and occupational photodermatosis of the face, *Hum. Hered.,* 35: 89–94, 1985.

61. Lunec, J. et al. Free-radical oxidation (peroxidation) products in serum and synovial fluid in rheumatoid arthritis, *J. Rheumatol.,* 8: 233–245, 1981.

62. Rantapaa-Dahlquist, S. and Beckman, L. Transferrin C subtypes in rheumatoid arthritis, *Hum. Hered.,* 35: 279–282, 1985.

63. Faulk, W.P. and Galbraith, G.M. Trophoblast transferrin and transferrin receptors in the host–parasite relationship, *Proc. R. Soc.,* 204: 83–97, 1979.

64. Ekblom, P. et al. Transferrin as fetal growth factor in acquisition of responsiveness related to embryonic induction, *Proc. Natl. Acad. Sci. U.S.A.,* 80: 2651–2655, 1983.

65. Sikstrom, C. et al. Transferrin C3 offers protection against smoking-associated lung cancer? *Carcinogenesis,* 17: 1447–1449, 1996.

66. Beckam, L.E. et al. Protective effect of transferrin C3 in lung cancer? *Oncology,* 56: 328–331, 1999.

67. Lineback-Zius, J. and Brew, K. Preparation and characterization of an NH_2-terminal fragment of human serum transferrin containing a single iron binding site, *J. Biol. Chem.,* 255: 708–713, 1980.

68. Harris, W.R. Thermodynamics of iron binding to human transferrin, *Biochemistry,* 24: 7412–7422, 1985.

69. Schlabach, M.R. and Bates, G.W. The synergistic binding of anions and Fe (3+) by transferrin, *J. Biol. Chem.,* 250: 2182–2190, 1975.

70. Folajtar, D. and Chasteen, D.N. Measurement of nonsynergistic anion binding to transferrin by EPR difference spectroscopy, *J. Am. Chem. Soc.,* 104: 5775–5780, 1982.

71. Kilar, F. and Simon, I. The effect of iron binding on the conformation of transferrin, *Biophys. J.,* 48: 799–785, 1985.

72. Nakazato, K., Yamamura, T., and Satake, K. Different stability of N- and C-domain of diferric ovotransferrin in urea and application to the determination of iron distribution between the two domains, *J. Biochem.,* 103: 823–832, 1988.

73. Azari, P.P. and Feeney, R.E. Resistance of metal complexes of conalbumin and transferrin to proteolysis and to thermal denaturation, *J. Biol. Chem.,* 232: 293–300, 1958.

74. Ulmer, D.D. Effect of metal binding on the hydrogen–tritium exchange of conalbumin, *Biochim. Biophys. Acta,* 181: 305–316, 1969.

75. Weaver, J. and Pollack, S. Iron binding to apotransferrin, *Acta Haematol.,* 84: 68–71, 1990.

76. Harrington, J.P., Stuart, J., and Jones, A. Unfolding of iron and copper complexes of human lactoferrin and transferrin, *Int. J. Biochem.,* 19: 1001–1008, 1987.

77. Harrington, J.P. Spectroscopic analysis of the unfolding of transition metal-ion complexes of human lactoferrin and transferrin, *Int. J. Biochem.,* 24: 275–280, 1992.

78. Bailey, S. et al. Molecular structure of serum transferrin at 3.3 Å resolution, *Biochemistry,* 27: 5804–5816, 1988.

79. Anderson, B.F. et al. Structure of human lactoferrin: crystallographic structure analysis and refinement at 2.8 Å resolution, *J. Mol. Biol.,* 209: 711–724, 1989.

80. Anderson, B.F. et al. Apolactoferrin structure demonstrates ligand-induced conformational change in transferrins, *Nature,* 34: 784–787, 1990.

81. Gorinsky, B. et al. Evidence for the bilobal nature of diferric rabbit plasma transferrin, *Nature,* 281: 157–159, 1979.
82. Evans, R.W. and Williams, J. Studies of the binding of different iron donors to human serum transferrin and isolation of iron binding fragments from the N- and C-terminal regions of the protein, *Biochem. J.,* 173: 543–552, 1978.
83. Aisen, P. and Listovsky, I. Iron transport and storage proteins, *Ann. Rev. Biochem.,* 49: 357–393, 1980.
84. Mecklenburg, S.L., Donohoe, R.J., and Olah, G.A. Tertiary structural changes and iron release from human serum transferrin, *J. Mol. Biol.,* 270: 739–750, 1997.
85. Zak, O. and Aisen, P. Evidence for functional differences between the two sites of rabbit transferrin: effects of serum and carbon dioxide, *Biochim. Biophys. Acta,* 1052: 24–28, 1990.
86. Rogers, T.B., Gold, B.A., and Feeney, R.E. Ethoxyformylation and photooxidation of histidines in transferrins, *Biochemistry,* 16: 2299–2305, 1977.
87. Gelb, M.A. and Harris, D.C. Correlation of proton release and ultraviolet difference spectra associated with metal binding by transferrin, *Arch. Biochem. Biophys.,* 200: 93–98, 1980.
88. Pecoraro, V.L. et al. Siderophillin metal coordination. Difference ultraviolet spectroscopy of di-, tri-, and tetravalent metal ions with ethylene bis (o-hydroxyphenyl) glycine, *Biochemistry,* 20: 7033–7039, 1981.
89. Koening, S.H. and Schillinger, W.E. Nuclear magnetic relaxation dispersion in protein solutions. II. Transferrin, *J. Biol. Chem.,* 244: 6520–6526, 1969.
90. Schlabach, M.R. and Bates, G.W. The synergistic binding of anions and Fe^{3+} by transferrin. Implications for the interlocking sites hypothesis, *J. Biol. Chem.,* 250: 2182–2188, 1975.
91. Campbell, R.C. and Chasteen, N.D. A anion binding study of vanadyl (IV) human serotransferrin. Evidence for direct linkage of the metal, *J. Biol. Chem.,* 252: 5996–6001, 1977.
92. Zweier, J.L. et al. Pulsed electron paramagnetic resonance studies of the copper complexes of transferrin, *J. Biol. Chem.,* 254: 3512–3515, 1979.
93. Kojima, N. and Bates, G.W. The formation of Fe^{3+}–transferrin–CO_3^{2-} via the binding and oxidation of Fe^{2+}, *J. Biol. Chem.,* 256: 12034–12040, 1981.
94. Cowart, R.E., Kojima, N., and Bates, G.W. The exchange of Fe^{3+} between acetohydroxamic acid and transferrin, *J. Biol. Chem.,* 257: 7560–7568, 1982.
95. Mizutani, K. et al. Alternative structural state of transferrin, *J. Biol. Chem.,* 274: 10190–10194, 1998.
96. Mac Gillivray, R.T. et al. The high-resolution crystal structures of the recombinant N-lobe of human transferrin reveal a structural change implicated in iron release, *Biochemistry* 37: 7919–7928, 1998.
97. Jeffrey, P.D. et al. Ligand-induced conformational change in transferrins: crystal structure of the open form of the N-terminal half-molecule of human transferrin, *Biochemistry,* 37: 13978–13986, 1998.
97a. Kurowaka, H. et al. Crystal Structure of hen apo-ovotransferrin, *J. Biol. Chem.,* 274: 28445–28452, 1999.
98. Grossmann, J.G. et al. X-ray solution scattering reveals conformational changes upon iron uptake in lactoferrin, serum and ovo-transferrins, *J. Mol. Biol.,* 255: 811–819, 1992.
99. Grossman, J.G. et al. Metal-induced conformational changes in transferrins, *J. Mol. Biol.,* 229: 585–590, 1993.

100. Grossman, J.G. et al. Asp ligand provides the trigger for closure of transferrin molecules, *J. Mol. Biol.*, 231: 554–558, 1993.
101. Lindley, P.F. et al. Transferrin: a study of the iron-binding sites during extended X-ray absorption fine structure and anomalous dispersion techniques, *Biochem. Soc. Trans.*, 12: 661–662, 1984.
102. Jhoti, H. et al. Crystallization and preliminary analysis of an 18,000 MW fragment of duck ovotransferrin, *J. Mol. Biol.*, 200: 423–425, 1988.
103. He, Q.Y. et al. Inequivalence of the two tyrosine ligands in the N-lobe of human serum transferrin, *Biochemistry*, 36: 14853–14860, 1997.
104. Grossman, J.G. et al. The nature of ligand-induced conformational change in transferrin in solution. An investigation using X-ray scattering, XAFS and site-directed mutants, *J. Mol. Biol.*, 279: 461–472, 1998.
105. He, Q.Y. et al. Mutations at nonliganding residues Tyr-85 and Glu-83 in the N-lobe of human serum transferrin, *J. Biol. Chem.*, 273: 17018–17024, 1998.
106. He, Q.Y. et al. Effects of mutations of aspartic acid 63 on the metal-binding properties of the recombinant N-lobe of human serum transferrin, *Biochemistry*, 36: 5522–5528, 1997.
107. He, Q.Y., Mason, A.B., and Woodworth, R.C. Iron release from recombinant N-lobe and single point Asp 63 mutants of human transferrin by EDTA, *Biochem. J.*, 328: 439–445, 1997.
108. Grossman, J.G. et al. Asp ligand provides the trigger for closure of transferrin molecules. Direct evidence for X-ray scattering studies of site-specific mutants of the N-terminal half-molecule of human transferrin, *J. Mol. Biol.*, 231: 554–558, 1993.
109. Mason, A.B. et al. Mutagenesis of the aspartic acid ligands in human serum transferrin: lobe–lobe interaction and conformation as revealed by antibody, receptor-binding and iron-release studies, *Biochem. J.*, 330: 35–40, 1998.
110. Steinlein, L.M. et al. Iron release is reduced by mutations of lysines 206 and 296 in recombinant N-terminal half-transferrin, *Biochemistry*, 37: 13696–13703, 1998.
110a. He, Q.Y. et al. Dual role of Lys 206–Lys 296 interaction in human transferrin N-lobe: iron-release trigger and anion-binding site, *Biochemistry*, 38: 9704–9711, 1999.
111. Sack, J.C., Saper, M.A., and Quiocho, F.A. Periplasmic binding protein structure and function, *J. Mol. Biol.*, 206: 171–190, 1989.
112. Nowalk, A.J., Tencza, S.B., and Miezner, T.A. Coordination of iron by the ferric iron-binding protein of pathogenic *Neisseria* is homologous to the transferrins, *Biochemistry*, 33: 12769–12777, 1994.
113. Brock, J.H. et al. The effect of trypsin on bovine transferrin and lactoferrin, *Biochim. Biophys. Acta*, 446: 214–228, 1976.
114. Nowalk, A.J. et al. Metal-dependent conformers of the periplasmic ferric ion binding protein, *Biochemistry*, 36: 13054–13059, 1997.
115. Kretchmar, S.A., Reyes, Z.E., and Raymond, K.N. The spectroelectrochemical determination of the reduction potential of diferric serum transferrin, *Biochim. Biophys. Acta*, 956: 85–94, 1988.
116. Zak, Z.O. and Aisen, P. Preparation and properties of a single-sited fragment from the C-terminal domain of human transferrin, *Biochim. Biophys. Acta*, 829: 348–353, 1985.
117. Kubal, G. et al. Uptake of Al^{3+} into the N-lobe of human serum transferrin, *Biochem. J.*, 285: 711–714, 1992.
118. Bali, P.K., Zak, O., and Aisen, P. A new role for the transferrin receptor in the release of iron from transferrin, *Biochemistry*, 30: 324–328, 1991.

119. Bali, P.K. and Aisen, P. Receptor-modulated iron release from transferrin: differential effects on N- and C-terminal sites, *Biochemistry,* 30: 9947–9952, 1991.
120. Lin, L.N. et al. Calorimetric studies of the binding of ferric ions to ovotransferrin and interactions between binding sites, *Biochemistry,* 30: 11660–11669, 1991.
121. Lin, L.N. et al. Calorimetric studies of the binding of ferric ions to human serum transferrin, *Biochemistry,* 32: 9398–9406, 1993.
122. Lin, L.N. et al. Calorimetric studies of serum transferrin and ovotransferrin. Estimates of domain interactions, and study of the kinetic complexities of ferric ion binding, *Biochemistry,* 33: 1881–1888, 1994.
123. Lin, L.N. et al. Calorimetric studies of the N-terminal half-molecule of transferrin and mutant forms modified near the Fe^{3+}-binding site, *Biochem. J.,* 293: 517–522, 1993.
124. Harris, W.R. et al. Binding of monovalent anions to human serum transferrin, *Biochim. Biophys. Acta,* 1383: 197–210, 1998.
125. Harris, W.R. and Nesset-Tollefson, D. Binding of phosphate chelating agents and pyrophosphate to apotransferrin, *Biochemistry,* 30: 6930–6936, 1991.
126. Harris, W.R. et al. Thermodynamic studies on anion binding to apotransferrin and to recombinant transferrin N-lobe half molecules, *Biochim. Biophys. Acta,* 1430: 269–280, 1999.
127. Workman, E.F. and Bates, G.W. Mobilization of iron from reticulocyte ghosts by cytoplasmic agents, *Biochim. Biophys. Res. Commun.,* 58: 787–794, 1974.
128. Morgan, E.H. Studies on the mechanisms of iron release from transferrin, *Biochim. Biophys. Acta,* 580: 312–326, 1979.
129. Pollack, S., Campana, T., and Weaver, J. Low molecular weight iron in guinea pig reticulocytes, *Am. J. Haematol.,* 19: 75–84, 1985.
130. Mansour, A.N. et al. Fe(III)–ATP complexes. Models for ferritin and other poly-nuclear iron complexes with phosphate, *J. Biol. Chem.,* 260: 7975–7979, 1985.
131. Li, C.Y. et al. Iron binding, a new function of the reticulocyte endosome $H^{(+)-}$ ATPase, *Biochemistry,* 34: 5130–5136, 1995.
132. Pollack, S. and Weaver, J. Iron release from transferrin: synergistic interaction between adenosine triphosphate and an ammonium sulfate fraction of hemolysate, *J. Lab. Clin. Med.,* 108: 411–414, 1986.
133. Carver, F.J. and Frieden, E. Factors affecting the adenosine triphosphate-induced release of iron from transferrin, *Biochemistry,* 17: 167–177, 1978.
134. Weaver, J. and Pollack, S. Low MW iron isolated from guinea pig reticulocytes as AMP and ATP-Fe complexes, *Biochem. J.,* 261: 787–796, 1989.
135. Kretchmar, S.A., Reyes, Z.E., and Raymond, K.N. The spectroelectrochemical determination of the reduction potential of diferric serum transferrin, *Biochim. Biophys. Acta,* 956: 85–94, 1988.
136. Egan, T.J., Barthakur, S.R., and Aisen, P. Catalysis of the Haber–Weiss reaction by iron diethylenetriaminepentaacetate, *J. Inorg. Biochem.,* 48: 241–249, 1992.
137. Egan, T.J., Zak, O., and Aisen, P. The anion requirement for iron release from transferrin is preserved in the receptor-transferrin complex, *Biochemistry,* 32: 8162–8167, 1993.
138. Montreuil, J. and Spik, G. in *Proteins of Iron Storage and Transport in Biochemistry and Medicine,* Crichton, R.R., Ed., Amsterdam, North Holland, 27–38, 1975.
138a. Li, Y. and Harris, W.R. Iron removal from monoferric human transferrin by 1,2-dimethyl-3-hydroxypyridin-4-one, 1-hydroxypyridin-2-one acetohydroxamic acid, *Biochim. Biophys. Acta,* 1387: 89–102, 1998.
139. Zak, O., Trinder, D., and Aisen, P. Primary receptor-recognition site of human transferrin is in the C-terminal lobe, *J. Biol. Chem.,* 269: 7110–7114, 1994.

140. Mason, A.B. et al. Receptor recognition sites reside in both lobes of human serum transferrin, *Biochem. J.,* 326: 77–85, 1997.

140a. Lawrence, C.M. et al. Crystal structure of the ectodomain of human transferrin receptor, *Science,* 286: 779–782, 1999.

141. Mac Gillivray, R.T. A. et al. The primary structure of human serum transferrin. The structures of seven cyanogen bromide fragments and the assembly of the complete structure, *J. Biol. Chem.,* 258: 3543–3553, 1983.

142. Spik, G. et al. Primary structure of two sialylated triantennary glycans from human serum transferrin, *FEBS Lett.,* 183: 65–69, 1985.

143. Léger, D. et al. Physiological significance of the marked inreased branching of the glycans of human serotransferrin during pregnancy, *Biochem. J.,* 257: 231–238, 1989.

144. Hoefkens, P. et al. Influence of transferrin glycans on receptor binding and iron donation, *Glyconj. J.,* 14: 289 295, 1997.

144a. Hoffman, A. et al. Brain-type N-glycosylation of asialo-transferrin from human cerebrospinal fluid, *FEBS Lett.,* 359: 164–168, 1995.

145. Mazurier, J. et al. Molecular modelling of glycans: three-dimensional structure and glycan protein interaction. The rabbit serotransferrin as a model, *Comptes Rendus Acad. Sci. Paris,* 313, Ser. 3, 7–14, 1991.

146. Campion, B. et al. Presence of fucosylated triantennary, tetraantennary and pentaantennary glycans in transferrin synthesized by the human hepatocarcinoma cell line Hep G2, *Eur. J. Biochem.,* 184: 405–413, 1989.

147. Yamashita, K. et al. Altered glycosylation of serum transferrin of patients with hepatocellular carcinoma, *J. Biol. Chem.,* 264: 2415–2423, 1989.

148. Ogata, S.I., Muramatsu, T., and Kobata, A. New structural characteristic of the large glycopeptides from transformed cells, *Nature,* 259: 580–582, 1976.

149. Coddeville, B. et al. Primary structure of horse serotransferrin glycans, *Eur. J. Biochem.,* 186: 583–590, 1989.

150. Spik, G. et al. Carbohydrate microheterogeneity of rat serotransferrin, *Eur. J. Biochem.,* 195: 397 405, 1991.

151. Stibler, H. Carbohydrate-deficient transferrin in serum: a new marker of potentially harmful alcohol consumption reviewed, *Clin. Chem.,* 37: 2029–2037, 1991.

152. Stibler, H., Jaeken, J., and Kristiansson, B. Biochemical characteristics and diagnosis of carbohydrate-deficient glycoprotein syndrome, *Acta Pediatr. Scand.,* Suppl. 375: 22–31, 1991.

153. Stibler, H. and Jaeken, J. Carbohydrate deficient serum transferrin in a new systemic hereditary syndrome, *Arch. Dis. Child.,* 65: 107–111, 1990.

154. Wada, Y. et al. Structure of serum transferrin in carbohydrate-deficient glycoprotein syndrome, *Biochim. Biophys. Res. Commun.,* 189: 832–836, 1992.

155. Van Schiftingen, E. and Jaeken, J. Phosphomannomutase deficiency is a cause of carbohydrate-deficient glycoprotein syndrome type I, *FEBS Lett.,* 377: 318–320, 1995.

156. Niehues, R. et al. Carbohydrate-deficient glycoprotein syndrome type Ib. Phosphomannose isomerase deficiency and mannose therapy, *J. Clin. Invest.,* 101: 1414–1420, 1998.

157. Coddeville, B. et al. Determination of glycan structures and molecular masses of the glycovariants of serum transferrin from a patient with carbohydrate-deficient syndrome type II, *Glyconj. J.,* 15: 265–273, 1998.

158. Burda, P. et al. A novel carbohydrate-deficient glycoprotein syndrome characterized by a deficiency in glycosylation of the dolichyl-linked oligosaccharide, *J. Clin. Invest.,* 102: 647–652, 1998.

159. Korner, C. et al. Carbohydrate-deficient glycoprotein syndrome type V-deficiency of dolichyl-P-Glc:Man9GlcNac2-PP-dolichyl glycosyltransferase, *Proc. Natl. Acad. Sci. U.S.A.,* 95: 13200–13205, 1998.

160. Charlwood, J. et al. Defective glycosylation of serum transferrin in galactosemia, *Glycobiology,* 8: 351–357, 1998.

161. Yamashita, K. et al. Altered glycosylation of serum transferrin of patients with hepatocellular carcinoma, *J. Biol. Chem.,* 264: 2415–2423, 1989.

162. Huerre, C. et al. The structural gene for transferrin (TF) maps to 3q21-3qter, *Ann. Genet.,* 27: 5–15, 1984.

163. Schaeffer, E. et al. Complete structure of human transferrin gene. Comparison with the analogous chicken gene and human pseudogene, *Gene,* 56: 109–116, 1987.

164. LeBeau, M.M. et al. Molecular analysis of transferrin (TF) and p97 antigen genes in INV(3) and T(3;3) in acute non-lymphocytic leukemia (ANLL), *Am. J. Hum. Genet.,* 37: 4–15, 1985.

165. McKnight, G.S. et al. Transferrin gene expression. Effect of nutritional iron deficiency, *J. Biol. Chem.,* 255: 144–147, 1980.

166. McKnight, G.S., Lee, D.C., and Palmiter, R.D. Transferrin gene expression. Regulation of mRNA transcription in chicken liver by steroid hormones and iron deficiency, *J. Biol. Chem.,* 255: 148–153, 1980.

167. Idzerda, R.L. et al. Rat transferrin gene expression: tissue specificity and regulation by iron deficiency, *Proc. Natl. Acad. Sci. U.S.A.,* 83: 2723–2727, 1986.

168. Skinner, M.K. and Griswold, M.D. Secretion of testicular transferrin by cultured Sertoli cells is regulated by hormones and retinoids, *Biol. Reprod.,* 27: 211–221, 1982.

169. Block, B. et al. Transferrin gene expression visualized in oligodendrocytes of the rat brain by using *in situ* hybridization and immunocytochemistry, *Proc. Natl. Acad. Sci. U.S.A.,* 82: 6706–6770, 1985.

170. Tsutsumi, M., Skinner, M.K., and Sanders-Bush, E. Transferrin gene expression and synthesis by cultured choroid plexus epithelial cells, *J. Biol. Chem.,* 264: 9626–9631, 1989.

171. Alred, A.R. et al. Distribution of transferrin synthesis in brain and other tissues in the rat, *J. Biol. Chem.,* 262: 5293–5297, 1987.

172. Grigor, M.R. et al. Transferrin gene expression in the rat mammary gland, *Biochem. J.,* 267: 815–819, 1990.

173. Meehan, R.R. et al. Pattern of serum protein gene expression in mouse visceral yolk sac and fetal liver, *EMBO J.,* 3: 1881–1885, 1984.

174. Ekblow, P. and Thesleff, I. Control of kidney differentiation by soluble factors secreted by the embryonic liver and the yolk sac, *Dev. Biol.,* 110: 29–38, 1985.

175. Levin, M.J. et al. Expression of the transferrin gene during development of non-hepatic tissues: high level of transferrin mRNA in fetal muscle and adult brain, *Biochem. Biophys. Res. Commun.,* 122: 212–217, 1984.

176. Kahn, A. et al. The transferrin gene, in *Oncogenes, Genes, and Growth Factors,* Guroff G., Ed., New York, Wiley-Interscience, 277–309, 1984.

177. Yang, F. et al. Tissue specific expression of mouse transferrin gene during development and aging, *Mechanisms Aging Dev.,* 56: 187–197, 1992.

178. Yang, F., Friedrichs, W.E., and Coalson, J.J. Regulation of transferrin gene expression during lung development and injury, *Am. J. Physiol.,* 273: L417-L426, 1997.

179. Morgan, E.H. and Peters, T. The biosynthesis of rat serum albumin. V. Effect of protein depletion and refeeding on albumin and transferrin synthesis, *J. Biol. Chem.,* 246: 3500–3507, 1971.

180. Morton, A.G. and Tavill, A.S. The role of iron in the regulation of hepatic transferrin synthesis, *Brit. J. Haematol.* 36: 383–394, 1977.

181. Vasay, J.M. et al. Ultrastructural indirect immuno-localization of transferrin in cultured rat hepatocytes permeabilized with saponin, *J. Histochem. Cytochem.*, 32: 538–540, 1984.

182. Schreiber, G. et al. The synthesis and secretion of rat transferrin, *J. Biol. Chem.*, 254: 12013–12019, 1979.

183. Thibodeau, S.N., Lac, D.C., and Palmiter, R.D. Identical precursors for serum transferrin and egg white conalbumin, *J. Biol. Chem.*, 252: 3771–3774, 1978.

184. Cassia, R. et al. Transferrin is an early marker of hepatic differentiation, and its expression correlates with the postnatal development of oligodendrocytes in mice, *J. Neurosci. Res.*, 50: 421–432, 1997.

185. Woodworth, C. and Isom, H.C. Tumorigenicity of simian virus 40 hepatocyte cell lines: effect of *in vitro* and *in vivo* passage on expression of liver-specific genes and oncogenes, *Mol. Cell. Biol.*, 7: 3740–3748, 1987.

186. Idzerda, R.L. et al. Expression from the transferrin gene promoter in transgenic mice, *Mol. Cell. Biol.*, 9: 5154–5162, 1989.

187. Srinivas, U.K., Revathi, C.J., and Das, M.R. Heat-induced expression of albumin during early stages of rat embryo development, *Mol. Cell. Biol.*, 7: 4599–4602, 1987.

188. Hon, S.L., Lin, Y.F., and Chou, C.K. Transcriptional regulation of transferrin and albumin genes by retinoic acid in human hepatoma cell line Hep 3B, *Biochem. J.*, 283: 611–615, 1992.

189. Heinrich, P.C., Castell, J.V., and Andres, T. Interleukin-6 and the acute-phase response, *Biochem. J.*, 165: 621–636, 1990.

190. Simpson, R.J. Effect of hypoxic exposure on iron absorption in heterozygous hypotransferrinaemic mice, *Ann. Hematol.*, 65: 260–264, 1992.

191. Rolfs, A. et al. Oxygen-regulated transferrin expression is mediated by hypoxia-inducible factor-1, *J. Biol. Chem.*, 272: 20055–20062, 1997.

192. Skinner, M.K. and Griswold, M.D. Sertoli cells synthesize and secrete transferrin like-proteins, *J. Biol. Chem.*, 255: 9523–9525, 1980.

193. Morales, C., Hughley, S., and Griswold, M.D. Stage-dependent levels of specific mRNA transcripts in Sertoli cells, *Biol. Reprod.*, 36: 1035–1046, 1987.

194. Huggenvik, J.I., Sylvester, S.R., and Griswold, M.D. Control of transferrin mRNA synthesis in Sertoli cells, *Ann. N.Y. Acad. Sci.*, 438: 1–7, 1985.

195. Perez-Infante, V. et al. Differential regulation of testicular transferrin and androgen-binding protein secretion in primary cultures of rat Sertoli cells, *Endocrinology*, 118: 383–392, 1986.

196. Huggenvik, I.J. et al. Transferrin messenger ribonuclei acid: molecular cloning and hormonal regulation in rat Sertoli cells, *Endocrinology*, 120: 332–340, 1987.

197. Hughly, S. and Griswold, M.D. Regulation of levels of specific Sertoli mRNAs by vitamin A, *Dev. Biol.*, 121: 316–324, 1987.

198. Hughly, S., Roberts, K.P., and Griswold, M.D. Transferrin and sulfated glycoprotein-2 messenger ribonucleic acid levels in the testes and isolated Sertoli cells of hypophysectomized rats, *Endocrinology*, 122: 1390–1396, 1988.

199. Fakunding, J.L. et al. Biochemical actions of follicle-stimulating hormone in the Sertoli cell of the rat testes, *Endocrinology*, 98: 392–402, 1976.

200. Boockfor, F.R. and Schwarz, L.K. Fibroblast growth factor modulates the release of transferrin from cultured Sertoli cells, *Mol. Cell. Endocrinol.*, 73: 187–197, 1990.

201. Spaliviero, J.A. and Handelsman, D.J. Effect of epidermal and insulin-like growth factors on vectorial secretion of transferrin by rat Sertoli cells *in vitro*, *Mol. Cell. Endocrinol.*, 81: 95–104, 1991.

202. Stallard, B.J. and Griswold, M.D. Germ cell regulation of Sertoli cell transferrin mRNA levels, *Mol. Endocrinol.*, 4: 393–401, 1990.

203. Grime, J. et al. Rat Sertoli cell clustering, α_2-macroglobulin, and testins: biosynthesis and differential regulation by germ cells, *Mol. Cell. Endocrinol.*, 89: 127–140, 1992.

204. Boockfor, F.R. and Schwaz, L.K. Effects of interleukin-6, interleukin-2, and tumor necrosis factor-α on transferrin release from Sertoli cells in culture, *Endocrinology*, 129: 256–262, 1991.

205. Hoeben, N. et al. Cytokines derived from activated human mononuclear cells markedly stmimulate transferrin secretion by cultured Sertoli cells, *Endocrinology*, 137: 514–521, 1996.

206. Hoeben, E. et al. Identification of IL-6 as one of the important cytokines responsible for the ability of mononuclear cells to stimulate Sertoli cell functions, *Mol. Cell. Endocrinol.*, 132: 149–160, 1997.

207. Sigillo, F. et al. *In vitro* regulation of rat Sertoli cell transferrin expression by tumor necrosis factor alpha and retinoic acid, *Mol. Cell. Endocrinol.*, 148: 163–170, 1999.

208. Boockfor, F.R., Schwarz, L.K., and Derrick, F.C. Sertoli cells in culture are heterogeneous with respect to transferrin release: analysis by reverse hemolytic plaque assay, *Endocrinology*, 125: 1128–1133, 1989.

209. Garza, M.M. et al. Sertoli cell function varies along seminiferous tubule: the proportion and response of transferrin secretors differ between stage-associated tubule segments, *Endocrinology*, 128: 1869–1874, 1991.

210. Skinner, M.K., Fetteroli, P.M., and Anthony, C.T. Purification of the paracrine factor P-Mod-S, produced by testicular peritubular cells that modulates Sertoli cell function, *J. Biol. Chem.*, 263: 2884–2890, 1989.

211. Anthony, C.T., Rosselli, M., and Skinner, M.K. Actions of the testicular paracrine factor (P-Mod-S) on Sertoli cell transferrin secretion throughout pubertal development, *Endocrinology*, 129: 353–360, 1991.

212. Hoeben, E. et al. Heregulins or Neu differentiation factors and the interactions between peritubular myoid cells and Sertoli cells, *Endocrinology*, 140: 2216–2223, 1999.

213. Edwards, R.G. Follicular fluid, *J. Reprod. Fertil.*, 37: 189–219, 1974.

214. Nagy, B. et al. The serum protein content of human follicular fluid and its correlation with the maturity of oocytes, *Acta Physiol. Hung.*, 73: 71–78, 1989.

215. Entman, S.S. et al. Follicular fluid transferrin levels in preovulatory human follicles, *J. In Vitro Embryo Transfer*, 4: 98–105, 1987.

216. Aleshire, S.L. et al. Localization of transferrin and its receptor in ovarian follicular cells: morphologic studies in relation to follicular development, *Fertil. Steril.*, 51: 444–449, 1989.

217. Orly, J., Sato, G., and Erickson, G.F. Serum suppresses the expression of hormonally induced functions in cultured granulosa cells, *Cell*, 20: 817–827, 1980.

218. Yu, J.H. and Findley, J.K. An inhibitory effect of transferrin on differentiation of rat granulosa cells *in vitro*, *Endocrinology*, 128: 1841–1848, 1991.

219. Lee, D.C., McKnight, G.S., and Palmiter, R.D. The action of estrogen and progesterone on the expression of the transferrin gene, *J. Biol. Chem.*, 253: 3494–3503, 1978.

220. Gerber, M.R. and Connor, J.R. Do oligodendrocytes mediate iron regulation in human brain? *Ann. Neurol.*, 26: 95–98, 1989.

221. Connor, J.R. et al. Cellular distribution of transferrin, ferritin, and iron in normal and aged human brains, *J. Neurosci. Res.,* 27: 595–611, 1990.
222. Dickson, P.W. et al. High prealbumin and transferrin mRNA levels in the choroid plexus of the rat brain, *Biochim. Biophys. Res. Commun.,* 890: 895–902, 1985.
223. Block, B. et al. Transferrin gene expression in choroid plexus of the adult rat brain, *Brain Res. Bull.,* 18: 573–576, 1987.
224. Connor, J.R. and Fine, R.E. The distribution of transferrin immunoreactivity in the rat central nervous system, *Brain Res.,* 368: 319–328, 1986.
225. Connor, J.R. and Fine, R.E. Development of transferrin-positive oligodendrocytes in the rat central nervous system, *J. Neurosci. Res.,* 17: 51–59, 1987.
226. Connor, J.R. and Menzies, S.L. Altered cellular distribution of iron in the central nervous system of myelin deficient rats, *Neuroscience,* 34: 265–271, 1990.
227. Connor, J.R. et al. Regional variation in the levels of transferrin in the CNS of normal and myelin-deficient rats, *J. Neurochem.,* 49: 1523–1529, 1987.
228. Tu, G.F. et al. The distribution of cerebral expression of the transferrin gene is species specific, *J. Biol. Chem.,* 266: 6201–6208, 1991.
229. Rothenberger, S. et al. Coincident expression and distribution of melanotransferrin and transferrin receptor in human brain capillary endothelium, *Brain Res.,* 712: 117–121, 1996.
230. Stamatos, C., Squicciarini, J., and Fine, R.E. Chick embryo spinal cord neurons synthesize transferrin-like myotrophic protein, *FEBS Lett.,* 153: 397–390, 1983.
231. Lin, H.H., Syndez, B.S., and Connor J.R. Transferrin expression in myelinated and non-myelinated peripheral nerves, *Brain Res.,* 526: 217–220, 1990.
232. Barakat-Walter, I. et al. Proliferation of chick embryo neuroblasts grown in the presence of horse serum requires exogenous transferrin, *J. Neurosci. Res.,* 28: 391–398, 1991.
233. Hyndman, A.G. High affinity binding of transferrin in cultures of embryonic neurons from the chick retina, *Brain Res.,* 564: 127–131, 1991.
234. Hyndman, A.G. et al. Transferrin can alter physiological properties of retinal neurons, *Brain Res.,* 561: 318–323, 1991.
235. De Los Monteros, A.E. et al. Transferrin gene expression and secretion by rat brain cells *in vitro, J. Neurosci. Res.,* 25: 576–580, 1990.
236. Ozawa, H. and Takashima, S. Immunocytochemical development of transferrin and ferritin immunoreactivity in the human pons and cerebellum, *J. Child Neurol.,* 13: 59–63, 1998.
237. Escobar-Cabrera, O.E. et al. Single intracranial injection of apotransferrin in young rats increases the expression of specific myelin protein mRNA, *J. Neurosci. Res.,* 47: 603–608, 1997.
238. Soltys, H.D. and Brody, J.I. Synthesis of transferrin by human peripheral blood lymphocytes, *J. Biol. Chem.,* 75: 250–258, 1969.
239. Nishiya, K., Chiao, J.W., and De Sousa, M. Iron-binding proteins in selected human peripheral blood cell sets: immunofluorescence studies, *Brit. J. Haematol.,* 46: 235–244, 1980.
240. Lum, J.B. et al. Transferrin synthesis by inducer T lymphocytes, *J. Clin. Invest.,* 77: 841–853, 1986.
241. Kondo, T., Sakaguchi, M., and Namba, M. Characteristics of intracellular transferrin produced by human fibroblasts: its post-transcriptional regulation and association with tubulin, *Exp. Cell. Res.,* 242: 38–44, 1998.
242. Schaeffer, E. et al. Cell type-specific expression of the human transferrin gene, *J. Biol. Chem.,* 264: 7153–7160, 1989.

243. Idzearda, R.L. et al. Expression from the transferrin gene promoter in transgenic mice, *Mol. Cell. Biol.,* 9: 5154–5162, 1989.

244. Adrian, G.S. et al. Human transferrin. Expression and iron modulation of chimeric genes in transgenic mice, *J. Biol. Chem.,* 265: 13344–13350, 1990.

245. Brunel, F. et al. Interactions of DNA-binding proteins with the 5′ region of the human transferrin gene, *J. Biol. Chem.,* 263: 10180–10185, 1989.

246. Mendelzon, D., Boissier, F., and Zakin, M.M. The binding sites for the liver-specific transcription factor Tf-LF1 and the TATA box of the human transferrin gene promoter are the only elements necessary to direct liver-specific transcription *in vitro*, *Nucleic Acid Res.,* 18: 5717–5721, 1990.

247. Guillo, F. et al. Sertoli cell-specific expression of the human transferrin gene, *J. Biol. Chem.,* 296: 9876–9884, 1991.

248. Cao, Z., Urnek, R., and McKnight, S.L. Regulated expression of three C/EBP isoforms during adipose conversion of 3T3-L1 cells, *Genes Dev.,* 5: 1538–1552, 1991.

249. Suire, S., Fontaine, I., and Guillou, F. Follicle-stimulating hormone (FSH) stimulates transferrin gene expression in rat Sertoli cells: *cis* and *trans*-acting elements involved in FSH action via cyclic 3′, 5′-monophosphate on the transferrin gene, *Mol. Endocrinol.,* 9: 756–766, 1995.

250. Chaudhary, J., Cupp, A.S., and Skinner, M.K. Role of basic helix–loop helix transcription factors in Sertoli differentiation: identification of an E-box response element in the transferrin promoter, *Endocrinology,* 138: 667–675, 1997.

251. Chaudhary, J. and Skinner, M.K. Comparative sequence analysis of the mouse and human transferrin promoters: hormonal regulation of the transferrin promoter in Sertoli cells, *Mol. Reprod. Devel.,* 50: 273–283, 1998.

252. Chaudhary, J. and Skinner, M.K. E-box and cyclic adenosine monophosphate response elements are both required for follicle-stimulating hormone-induced transferrin promoter activation in Sertoli cells, *Endocrinology,* 140: 1262–1271, 1999.

253. Petropolos, I., Augé-Gouillou, C., and Zakin, M.M. Characterization of the active part of the human transferrin gene enhancer and purification of two liver nuclear factors interacting with the TGTTTGC motif present in this region, *J. Biol. Chem.,* 266: 24220–24225, 1991.

254. Lee, Y.H. et al. Disruption of the C/EBP alpha gene in adult mouse liver, *Mol. Cell. Biol.,* 17: 6014–6022, 1997.

255. Brunel, F. et al. Cloning and sequencing of PYBP, a pyrimidine-rich specific single-strand DNA-binding protein, *Nucleic Acid Res.,* 19: 5237–5245, 1991.

256. Gil, A. et al. Characterization of cDNAs encoding the polypyrimidine tract-binding protein, *Genes Dev.,* 5: 1224–1236, 1991.

257. Sladek, F.M. et al. Liver-enriched transcription factor HNF-4 is a novel member of the steroid hormone receptor superfamily, *Genes Dev.,* 4: 2353–2365, 1990.

258. Hertz, R. et al. Transcriptional suppression of the transferrin gene by hypolipidemic peroxisome proliferators, *J. Biol. Chem.,* 271: 218–224, 1996.

259. Schaeffer, E. et al. A different combination of transcription factors modulates the expression of the human transferrin promoter in liver and Sertoli cells, *J. Biol. Chem.,* 268: 23399–23408, 1993.

260. Zakin, M.M. Regulation of transferrin gene expression, *FASEB J.,* 6: 3253–3258, 1992.

261. Boissier, F. et al. The enhancer of the human transferrin gene is organized in two structural and functional domains, *J. Biol. Chem.,* 266: 9822–9828, 1991.

262. Lai, E. et al. HNF-3A, a hepatocyte-enriched transcription factor of novel structure is regulated transcriptionally, *Genes Dev.,* 4: 1427–1436, 1990.

263. Lai, E. et al. Hepatocyte nuclear factor 3α belongs to a gene family in mammals that is homologous to the *Drosophila* homeotic fork head, *Genes Dev.*, 5: 416–427, 1991.
264. Vallet, V. et al. Overproduction of a truncated hepatocyte nuclear factor 3 protein inhibits expression of liver-specific genes in hepatoma cells, *Mol. Cell. Biol.*, 15: 5453–5460, 1995.
265. Kaestener, K.H., Hiemish, H., and Schutz, G. Targeted distruption of the gene encoding hepatocyte nuclear factor 3 gamma results in reduced transcription of hepatocyte-specific genes, *Mol. Cell. Biol.*, 18: 4245–4251, 1998.
266. Espinosa de los Monteros, A. et al. Brain-specific expression of the human transferrin gene, *J. Biol. Chem.*, 269: 24504–24510, 1994.
267. Sawaya, B.E. and Schaeffer, E. Transcription of the human transferrin gene in neuronal cells, *Nucleic Acid Res.*, 23: 2206–2211, 1995.
268. Bowman, B.H. et al. Discovery of a brain promoter from the human transferrin gene and its utilization for development of transgenic mice that express human apolipoprotein E alleles, *Proc. Natl. Acad. Sci. U.S.A.*, 92: 12115–12119, 1995.
269. Sawaya, B.E., Aunis, D., and Schaeffer, E. Distinct positive and negative regulatory elements control neuronal and hepatic transcription of the human transferrin gene, *J. Neurosci. Res.*, 43: 261–272, 1996.
270. Lescoat, G. et al. Transferrin gene modulation by iron, *Biol. Cell.*, 65: 221–229, 1989.
271. Cox, L.A. and Adrian, G.S. Post-transcriptional regulation of chimeric human transferrin genes by iron, *Biochemistry*, 32: 4738–4745, 1993.
272. Semenza, G.L. Perspectives on oxygen sensing, *Cell*, 98: 281–284, 1999.
273. Rolfs, A. et al. Oxygen-regulated transferrin expression is mediated by hypoxia-inducible factor-1, *J. Biol. Chem.*, 272: 20055–20062, 1997.
274. Gu, H., Gupta, J., and Schoenberg, D.R. The poly(A)-limiting element is a conserved *cis*-acting sequence that regulates poly(A) tail length on nuclear pre-mRNAs, *Proc. Natl. Acad. Sci. U.S.A.*, 96: 8943–8948, 1999.
275. Braun, V. and Killman, H. Bacterial solutions to the iron-supply problem, *TIBS*, 24: 104–109, 1999.
276. Ferguson, A.D. et al. Siderophore-mediated iron transport: crystal structure of FhuA with bound lipopolysaccharide, *Science*, 282: 2215–2220, 1998.
277. Locher, K.P. et al. Transmembrane signaling across the ligand-gated FhuA receptor: crystal structures of free and ferrichrome-bound states reveal allosteric changes, *Cell*, 95: 771–778, 1998.
277a. Arnoux, P. et al. The crystal structure of Has-A, a hemophore secreted by *Serratia marcescens*, *Nature Struct. Biol.*, 6: 516–520, 1999.
278. Crosa, J.H. Genetics and molecular biology of siderophore-mediated iron transport in bacteria, *Microbiol. Rev.*, 53: 517–530, 1989.
279. Weinberg, E.D. Cellular regulation of iron assimilation, *Q. Rev. Biol.*, 64: 261–290, 1989.
280. Martinez, J.L., Delgado-Iribarren, A., and Baquero, F. Mechanisms of iron acquisition and bacterial virulence, *FEMS Microbiol. Rev.*, 75: 45–56, 1990.
281. Griffiths, E. The iron-uptake systems of pathogenic bacteria, in *Iron and Infection: Molecular, Physiological and Clinical Aspects*, Bullen, J.J. and Griffiths, E., Eds., Chichester, John Wiley, 69–137, 1995.
282. Morton, D.J. and Williams, P. Utilization of transferrin-bound iron by *Haemophilus* species of human and porcine origins, *FEMS Microbiol. Lett.*, 65: 123–128, 1989.
283. Morton, D.J. and Williams, P. Siderophore-independent acquisition of transferrin-bound iron by *Haemophilus influenzae* type B, *J. Gen. Microbiol.*, 136: 927–933, 1990.

284. Schryvers, A.B. and Morris, L.J. Identification and characterization of the transferrin receptor from *Neisseria meningitidis, Mol. Microbiol.,* 2: 281–288, 1988.

285. Schryvers, A.B. and Morris, L.J. Identification and characterization of the human lactoferrin-binding protein from *Neisseria meningitidis, Infect. Immun.,* 56: 1144–1149, 1988.

286. Schryvers, A.B. and Gonzales, G.C. Comparison of the abilities of different protein sources of iron to enhance *Neisseria meningitidis* infection in mice, *Infect. Immun.,* 57: 2425–2429, 1989.

287. Blanton, K.J. et al. Genetic evidence that *Neisseria gonorrhoeae* produces specific receptors for transferrin and lactoferrin, *J. Bacteriol.,* 172: 5225–5235, 1990.

288. McKenna, W.R. et al. Iron uptake from lactoferrin and transferrin by *Neisseria gonorrhoeae, Infect. Immun.,* 56: 785–791, 1988.

289. Niven, D.F., Donga, J., and Archibald, F.S. Responses of *Haemophilus pleuropneumoniae* to iron restriction: changes in the outer membrane protein profile and the removal of iron from porcine transferrin, *Mol. Microbiol.,* 3: 1083–1089, 1989.

290. Tsai, J., Dyer, D.W., and Sparling, P.F. Loss of transferrin receptor activity in *Neisseria meningitidis* correlates with inability to use transferrin as an iron source, *Infect. Immun.,* 56: 3132–3138, 1988.

291. Banerjee-Bhatnagar, N. and Frasch, C.E. Expression of *Neisseria meningitidis* iron-regulated outer membrane proteins, including a 70-kilodalton transferrin receptor, and their potential for use as vaccines, *Infect. Immun.,* 58: 2875–2881, 1990.

292. Schryvers, A.B. and Lee, B.C. Comparative analysis of the transferrin and lactoferrin binding proteins in the family Neisseriaceae, *Can. J. Microbiol.* 35: 409–415, 1989.

293. Holland, J., Towner, K.J., and Williams, P. Isolation and characterization of *Haemophilus influenzae* type B mutants defective in transferrin-binding and iron assimilation, *FEMS Microbiol. Lett.,* 77: 283–288, 1991.

294. Schryvers, A.B. Identification of the transferrin- and lactoferrin-binding proteins in *Haemophilus influenzae. J. Med. Microbiol.,* 29: 121–130, 1989.

295. Blanton, K.J. et al. Genetic evidence that *Neisseriae gonorrhoeae* produces specific receptors for transferrin and lactoferrin, *J. Bacteriol.,* 172: 5225–5235, 1990.

296. Lee, B.C. and Schryvers, A.B. Specificity of the lactoferrin and transferrin receptors in *Neisseria gonorrhoeae, Mol. Microbiol.,* 2: 827–829, 1988.

297. Pintor, M., Ferreiros, C.M., Criado, M.T. Characterization of the transferrin-iron uptake system in *Neisseria meningitidis, FEMS Microbiol. Lett.,* 112: 159–166, 1993.

298. Simonson, C., Brener, D., and DeVoe, I.W. Expression of a high-affinity mechanism for acquisition of transferrin iron by *Neisseria meningitidis, Infect. Immun.,* 36: 107–113, 1982.

299. West, S.E.H. and Sparling, P.F. Response of *Neisseria gonorrhoeae* to iron limitation: alterations in expression of membrane proteins without apparent siderophore production, *Infect. Immun.,* 47: 388–394, 1985.

300. Dyer, D.W. et al. Isolation by streptonigrin enrichment and characterization of a transferrin-specific iron uptake mutant of *Neisseria meningitidis, Microb. Pathog.,* 3: 351–363, 1987.

301. Ward, J.H. The structure, function, and regulation of transferrin receptors, *Invest. Radiol.,* 22: 74–83, 1987.

302. Heller, K. and Kadner, R.J. Nucleotide sequence of the gene for the vitamin B_{12} receptor protein in the outer membrane of *Escherichia coli, J. Bacteriol.,* 161: 904–908, 1985.

303. Padda, J.S. and Schryvers, A.B. N-linked oligosaccharides of human transferrin are not required for binding to bacterial receptors, *Infect. Immun.,* 58: 2972–2976, 1990.

304. Gotschlich, E.C. et al. The mechanisms of genetic variation of gonococcal pili. Iron-inducible proteins of *Neisseria*. A consensus, in *Neisseriae 1990*, Achtman, M. et al., Eds., New York, Walter de Gruyter, 405–414, 1991.

305. Powell, N.B. et al. Differential binding of apo and holo human transferrin to meningococci and co-localisation of the transferrin-binding proteins (TBPA and TBPB), *J. Med. Microbiol.*, 17: 257–264, 1998.

306. Cornelissen, C.N. et al. Gonococcal transferrin-binding protein 1 is required for transferrin utilization and is homologous to TonB-dependent outer membrane receptors, *J. Bacteriol.*, 174: 5788–5797, 1992.

307. Archibald, F.S. and DeVoe, I.W. Removal of iron from human transferrin by *Neisseria meningitidis*, *FEMS Microbiol. Lett.*, 6: 159–162, 1979.

308. Genco, C.A. et al. Isolation and characterization of a mutant of *Neisseria gonorrhoeae* that is defective in the uptake of iron from transferrin and haemoglobin and is avirulent in mouse subcutaneous chambers, *J. Gen. Microbiol.*, 137: 1313–1321, 1991.

309. Alà Aldeen, D.A. et al. Localization of the meningococcal receptors for human transferrin, *Infect. Immun.*, 61: 751–759, 1993.

310. Griffiths, E. et al. Antigenic relationship of transferrin-binding proteins from *Neisseria meningitidis*, *N. gonorrhoeae* and *Haemophilus influenzae*: cross-reactivity of antibodies to NH$_2$-terminal peptides, *FEMS Microbiol. Lett.*, 109: 85–92, 1993.

311. Cornelissen, C.N., Biskwas, G.D., and Sparling, P.F. Expression of gonococcal transferrin-binding protein 1 of *Escherichia coli* to bind human transferrin, *J. Bacteriol.*, 175: 2448–2450, 1993.

312. Alà Aldeen, D.A., Wall, R.A., and Borriello, S.P. The 70-kilodalton iron-regulated protein of *Neisseria meningitidis* is not the human transferrin receptor, *FEMS Microbiol. Lett.*, 69: 37–42, 1990.

313. Ferreiros, C.M. et al. Analysis of the molecular mass heterogeneity of the transferrin receptor in *Neisseria meningitidis* and commensal *Neisseria*, *FEMS Microbiol. Lett.*, 83: 247–254, 1991.

314. Cornelissen, C.N., Anderson, J.E., and Sparling, P.F. Characterization of the diversity of the transferrin-binding protein 2, *Infect. Immun.*, 65: 822–828, 1997.

315. Stevenson, P., Williams, P., and Griffiths, E. Common antigenic domains in transferrin-binding protein 2 of *Neisseria meningitidis*, *Neisseria gonorrheae* and *Haemophilus influenzae* type B, *Infect. Immun.*, 60: 2391–2396, 1992.

315a. Retzer, M.D., Yu, R., and Schryvers, A.B. Identification of sequences in human transferrin that bind to the bacterial receptor protein, transferrin-binding protein B, *Mol. Microbiol.*, 32: 111–121, 1999.

316. Otto, B.R., van Verweif, V., and Mac Laren, D.M. Transferrin and heme-compounds as iron sources for pathogenic bacteria, *Crit. Rev. Microbiol.*, 18: 217–233, 1993.

317. Boulton, I.C. et al. Transferrin binding protein B isolated from *Neisseria meningitidis* discriminates between apo and diferric human transferrin, *Biochem. J.*, 334: 269–273, 1998.

318. Renauld-Mongénie, G. et al. Identification of human transferrin-binding sites within meningococcal transferrin-binding protein B, *J. Bacteriol.*, 179: 6400–6407, 1997.

319. Alcantara, J., Yu, R.H., and Scryvers, A.B. The region of human transferrin involved in binding to basolateral transferrin receptors is localized in the C-lobe, *Mol. Microbiol.*, 8: 1135–1143, 1993.

320. Retzer, M.D. et al. Production and characterization of chimeric transferrins for the determination of the binding domains for bacterial transferrin receptors, *J. Biol. Chem.*, 271: 1166–1173, 1996.

321. Boulton, C. et al. Purified meningococcal transferrin-binding protein B interacts with a secondary strain-specific binding site in the N-terminal lobe of human transferrin, *Biochem. J.,* 339: 143–149, 1999.

322. Retzer, M.D. et al. Discrimination between apo and iron-loaded forms of transferrin by transferrin binding protein B and its N-terminal subfragment, *Microbiol. Pathog.,* 25: 175–180, 1998.

323. Cornelissen, C.N. and Sparling, P.F. Binding and surface exposure characteristics of the gonococcal transferrin receptor are dependent on both transferrin binding proteins, *J. Bacteriol.,* 178: 1437–1444, 1996.

324. Coulton, J.W. and Pang, J.C.S. Transport of hemin by *Haemophilus influenzae* type B, *Current Microbiol.,* 9: 93–98, 1983.

325. Kilian, M. and Biberstein, E.L. *Haemophilus,* in *Bergey's Manual of Systematic Bacteriology,* Vol. 1, Krieg, N.R., Ed., Baltimore, Williams & Wilkins, 558–569, 1992.

326. Stull, T.L. Protein sources of heme for *Haemophilus influenzae, Infect. Immun.,* 55: 148–153, 1987.

327. Pidcock, K.A. et al. Iron acquisition by *Haemophilus influenzae, Infect. Immun.,* 56: 721–725, 1988.

328. Herrington, D.A. and Sparling, P.F. *Haemophilus influenzae* can use transferrin as a sole source for required iron, *Infect. Immun.,* 48: 248–251, 1985.

329. Morton, D.J. and Williams, P. Utilization of transferrin-bound iron by *Haemophilus* species of human and porcine origins, *FEMS Microbiol. Lett.,* 65: 123–128, 1989.

330. Morton, D.J. and Williams, P. Siderophore-independent acquisition of transferrin-bound iron by *Haemophilus influenzae* type B, *J. Gen. Microbiol.,* 136: 927–933, 1990.

331. Gray-Owen, S.D., Loosmore, S., and Schryvers, A.B. Identification and characterization of genes encoding the human transferrin-binding proteins from *Haemophilus influenzae, Infect. Immun.,* 63: 1201–1210, 1998.

332. Loosmore, S.M. et al. Cloning and expression of the *Haemophilus influenzae* transferrin receptor genes, *Mol. Microbiol.,* 19: 575–586, 1996.

333. Hasan, A.A. et al. Elemental iron does repress transferrin, hemopexin and hemoglobin receptor expression in *Haemophilus influenzae, FEMS Lett.,* 150: 19–26, 1997.

334. Fleishmann, R.D. Whole genome random sequencing and assembly of *Haemophilus influenzae, Science,* 269: 496–512, 1995.

335. Holland, J. et al. Evidence for *in vivo* expression of transferrin-binding proteins in *Haemophilus influenzae* type B, *Infect. Immun.,* 60: 2986–2991, 1992.

336. Hardie, K.R., Adams, R.A., and Towner, K.J. Transferrin-binding ability of invasive and commensal isolates of *Haemophilus* spp., *J. Med. Microbiol.,* 39: 218–224, 1993.

337. Whitby, P.W. et al. Transcription of genes encoding iron and heme acquisition proteins of *Haemophilus influenzae* during acute otitis media, *Infect. Immun.,* 65: 4696–4700, 1997.

338. Bruns, C.M. et al. Structure of *Haemophilus influenzae* Fe (+3)-binding protein reveals convergent evolution within a superfamily, *Nature Struct. Biol.,* 4: 919–924, 1997.

339. Redhead, K. and Hill, T. Acquisition of iron from transferrin by *Bordetella pertussis,* *FEMS Microbiol. Lett.,* 77: 303–308, 1991.

340. Yu, R.H. et al. Interaction of ruminant transferrins with transferrin receptors in bovine isolates of *Pasteurella haemolytica* and *Haemophilus somnus, Infect. Immun.,* 60: 2992–2994, 1992.

341. Ogunnariwo, J.A. et al. Characterization of the *Pasteurella haemolytica* transferrin receptor genes and the recombinant receptor proteins, *Microb. Pathog.,* 23: 273–284, 1997.

342. Konetschny-Rapp, S. et al. Staphyloferrin A: a structurally new siderophore from staphylococci, *Eur. J. Biochem.*, 191: 65–74, 1991.

343. Courcol, R.J. et al. Siderophore production by *Staphylococcus aureus* and identification of iron-regulated proteins, *Infect. Immun.*, 65: 1944–1948, 1997.

344. Lindsay, J.A., Riley, T.V., and Mee, B.J. *Staphylococcus aureus* but not *Staphylococcus epidermidis* can acquire iron from transferrin, *Microbiology*, 141: 197–203, 1995.

345. Modun, B., Kendall, D., and Williams, P. Staphylococci express a receptor for human transferrin: identification of a 42-kilodalton cell wall transferrin-binding protein, *Infect. Immun.*, 62: 3850–3858, 1994.

346. Modun, B. et al. The *Staphylococcus aureus* and *Staphylococcus epidermidis* transferrin-binding proteins are expressed *in vivo* during infection, *Microbiology*, 144: 1005–1012, 1998.

347. Modun, B. et al. Receptor-mediated recognition and uptake of iron from human transferrin by *Staphylococcus aureus* and *Staphylococcus epidermidis*, *Infect. Immun.*, 66: 3591–3596, 1998.

348. Lim, Y. et al. A human transferrin-binding protein of *Staphylococcus aureus* is immunogenic *in vivo* and has an epitope in common with human transferrin receptor, *FEMS Microbiol. Lett.*, 166: 225–230, 1998.

349. Duchesne, P., Grenier, D., and Mayrand, D. Binding and utilization of human transferrin by *Prevotella nigrescens*, *Infect. Immun.*, 67: 576–580, 1999.

350. Lalounde, R.G. and Holbein, B.E. Role of iron in *Trypanosoma cruzi* infection in mice, *J. Clin. Invest.*, 73: 470–476, 1984.

351. Loo, V.G. and Lalounde, R.G. Role of iron in intracellular growth of *Trypanosoma cruzi*, *Infect. Immun.*, 45: 726–730, 1984.

352. Lima, M.F. and Villalta, F. *Trypanosoma cruzi* receptors for human transferrin and their role, *Mol. Biochem. Parasitol.*, 38: 245–252, 1990.

353. Voyiatzaki, C.S. and Seteriadan, K.P. Evidence of transferrin binding sites on the surface of *Leishmania* promastigotes, *J. Biol. Chem.*, 265: 22380–22385, 1990.

354. Clemens, L.E. and Bash, P.F. *Schistosoma mansoni*: effect of transferrin and growth factors on the development of schistosomula *in vitro*, *J. Parasitol.*, 75: 417–421, 1989.

355. Schell, D. et al. A transferrin binding protein of *Trypanosoma brucei* is encoded by one of the genes in the variant surface glycoprotein expression site, *EMBO J.*, 10: 1061–1066, 1991.

356. Steverding, D. et al. ESAG6 and 7 products of *Trypanosoma brucei* form a transferrin-binding protein complex, *Eur. J. Cell Biol.*, 64: 78–87, 1994.

357. Salmon, D. et al. A novel heterodimeric transferrin receptor encoded by a pair of VSG expression site-associated genes in *T. brucei*, *Cell*, 78: 75–86, 1994.

358. Janssen, H., Calafat, J., and Borst, P. Reconstitution of a surface transferrin binding complex in insect from *Trypanosoma brucei*, *EMBO J.*, 13: 2565–2573, 1994.

359. Steverding, D. et al. Transferrin-binding protein complex is the receptor for transferrin uptake in *Trypanosoma brucei*, *J. Cell Biol.*, 131: 1173–1182, 1995.

360. Steverding, D. Bloodstream forms of *Trypanosoma brucei* require only small amounts of iron for growth, *Parasitol. Res.*, 84: 59–62, 1998.

361. Bitter, W. et al. The role of transferrin-receptor variation in the host range of *Trypanosoma brucei*, *Nature*, 391: 499–502, 1998.

362. Voyiatzaki, C.S. and Soteriadou, K.P. Identification and isolation of the *Leishmania* transferrin receptor, *J. Biol. Chem.*, 267: 9112–9117, 1992.

363. Wilson, M.E. et al. Acquisition of iron from transferrin and lactoferrin by the protozoan *Leishmania chagasi*, *Infect. Immun.*, 62: 3262–3269, 1994.

364. Britigan, B.E. et al. Evidence for the existence of a surface receptor for ferrilactoferrin and ferrictransferrin associated with the plasma membrane of the protozoan parasite *Leishmania donovani, Adv. Exp. Med. Biol.,* 443: 135–140, 1998.

365. Rodriguez, M.H. and Jurgery, M. A protein on *Plasmodium falciparum*-infected erythrocytes functions as a transferrin receptor, *Nature,* 324: 388–391, 1986.

366. Gobin, J. and Horwitz, M.A. Exochelins of *Mycobacterium tuberculosis* remove iron from human iron-binding proteins and donate iron to mycobactins in the *M. tuberculosis* cell wall, *J. Exp. Med.,* 183: 1527–1532, 1996.

367. Newman, S.L. et al. Chloroquine induces human macrophage killing of *Histoplasma capsulatum* by limiting the availability of intracellular iron and is a therapeutic in a murine model of histoplasmosis, *J. Clin. Invest.,* 93: 1422–1431, 1994.

368. Strosser, J.E. et al. Regulation of the macrophage vacuolar ATPase and phagosome–lysosome fusion by *Histoplasma capsulatum, J. Immunol.,* 162: 6148–6154, 1999.

369. Danve, B.L. et al. Transferrin-binding proteins isolated from *Neisseria meningitidis* elicit protective and bactericidal antibodies in laboratory animals, *Vaccine,* 11: 1214–1220, 1993.

370. Potter, A. et al. Protective capacity of the *Pasteurella haemolytica* transferrin-binding proteins TBPA and TBPB in cattle, *Microb. Pathog.,* 27: 197–206, 1999.

371. Chen, D. et al. Evaluation of a 74-kDa transferrin-binding protein from *Moraxella catarrhalis* as a vaccine candidate, *Vaccine,* 18: 109–118, 1999.

372. Holland, J.P. et al. Evidence for *in vivo* expression of transferrin-binding proteins in *Haemophilus influenzae* type B, *Infect. Immun.,* 60: 2986–2991, 1992.

373. Ala'Aldeen, D.A. et al. Immune responses in humans and animals to meningococcal transferrin-binding proteins: implications for vaccine design, *Infect. Immun.,* 62: 2984–2990, 1994.

374. Yu, R. et al. Analysis of the immunoglobulin response to transferrin and lactoferrin receptor proteins from *Moraxella catarrhalis, Infect. Immun.,* 67: 3793–3799, 1999.

375. Lissolo, L. et al. Evaluation of transferrin-binding protein 2 within the transferrin binding protein complex as a potential antigen for future meningococcal vaccines, *Infect. Immun.,* 63: 884–890, 1995.

376. Myers, L.E. et al. The transferrin binding protein B of *Moraxella catarrhalis* elicits bactericidal antibodies and is a potential vaccine antigen, *Infect. Immun.,* 66: 4183–4192, 1998.

377. Webb, D.C. and Cripps, A.W. Immunization with the recombinant binding protein B enhances clearance of nontypeable *Haemophilus influenzae* from the rat lung, *Infect. Immun.,* 67: 2138–2144, 1999.

378. Robbins, E. and Pederson, T. Iron: its intracellular localization and possible role in cell division, *Proc. Natl. Acad. Sci. U.S.A.,* 66: 1244–1259, 1970.

379. Lederman, H.M. et al. Deferoxamine: a reversible S-phase inhibitor of human lymphocyte proliferation, *Blood,* 64: 748–754, 1984.

380. Eriksson, S. et al. Cell-cycle dependent regulation of mammalian ribonucleotide reductase. The S-phase-correlated increase in subunit M2 is regulated by *de novo* protein synthesis, *J. Biol. Chem.,* 259: 11695–11703, 1984.

381. Trowbridge, I.A. and Lopez, F. Monoclonal antibody to transferrin receptor blocks transferrin binding and inhibits human tumor growth *in vitro, Proc. Natl. Acad. Sci. U.S.A.,* 79: 1175–1180, 1982.

382. Teatle, R., Castanola, J., and Mendelsohn, J. Mechanisms of growth inhibition by anti-transferrin receptor monoclonal antibodies, *Cancer Res.,* 46: 1759–1766, 1986.

383. Chitambar, C.R. and Zivkovic, Z. Uptake of gallium-67 by human leukemic cells: demonstration of transferrin receptor-dependent and transferrin-independent mechanisms, *Cancer Res.*, 47: 3929–3937, 1987.

384. Moran, P.L. and Seligman, P.A. The effects of transferrin–indium on cellular proliferation of a human leukemic cell line, *Cancer Res.*, 49: 4237–4244, 1989.

385. Blatt, J. and Stitely, S. Anti-neuroblastoma activity of desferrioxamine in human cell lines, *Cancer Res.*, 47: 1749–1756, 1987.

386. Breuer, W., Epsztejn, S., and Cabantchik, Z.I. Iron acquired from transferrin by K562 cells is delivered into a cytoplasmic pool of chelatable iron (II), *J. Biol. Chem.*, 270: 24209–24216, 1995.

387. Konijn, A. et al. The cellular labile iron pool and intracellular ferritin in K562 cells, *Blood*, 94: 2128–2134, 1999.

388. de Jong, G., Van Dijk, J.P., and Van Dijk, H.G. The biology of transferrin, *Clin. Chim. Acta*, 190: 1–46, 1990.

389. Buys, S.S., Keogh, E.A., and Kaplan, J. Fusion of intracellular membrane pools with the cell surfaces of macrophages stimulated by phorbol esters and calcium ionophores, *Cell*, 38: 569–576, 1984.

390. Wiley, H.S. and Kaplan, J. Epidermal growth factor rapidly induces a redistribution of transferrin receptor pools in human fibroblasts, *Proc. Natl. Acad. Sci. U.S.A.*, 81: 7456–7460, 1984.

391. McKey, W.D. and Kaplan, J. Mitogenic agents induce redistribution of transferrin receptors from internal pools to the cell surface, *Biochem. J.*, 238: 721–728, 1986.

392. Sun, I.L., Garcia-Carrero, R., and Lin, W. Diferric transferrin reduction stimulates the Na^+/H^+ antiport of HeLa cells, *Biochem. Biophys. Res. Commun.*, 145: 467–473, 1987.

393. Ellem, R.A.O. and Kay, G.F. Ferricyanide can replace pyruvate to stimulate growth and attachment of serum restricted human melanoma cells, *Biochem. Biophys. Res. Commun.*, 112: 183–190, 1983.

394. Sun, I.L. et al. Growth stimulation by impermeable oxidants through plasma membrane redox, *J. Cell. Biol.*, 99: 293–303, 1984.

395. Morré, D.J. et al. NADH oxidase of liver plasma membrane stimulated by diferric transferrin and neoplastic transformation induced by the carcinogen 2-acetylaminofluorene, *Biochim. Biophys. Acta*, 1057: 140–146, 1991.

396. Sun, J.L. et al. NADH diferric transferrin reductase in liver plasma membrane, *J. Biol. Chem.*, 262: 15915–15921, 1987.

397. Sun, J.L. et al. Diferric transferrin reduction stimulates the Na^+/H^+ antiport of HeLa cells, *Biochim. Biophys. Res. Commun.*, 145: 467–473, 1987.

398. Sun, J.L. et al. Reduction of diferric transferrin by SV-40 transformed pineal cells stimulates Na^+/H^+ antiport activity, *Biochim. Biophys. Acta*, 938: 17–23, 1988.

399. Low, H. et al. Transplasmalemma electron transport from cells is part of a ferric transferrin reductase system, *Biochim. Res. Commun.*, 139: 1117–1123, 1986.

400. Sun, J.L. et al. Transmembrane redox in control of cell growth. Stimulation of HeLa cell growth by ferricyanide and insulin, *Exp. Cell Res.*, 156: 528–536, 1985.

401. Draz-Gil, J.J. et al. Purification of a liver DNA-synthesis promoter from plasma of partially hepatectomized rats, *Biochem. J.*, 235: 49–55, 1986.

402. Sun, J.L. et al. Inhibition of transplasma membrane electron transport by transferrin–adriamycin conjugates, *Biochim. Biophys. Acta*, 1105: 84–88, 1992.

403. Brightman, A.O. et al. A growth factor- and hormone-stimulated NADH oxidase from rat liver plasma membrane, *Biochim. Biophys. Acta*, 1105: 109–117, 1992.

404. Sun, I.L. et al. Requirement for coenzyme Q in plasma membrane electron transport, *Proc. Natl. Acad. Sci. U.S.A.,* 89: 11126–11130, 1992.

405. Thorstensen, K. Hepatocytes and reticulocytes have different mechanisms for the uptake of iron from transferrin, *J. Biol. Chem.,* 263: 16837–16841, 1988.

406. Thorstensen, K. and Romslo, I. Uptake of iron from transferrin by isolated rat hepatocytes. A redox-mediated plasma membrane process, *J. Biol. Chem.,* 263: 8844–8850, 1988.

407. Thorstensen, K. and Aisen, P. Release of iron from diferric transferrin in the presence of rat liver plasma membranes: no evidence of a plasma membrane diferric transferrin reductase, *Biochim. Biophys. Acta,* 1052: 29–35, 1990.

408. Goldenberg, H., Dodel, B., and Seidl, D. Plasma membrane Fe_2-transferrin reductase and iron uptake in K562 cells are not directly related, *Eur. J. Biochem.,* 192: 475–480, 1990.

409. Oshiro, S. et al. Redox, transferrin-independent, and receptor-mediated endocytosis iron uptake systems in cultured human fibroblasts, *J. Biol. Chem.,* 268: 21586–21591, 1993.

410. Van Duijn, M.M. et al. Ascorbate stimulates ferricyanide reduction in HL-60 cells through a mechanism distinct from the NADH-dependent plasma membrane reductase, *J. Biol. Chem.,* 273: 13415–13420, 1998.

411. May, J.M. Is ascorbic acid an antioxidant for the plasma membrane? *FASEB J.,* 13: 995–1006, 1999.

412. Han, O. et al. Reduction of Fe (III) is required for uptake of nonheme iron by Caco-2 cells, *J. Nutr.,* 125: 1291–1299, 1995.

413. Philips J.L., Boldt D.H., Harper J. Iron transferrin induces protein kinase C activity in CCRF-CEM cells, *J. Cell. Physiol.,* 132: 349–358, 1987.

414. Alcantara, O., Javors, M., and Boldt, D.H. Induction of protein kinase C mRNA in cultured lymphoblastoid T cells by iron-transferrin, but not by soluble iron, *Blood,* 77: 1290–1297, 1991.

415. Thelander, L. and Reichard, P. Reduction of ribonucleotides, *Ann. Rev. Biochem.,* 48: 133–158, 1979.

416. Graslund, A., Sahlin, M., and Sjoberg, B.M. The tyrosyl free radical in ribonucleotide reductase, *Environ. Health Perspect.,* 64: 139–149, 1985.

417. Atkin, C.L. et al. Iron and free radical in ribonucleotide reductase. Exchange of iron and Mossbauer spectroscopy of the protein B2 subunit of the *Escherichia coli* enzyme, *J. Biol. Chem.,* 248: 7464–7472, 1973.

418. Terada, N. et al. Definition of the role for iron and essential fatty acids in cell cycle progression of normal human T lymphocytes, *Exp. Cell. Res.,* 204: 260–267, 1993.

419. Trowbridge, I.S. and Omary, M.B. Human cell surface glycoprotein related to cell proliferation is the receptor for transferrin, *Proc. Natl. Acad. Sci. U.S.A.,* 78: 3039–3043, 1981

420. Wu, K.J., Polack, A., and Dalla-Favera, R. Coordinated regulation of iron-controlling genes, H-ferritin and IRP2 by c-*myc*, *Science,* 283: 676–679, 1999.

421. Loulegue, C., Lebruss, M., and Briat, J.F. Expression cloning in Fe^{2+} transport defective yeast of a novel maize MYC transcription factor, *Gene,* 225: 47–57, 1998.

422. Chitambar, C.R., Massey, E.J., and Seligman, P.A. Regulation of transferrin receptor expression on human leukemic cells during proliferation and induction of differentiation: effects of gallium and dimethyl sulfoxide, *J. Clin. Invest.,* 72: 1314–1325, 1983.

423. Pelosi-Testa, E. et al. Expression of transferrin receptors in human erythroleukemic lines: regulation in the plateau and exponential phase of growth, *Cancer Res.,* 46: 5330–5334, 1986.

424. Neckers, L.M. and Cossman, J. Transferrin receptor induction in mitogen stimulated human T lymphocytes is required for DNA synthesis and cell division and is regulated by interleukin-2, *Proc. Natl. Acad. Sci. U.S.A.,* 80: 3494–3498, 1983.

425. Kronke, M. et al. Sequential expression of genes involved in human T lymphocyte growth and differentiation, *J. Exp. Med.,* 161: 1593–1598, 1985.

426. Pelosi, E. et al. Expression of transferrin receptors in phytohemagglutinin-stimulated human T-lymphocytes. Evidence for a three-step model, *J. Biol. Chem.,* 261: 3036–3041, 1986.

427. Brock, J.H. and Rankin, M.C. Transferrin binding and iron uptake by mouse lymph-node cells during transformation in response to concavalin A, *Immunology,* 43: 393–402, 1981.

428. Neckers, L.M. et al. Diltiazem inhibits transferrin receptor expression and causes G1 arrest in normal and neoplastic T cells, *Mol. Cell. Biol.,* 6: 4244–4250, 1986.

429. Trowbridge, I.S. and Lopez, F. Monoclonal antibody to transferrin receptors blocks transferrin binding and inhibits human tumor cell growth *in vitro, Proc. Natl. Acad. Sci. U.S.A.,* 79: 1175–1179, 1982.

430. Mendelsohn, J., Trowbridge, I.S., and Castagnola, J. Inhibition of human lymphocyte proliferation by monoclonal antibody to transferrin receptor, *Blood,* 62: 821–826, 1983.

431. Taetle, R. et al. Role of transferrin, Fe, and transferrin receptors in myeloid leukemia cell growth. Studies with anti-transferrin receptor monoclonal antibody, *J. Clin. Invest.,* 75: 1061–1067, 1985.

432. Trowbridge, I.S. and Shackelford, D.A. Structure and function of transferrin receptors and their relationship to cell growth, *Biochem. Soc. Symp.,* 51: 117–119, 1986.

433. Lesley, J.F. and Schulte, R.J. Inhibition of cell growth by monoclonal anti-transferrin receptor antibodies, *Mol. Cell. Biol.,* 5: 1814–1821, 1985.

434. Lesley, J.F., Schulte, R.J., and Woods, J. Modulation of transferrin receptor expression and function by anti-transferrin receptor antibodies and antibody fragments, *Exp. Cell Res.,* 182: 215–233, 1989.

435. White, S. et al. Combinations of anti-transferrin receptor monoclonal antibodies inhibit human cell growth *in vitro* and *in vivo*: evidence for synergistic antiprolifer-ative effects, *Cancer Res.,* 50: 6295–6301, 1990.

436. Taetle, R., Honeysett, J.M., and Bergeron, R. Combination iron depletion therapy, *J. Natl. Cancer Inst.,* 81: 1229–1235, 1989.

437. Taetle, R. et al. Effects of combined antigrowth receptor treatment on *in vitro* growth of multiple myeloma, *J. Natl. Cancer Inst.,* 86: 450–455, 1994.

438. Hsi, B.I., Yeh, C.J., and Faulk, W.P. Human amniochorion: tissue-specific markers, transferrin receptors and histocompatibility antigens, *Placenta,* 3: 1–12, 1982.

439. Bulmer, J.N. and Johnson, P.M. Antigen expression by trophoblast populations in the human placenta and their possible immunological relevance, *Placenta,* 6: 127–140, 1985.

440. Bulmer, J.N., Morrison, L., and Johnson, P.M. Expression of the proliferation markers Ki 67 and transferrin receptor by human trophoblast populations, *J. Reprod. Immu-nol.,* 14: 291–302, 1988.

441. Bierings, M.B. et al. Pregnancy and guinea pig isotransferrin isolation, *Clin. Chim. Acta,* 165: 141–145, 1987.

442. Sorokin, L.M., Morgan, E.H., and Yech, G.C.T. Transferrin receptor number and transferrin and iron uptake in cultured chick muscle cells at different stages of development, *J. Cell. Physiol.,* 131: 342–353, 1987.

443. Ozawa, E. and Hagiwaza, Y. Degeneration of large myotubes following removal of transferrin from cultured medium, *Biomed. Res.,* 3: 16–23, 1982.

444. Robbins, E. and Pederson, T. Iron: its intracellular localization and possible role in cell division, *Proc. Natl. Acad. Sci. U.S.A.,* 66: 1244–1251, 1970.

445. Vogt, A., Mishell, R.I., and Dutton, R.W. Stimulation of DNA synthesis in cultures of mouse spleen suspensions by bovine transferrin, *Exp. Cell Res.,* 54: 195–200, 1969.

446. Tormey, D.C., Imrie, R.C., and Mueller, G.C. Identification of transferrin as a lymphocyte growth promoter in human serum, *Exp. Cell Res.,* 74: 163169, 1972.

447. Tormey, D.C. and Mueller, G.C. Biological effects of transferrin on human lymphocytes *in vitro*, *Exp. Cell Res.,* 74: 220–226, 1972.

448. Messmer, T.O. Nature of the iron requirement for Chinese hamster V-79 cells in tissue culture medium, *Exp. Cell Res.,* 77: 404–408, 1973.

449. Guilbert, L.J. and Iscove, N.N. Partial replacement of serum by selenite, transferrin, albumin and lecithin in hemopoietic cell cultures, *Nature,* 263: 594–595, 1976.

450. Barnes, D. and Sato, G. Serum-free cell culture, *Cell,* 22: 649–655, 1980.

451. Ponka, P., Schulman, H.M., and Wilczynska, A. Ferric pyridoxal isonicotinoyl hydrazone can provide iron for heme synthesis in reticulocytes, *Biochim. Biophys. Acta,* 718: 151–156, 1982.

452. Ponka, P. and Schulman, H.M. Acquisition of iron from transferrin regulates reticulocyte heme synthesis, *J. Biol. Chem.,* 260: 14717–14721, 1985.

453. Titeux, M. et al. The role of iron in the growth of human leukemic cell lines, *J. Cell. Physiol.,* 121: 251–256, 1984.

454. Landschulz, W., Thesleff, I., and Ekblom, P. A lipophilic iron chelator can replace transferrin as a stimulator of cell proliferation and differentiation, *J. Cell. Biol.,* 98: 596–601, 1984.

455. Landschulz, W. and Ekblom, P. Iron delivery during proliferation and differentiation of kidney tubules, *J. Biol. Chem.,* 260: 15580–15584, 1985.

456. Ekblom, P., Landschulz, W., and Andersson, L.C. A lipohilic iron chelator induces an enhanced proliferation of human erythroleukemia (HEL) cells. *Scand. J. Haematol.,* 36: 258–262, 1986.

457. Partanen, A.M. and Thesleff, I. Transferrin and tooth morphogenesis: retention of transferrin by mouse embryonic teeth in organ culture, *Differentiation,* 34: 25–31, 1987.

458. Thesleff, I. et al. The role of transferrin receptors and iron delivery in mouse embryonic morphogenesis, *Differentiation,* 30: 152–158, 1985.

459. Forsbeck, K., Bjelkenkrantz, K., and Nilsson, K. Role of iron in the proliferation of the established human cell lines U-937 and K-562: effects of suramin and a lipophilic iron chelator (PIH), *Scand. J. Haematol.,* 37: 429–437, 1986.

460. Tsao, M.S., Sanders, G.H. S., and Grisham, J.W. Regulation of growth of cultured hepatic epithelial cells by transferrin, *Exp. Cell Res.,* 171: 52–62, 1987.

461. Flynn, T.J. et al. Methionine and iron as growth factors for rat embryos cultured in canine serum, *J. Exp. Zool.,* 244: 319–324, 1987.

462. Brock, J.H. and Stevenson, J. Replacement of transferrin in serum-free cultures of mitogen-stimulated mouse lymphocytes by a lipophilic iron chelator, *Immunol. Lett.,* 15: 23–27, 1987.

463. Sanders, E.J. and Cheung, E. Transferrin and iron requirements of embryonic mesoderm cells cultured in hydrated collagen matrices, *In Vitro Cell Dev. Biol.,* 24: 581–587, 1988.

464. Sirbasku, D.A. et al. Purification of an equine apotransferrin variant (thyromedin) essential for thyroid hormone dependent growth of GH1 rat pituitary tumor cells in chemically defined culture, *Biochemistry,* 30: 295–304, 1991.

465. Sirbasku, D.A. et al. Thyroid hormone regulation of rat pituitary tumor cell growth: a new role for apotransferrin as an autocrine thyromedin, *Mol. Cell. Endocrinol.*, 77: c47–c55, 1991.

466. Sirbasku, D.A. et al. Thyroid hormone dependent pituitary tumor cell growth in serum-free chemically defined culture. A new regulatory role for apotransferrin, *Biochemistry*, 30: 7466–7477, 1991.

467. Sirbasku, D.A. et al. Thyroid hormone dependent pituitary tumor cell growth in serum-free chemically defined culture. A new regulatory role for apotransferrin, *Biochemistry*, 30: 7466–7477, 1991.

468. Tilemans, D. et al. Production of transferrin-like monoreactivity by rat anterior pituitary and intermediate lobe, *J. Histochem. Cytochem.*, 43: 657–664, 1995.

469. Tampanany-Sarmesin, A. et al. Transferrin and transferrin receptor in human hypophysis and pituitary adenomas, *Am. J. Pathol.*, 152: 413–422, 1998.

470. Lombardi, A. et al. Transferrin in FRTL5 cells: regulation of its receptor by mitogenic agents and its role in growth, *Endocrinology*, 125: 652–658, 1989.

471. Fayada, L. et al. Role of heme in intracellular trafficking of thyroperoxidase and involvement of H_2O_2 generated at the apical surface of thyroid cells in autocatalytic covalent heme binding, *J. Biol. Chem.*, 274: 10553–10538, 1999.

472. Kovar, J. and Franek, F. Growth-stimulating effect of transferrin on a hybridoma cell line: relation to transferrin iron-transporting function, *Exp. Cell Res.*, 182: 358–369, 1989.

473. Chackal-Roy, M. et al. Stimulation of human prostatic carcinoma cell growth by factors present in human bone marrow, *J. Clin. Invest.*, 84: 43–50, 1989.

474. Chackal-Roy, M. and Zetter, B.R. Selective stimulation of prostatic carcinoma cell proliferation by transferrin, *Proc. Natl. Acad. Sci. U.S.A.*, 89: 6197–6201, 1992.

475. Bhatti, R.A., Gadarowski, J.J., and Ray, P.S. Metastatic behavior of prostatic tumor as influenced by the hematopoietic and hematogenous factors, *Tumor Biol.*, 18: 1–5, 1997.

476. Nguyen-Lee, X.K., Briere, N., and Carcos, J. The effects of insulin, transferrin and androgens on rat prostate explants in serum-free organ culture, *Biofactors*, 6: 339–349, 1997.

477. Douat, S.M. et al. Reversal by transferrin of growth-inhibitory effect of suramin on hormone-refractory human prostate cancer cells, *J. Natl. Cancer Inst.*, 87: 41–46, 1995.

478. Vostreys, M., Morau, P.L., and Seligman, P.A. Transferrin synthesis by small lung cancer cells acts as an autocrine regulator of cellular proliferation, *J. Clin. Invest.*, 82: 331–339, 1988.

479. Denstam, S. et al. Identification of transferrin as progression factor for ML-1 human myeloblastic leukemia cell differentiation, *J. Biol. Chem.*, 266: 14876–14993, 1991.

480. Trayner, I.D. and Clemens, M.J. Phorbol-ester induced macrophage-like differentiation of human promyelocytic leukemia (HL-60) occurs independently of transferrin availability, *Cancer Res.*, 50: 7221–7225, 1990.

481. Oh, T.H. and Markelonis, G.J. Chicken serum transferrin duplicates the myotrophic effects of sciatin on cultured muscle cells, *J. Neurosci. Res.*, 8: 535–545, 1982.

482. Beach, R.L., Popoeila, H., and Festoff, B.W. Specificity of chicken and mammalian transferrins in myogenesis, *Cell Differentiation*, 16: 93–100, 1985.

483. Festoff, B.W. et al. Monoclonal antibody detects embryogenic epitope specific for nerve-derived transferrin, *J. Neurosci. Res.*, 22: 425–438, 1989.

484. Ghigo, D. et al. Chloroquine stimulates nitric oxide synthesis in murine, porcine, and human endothelial cells, *J. Clin. Invest.*, 102: 595–605, 1998.

485. Fairbanks, V.F. and Beutler, E. Congenital atransferrinemia and idiopathic pulmonary hemosiderosis, in *Hematology,* Williams W.J. et al., Eds., New York, McGraw-Hill, 506–510, 1990.
486. Heilmeyer, L. et al. Kongenitale atransferrinemie bei einem sieben Jahre anten Kind, *Dtsch. Med. Wochenschr.,* 86: 1745–1751, 1961.
487. Heilmeyer, L. Die Atransferrenamien, *Acta Haematol.,* 36: 40–49, 1966.
488. Cap, J., Lehotska, V., and Mayerova, A. Kongenitalna atransferrinemia U 11-mesacneho dietata, *Cesk. Pediatr.,* 23: 1020–1025, 1968.
489. Sakata, T. A case of congenital atransferrinemia, *J. Pediatr. Pract.,* 32: 1523–1528, 1969.
490. Walbaum, R. Déficit congénital en transferrine, *Lille Méd.,* 16: 1122–1124, 1971.
491. Goya N. et al. A family of congenital atransferrinemia, *Blood,* 40: 239–245, 1972.
492. Loperena, L. et al. Atransferrinemia hereditaria, *Bol. Med. Hosp. Inf.,* 31: 519–535, 1974.
493. Gaston-Morata, J.L., Rodriguez, C.A., and Urbano, J.F. Atransferrinemia secondary to hepatic cirrhosis, hemochromatosis, and nephrotic syndrome, *Rev. Esp. Enferm. Apar. Dig.,* 62: 491–495, 1982.
494. Hamill, R.L., Woods, J.C., and Cook, B.A. Congenital atransferrinemia: a case report and review of the literature, *J. Clin. Pathol.,* 96: 215–218, 1991.
495. Hayashi, A. et al. Studies on familial hypotransferrinemia: unique clinical course and molecular pathology, *Am. J. Hum. Genet.,* 33: 201–213, 1993.
495a. Raja, K.B. et al. Importance of anemia and transferrin levels in the regulation of intestinal iron absorption in hypotransferrinemic mice, *Blood,* 94: 3185–3192, 1999.
496. Dorantes-Mesa, S., Marquez, J.L., and Valencia-Mayotal, P. Iron overload in hereditary atransferrinemia, *Bol. Med. Hosp. Infant. Mex.,* 43: 99–101, 1986.
497. Raso, V. and Basala, M. A highly cytotoxic human transferrin–ricin A-chain conjugate used to select receptor-modified cells, *J. Biol. Chem.,* 259: 1143–1149, 1984.
498. Recht, L.D. et al. Potent cytotoxicity of an anti-human transferrin receptor–ricin A-chain immunotoxin on human glioma cells *in vitro*, *Cancer Res.,* 50: 6696–6700, 1990.
499. Martell, L.A. et al. Efficacy of transferrin receptor-targeted immunotoxins in brain tumor cell lines and pediatric brain tumors, *Cancer Res.,* 531348–1353, 1993.
500. Laske, D.W. et al. Efficacy of direct intratumoral therapy with targeted protein toxins for solid human gliomas in nude mice, *J. Neurosurg.,* 80: 520–526, 1994.
501. Murasko, K. et al. Pharmakinetics and toxicology of immunotoxins administered into the subarachnoid space in nonhuman primates and rodents, *Cancer Res.,* 53: 3752–3757, 1993.
502. Laske, D.W. et al. Chronic interstitial infusion of protein to primate brain: determination of drug distribution and clearance with single-photon emission computerized tomography imaging, *J. Neurosurg.,* 87: 586–594, 1997.
503. Laske, D.W., Youle, R.J., and Oldfield, E.H. Tumor regression with regional distribution of the targeted toxin TF-CRM107 in patients with malignant brain tumors, *Nat. Med.,* 3: 1362–1368, 1997.
504. Friden, P.M. et al. Anti-transferrin antibody and antibody–drug conjugates cross the blood-brain barrier, *Proc. Natl. Acad. Sci. U.S.A.,* 88: 4771–4775, 1991.
505. Wu, D. and Pardridge, W.M. Neuroprotection with noninvasive neurotrophin delivery to the brain, *Proc. Natl. Acad. Sci. U.S.A.,* 95: 254–259, 1999.
506. Kordower, J.H. et al. Intravenous administration of a transferrin receptor antibody-nerve growth factor conjugate prevents the degeneration of cholinergic striatal neurons in a model of Huntington disease, *Proc. Natl. Acad. Sci. U.S.A.,* 91: 9077–9082, 1994.

507. Li, J.Y. et al. Genetically engineered brain drug delivery vectors: cloning, expression and *in vivo* application of an anti-transferrin receptor single chain antibody–streptavidin fusion gene and protein, *Prot. Eng.*, 12: 787–796, 1999.

508. Penichet, M. et al. An antibody–avidin fusion protein specific for transferrin receptor serves as a delivery vehicle for effective brain targeting: initial applications in anti-HIV antisense drug delivery to the brain, *J. Immunol.*, 163: 4421–4429, 1999.

509. Ali, S.A. et al. Transferrin Trojan horses as a rational approach for the biological delivery of therapeutic peptide domains, *J. Biol. Chem.*, 274: 24066–24073, 1999.

510. Wagner, E. et al. Transferrin–polycation conjugates as carriers for DNA uptake into cells, *Proc. Natl. Acad. Sci. U.S.A.*, 87: 3410–3414, 1990.

511. Zenke, M. et al. Receptor-mediated endocytosis of transferrin-polycation conjugates: an efficient way to introduce DNA into hematopoietic cells, *Proc. Natl. Acad. Sci. U.S.A.*, 87: 3655–3659, 1990.

512. Cotton, M. et al. Transferrin–polycation-mediated introduction of DNA into human leukemic cells: stimulation by agents that affect the survival of transfected DNA or modulate transferrin receptor levels, *Proc. Natl. Acad. Sci. U.S.A.*, 87: 4033–4037, 1991.

513. Wagner, E. et al. Coupling of adenovirus to transferrin-polylysine/DNA complexes greatly enhances receptor-mediated gene delivery and expression of transfected genes, *Proc. Natl. Acad. Sci. U.S.A.*, 89: 6099–6103, 1992.

514. Wagner, E. et al. Influenza virus hemagglutinin HA-2 N-terminal fusogenic peptides augment gene transfer by transferrin-polylysine-DNA complexes: toward a synthetic virus-like gene-transfer vehicle, *Proc. Natl. Acad. Sci. U.S.A.*, 89: 7934–7938, 1992.

5 Transferrin Receptor

CONTENTS

5.1 Introduction 249
5.2 General Features 251
5.3 Molecular Biology 254
5.4 Transferrin Receptor Glycosylation..258
5.5 Transferrin Receptor Endocytosis...262
    5.5.1 Basic Features 262
    5.5.2 Structural Requirement for Efficient Transferrin Receptor
          Internalization...272
5.6 Transferrin Receptor Phosphorylation...275
5.7 Intracellular Pools of Transferrin Receptors 277
5.8 Effect of CA^{2+} and Oxidation on Transferrin Receptor Internalization 279
5.9 Tissue Distribution and Expression in Selected Cell Types 281
5.10 Regulation of Transferrin Receptors...294
    5.10.1 The Expression of Transferrin Receptors is Correlated with
           Cell Proliferation...294
    5.10.2 Iron-Dependent Regulation of Transferrin Receptors 297
    5.10.3 Gene Elements Required for Receptor Expression and
           Regulation..301
5.11 Expression and Regulation in Different Cell Types.....................................305
    5.11.1 Expression in Activated T Lymphocytes 305
    5.11.2 Expression in Monocytes–Macrophages 317
    5.11.3 Expression of Transferrin Receptor during Erythropoietic
           Differentiation and Maturation 324
5.12 Interaction between Transferrin Receptors and HFE: The Class I
    MHC-Related Protein Mutated in Hereditary Hemochromatosis.............332
5.13 A General Overview of Iron Uptake Mechanisms in Various Cell
    Types 334
References...339

5.1 INTRODUCTION

Iron is an essential element for all living cells and takes part in several metabolic pathways. The strict dependence of all forms of life on iron may be related to the well-known capacity of this element to be reversibly oxidized and reduced, and to its abundance in soil and water. Primitive organisms required iron in their energy-generating systems.[1] In prehistoric times, the oxygen concentration in the atmosphere was extremely low, and iron was present mostly in the reduced ferrous form. Following the passage to an aerobic environment, iron was oxidized to the ferric form and soluble iron became scarce.[1,2] Ferrous iron is soluble up to 0.1 mol/L even at physiologic pH levels, while ferric iron is virtually insoluble at neutral pH; its solubility is limited to acid pH. Thus, at neutral pH, the formulation of soluble iron hydroxides limits the formation of ferric iron to very small amounts corresponding to about 10^{-18} mol/L[3]; this concentration is too low to provide a sufficient iron supply to sustain all metabolic activities that require this metal.

In addition, ferric iron may undergo cycles of reduction and oxidation that may generate the formation of dangerous oxidative radicals. To bypass this diminished iron supply, some bacteria and plants developed the capacity to synthesize low molecular iron chelators or siderophores, which are able to bind iron present in their immediate vicinity. They then developed biochemical mechanisms to improve their iron uptake from food and to transport iron into the body.[4] In the complex domain of invertebrates and vertebrates, both of these systems are based on the capacity of iron to bind to a high molecular weight transporter known as transferrin.[5,6]

Inside the body, three proteins, the transferrin iron transport protein, the transferrin receptor on the cell surface, and the intracellular iron storage protein (ferritin) provide an essential sequence for making iron available for cellular metabolism.

Iron participates in several metabolic activities. Iron-containing proteins of the respiratory chain are involved in electron transport, a process that provides the energy necessary for cellular metabolism.[7] Iron is also involved in the regulation of cell proliferation through the control of several metabolic pathways, in particular ribonucleotide reductase activity. Ribonucleotide reductase is the only enzyme that catalyzes the conversion of ribonucleotides to deoxyribonucleotides, a key step in DNA synthesis.[8]

Transferrin receptor is a membrane receptor that plays a key role in the control of iron uptake for specific cellular functions. Polypeptide receptors on mammalian plasma membranes can be classifed on the basis of their function in two different categories: class I receptors essentially transmit information, while class II receptors internalize the ligand,[9] thus providing cells with essential metabolic factors (e.g., iron, cobalamin,[10] and cholesterol.[11]

Upon exposure to ligand, class I receptors are down-regulated (i.e., the number of surface receptors is reduced). In contrast, binding of ligand by class II receptors does not usually significantly alter their number. The transferrin receptor is a typical example of a class II membrane receptor.

The function of transferrin receptor is mediating iron uptake from circulating transferrin. The receptor is strictly required for mediating iron uptake at the levels of some tissues, as shown by studies of gene knockout mice. In fact, embryonic mice rendered deficient in both transferrin receptor alleles (TfR$^{-/-}$) died before embryonic day 12.5.[12] Histology showed edema and diffused necrosis in all tissues as a consequence of anemia and consequent tissue hypoxia. These TfR$^{-/-}$ animals also showed marked abnormalities of nervous system development, with morphological evidence of apoptosis at the level of neurons.[12]

In spite these marked defects in erythropoiesis observed in TfR$^{-/-}$ mice at day 12.5, at day 10, these animals showed apparently normal red cell masses, thus suggesting that the transferrin receptor is critical for erythropoiesis only after a certain stage of embryonic development; before this stage, its function may be handled by other iron uptake mechanisms.

In conclusion, transferrin receptor knockout studies clearly indicate that transferrin receptors play a key role in erythropoiesis and neurologic development. The requirement for transferrin receptors to aid erythropoiesis is not surprising since iron is required for hemoglobin synthesis. The need of transferrin receptors for neurologic development implies that an optimal iron supply is required for neuronal development.

5.2 GENERAL FEATURES

The structures, properties, and functions of transferrin receptors have been studied in detail. Table 5.1 lists the main biochemical and functional properties of transferrin receptors, and Figure 5.1 shows the basic architecture of the transferrin receptor molecule. Cloning and sequencing of the cDNA encoding the transferrin receptor allowed us to establish the primary structure and predict a model of receptor localization at the level of cell membrane.[13,14]

The amino acid sequence of this receptor is shown in Figure 5.2, and the structure is summarized as follows:

1. It consists of two identical subunits of about 90 kDa each, organized into three domains, extracellular, transmembrane, and intracellular.
2. It has a transmembrane domain of 27 amino acid residues that contains two cysteine residues at positions 62 and 67 which serve as sites for covalent attachment of palmitic acid.[15]
3. It has an intracellular segment of 62 amino acid residues, with a serine residue at position 24 which represents the site for receptor phosphorylation triggered by protein kinase C activators,[16] but is not required for receptor internalization.
4. The long extracellular C terminal domain is involved in ligand binding.

The extracellular domain contains three cysteine residues involved in the binding of a fatty acid chain and the formation of two subunit-linking disulfide bridges; the sites of intermolecular disulfide binding are Cys 89 and Cys 98.[17] The extracellular domain also contains three potential sites for N-linked glycosylation at Asn residues 251, 317 and 727. The glycosylation of the last residue is the most critical to the structure of the receptor (for a detailed analysis of glycosylation see Section 5.4).

Functional analysis of deletion mutants and a series of mutant human transferrin receptors has shown that the cytoplasmic domain of the receptor is essential for the receptor internalization signal required for spontaneous clustering of receptors in clathrin-coated pits and promotes high-efficiency endocytosis (for details see Section 5.5).

The transferrin receptor undergoes several co- and post-translational modifications during its synthesis and progressive localization at the level of the cell membrane. The modifications include dimer formation, intersubunit disulfide bond formation, acylation with palmitate, phosphorylation of a serine residue, N-linked glycosylation, and O-linked glycosylation.

TABLE 5.1
Biochemical Characteristics of Transferrin Receptor

Parameter	Characteristic
Molecular weight (kDa)	180
No. of subunits	2
Molecular weight/subunit (kDa)	90
No. of amino acids/subunit	760
No. of amino acids in cytoplasmic tail	62
No. of amino acids in cytoplasmic segment	26
No. of amino acids in extracellular segment	672
N terminal	Cytoplasm
COOH terminal	Extracellular
No. of carbohydrate chains/subunit	3
N-glycosylation sites	Asn 251, Asn 317, Asn 727
O-glycosylation sites	Thr 104
No. of fatty acid chains/subunit	1
Phosphorylation site	Ser 24
Internalization sequence	Cytoplasm (Tyr Gln Arg Phe)
Site of proteolytic cleavage	Extracellular (Arg 100, Leu 102)
Sites of intermolecular disulfide bridges	Cys 89, Cys 98
Gene (Kb)	31
No. of exons	19
Gene localization (chromosome)	3 (q26.2-qter)
Transcript (Kb)	4.9
Iron loading effect on mRNA level	Negative
Iron loading effect on protein level	Negative
Iron deprivation effect on mRNA level	Positive
Iron deprivation effect on protein level	Positive
Three-dimensional structure	Butterfly-like
Transferrin-binding capacity (Tf molecules per receptor)	2
Main functions	Tf binding and internalization
Effect of gene knockout	Anemia, abnormalities of central nervous sytem

Analysis of the biosynthesis of the receptor and its post-translational modifications showed the following sequence of events. The nascent transferrin receptors contain core-glycosylated, asparagine-linked oligosaccharides; they do not possess intersubunit disulfide bonds, exist predominantly in the form of monomers, and show little capacity to bind transferrin.[18] Within 20 to 30 minutes after synthesis, the transferrin receptors acquire the ability to bind transferrin, and intersubunit disulfide bond formation occurs slowly during the transit of the receptors to the cell surface.[18]

Recent studies suggest that the carboxy terminal region of the receptor is important for ligand binding.[19] This conclusion cannot be derived from the analysis of the human transferrin receptors with truncations of the carboxy terminal region because these mutants failed to be expressed after transfection.

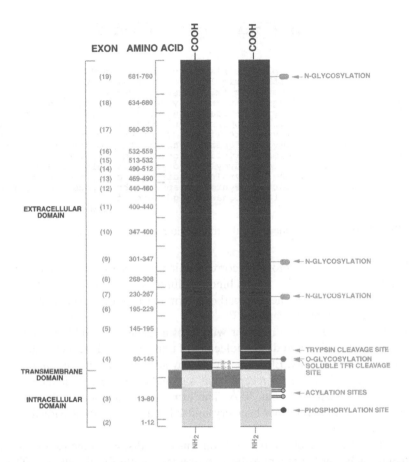

EXON   AMINO ACID

(19)	681-760	N-GLYCOSYLATION
(18)	634-680	
(17)	560-633	
(16)	532-559	
(15)	513-532	
(14)	490-512	
(13)	469-490	
(12)	440-460	
(11)	400-440	
(10)	347-400	
(9)	301-347	N-GLYCOSYLATION
(8)	268-308	
(7)	230-267	N-GLYCOSYLATION
(6)	195-229	
(5)	145-195	

EXTRACELLULAR DOMAIN

TRYPSIN CLEAVAGE SITE
O-GLYCOSYLATION
SOLUBLE TFR CLEAVAGE SITE

| (4) | 80-145 |

TRANSMEMBRANE DOMAIN

ACYLATION SITES

INTRACELLULAR DOMAIN   (3)   13-80

PHOSPHORYLATION SITE

(2)   1-12

FIGURE 5.1 Schematic representation of the transferrin receptor molecule formed by two identical subunits linked by two disulfide bridges. Each subunit contains three domains. The extracellular domain is the COOH terminal part and contains 672 amino acids. The glycan chains are located on this domain, which also contains the transferrin-binding site. Located in the segment of this domain near the cell membrane is a site involved in proteolytic cleavage and consequent generation of the soluble transferrin receptor. The hydrophobic transmembrane domain contains 28 amino acids. The hydrophobic intracellular domain is the N terminal part of the molecule containing 61 amino acids, a fatty acid, and a phosphate group bound to a serine residue. The correspondence between exon numbers of the transferrin receptor gene and the amino acid sequence is shown on the left.

An alternative approach involved the use of chimeric human/chicken transferrin receptors. The regions of the human transferrin receptor's external domain were replaced with the corresponding chicken transferrin receptor sequences. Since the chicken transferrin receptors do not bind human transferrin, these chimeric receptors can be used to identify the regions of the human receptor involved in ligand binding. The functional analysis of the chimeric receptors provided evidence that the carboxy terminal region of the external domain of the human transferrin receptor is critical for transferrin binding. More particularly, replacement of the 72 carboxy terminal

HUMAN TRANSFERRIN RECEPTOR

Origin

```
  1 mmdqarsafs nlfggeplsy trfslarqvd gdnshvemkl avdeeenadn ntkanvtkpk
 61 rcsgsicygt iavivfflig fmiqylgyck gvepktecer lagtespvre epgedfpaar
121 rlywddlkrk lsekldstdf tstikllnen syvpreagsq kdenlalyve nqfrefklsk
181 vwrdqhfvki qvkdsaqnsv iivdkngrlv ylvenpggyv ayskaatvtg klvhanfgtk
241 kdfedlytpv ngsivivrag kitfaekvan aeslnaigvl iymdqtkfpi vnaelsffgh
301 ahlgtgdpyt pgfpsfnhtq fppsrssglp nipvqtisra aaeklfgnme gdcpsdwktd
361 stcrmvtses knvkltvsnv lkeikilnif gvikgfvepd hyvvvgaqrd awgpgaaksg
421 vgtalllkla qmfsdmvlkd gfqpsrsiif aswsagdfgs vgatewlegy lsslhlkaft
481 yinldkavlg tsnfkvsasp llytliektm qnvkhpvtgq flyqdsnwas kveklt1dna
541 afpflaysgi pavsfcfced tdypylgttm dtykelieri pelnkvaraa aevagqfvik
601 lthdvelnld yerynsqlls fvrdlnqyra dikemglslq wlysargdff ratsrlttdf
661 gnaektdrfv mkklndrvmr veyhflspyv spkespfrhv fwgsgshtlp allenlklrk
721 qnngafnetl vrnqlalatw tiqgaanals gdvwdidnef
```
                                                  End

FIGURE 5.2 Amino acid sequence of the transferrin receptor.

residues of the human receptor with corresponding sequences of the chicken receptor resulted in a marked decline of the binding affinity of the receptor for transferrin, while the replacement of the 192 carboxy terminal residues resulted in complete loss of transferrin-binding capacity.[19]

Treatment of transferrin receptor with trypsin releases a soluble fragment (residues 121 to 760) able to bind two molecules of transferrin and the membrane-bound receptor with high affinity. Electron microscopy studies have shown that this extracellular domain of transferrin receptor has a globular structure, separated from the membrane by a stalk of about 30 Å. The extracellular domain of the transferrin receptor in its dimeric form has been crystallized and its three-dimensional structure has been elucidated.[20]

The transferrin receptor monomer contains three distinct domains, organized in a structure such that the receptor dimer possesses a butterfly-like shape.[21] The first domain (domain A) is a protease-like domain, containing residues 122 through 188 and 384 through 606, and is located laterally with respect to the cell membrane (Figure 5.3). The second domain (domain B), containing residues 189 to 383, is located in an apical position and resembles a β sandwich in which the two sheets are splayed apart. The third domain (domain C), containing residues 607 to 760, has the shape of a four-helix bundle formed by a pair of parallel α-hairpins (Figure 5.3). This three-dimensional structure of the transferrin receptor also has important implications for the understanding of the mechanisms underlying transferrin–transferrin receptor interactions.

5.3 MOLECULAR BIOLOGY

The gene encoding the human transferrin receptor has been located on chromosome 3.[22-25] Detailed analysis has shown that the gene encoding the transferrin receptor may be localized to region q12 to qh, q22 to ter22, or q21 to 25.[26] Using a chromosome pumping technique made it possible to show that the human transferrin receptor gene maps to region 3q29.[27]

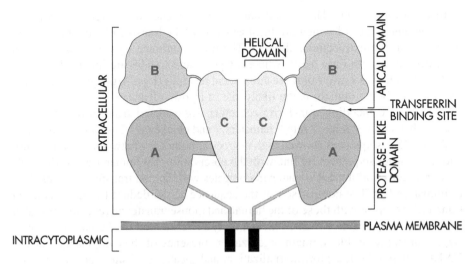

FIGURE 5.3 Schematic representation of the three-dimensional structure of the transferrin receptor derived from crystallographic studies. Three domains are present in the extracellular fragment of a single subunit forming the homodimeric transferrin receptor molecule: domain A (protease-like domain); domain B (apical domain); and domain C (helical domain where the two identical subunits of the receptor interact). The receptor exhibits a butterfly-like shape. (Lawrence, C.M. et al., *Science*, 286: 779–782, 1999)

A single gene apparently exists for the haploid genome. Note that chromosome 3 also encodes the genes responsible for synthesis of plasma transferrin and melanotransferrin.[28,29]

Cloning of the transferrin receptor cDNA[13,14] led to the elucidation of the amino acid sequence and understanding of the general structure of the receptor (see Figures 5.1 and 5.2).

The mRNA corresponding to the cDNA encodes a peptide of 760 residues.[13,14] The receptor lacks the N terminal hydrophobic signal sequence present in most membrane-bound and secreted proteins. However, a cluster of basic hydrophilic residues (KPKRs) exists at a brief distance from the N terminus, followed by a highly hydrophobic region, which, in addition to anchoring the molecule to the membrane, may also serve as an internal signal sequence.

The molecule is divided into three different structural and functional domains:

1. The cytoplasmic domain consists of the first 61 amino acids of the N terminal sequence.
2. The transmembrane domain is constituted by 27 hydrophobic amino acids.
3. The extracellular domain is composed of 672 amino acids on the C terminus.[13,14]

The orientation of the transferrin receptors is unusual in that a large majority of other membrane proteins are positioned in the apposite way, i.e., with the C terminus on the cytoplasmic site. The extracellular domain contains three possible sites for

N-linked glycosylation. The overall amino acid sequence of the receptor contains eight cysteines, four of which are localized around the transmembrane domain and the others in the extracellular domain. The transmembrane Cys 62 is involved in thioester bonds with fatty acids, whereas Cys 89 and Cys 98 are involved in intermolecular bonds. A cluster of three amino acid residues (Lys–Arg–Lys) located near the transmembrane domain is the likely site of trypsin cleavage.

The cDNA encoding transferrin receptors was isolated from different animal species: humans,[13,14] mice,[30] rats,[31] hamsters,[32] and chickens.[33] Comparative analysis of these cDNA sequences showed high degrees of homology between human–mouse and rat–hamster sequences, in line with the observation that human transferrin binds to murine cells and murine transferrin competes with human transferrin for binding to human cells.[34] The comparison of the sequence of the deduced primary chicken transferrin receptor with those of the human and mouse transferrin receptors revealed some interesting features.[33] The chicken receptor resembled the human receptor in size, hydropathy profile, domain organization, presence of the cytoplasmic domain YXRF required for receptor internalization, and locations of sites for post-translational modifications. However, the extracellular domain of chicken transferrin receptor is only 53% identical to the extracellular domains of human and mouse transferrin receptors; these differences account for inability of chicken transferrin receptors to bind human transferrin.

The gene encoding the human transferrin receptor has been identified and almost entirely cloned.[35] An electron microscopy study of the cDNA/genomic DNA heteroduplex showed 19 exons and 18 introns.[35] The human transferrin receptor gene spans more than 31 Kb and is comprised of at least 19 exons of about 70 to 650 bases. Using a PCR-based intron jumping strategy the exon–intron structure of the human transferrin receptor gene was recently reinvestigated.[36] The study confirmed the previous observations of McClelland et al.[35] and provided further detail. Evans and Kemp proposed a functional map of the different exons:[36]

Exon 1 and part of exon 2 are untranslated and encode for the 5′ non-coding region of the receptor cDNA; all the translated segments are encoded by exons 2 to 19.

Exon 2 contains the start codon and the first 12 amino acid residues of the receptor.

Exon 3 encodes the four amino acids YXRF that constitute the high-turn structural motif essential for receptor endocytosis; the serine residue at position 24 phosphorylates following protein kinase C activation; the cysteine residues at positions 62 and 67 serve as sites for covalent attachment of palmitic acid.

Exon 4 encodes the extracellular cysteines at position 89 and 98 which form disulfide bonds in the receptor homodimer, the arginine and leucine at positions 100 and 101 representing the proteolytic site which generates the soluble transferrin receptor, the threonine at position 104 which forms a site for O-linked oligosaccharides, and the Lys–Arg–Lys residues at positions 128 to 130 which form a trypsin cleavage-sensitive site; N-glycosylation sites are encoded by exon 7 (Asn–Gly–Ser at positions 251 to 253), exon 9 (Asn–Gly–Thr at positions 317 to 319), and exon 19 (Asn–Gly–Thr at positions 727 to 729), respectively.

Exons 5, 6, 8, and 10 to 18 encode amino acid sequences not involved in post-translation modifications.

Exons 17, 18, and 19 encode the extracellular region of the receptor involved in ligand binding.

Exon 19 encodes the last translated segment and a sizeable 3′ untranslated segment (see Figure 5.1).

Several polymorphisms were described at the level of the transferrin receptor gene; particularly, a *Hin* 6I polymorphism in intron 7 and a *Bam* I polymorphism in exon 1 have been reported.[37] The DNA polymorphisms are of value in defining ethnically distinct haplotypes as markers of some populations.

Recently, a human gene homologous to transferrin receptor, termed transferrin receptor 2, and mapping on chromosome 7q22 was revealed. The transferrin receptor 2 gene generates two different transcripts: an alpha transcript of about 2.9 Kb and a beta transcript of about 2.5 Kb (Figure 5.4). The deduced amino acid sequence corresponding to the alpha transcript was consistent with a type II transmembrane protein exhibiting 45% homology in its extracellular domain with transferrin receptor.[38] The beta transcript encoded for a protein lacking the NH_2-terminal portion of the alpha transcript, including the putative transmembrane region.[38]

The transferrin receptor 2α protein exhibits an apparent molecular weight of about 105 kDa and possesses four potential N-glycosylation sites. Finally, transferrin receptor 2α protein is able to bind transferrin and mediate iron uptake. The exact biological role of the transferrin receptor 2 is unknown.

Transferrin receptor 2 was cloned also in the mouse, with several interesting findings.[38a] Transferrin receptor 2, although homologous to the classical murine transferrin receptor in the coding region (53% similarity between the two receptors was observed), did not contain in its 3′ untranslated region the iron regulatory elements (IREs), found in the 3′ untranslated region of transferrin receptor mRNA. IREs will be discussed in detail in Chapter 8.

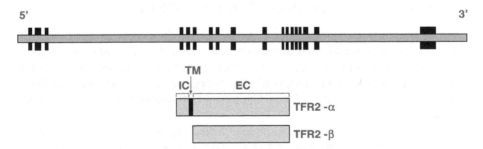

FIGURE 5.4 Genomic structure of transferrin receptor 2. The gene is composed of 18 exons represented by black boxes, and generates two different transcripts: TFR2α and TFR2β. Only the TFR2α mRNA encodes for the full-length protein with an extracellular domain (EC), a transmembrane domain (TM), and an intracytoplasmic domain (IC). Exons and protein structure are indicated according to Kawabata, H. et al., *J. Biol. Chem.*, 274: 20826–20832, 1999. With permission.

These IREs are required to produce iron-dependent regulation of transferrin receptor mRNA levels. According to the structural findings, transferrin receptor 2 mRNA is not modulated by iron, while transferrin receptor mRNA is modulated by iron.[38a] Furthermore, Northern blot analysis of transferrin receptor 2 mRNA expression in various murine tissues showed that this mRNA species is particularly expressed in hepatic tissue.[38a] Given the peculiar expression of transferrin receptor 2 at the level of the liver and the lack of down-modulation of this receptor in response to iron overload, it was suggested that this phenomenon may explain the increased susceptibility of the liver to iron loading in hereditary hemochromatosis.

Additional studies have provided evidence that transferrin receptor 2 exhibits a lower affinity of binding for transferrin than transferrin receptor and thus may represent a low-affinity iron uptake system not modulated by cellular iron status.[38b] Furthermore, transferrin receptor 2 expression is modulated by the cell cycle, with highest expression in the late G1 phase and virtually no expression in phases G0/G1.[38b]

It is of interest that the amino acid sequence of the extracellular domain of transferrin receptor exhibits a significant homology (28% of identity) with those of membrane glutamate carboxypeptidase II and prostate-specific membrane antigen.[39,40] These two highly homologous proteins exhibit peptidase activity specific for N-acetylaspartylglutamate, the most prevalent peptide neurotransmitter in the mammalian nervous system. This ligand activates the glutamate receptor inactivated by peptidase activity (membrane glutamate carboxypeptidase II) present on the extracellular faces of the membranes of neurons and glia.

On the basis of these structural similarities, it was suggested that transferrin receptor may have evolved from a peptidase related to membrane glutamate carboxypeptidase II.[41,42] In the transferrin receptor molecule, the Zn ligands of the protease-like domain are lost and, consequently, the protein lacks peptidase activity.

Since transferrin receptor 2 exhibits higher homology than transferrin receptor for prostate-specific membrane antigen, it was suggested that it may represent a more primitive form of transferrin receptor.[38b]

5.4 TRANSFERRIN RECEPTOR GLYCOSYLATION

Like all integral plasma membrane proteins, transferrin receptor is synthesized in the endoplasmic reticulum and undergoes many co- and post-translation modifications during its transit to the cell surface. These modifications include dimer formation, acylation with palmitate on Cys 62, phosphorylation of a serine (Ser 24) residue, N-linked (asparagine-linked) glycosylation, and O-linked (serine/threonine-linked) glycosylation.

Transferrin receptor contains a single O-linked oligosaccharide,[43,44] localized to the threonine residue at position 104 (Thr 104).[45] O-linked glycosylation is characteristic of a number of cell surface receptors, including the low-density lipoprotein (LDL) receptor, the insulin receptor, and the interleukin-2 receptor. The O-linked oligosaccharides of the LDL receptor play a role in receptor stability and function. Additionally, O-glycocosylation of the interleukin-2 receptor is important for expression at the cell surface.

The structures of O-linked oligosaccharides were determined by studies in which three human cell lines were metabolically labeled with ^{3}H-mannose and ^{3}H-glucosamine; the radiolabeled transferrin receptors were isolated and treated with a mild base borohydride to release oligosaccharides.[33] Analysis of the released oligosaccharides showed that they were composed of sialic acid, galactose-N-acetylgalactosamine, and N-acetylgalactosamine.[44]

O-linked oligosaccharides are linked through N-acetyl-galactosamine to peptide. There are structural differences in the O-linked oligosaccharides in transferrin receptors between different human cell lines.[44] The receptor isolated from K-562 human erythroleukemia cells contains at least one O-linked oligigosaccharide having two sialic acid residues and a core structure of the disaccharide galactose-N-acetylgalactosamine. In contrast, the O-linked oligosaccharides in the transferrin receptors from both A-431 and BeWo cell lines are not as highly sialylated and were identified as the neutral disaccharide galactose-N-acetylgalactosamine. These differences in O-glycosylation of the transferrin receptor between different cell lines may affect receptor function and/or binding activity.

Analysis of the peptides originated by the tryptic digestion of the transferrin receptor determined that the O-glycosylation chain is located near the transmembrane domain at the Thr 104 residue.[45,46]

To elucidate the effect of the O-linked carbohydrate on transferrin receptor function, the oligosaccharide was eliminated by mutation of Thr 104 to Asp and the mutated cDNA was expressed in a cell line lacking endogenous transferrin receptor.[47] The mutant receptor does not show important abnormalities in its main biochemical characteristics in that binds transferrin with the same affinity as wild-type transferrin receptor and possesses a normal cellular distribution.

The only known effect of this mutation consisted of potentiating the clearance in CHO cells of the human transferrin receptor at the level of the Arg 100 site,[47] which is the identical site reported for the circulating soluble transferrin receptor isolated from human blood. This observation suggests that elimination of the O-linked carbohydrate at position 104 may represent one of the biochemical mechanisms underlying the generation of soluble human transferrin receptor.

The extracellular domain of each subunit of transferrin receptor contains three potential N-glycosylation sites. Digestion of the transferrin receptor with endoglycosidases has suggested the presence of three N-linked oligosaccharides on each receptor subunit, indicating that each site is glycosylated.

The structure of the N-linked oligosaccharide of the human transferrin receptor isolated from placenta and hepatocarcinoma cell line HepG2 was determined.[48] Transferrin receptor from placenta predominantly carries diantennary and triantennary N-acetyllactosaminic glycans as well as hybrid-type species. Distinct from placental transferrin receptor, the receptor from hepatocarcinoma cells contained larger amounts of oligomannosidic glycans with six to nine mannose residues and tetrasialylated complex-type oligosaccharides apart from mono-, di-, and trisialylated species.

Through a combination of oligosaccharide analysis and site-directed mutagenesis, the oligosaccharide composition of each N-linked glycosylation site was determined. The Asn 251 site contains complex oligosaccharides, the Asn 317 site contains

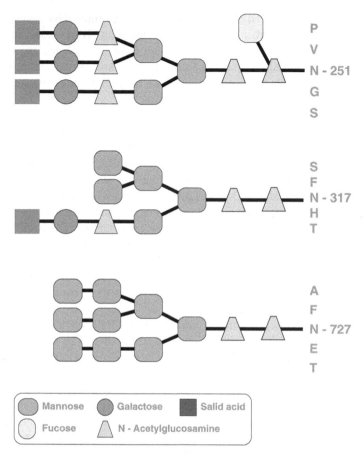

FIGURE 5.5 The structures of the three N-linked oligosaccharide chains of the transferrin receptor are shown. The Asn 251 site contains complex oligosaccharides, the Asn 317 site contains hybrid oligosaccharides, and the Asn 727 site is totally mannose in character.

hybrid oligosaccharides,[49] and the Asn 727 site is entirely high mannose in character.[50] (See Figure 5.5.) Thus, it is clear that the types of oligosaccharides are not randomly distributed at each glycosylation site; rather, each site has its own distinctive oligosaccharide array.

Several studies have shown that asparagine-linked oligosaccharides play an important role in conferring function to the transferrin receptor during its biosynthesis. In this context, the properties of the unglycosylated transferrin receptor in tunicamycin-treated cells were examined. In human epidermoid A-431 cell line treated with tunicamycin, an inhibitor of N-linked glycosylation, the unglycosylated transferrin receptor was not transported to the cell surface, was unable to bind transferrin, and was unable to form dimers.[51]

These observations were confirmed by studies of the properties of the nascent form of the transferrin receptor which is a monomer and is uncapable of binding to transferrin.[18] Treatment of cells with swainsonine or deoxynojirimycin, two inhibitors of N-linked complex carbohydrate formation, does not affect the ability of transferrin receptor to dimerize, form intersubunit disulfide bonds, or be transported to the cell surface; however, it does reduce the affinity of the receptor for transferrin.[18] Because of the pleiotropic natures of tunicamycin and the other glycosylation inhibitors, however, the putative role of the N-linked glycosyl groups was investigated by more specific means involving site-directed mutagenesis of the consensus sites for N-linked glycosylation.

Site-directed mutagenesis was performed to abolish the three asparagine-linked glycosylation sequences of the human transferrin receptor.[52] The DNA encoding the mutated transferrin receptor was transfected into mouse fibroblasts lacking endogenous transferrin receptor. This mutant of transferrin receptor showed reduced transferrin-binding capacity, reduced intersubunit bonding function, and reduced cell surface expression, thus indicating that the transferrin receptor lacking asparagine-linked glycosylation is not fully functional.[52]

Subsequent studies focused on evaluating the role of the single asparagine residues. The analysis of the functional role of Asn 251 residue generated conflicting evidence. Asn 251 is the site nearest to plasma cell membrane where oligosaccharides are linked, and is also the first to be glycosylated when the nascent protein emerges into the lumen of the endoplasmic reticulum.

One study showed that loss of the sugar chain attached to this Asn residue leads to a retention of the mutated receptor in the endoplasmic reticulum, failure to form dimers, and site-specific proteolysis that gave rise to 73-kDa monomers.[53] When this cleavage was inhibited by the introduction of a second mutation into the molecule, the resulting transferrin receptors lacking glycosylation at Asn 251 functioned normally.[54]

However, in another study, the generation of single mutants of transferrin receptors lacking Asn 251 glycosylation sites did not affect the processing and surface localization of the mutant receptor.[55]

It must be noted that the apparent discrepancy between these results may be related to the fact that in the first study,[52] a glutamine was substituted for the Asn 251. In the second study,[55] a glutamine was substituted for the serine 253. Mutation of the C terminal glycosylation site (Asn 727) has negative effects on the appearance of the receptor at the cell surface.[54,55] Thus, mutants lacking Asn 727 glycosylation sites appear to be retained in the endoplasmic reticulum. Addition of new glycosylation sites in the C terminal region of the unglycosylated mutated transferrin receptor restores the cell surface. Finally, mutations of the Asn 217 glycosylation site (between the N terminal Asn 251 and the C terminal Asn 727 glycosylation sites) did not affect the processing and surface localization of the transferrin receptor.[54,55]

The studies of the glycosylation of transferrin receptors suggest that the three-dimensional structures of the different domains of the neosynthesized transferrin receptor molecule determine the types of modifications caused by the oligosaccharide-processing machinery following a pattern that is site-specific.[49] Thus, the N-glycosylation sites Asn 251 and Asn 317 that are less important for receptor function (processing and expression at the level of cell membrane) contain highly processed oligosaccharides, probably because they are fully accessible to the Golgi glycosylation machinery, while the N-glycosylation site (Asn 727), that is of fundamental importance for receptor folding and transport to the cell membrane, contains high-mannose oligosaccharides, probably because the steric structure of the protein at the level of this residue impairs more complex glycosylation.

Recent studies have shown that human transferrin receptor expressed in the yeast *Saccharomyces cervisiae* is post-translationally modified in a way comparable to that observed for the native receptor with respect to both glycosylation and dimer formation.[56] In some pathological conditions, the glycosylation pattern of the transferrin receptor may be modified and this may lead to a change in transferrin binding capacity. Thus, increased N-glycosylation and reduced transferrin-binding capacity of transferrin receptor isolated from placentae of diabetic women were observed. This phenomenon may help to explain the origin of the iron deficiency frequently observed in infants of diabetic women.[57]

5.5 TRANSFERRIN RECEPTOR ENDOCYTOSIS

5.5.1 Basic Features

Receptor-mediated endocytosis is the process by which integral membrane proteins are selectively internalized from the plasma membrane to the endocytic pathway. This mechanism represents an efficient pathway for mediating the cellular uptake of nutrients transported by specific plasmatic carriers and bound to their specific membrane receptors. During the process of receptor-mediated endocytosis, membrane proteins are concentrated in specialized regions of the plasma membrane known as clathrin-coated pits. Invagination of coated pits into the cytoplasma generates transport vesicles that deliver their content to the endosomal compartment. Some proteins then recycle back to the cell surface, while others are sorted and sent to different destinations.

Receptor-mediated endocytosis of transferrin is an endocytic pathway investigated in recent years in detail. The basic features of the process may be outlined as follows.[58] Following the interaction of the ligand with its receptor at level of the plasma membrane, transferrin–transferrin receptor complexes are rapidly internalized via clathrin-coated pits generating coated pit vesicles that rapidly lose their coats and undergo fusion with a heterogeneous tubulovesicular network collectively defined as early endosomes.

Within the early endosomes, the pH is acidic and this chemical condition mediates the dissociation and the sorting of ligands and receptors. In the case of transferrin receptors, within early endosomes, Fe^{3+} dissociates from transferrin and then diffuses out of endosomes, thus becoming available for cellular metabolism. Transferrin remains bound to its receptor; the apotransferrin–transferrin receptor complex remains within the endosomes and is recycled back to the cell surface through a system of tubular extensions of endosomes that form a population of transport vesicles. Finally, when the transferrin–transferrin receptor complex is recycled back to the cell surface and the neutral pH of the cell surface favors the detachment of apotransferrin from transferrin receptor. After binding of iron, apotransferrin is available to bind its receptor and start a new cycle of endocytosis (Figure 5.6).

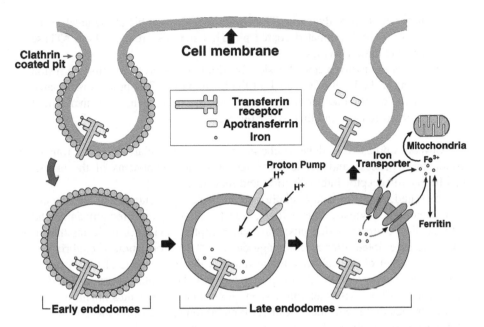

FIGURE 5.6 Outline of the receptor-mediated endocytosis of transferrin. Following the inter-action of the ligand with its receptor at the plasma membrane, transferrin–transferrin receptor complexes are internalized via clathrin-coated pits that rapidly lose their coats and undergo fusion with a heterogeneous tubulovesicular network collectively known as early endosomes. The pH of the early endosomes is acidic and the acidity mediates the dissociation of iron from transferrin, while apotransferrin remains bound to its receptor. The apotransferrin–trans-ferrin receptor complex initially remains within the endosomes and is then recycled back to the cell surface through a system of tubular extensions of endosomes that form transport vesicles. The neutral pH on the cell surface favors the detachement of apotransferrin from transferrin receptor, and apotransferrin, after conversion to transferrin through binding of iron, is available to complex with its receptor and start a new cycle of endocytosis.

A recent view of the mechanism of receptor endocytosis through clathrin-coated endocytosis includes several steps[59]:

1. Formation of clathrin-coated vesicles at the level of plasma membrane through the recruitment of adaptor protein 2 (AP-2). This adaptor protein interacts with endocytosis signal sequences present at the level of the cytoplasmic tails of transferrin receptors and similar receptors endocy-tosed through coated pits (see Section 5.5.2).
2. Following the recruitment of AP-2 at the level of the plasma membrane, the docked AP-2 complexes promote the assembly of clathrin molecules at the level of the plasma membrane.
3. Invagination of coated pits, a process requiring ATPase activity occurs; during this stage, dynamin, a GTPase, associates with coated pits.
4. The next step involves the formation of coated pits that remain attached to the plasma membrane via a narrow neck. The pits are associated with

redistribution of dynamin. They migrate and assemble at the level of the neck of the invaginated coated pit. This step also requires both ATPase and GTPase activities.

5. This step involves coated vesicle budding, a process that requires a fusion event at the level of the neck of the invagination. The process is initiated by the external leaflet of the membrane; the subsequent steps involve first clathrin release and then AP-2 release.

Several studies were done to characterize some peculiarities of the transferrin endocytosis process and define the roles of some components of the endocytic machinery in receptor internalization and recycling.

As stated, the formation of clathrin-coated pits represents one of the first events of receptor endocytosis. Interestingly, transferrin receptors predominantly cluster into existing clathrin-coated pits; for other receptors, evidence indicates that these coats are assembled after receptor aggregation.[60] The formation of clathrin-coated pits at the level of the cell membrane requires the interactions of the cytoplasmic tails of the receptors with adaptor proteins (Aps). The clathrin-associated adaptors AP-1 and AP-2 are heterotetrameric complexes that mediate attachment of clathrin to membranes and recruitment of integral membrane proteins for incorporation into clathrin-coated vesicles.

AP-1 is localized to the trans-Golgi network while AP-2 is localized to the plasma membrane and mediates rapid internalization of endocytic receptors. AP-2 is a heterodimer composed of two large chains, the α- and β2-adaptins (100 kDa), a medium chain μ2 (50 kDa), and a small chain σ2. There is evidence that the μ2 subunit of AP-2 interacts with the internalization YTRF motif of the transferrin receptor,[61] while AP-1 is unable to interact with the transferrin receptor.

Both the position and the exact sequence of the tyrosine internalization motif are major determinants for the selectivity of interaction of these sequences with the Aps. Other receptors mediate clathrin coat formation through interaction with dileucine-based motifs. The internalization process via dileucine- and tyrosine-based motifs is a saturable process that implies the utilization of different cellular components. In fact, overexpression of a protein containing the tyrosine-based motif slowed clearance of other proteins that contain similar motifs (such as the transferrin receptor) from the cell surface, while the clearance of molecules containing dileucine motifs was not affected.[62]

Recent studies showed that AP-3, a recently identified adaptor-like protein, co-localizes with transferrin receptor in early endosomes.[63] This observation suggests that AP-3 and AP-2 interact with the tyrosine motif of the transferrin receptor and may play a role in endocytosis and/or recycling of this receptor (Figure 5.7).

The direct involvement of AP-2 in transferrin receptor endocytosis was directly supported by experiments using a mutant clathrin adaptor protein AP-2 acting as a dominant negative. In fact, overexpression of a mutated μ2 subunit of the AP-2 protein, unable to interact with the tyrosine-based internalization motif of the transferrin receptor, elicited a marked inhibition of transferrin endocytosis.[64]

Clathrin adaptor proteins also mediate the basolateral targeting of the transferrin receptor in polarized epithelial cells. The localization of the transferrin receptor at

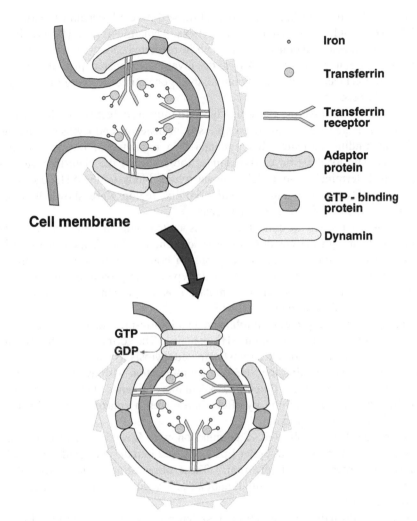

Cell membrane

Iron

Transferrin

Transferrin
receptor

Adaptor
protein

GTP - binding
protein

Dynamin

GTP
GDP

FIGURE 5.7 Schematic outline of the early steps of transferrin–transferrin receptor internal-ization.

the basolateral surfaces in polarized epithelial cells depends on the expression of μ1B, an epithelial cell-specific component of the AP-1 complex.[65] This AP-1B adaptor complex assembles with membranes at the level of the basal membrane, together with clathrin, to form nascent secretory vesicles that selectively accumulate proteins destined for the basolateral surface.

The α subunit of the endocytic AP-2 complex contains a 30-kDA appendage domain, joined to the rest of the molecule through a flexible linker. Several proteins may bind to this appendage domain, including dynamin, amphiphysin heterodimer, Eps 15, and epsin. Overexpression of this domain in fibroblasts leads to a significant inhibition of transferrin uptake; this effect requires the integrity of amino acid residues involved in binding of its different biochemical partners.[66]

Other studies have evaluated the roles of these different biochemical partners of the AP-2 α subunit on transferrin endocytosis. Dynamin is a GTP-hydrolyzing protein that is an essential constituent in clathrin-mediated endocytosis by cells. This protein self-assembles at the necks of the invaginated coated pits; this process stimulates the GTPase activity of dynamin, thus generating the energy required for the endocytic process.

In contrast to this assumption, when dynamin mutants defective in self assembly-stimulated GTPase activity are overexpressed, transferrin endocytosis is accelerated.[67] This finding indicates that dynamin is a regulator of receptor-mediated endocytosis. The localization of dynamin at the levels of the clathrin-coated pits requires its C terminal proline-rich domain which binds to the SH3 domain of amphiphysin. Dynamin mutants defective in this region were shown to inhibit clathrin-dependent endocytosis of transferrin.

The Eps 15 protein is constitutively associated with the plasma membrane adaptor AP-2 through the ear domain of α-adaptin. The localization of Eps 15 at the levels of the clathrin-coated pits depends on both N terminal domains and the AP-2 binding site. Overexpression of the wild-type Eps 15 protein strongly inhibited endocytosis of transferrin, whereas a mutant Eps 15 protein in which the AP-2 binding sites were deleted had no effect.[68]

Amphiphysins are proteins implicated in clathrin-mediated endocytosis that interact with dynamin through their SH3 domains. The formation of the dynamin self-assembly into rings prevents transferrin endocytosis.[69] This effect was lost in amphiphysin mutants lacking the SH3 domain interaction with dynamin. Assays on permeabilized cells confirmed that the SH3 domains of a series of proteins, including amphiphysins, exerted an inhibitory effect on transferrin uptake and that this effect is due to the binding to dynamin.[70]

Epsins have been implicated in conjunction with Eps 15 in clathrin-mediated endocytosis. Overexpression of either Eps 1 or Eps 2 into the cells leads to a marked inhibition of transferrin internalization.[71] These observations indicate that all the proteins capable of interacting with dynamin exert regulatory control on transferrin receptor endocytosis.

Recent studies indicate that actin cytoskeleton is required for the initial phases of transferrin receptor endocytosis. In fact, agents that specifically sequester actin monomers such as latrunculin A cause a marked inhibition of transferrin receptor endocytosis, without concomitant effect on receptor recycling.[72] Other studies also suggest a role for microtubules in transferrin receptor endocytosis; based on the observation that agents such as nocodazole that disrupt microtubules markedly inhibit transferrin receptor endocytosis.[73,74]

At steady state, most transferrin receptors on the plasma membrane are outside the coated pits.[75,76] It is therefore evident that these receptors must be mobile on the plasma membrane to reach coated pits; this process seemingly is to a large extent inhibited by microtubule depolymerizing agents. However, it must be emphasized that microtubule disruption affected endocytosis in hemopoietic cell lines growing in suspension, while it had no effect on adherent cells of epithelial origin. These differences may be related to differences in the cytoskeletal organization as well as to the short half-lives of microtubules in cells of hemopoietic origin.

Other studies suggest that the routing of the transferrin receptor through early endocytic compartments is independent of the actin filament systems, as shown by experiments with cytochalasin D[77] and studies on cells deficient in the cytoskeletal protein ABP-280.[78] However, other studies suggest an involvement of actin filaments in clathrin-coated pit mobility.[79]

Particular attention was focused to the study of Rab proteins, members of a large GTPase family with homology to *ras*. These proteins are important regulators of several intracellular membrane transport steps that mediate their function through cyclical binding and hydrolysis of GTP. Rab proteins are localized to the cytoplasmic faces of all organelles involved in intracellular transport, and are associated with vesicle membranes in their activated GTP-bound form.

After GTP hydrolysis, GDP-bound Rab proteins detach from the membrane and become associated with the GDP dissociation inhibitor in the cytosol. Some Rab proteins are associated with the endocytic pathway. Rab 4 and Rab 5 proteins co-localize with the transferrin receptor in early endosomes.[80,81] Others, such as Rab 7, are localized in late endosomes and do not co-localize with transferrin receptors. Rab 5 promotes the fusion of early endosomes, while Rab 4 seems involved in receptor recycling. Rab 4 overexpression leads to an inhibition of iron release from transferrin, a phenomenon seemingly related to a dislocation of internalized transferrin to a population of non-acidic endosomes.[82]

Recent studies have shown that shortly after internalization, transferrin and transferrin receptors enter a compartment of early endosomes that contain both Rab 4 and cellubrevin. Transferrin then accumulates at the recycling vesicles accumulated around the microtubule organizing center. These pericentriolar endosomes contain cellubrevin, but are totally depleted of Rab 4.[83]

Stimulation of Rab 5 activity results in formation of enlarged early endosomes, a finding consistent with the role for Rab 5 in fusion of endosomal structures. Furthermore, Rab 5 overexpression determines redistribution of endosomes from a punctuate pattern dispersed throughout the cytoplasm to a juxtanuclear localization. Recent observations indicate that Rab 5 plays an important role in promoting the movement of endosomes toward microtubules.[84] Thus, Rab 5 is involved in the regulation of the transferrin receptor traffic at the level of the early endosomal compartment.

The molecular mechanisms by which Rab proteins regulate transferrin receptor traffic are not well understood. However, recent studies shed light on this issue by showing that Rab proteins exert their effects on membrane traffic through interaction with other proteins. In the case of Rab 5, rabaptin 5 acts as an effector protein that binds to GTP–Rab 5,[85] while rabphillin 3 interacts with GTP–Rab 3. Rabaptin 5 and rabphillin 3 form a complex that is required for transferrin receptor endocytosis.[86] The stimulatory effect of this complex is inhibited by Rab 3 which competes with rabphillin 3 for the binding of rabaptin 5.

Recent observations suggest that, in addition to Rab 4 and Rab 5, Rab 15b is localized at the levels of early endosomes where it may act as an inhibitory GTPase in early endocytic trafficking,[87] as suggested by enforced gene expression.

Other Rab proteins may play a role in late events related to receptor recycling and not in the initial steps of transferrin receptor endocytosis. In fact, Rab 11 appears to be associated predominantly with the receptor recycling compartment.[88] Recent studies indicate that Rab 11 is strictly associated in its cellular distribution with transferrin-containing recycling compartments in K-562 cells, thus suggesting that it may play a role in the trafficking and recycling of internalized transferrin receptors.[89] Rab 11 exerts its effect on receptor recycling through interaction with the target protein rabphillin.[90] It can be concluded that endocytosed transferrin–transferrin receptor moves sequentially through endosome compartments that can be immunoadsorbed with anti-Rab 5 and anti-Rab 11 antibodies, consistent with the theory that Rab 5 and Rab 11 localize to sorting and recycling endosomes, respectively.

Another category of coating proteins different from clathrins are called coatomer proteins (COPs). These proteins were mainly implicated in forming a coat for transport vesicles associated with the Golgi complex. However, recent studies have shown that some components of COPs associate with early endosomes.[91,92] A recent study showed that Chinese hamster ovary (CHO) cells bearing mutations of the ε-COP subunit do not display defects in transferrin–transferrin receptor internalization, but exhibit inhibition of transferrin receptor recycling.[93] More particularly, CHO cells harboring mutated ε-COPs displayed a transferrin recycling time with an overall $t_{1/2}$ of ~11.5 minutes, compared to a value of ~3.5 minutes observed in controls.[93] These observations suggest that loss of ε-COP function may inhibit the recycling activities of one or more populations of early endosomes.

The above-mentioned study documented the expression of cellubrevin in both early and pericentriolar endosomes. Other studies investigated the role of this protein in transferrin receptor endocytosis. Cellubrevin is a synaptobrevin homologue present in all cells and tissues. Its pattern of cellular expression strictly resembles that of recycling vesicles. The intracellular expression of cellubrevin showed a localization at the levels of small vesicles present in all parts of the cell and particularly concentrated around the Golgi complex. All transferrin-recycling vesicles also stain for cellubrevin; however, cellubrevin antibodies stain more vesicles than transferrin.[94] The function of cellubrevin is unknown; however, experiments on permeabilized CHO cells treated with tetanus toxin, a protease that cleaves cellubrevin and other synaptobrevin-like proteins, showed an inhibition of transferrin receptor recycling.[95]

Other studies have shown that cellubrevin interacts with Bap 31, a representative of a highly conserved family of integral membrane proteins. Bap 31 has an intracellular distribution different from that observed for transferrin receptor (the two proteins co-localize only at vesicles accumulating around the microtubule organizing center) and controls anterograde transport of certain membrane proteins from the endoplasmic reticulum to the Golgi complex.[96]

Recent studies involved a novel technique that allows visualization and immunolabeling of the endocytic system in non-sectioned cells at the electron microscopic level. The studies showed the existence of a novel class of clathrin-coated vesicles budding from endosomes and involved in transferrin receptor recycling.[97]

These endosome-associated clathrin-coated buds were discerned from both plasma membrane-derived clathrin-coated vesicles and trans-Golgi network-derived clathrin-coated vesicles by their location, size (100 nm), and lack of labeling for β-adaptin. These buds are vesicle structures seemingly involved in transferrin recycling, as suggested by several findings, including their positivity for transferrin receptor, their localization in periphery near to early endosomes, and the inhibition of their formation by brefeldin.[97]

This last observation is particularly relevant to understanding the physiological role of these endosomic structures in transferrin receptor recycling. Brefeldin treatment of the cells leads to a two- to three-fold reduction of receptor expression at the cell surface, a phenomenon related to a reduced rate of receptor externalization.[98] It also produces a complete disappearance of clathrin-coated endosome buds.

Other factors involved in the control of transferrin receptor recycling are represented by a small group of *ras*-like guanosine triphosphates (GTPases) called adenosine diphosphate (ADP) ribosylation factors (ARFs).

ARFs form a group of six, small (about 20 kDa), ubiquitous GTP-binding proteins required for maintaining the integrity of organelle structure and intracellular transport. ARF6 has been implicated in the transport between the plasma membrane and the endosomes.[99,100] ARF6 localizes at the cytoplasmic side of the plasma membrane and to an internal, tubulovesicular compartment.

Overexpression of a dominant negative ARF6 mutant defective in GTP binding resulted in impaired recycling of endocytosed transferrin[99] and accumulation of 100- to 300-nm vesicles with unidentified coats that lacked clathrin.[100] The vesicles may represent the equivalent of clathrin-coated endosome buds.

Overexpression of wild-type ARF6 or its constitutively activated mutant inhibited transferrin uptake into Chinese hamster ovary (CHO) cells.[99] Constitutively activated ARF6 accumulates at the plasma membrane where it induces extensive membranes in CHO cells invaginations. In contrast, ARF6 and GTP-binding defective mutant co-localize with the transferrin receptor to the pericentriolar recycling compartment.[101] In these cells, ARF6 and transferrin receptors co-localize. However, in other cell types, such as HeLa cells, the cellular compartments where ARF6 localizes are distinct in their morphology and behavior from the transferrin receptor compartments.[102] Thus, the effect of ARF6 on transferrin receptor endocytosis is cell type-specific.

Activation of the low molecular weight GTP-binding proteins is mediated by guanine nucleotide exchange factors (GEFs), which are involved in the catalysis of the replacement of bound GDP with GTP. A GEF protein, designated EFA6 (exchange factor for ARF6) promotes efficient guanine exchange on ARF6.[103]

Studies of enforced expression of EFA6 in a CHO-derived cell line overexpressing the human transferrin receptor have shown that EFA6 protein localizes to the plasma membrane and induces membrane invaginations.[103]

From a functional point of view the overexpression of EFA6 leads to an increase in the pool of ARF6 in a GTP-bound state and a consequent inhibition of transferrin internalization.[103] Experiments with different EFA6 mutants determined that the effects of EFA6 on transferrin endocytosis depend on the capacity to activate ARF6 and not on its effects on the cytoskeleton.

Other studies investigated the role of the Rho-family GTPases in transferrin receptor endocytosis and recycling. Transfection of constitutively activated Rac induced a marked inhibition of intracellular transferrin receptor accumulation, a phenomenon due to inhibition of receptor endocytosis.[104]

Experiments with agents that either inhibit actin assembly or stabilize actin filaments exclude the possibility that the effects of Rho and Rac on receptor endocytosis are due to modifications of the actin network, and suggest that these two GTPases directly control a critical step in transferrin receptor endocytosis.[105] The effects of Rac and Rho on receptor endocytosis may be related to the activization of phosphatidylinositol 3′-kinase or PI(3)K. In fact, wortmannin, a potent inhibitor of PI(3)K, leads to down-regulation of cell surface transferrin receptors associated with accumulation of the receptor at the intracellular sites.[105] Furthermore, wortmannin-treated cells display an increased endocytic rate constant for transferrin internalization and decreased exocytic rate constants for transferrin recycling.[106]

Acidification is an important property for endosomes. Studies of CHO cells have shown that sorting endosomes have a pH of about 6.0, and the recycling tubulovesicular system has a pH of 6.4. In several receptorial systems, one of the major functions of endosomes is the dissociation of ligands from their receptors. The situation is different with the transferrin receptor in that at the acidic pH of sorting endosomes, iron is released from transferrin, and apotransferrin remains bound to its receptor in the acidic endosomes.[107] The acidic pH of endosomes is maintained by an ATP-dependent proton pump called vacuolar H^+-ATPase, part of the family of multi-subunit proton pumps.

Treatment of Chinese hamster ovary cells with the vacuolar proton pump inhibitor bafilomycin A_1 causes a marked delay in the rate of movement of transferrin receptors back to the cell surface.[108] A detailed analysis of the different steps of receptor recycling showed that in bafilomycin A_1-treated cells, removal of transferrin from sorting endosomes and accumulation in the pericentriolar endocytic recycling compartment occurred normally, but the rate constant for exit of transferrin receptors from recycling endosomes is significantly reduced.[107]

Similar effects were observed when the cells were treated with tamoxifen[109] or transfected with the gene encoding influenza M2 protein.[110] Both agents produced neutralization of the endosomal pH and, as a consequence, inhibition of transferrin–transferrin receptor recycling.

Recent studies suggest that the acidification of endosomes may imply a G protein-dependent mechanism. A short region of the cytosolic domain of the transferrin receptor containing a KPKR sequence induced a G-mediated signal, as evidenced by stimulation of the rate of GTP-γ-S binding.[107]

A cytoplasmic fragment of the transferrin receptor lacking the KPKR sequence failed to activate the pathway that contributes to maintaining the acidification of endosomes.

Recent studies have raised the problem of determining the importance of the externalization phase (recycling phase) during the process of transferrin receptor endocytosis. Initial studies carried out in hepatoma HepG2 cells have shown that endocytosed transferrin receptors exit the sorting endosome within 1 to 2 minutes after uptake, whereas the entire endocytic cycle takes about 10 minutes.[112] Thus, the initial steps of endocytosis as well as the exit of transferrin receptors from endosomes do not constitute the rate-limiting steps of the transferrin receptor cycle through the cell.

In line with these observations, recent studies carried out in cell lines expressing supranormal levels of transferrin receptors have shown that the rate of iron uptake is not linearly related to the number of transferrin receptors expressed on the cell surface.[113] An obvious interpretation is that cellular components required for the endocytosis of the transferrin receptor are not saturated at the levels of receptors normally observed in the cells, but one or more molecules that form the exocytic machinery are present in amounts sufficient to handle the normal levels of receptor expression, but become rate limiting at high levels of transferrin receptor expression.[113]

The reduced rate of transferrin receptor recycling observed in cells overexpressing transferrin receptor is a phenomenon specific to transferrin receptors and does not affect the rates of endocytosis and recycling of other receptors.[114]

Studies using multivalent transferrin (prepared by chemical cross-linking) also provided evidence about a sorting function of the recycling compartment of the transferrin receptor.[115] The multivalent ligand determined cross-linking of transferrin receptors and both multivalent transferrin and oligomerized receptor were retained within lumenal sorting tubulovesicular elements localized in the pericentriolar area. This organelle remains accessible to subsequent endocytosed transferrin.[115]

Few studies have addressed iron mobilization from endocytic vesicles. Some of these studies have clarified the mechanisms responsible for iron removal from endocytic vesicles for reticulocytes.[116] Using purified preparations of reticulocyte endocytic vesicles, it was shown that the cellular incorporation of iron through the endocytic pathway involves three sequential biochemical components: an acidification pathway; an iron reduction system; and an iron transporter system.[116]

Finally, studies of polarized intestinal Caco-2 cells provided evidence that apotransferrin and diferric transferrin undergo different endocytic cycles. After internalization, both compounds recycled to the cell surface, but apotransferrin exhibited a much slower recycling time than diferric transferrin.[117] As a consequence of this delayed recycling rate, a significant proportion of cellular transferrin receptors became sequestered in the cell interior. This mechanism seems to be peculiar to polarized intestinal cells since it was not observed in non-polarized hematopoietic cells.[117]

Studies of a recently identified iron transporter may offer some understanding of the system responsible for the transport of iron from the endosomes. Studies on mice with a hereditary form of microcytic anemia led to the identification of an iron transporter called Nramp2 or DMT1 (divalent metal transporter 1). Animals with defective Nramp2 protein also have defective intestinal iron absorption and reduced cellular iron uptake from plasma transferrin into various tissues. Animals with nonfunctional Nramp2 have a major defect in transferrin recycling, thus suggesting a role for this protein in endosomal iron transport.[117a] This suggestion was directly supported by studies of cellular localization of Nramp2 protein.[117b] Immunofluorescence and confocal microscopy studies showed that Nramp2 is primarily expressed on recycling endosomes, and to a lesser extent, at the plasma membrane, co-localizing at both sites with transferrin.[117b] According to these findings, Nramp2 may play a key role in iron metabolism at the level of the transport of free Fe^{2+} across the endosomal membrane and into the cytoplasm.

The understanding of the mechanism of transferrin receptor internalization led also to some interesting applications. Bispecific monoclonal antibodies were assembled by coupling monoclonal antibodies that recognize transferrin receptors to acid-sensitive epitopes on mutant diphtheria toxin.[118] The mutant toxin was unable to bind to membrane receptors and thus does not mediate cytotoxic effects. However, the cytotoxic activity of the mutant diphtheria protein was restored via bispecific antibody-mediated delivery and release within low pH intracellular sites.[118]

5.5.2 Structural Requirement for Efficient Transferrin Receptor Internalization

Receptors such as the transferrin and low-density lipoprotein (LDL) receptors are constitutively clustered in coated pits and undergo rapid internalization in the absence of ligand. Initial studies have shown the existence of internalization signals in the cytoplasm domains of constitutively recycling receptors that are believed to interact with adaptor proteins of coated pits and promote high efficiency endocytosis.[119,120] These signals are represented by a contiguous short stretch of four to six amino acids.[58,121] The internalization signal sequences are self-determined three-dimensional structures in which a tight turn is implicated as the conformation determinant for high efficiency endocytosis.[58,121]

The amino acid sequences responsible for the internalization signals of the LDL and transferrin receptors have been elucidated in experiments in which mutant receptors were expressed in recipient cells lacking functional endogenous receptors; their internalization efficiencies were then determined using quantitative assays.

Using this approach, the LDL receptor internalization signals localized to the first 22 amino acids of the 50-residue carboxy-terminal domain, including Tyr 807, the amino acid mutated in familial hypercholesterolemia.[122]

The exact signal sequence of LDL receptor was represented by a six-nucleotide sequence: Phe–Asp–Asn–Pro–Val–Tyr.[122] The internalization signal of transferrin receptor was identified by functional analysis of mutant human transferrin receptors expressed in chick embryo fibroblasts using a helper-independent retroviral vector.[123] Mutant tailless human transferrin receptors, in which all but 4 of the 61 amino acids of the cytoplasmic domain were deleted, were not excluded from coated pits, but were endocytosed with an efficiency of approximately 10% of wild-type receptors.[124]

Functional analysis of a series of mutant transferrin receptors with different regions of the cytoplasmic domain deleted indicated that residues within a 10-amino acid region (amino acids 19 through 28) of the human transferrin receptor cytoplasmic domain were sufficient for rapid internalization; more particularly, Tyr 20, the only Tyr residue in the transferrin receptor cytoplasmic tail, was a fundamental element of the signal sequence. An aromatic, but not hydrophobic, amino acid could substitute for Tyr.[124-126]

Subsequent work on functional analysis of mutant human transferrin receptors expressed in chicken embryo fibroblasts showed that the critical sequence required for receptor internalization consists of the tetrapeptide sequence Tyr–Gln–Arg–Phe,[127] residues 20 through 23.

The most critical residues in the transferrin receptor internalization signal appear to be the two aromatic residues. A large hydrophobic residue can substitute for Phe 23 without loss of activity.[128]

Deletional analysis has shown that the activity of the YTRF signal is relatively independent of its position within the transferrin receptor cytoplasmic domain, provided it is separated from the transmembrane region by at least seven amino acids.[127] Comparison of this internalization signal with that of the LDL receptor signal indicated that the only similarity in primary structure between the two signals was the presence of a Tyr residue.[128] Analysis of the possible three-dimensional structures of these two different signal sequences revealed that they have the propensity to form tight turns in proteins of known three-dimensional structure.[127]

The conclusion that Tyr–Gln–Arg–Phe represents the signal sequence for internalization of human transferrin receptors has been questioned.[129,130] Studies showed that the transferrin receptor internalization signal included residues in the amino terminal 16 residues of the transferrin receptor cytoplasmic tail because amino acid substitutions or deletions in this region reduced internalization activity.[129,130] However, since residues 19 through 28 are sufficient for endocytic activity,[123] the effects of mutations in the amino terminal 18 residues of the transferrin receptor tail must be indirect; in fact, they may change the conformation of the internalization signal or, alternatively, may sterically block its interaction with coated pits.

The role of the YTRF sequence in mediating receptor internalization is further supported by studies in which this sequence was inserted into the cytoplasmic tail of the human IL-2 receptor α chain. The wild-type IL-2R α chain is internalized only when it forms a complex with IL-2R β and γ chains, while the mutant IL-2R α chain internalizes with high efficiency.[131]

In another set of studies,[132] the effect of insertion of an extra copy of a Tyr–Gln–Arg–Phe sequence at three different locations within the human transferrin receptor cytoplasmic domain was examined. The insertion of another Tyr–Gln–Arg–Phe signal at position 31 to 34 in the wild-type transferrin receptor, but not at the other two locations (9 to 12 and 47 to 50), increases the rate of endocytosis two-fold. Furthermore, the insertion of a Tyr–Gln–Arg–Phe signal at position 31 to 34 in an internalization-defective mutant receptor restores endocytosis to wild-type levels, thus indicating that a Tyr–Gln–Arg–Phe signal at position 20 to 23 or position 31 to 34 is necessary and sufficient to promote transferrin receptor internalization and function in an independent and additive manner.[132]

An internalization motif may be created in the cytoplasmic domain of the transferrin receptor by substitution of a tyrosine at the first position of a predicted tight turn at the level of residues 31 to 34. Such substitutions restore wild-type levels of internalization to transferrin receptors made defective in endocytosis capacity by a missense mutation in the YTRF motif.[133]

The endocytic machinery can recognize receptors destined for internalization through short stretches of amino acids present in their cytoplasmic tails. Two types of internalization signals have been described: the tyrosine-based motif and the dileucine-based motif. The recycling receptors that utilize tyrosine-based motif internalization have some important differences in that some (such as transferrin and

LDL receptors) are constitutively clustered in coated pits, while others (such as EGF and insulin receptors) require ligand binding before being concentrated into clathrin-coated pits.

It was of interest to evaluate the effect of overexpression of one of the receptors on the internalization and recycling of the other receptors. The overexpression of the transferrin receptor leads to saturation of endocytosis for the receptor associated with increased membrane expression, but does not affect the rate of internalization of EGF receptors.[134] Conversely, overexpression of the EGF or LDL receptors leads to saturation of their own endocytic mechanisms, but does not alter their internalization of each other or of the transferrin receptor.[135] These observations indicate that multiple distinct saturable components exist for clathrin-mediated endocytosis.

Somewhat different conclusions were reached in another study in which several transmembrane proteins internalized through the tyrosine-based or leucine-based motifs were overexpressed in Madin–Darby canine kidney cells. Overexpression of proteins containing tyrosine- or leucine-based sorting signals resulted in reduced internalization of the transferrin receptor, while recycling and polarized distribution were not influenced.[136]

Additional studies were carried out to evaluate possible roles of other sequences of transferrin receptor molecules in modulating the rate of endocytosis of the receptor. These analyses focused on the sequences involved in receptor phosphorylation and palmitoylation.

Mutant transferrin receptors, obtained by replacing the phosphorylation site, Ser 24, with Ala through site-directed mutagenesis, and expressed in mouse 3T3 fibroblasts showed lack of phosphorylation and efficient endocytosis and recycling.[137] This observation was subsequently confirmed in experiments in which the same transferrin receptor mutant was transfected in Madin–Darby canine kidney cells.[138]

The role of palmitoylation in the control of receptor internalization was analyzed by investigating the rate of receptor endocytosis in Chinese hamster ovary cells[139] and chick embryo fibroblasts[140] transfected with mutant human transferrin receptors lacking both Cys 62 and Cys 67 residues. The mutant receptors were found to be palmitoylation-defective, but were internalized efficiently and mediated iron uptake from transferrin at rates similar to those of wild-type receptors in chick embryo fibroblasts.[140] However, experiments on Chinese hamster ovary cells showed that mutant transferrin receptors defective in palmitoylation were internalized more efficiently than wild-type receptors,[139] but had no effect on receptor recycling.

These observations indicated that acylation of the human transferrin receptor may not be essential for endocytosis and recycling; in some cell types, transferrin receptor palmitoylation may exert an inhibitory effect on the rate of receptor internalization.

Polarized epithelial cells possess the unique ability to specialize their membranes into structurally and functionally distinct domains. The apical membrane is the site of specialized cellular functions peculiar to each cell type and is separated by tight junctions from the basolateral membrane, which performs non-specialized cellular functions and is in contact with the underlying tissue. In the majority of polarized epithelial cells, transferrin receptors are preferentially expressed at the level of the basolateral membrane.

Recent studies carried out in Madin–Darby canine kidney cells showed that signal-dependent sorting mechanisms operate in the endocytic pathways of cells that selectively deliver transferrin receptors to the basolateral border regardless of the surface from which they are internalized.[141]

Subsequent studies have shown that the basolateral sorting of transferrin receptors depends on a signal present in the cytoplasmic tail of the receptor.[142] Residues 29 to 35 (VDGDNSH) are the most important for transferrin receptor basolateral sorting from the biosynthetic pathway.

The differential localization (a peculiar type of polarization) of transferrin receptors in neuronal cells represents the basis for transcytosis of transferrin observed in neuronal cells. Transcytosis can be defined as the transport of molecules endocytosed at one plasma membrane domain to the opposite domain. Studies of cultured hippocampal cells have shown dendritic uptake, transcytotic transport, and axonal release of transferrin, whose extent was greatly enhanced by excitatory neurotransmitters.[143]

Transferrin receptors are localized in dendritic endosomes and are strictly excluded from axons, even when the wild-type receptor is overexpressed in neuronal cells through gene transfer.[143] The neuronal polarized targeting of the transferrin receptor is abolished by deletion of amino acids 7 through 10, 11 through 14, or 19 through 28.[144]

Transferrin receptors were absent at the synoptic vesicles[145]; addition of the cytoplasmic domain of synaptobrevin onto human transferrin receptor was sufficient to retarget the transferrin receptor from dendrites to presynaptic sites in the axon.[146]

5.6 TRANSFERRIN RECEPTOR PHOSPHORYLATION

Studies carried out in HL-60 and K-562 leukemic cell lines showed that treatment with phorbol esters resulted in a rapid down-modulation of surface transferrin receptors, a phenomenon due to receptor internalization.[147,148] Subsequent studies showed that phorbol esters mediate receptor phosphorylation through activation of protein kinase C.[149] Colchicine inhibited in a dose-dependent manner phorbol ester-induced transferrin receptor down-modulation, but had no effect on surface receptor hyperphosphorylation, thus suggesting the requirement for an intact cytoskeleton.[149]

The decreased surface expression of transferrin receptors induced by phorbol esters was accompanied by a parallel decrease of the rate of iron uptake.[150] Tryptic phosphopeptide mapping of transferrin receptors isolated from phorbol ester-treated cells indicated that the major site of protein kinase C phosphorylation of the transferrin receptor in vivo and in vitro is Ser 24, an amino acid residue located within the intracellular domain of the receptor, 38 residues away from the predicted transmembrane domain.[151]

Subsequent experiments showed that the transfection of phosphorylation site mutant receptors and wild-type receptors mediates internalization, thus indicating that receptor phosphorylation does not play an important role in receptor endocytosis.[152,153]

It must be pointed out that the down-modulation of surface transferrin receptors observed in leukemic cell lines is associated with inhibition of cell growth and induction of cell differentiation.

In other cellular systems, however, phorbol esters increase surface transferrin receptor expression and stimulate cell growth.[153] Thus, treatment of Swiss 3T3 fibroblasts with phorbol diester or with platelet-derived growth factor caused the phosphorylation of transferrin receptor by protein kinase C and increased the cell surface expression of the transferrin receptor. The increased expression of surface transferrin receptor depends on an increased rate at the exocytic pathway of the receptor.[154] Experiments with receptor mutants showed that receptor phosphorylation is not required for phorbol ester-induced receptor distribution.[154,155]

These observations suggest that the phorbol ester effect on transferrin receptor expression may result from a general perturbation of membrane trafficking rather than a specific modulation of the transferrin receptor.

The fact that phosphorylation does not play a role in phorbol ester-induced redistribution of transferrin receptors is directly supported by a study in which phosphorylated and non-phosphorylated transferrin receptor protein species were separated by high resolution isoelectric focusing. Phosphorylated transferrin receptor behaved in a manner identical to non-phosphorylated transferrin receptor in terms of internalization.[156]

The mechanisms responsible for receptor down-modulation induced by phorbol esters were investigated in detail in erythroleukemic K-562[157] cells showing that:

1. Phorbol esters significantly enhanced (almost doubled) the endocytic rate constant for transferrin internalization compared to control cells.
2. Phorbol esters induced the internalization of transferrin receptors at the level of the intracellular compartments normally involved in receptor endocytosis and recycling.
3. Phorbol esters did not modify receptor synthesis; during short incubation times incubation receptor transcription was markedly inhibited during longer incubation times.
4. Down-modulation of transferrin receptors by phorbol esters is not due to induction of cell differentiation because other protein kinase C activators, such as bryostatin-1, also down-regulate transferrin receptors.

More recent studies evaluated the role of protein kinase and/or phosphatases in transferrin receptor endocytosis. Treatment of hepatoma HepG2 cells with tyrosine kinase inhibitors (such as thryphostin, genistein, and staurosporine) leads to inhibition of transferrin receptor endocytosis; this inhibition is limited to the early phases of receptor internalization.[158] Asialoglycoprotein receptor endocytosis was similarly inhibited by tyrosine kinase inhibitors. These observations suggest that tyrosine kinase activity modulates the rate of receptor endocytosis early in the internalization process.[158]

Furthermore, studies carried out in lymphoid cell lines have shown that stimulation of the transferrin receptor with specific monoclonal antibodies resulted in the tyrosine phosphorylation of the T cell receptor ζ-chain.[159]

In another study, the effects of a series of tyrosine kinase, protein kinase C, and serine/threonine kinase inhibitors on transferrin receptor internalization in HeLa and chicken embryonic fibroblasts were assayed. Transferrin receptor endocytosis was not affected by tyrosine kinase C inhibitors, but was inhibited by serine/threonine kinase inhibitors.[160] Studies with peptide inhibitors on permeabilized cells provided evidence that the inhibitory effects of the protein kinase inhibitors on transferrin receptor internalization may be ascribed to inhibition of casein kinase II.[160]

Finally, another set of studies focused on the effects of phosphatase inhibitors (i.e., inhibitors of the serine–theonine phosphatases, microcystin and okadaic acid) on transferrin receptor endocytosis. These inhibitors caused a significant reduction of transferrin and transferrin receptor uptake.[161,162] These observations suggest that transferrin uptake by receptor-mediated endocytosis involves phosphatase 2A, probably through regulation of microtubule-based vesicle transport.

5.7 INTRACELLULAR POOLS OF TRANSFERRIN RECEPTORS

For many receptors involved in ligand accumulation, the bulk of their content is intracellular. Thus, in HeLa cells, only 20% of the transferrin receptors are on the cell surface, with 60% residing in the endosome and the remaining 20% in the Golgi complex and the biosynthethic pathway.[163] Similar observations were made for other receptors. The presence of this pool of internal receptors may be explained on the basis of two general models for receptor internalization:

1. The pool of internal receptors may include receptors in membrane vesicles that are in transit through the cell.
2. The pool of internal receptors is mainly represented by transferrin receptor–transferrin complexes present in the endosomes.

Studies using transferrin–horseradish peroxidase conjugates showed that the intracellular pool of transferrin receptors results from continual internalization of unoccupied transferrin receptors.[164]

In addition to the main route of receptor recycling, 5 to 15% of the surface transferrin receptor that enters the cell is transported through the Golgi complex. Initial studies were consistent with the transport of transferrin receptor from the cell surface to the juxtanuclear region that contains the Golgi complex. In electron microscopic studies, the transferrin receptor bound to surface receptor was transported to this juxtanuclear region in cultured cells.[165,166]

Furthermore, Willingham[167] and Yamashiro et al.[168] found that transferrin bound to surface transferrin receptor moves into vesicles in the Golgi region. Finally, when cells are incubated with transferrin for long periods, material is accumulated in intracellular compartments and exchanged slowly with the cell surface.[169]

None of these initial studies provided direct evidence about the movement of transferrin–transferrin receptor complexes into Golgi cisternae. This phenomenon was demonstrated for the first time in human erythroleukemia K-562 cells where the sialic residues of transferrin receptor glycoprotein were used to monitor transport to the Golgi complex, the cellular site of sialyltransferases. Surface-labeled cells were treated with neuraminidase, and sialic acid residues were added. Isoelectric focusing of immunoprecipitated transferrin receptors was used to assess the movement of receptor to sialyltransferase-containing compartments.[170] Asialo-transferrin receptor was resialylated by the cells with a half-life of 2 to 3 hours, a result at variance with the very rapid rate of transferrin uptake and release (the ligand is bound to the cell surface, and released into medium within 10 to 20 minutes).

This observation suggests that not every transferrin–transferrin receptor complex moves through the Golgi complex during iron uptake. This is supported by a subsequent study showing that transferrin receptors with immature oligosaccharides (as a consequence of treatment with dioxy-mannojirimycin, an inhibitor of mannosidase I) are transported to the Golgi complex.[171] Via electron microscopy, internalized transferrin receptors were detected in all cisternae (cis, middle, and trans) of the Golgi stacks (as well as in endosomes and trans-Golgi reticular elements).[172] The finding that recycling membrane transferrin receptors can visit all or most Golgi subcompartments raises the possibility that any Golgi-associated post-translational modification can occur during recycling. These results can be explained by assuming that each time transferrin and its receptors are internalized, there is some probability that they will be diverted from the normal short-term recycling pathway and will traverse the long-term recycling pathway that involves the Golgi complex.[173]

Using a cholinesterase-mediated density shift technique that allows the separation of rat liver endocytic- from exocytic-coated vesicles, Fishman and Fine[174] demonstrated in the perfused rat liver that ^{125}I-transferrin is initially associated with an endocytic-coated vesicle upon internalization, and a significant fraction is subsequently transferred to an exocytic-coated vesicle. Asialo-transferrin in the exocytic-coated vesicles (but not in the endocytic-coated vesicles) is partially resialylated, thus indicating that the trans-Golgi network is the origin of coated vesicles that carry newly synthesized and recycled material to the cell surface.[174]

Similar cell surface-to-Golgi traffic has also been described for other cell surface receptors, including asialo-glycoprotein receptor, mannose 6-phosphate receptor, and low density lipoprotein receptor.

Studies performed in hepatoma HepG2 cells have shown that endocytosed transferrin (supplied as ^{125}I-transferrin or a conjugate of transferrin and horseradish peroxidase) recycled and was released in the culture medium with a half-life of ~10 minutes. During that time, endocytosed transferrin reached the trans-Golgi reticulum which connections the endocytosis and exocytosis pathways.[175] The endocytosed conjugate of transferrin and horseradish peroxidase was found only in a reticular network at the trans-Golgi region, and not in the Golgi stack.[175]

Experiments with anti-clathrin antibodies provided evidence that the antibodies blocked receptor-mediated endocytosis, but had no effect on receptor transport from the endosomes to the trans-Golgi network.[176] This observation was strengthened by experiments carried out by Robertson et al.[177] in K-562 cells. Depletion of cellular potassium, which blocks formation of coated vesicles at the cell surface, stimulated asialo-transferrin receptor resialylation by 60% over controls, suggesting that coated vesicle formation is not the rate-limiting step in cell surface-to-Golgi transport. Similarly, culture in sodium-free medium, which blocks transport from endosomes to lysosomes, increased asialo-transferrin receptor resialylation by 40%, indicating that lysosomes do not lie on the transport pathway.

In contrast, incubation of cells in hypertonic medium, which blocks many vesicular transport steps, inhibited transferrin receptor resialylation by 40%, thus suggesting an important role of vesicular traffic in transport of asialo-transferrin receptor from the cell surface to the Golgi complex. Finally, additional experiments using microtubule inhibitors provided other important details on the mechanism of transferrin receptor transport from the cell surface to endosomes and the Golgi complex. Nocodazole, an agent capable of depolymerizing microtubules, caused:

1. A decrease of the rate of receptor endocytosis, whereas the recycling of internalized receptors to the cell surface was unaffected.
2. No significant changes in the transport of transferrin receptors from the cell surface to the Golgi complex, thus indicating that the fragmentation of the Golgi complex caused by microtubule depolymerization does not block recycling through the Golgi complex.[178]

These observations point to two possible routes of surface-to-Golgi transport:

1. Transferrin receptors are transported from the cell surface to the Golgi complex by a route that does not involve endosomes (the rate-limiting step of this route cannot involve coated-vesicle formation).
2. Transferrin receptors are transported to the Golgi complex via endosomal intermediates (the rate-limiting step of this route cannot involve coated-vesicle formation, but must occur between endosomes and the Golgi complex).

A connection between the biosynthetic and secretory pathways of transferrin receptor was also observed. In the epithelioid Hep2 cell line, a significant proportion of newly synthesized transferrin receptors are delivered to endosomes before reaching the cell surface.[179] It is hypothesized that the internalization signal present in the cytoplasmic tail of the transferrin receptor is also recognized by clathrin lattices in the trans-Golgi region and delivered to sorting endosomes through this structure.[179]

The transfer of newly synthesized transferrin receptor from the rough endoplasmic reticulum to the Golgi complex was also demonstrated by electron microscopy studies in Hep2 cells transfected with cDNA encoding a chimeric protein composed of a reporter enzyme (horseradish peroxidase) anchored to the transmembrane domain of the transferrin receptor.[180] The chimeric protein was present on the plasma membrane and in cytoplasmic vesicles, vacuoles, and tubules, and to a lesser extent in the Golgi complex. In cells incubated at 20°C to inhibit transport from the trans-Golgi network, localization of the chimeric protein at the rough endoplasmic reticulum and at the Golgi cisternae was observed.[180] The transferrin receptor observed in these early exocytic compartments may be ascribed to newly synthesized protein in transit since it can be chased away by a temperature shift to 37°C.[180]

5.8   EFFECT OF CA²⁺ AND OXIDATION ON TRANSFERRIN RECEPTOR INTERNALIZATION

Several lines of evidence indicate that Ca^{2+} plays a role in transferrin receptor internalization. Studies performed in K-562 cells,[181] reticulocytes,[182] and Madin–Darby canine kidney (MDCK) cells[183] indicate the existence of a calmodulin-dependent pathway in transferrin receptor recycling. Other studies have shown that modulation of either intracellular or extracellular Ca^{2+} levels affects transferrin receptor recycling. Increasing intracellular Ca^{2+} increases the rate of receptor recycling[184] and stimulates apical transcytosis of transferrin.[185] Removal of Ca^{2+} from the incubation medium causes inhibition of transferrin receptor recycling.[186] Reticulocyte endocytic vesicles contain Ca^{2+}-ATPase activity whose function is under the control of a trimeric GTP-binding regulatory protein[187].

A recent study on T and B lymphoid cell lines indicates a direct link between transferrin internalization and Ca^{2+} mobilization. Transferrin addition to either Jurkat or L2C cells caused a moderate increase in intracellular Ca^{2+} concentration.[188] This mechanism may be responsible for the increased rate in transferrin receptor internalization caused by transferrin. Transferrin receptor internalization and recycling are constitutive processes, whose rates depend on the presence of the ligand.

In the absence of exogenous transferrin, the rate of transferrin receptor endocytosis and recycling markedly diminishes, particularly the rate of recycling of unoccupied receptors. Additional experiments have further supported the role of Ca^{2+} in the control of transferrin receptor recycling in these cells. Perhexiline, a Ca^{2+} antagonist, inhibited the rise of Ca^{2+} induced by transferrin and elicited a marked inhibition of transferrin recycling following its internalization. Ca^{2+} chelators and calmodulin inhibitors also markedly inhibited transferrin receptor recycling.

The mechanisms through which Ca^{2+} controls transferrin receptor recycling are largely unknown, but it is reasonable to hypothesize that tyrosine phosphorylation may be involved. On the other hand, the effect of calmodulin on receptor recycling may be related to the interaction of this protein with endosomal-associated proteins.

As will be discussed in detail in Chapter 8 covering iron regulatory proteins, an important link exists between iron metabolism and oxidative stress. The clearest evidence of this link is the observation of the toxicity of reactive oxygen, which reacts with H_2O_2 by a Fenton chemical reaction and gives rise to a highly toxic hydroxyl radical (OH).

Several proteins involved in iron metabolism represent cellular targets for oxidative stress. H_2O_2 markedly activates iron regulatory protein (IRP) activity, along with a corresponding increase in transferrin receptor mRNA levels and repressed ferritin synthesis;[189] oxidative-stress induces a dose-dependent increase of heme oxygenase, the enzyme that catabolizes heme to biliverdin, carbon monoxide, and free iron.[190]

In addition to these effects on IRP and heme oxidase, which will be analyzed in Chapter 8, oxidative stress leads to a rapid down-modulation of surface transferrin receptor.[191,192] Addition of agents that generate extracellular (H_2O_2) or intracellular (menadione) oxidative stress leads to a pronounced decline of membrane transferrin receptors.[191,192] This effect is dependent upon receptor redistribution, and not on receptor loss. Internalization studies showed that membrane receptor down-modulation was due to a block of receptor recycling on the cell surface.[191,192] Pretreatment of the cells with the antioxidant thiol supplier, N-acetyl-cysteine, inhibits the down-modulation of transferrin receptor expression elicited by menadione or hydrogen peroxide, thus suggesting that intracellular thiol redox status is a major determinant of receptor down-modulation induced by oxidative stress.[192]

Immunocytochemical studies with anti-transferrin receptor monoclonal antibodies showed that in menadione-treated cells, a large number of transferrin receptors are localized in an intracellular area corresponding to the Golgi complex. Okadaic acid, an inhibitor of type I and II phosphatases, prevented the down-modulation of transferrin receptors mediated by hydrogen peroxide, but did not inhibit the effect of menadione on surface transferrin receptors.[192] This observation suggests that extracellular and intracellular oxidants modulate membrane transferrin receptor expression through different molecular mechanisms.

These observations have interesting physiological implications for understanding the effects of oxidative stress on iron metabolism. The marked down-modulation of transferrin receptors following oxidative stress may help prevent an increase in reactive intracellular iron, which would otherwise aggravate the oxidative stress condition. Thus, the down-modulation of surface transferrin receptors induced by oxidative stress represents a protective mechanism against the toxicity mediated by oxidative radicals.

5.9 TISSUE DISTRIBUTION AND EXPRESSION IN SELECTED CELL TYPES

Transferrin receptors (TfRs) have been studied in a wide variety of human cells. Red blood cell precursors, which require large amounts of iron for hemoglobin production, were the first human cells shown to possess high numbers of specific membrane receptors for TfRs. Other tissues that require iron for incorporation into enzymes involved in oxidative metabolism have been found to possess these receptors. Thus, a high number of specific membrane receptors were found on the microvillus surface membranes of human placental syncytiotrophoblasts. This is in keeping with the finding that as fetal requirements for iron increase toward the end of pregnancy, daily placental uptake of iron may represent 90% of total maternal iron delivered to tissues.

Using several different monoclonal antibodies against the human TfR, Gatter et al. examined a variety of normal human tissues by the antiperoxidase–immunoperoxidase method. The receptors exhibited restricted expression patterns in normal human tissues. They were present in the basal layers of squamous epithelium but not in more superficial cell layers. Furthermore, they were found in the endocrine portion, but not the exocrine portion, of pancreas, in seminiferous tubules of testes, in the pituitary, in hepatic Kupffer cells, and in hepatocytes.[193] Normal mature monocytes did not possess TfR; however, receptors specific for transferrin appeared on the surfaces of these cells when they differentiated into macrophages.

The explanation for the pattern of TfR distribution in human tissues is complex. Their presence in the basal layers of squamous epithelium and in seminiferous tubules is in agreement with reports indicating that TfRs are highly expressed on rapidly proliferating cells. For other sites, for example, liver, this explanation is not tenable. Gatter et al. speculated that the tissues expressing TfRs also represent the sites at which excessive iron deposition occurs in primary hemochromatosis.[193]

The expression of transferrin receptors was investigated in detail in virtually all tissues of the body. Studies have shown that normal adult lung tissue is scarcely reactive with monoclonal antibodies to transferrin receptors. Significant reactivity was observed only at the level of macrophages; interstitial macrophages possess more transferrin receptors than alveolar macrophages.[194]

These observations were confirmed by *in vitro* culture studies of bronchial epithelium. These cells, in spite of their proliferative status, only weakly expressed transferrin receptors, while they were strongly positive for epidermal growth factor receptor expression.[195]

In contrast to the findings observed in normal lung, small lung cancer cells are usually positive for transferrin receptors. Transferrin receptors are preferentially expressed on squamous cell carcinomas and to a lesser extent in adenocarcinomas.[196] This observation explains the positivity of the majority of lung cancers after *in vivo* inoculation of ^{67}Ga which accumulates in malignant tissues via the transferrin receptor. Recent studies have shown gene amplifications at 3q26 in squamous cell lung carcinoma; among the genes present in this chromosome region, the transferrin receptor gene was amplified in a high percentage of cases.[197]

Particular attention was focused on transferrin receptor in hepatic tissue. Studies of their distribution in human normal hepatic tissue have shown that these receptors are preferentially expressed on parenchymal cells.[198,199] This observation is consistent with the fact that about 10% of the total iron body content is present in hepatic stores, mainly in hepatocytes, in the form of ferritin.[200]

Using an optimized alkaline phosphatase detection system, widespread expression of transferrin receptor with both membrane and cytoplasmic staining was found in all parenchymal and non-parenchymal cells.[201] The system of iron uptake by hepatic cells is complex in that these cells internalize iron by three different iron uptake mechanisms: (1) from transferrin by the transferrin receptor-mediated pathway; (2) from transferrin by the transferrin receptor-independent pathway; and (3) from nontransferrin-bound iron.[202,203]

Uptake of transferrin-bound iron by a nontransferrin receptor-mediated process involves transferrin binding to membrane proteins, internalization, release of iron from transferrin by a pH-dependent mechanism, and intracellular transport of iron into ferritin and heme.[202] Uptake of nontransferrin-bound and transferrin-bound iron by a nontransferrin receptor-mediated process occurs through a common cellular pathway involving the same iron carrier.[203]

The modifications transferrin receptor expression in four pathological conditions were evaluated in detail: (1) liver regeneration; (2) iron overload; (3) liver tumor; and (4) inflammation.

Initial studies in 1984 showed that a two- to three-fold increase in receptor numbers is observed during liver regeneration,[204] and this phenomenon is regarded as an index of increased proliferation. Subsequent studies have in part clarified the mechanisms responsible for the receptor up-modulation observed in liver regeneration. One study suggested that the increase in surface receptor numbers is due to the translocation of intracellular transferrin receptor to the cell surface.[205] A second study showed that an increase in transferrin receptor mRNA was observed in regenerating liver as a consequence of mRNA stabilization induced by iron regulatory protein 2 activation.[206]

This last observation was unexpected because the activation of iron regulatory protein observed during liver regeneration should lead to an increase of transferrin receptor levels associated with a decrease of ferritin content. A possible explanation for this apparent discrepancy is that growth-dependent signals stimulate ferritin synthesis mediated by IRP activation during liver regeneration.[207] Surprisingly, the receptor increase in regenerating liver is accompanied by an increase of ferritin content.

Expression of transferrin receptors in conditions associated with iron overload was carefully investigated. Transferrin receptor expression is virtually absent on hepatic cells derived from liver biopsies of most primary hemochromatosis patients.[208] However, in secondary hemochromatosis, transferrin receptor expression is detectable in most cases; the positivity is predominantly localized on hemosiderin-free hepatocytes.[209]

Immunohistochemical studies of rats loaded with iron revealed decreased transferrin receptors at the level of liver parenchymal cells, and increased receptor expression at the reticuloendothelial cells.[210] These observations were confirmed by *in vitro* studies on human hepatocyte cultures; they showed that iron loading was associated with transferrin receptor down-modulation.[211]

Studies carried out on liver tumors provided clear evidence that hepatocellular carcinomas express large numbers of transferrin receptors, as demonstrated by the intense receptor immunostaining observed on liver biopsies.[212] Interestingly, histological sections of tumoral and non-tumoral tissue showed that tumor tissue is more intensely stained than the surrounding liver parenchyma.[212]

Studies evaluating receptor expression in benign liver tumors showed a lower expression of transferrin receptors compared to expression observed in malignant tumors. In cases of focal nodular hyperplasia, nodular regeneration hyperplasia, and adenoma, liver cells displayed weak to moderate transferrin receptor staining.[213] Other malignant liver tumors, such as hepatoblastomas, frequently displayed intense staining of tumor cells with anti-transferrin receptor monoclonal antibodies.[213] Studies performed in models of hepatocellular carcinoma confirmed that a rise in transferrin receptor expression is a marker of preneoplastic progression to malignancy.[214]

It is well known that the liver is involved in the synthesis of acute inflammatory proteins and it is clear that chronic inflammatory diseases are associated with tissue accumulation of iron. These findings stimulated several studies done to evaluate the effects of cytokines involved in the inflammatory response (i.e., interleukin-1, interleukin-6 and tumor necrosis factor) on the synthesis of transferrin receptors in hepatic cells. Studies of the hepatoma HepG2 cell line have shown that these three cytokines have pronounced effects on ferritin synthesis, but only moderately affect transferrin receptor expression.[215,216]

The administration of IL-1, IL-6, or tumor necrosis factor (TNF) for 24 hours elicited a marked increase in ferritin synthesis associated with a moderate decrease of transferrin release, a slight increase in transferrin receptor expression, and a concomitant increase in ^{59}Fe-transferrin uptake.[216]

Transferrin receptor may have other biological functions in hepatocytes in addition to mediating the binding of transferrin. Franco et al. used the antigen-presenting

T cell system to identify the viral receptor employed by hepatitis B to enter cells and the sequence of hepatitis B envelope antigen (HBenvAg) involved in this inter-action. Results show that both CD4$^+$ and CD8$^+$ T clones can process and present HBenvAg to class II restricted cytotoxic T cells (CTLs), and that transferrin receptor is involved in efficient HBenvAg uptake by T cells.[217] Since transferrin receptor is expressed on hepatocytes, it may serve as a portal of cellular entry for hepatitis B virus infection.

Early studies of the distribution of transferrin receptors in the brain produced a peculiar finding: in humans and rats, brain monoclonal antibodies against transferrin receptors label blood capillaries in the brain.[218] It is of interest that anti-transferrin receptor monoclonal antibodies do not label blood capillaries in other tissues. This study also showed that this labeling occurs after injection of the antibody into the blood in the rat, thus indicating that transferrin receptors are accessible at the endothelial surface and allow transport into brain tissues.

Postmortem studies of human brain tissue confirmed that transferrin receptors are mostly localized at the blood capillaries. Some labeling was also observed in the brain tissue, with the highest receptor density in cortical and brain stem structures, apparently related to the requirement of neurons for iron for mitochondrial respira-tory activity.[219]

This hypothesis is directly supported by a study investigating the distribution of transferrin receptors in relation to cytochrome oxidase activity in the human spinal cord, lower brain stem and cerebellum.[220] All three markers showed similar patterns of distribution, with the highest levels of labeling in the substantia gelatinosa and motor neurons in the spinal cord and the nerve nuclei in the medulla and pons. A recent study carried out in the murine central nervous system showed that transferrin receptors are localized at two different types of structures:

1. Brain capillary endothelial cells, with the exception of circumventricular organs and choroid plexus epithelial cells
2. Intraneuronally in several brain regions without access to peripheral blood, including cerebral cortex, hippocampus, red nucleus, sustantia nigra, pon-tine nuclei, reticular formation, deep cerebellar nuclei, cerebellar cortex, and several cranial nerve nuclei[221]

Studies on cultured neuronal cells have shown that transferrin receptors are selectively expressed at the level of the somatodendritic domain, while no expression was observed in the axonal domain.[143,144] Transferrin is taken up at the dendritic extremities, transported into the cytoplasm, and accumulated and secreted at the level of the axonal domain.

This distribution of transferrin receptors in brain parallels iron distribution, with the highest concentrations observed in the basal ganglia, sustantia nigra, and deep cerebellar nuclei.

The localization of transferrin receptors at the blood–brain barrier sites and at some intraneuronal areas suggests that plasma transferrin is transported through cap-illaries and the choroid plexus into brain interstitial space. Iron is then bound by transferrin, synthesized by oligodendrocytes and choroid plexus epithelial cells, and subsequently taken up by neurons possessing transferrin receptors on their membranes.

These observations indicate that transferrin receptors in the brain are primarily located in gray matter areas, and are absent in the iron-rich white matter tracts. However, ferritin-binding sites are observed in the white matter tracts, while they are absent in gray matter nuclei.[222] This observation suggests that ferritin may act as an iron delivery system to the white brain matter.

The observation that transferrin receptors are localized at blood–brain barrier has some important implications.[223] Studies of *in vivo* intravenous injection of anti-transferrin receptor monoclonal antibodies have shown selective localization in the brain and not in other organs or tissues. Importantly, the antibodies were found in the brain parenchyma, rather than in the capillaries, thus indicating that they trans-cytosed the blood–brain barrier. This means that the transferrin receptor may be used to deliver nonlipophilic compounds to the brain. In fact, using anti-transferrin receptors as carriers, it was possible to transport nerve growth factor across the blood–brain barrier in a biologically active form and at levels sufficient to produce significant biological effects.[224,225]

An *in vitro* model of the blood–brain barrier (brain capillary endothelial cells) provided evidence for transcytosis of iron-loaded transferrin across the cerebral endothelium by means of a specific transferrin receptor-mediated pathway.[226] No intracellular degradation of transferrin was observed, indicating that a pathway through the cultured endothelium bypasses the lysosomal compartment.[226] Studies on cultures of brain capillary endothelium cells demonstrated that these cells also possess receptors for low density lipoproteins.[227] This receptor system may be used in addition to the transferrin receptor to transport nonlipophilic molecules to the brain.

Experiments involving intravenous ^{59}Fe-transferrin injections in mice have shown that endothelial blood–brain barrier cells mediate iron uptake into the brain[228]; the transferrin receptors present on the blood–brain barrier mediate brain iron uptake as shown by the observation that pre-treatment with mAb to transferrin receptor prevented transferrin-mediated iron uptake. Studies carried out on primary cultures of blood–brain barrier endothelial cells have shown that these cells possess, in addition to membrane transferrin receptors, large pools of intracellular receptors that represent 90% of the total cellular receptor pool. This large intracellular transferrin receptor pool may function as a storage site for spare receptors or be activated by the cell to increase its capacity to transport iron.[229]

It is of interest that brain capillary endothelium possesses, in addition to trans-ferrin receptors, additional membrane proteins capable of mediating iron uptake. Brain capillary endothelium membrane expresses melanotransferrin, a glycosylphos-phatidylinositol-anchored molecule that provides a route for cellular iron uptake independent of transferrin and its receptor.[230]

It is possible to outline a general view of iron transport and metabolism within the central nervous system.[231] Iron is an important element and one of the main mineral constituents of the brain, reaching the highest concentrations at the basal nuclei. In addition to its typical metabolic functions, iron plays an important role in cathecolamine and myelin synthesis. Iron is absorbed at the level of the blood–brain barrier. Within the endothelium, iron is separated from transferrin in the endosomes of the endothelial cells. It is then transported through the vesicular system and across the abluminal plasma membrane into the interstitial fluid. Fe^{2+} ions cross the blood–brain barrier and reach the interstitial fluid where they are complexed with unsaturated transferrin synthesized by oligodendrocytes, then transported in the form of transferrin to various areas of the central nervous system.

Transferrin is taken from interstitial fluid by neurons and glial cells through a process involving receptor-mediated endocytosis. At the level of neuronal cells, transferrin receptors are mainly localized at the somatodendritic extremities where they mediate the uptake of transferrin, which is then transported and released to the axonal domain through a process known as transcytosis.

Initial studies of transferrin receptor expression in brain tumors have shown staining for transferrin receptors in variable fractions of neoplastic cells of all histological tumor types.[232] Immunoreactivity was found only in glioblastomas. These observations were confirmed in another study, showing that: (1) transferrin receptor positivity in astrocytomas correlates with tumor grade; and (2) meningiomas showed an identical and characteristic focal staining pattern.[233]

The pronounced expression of transferrin receptors in brain tumors stimulated the development of experimental therapeutic approaches based on the use of monoclonal antibody anti-transferrin receptors. In a recent study, transferrin–CRM107, a conjugate of human transferrin and a genetic mutant of diphtheria toxin (CRM107) that lacks native binding, was infused through high-flow interstitial microinjection into brain tumor lesions. Nine of 15 patients treated with this approach presented >50% reduction in tumor volume, including two complete responses.[234]

A second clinical trial was based on intraventricular infusion of an immunotoxin composed of a conjugate of a monoclonal antibody against the human transferrin receptor and recombinant ricin A-chain. This immunotoxin was relatively well tolerated and induced in half the patients greater than 50% reductions of tumor cell counts in lumbar cerebrospinal fluid.[235]

Recently, a new strategy was proposed to deliver anti-tumor drugs to the brain via the transferrin receptor, based on the construction of a fusion protein composed of anti-transferrin receptor single-chain antibody and streptavidin.[236,237] This fusion protein, when injected *in vivo*, binds to the brain capillaries and is then transcytosed into the brain. The advantage of this fusion protein is its capacity to bind biotin-conjugated drugs that can be then delivered to the brain.

Brain cells also possess the biochemical machinery required for the post-transcriptional modulation of transferrin receptor expression according to cellular iron status. Brain cells possess iron regulatory protein 1 (IRP-1), capable of forming two RNA–protein complexes with ferritin or transferrin receptor RNA.[238]

The initial study of the distribution of transferrin receptors in normal human tissues indicated a positivity at the level of basal epidermis.[239] Subsequent studies focused on evaluating the expression of this receptor at skin level in normal and pathological conditions. In normal human skin, the transferrin receptor shows characteristic tissular distribution in the basal cell layer, at the level of the dermal–epidermal interface.[239] An identical pattern was observed in embryonic and fetal epidermis.[240]

Cultured keratinocytes are only weakly positive for transferrin receptor; their synthesis is markedly increased by retinoic acid treatment.[241] In epidermal fibroblasts, the expression of transferrin receptors is strictly related to cell proliferation. Treatment of quiescent fibroblasts with mitogenic growth factors led to a rapid redistribution of transferrin receptors from an intracellular pool to the cell membrane.[242] The majority of skin tumors exhibit positivity for transferrin receptors:

1. In cutaneous lymphomas of high-grade malignancy, transferrin receptor was expressed in most of the cells, compared to a variable proportion (25 to 75%) of positive cells in low-grade lymphoma.[239]
2. Malignant melanomas were totally positive, whereas benign melanocytic nevi were largely negative.[239]
3. Transferrin receptor positivity was frequently observed in cutaneous papillomas and maximally in lesions in which viral DNA was clearly detected.[243]

After an initial study describing localization of transferrin receptors at the level of pancreas,[244] few further studies were conducted. In iron-deficient and iron-loaded rats, transferrin receptor positivity occurs at the acinar and Langherans islet cells of the pancreas.[244] The expression of transferrin receptors and the extent of iron uptake are controlled by the level of intracellular iron determined through a negative feedback. Maximal pancreatic iron uptake was observed in iron-deficient or hypotransferrinemic mice.[245]

A recent study indicated that the capacity of pancreatic acinar cells to express and internalize transferrin receptors is acquired during ontogenesis; this phenomenon parallels the maturation of various regulatory secretory pathways in pancreatic acinar cells.[246]

In vitro-grown kidney cells were one of the first cell types in which the presence of transferrin receptors was demonstrated.[247] The expression of transferrin receptors on kidney cells was confirmed by other studies, particularly at the level of the distal convoluted tubular epithelium of the kidney.[248]

The expression and possible role of transferrin receptor during renal tubulogenesis were also investigated. Examination of tissue sections of human fetal kidneys showed that the receptor appears during nephron induction, is then strongly expressed during the early stages of renal tubulogenesis, and its expression declines with progressive maturation.[248] *In vitro* studies using murine explants of embryonic kidneys showed that addition of blocking anti-transferrin receptor antibodies completely inhibits kidney morphogenesis.[249]

In vitro studies on monolayer cultures of canine kidney epithelial cells have shown that transferrin receptors have a polarized distribution, with a ratio of basolateral to apical receptors of approximately 800:1.[141] The transferrin receptor is selectively delivered to the basolateral surface, where it internalizes transferrin via clathrin-coated pits and recycles it back to the basolateral border.

Studies based on the transfection of transferrin receptor mutant cDNA showed that residues 29 to 35 contained in the cytoplasmic tail of the receptor play a key role in the basolateral sorting of this membrane protein.[142] Specific sorting mechanisms based on a clathrin-dependent mechanism retrieve transferrin receptors from the pathways that deliver other membrane receptors either to the apical surface or to the lysosomes.[250] These cells represent an important model for the study of transferrin endocytosis in polarized cells, allowing the identification of specific components of the endocytic machinery such as synthenin and SNAP-23.[251,252]

Transferrin receptor expression has been extensively investigated in both normal and neoplastic mammary tissue. Although initial studies claimed that transferrin receptor expression was limited to neoplastic tissue, subsequent studies have shown that normal mammary cells also express transferrin receptors. Significant receptor expression was observed in pregnant breast[253] and in developing murine mammary gland.[254] Compared to virgin controls, the total number of mammary transferrin receptors markedly increased during pregnancy and particularly during lactation. Parallel measurement of ferritin content showed that the level of this iron storage protein moderately but significantly decreased in these conditions, indicating that the increased iron taken up by mammary cells is utilized to sustain increased proliferation and metabolism. In vitro studies have shown that transferrin receptor expression correlated in both pregnant and lactating mammary glands with cell proliferation.

Many studies provided evidence that transferrin receptors are highly expressed in human breast cancer cells. These studies showed that an inverse relationship between transferrin receptor and estrogen expression correlated with menopausal status, with tumors from premenopausal patients exhibiting higher levels[255]; estrogen stimulated in vitro transferrin receptor expression in estrogen receptor-positive human breast cancer cells[256]; the number of transferrin receptors present on the surfaces of sublines derived from breast cancer MCF-7 cell lines correlated with metastatic and invasive properties of these cells[257]; and anaplastic breast cancers exhibit the highest levels of transferrin receptor and ferritin.[258]

Recent observations support the existence of a link between the number of transferrin receptors expressed by breast cancer cells and their metastatic potential. The selection of cells of the mammary adenocarcinoma cell line MTLn2 with high levels of transferrin receptors and a higher proliferative response to transferrin results in sublines with greater potential for spontaneous metastasis.[259,260]

Preliminary observations suggest that breast cancer cells may express, in addition to the canonical transferrin receptor, a peculiar form of receptor described in estrogen-responsive chicken cells.[261] This protein was initially described as a 95-kDa membrane protein induced in chicken oviduct by estrogen treatment. It has a high degree of sequence homology with chicken heat shock protein 108, mouse endoplasmic reticulum protein p99, and human tumor rejection antigen gp96; it also possesses the important property of binding transferrin.[261]

Antibodies raised against this purified protein showed an intracellular localization; this antiserum also reacts with human breast cancer cell lines and selectively immunoprecipitates an antigen of 104 kDa.[261] The role, if any, of this protein in normal cellular physiology and in breast cancer cells remains to be evaluated.

Particular attention focused on the study of transferrin receptors in the urogenital system. Studies have shown that transferrin receptor expression in normal bladder mucosa was negative except for the proliferating cells of the basal layer.[262] However, the majority of bladder carcinomas are clearly positive for transferrin receptor; the receptor positivity is correlated with tumor proliferation index[263] or tumor grade and stage.[264]

In another study of 55 bladder cancer patients, transferrin receptor positivity correlated with tumor aneuploidity; receptor expression was significantly higher in aneuploid tumors than in euploid tumors.[265]

Other studies also showed that positivity for transferrin receptor in bladder carcinoma had a prognostic value. Superficial bladder tumors positive for transferrin receptor had recurrences in a higher proportion (74% of cases) than tumors negative for this receptor (37% of cases).[266] Interestingly, a recent study has shown that various bladder carcinoma cell lines have the capacity to synthesize a transferrin-like 70-kDa protein, that represents the main growth-promoting factor present in the conditioned media of these lines.[267] According to these observations, the synthesis of transferrin (or of a transferrin-like molecule) may represent a mechanism for providing iron required to sustain the growth of these cells; this mechanism may be particularly important in areas that are poorly vascularized.[267]

Based on these observations, the potential of transferrin–adriamycin conjugates in the treatment of bladder cancer was evaluated. Incubating bladder cancer cell lines in the presence of these conjugates did not provide a system to prevent cyto-toxicity to transferrin receptor-negative cells or overcome the multi-drug resistance of tumor cells.[268]

Similar observations were also made through the analysis of transferrin receptor expression in the prostate. Initial studies showed that human prostate cancer cells express elevated levels of transferrin receptors, while only a low rate of expression of these receptors was observed in normal tissue or in benign prostatic hyperplasia. The expression of the transferrin receptor was observed only at the level of epithelial cells, while stromal cells were negative.[269] This increased expression of transferrin receptors on cancer prostate cells seems relevant to the growth of these tumor cells. Several lines of evidence indicate that prostate cancer cells possess the capacity to synthesize and release transferrin:

1. Prostatic fluid secreted by adenocarcinoma cells contains high levels of transferrin, and urine transferrin levels are significantly increased in patients with prostate cancer in comparison to age-matched controls.[270]
2. Studies in rats have shown that staining for transferrin was found in the interstitia of all regions of the normal prostate; epidermal cells of rat prostate Dunning tumor possess the capacity to synthesize high levels of transferrin.[271]
3. The growth of prostate cancer cell lines is greatly inhibited by monoclonal antibody anti-transferrin receptor.[272]
4. The metastatic behavior of prostatic cancer cells is influenced by factors present in the hematopoietic tissue.[273]

The last point is particularly important because it is well known that metastasis occurring in the vertebrae is rapid. The difference between the growth rates of primary and secondary tumors is seemingly related to a stimulatory effect of the bone marrow milieu. In line with this interpretation, bone marrow contains factors that are highly stimulatory for sustaining the growth of prostatic carcinoma cells. The main factor is transferrin.[274]

More particularly, the stimulatory effect of transferrin is limited to hormone-insensitive prostatic carcinoma cells; hormone-dependent cells were not stimulated by transferrin.[275] The inhibitory effect of suramin on the growth of prostatic carcinoma cells is counteracted by transferrin.[275]

Prostate cells express on their membranes an antigen, called prostate-specific membrane antigen, that exhibits moderate homology with the transferrin receptor (54% homology of nucleic acid sequence).[276] Normal prostate cells produce a short mRNA species encoding a cytosolic form of the protein, while cancer prostate cells produce a long mRNA species encoding a membrane-bound form of the protein.[276] This protein possesses folate hydrolase activity.[277]

At the level of testis, the pattern of transferrin–transferrin receptor expression is peculiar in that a cell type involved in transferrin production (Sertoli cells) is utilized by spermatocytes. Within the seminiferous tubules, the Sertoli cells create an impermeable blood–testis barrier. (The barrier is formed by tight functions between Sertoli cells in seminiferous tubules and by myoid cells that encircle these tubules.)[278]

Transferrin is required to sustain spermatogenesis, as supported by the observation that a mouse mutant model that lacks the ability to synthesize transferrin is defective in spermatogenesis. To meet the iron requirement of spermatogenesis, Sertoli cells act as "nurses" that provide transferrin to developing spermatocytes.[279] They deliver transferrin to the adluminal compartment of the seminiferous epithelium. This transferrin is taken up by Sertoli cells from the blood synthesized by these cells.[280]

Initial studies of immunolocalization of transferrin and transferrin receptors in human testes have shown that transferrin is mainly found in Sertoli cells and in lesser amounts in spermatocytes and early spermatids. Transferrin receptors are found only in spermatocytes and early spermatids.[281] However, studies of rat testes have shown that Sertoli cells also possess transferrin receptors, and provided detailed information on the distribution of transferrin receptors in the spermatocytes.

Intratesticular injection of radioactive transferrin results in rapid, strong labeling of the basal compartments of the seminiferous epithelium, thus indicating a high density of transferrin receptors in this site.[280] Northern blot analysis of RNA isolated from adluminal germ cells showed that transferrin receptor mRNA levels are inversely related to the stage of maturation of spermatocytes, with highest levels observed in round spermatids.[282] Comparison of the levels of transferrin receptor mRNA in various cell types showed that the highest levels were observed in Sertoli cells and were not influenced by hormones.[283]

According to all these observations, diferric transferrin secreted by Sertoli cells in the testes binds to the transferrin receptors on the surfaces of adluminal germ cells; is then internalized by receptor-mediated endocytosis. Iron is utilized by germ cells, while the apotransferrin–transferrin receptor complex is recycled back to the cell surface where apotransferrin is released into the adluminal fluid.

Few studies have evaluated the expression of transferrin receptors in testes cancer. One study showed that the greatest expression of transferrin receptors was observed in seminoma, while other testicular tumors such as intratubular carcinoma and teratoma are less positive for transferrin receptor.[284]

Some studies characterized transferrin receptor expression in normal and neoplastic ovary cells. Studies of freshly isolated tissue and in vitro-grown primordial follicles have shown that pre-granulosa cells and oocytes of the quiescent primordial follicles do not express proliferation markers, including transferrin receptors, while granulosa cells and oocytes of the growing primary follicles are positive for proliferation markers, including transferrin receptors.[285]

Studies in chickens showed that transferrin receptors are highly expressed in oviducts and ovaries, thus suggesting a major role for the oviduct transferrin receptor in oogenesis.[286] Studies in other animal species have shown that transferrin is required to sustain the differentiation and growth of primary follicles from primordial follicles.[287-289]

Ovarian cancer cells express markedly higher levels of transferrin receptors than normal ovary cells.[290,291] This observation prompted the development of experimental therapeutical approaches based on the use of immunotoxins composed of either Pseudomonas toxin or ricin-A toxin coupled to an antibody to the human transferrin receptor. The injection of this immunotoxin greatly improved the survival of nude mice intraperitoneally injected with ovarian cancer cells.[290,291] Finally, in vitro studies indicate that ovarian cancer cells grown under appropriate conditions acquire the ability to endogenously synthesize transferrin.[292]

Few studies focused on the expression of transferrin receptors in endocrine glands. Studies on thyroid glands are limited to the analysis of the effect of transferrin on the growth of thyroid cell lines.[293] These studies showed that FRTL5 rat thyroid cells possess elevated numbers of transferrin receptors whose expression is regulated by thyroid-stimulating hormone.[293] Transferrin enhances the effects of the cAMP-dependent mitogens, such as thyroid-stimulating hormone, without modifying basal or stimulated cAMP production.

Few studies investigated the localization of transferrin receptors in epithelial thyroid cells. Studies performed in Fisher rat thyroid cell (FRT) line have shown the basolateral distribution of transferrin receptors.[294]

Similarly, studies on pituitary cells were virtually limited to in vitro-grown cell lines, with the exception of one study focused on the investigation of pituitary tumors. Most of these studies were performed on GH1 rat pituitary tumor cells which require thyroid hormones for growth. Initial studies on the in vitro growth of these cells under serum-free conditions showed a horse serum requirement for a protein of about 70 kDa identified as equine transferrin. Interestingly, the growth-enhancing activity of this protein was observed only when the protein was devoid of iron (apotransferrin), while iron saturation of equine transferrin caused total loss of its biological activity as a pituitary growth stimulator.[295]

Subsequent studies have clarified that the stimulatory role of apotransferrin is not dependent on the triggering of a growth-stimulating signal; it depends on chelation of growth-inhibitory Fe^{3+} levels.[296]

Normal rat pituitary cells also possess the capacity to synthesize transferrin. Immunostaining of pituitary gland showed that in the anterior and intermediate lobes, some cells displayed reactivity with anti-transferrin antibody. This reactivity was limited to gonadotrophs and somatotrophs, and no co-localization was seen with cells producing TSH, MSH, ACTH, and prolactin.[297] These observations suggest that locally-produced transferrin may act as a growth factor for some cells of the pituitary gland. Autopsy studies of human pituitaries showed that the majority of the adenohypophyseal cells were transferrin receptor-positive.[298] Another study of the expression of transferrin receptors in pituitary tumors showed that 8 of 13 clinically non-functioning adenomas showed significant levels of transferrin receptor expression, while adenomas associated with acromegaly and prolactinomas were receptor negative.[299] All eight transferrin receptor-positive tumors were gonadotrophinomas. These observations suggest that gonadotrophin, at variance with other pituitary cell hormones, has a peculiar requirement for iron; this may explain the high susceptibility of these cells to malfunction in primary and secondary iron overload syndromes.[299]

Some studies analyzed the expression of transferrin receptors in muscle cells. Initial studies on primary cultures of chick embryo muscle cells showed that myoblasts during exponential growth clearly express transferrin receptors. Quiescent myoblasts express lower numbers of receptors than proliferating myoblasts.[300]

Differentiated myotubes express transferrin receptors at levels comparable to those observed in proliferating myoblasts. However, the level of iron uptake in myotubes is higher than in myoblasts.[300]

Transferrin receptor expression increases during skeletal regeneration. Since the rise of transferrin receptor expression is preceded by an increase of IGF-I receptor, it was suggested that IGF-I may have a regulatory role on transferrin receptor expression in muscle cells.[301]

Studies of intracellular localization of the transferrin receptor, both at protein and mRNA levels, showed distribution in a narrow band around the nuclei of the myotubes.[302] Analysis of the distribution in intracellular vesicular structures showed a partial co-localization with GLUT4, a glucose transporter recruited by insulin from intracellular vesicular structures to the plasma membrane.[303]

Few studies have characterized the pattern of transferrin receptor expression in cartilaginous tissue. This tissue is normally avascular, but after differentiation in hypertrophic cartilage, blood vessel invasion is required for bone formation. Transferrin receptors are expressed on the surfaces of hypertrophic chondrocytes and at high levels on the surfaces of adjacent undifferentiated cells in the diaphyseal collar.[304] The absence of vascularization in resting cartilage is due to the presence of angiogenesis inhibitors, while the vascularization of hypertrophic chondrocytes is promoted by transferrin produced and released by these cells.[305]

Particular attention was given to studies of transferrin receptor in the placenta, which mediates and controls the supply of all nutrients to the fetus, where an important role is played by iron. The placenta behaves like a low-permeability membrane which contains specific transport systems for the different nutrients. To satisfy the high fetal iron requirement, trophoblastic cells of the placenta possess high numbers of transferrin receptors. Transferrin receptors can be isolated from placenta and they possess biochemical properties identical to those of the receptors isolated from other cell types.[306,307]

Immunohistochemical studies showed that transferrin receptors were present on the apical surfaces of the syncytiotrophoblasts.[308,309] Studies of rat placenta during the different phases of gestation showed that transferrin binding by the placenta increased progressively from day 14 to day 21 of gestation, associated with corresponding changes in the rate of iron transfer to the fetus.[310] This observation indicates that increased iron transfer to the fetus is mainly related to an increase of the number of transferrin receptors present on the maternal surface of the placenta. It must be emphasized that the increase of the total number of placental transferrin receptors is accompanied by a decrease of receptor density because the placental size increases during pregnancy. Studies on cultured trophoblasts isolated from human term placentas have shown that transferrin uptake is more predominant at the microvillus membrane than at the basal one.[311]

In addition to apical receptors, syncytiotrophoblasts also possess transferrin receptors at the level of the basal membrane.[312] The number of receptors at the basal membrane is about half the number present on the microvillus membrane. The biochemical properties of the transferrin receptors located at the basal membranes of the syncytiotrophoblasts are identical to those of the apical receptors. It was suggested that basal receptors participate in iron transfer to the fetus.[312]

The different endocytic pathways displayed by apical and basal transferrin receptors of the syncytiotrophoblasts were elucidated in the polarized trophoblast-like lice BeWo cells. These cells possess apical and basolateral receptors in a 2:1 ratio, and after binding and internalization, transferrin recycles predominantly to the domain of administration.[313]

This conclusion was reached through the study of the intracellular distribution of ^{125}I-transferrin and peroxidase-labeled transferrin simultaneously internalized from opposite cellular domains (basal and apical).[313] Subsequent studies on BeWo cells have shown that transferrin receptors migrate from one cellular domain to the other domain, a phenomenon known as transcytosis.[314] Newly synthesized transferrin receptors preferentially (>80%) localize at the level of the basal membrane domain; after arrival at the cell surface, the newly synthesized receptors equilibrate between the apical and basolateral cell surfaces, as a consequence of a transocytotic transport occurring in both directions.[315]

More recently, another trophoblast cell line (HRP-1) was isolated from normal rat placenta. These cells express transferrin receptors on both apical and basolateral membrane domains, and bind and internalize transferrin on both sites.[316]

Studies performed on microvillus and basal membrane vesicles isolated from human term placenta have shown that the ratio between apical and basolateral receptors is higher than that observed in trophoblast cell lines. Thus, the number of transferrin receptors on basal membranes was four times higher than the number present on microvillus membranes.[316]

The multinucleated syncytiotrophoblasts in the human placenta are derived from the differentiation of mononucleated cytotrophoblasts and are in direct contact with the maternal blood in the placenta. Human cytotrophoblasts cultured under appropriate conditions differentiate to syncytiotrophoblasts; during this differentiation, a five-fold increase in receptor numbers is observed.[317] Syncytiotrophoblasts exhibit a higher affinity for transferrin than cytotrophoblasts.

The expression of placental transferrin receptors is controlled in a manner similar to that observed in other cell types in that iron loading leads to a down-modulation of transferrin receptors, while iron deprivation leads to an opposite effect.[318] Modulation of transferrin receptors by iron is mediated mainly by diferric transferrin, and not by iron soluble salts.[318] In line with this observation, placental transferrin receptor expression is increased in diabetic pregnancies complicated by fetal iron deficiency. This phenomenon is mediated by a rise of iron regulatory protein 1 activity.[319]

The expression of transferrin receptors on trophoblasts according to environmental iron availability has some obvious physiological implications in that it provides a way to modulate placental iron uptake based on maternal iron status. Malignant transformation of trophoblast cells is associated with a marked increase of transferrin receptor expression.[320]

In vitro studies indicate that stimulation of cell proliferation enhances transferrin receptor expression on trophoblasts. However, this does not reflect the in vivo situation where proliferative cytotrophoblasts and quiescent syncytiotrophoblasts possess transferrin receptors.[321] This observation indicates that normal trophoblasts differ from other cell types in the relationship of proliferation to transferrin receptor expression.

In addition to the ability to express transferrin receptors, trophoblasts also possess the capacity to synthesize transferrin.[322] Human cytotrophoblasts and syncytiotrophoblasts both possess the capacity to synthesize transferrin. Transferrin isolated from trophoblasts was different in maternal and fetal sera with respect to the distribution of isoforms, a phenomenon seemingly related to a higher number of sialic acid residues in trophoblast transferrin in maternal serum.[322] The function of trophoblast transferrin remains to be evaluated.

Finally, a recent study provides evidence of an association between transferrin receptor and HLA-A (also known as HFE), the candidate gene for hemochromatosis.[323] The presence of the HFE protein in association with the transferrin receptor at the site of contact with maternal blood raises the posssibility that the HFE–transferrin receptor complex may be relevant for the control of maternal–fetal iron metabolism.

5.10 REGULATION OF TRANSFERRIN RECEPTORS

5.10.1 THE EXPRESSION OF TRANSFERRIN RECEPTORS IS CORRELATED WITH CELL PROLIFERATION

The relationship between iron and cell proliferation stems from a series of observations indicating that:

1. Iron deprivation prevents cell multiplication.
2. Transferrin or soluble iron salts are essential for in vitro cell proliferation in serum-free systems.

3. Iron is required to sustain the activity of ribonucleotide reductase, a key enzyme in the control of DNA synthesis.
4. Acidic isoferritins suppress the *in vitro* growth of normal hematopoietic progenitors.[324-327]

Studies performed on several continuous cell lines have clarified the relationship between the expression of transferrin receptors (TfRs) and cell proliferation. Non-replicating cells generally have stable iron balances and low numbers of TfRs, but cells undergoing multiplication have increased needs for iron uptake and markedly increase their receptor numbers. In cultured cells, transferrin-binding sites markedly increase soon after these cells are exposed to tissue culture conditions that stimulate proliferation.[328-331] Immunoassay studies of transferrin binding indicated that this increase results from an augmentation of total cellular immunoreactive receptor molecules, and not from a change in receptor affinity or availability.[332,333] Other observations suggest that TfRs are preferentially expressed on actively proliferating cells. When normal lymphocytes are transformed by mixed lymphocyte cultures or mitogens, they rapidly undergo a marked increase in the number of binding sites.[326,331]

The number of TfRs on continuous cell lines and normal fibroblasts is down-regulated during the transition from the log phase of growth to the stationary phase.[328,332] Finally, in different types of human leukemias, a correlation between receptor expression and proliferation exists.

Studies of the regulation of receptor expression in fibroblasts revealed additional mechanisms of control related to cell proliferation. Addition of growth factors to quiescent fibroblasts elicited a rapid increase of surface transferrin-binding capacity, a phenomenon related to the redistribution of transferrin receptors from intracellular pools to the cell surface.[334,335]

Epidermal growth factor caused only a transient increase in transferrin binding with peak effect at 5 minutes and return to control values within 25 minutes. In contrast, platelet-derived factor and insulin-like growth factor I induced a prolonged stimulation of surface transferrin binding up to 2 hours.[335] The rise of transferrin receptors following growth factor addition was calcium-dependent.[336] Subsequent studies have shown that some cytokines, such as tumor necrosis factor, which stimulates fibroblast proliferation, induce a rapid rise of surface transferrin receptors.[337]

Interestingly, the increase of surface transferrin receptors following growth factor stimulation does not represent a generalized phenomenon because other membrane receptors and epidermal growth factor receptors do not change following proliferation stimulus.[338]

Analysis of DNA content of transferrin receptors in activated lymphocytes revealed that cells in the G1, G2, S, and M phases of the cell cycle express transferrin receptor,[325] but cells in the active phases of the cycle express the highest receptor density. Thus, in HL-60 cells, TfRs show increasing expression from the late G1 phase to the onset of the S phase of the cell cycle, indicating a close linkage with the initiation of DNA synthesis.[339] However, the level of transferrin receptor mRNA seems to be the same in HL-60 cells corresponding to different phases of the cell cycle.[340]

This conclusion has been supported by modulation studies of transferrin receptor expression by agents that inhibit DNA synthesis.[341] However, during mitosis, there is a marked reduction in the number of surface transferrin receptors.[342] This finding has been explained by inhibition of recycling of receptors back to the cell surface. In telophase, recycling resumes before externalization, resulting in a temporary excess of transferrin receptors at the cell surface.[342] Finally, transferrin receptor gene expression during the cell cycle differs from expression of c-*myc*, a gene closely linked to cellular proliferation. Expression of c-*myc* appears to be highest in the early G1 phase, whereas transferrin receptor gene expression is very low in the early G1 phase and high in the late G1 phase.[343]

The linkage between cell proliferation and transferrin receptor expression is supported by studies of the effects of cell growth inhibitors. Interferon α and interferon β concurrently inhibited both cell growth and the rise in transferrin receptor expression observed when cells were exposed to culture conditions that stimulated proliferation.[344,345] Evidence suggested that the interferon-induced transferrin receptor inhibition is not a consequence of cell growth arrest, but rather is one of the causes of the anti-proliferative effects of interferon.[345]

The correlation between transferrin receptor expression and cell proliferation is confirmed by studies on differentiating cells. These studies indicate that induction of differentiation of leukemic cell lines with chemical inducers is associated with a marked decline of transferrin receptor expression.[346]

In other cell types, such as in adipocytes differentiating from fibroblasts, a redistribution of transferrin receptors from intracellular sites to cell membrane was observed, thus leading to an increase in surface expression of the receptor.[348] More recent studies have shown that in HL-60 cells induced to granulocytic maturation, negativity for transferrin receptors correlates with commitment to cell differentiation, while the cells remaining positive for transferrin receptor tend to proliferate.[347]

Similarly, in some tissues, transferrin receptor expression was observed only during embryonic or fetal life, when these tissues were actively proliferating. Thus, transferrin receptors appear during nephron induction, are expressed during the early stages of renal tubulogenesis, and are lost with progressive maturation.[349]

These observations suggest that the expression of transferrin receptors is dependent on the rate of cellular proliferation and that these receptors may represent a good marker for cells proliferating at a high rate. Accordingly, it was hypothesized that events preceding cell division provide the regulatory stimulus for the synthesis and subsequent appearance of the transferrin receptor on the cell surface. The presence of these receptors seems necessary to sustain cellular proliferation because it allows the high rate of iron uptake required for cell multiplication.[350]

Transferrin and cell surface receptors are required for cell proliferation, as shown by *in vitro* experiments using antibodies that block the binding of transferrin by its receptor.[351] As a result of inhibiting transferrin binding, the antibody had a profound effect on iron uptake and inhibited cell growth. The inhibitory effect of these blocking anti-transferrin receptor monoclonal antibodies may be alleviated by iron chelates.[352]

Cells grown in the presence of this type of monoclonal antibody accumulated in the S phase. This suggests that the major effect of iron deprivation is on metabolic steps associated with DNA synthesis.

One candidate may be the activity of the enzyme ribonucleotide reductase, which plays a key regulatory role in DNA synthesis and requires iron for the catalytic reduction of ribonucleotides. This enzyme contains a unique binuclear ferric iron center that stabilizes a tyrosyl-free radical essential for enzyme activity. Different types of ribonucleotide reductase enzymes are present within the cells, some of which are involved in DNA synthesis and are required when the cells proliferate. Others, such as p53 ribonucleotide reductase 2, are specifically involved in the p53 checkpoint for repair of damaged DNA.

Treatment of the cells with iron chelator leads to removal of the tyrosyl radical and to the consequent inactivation of ribonucleotide reductase, inhibition of DNA synthesis, and cell proliferation.[353] However, some studies have shown that not all cells are arrested in the S phase after exposure to antibodies inhibiting transferrin binding. Different cell types treated with picolinic acid, a component capable of inhibiting cellular incorporation of iron,[354] are arrested at the S phase[355] and at other phases of the cell cycle.[356]

Several studies have confirmed this observation. The addition of deferoxamine to normal human T lymphocytes,[357] neuroblastoma,[358] and bladder cancer[359] cell lines arrested cells at a point in the G1 phase, earlier than the G1/S boundary. This finding implies that the need for iron in the control of cell proliferation is not restricted to events occurring in the S phase of the cell cycle (ribonucleotide reductase activity), but also concerns events occurring during the G1 phase. Studies carried out in T lymphocytes[360] have shown that deferoxamine induces a series of events: (1) p34 cdc2, which normally appears just before the S phase entry, is markedly inhibited; (2) the retinoblastoma protein (Rb 110), which plays a key role in G1 phase progression, is only partially phosphorylated; and (3) marked inhibition of the synthesis of the cyclin A-associated component of p33 cdk2 kinase, associated with a moderate decrease of cyclin E Cdk2 kinase activity occurs.

TfRs are also expressed on several types of nondividing cells, such as reticulocytes, trophoblasts, hepatocytes, and tissue macrophages; this suggests that the relationship between TfR expression and cell activation is not a generalized phenomenon.

Other agents capable of inhibiting the binding of transferrin to its receptors, such as the acute-phase protein, α-1-anti-trypsin, lead to inhibition of cell proliferation.[361]

5.10.2 IRON-DEPENDENT REGULATION OF TRANSFERRIN RECEPTORS

Precise modulation of cellular iron is required to produce the synthesis of iron proteins, such as ribonucleotide reductase and cytochromes, and to prevent damage from free radicals induced by iron–dioxygen formation. Cellular iron metabolism is self-regulated through iron-dependent changes in the abundance of transferrin receptors, which control iron uptake, and ferritin, which sequesters iron within the cell.

Several studies have provided evidence that expression of transferrin receptors is modulated through a negative feedback mechanism dependent on the intracellular levels of iron and heme. The transferrin-binding capacity of human leukemic cell lines (K-562, HL-60, U-937, and HEL) grown in the presence of hemin is greatly reduced.[362,363] The addition of hemin to the culture medium did not modify the growth rate. Thus, the inhibitory effect of hemin on transferrin receptor expression cannot be related to inhibition of cell proliferation.

The down-modulation of transferrin receptors induced by hemin was accompanied by an opposite effect on ferritin whose synthesis was enhanced.[362,363] The heme-dependent down-modulation of transferrin receptor expression is observed when the cells are incubated either with free hemin or hemopexin. Hemopexin is a serum glycoprotein that binds heme with a high affinity ($K_d < 1$ pM) and transports heme into circulation, thus serving as an important link between heme and iron metabolism.

Heme derived from hemolysis and damaged tissues is bound by hemopexin and transported to hepatic cells that possess high-affinity receptors for this protein. Taketani et al.[364] showed that the uptake of heme by the cells was mediated by the heme–hemopexin complex system, via hemopexin receptors, and that treatment of cells with hemopexin leads to a decrease in iron uptake concomitant with down-regulation of transferrin receptor expression, the induction of heme oxygenase, and the degradation of heme.

However, subsequent studies have shown that the effect of heme on transferrin receptor expression can be related to a large extent to the capacity of heme to donate iron to the chelatable iron pool through the action of heme oxygenase.[365] In fact, the effect of heme on both transferrin receptors and ferritin was inhibited by addition of an iron chelator. This implies that heme taken up by the cells must be degraded through the action of heme oxygenase; free iron thus generated down-modulates transferrin receptors.

Heme oxygenase catabolizes cellular heme to biliverdin, carbon monoxide, and free iron. Two different isoforms of heme oxygenase, heme oxygenase 1 and heme oxygenase 2, encoded by separate genes, are present in mammalian cells. Heme oxygenase 1 plays a role in antioxidant defense mechanisms, and participates in the control of iron homeostasis.

A recent study showed that mice with targeted heme oxygenase null mutations, developed serum iron deficiency associated with pathological tissue iron-loading. Iron accumulation was particularly evident at the renal cortical tubules, hepatocytes, and Kupffer cells.[366] This plasmatic iron deficiency associated with tissue iron loading resembles the situation in patients with anemia associated with chronic inflammation.

These observations indicate that heme oxygenase 1 is crucial for the expulsion of iron from tissue stores. Furthermore, studies carried out in fibroblasts derived from animals genetically deficient in heme oxygenase 1 have shown that these cells tend to accumulate iron and are particularly sensitive to stress conditions, such as serum deprivation.[367]

Conversely, overexpression of heme oxygenase 1 in normal fibroblasts reduced their intracellular iron content and exerted a protective effect against cell death induced by serum deprivation.[367] On the other hand, mice lacking heme oxygenase 2 showed no disturbances in iron metabolism, in line with recent studies showing that the function of this enzyme is related to the regulation of intracellular heme concentration, and not to heme degradation.[368]

Similarly, HeLa cells,[369] human fibroblasts,[370] human leukemic cell lines,[371,372] mitogen-activated T lymphocytes,[373] T-acute lymphocyte leukemic cells,[374] and hepatocytes[375] grown in the presence of iron salts showed concentration- and time-dependent decreases of their transferrin binding capacities. The decrease reflected a reduction in receptor numbers elicited by iron load and was associated with enhanced intracellular ferritin content.[372,376]

The study of the rate of transferrin receptor biosynthesis in cells grown in the presence of iron salts or hemin showed that both these compounds elicited marked reductions of the synthesis rate of the transferrin receptors, without modification of their half-lives.[372] Conversely, incubation of the cells with iron chelators caused an increase in the number of transferrin receptors, which is dependent on enhancement of the rate of receptor synthesis.[377-381] Only chelators that reduced the intracellular content of iron, and hence of ferritin (e.g., deferoxamine) enhance the rate of biosynthesis of transferrin receptors.[380]

The regulation of transferrin receptor expression observed after addition of agents that modify intracellular iron concentration was attributed to a cytoplasmic pool of chelatable iron. This poorly defined intracellular iron pool serves as a transit pool where iron is delivered after it is sorted out of endosomes.[382]

It was suggested that this pool is composed of low-molecular weight ligands such as amino acids, ascorbate, sugars, riboflavin, ATP, and other nucleotides that bind iron.[383] However, studies using a calcein metal-sensitive probe and extraction of ^{55}Fe loaded-cells with benzyl alcohol showed that a significant proportion of *in situ* cell chelatable iron is represented by ferrous iron. The cytoplasmic concentration of Fe^{2+} fluctuated in K-562 cells between 0.3 and 0.5 μM.[383]

The intracellular iron pool is depleted following treatment of the cells with an iron chelator; after depletion of the iron pool, iron can be mobilized into the intracellular iron pool from intracellular sources and slow release from ferritin via proteolysis.[384] Although the definition of the transit iron pool remains vague, it is conceivable that it contains a form of iron that is sensed by iron-responsive proteins. Interestingly, the addition of protoporphyrin IX into the culture medium elicited a marked up-modulation of surface transferrin receptors associated with a pronounced inhibition of ferritin synthesis.[372,385] Both these phenomena are seemingly related to intracellular iron chelation induced by protoporphyrin IX, as suggested by the activation of iron regulatory protein.[386]

The availability of cDNA probes encoding for human transferrin receptor facilitated the investigation of the molecular mechanisms underlying the regulation of transferrin receptor gene by iron. Northern blot experiments showed that treatment with an iron chelator induces a four- to six-fold increase of transferrin receptor RNA levels. Conversely, a reduction of 50 to 75% is observed after iron load.[379,387] The effect of iron chelators is rapid. Transferrin receptor mRNA is enhanced as early as 30 minutes after addition of an iron chelator, and up to six-fold over control values at 20 hours.[388]

Additional experiments showed that the simultaneous addition of a transcriptional inhibitor, such as actinomycin D, and an iron chelator blocked the rise of transferrin receptor RNA level and biosynthesis,[389] which the chelator alone induced. Furthermore, cells grown under standard conditions or in the presence of either a chelator or iron salts showed similar rates of transferrin receptor gene transcription.[388] On the basis of these findings, it was suggested that:

1. Iron modulates the expression of transferrin receptor gene at the post-transcriptional level.
2. At least one step in the regulatory process is mediated by a transcriptional event, as suggested by experiments with transcription inhibitors.
3. The regulatory mechanism does not operate at the transcriptional level.

Subsequent studies, fully detailed in Chapter 8 on iron regulatory protein, clarified the mechanisms underlying these observations. These studies are briefly summarized as follows. During iron deprivation, the iron regulatory protein stabilizes transferrin receptor mRNA by binding to stem–loop structures in its 3′-untranslated region; in the presence of iron loading, the iron regulatory protein is inactive and does not bind to transferrin receptor mRNA. Rapid degradation of transferrin receptor mRNA follows.

The transcriptional requirement for the modulation of transferrin receptor expression mediated by iron is dependent upon the transcription of an unstable protein that is an important component of the transferrin receptor mRNA decay machinery.[390]

Recent studies indicate that oxygen tension may represent an important mechanism of modulation of transferrin receptor expression. These studies suggest a relationship between transferrin receptor modulation by iron and by oxygen. Initial studies of human hepatoma and erythroleukemia cell lines have shown that a hypoxic stimulus induced an increase of transferrin receptor expression, a phenomenon related to an increase in the level of mRNA encoding this receptor.[391]

Subsequent studies have somewhat clarified the molecular mechanisms responsible for this phenomenon, and indicate that both post-transcriptional and transcriptional mechanisms are involved. Post-transcriptional mechanisms are mediated by activation of iron regulatory protein, thus leading to a stabilization of transferrin receptor mRNA. Transcriptional mechanisms are mediated through induction of hypoxia-inducible factor 1 (HIF-1), which stimulates the transcriptional rate of the transferrin receptor gene.[392,393]

The activity of hypoxia-regulated genes is mediated through transcriptional activation by HIF-1, which binds to particular sites present in the promoters of responsive genes. Examples of hypoxia-regulated genes are erythropoietin, vascular endothelial growth factor, glucose transporter 1, and several glycolytic enzymes.

All these genes are activated by hypoxia, iron chelators, and cobalt ions. The molecular mechanism of HIF triggered by hypoxia is related to stabilization of HIF-1-α which is, under normoxic conditions, rapidly degraded by the ubiquitin–proteosome system.[394]

A similar mechanism seems to mediate HIF-1 activation by iron chelation. In normal iron conditions, HIF-1-α forms a complex with the Von Hippel–Lindau (VHL) tumor suppressor gene product, and this complex is rapidly degraded. After iron chelation, HIF-1-α dissociates from pVHL and becomes resistant to proteosome-mediated degradation.[395]

Experiments on cells exposed to an iron chelator showed a significant increase in the rate of transferrin receptor gene transcription.[396] Following treatment with an iron chelator, a significant enhancement of HIF-1 binding to the hypoxia-sensitive site of the transferrin receptor gene occurred.[396] Furthermore, experiments in HIF-deficient cells showed that the induction of transferrin receptor mRNA elicited by iron chelators in these cells was significantly lower than the induction observed in control cells.[396] These observations provide clear evidence that HIF is involved in iron-dependent modulation of transferrin receptors.

5.10.3 Gene Elements Required for Receptor Expression and Regulation

In order to identify the gene elements required for receptor expression and regulation, transfection experiments were performed with different constructs of the transferrin receptor gene.[397,398]

A full-length human cDNA clone (designated pcD-TR1 and comprised of 96 bp upstream from the initiation codon, 2280 bp of coding sequences, and a 2.6-Kb 3' non-coding region) containing either the natural transferrin receptor promoter (a TATA box is located –27 bp from the cap site) or the SV-40 promoter was transfected to LtK⁻ cells.

These studies made important contributions to the understanding of the molecular mechanisms involved in the control of transferrin receptor expression:[398]

1. Expression of intact full-length transferrin receptor cDNA is regulated by iron when transcription is controlled by the SV-40 promoter or by the transferrin receptor promoter with as few as 14 bp upstream from the TATA box.
2. Deletion of a 2.3-Kb fragment within the 2.6-Kb 3' non-coding region of the cDNA abolishes iron-dependent regulation and increases the constitutive level of receptor expression.
3. The 3' deletion does not affect the decrease in receptors observed in response to growth arrest or the increase in receptor expression following stimulation of cell proliferation.

These findings indicate that the transferrin receptor gene is controlled by at least two distinct mechanisms:

1. Iron-dependent regulation requires sequences contained in the 3' region and operates mainly through post-transcriptional mechanisms.
2. Proliferation-dependent regulation of the transferrin receptor requires 5' sequences and operates presumably through transcriptional mechanisms.

Other studies focused on the role of the 5' flanking region in the control of transferrin receptor gene expression. The analysis of 115 bp of the 5' flanking sequence showed that this region contains a TATA box and several GC-rich regions (GC box containing the sequence GGGGCGG)[399] (Figure 5.8).

This DNA segment exhibits homology to the promoter regions of other genes such as dihydrofolate reductase and interleukin-3.[400] These regions have sequences

TRANSFERRIN RECEPTOR PROMOTER

5'   HIF-1   Ets   AP-1   Sp1   3'
gggcgatctgtcagagagcacctcgcgagcg acgtgcc tca gaaag   cagcccccctgggggccg gggcggcg gccaggc
-120

FIGURE 5.8 DNA binding sites present in the proximal region of the promoter of the human transferrin receptor gene.

similar to the consensus sequence identified as the binding site for transcription factor Sp 1. However, these sequences do not seem to be involved in the control of transferrin receptor gene expression because removal within the receptor promoter of three potential binding sites for the transcription factor Sp 1 did not decrease the promoter activity.[401]

Gel shift experiments showed that several high molecular weight proteins (88, 95, 105, and 120 kDa, respectively) can interact with the transferrin receptor gene promoter.[399,400] Each of these proteins is expressed at a higher level in proliferating rather than in quiescent Swiss/3T3 fibroblasts.

Particularly interesting is the kinetics of the binding activity of an 88-kDa protein following stimulation of cell proliferation. The binding activity was virtually absent in the nuclei of quiescent 3T3 fibroblasts and was rapidly stimulated following serum addition. This increase preceded a rise in transferrin receptor mRNA levels in the cytoplasm.[399]

The observation of a conserved sequence homology in the promoters of the transferrin receptor gene and several other mitogen-responsive genes, along with the demonstration that these conserved sequences serve as sites of interaction with regulatory proteins in the transferrin receptor promoter, suggest a mechanism for coordinate gene expression during cell proliferation.

Subsequent studies have shown that the 5' region flanking the transferrin receptor promoter plays a key role in the control of the transcriptional activity of this gene. Casey et al.[401] cloned fragments of human genomic DNA corresponding to the promoter region of the gene of transferrin receptor upstream of the bacterial gene for chloramphenicol acetyltransferase, and used these constructs to assess promoter activity following transfection into a recipient human cell line. Progressive 5' deletions and internal linker-substitution constructs support a critical role in gene expression of a sequence element approximately 70 bp upstream of the mRNA start site.[401] The receptor gene in this region was found to contain 11 bp that are identical to segments of the enhancers of polyoma virus and adenovirus.

Other studies determined that the promoter of the transferrin receptor gene contains a sequence similar to the cAMP- and phorbol ester-responsive elements.[402] Subsequent studies more clearly defined a sequence critical for the control of transferrin receptor promoter activity.

The transferrin receptor promoter contains the sequence 5'-GTGACGCAC (located from −74 to −66 bp upstream of the mRNA start site), which is closely related to, but not identical to, the CREB/ATF binding site central motif GTGACGTCA and that of AP-1 GTGACTCA[403] (Figure 5.8).

Oligonucleotides containing this sequence form a complex with two nuclear proteins with sizes of 55 and 47 kDa, respectively. This protein–DNA complex was efficiently reduced by competition with oligonucleotides containing the binding sequences of AP-4 or Sp 1 proteins.

This finding strongly suggests that the protein able to interact with this regulatory sequence of the transferrin receptor promoter is part of the family of proteins that recognize AP-1- or CRE-related sequences, and includes the *jun*, *fos*, CREB, and ATF proteins. All these findings suggest that:

1. The transcription rate of transferrin receptor gene is controlled by a transcriptional control element located about 75 bp upstream of the mRNA start site and interacts with a CRE/AP-1 protein.
2. The interaction between this regulatory protein and the TfR transcriptional element is seemingly involved in the activation of TfR transcription observed when quiescent G0-arrested cells are activated to proliferate. Consistent with this hypothesis is the observation that AP-1 activity is stimulated in cells triggered to proliferate.[404]

More recent studies have in part clarified these issues. Ouyang et al. defined the functional role of this CREB/AP1 site (also called the TRB box) and showed that it acts in cooperation with an Sp 1 site (also called the TRA box) located downstream 48 to −38 bp upstream of the mRNA start site.[405] A CAT assay using a minipromoter of the transferrin receptor gene that contains the first 78 nucleotides of the promoter region showed that:

1. This promoter region is sufficient to support growth factor-dependent transcriptional activity.
2. Deletion of the TRB box (CREB/AP 1 site) from this promoter region resulted in a marked decline of the transcriptional activity in response to a growth factor stimulus.
3. Full responsiveness of the promoter also requires the TRA box.[405]

In a subsequent study the same authors characterized the nuclear proteins interacting with TRA and TRB boxes:

1. Box A, which is composed of an unusual GC-rich sequence, forms several DNA–protein complexes progressively induced during stimulation of cell growth. One of these complexes contains a nuclear protein supershifted by anti-Sp 1 antibodies.
2. Box B is composed of two overlapping sequences, one corresponding to the previously identified CREB/AP-1 site and the other to a GC-rich region, and c-jun/c-fos at the AP-1 region.[406] Inhibition of Sp 1 binding at the GC-rich region enhanced the formation of the AP-1 complexes, while inhibition of c-*jun*/c-*fos* at the level of AP-1 region caused a marked increase of Sp 1 binding at the GC region of the TRB box.

Other studies have shown that other nuclear proteins may interact with the TRA box of the transferrin receptor promoter. Gel shift assays using nuclear extracts from proliferating HeLa cells in the absence of the copolymer poly (dI-dC), which is usually added in this type of assay, revealed the binding of an apparently unique protein complex called TRAC.[407]

Recent studies suggest that one component of the TRAC complex may be the PIF (parvovirus initiation factor) transcription factor composed of two subunits, p95 and p79.[408] The possible function of this factor and its biochemical structure remain to be determined. The TRB box specifically binds to the transcription factor ATF-1 in gel shift assays using cell extracts prepared from mouse melanoma cells, but does not bind to other members of the CREB/ATF family of transcription factors.[409]

Other experiments have provided evidence that phosphorylation events are required for the stimulation of transferrin receptor transcription during induction of cell proliferation. An initial study showed that the transferrin receptor promoter is responsive to treatment with sodium orthovanadate, an inhibitor of tyrosine phosphatases.[405] This observation suggests that a tyrosine phosphorylation event is implicated in the transcriptional activation of the transferrin receptor. This conclusion was supported by a subsequent study showing that wortmannin, a specific and potent inhibitor of phosphatidylinositol 3-kinase, blocks the mitogenic activation of the transferrin receptor gene.[409]

Transferrin receptor expression usually decreases during cell differentiation; however, a notable exception is the erythroid cell in which receptor expression increases during cell differentiation.

The molecular basis for these two patterns of receptor expression observed during cell differentiation has been investigated. The marked decline of transferrin receptor expression coupled with induction of cell differentiation is mainly related to a transcriptional mechanism.

The reduced transcription of transferrin receptor occurring during cell differentiation was investigated in a model of myeloblastic HL-60 cells induced to monocytic differentiation by phorbol ester addition.[410] Analysis of the protein complexes binding to the TRB box of the transferrin receptor promoter showed that phorbol treatment markedly induced a fast-migrating band that interacts with the AP-1 component of this box. It was suggested that the binding of this complex to the TRB box is responsible for the inhibition of receptor transcription.

Taking into account the studies of proliferating cells and HL-60 differentiating cells, it was suggested that the TRB box is able to mediate both activation and inhibition of transcription of the transferrin receptor gene, depending on the proliferation and differentiation status.[410] This dual capacity mediated by the same regulatory element may imply either the binding of different transcription factors or the binding of the same transcription factors present in different molecular forms related to post-translational modifications.

Transferrin receptor is highly expressed in erythroid cells where elevated levels of the receptor are required to allow a high rate of iron uptake to sustain hemoglobin synthesis. The hyperexpression of transferrin receptors observed in the erythroid lineage is mainly related to the high rate of transcription. The molecular basis of this phenomenon remains unknown.

A recent study provided some important findings for the understanding of the high rate of transcription of transferrin receptor in erythroid cells.[411] Sieweke et al. showed that *Ets-1* potentiated the transcription of transferrin receptor.[411] Mutation of the *Ets* binding site, present in the transferrin receptor promoter just upstream of the AP-1 site, markedly decreases the promoter activity observed in erythroid cells.[411]

Interestingly, in other hemopoietic cell lineages (i.e., myelomonocytic cells), the binding and transactivation of the transferrin receptor promoter is prevented by MafB, an AP-1 like protein specifically expressed in myelomonocytic cells, and protein with *Ets-1* inhibiting transactivation by *Ets-1*.

A possible role for *Ets-1* in the control of transferrin receptor expression was also confirmed in Friend erythroleukemia cells induced to differentiate by chemical inducers.[412] A minimal promoter region composed of the first 120 base pairs of the promoter from the transcription starting codon was necessary to produce transcriptional regulation in erythroleukemia cells induced to differentiate by dimethyl sulfoxide.

This region comprises partially overlapping consensus recognition sequences for both AP-1 and CREB/ATF transcription factors and for proteins of the *Ets* family and a GC-rich region downstream of the *Ets* binding site and capable of binding Sp 1 (Figure 5.8). Mutation of either the *Ets-1* or Sp1 binding site alone or in combination elicited a marked decrease of the promoter response to dimethyl sulfoxide.[412]

According to these observations, it was concluded that transcriptional regulation of transferrin receptors in differentiating erythroleukemia cells is dependent on cooperation between *Ets-1* and Sp 1-binding sites.

Recent studies have provided evidence that hypoxia acts as a modulator of transferrin receptor expression, as noted above. Since the biological effects of hypoxia are mediated through the activation of the HIF transcription factor, it was logical to check for the existence of a hypoxia-responsive element at the level of the transferrin receptor promoter. The inspection of the transferrin receptor promoter sequence showed at position −89 to −81 the putative hypoxia-responsive element sequence TACGTGCC.[392] This promoter element binds the HIF in extracts derived from cells exposed to hypoxia. Mutation of the CGTG core abolishes the hypoxia-mediated transcriptional activation of the transferrin receptor.

5.11 EXPRESSION AND REGULATION IN DIFFERENT CELL TYPES

This section will discuss the different cell types that utilize iron to sustain their proliferation (activated lymphocytes), those involved in iron storage (monocytes–macrophages), and those requiring iron to synthesize large amounts of structural and functional proteins (erythroid cells).

5.11.1 EXPRESSION IN ACTIVATED T LYMPHOCYTES

Interaction of antigen in the proper histocompatibility context with the T lymphocyte antigen receptor leads to an orderly series of events resulting in morphologic change, proliferation, and the acquisition of immunologic function. Two signals are required to initiate this process, one supplied by the antigen receptor and the other by accessory cells or agents that activate protein kinase C.

The mechanism of T lymphocyte activation and mitogenesis may be outlined as follows. Quiescent T lymphocytes are activated by antigen or mitogen (i.e., a mitogenic lectin such as phytohemagglutinin [PHA]) in the presence of accessory cells (i.e., monocytes), thus entering into and progressing through the cell cycle to mitosis. Some mechanisms underlying the activation of T cells have been delineated. Interaction of the mitogenic lectin with the cell membrane or the antigen with the T cell receptor induces a Ca^{2+} influx,[413-415] thus activating protein kinase C. This process plays a key role in the early programming of T cell activation.[416-418]

One of the first events appears to be phospholipase C catalyzed hydrolysis of the membrane component phosphatidyl inositol 4,5-biophosphate (PIP_2) and other phosphoinositides. The hydrolysis of PIP_2 yields inositol 1,4,5-triphosphate (PIP_3) and diacylglycerol (DAG), both of which play important roles as second messengers in the activation process.

The cascade of events initiated by the Ca^{2+} influx, protein kinase C activation, and other signals elicited during T cell activation ultimately leads to transcriptional activity in a variety of genes required for proliferation, differentiation, and effector activity of the cell.

Some studies suggest that transferrin receptor may be involved also in T cell activation by antigens. Studies of T cell lines have shown that transferrin receptors may mediate T cell activation independently of their role as iron transporters.[419-421] Thus, stimulation of T lymphoblastoid cells with iron transferrin or anti-transferrin receptor monoclonal antibody, but not with free iron, elicited the activation of protein kinase C[422] and the release of cytokines[423] and cell cytokines.[424]

In the Jurkat T cell line, the addition of anti-transferrin receptor monoclonal antibodies induced cell proliferation through induction of IL-2 release.[424] Subsequent studies provided evidence of a possible connection between the transferrin receptor and the T cell receptor (TCR). The TCR of the T cell is composed of a disulfide-linked TCR-$\alpha\beta$ heterodimer that recognizes the antigen as a peptide presented by the MHC complex and is non-covalently associated with the CD3 complex.

The complex is involved in intracellular signal transduction and is composed of paired CD3γε-chains and CD3γε-complex, associated to disulfide-linked dimers composed of ζ-chain homodimers. Following TCR activation, the CD3ζ chain becomes tyrosine-phosphorylated and binds the tyrosine kinase ZAP70.

Studies performed on Jurkat cells have shown that incubation with an anti-transferrin receptor monoclonal antibody leads to a stimulation of tyrosine kinase activity, while stimulation of the TCR leads to tyrosine phosphorylation of the transferrin receptor.[422] More importantly, in both Jurkat cells and lectin-stimulated T cells, the transferrin receptor is associated with the TCR ζ-chain.[422] The TCR is internalized following interaction with specific antigen and, interestingly, the internalized TCR complex is found within the early endosome associated with the transferrin receptor.[423]

However, it must be noted that potential cooperation of the transferrin receptor and TCR may operate only in activated T lymphocytes and not in quiescent lymphocytes since the activated but not the quiescent ones possess on their surfaces the transferrin receptor in addition to the TCR.

The initial events of T lymphocyte activation include a very early transcription of genes coding for RNA molecules involved in protein synthesis such that newly transcribed mRNA molecules can be efficiently translated into proteins. As a consequence, T lymphocytes secrete a 15,500 molecular weight glycoprotein, interleukin-2 (IL-2).[424] They also express specific high-affinity membrane receptors for this lymphokine[425-428] that are not detected in resting T cells.[429] Exogenous IL-2 produced through an autocrine pathway is clearly required for optimal expression of these receptors.[417]

Following stimulation with PHA, IL-2 receptors (IL-2Rs) are first detected at 4 to 8 hours, peak at day 2 or day 3, and then decline to 5 to 20% of maximum levels at days 7 through 12.[424] After IL-2R expression, TfRs are induced and T lymphocytes enter the S phase.[430-433] If IL-2 or the TfRs are blocked, DNA synthesis and cell growth are inhibited.[432]

Before the discovery that TfRs are induced during T cell activation, several studies established that iron is essential to sustain the proliferation of T lymphocytes. Early observations showed that a protein factor from serum is required for optimal DNA synthesis in response to mitogenic lectins.[434] Later studies clearly demonstrated that transferrin is the main serum protein component required for *in vitro* activation of T lymphocytes by mitogen.[435,436]

In T cell proliferation, the major function of transferrin is to supply iron. Indeed, the effect of holo-transferrin is completely eliminated by an iron chelator.[437] Similarly, gallium, which binds to transferrin, inhibits mitogen-induced proliferative responses of T lymphocytes.[438] This conclusion is confirmed by studies showing that iron-saturated transferrin may be replaced by iron salts to sustain T lymphocyte proliferation *in vitro*.

Although many studies have reported that iron donors such as chelates either cannot support proliferation or are less effective than transferrin,[436,439,440] others have clearly shown proliferation in the absence of transferrin.[433,441,442] These differences may be related to the different iron salts used in these studies and also to the fact that CD4+ T lymphocytes can themselves synthesize transferrin.[443]

The capacity of CD4+ T lymphocytes to synthesize transferrin is stimulated by mitogenic lectins, (thus proliferating CD4+ T lymphocytes synthesize more transferrin than quiescent CD4+ T lymphocytes)[444] and by some cytokines.[445] Furthermore, the fact that the major function of transferrin in T cell proliferation is to supply iron is proven directly by experiments showing that blockade of transferrin-mediated iron uptake inhibits lymphocyte proliferation. Anti-transferrin receptor monoclonal antibodies that block transferrin binding and iron uptake in activated lymphocytes inhibit proliferation, whereas antibodies that bind to the transferrin receptor, but do not block transferrin binding, have no effect.[440,446-448]

The absolute need of a transferrin-mediated iron supply to sustain T lymphocyte proliferation, as well as the proliferation of virtually all cell types, deserves comment. Iron is required for cell growth in view of its well-known role in enzyme ribonucleotide reductase activity. Ribonucleotide reductase is the only enzyme controlling the conversion of ribonucleotides to their deoxy derivatives.[449] It has a very rapid turnover and needs a continuous supply of iron to remain in its active state.[449] In this context, studies on lymphoid leukemic cell lines have shown that optimal iron uptake by these cells is required to sustain high ribonucleotide reductase activity.

Recent studies suggest that iron is required for T lymphocyte cell proliferation, not only for sustaining ribonuclease reductase activity (i.e., for DNA synthesis), but also for biochemical events required in the G1/S transition. Studies of *in vitro* growth of normal human T lymphocytes under serum-free conditions showed that iron is required for the synthesis of p34 Cdc2.[450] Furthermore, iron chelators blocked the growth of activated T lymphocytes before S phase entry as a consequence of marked inhibition of Cdc2 synthesis, moderate decreases of cyclin A and Cdk2, and reduced phosphorylation of p110 Rb.[451] These studies thus indicate that iron is required for the synthesis of some cyclins and cyclin kinases that play key roles in the control of cell cycle progression.

An additional hypothesis was proposed to explain the role of transferrin in cell proliferation control, based on the observation that modulation of the ability of the cell to transfer reducing equivalents from cytosol to extracellular electron acceptors controls cell growth.[452] Based on this hypothesis, transferrin iron may be one of the physiological electron acceptors. This theory suggests that the modulation of the number of transferrin receptors present on the cell surface regulates its ability to internalize iron from transferrin and also its capacity to donate electrons to transferrin iron.

Ellem and Kay[453] showed that ferricyanide may sustain the growth of human melanoma cells under serum-free conditions. The growth-promoting activity of ferricyanide cannot be ascribed to its iron content because non-iron extracellular oxidants were shown to have similar effects. Thus, in this cell system, the effect of ferricyanide is apparently related to its capacity to function as a sink for electrons furnished through the plasma membrane NADH ferricyanide oxidoreductase.

This hypothesis on the mechanism of the growth-promoting activity of iron is also indirectly supported by the observation that all the typical intracellular signals and responses associated with cell growth, i.e., alkalinization of the cytosol through proton extrusion via the Na^+/H^+ antiport, increase in cytosolic free calcium concentration, changes in intracellular $NAD^+/NADH$ ratio, and activation of immediate early genes such as c-*myc* and c-*fos*, may be induced by stimulation of the NADH ferricyanide oxidoreductase.[454,455] Several growth inhibitors are also inhibitors of the redox system,[456] whereas cell growth promoters stimulate the NADH ferricyanide oxidoreductase.[457,458]

Other studies have shown that thiol-mediated redox regulation plays an important role in T lymphocyte proliferation and, particularly, in transferrin receptor expression. Observations suggest that reduced thiol groups are important for human T lymphocyte proliferation:

1. The capacity of human T lymphocytes to enter the cell cycle correlates with the intracellular GSH content and this capacity is enhanced by β-mercaptoethanol.
2. Inhibitors of GSH synthesis cease lectin- or anti-CD3-induced proliferation of purified human T lymphocytes.
3. L-cysteine, β-mercaptoethanol, and other thiols and disulfides enhance the proliferation of T lymphocytes induced by mitogens.[459]

Growth of phytohemagglutinin (PHA)-stimulated human T lymphocytes in GSH and L-cysteine-deficient medium resulted in a complete inhibition of transferrin receptor expression, while interleukin-2 receptor expression was not affected.[460] The requirement of normal intracellular GSH levels for optimal induction of transferrin receptors is seemingly related to the activity of iron regulatory protein, which requires that thiol-related active sites (iron–sulfur centers) must be maintained in a reduced form for IRP activity.

Other authors suggest that the stimulatory role of iron[461] or hemin[462] in lymphocyte mitogenesis may be related to the generation of hydroxyl free radicals. This hypothesis is based on the observation that a variety of agents known to be scavengers of free radicals inhibited lymphocyte proliferation[463,464] through the mechanisms responsible for the inhibition of lymphocyte proliferation mediated by monoclonal antibody anti-transferrin receptor. This suggests that the growth inhibition caused by these antibodies cannot be attributed simply to iron starvation, but rather to other mechanisms, such as interference with receptor-mediated uptake of other nutrients, growth factors, or inhibition of the generation of intracellular signals transduced by the transferrin receptor.

Under physiological conditions, free heme is not found because this metalloporphyrin is bound with high affinity by hemopexin, a 60-kDa plasma glycoprotein which represents an important link between heme and iron metabolism. Hemopexin acts as a transporter of heme derived from lysis of senescent red blood cells and damaged tissues through specific receptors to hepatic cells. A recent study indicates that cell lines possess high-affinity hemopexin receptors; the addition of heme–hemopexin to these cells stimulates cell proliferation that is not inhibited by the contemporaneous addition of the iron chelator desferrioxamine.[465]

Several studies were performed to establish the kinetics of transferrin receptor gene activation and expression following PHA activation. The expression of transferrin receptor was compared with that of other activation markers. Particularly important are IL-2 and IL-2R in that a relationship exists between the induction of IL-2, IL-2R and the expression of transferrin receptors during lymphocyte activation.[432,466,467] *In vitro* run-off transcription experiments with nuclei isolated from quiescent and activated T lymphocytes showed that neither IL-2, IL-2R, nor transferrin receptor genes are transcribed in uninduced T cells.[441] However, after stimulation with PHA, IL-2 and IL-2R gene expression was rapidly induced, with peak transcriptional activity occurring at 24 and 9 hours, respectively. Induction of transferrin receptor transcription occurred more slowly, with initiation between 15 and 24 hours and peak expression at 48 hours.[468]

The addition of a protein synthesis inhibitor (cycloheximide) did not alter IL-2 and IL-2R gene expression but completely blocked transferrin receptor gene expression. Similar kinetics of transferrin receptor mRNA expression during T lymphocyte mitogenesis was observed in Northern blot experiments.[469] Experiments of ^{125}I-transferrin binding to activated T cells showed that the level of specific binding is virtually absent on resting T lymphocytes, is very low after 24 hours of stimulation, becomes clearly detectable at 48 hours, and usually attains peak level at 72 hours.

It is of some interest that the kinetics of transferrin-binding capacity of PHA-stimulated T lymphocytes parallels the kinetics of ^{3}H-thymidine incorporation, i.e.,

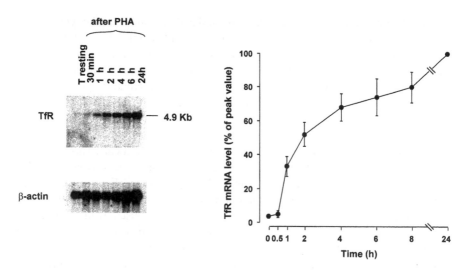

FIGURE 5.9 Kinetics of transferrin receptor mRNA induction in T lymphocytes stimulated with a mitogenic lectin. T lymphocytes were stimulated with phytohemagglutinin; cell aliquots were harvested at different time points; and poly-A⁺ mRNA was purified by affinity chromatography and analyzed by Northern blot.

the level of DNA synthesis. Similar findings were observed in experiments where the level of transferrin receptor synthesis was studied at different times after PHA stimulation.[470]

In mixed lymphocyte cultures, the kinetics of transferrin receptor expression on T cells is significantly retarded when compared with that observed in PHA-stimulated cultures.[471] This phenomenon can be easily explained by the fact that in mixed lymphocyte cultures, the activation of T lymphocytes, as evaluated by ^{3}H-thymidine incorporation, occurs more slowly than in corresponding PHA-stimulated cultures.

The kinetics of transferrin receptor expression at early time points after stimulation with a mitogenic lectin was recently investigated in detail (unpublished observations). Using a sensitive method to quantitate transferrin receptor mRNA by Northern blot on poly-A⁺ preparations, this mRNA species was induced few hours after lectin addition (Figure 5.9).

The kinetics of transferrin receptor mRNA induction parallels IL-2 and IL-2Rα chain mRNA induction, but it is clearly preceded by c-*fos* mRNA induction (Figure 5.10). The activity of iron regulatory protein, the protein that stabilizes transferrin receptor mRNA, is induced only at later times.

These observations clearly suggest that transferrin receptor can be regarded as an early-activated gene during the process of T cell activation. The induction of transferrin receptor during T lymphocyte activation is inhibited both by transcription and phosphorylation inhibitors, thus suggesting that transcriptional and phosphorylation events are required for the initial induction of the transferrin receptor gene during T lymphocyte activation.

The cellular and molecular mechanisms responsible for the induction and optimal expression of transferrin receptors during the mitogenic activation of T lymphocytes

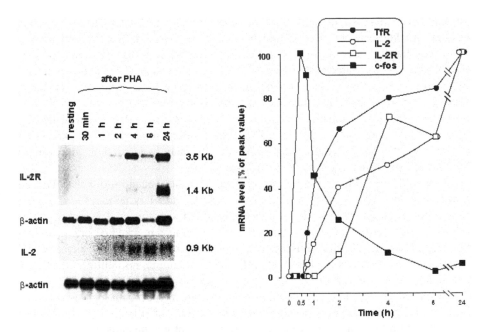

FIGURE 5.10 Comparative kinetics of transferrin receptor, interleukin-2, interleukin-2 receptor, and c-*fos* mRNA in T lymphocytes activated by a mitogenic lectin.

were investigated, leading to the discovery of several pathways involved in the activation and modulation of the expression of this gene in these cells. Initial studies analyzed the relationship between the activation of IL-2 and IL-2R synthesis and transferrin receptor expression. Thus, Neckers and Cosmann[432] showed that:

1. If IL-2Rs are blocked by addition of a specific monoclonal antibody, the appearance of transferrin receptor and induction of DNA synthesis are inhibited.
2. If transferrin receptors are blocked by addition of a specific monoclonal antibody that recognizes the epitope of the receptor involved in transferrin binding, DNA synthesis is inhibited, but IL-2R expression is virtually unmodified.

These results led to the important conclusion that IL-2 and IL-2R are necessary for transferrin receptor induction in activated T lymphocytes.

Other studies evaluated the role of iron in the regulation of transferrin receptor expression in mitogen-activated human T lymphocytes. Pelosi et al.[467] showed that resting human T lymphocytes possess an elevated intracellular ferritin concentration. In contrast, transferrin receptors are undetectable. Addition of an iron chelator to resting T lymphocytes induced a marked decrease of their ferritin content, but not the appearance of transferrin receptors.

Stimulation by mitogenic lectins, such as PHA or concanavalin A (Con A), markedly lowered their ferritin content and induced the synthesis of transferrin receptors.

Addition of iron salts (ferric ammonium citrate) to activated T lymphocyte cultures caused a marked enhancement of both ^{3}H-uridine and ^{3}H-thymidine incorporation.

It also induces a concentration-dependent decrease in transferrin receptor synthesis, which is associated with a marked rise of ferritin production. Hemin treatment exerts the same effects. Addition of an iron chelator is specific for transferrin receptors because the expression of other membrane markers of activated T lymphocytes (e.g., IL-2R, insulin receptor, and HLA-DR antigen) is not modified by treatment with iron salts or chelators. These observations suggest that the expression of TfRs in activated T lymphocytes is specifically modulated by their intracellular iron levels rather than their proliferative rates.

This study used an iron chelator (picolinic acid) at concentrations that induced a moderate decrease of the iron available to the activated T lymphocytes. If a more potent iron chelator, such as deforoxamine, is added to activated T lymphocytes at relatively high doses (300 μM), T-lymphocyte activation is blocked before the induction of IL-2 and IL-2R and, consequently, transferrin receptors are not expressed.[472] However, at lower concentrations (15 μM) of deforoxamine, no inhibition of IL-2R expression was observed in PHA-activated T lymphocytes.[473]

A second study, in which the expression of transferrin receptors was analyzed in T lymphocytes pulsed for 3 hours with a mitogenic lectin (Figure 5.11), showed three phases underlying the process of transferrin receptor gene induction and expression during T-lymphocyte mitogenesis (Figure 5.12):

Phase I: in resting T lymphocytes, the gene encoding TfR is in a transcriptionally inactive configuration.

Phase II: even in the absence of exogenously or endogenously synthesized IL-2, a mitogen pulse is sufficient to initiate the expression of transferrin receptors via transcriptional activation of transferrin receptor gene.

Phase III: transferrin receptor synthesis is then amplified by IL-2 via a decrease of the size of a regulatory intracellular iron pool.

It remains to be proven whether IL-2 exerts its effect on transferrin receptor mRNA levels only indirectly by influencing the rate of transferrin receptor gene transcription. More recent studies have provided further evidence in favor or this model.[474] One study evaluated levels of transferrin receptor expression, ferritin, and iron regulatory protein (IRP) activity. IRP activity, already present at very low levels in quiescent T lymphocytes, shows a gradual and marked rise in PHA-stimulated T lymphocytes from day 1 of culture onward, reaches a peak on day 3 of culture, and then declines at day 4 or day 5 when T lymphocytes again become quiescent.[474]

This increase is directly correlated with the initiation and gradual rise of transferrin receptor expression, which is in turn associated with a decrease of ferritin content. Both the rise of transferrin receptor and high-affinity IRP activity are completely inhibited in iron-supplemented T-cell cultures.

In contrast, low-affinity IRP level progressively declines when T lymphocytes are activated to proliferate by PHA. These results suggest that the amplification of transferrin receptor expression occurring in phase III of the above-mentioned model is mediated via activation of IRP (Figure 5.12).

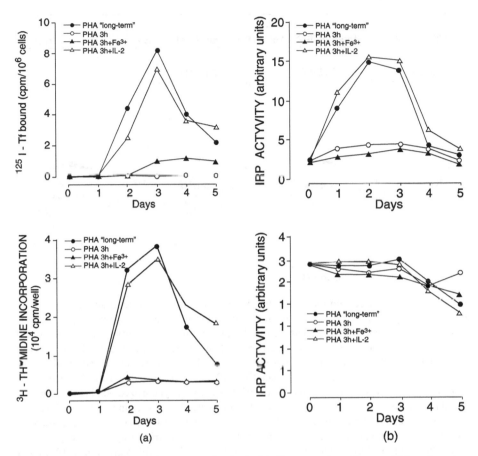

FIGURE 5.11 (a) DNA synthesis and transferrin-binding capacity of human T lymphocytes stimulated for 3 hours in the presence of PHA, washed to remove unbound PHA, and then grown in the absence or in the presence of interleukin-2. (b) Iron regulatory protein activity in T lymphocytes grown as above, measured in the absence of β-mercaptoethanol or in the presence of the reducing agent.

Other studies were performed to identify the intracellular messengers involved in the activation of transferrin receptor expression during T lymphocyte mitogenesis.[475,476] These studies demonstrated that Ca^{2+} ionophores and protein kinase C (PKC) activators are both inducers of transferrin receptor mRNA in quiescent T lymphocytes. However, when these agents are individually added, a low and late expression of transferrin receptor is observed. In contrast, when they are added together, an early and sustained expression of transferrin receptor is observed that is comparable or even higher than that detected in PHA-stimulated T lymphocytes.[476]

On the basis of these findings, it is reasonable to suggest that the rise in intracellular Ca^{2+} concentration elicited directly by the antigen or the mitogen and the stimulation of PKC activity induced by IL-2 following the binding to its receptor may represent the two signals involved in the activation of transferrin receptor expression during T lymphocyte mitogenesis.

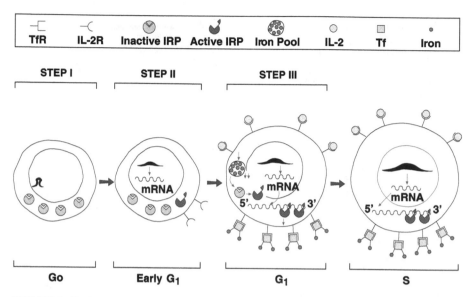

FIGURE 5.12 A three-step model for the expression of transferrin receptors during T lymphocyte mitogenesis. Step I shows resting T lymphocytes; the transferrin receptor gene is transcriptionally inactive and virtually all IRP is present in a low-affinity state, unable to bind to transferrin receptor mRNA. Step II corresponds to the G1 phase of the cell cycle, when transcription of the transferrin receptor and transcription of interleukin-2 and interleukin-2 receptor genes are activated. Concomitantly, IRP is activated and transferrin receptor is induced. At this stage transferrin receptor mRNA is actively synthesized but, because the IRP is mainly in the low-affinity state, it rapidly degrades, thus generating only low levels of transferrin receptor synthesis. Phase III corresponds to a late G1 stage when IL-2 secreted by activated T lymphocytes stimulates the synthesis of transferrin receptors via depletion of a regulatory intracellular iron pool, which in turn gives rise to full activation of IRP. After phase III, T lymphocytes can actively take up iron required to sustain the activity of ribonucleotide reductase and enter the S phase of the cell cycle.

Transferrin receptors are also expressed on leukemic T cells, particularly T-ALL-blasts.[477] Recent studies showed that transferrin receptors expressed on these cells exhibit some relevant abnormalities. Infection of normal T cell clones with HTLV-1 virus causes a marked overexpression of membrane transferrin receptor (as well as of IL-2R), not accompanied by an increase growth rate.[478] Surface transferrin receptor overexpression by these cells is associated with faulty internalization of the receptor and a resultant redistribution of receptor from internal stores to the plasma membrane.[478]

Another study[479] carried out on T-ALL blasts showed that transferrin receptors are constitutively expressed in 100% of the cases of T-ALL leukemia and exhibit some immunological and biochemical abnormalities consisting of:

1. Preferential reactivity with anti-transferrin receptor monoclonal antibodies (mAbs) that recognize ligand-binding domains of transferrin receptor.
2. Reduced molecular weight, when compared with transferrin receptor expressed on normal immature thymocytes or activated T lymphocytes, which is apparently related to a defective receptor glycosylation.

It was suggested that the constitutive high level of transferrin receptors on T-ALL blasts may play a key role in the stepwise progression of this malignancy and, in particular, provide a proliferative advantage to T-ALL blasts vs. normal T lymphocytes.[470] Because both normal and malignant T lymphocytes exhibit a wide range of proliferative responses, it would not be surprising to detect other adaptations that would improve the use of nutrients such as Fe.

An example of this phenomenon is transferrin synthesis by CD4$^+$ helper T lymphocytes, an event that seems to facilitate T lymphocyte proliferation.[480] It was suggested that this adaptation may result because areas of the lymph nodes are poorly vascularized, and these cells cannot obtain enough transferrin from serum.[481] Other reports suggest that some malignant lymphocyte lines synthesize "transferrin-like" factors that allow for continued proliferation *in vitro.*[482,483]

Transferrin receptor expression on T lymphocytes was analyzed in other pathological conditions, including human immunodeficiency virus type 1 (HIV-1) infection and allograft rejection. HIV infection of T lymphocytes causes a down-modulation of transferrin receptor expression and this phenomenon is correlated with the cytopathic effects of the virus.[484]

Graft rejection after transplantation is related to the capacity of T cells to recognize non-self host antigens. During this process, T lymphocytes are activated and induced to express transferrin receptors on their membranes.

Based on these observations, experimental models were used to determine whether anti-transferrin receptor monoclonal antibody may prolong graft survival by inhibiting T cell responses. Administration of anti-transferrin receptor antibodies to animals undergoing allograft transplantations elicited marked inhibition of T cell responses and significant prolongation of allograft survival. The antibody treatment induced a marked inhibition of the production of cytokines by lymphocytes. Particularly relevant is the inhibition of production of interleukin-12, a cytokine required for the development of specific T helper cells.[485]

The expression of TfRs was also investigated in B lymphocytes activated to proliferate. Particularly relevant is a study by Teixeira and Kühn[486] concerning the expression of TfR in mouse spleen cells after stimulation with 4β-phorbol-12-myristate-13-acetate (PMA) and Ca^{2+}-ionophore, ionomycin, and recombinant IL-2.

TfR mRNA was barely detectable in resting cells but markedly increased within a few hours of growth stimulation.[486] In run-on transcription assays, TfR gene was transcribed to almost the same extent in resting and growth-stimulated cell populations.[486] This observation suggests that post-transcriptional control mechanisms in splenocytes are mainly responsible for the accumulation of transferrin receptor mRNA occurring at the onset of cell proliferation.

In addition to the studies performed in T lymphocytes activated to proliferate by the antigen of a mitogenic lectin, other investigations focused on the pattern of expression of transferrin receptor during the early intrathymic stages of T cell differentiation. Phenotypic analysis of the thymic cells showed that only large, immature CD4$^-$/8$^-$/3$^-$ and CD4$^+$/8$^+$/3$^-$ cells expressed the transferrin receptor.[487] In the human fetal thymus, three stages of differentiation were identified:

1. The earliest stage was characterized by the positivity of both interleukin-2 and transferrin receptors, while CD3 and CD6 antigens were not expressed.
2. The second stage was characterized by expression of typical T lymphocyte markers and positivity for transferrin receptor.
3. The third stage was characterized by full positivity for CD3 antigen and loss of the transferrin receptor.[488]

In another set of experiments, the effect of the addition of an anti-transferrin receptor monoclonal antibody to thymus organ culture was investigated. The anti-transferrin receptor antibody caused decreased iron uptake which abrogated the terminal differentiation of thymocytes into T cell receptor α/β positive cells, while the development of T cell receptor γ/δ positive cells was not affected.[489] This observation suggests that T cells expressing the T cell receptor γ/δ may have mechanisms of iron uptake alternative to the transferrin receptor.

Studies performed on B lymphocytes showed that in these cells, as well as in T lymphocytes, transferrin receptor expression is correlated with induction of cell proliferation. Resting B lymphocytes are activated *in vitro* by polyclonal B cell stimulators to enter into the cell cycle and proliferate. A polyclonal B cell activator, *Staphylococcus aureus* Cowan strain I (SAC) cross-links surface-membrane Ig in a way similar to anti-IgM antibody, resulting in cellular activation and proliferation.

Alternatively, B lymphocytes can be stimulated to proliferate by anti-IgM antibodies. Addition of anti-IgM antibodies to human B lymphocytes induces an activation of transferrin receptor which starts sometime after these cells enter the G1 phase of the cell cycle. Similarly *in vivo* injection of anti-IgD antibodies elicited a stimulation of transferrin receptor expression at the splenic B lymphocytes.[490]

Run-on transcription experiments on nuclei isolated from SAC-stimulated B lymphocytes showed that transferrin receptor gene transcription is induced in response to the mitogen with peak expression 30 minutes after stimulation.[491]

The activation of B lymphocytes through B cell antigen receptor involves two transduction pathways: the tyrosine kinase pathway and the phosphatidylinositol pathway. Only the phosphatidylinositol pathway is involved in the induction of transferrin receptors during B lymphocyte activation.[492,493]

In B lymphocytes, MHC class II molecules acquire antigenic peptide in lysosome-related intracellular compartments. Experiments investigating inhibition of early and recycling endosome compartments have shown that:

1. Inhibition of the endosome compartment inhibited transferrin recycling and antigen processing. This observation implies that antigen is processed

by human B lymphocytes involving the same pathway used for transferrin endocytosis and recycling.

2. Endosome ablation did not modify the capacity of B lymphocytes to deliver HLA class II–peptide complexes to the cell surface if these cells have already taken up antigen and assembled intracellular class II–peptide complexes before endosome blockade.[494,495]

These observations indicate that the endocytic machinery is required for antigen delivery to HLA class II, but is not involved for functional class II–peptide complexes to reach the cell surface.[495]

5.11.2 EXPRESSION IN MONOCYTES–MACROPHAGES

It is well-known that macrophages play a key role in the storage of iron. This is related to the destruction of aged erythrocytes and hemoglobin degradation occurring within macrophages, chiefly in liver and spleen. Alterations of iron handling by monocytes–macrophages also may be involved in the pathogenesis of disorders of iron metabolism, such as genetic hemochromatosis. Furthermore, iron metabolism in monocyte–macrophages plays a key role in the inflammatory response. In fact, uptake of iron from the circulation to iron storage compartments, mainly represented by tissue macrophages, is required for the production of reactive oxygen species essential for macrophage cytotoxic activity and also diminishes iron availability to pathogenic bacteria.

Several studies firmly established that TfRs are not expressed on circulating monocytes, and are clearly expressed on macrophages. These conclusions were derived from studies in which monocytes isolated from peripheral blood mononuclear cells by plastic adherence were grown either in Teflon® membranes[496] or in Petri dishes.[497,498] Monocytes grown under these conditions undergo spontaneous *in vitro* maturation to macrophages, and progressively acquire optimal transferrin receptor expression.[496-498]

However, macrophages generated *in vivo* differed in their expression of transferrin receptors, depending on the anatomical site from which they were obtained. Although the majority of alveolar macrophages reacted with anti-transferrin receptor mAbs, pleural and peritoneal macrophages were mainly negative.[496] However, when pleural and peritoneal macrophages are grown *in vitro*, they express transferrin receptors as effectively as alveolar macrophages.[496] Note that during the process of monocyte maturation to macrophages, in parallel with the induction of transferrin receptors, a marked increase of ferritin content was observed.

Because macrophages acquire iron from the breakdown of senescent red blood cells and rapidly recycle it to plasma, it is difficult to understand the physiological significance of this receptor on the membranes of these cells. It can be argued that these receptors are necessary for macrophages to provide iron for their intracellular stores and thus meet the needs of the erythropoietic system.[499-502]

Furthermore, transferrin receptors may be necessary to sustain the proliferation of macrophages.[503] Finally, macrophages are capable of producing oxidants, such as superoxide anion, hydrogen peroxide, and hydroxyl radical, a function that allows macrophages to kill infectious organisms. Production of hydroxyl radicals requires iron[504] and thus the presence of transferrin receptors on the cell surface probably plays an important role in helping these cells to function effectively in host defense.

A recent study[505] suggests that transferrin and lactoferrin may be involved directly in the biochemical mechanisms that lead to the formation of cytotoxic radicals. In fact, apotransferrin and apolactoferrin can accelerate the autoxidation of Fe^{2+} at acid pH by the formation of oxidants with cytotoxic properties. The formation of these oxidants is inhibited by catalase, whereas superoxide dismutase had no effect. On the basis of these findings it was proposed that Fe^{2+} and apotransferrin or apolactoferrin can generate OH^- via H_2O_2 intermediate with marked toxicity to microorganisms, and thereby contribute to the microbicidal actvity of phagocytes.

It was also shown that the levels of intracellular iron are important regulators of macrophage listericidal activity.[506] These studies also demonstrate that:

1. Only macrophage populations expressing significant levels of transferrin receptors are able to both phagocytose and kill *Listeria monocytogenes* bacteria, whereas macrophages exhibiting low levels of transferrin receptors are able to phagocytose but not kill these bacteria.
2. Antibody to transferrin, which prevents binding to its receptor, inhibits listericidal cells from killing this bacterium.

The situation is different for other bacteria, such as *Legionella pneumophila*, whose growth is inhibited by intracellular iron deprivation, and as a consequence, alterations in human monocyte iron metabolism induced by interferon-γ influence the growth of these bacteria.[507,508]

Similarly, chloroquine inhibits the intracellular multiplication of *Legionella pneumophila* by limiting the availability of iron.[509] Comparable observations were also made for *Histoplasma capsulatum* yeast, whose growth within human macrophages depends upon intracellular iron levels. Chloroquine, which prevents release of iron from transferrin within the phagosome by raising endocytic and lysosomal pH, induced human macrophages to kill *Histoplasma capsulatum*.[510]

Similarly, *Francisella tularensis* is internalized by macrophages and confined to a vacuolar endosomal compartment where iron is available for its growth. Treatment of macrophages with agents that block endosome acidification leads to an inhibition of the growth of this bacterium through iron depletion at the levels of these endosomal compartments.[511]

Chloroquine inhibits the growth of *Cryptococcus neoformans* within macrophages by a mechanism independent of iron deprivation and related to the rise of intracellular pH induced by the drug.[512]

Interestingly, the growth of *Mycobacterium avium* within human macrophages is inhibited by apotransferrin, but is enhanced by iron or by iron-saturated transferrin.[513] *Mycobacterium tubercolosis* is taken up by macrophages through a phagocytosis process and is found within early and late endosomes where it is in contact with internalized transferrin.[514] However, at variance with *Mycobacterium avium*, the growth of *Mycobacterium tuberculosis* is inhibited by iron and this effect seems to be mediated through inhibition of TNF-α release required to render monocytes–macrophages permissive to the infection of this bacterium.[515]

These observations indicate that iron and transferrin receptors play a key role in the control of the growth of bacteria within macrophages. However, another protein capable of binding iron plays an important role in the control of macrophage susceptibility to these infectious agents. Studies performed in mice have shown that natural resistance of Bcg mice to infection with *Mycobacterium* and *Leishmania* is dependent upon expression of the Nramp1 gene.[516] This gene encodes for a membrane protein with a structural organization typical of families of divalent cation transporters. This protein is able to bind iron and other divalent cations. It has a selective pattern of expression in professional phagocytes, such as monocytes–macrophages.[516]

Nramp1 protein is not localized at the level of the cell membrane, but is found at the level of the late endocytic compartments. Upon phagocytosis, Nramp1 is recruited to the membrane of the phagosome and remains associated with this structure during its maturation to phagolysosome.[517]

This pattern of localization is compatible with the hypothesis that Nramp1 controls the replication of intracellular parasites by altering the intravacuolar environment of the phagosomes containing bacteria.[517] In line with this interpretation, recent studies of West African populations indicate that genetic variation of Nramp1 affects susceptibility to tuberculosis in humans.[518]

Recent studies directly support the concept that Nramp1 exerts its anti-microbial activity by controlling iron availability for pathogens. In fact, it was suggested that Nramp1 may act as an iron pump that depletes the phagosomal compartments of iron, thus leading to a deprivation for the pathogen of this essential element.[519]

Expression of the Nramp1 protein in macrophage cell lines through gene transfer showed that this protein exerted some effects on iron traffic within these cells. Nramp1 expression was associated with reduced cellular iron load following challenge with iron/nitrolotriacetate and with increased iron flux into the cytoplasm from a cellular compartment seemingly represented by lysosomes.[520]

These observations indicate that Nramp1 mobilizes iron from intracellular vesicles. In spite of these findings, it was observed that low amounts of iron stimulate Nramp1 function, a phenomenon presumably related to the capacity of Nramp1 to facilitate an Fe-dependent Haber–Weiss reaction with consequent hydroxyl radical production and inhibition of bacterial growth.[521]

Other bacteria that multiply within monocytes–macrophages developed a peculiar strategy to obtain optimal amounts of iron required to sustain their growth within cells. Macrophage cultures infected with *Coxiella burnetii* displayed a significant enhancement of transferrin receptor expression, a phenomenon associated with a 2.5-fold increase in iron uptake and accumulation within macrophages.[522]

A peculiar strategy for survival and proliferation within macrophages was developed by another bacterium, *Salmonella enterica*. This strategy is based on a type III secretion system, designated Spi/Ssa, that allows the synthesis and release within the cytosol of the macrophage of the host of a protein, SpiC, which interferes with intracellular membrane trafficking and particularly, with normal trafficking of transferrin.[523]

Several studies performed in different cell types have shown that these cells, in addition to the well-known mechanism of iron uptake through the transferrin–transferrin receptor pathway, also possess the capacity to acquire substantial amounts of iron from low moleclar weight chelates. Under normal conditions, low levels of non-transferrin-bound iron are detected in serum, but these levels markedly increase in some pathological conditions including idiopathic hemochromatosis, thalassemia, aplastic anemia, and during myeloablative chemotherapy.[524]

Thus, acquisition of iron in the non-transferrin-bound form is likely to be relevant *in vivo* in that macrophages have a particularly pronounced capacity to take up iron from a variety of low molecular weight iron complexes.[525] This mechanism is particularly important from a physiological view in that the removal of this form of iron by macrophages limits the potential for iron to act as a hydroxyl radical catalyst. In fact, studies performed on human alveolar macrophages and on *in vitro* monocyte-derived macrophages have shown that these cells have the capacity to incorporate large amounts of iron from low molecular weight chelates and to incorporate it into ferritin, thus avoiding the opportunity for iron to interact with H_2O_2 or O_2^- and to generate hydroxyl (OH) radicals via the Haber–Weiss reaction.[525]

Iron sequestration by macrophages may protect the cells located in close proximity from exposure to potentially cytotoxic iron-catalyzed oxidants such as hydroxyl radicals. The mechanism of iron uptake by low molecular weight chelates is poorly characterized; however, it is known that:

1. Iron acquisition is influenced by the nature of the iron chelate.
2. Iron acquisition from low molecular weight chelates is greater than from diferric transferrin.
3. This type of iron uptake is temperature-dependent and pH-independent.
4. Treatment of macrophages with different types of proteases does not modify the iron uptake from low molecular weight chelates.[526]

Interestingly, studies performed on rat bone marrow macrophages have shown that most of their transferrin uptake is mediated by a non-receptor mechanism that is independent of the degree of transferrin saturation.[527]

It must be emphasized that the induction of transferrin receptors on *in vitro*-grown human monocytes represents an event not related to proliferation because these cells remain quiescent during culture.[498] Thus, the expression of transferrin receptor on macrophages seems to be related to the differentiation of monocytes to macrophages rather than induction of cell proliferation.

However, in certain cell culture conditions, monocytes are induced to proliferate. Thus, human monocytes isolated from peripheral blood are quiescent and remain in this quiescent state if no growth factor is added in the culture medium. In fact, the addition of M-CSF induced DNA synthesis of human monocytes, a phenomenon potentiated by other CSFs such as GM-CSF and IL-3, but inhibited by interferon-γ or interleukin-4.[528] Addition of M-CSF to quiescent bone marrow macrophages induced these cells to proliferate and induced a marked increase of transferrin receptor expression.[529]

Several factors that can modulate the expression of transferrin receptor on macrophages were identified and may be divided into two categories: agents that activate or stimulate the proliferation of macrophages, and agents that modify the intracellular iron concentration of macrophages. The study of the effects of agents activating monocytes–macrophages on transferrin receptor expression was stimulated by initial observations by Hamilton et al.[530] The studies indicate that the expression of transferrin receptor is different in macrophages exhibiting qualitatively and quantitatively the different degrees of functional activation. Thus, resident peritoneal macrophages, exudate macrophages primed by elicitation with a pyran copolymer, and activated macrophages induced by chronic infection of mice with *Bacillus Calmette-Guerin* (BCG) or elicitation with heat-killed *Propionibacterium acnes* had low numbers of transferrin-binding sites.[529]

In contrast, macrophages elicited by sterile inflammatory agents (thioglycollate broth, fetal bovine serum, or casein) all exhibited greater numbers of transferrin receptors.[527] These observations suggest that the level of transferrin receptor expression varies in response to discrete extracellular signals known to induce macrophage activation.

In order to test this concept directly, further studies were performed to evaluate the effect of the main cytokine involved in macrophage activation, i.e., interferon-γ (IFN-γ). A series of studies showed that IFN-γ down-modulates the expression of transferrin receptors in both murine[531] and human[532] monocytes. This effect is relatively rapid in that it requires 8 to 10 hours of exposure before substantial reduction in receptor expression becomes apparent, and induces maximal reduction 16 to 20 hours after addition of the cytokine.[531]

The simultaneous addition of bacterial endotoxin (LPS) does not produce a further inhibition of transferrin receptor expression by macrophages. The reduced expression elicited by IFN-γ seems to have an important effect on the anti-microbial activities of macrophages, which leads to a limited availability of intracellular iron. That in turn inhibits the intracellular multiplication of parasites.[531]

A large number of studies have shown that the intracellular multiplication of several parasites, such as *Legionella pneumophila*,[532] *Trypanosoma cruzi*,[533] and *Plasmodium falciparum*,[534] is strictly iron-dependent. Consistent with this interpretation, other studies clearly showed that activated macrophages release much less iron than resting ones.[535-537]

In contrast to these observations, Taetle et al. reported the increased expression of transferrin receptor following overnight exposure of *in vitro*-grown human monocytes to IFN-γ.[538] The contradiction of these results may be related to both the dose of IFN-γ and the monocyte–macrophage population used to perform the assay.

IFN-γ, whose synthesis is induced in monocytes–macrophages by LPS,[539] even at low doses, can elicit a marked down-modulation of transferrin receptor in monocytes differentiating to macrophages.[540] This inhibitory effect was not specific for transferrin receptor, and was also observed with other membrane antigens whose expression greatly increased during the process of maturation of monocytes to macrophages (i.e., FcRI), and was due to a post-translational mechanism involving sequestering of the transferrin receptor at the intracellular compartments.[540]

Other studies investigated the effect of iron on the expression of transferrin receptors and ferritin[498] chains in cultures of human peripheral blood monocytes maturing to macrophages.[498] At all culture times, treatment with iron salts induced a dose-dependent rise of transferrin-binding capacity, which is dependent on an increase in transferrin receptor synthesis associated with a sustained elevation of the transferrin receptor RNA level.[498]

In parallel, the addition of iron salts to monocyte–macrophage cultures sharply stimulated ferritin synthesis, but only slightly enhanced the level of ferritin RNA, thus indicating a modulation at translation level.[498] These findings suggest that in cultured human monocytes–macrophages, iron up-regulates transferrin receptor expression. This is in sharp contrast to the negative feedback reported in a variety of other cell types.

Studies on the activity of iron regulatory protein (IRP) in monocytes maturing *in vitro* to macrophages[474] further supported the concept that in these cells iron up-regulates transferrin receptor expression. Monocytes freshly isolated from peripheral blood exhibit very low levels of high-affinity IRP and do not express transferrin receptors.

Monocytes maturing *in vitro* to macrophages show a sharp increase of high-affinity IRP and, to a lesser extent, of total IRP. The addition of iron salts moderately stimulates the high-affinity IRP but not the total IRP.[474]

The progressive rise of high-affinity IRP from very low to high activity is strictly related to the parallel increase of transferrin receptor expression and, surprisingly, to a very pronounced rise of ferritin expression observed both at the mRNA and the protein levels.

Further studies are necessary to clarify the apparent dichotomy of the effects of IRP in monocytes–macrophages, i.e., the rise of the high-affinity is linked to a rise of transferrin receptor expression at the RNA and protein levels, and to an increase of ferritin expression at the mRNA and protein levels. It is conceivable that the remarkable increase in ferritin gene expression observed during maturation of monocytes to macrophages may be related to moleclar mechanisms other than those mediated by IRP. Similarly, the stimulation of ferritin synthesis elicited in these cells by iron load cannot be accounted for by IRP.

Recent studies have shown that nitric oxide (NO) produced by activated macrophages may play an important role in the control of transferrin receptor expression and, more generally, of iron metabolism in these cells. NO is a short-lived messenger molecule involved in neurotransmission, regulation of blood pressure, and cytotoxicity.

Many biological effects of NO depend on its capacity to interact with iron; the production of NO depends on the activity of a peculiar group of enzymes known as NO synthases. Three isoforms of NO synthases (NOSs) are observed in mammallian cells, the NOS2 isoform being specific to monocytes–macrophages.[541] These three isoforms of NOS catalyze the production of NO through a similar biochemical pathway, which involves two sequential monooxygenase reactions. The first reaction leads to oxidation of L-arginine to N^w-OH-L arginine, which is further oxidized to yield one molecule of NO and L-citrulline.[541] The production of NO and the activation of inducible NOS in human monocytes–macrophages were difficult to demonstrate, but it is now clear that:

1. Human macrophages do not produce measurable levels of NO when induced by inflammatory cytokines.
2. Some tumor cells, bacteria, and viruses are able to activate NOS in human macrophages.
3. NO synthase inhibitors decrease the release of proinflammatory cytokines by human monocytes–macrophages stimulated by bacterial lipolysaccharide.

A series of studies have shown that NO may play a key role in the control of iron metabolism, and iron itself controls the expression of NOS. These observations provided evidence of an autoregolatory feedback loop in macrophages that links maintenance of iron homeostasis with optimal formation of NO for host defense.

In response to NO production, several cell types including macrophages and neurons, exhibit an activation of IRP activity which leads to ferritin repression and an increase in transferrin receptor expression, thus showing a correction between the L-arginine–NO pathway and iron metabolism.[542]

Inducers of NOS activity of murine macrophages, such as interferon-γ, elicited a significant rise of IRP activity. The rise is specifically inhibited by NOS inhibitors, thus showing the connection between NO and modulators of IRP activity.[542] Recent studies performed on the murine macrophage J774 cell line have shown that the effect of NO on IRP may be more complex than previously reported. Inducers of NO production such as IFN-γ, elicited a rise of IRP1 activity, associated with a marked down-modulation of IRP2 activity.[543,544]

The increase in IRP1 activity elicited by IFN-γ seems to depend largely on NO production, as the increase was largely reversed by NOS inhibitors, such as N-monomethylarginine.[544,545] In contrast, the decrease in IRP2 activity was not affected by addition of NOS inhibitors. Furthermore, the effect of IFN-γ on IRP1 activity, but not on IRP2 activity, requires *de novo* protein synthesis. Cycloheximide blocked the rise in IRP1 activity elicited by IFN-γ, while it had no effect on the decrease of IRP2 activity induced by this cytokine.[544]

Interestingly, in this model, these changes in IRP activity are accompanied by a rise in ferritin synthesis and a decrease in transferrin receptor expression, a condition that mimics the regulation of iron homeostasis observed in tissue macrophages during inflammation.[546]

On the other hand, there is evidence that iron modulates NOS activity by controlling nuclear transcription; in fact, in murine macrophage J774 cells, the induction of NOS activity elicited by interferon-γ is markedly inhibited by iron loading, and inhibited by iron deprivation.[546]

Anti-inflammatory cytokines, such as interleukin-4 and interleukin-13, modulate the responses of macrophages to interferon-γ in terms of NO synthesis and changes in iron metabolism. These cytokines:

1. Inhibit the NO-mediated rise of IRP activity.
2. Through an IRP-independent mechanism, induce an augmentation of transferrin receptor mRNA levels.[547]

Finally, recent studies suggest that iron may play an important role in mediating NF-kB activation in macrophage.[548] NF-kB is a nuclear transcription factor that induces in activated macrophages the transcription of cytokines such as TNF-α and IL-6, as well as adhesive membrane molecules, all of which are involved in inflammation. Supplementation of iron increases activation of NF-kB elicited by lipopolysaccharides, while an opposite effect is induced by iron deprivation.[548] These observations indicate a further link between the inflammatory responses of macrophages and iron metabolism.

A peculiar aspect of macrophage physiology is related to their ability to exhibit a pronounced capacity for phagocytosis. During phagocytosis these cells internalize significant proportions of their plasma membranes, and must possess a mechanism for rapid renewal of their membranes. Since macrophages lack distinct pericentriolar recycling compartments, unlike the Golgi apparatus, their capacity to rapidly renew the phagocytosed plasma membrane must be related to the development of an extensive network of endocytic vesicles and tubules. Macrophages display networks of transferrin receptor-positive endocytic vesicles and tubules. This endocytic apparatus accelerates the rate of transferrin exocytosis by these cells, a phenomenon further potentiated by phagocytosis.

5.11.3 Expression of Transferrin Receptor during Erythropoietic Differentiation and Maturation

The cellular requirements for iron change greatly during differentiation and maturation of erythroid cells, largely due to the need for high amounts of iron to sustain heme and hemoglobin synthesis.

Studies analyzed the expression of transferrin receptors in erythroid cells during the different steps of hemopoietic differentiation and maturation, starting with the most undifferentiated hemopoietic progenitors through committed progenitors to the late stages of maturation.

Hemopoietic differentiation starts from pluripotent hemopoietic stem cells that have the capacity to self-renew and generate differentiated progeny. This definition implies that at each division a stem cell is able to pass its properties to at least one of its two daughters.[549]

Some cells identified *in vitro* as the hemopoietic progenitors capable of initiating long-term cultures (LTC-IC) correspond to true hematopoietic stem cells. The first steps of hematopoietic stem cell differentiation generate early hematopoietic progenitors defined by *in vitro* assay as progenitors generating blast cell colonies (BCCs), highly proliferative potential colony-forming cells (HPPCFCs), and mixed granulo–erythroid–monocytic–megakaryocytic colony-forming units (CFU-GEMMs).

The subsequent step of hemopoietic differentiation is the generation of committed progenitors defined by *in vitro* assays as burst-forming and colony-forming unit–erythroid (BFU-E and CFU-E), colony-forming unit–granulomonocytic (CFU-GM), and colony-forming unit–megakaryocytic (CFU-Mk) progenitors. These committed hemopoietic progenitors progressively undergo differentiation and maturation until they become mature blood elements circulating in the blood.

Cell survival, proliferation, and differentiation are controlled by hemopoietic growth factors (HGFs) which, according to their level of action, may be subdivided into early-acting HGFs (kit ligand and flt3 ligand), multilineage HGFs (IL-3 and GM-CSF), and late-acting unilineage HGFs (erythropoietin, thrombopoietin, M-CSF, and G-CSF).

A number of experiments focused on evaluating the expression of transferrin receptors on early human progenitors. Initial studies by Brandt et al. showed that human marrow cells selected according to positivity for CD34 antigen and negativity for CD15, HLA-DR, and CD71 (transferrin receptor) contained the most immature populations of hemopoietic progenitors.[550] Subsequent studies on human marrow cells confirmed these observations and showed also that CD34$^+$/CD71$^-$ fraction contained cells capable of initiating and maintaining *in vitro* long-term cultures. After *in vitro* culture in the presence of kit ligand, IL-3, IL-6, and erythropoietin, these cells became transferrin receptor-positive.[551]

Similar observations were also made on hemopoietic progenitors purified from cord blood[552] and fetal liver.[553] The fraction of CD34$^+$/45RA^{low}/71^{low} cord blood cells contains the most immature progenitor cells.[552] CD34$^+$/71^{low} cells isolated from fetal liver when grown *in vitro* in the presence of appropriate cytokines exhibit a proliferative potential markedly higher than the corresponding cells isolated from adult bone marrow.[553]

Subsequent studies of fractionation of cord blood mononuclear cells showed that: (1) 34$^+$/45Ra^{low}/71^{low} cells are predominantly composed of multipotent progenitors and generate mixed hematopoietic populations when grown *in vitro*; (2) 34$^+$/45RA$^+$/71^{low} cells are mainly composed of myeloid progenitors; and (3) 34$^+$/45Ra^{low}/71$^+$ cells predominantly (70%) contain erythroid progenitors[554].

The low or absent expression of transferrin receptor on very primitive human hemopoietic cells was confirmed by cell fractionation focused on isolating primitive human stem cells. The results were as follows: (1) 34$^+$/Thy-1$^+$ cells express very low levels of CD71[555]; (2) 34$^+$/DR$^-$/38$^-$ cells exhibit very low or no CD71 expression[556]; and (3) 34$^+$/flt3$^+$ cells are CD71$^-$.[557]

Similar studies carried out on acute leukemia blasts generally showed that it is possible to isolate from leukemic cells a population endowed with a high proliferative potential, characterized by a 34$^+$/38$^-$ membrane phenotype; those exhibiting a long-term culture-initiating cell capacity have lower CD71 expression as compared to those able to originate leukemic blast cell colonies.[558]

Studies of the expression of transferrin receptors on committed hemopoietic progenitors produced conflicting results. Sieff et al.[559] investigated the expression of transferrin receptors on erythroid progenitors (BFU-E and CFU-E) by fractionating human bone marrow cells on a fluorescence-activated cell sorter according to their binding of either anti-transferrin receptor OKT9 monoclonal antibody or FITC-labeled transferrin. The positive and negative fractions were then cloned in methylcellulose for BFU-E and CFU-E. Using the OKT9 monoclonal antibody, a wide range of positivity was found in erythroid progenitors (BFU-E, 1 to 78%; CFU-E, 14 to 96%). In contrast, using FITC-labeled transferrin, virtually all erythroid progenitors were positive (BFU-E, 60 to 95%; CFU-E, 99.5 to 99.8%). With either OKT9 or transferrin, a consistently lower percentage of myeloid progenitors was found among the positive fraction. The discordance in reactivity of erythroid progenitors with transferrin as compared to OKT9 monoclonal antibody may be related to the fact that receptors expressed on these cells expose the transferrin-binding domain more than the OKT9 epitope.[559]

Using a different monoclonal antibody anti-transferrin receptor, Loken et al.[560] found that a majority of bone marrow BFU-E and CFU-E progenitors express transferrin receptors. They also performed double-labeling experiments using anti-glycophorin and anti-transferrin receptor antibodies. Erythroid progenitors in the fractions exhibited low or intermediate positivity for transferrin receptor and negativity for glycophorin.

Studies carried out on normal human bone marrow indicate that transferrin receptors are expressed on a majority of erythroid progenitors, and their expression apparently increases from the BFU-E stage to the CFU-E stage of erythroid progenitor differentiation.

Using a method that produces virtually pure human hemopoietic progenitors from peripheral blood, Gabbianelli et al.[561] investigated the expression of transferrin receptors on these cells. They used different types of monoclonal antibodies to transferrin receptor (OKT9, B3/25, and 42/6). Most erythroid and myeloid progenitors were observed among transferrin receptor-negative cells.

However, purified progenitors grown in liquid suspension in the presence of IL-3 became transferrin receptor-positive and proliferated. Thus, the fact that circulating erythroid progenitors are basically transferrin receptor-negative, and those found in bone marrow are predominantly transferrin receptor-positive, may be explained by the observation that peripheral blood progenitors are subject to significantly less cycling than those found in bone marrow.

The difference in CD71 expression between bone marrow and peripheral blood progenitors was confirmed by other studies. Bender et al.[562] found that >90% of bone marrow CD34$^+$ cells express transferrin receptors, while only 17% of peripheral blood CD34$^+$ cells express these receptors.

Other experiments analyzed the effect of the addition of anti-transferrin receptor mAbs to the culture medium on the growth and differentiation of hemopoietic progenitors. Initial experiments[563] showed that different anti-transferrin receptor mAbs (B3/25, 42/6 and 43/31), that either inhibit or do not inhibit the binding of transferrin to its receptor, caused a dose-dependent inhibition of granulocyte–macrophage progenitors (CFU-GM). Further studies[564] focused on the effect of anti-transferrin receptor on erythroid progenitors determined that:

1. Monoclonal antibodies against transferrin receptor abrogate the growth of erythroid progenitor colonies derived from BFU-E at much lower concentrations than those required to inhibit CFU-GM development.
2. The requirement of developing bursts for transferrin extends into the second week of *in vitro* growth, at which time erythroid precursors develop within each BFU-E colony.

These observations are in line with *in vivo* studies showing that a patient who developed autoantibody against transferrin receptor exhibited severe microcytic anemia associated with normal white blood cells and platelets.[565]

The differential sensitivities of erythroid and myeloid progenitors to the inhibitory effect of anti-transferrin receptor monoclonal antibody seem to be related more to quantitative than qualitative differences in the levels of expression of transferrin receptors. Some studies[566,567] suggest that transferrin receptors expressed on erythroid cells may exhibit some differences when compared with receptors observed on other cell types. This conclusion is based on studies in which new anti-transferrin receptor monoclonal antibodies were developed and characterized.

In one study,[566] the L5.1 monoclonal antibody reacted preferentially with transferrin receptor expressed on erythroid cells. A later study[567] indicated that five different anti-transferrin receptor mAbs obtained by immunizing against erythroid precursors derived from fetal liver BFU-E recognize a form of the transferrin receptor different from that reacting with OKT9 or 5E9 mAbs. On the basis of these findings, the authors proposed that cells with high iron requirements, such as erythroid cells, synthesize two forms of transferrin receptors, possibly by means of differential mRNA splicing or by post-translational modification of the transferrin receptor.[567]

Experiments analyzed the expression of transferrin receptors on erythroid precursors from immature erythroid cells (proerythroblasts) to mature erythrocytes. These studies showed that immature erythroid cells express a high number of transferrin receptors[568] and actively synthesize ferritin.[569,570]

As erythroid cells differentiate further, the rate of hemoglobin synthesis begins to rise. Concurrently, the number of transferrin receptors displayed by erythroid cells decreases[568] and ferritin synthesis declines.[569,570] In early normoblast stages in the rat, the receptor number per cell has been reported as ~300,000,[571] increasing to a maximum of 800,000 per cell on the intermediate normoblast. During further maturation, this number declines to ~100,000 in the reticulocyte.

Transferrin receptors are present on all reticulocytes, but their number decreased markedly during reticulocyte maturation and no receptors were detectable on mature erythrocytes. Transferrin receptors are lost from reticulocytes by externalization of membrane vesicles.[572]

Using a method that allows a virtually complete purification of hemopoietic progenitors[561] and a culture method allowing the selective growth of erythroid cells, it was possible to evaluate transferrin receptor expression during all stages of differentiation and maturation.

Studies showed that transferrin receptors are initially expressed at low levels in quiescent hemopoietic progenitors, but their expression markedly increases when these cells are triggered to proliferate by hemopoietic growth factors. Starting at day 5 of culture, when the cells reach a differentiation stage corresponding to CFU-E, transferrin receptor expression increases, and remains at high levels through the terminal stages of erythroid maturation (Figures 5.13A and B).

During the differentiation of hemopoietic progenitors to granulocytic, monocytic, or megakaryocytic cells an increase in transferrin receptor was observed only during initial stages of progenitor cell differentiation. At later stages of differentiation and maturation, receptor expression progressively declines.

Studies of labeling of bone marrow cells with anti-transferrin receptor monoclonal antibodies showed that: (1) transferrin receptor expression is largely confined to erythroid cells in bone marrow[573]; (2) the reactivity of erythroid cells with anti-transferrin receptor monoclonal antibodies decreases with erythroid maturation[571]; and (3) transferrin receptor expression on bone marrow erythroid cells precedes the onset of glycophorin synthesis.[571-574]

The expression of transferrin receptors and other proteins involved in iron metabolism was carefully investigated in Friend erythroleukemia cells (FLCs). Friend cells are a transformed, erythroblastic cell line that can be induced to differentiate in vitro by dimethyl sulfoxide (DMSO). DMSO induces formation of heme biosynthetic enzymes,[575] increased iron uptake,[576] and formation of hemoglobin.[577]

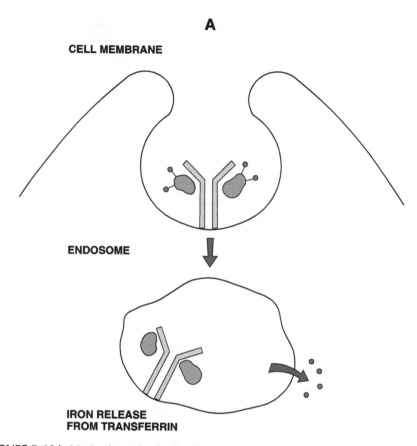

FIGURE 5.13A Mechanism of control of iron uptake by HFE. When HFE is absent transferrin interacts with its receptor at the cell membrane, the transferrin–transferrin receptor complex is endocytosed; when it reaches the acidic pH of the endosome, iron is released from transferrin and becomes available.

Investigation of the transferrin receptors in FLCs showed that the induction of differentiation is accompanied by a marked increase in the rate of iron uptake and in the number of transferrin receptors.[576] Activation of iron transport and heme biosynthesis occurs before hemoglobinization of the cells and seems to be required for the hemoglobinization to take place during erythroid maturation.[578] Increasing levels of intracellular heme are required to sustain optimal transferrin receptor synthesis in FLCs induced to terminally differentiate by DMSO.[579,580] This conclusion is supported by experiments showing that, in DMSO-induced FLCs, inhibition of heme biosynthesis by succinyl acetone leads to a decrease of transferrin receptor expression, while an opposite phenomenon is observed when heme content is increased by the addition of exogenous hemin.[579,580]

However, this positive feedback on transferrin receptor expression mediated by heme is observed only in DMSO-induced FLCs; uninduced cells exhibited the usual negative feedback observed in the majority of other cell types.[580]

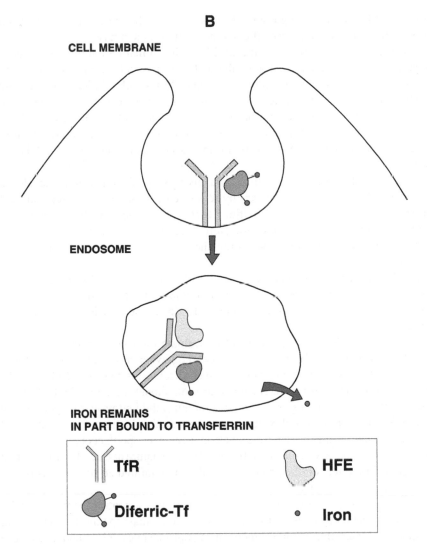

FIGURE 5.13B Mechanism of control of iron uptake by HFE. When HFE is present, it interferes with iron uptake through two different mechanisms: (1) it competes with transferrin at the cell membrane for binding to the transferrin receptor; and (2) it inhibits the release of iron from transferrin in the endosome.

Differentiating FLCs, in addition to exhibiting the peculiar mechanism of transferrin receptor control, also show mechanisms of heme synthesis control. Thus, when FLCs are induced to differentiate by DMSO, they show a marked increase in the level of all the enzymes involved in the heme biosynthesis pathway.[575]

Treatment of FLCs with exogenous hemin can cause not only increased synthesis of hemoglobin, but also increased levels of different enzymes of the heme pathway, including ALA synthase, ALA dehydradatase, PBG deaminase, and ^{59}Fe incorporation into heme.[581,582] Increases in ALA synthase activity in response to hemin treatment in erythroid cells are in striking contrast to the effect of hemin in the liver and other tissues, where hemin suppresses the synthesis of ALA synthase.

The expression of ferritin during DMSO-induced maturation of FLCs was also investigated.[583,584] Glass et al. found that cellular iron uptake and deposition into ferritin were greatly stimulated after treatment of FLCs with DMSO.[583] These results suggest that ferritin plays a key role in supplying iron required to sustain heme synthesis. Another study showed that in DMSO-induced FLCs, ferritin mRNA and synthesis were increased before induction of globin synthesis.[584] Furthermore, it was suggested that ferritin mRNA levels are increased before the rise in cellular iron uptake.[584]

These findings suggest that the initial increase in ferritin mRNA levels may allow the cells to respond more rapidly to subsequent increases in iron uptake by synthesizing large amounts of ferritin. These studies indicate that a coordinated pathway of modifications involves transferrin receptor expression, iron uptake, and ferritin synthesis, and all these events together allow the synthesis of optimal levels of heme and hemoglobin.

Studies performed in other erythroid cell lines, such as the erythropoietin-dependent J2E cell line, showed that erythropoietin exerted a stimulatory role on transferrin receptor transcription and on transferrin receptor mRNA stabilization via IRP activation.[585]

Additional studies were carried out to evaluate the role of transferrin receptor in erythropoiesis in a different *in vitro* model, i.e., avian erythroid cells transformed with temperature-sensitive mutants of avian erythroblastosis virus or S/3 virus.[586,587] The addition of an anti-transferrin receptor monoclonal antibody (JS-8)[588] had no effect on transferrin binding or internalization but inhibited proper recycling of the apotransferrin-receptor complex, leading to a marked reduction of iron uptake. The antibody caused an arrest of temperature-induced and spontaneous differentiation of transformed avian erythroid cells, followed by premature cell death. This antibody, however, had no effect on the proliferation of noncommitted erythroid cells or other types of hemopoietic cells.[589]

To explain all these findings, it was suggested that a control mechanism which ensures coordinated synthesis of erythrocyte proteins may operate in erythroid differentiation. To sustain this control mechanism, optimal rates of iron uptake and levels of heme synthesis are absolutely required during all steps of erythroid maturation.[589]

The development of a unilineage cell culture system allowing the selective growth of erythroid cells from highly purified preparations of human hematopoietic progenitors made it possible to investigate the expression of transferrin receptors during all stages of differentiation and maturation from undifferentiated erythroid progenitor cells (BFU-E) to mature erythroblasts. This technique showed that:

1. Transferrin receptor is scarcely expressed on quiescent hemopoietic progenitor cells, and its expression is markedly enhanced during initial stages of progenitor proliferation/differentiation (day 0 to 3 of culture).
2. At day 5 of culture, when the majority of erythroid progenitor cells reach the CFU-E stage, a further marked increase in transferrin receptor expression

is observed, and this increase precedes the appearance of the first detectable membrane erythroid markers (i.e., glycophorin A).

3. At later stages of erythroid maturation, corresponding to the steps progressively leading to the formation of mature erythroblasts from proerythroblasts through intermediate erythroblasts (polychromatophilic erythroblasts), a high level of receptor expression is maintained (unpublished observations).

Both transcriptional and post-transcriptional mechanisms contribute to maintaining elevated levels of transferrin receptor expression during all stages of erythroid differentiation/maturation, as suggested by two observations:

1. A high rate of transferrin receptor gene transcription is observed in erythroid cells.
2. Erythroid cells express high levels of iron regulatory protein 1 which is present in a highly active state (unpublished observations).

In erythroid cells, a strict linkage exists between iron and heme metabolism.[590,591] Evidence indicates that the rate of iron uptake controls the rate of heme synthesis, while heme controls the rate of iron uptake from transferrin either by inhibiting transferrin endocytosis or iron release from transferrin. This double control mechanism allows fine tuning and coupling of iron uptake with heme synthesis in order to sustain an optimal level of hemoglobin synthesis.

Several peculiarities of iron–heme metabolism in erythroid cells have been shown:[590,591]

1. In erythroid, but not in non-erythroid cells, iron uptake stimulates heme synthesis.
2. As mentioned earlier, the level of transferrin receptor expression in erythroid cells is much higher than in all other tissues, a phenomenon mainly related to the high rates of transcription of the transferrin receptor gene observed in these cells.[592,593]
3. IRP1 activity in erythroid cells is elevated under the control of erythropoietin. The elevated IRP1 activity and the high rate of transferrin receptor gene transcription cooperate to induce high levels of transferrin receptor mRNA in erythroid cells.[594]
4. In erythroid cells, but not in other cell types, during the transferrin endocytosis cycle, iron, after sorting from endosomes to the cytosol, is rapidly targeted to mitochondria.
5. Free heme inhibits iron uptake from transferrin and, through this mechanism, regulates the rate of heme synthesis.
6. In erythroid cells, but not in non-erythroid cells, protoporphyrin IX levels are finely tuned to the level of iron uptake as a consequence of a biochemical mechanism involving control of aminolevulinic acid synthetase activity by iron through a post transcriptional mechanism.

The linkage between heme and iron metabolism in erythroid cells is reinforced by recent studies of knockout of the genes encoding heme synthesis enzymes. Disruption of the gene encoding murine erythroid δ-aminolevulinate synthase (an erythroid-specific isozyme encoded by a specific gene which greatly contributes to the high levels of heme synthesis in erythroid cells) leads to a pronounced heme deficiency in erythroid cells associated with a cytoplasmic iron overload.[595]

Although erythroid cells have the ability to immediately utilize the majority of endocytosed iron for heme synthesis, a low amount of endocytosed iron is incorporated into ferritin. However, intracellular ferritin may also donate iron for heme synthesis; proteolytic degradation of ferritin in a lysosomal compartment is required for iron release and its subsequent transfer to the heme biosynthetic pathway.[596] Ferritin may also donate iron to erythroid precursors through a receptor-mediated process. Erythroblasts possess ferritin receptors, whose expression is regulated through an iron-dependent process.[597]

In addition to transferrin receptor-mediated iron uptake, normal erythroid cells possess the capacity to incorporate non-transferrin bound iron (NTBI).[598] The uptake of NTBI does not require transferrin receptors since it is unaffected by pronase treatment of the cells. Uptake of iron from NTBI requires passage through an endosomal compartment, as shown by the inhibitory effect on this process displayed by agents that alkalinize the endosomal pH. However, iron incorporated into erythroid cells from NTBI predominantly moves to stromal and non-heme fractions, while iron derived from a transferrin source is predominantly incorporated into heme.[598]

5.12 INTERACTION BETWEEN TRANSFERRIN RECEPTORS AND HFE: THE CLASS I MHC-RELATED PROTEIN MUTATED IN HEREDITARY HEMOCHROMATOSIS

Recent studies based on positional cloning have led to the identification of the gene that is mutated in hereditary hemochromatosis. The gene encoding the hereditary hemochromatosis protein maps within the locus first described as HLA-H. It was renamed HFE and encodes a protein that typically resembles class I major histocompatibility complex (MHC) molecules, both at its primary sequence and at the three-dimensional structure.

HFE shows a typical class I architecture in which each protein domain is encoded by a different exon, including a leader peptide; α1, α2, and α3 extracellular domains; the transmembrane, and the cytoplasmic 3' untranslated region. However, an important difference between HFE and class I HLA molecules was observed at the three-dimensional level. In MHC molecules, a groove between two helices of the molecule is involved in antigen binding, whereas the analogous groove in the HFE molecule is markedly narrowed, thus impeding the binding and therefore the presentation of the antigen to T cells.[599]

The HFE gene encodes a 2.7-Kb transcript predominantly expressed at the level of epithelial cells present in various tissues. This transcript generates a translation product of about 48 kDa that interacts noncovalently with a β2-microglobulin molecule. Immunohistochemical labeling with a specific antibody showed a pattern of tissue distribution with preferential expression at the sites of major iron absorption.

Most hereditary hemochromatosis patients are homozygous for a mutation (845 G to A) that converts Cys 260 to a tyrosine, preventing the formation of a disulfide band at the α3 domain, interaction with β2-microglobulin, and expression at the cell surface. This defect results in an increased rate of iron absorption at the intestinal crypts.

The biological basis of the role of HFE in the control of iron absorption derives from studies showing that this protein is able to interact with transferrin receptor. HFE can associate in the placenta with transferrin receptor,[600,601] and decrease the affinity for diferric transferrin.[601]

The HFE protein in crypt enterocytes is physically associated with transferrin receptor and β2-microglobulin.[602] Interestingly, the crypt cell fraction exhibits dramatically higher transferrin-bound uptake than villus cells, but villus cells display higher iron uptake than crypt cells. On the basis of these observations, it was suggested that HFE modulates the uptake of transferrin-bound iron from plasma.[602]

In parallel, the association of the HFE protein with the transferrin receptor was observed both at the level of the cell membrane and at the intracellular sites occupied by the transferrin receptor during its trafficking.[603]

Thus, the HFE protein endocytoses together with the transferrin–transferrin receptor complex and then co-localizes with the transferrin receptor during endosome trafficking. Free HFE is rapidly degraded, whereas the HFE–transferrin receptor complex is stable.[603]

To more carefully evaluate the mechanism through which HFE modulates transferrin receptor function, additional experiments were performed using human cell lines (HeLa) transfected with the human HFE gene. One study showed that the expression of the HFE protein in these cells elicited a significant reduction of transferrin-dependent iron uptake, without affecting the endocytic or exocytic rate of transferrin receptor cycling.[604] A second study based on the same experimental model provided evidence that HFE reduces the number of functional transferrin binding sites and reduces the rate of transferrin–transferrin receptor internalization; these two events lead to a decrease in iron uptake.[605]

The use of this cell line expressing HFE determined that the interaction between HFE and the transferrin receptor is strong and occurs during their biosynthesis, before they reach the cell surface.[605] Finally, it was possible to demonstrate that transferrin receptor is required for the intracellular transport and surface expression of HFE; in fact, the overexpression of the transferrin receptor in cells expressing HFE leads to an increase of the HFE levels on the cell surface.[605]

A peculiar feature of the HFE-transferrin receptor interaction is its strong pH dependence; in fact, HFE binds tightly to transferrin receptor at pH 7.50, but weakly or not at all at pH 6.0. It was proposed that histidine clusters present on the HFE molecule may be responsible for this sharp pH dependence in the HFE–transferrin receptor interaction.

However, studies on mutants of HFE in which the histidine residues were mutated ruled out a direct involvement of these amino acid residues in mediating HFE–transferrin receptor interaction.[606] The analysis of a series of HFE mutants showed that the transferrin receptor binding site on HFE is localized at the level of the C terminal portion of the α1 domain helix.[606] Finally, using a preparation of soluble HFE, it was possible to demonstrate that HFE competes with transferrin for binding to the transferrin receptor.[607] The mechanism for this competition between HFE and transferrin for binding to the transferrin receptor depends on the binding of HFE to a region of the transferrin receptor near the transferrin binding site.[607]

HFE also exerts a modulatory activity on iron regulatory protein activity. In fact, expression of HFE protein in HeLa cells stimulated IRP activity, and that in turn induced an increase in transferrin receptor synthesis associated with decreased ferritin synthesis.[608]

HFE is a competitive inhibitor of the interaction between transferrin and its receptor. The stoichiometry of the ternary complex formed by HFE, transferrin receptor, and transferrin is 1:2:1 (Figure 5.13B).

These observations stimulated the analysis of the three-dimensional structure of the HFE–transferrin receptor complex. The 2.8 Å crystal structure of the complex of HFE and the transferrin receptor has been recently reported.[608] The complex exhibits a two-fold symmetry with a 2:2 stoichiometry, such that one HFE molecule contacts each polypeptide chain of the transferrin receptor homodimer.[609]

Whereas the structure of HFE changes only slightly following formation of the complex, the three-dimensional structure of the transferrin receptor significantly changes following complex formation, particularly at the level of the dimer interface. The formation of the HFE–transferrin receptor complex occurs only at neutral pH; at acidic pH, the complex dissociates. This phenomenon is seemingly due to protonation of His residues which interferes with the three-dimensional structure of the transferrin receptor at the level of a domain critical for the interaction with HFE.

A recent study addressed the role of HFE in iron metabolism in intestinal cells through breeding of HFE knockout with mice strains carrying other mutations that affect iron metabolism.[609a] Interestingly, HFE knockout mice carrying mutations in the iron transporter Nramp2 failed to accumulate iron, thus indicating that in hereditary hemochromatosis, iron flux mostly occurs through the Nramp2 iron transporter. Mice deficient in both HFE and hephaestin exhibited a reduced iron loading as compared to animals only deficient in HFE, suggesting that these two proteins contribute to the control of iron absorption within duodenal cells. Finally, HFE knockout mice lacking transferrin receptors showed significantly higher iron loading than levels observed in animals with HFE deficiency alone. This observation further reinforces the idea that the interaction between HFE and transferrin receptor exerts a negative control on iron absorption.[609a]

5.13 A GENERAL OVERVIEW OF IRON UPTAKE MECHANISMS IN VARIOUS CELL TYPES

The first cell types involved in iron metabolism are the intestinal villus cells present in the duodenum, where iron absorption occurs. Recent studies have in part clarified the mechanisms of intestinal iron absorption. This advance was mainly due to the cloning of several iron transporters localized at the levels of different cellular domains. Using an expression cloning approach, Gunshin et al.[610] identified the divalent cation/metal ion transporter (DCT1) from a duodenal cDNA library prepared from mRNA from rats fed a low-iron diet. This metal transporter localized at the level of the cell membrane and of recycling, and was found identical to the human homologue gene Nramp2 encoding one of the natural resistance-associated macrophage proteins. It was also observed that a mutation of the DCT1 gene was responsible for murine microcytic anemia.

Other studies led to the identification of iron transporters localized at the basolateral membranes of the intestinal villus cells. These transporters are mostly involved in mediating iron efflux from intestinal cells to the blood. One such iron transporter was identified through the study of *sla* mice exhibiting microcytic anemia due to a defect in the release of iron into blood from enterocytes. The gene defective in this mouse strain encodes a transmembrane protein homologue of ceruloplasmin called hephaestin.[611] This protein functions as multi-copper ferrioxidase and seemingly is not the basolateral transporter. It interacts with the true iron transporter, facilitating iron release into the blood by its ferrioxidase activity.

Several studies support a role for ceruloplasmin in iron metabolism. In fact, ceruloplasmin increases ^{55}Fe-uptake by iron-deficient HepG2 cells by a transferrin-independent pathway.[612] Similar observations were also made in erythroleukemia K-562 cells. Two important additional observations were made in these cells: (1) ceruloplasmin-stimulated iron uptake was completely inhibited by trivalent, but not divalent cations; and (2) ceruloplasmin ferrioxidase activity was required for the stimulation of iron uptake.[613]

According to these observations, extracellular iron is first reduced by a membrane ferrireductase re-oxidized through the action of ceruloplasmin, and then transported into the cell via a trivalent cation-specific membrane transporter. A role for ceruloplasmin in iron metabolism is also supported by studies of patients with hereditary aceruloplasminemia and mice with targeted disruptions of the ceruloplasmin gene. In both these conditions, increased accumulations of parenchymal iron associated with impaired mobilization of iron from the iron storage tissues was observed.[614]

Thus, the major function of ceruloplasmin in iron metabolism is facilitating the flux of iron between cells and tissues. The synthesis of ceruloplasmin is stimulated by iron deficiency through a transcriptional mechanism dependent upon the binding of hypoxia-inducible factor to the ceruplasmin gene promoter.[614a]

A second iron transporter of the basolateral membranes of enterocytes was found by positional cloning to identify the gene responsible for the hypochromic anemia of the zebrafish Weissherbst mutant. The identified gene was named ferroportin 1. It encodes a multiple-transmembrane domain protein that functions as an iron exporter when expressed in *Xenopus* oocytes.[615]

Human ferroportin 1 was found at the basolateral membranes of duodenal enterocytes and at the basal surfaces of placental syncytiotrophoblasts. According to these observations, ferroportin 1 may represent the iron exporter of the duodenal enterocytes.

This same gene was cloned by another group of investigators and called IREG-1. This study showed that IREG-1/ferroportin 1 expression increased under conditions of increased iron absorption, a phenomenon due to the presence in the 5′ untranslated region of the IREG-1 mRNA of a functional iron-responsive element.[616]

In mice fed iron-deficient diets, duodenal IREG-1 mRNA levels significantly increased, compared to levels in mice fed the iron-replete diet. Another modulator of IREG-1 expression is hypoxia. During hypoxia, the rate of iron absorption and, more importantly, the rate of iron efflux from duodenal cells to circulation, significantly increase with kinetics mimicking the kinetics of IREG-1 under these conditions.

The IRE element of IREG-1 possesses an affinity for IRP comparable to that of ferritin IRE. The IREG-1 protein was conserved during evolution. Finally, the expression of IREG-1 is developmentally regulated, reaching adult levels after birth. This observation further suggests that IREG-1 expression is related to the absorption of iron from foods and is less important in neonates who absorb iron from all regions of the small intestine.

This same gene was cloned by a third group of investigators.[616a] In addition to confirming the findings of the other groups, they studied the pattern of expression of this protein in macrophagic cells. IREG-1 was localized at the level of its cytoplasm and its expression is modulated by iron levels in a way opposite to that observed in duodenal cells (iron loading up-modulates macrophage IREG-1 levels, while iron deprivation down-modulates this protein).[616a]

This pattern of expression and localization of IREG-1 within macrophages suggests that IREG-1 is likely involved in the reutilization and storage of iron derived from the destruction of senescent erythrocytes.

Based on these new findings it is now possible to describe the patterns of expression and functions of the different iron import and export proteins differentially expressed at the different cellular compartments. The most relevant cellular compartments for iron metabolism are the cells involved in iron absorption (enterocytes of the duodenum) or iron storage (monocytes–macrophages). Erythroid cells also play a role by utilizing large amounts of iron to sustain heme and hemoglobin synthesis.

Iron absorption at the level of enterocytes occurs as a three-step process, involving (1) uptake by the enterocytes from the intestinal lumen across the intestinal brush border; (2) intracellular transport; and (3) transfer across the basolateral membrane into plasma.[617]

The initial phase of iron entry at the apical membranes of enterocytes requires first the reduction of Fe^{3+} to Fe^{2+}, and then its transport across the membrane is mediated by the Nramp2 iron transporter whose expression is regulated by iron status. More particularly, Nramp2 produces two alternatively spliced transcripts generated by alternative use of 3′ untranslated region.

The mRNA transcript containing an IRE element in its 3′ untranslated region is the major transcript of Nramp2 and is modulated according to the iron status in a manner similar to that observed for transferrin receptors. Thus, iron deprivation elicited a significant increase in the level of Nramp2 mRNA, with a consequent increase in the amount of Nramp2 protein at the apical membranes of enterocytes.[618]

Since the villus cells originate from the differentiation of progenitor cells at the crypts, it was hypothesized that the crypt cells sense body iron stores through the transferrin receptor–HFE complex present at their basolateral membranes. The amount of iron adsorbed by these cells at the basolateral membranes determines the expression of Nramp2 along the crypt-to-villus axis and the amount of this iron transporter expressed on the apical surfaces of villus cells (Figure 5.14).

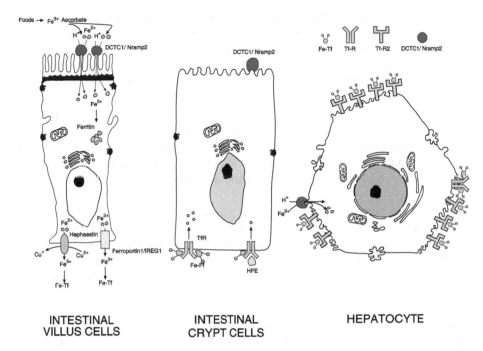

FIGURE 5.14 Iron import and export mechanisms of enterocytes and hepatocytes. Left: Dietary ferric iron (Fe^{3+}) in the intestinal lumen is first reduced to ferrous iron (Fe^{2+}) and then absorbed at the apical sides of the villus cells through the action of the cation transporter DCTC1/Nramp2. At the basolateral membrane, Fe^{2+} is released from villus cells into plasma by an iron transporter, ferroportin 1/IREG-1, which requires a multicopper oxidase (hephaestin). Middle: The crypt cells (progenitors of villus cells) sense iron stores through the transferrin receptor–HFE complex at the basolateral membranes. If the body iron stores are low, most of the transferrin receptor present at the basolateral membranes of the crypt cells is present as a complex with HFE, producing a decrease of iron uptake. The decrease induces activation of the IRP in the cells; the activated IRP increases DCTC1/Mramp2 mRNA stability, thus leading to a significant increase of DCTC1/Nramp2 protein at the villus cells. An opposite effect is observed when crypt cells sense increased iron stores. Right: The hepatic cells and reticuloendothelial cells play an important role in body iron storage. The pattern of expression of transferrin receptors on their membranes is unique: transferrin receptor 2 expression is predominant over transferrin receptor expression. Since the expression of transferrin receptor 2 is modulated by body iron stores, the capacity of hepatic cells to store iron remains unmodified even in the presence of excessive iron loading, thus exposing these cells to the danger of tissue damage. This phenomenon is observed in primary and secondary hemochromatosis.

The liver plays an important role in iron metabolism. On the membranes of hepatic cells, three types of iron transporters are expressed: transferrin receptor, transferrin receptor 2, and Nramp2. The first two molecules are responsible for transferrin-dependent iron uptake by hepatic cells. Interestingly, on the membranes of hepatic cells, transferrin receptor 2, whose expression is not modulated by iron levels, is much more expressed than transferrin receptors.

This observation has many important implications, particularly concerning the role played by the liver in iron storage. Thus, transferrin receptor 2 significantly contributes to the liver's ability to uptake and store transferrin-bound iron. However, when body iron stores are supranormal, hepatic cells do not down-modulate transferrin receptor 2 expression and cannot be protected from excessive iron loading (Figure 5.14).

The two other cell compartments that have particular functions in iron metabolism are reticuloendothelial cells and erythroid cells.

Macrophages play a key role in iron storage, an important function required to avoid potential toxicity and make available the iron derived from the destruction of senescent red blood cells and enzymatic digestion of hemoglobin for cellular metabolism.

These cells express on their membranes transferrin receptor, transferrin receptor 2, and Nramp2 (Figure 5.15). However, the majority of the iron incorporated into macrophages does not derive from uptake of exogenous iron, but from phagocytosis and destruction of senescent red blood cells. To avoid the toxicity related to the large amounts of iron, these cells developed the capacity to synthesize large amounts of ferritin.

In addition to their role in iron storage, the main function of macrophages is host defense against certain bacterial infections. This function depends in part on the capacity of macrophages to limit divalent cation concentration, and particularly of iron, through the action of a cation exporter (Nramp1) located at the phagosome membranes.

Macrophages, in addition to their capacity to store iron, must possess the capacity to release this stored iron to make it available for cellular metabolism. This iron export function is probably mediated by the iron exporter ferroportin 1/IREG-1, whose expression is particularly pronounced in macrophagic cells.

The situation is different in erythroid cells where the pattern of erythroid import proteins is finalized to allow high rates of iron uptake into these cells. Erythroid cells express high levels of both transferrin receptor and transferrin receptor 2 on their membranes, thus allowing high rates of iron uptake into the cells (Figure 5.15). These high amounts of iron incorporated into erythroid cells are used to sustain the high level of heme synthesis required for hemoglobin synthesis.

The precise mechanism of iron uptake into mitochondria is still unknown. However, recent studies led to the identification of a mammalian protein involved in mitochondrial iron metabolism. This protein, called frataxin (named for a mutation of this protein responsible for Friedreich's ataxia), encodes a 210-amino acid protein localized at the mitochondria and is a homologue of yeast YFH1 protein.[619]

The knockout in yeast of YFH1 genes and the mutation of frataxin gene in Fridreich's ataxia produce similar phenotypes characterized by severe defects in mitochondrial respiration and loss of mitochondrial DNA associated with elevated intramitochondrial iron.

These studies and studies of other mitochondrial proteins suggest potential cross-talk between mitochondria and cytosol to couple the need of mitochondria for iron to the general mechanism of cellular iron uptake.[620] This cross-talk is based on the capacity of some mitochondrion proteins to generate mitochondrial and cytosolic

MACROPHAGE　　　ERYTHROID CELLS

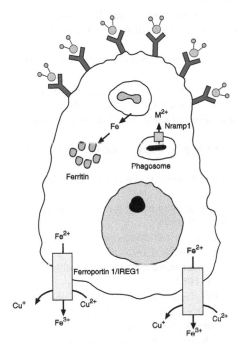

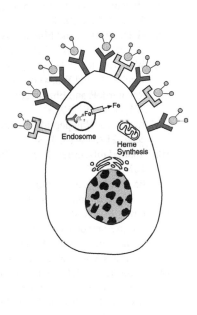

FIGURE 5.15 Left: Iron import and export mechanisms of macrophages. Macrophages import iron through their membranes by using different iron uptake molecules, including transferrin receptor, transferrin receptor 2, and Nramp2. However, they acquire most of their iron through a different mechanism, the phagocytosis of senescent red blood cells. Macrophages must have efficient mechanisms of iron efflux to make the stored iron available for metabolism. Iron efflux is probably mediated by the ferroportin 1/IREG-1 iron exporter. Right: Iron import mechanisms in erythroid cells. These cells endocytose elevated levels of iron via endocytosis of transferrin mediated by transferrin receptor and transferrin receptor 2, both highly expressed on the membranes of erythroid cells. The large amount of iron endocytosed into erythroid cells is incorporated into mitochondria, where iron is used for heme synthesis.

Fe–S proteins. This function is mediated by some mitochondrial proteins, such as Nfs1, that initiate biogenesis of Fe–S; other proteins, such as the ABC transporter Atm1, mediate the export of Fe–S cluster precursors.[621] Cytosolic Fe–S clusters play a key role in the control of iron metabolism through modulation of iron regulatory protein activity.

REFERENCES

1. Frieden, E. The biochemical evolution of the iron and copper proteins, in *Trace Element Metabolism in Animals*, Hoekstra, E.E., Ed., Baltimore, University Park, 105, 1974.
2. Neilands, J.B. Microbial metabolism of iron, in *Iron in Biochemistry and Medicine*, Jacobs, A. and Worwood, M., Eds., London, Academic Press, 329, 1980.

3. McClendon, J.H. Elemental abundance as a factor in the origin of mineral nutrient requirements, *J. Mol. Evol.*, 8: 175–191, 1976.

4. Sylva, R.N. The hydrolysis of iron (III). *Rev. Pure Appl. Chem.*, 22: 115–132, 1972.

5. Huebers, H. et al. The significance of transferrin for intestinal iron adsorption, *Blood*, 61: 283–290, 1983.

6. Huebers, H.A. and Finch, C.A., Transferrin: physiologic behaviour and clinical implications, *Blood*, 64: 763–790, 1984.

7. Williams, R.J.P. Haeme protein and oxygen, in *Iron in Biochemistry and Medicine*, Jacobs, A. and Worwood, M., Eds., London, Academic Press, 183, 1980.

8. Reichard, P. and Ehzenbergn, A. Ribonucleotide reductase: a radical enzyme, *Science*, 221, 514–518, 1983.

9. Kaplan, J. Polypeptide binding membrane receptor: analysis and classification, *Science*, 212, 14–20, 1981.

10. Youngdahl-Turner, P. et al. Protein mediated vitamin up-take, *Exp. Cell Res.*, 118, 127–137, 1979.

11. Brown, M.S. and Goldstein, J. Analysis of the activity of the low density lipoprotein receptor in human fibroblasts, *Cell*, 6: 307–320, 1975.

12. Levy, J.E. et al. Transferrin receptor is necessary for development of erythrocytes and the nervous system, *Nature Genet.*, 21: 396–399, 1999.

13. Schneider, C. et al. Primary structure of human transferrin receptor deduced from the mRNA sequence, *Nature*, 311: 675–678, 1984.

14. Kühn, L.C., McClelland, A., and Ruddle, F.H. Gene transfer, expression, and molecular cloning of the human transferrin receptor gene, *Cell*, 37: 95–103, 1984.

15. Alvarez, E., Girones, N., and Davis, R.J. inhibition of the receptor-mediated endocytosis of diferric transferrin is associated with the covalent modification of the transferrin receptor with palmitic acid, *J. Biol. Chem.*, 265: 16644–16655, 1990.

16. Davis, R.J. et al. Identification of serine 24 as the unique site on the transferrin receptor phosphorylated by protein kinase C, *J. Biol. Chem.*, 261: 9034–9041, 1986.

17. Jing, S. and Trowbridge, I.S. Identification of the intermolecular disulfide of the human transferrin receptor and its lipid-attachment site, *EMBO J.*, 6: 327–333, 1987.

18. Enns, C.A. et al. Acquisition of the functional properties of the transferrin receptor during its biosynthesis, *J. Biol. Chem.*, 266: 13272–13277, 1991.

19. Buccheger F. et al. Functional analysis of human/chicken transferrin receptor chimeras indicates that the carboxy-terminal region is important for ligand binding, *Eur. J. Biochem.*, 235: 9–17, 1996.

20. Fuchs, H. et al. Structural model of phospholipid-reconstituted human transferrin receptor derived by electron microscopy, *Structure*, 6: 1235–1249, 1998.

21. Lawrence, C.M. et al. Crystal structure of the ectodomain of human transferrin receptor, *Science*, 286: 779–782, 1999.

22. Enns, C.A. et al. Human transferrin receptor: expression of the receptor is assigned to chromosome 3, *Proc. Natl. Acad. Sci. U.S.A.*, 79: 3241–3245, 1982.

23. Goodfellow, P. et al. Expression of human transferrin receptor is controlled by a gene on chromosome 3: assignment using specificity of a monoclonal antibody, *Somatic Cell. Genet.*, 8: 197–205, 1982.

24. Van di Rign, M. et al. Localization of a gene controlling the expression of the human transferrin receptor to the region q12-qter of chromosome 3, *Cytogenet. Cell. Genet.*, 36: 525–531, 1983.

25. Miller, Y.E. et al. Chromosome 3q(22-ter) encodes the human transferrin receptor, *Am. J. Hum. Genet.*, 35, 573–580, 1983.

26. Rabin, M. et al. Regional localization of human transferrin receptor gene to 3q26.2-ter, *Am. J. Hum. Genet.*, 37, 1112–1118, 1985.

27. Kashuba, V.I. et al. Not-I linking/jumping clones of human chromosome 3, *FEBS Lett.*, 419: 181–185, 1997.

28. Yang, F. et al. Human transferrin cDNA characterization and chromosomal localization, *Proc. Natl. Acad. Sci. U.S.A.*, 81: 2752–2576, 1984.

29. Brown, J.P. et al. Human melanoma-associated antigen p97 is structurally and functionally related to transferrin, *Nature*, 296: 171–173, 1982.

30. Grego, B. et al. A microbore high-performance liquid chromatography strategy for the purification of polypeptides for gas-phase sequence analysis. Structural studies on the murine transferrin receptor, *Eur. J. Biochem.*, 148: 485–491, 1985.

31. Roberts, K.P. and Griswold, M.D. Characterization of rat transferrin receptor cDNA: the regulation of transferrin receptor mRNA in testes and in Sertoli cells in culture, *Mol. Endocrinol.*, 4: 531–542, 1990.

32. Collawn, J.F. et al. YTRF is the conserved internalization signal of the transferrin receptor, and a second YTRF signal at position 31–34 enhances endocytosis, *J. Biol. Chem.*, 268: 21686–21692, 1993.

33. Gerhardt, E.M. et al. The cDNA sequence and primary structure of the chicken transferrin receptor, *Gene*, 102: 249–254, 1991.

34. Titeux, M. et al. The role of iron in the growth of human leukemic cell lines, *J. Cell. Physiol.*, 121: 251–261, 1984.

35. McClelland, A., Kühn, L.C., and Ruddle, F.H. The human transferrin receptor gene: genomic organization, and the complete primary structure of the receptor deduced from a cDNA sequence, *Cell*, 39: 267–274, 1984.

36. Evans, P. and Kemp, J. Exon/intron structure of the human transferrin receptor gene, *Gene*, 199: 123–131, 1997.

37. Van Landeghem, G.F. et al. New DANN polymorphisms define ethnically distinct hapoplytes in the human transferrin receptor gene, *Hum. Hered.*, 48: 245–250, 1999.

38. Kawabata, H. et al. Molecular cloning of transferrin receptor 2, *J. Biol. Chem.*, 274: 20826–20832, 1999.

38a. Fleming, R.E. et al. Transferrin receptor 2: continued expression in mouse liver in the face of iron overload and in hereditary hemochromatosis, *Proc. Natl. Acad. Sci. U.S.A.*, 97: 2214–2219, 2000.

38b. Kawabata, H. et al. Transferin receptor-2 supports cell growth both in iron-chelated cultured cells and *in vivo*, *J. Biol. Chem.*, 275: 16618–16625, 2000.

39. Carter, R.E., Feldman, A.R., and Coyle, J.T. Prostate-specific membrane antigen is a hydrolase with substrate and pharmacologic characteristics of a neuropeptidase, *Proc. Natl. Acad. Sci. U.S.A.*, 93: 749–753, 1996.

40. Luthi-Carter, R. et al. Isolation and expression of a rat brain cDNA encoding glutamate carboxypeptidase II, *Proc. Natl. Acad. Sci. U.S.A.*, 95: 2315–3220, 1998.

41. Bzdega, T. et al. Molecular cloning of a peptidase N-acetylaspartylglutamate from a rat hippocampal cDNA library, *J. Neurochem.*, 69: 2270–2277, 1997.

42. Rawlings, N.D. and Barret, A.J. Structure of membrane glutamate carboxypeptidase, *Biochim. Biophys. Acta*, 1339: 247–252, 1997.

43. Neefjes, J.J. et al. Recycling glycoproteins do not return to the *cis*-Golgi, *J. Cell. Biol.*, 107: 79–87, 1988.

44. Do, S.I., Enns, C., and Cummings, R.D., Human transferrin receptor contains O-linked oligosaccharides, *J. Biol. Chem.*, 265: 114–125, 1990.

342 Proteins of Iron Metabolism

45. Hayes, G.R., Enns, C.A., and Lucas, J.J. Identification of the O-linked glycosylation site of the human transferrin receptor, *Glycobiology*, 2: 355–359, 1992.
46. Do, S.I. and Cummings, R.D. Presence of O-linked oligosaccharide on a threonine residue in the human transferrin receptor, *Glycobiology*, 2: 345–351, 1992.
47. Rutledge, E.A. et al. Elimination of the O-linked glycosylation site at the 104 results in the generation of a soluble human-transferrin receptor, *Blood*, 83: 580–586, 1994.
48. Orberger, G. et al. Structure of the N-linked oligosaccharides of the human transferrin receptor, *Eur. J. Biochem.*, 205: 257–267, 1992.
49. Hayes, G.R. et al. Structure of human transferrin receptor oligosaccharides: conservation of site-specific processing, *Biochemistry*, 36: 5276–5284, 1997.
50. Hayes, G.R. et al. The critical glycosylation site of human transferrin receptor contains a high-mannose oligosaccharide, *Glycobyology*, 5: 227–232, 1995.
51. Reckhow, C.L. and Enns, C.A. Characterization of the transferrin receptor in tunicamycin-treated A431 cells, *J. Biol. Chem.*, 263: 7297–7301, 1988.
52. Williams, A.N. and Enns, C.A. A mutated transferrin receptor lacking asparagine-linked glycosylation sites shows reduced functionality and an association with binding immunoglobulin protein, *J. Biol. Chem.*, 266, 17648–17654, 1991.
53. Hoe, M.H. and Hunt, R.C. Loss of one asparagine-linked oligosaccharide from human transferrin receptors results in specific cleavage and association with the endoplasmic reticulum, *J. Biol. Chem.*, 267: 4916–4923, 1992.
54. Yang, B. et al. Role of oligosaccharides in the processing and function of human transferrin receptors, *J. Biol. Chem.*, 268: 7435–7441, 1993.
55. Williams, A.M. and Enns, C.A. A region of the C-terminal position of the human transferrin receptor contains an asparagine-linked glycosylation site critical for receptor structure and function, *J. Biol. Chem.*, 268: 12780–12786, 1993.
56. Ternh, H.J. et al. Human transferrin receptor is active and plasma membrane-targeted in yeast, *FEMS Microbiol. Lett.*, 160: 61–67, 1998.
57. Georgieff, M.K. et al. Increased N-glycosylation and reduced transferrin-binding capacity of transferrin receptor isolated from placentae of diabetic women, *Placenta*, 18: 563–568, 1997.
58. Trowbridge, I.S., Collwan, J.F., and Hopkins, C.R. Signal-dependent membrane protein trafficking in the endocytic pathway, *Annu. Rev. Cell. Biol.*, 9: 129–162, 1993.
59. Schmid, S.L. Clathrin-coated vesicle formation and protein sorting, *Annu. Rev. Biochem.*, 66: 511–548, 1997.
60. Hopkins, C.R., Miller, K., and Beardmore, J.M. Receptor-mediated endocytosis of transferrin and epidermal growth factor receptors: a comparison of constitutive and ligand-induced uptake, *J. Cell Sci.*, 3: 173–186, 1985.
61. Mukherjee, S., Ghosh, R.N., and Maxfield, F.R. Endocytosis, *Physiol. Rev.*, 77: 759–803, 1997.
62. Marks, M.S. et al. Protein targeting by tyrosine- and di-leucine-based signals: evidence for distinct saturable components, *J. Cell Biol.*, 135: 341–354, 1996.
63. Dell'Angelica, E.C. et al. AP-3: an adaptor-like protein complex with ubiquitous expression, *EMBO J.*, 16: 917–928, 1997.
64. Nesterov, A. et al. Inhibition of the receptor-binding function of clathrin adaptor AP-2 by dominant-negative mutant μ2 subunit and its effects on endocytosis, *EMBO J.*, 18: 2489–2499, 1999.
65. Folsch, H. et al. A novel clathrin adaptor complex mediates basolateral targeting in polarized epithelial cells, *Cell*, 99: 189–198, 1999.
66. Owen, D.J. et al. A structural explanation for the binding of multiple ligands by the α-adaptin appendage domain, *Cell*, 97: 805–815, 1999.

67. Sever, S., Muhlberg, A.B., and Schmid, S.L. Impairment of dynamin's GAP domain stimulates receptor-mediated endocytosis, *Nature,* 398: 481–486, 1999.

68. Benmerak, A. et al. AP-2/Eps15 interaction is required for receptor-mediated endocytosis, *J. Cell. Biol.,* 140: 1055–1062, 1998.

69. Owen, D.J. et al. Crystal structure of the amphiphysin-2 SH3 domain and its role in the prevention of dynamin ring formation, *EMBO J.,* 17: 5273–5285, 1998.

70. Simpson, F. et al. SH3-domain-containing proteins function at distinct steps in clathrin-coated vesicle formation, *Nature Cell. Biol.,* 1: 119–124, 1999.

71. Rosenthal J.A. et al. The epsins define a family of proteins that interact with components of the clathrin coat and contain a new protein module, *J. Biol. Chem.,* 274: 33959–33965, 1999.

72. Lamaze, C. et al. The actin cytoskeleton is required for receptor-mediated endocytosis in mammalian cells, *J. Biol. Chem.,* 272: 20332–20335, 1997.

73. Thatte, H.S., Bridges, K.R., and Golan, D.E. Microtubule inhibitors differentially affect translation movement, cell surface expression and endocytosis of transferrin receptors in K562 cells, *J. Cell Physiol.,* 160: 345–357, 1994.

74. Subtil, A. and Dantry-Versat, A. Microtubule depolymerization inhibits clathrin coated-pit internalization in non-adherent cell lines while interleukin-2 endocytosis is not affected, *J. Biol. Chem.,* 272: 20332–20335, 1997.

75. Ohno, H. et al. Structural determinants of interaction of tyrosine-based sorting signals with the adaptor medium chains, *J. Biol. Chem.,* 271: 29009–29015, 1996.

76. Hansen, S.H., Sandvig, K., and Van Deurs, B. Internalization efficiency of the transferrin receptor, *Exp. Cell Res.,* 199: 19–29, 1992.

77. Sandvig, K. and Van Deurs, B. Selective modulation of the endocytic uptake of ricin and fluid phase markers without alteration in transferrin endocytosis, *J. Biol. Chem.,* 265: 6382–6388, 1990.

78. Liu, G. et al. Cytoskeletal protein ABP-280 directs the intracellular trafficking of furin and modulates protein processing in the endocytic pathway, *J. Cell. Biol.,* 139: 1719–1733, 1997.

79. Gaidarov, I. et al. Spatial control of coated-pit dynamics in living cells, *Nature Cell. Biol.,* 1: 1–7, 1999.

80. van der Sluijs, P. et al. The small GTP-binding protein rab4 is associated with early endosomes, *Proc. Natl. Acad. Sci. U.S.A.,* 88: 6313–6317, 1991.

81. Bucci, C. et al. The small GTPase rab5 functions as a regulatory factor in the early endocytic pathway, *Cell,* 70: 715–728, 1992.

82. van der Sluijs, P. et al. The small GTP-binding protein rab4 controls an early sorting event on the endocytic pathway, *Cell,* 70: 729–740, 1992.

83. Daro, E. et al. Rab4 and cellubrevin define different early endosome populations on the pathway of transferrin receptor recycling, *Proc. Natl. Acad. Sci. U.S.A.,* 93: 9559–9566, 1996.

84. Nelsen, E. et al. Rab5 regulates motility of early endosomes on microtubules, *Nature Cell. Biol.,* 1: 376–382, 1999.

85. Stenmark, H. et al. Rabaptin-5 is a direct effector of the small GTPase rab5 in endocytic membrane fusion, *Cell,* 83: 423–432, 1995.

86. Ohya, T. et al. Involvement of rabphillin 3 in endocytosis through interaction with rabaptin 5, *J. Biol. Chem.,* 273: 613–617, 1998.

87. Zuk, P.A. and Elferink, L.A. Rab15 mediates an early endocytic event in Chinese hamster ovary cells, *J. Biol. Chem.,* 274: 22303–22312, 1999.

88. Ullrich, O. et al. Rab11 regulates recycling through the pericentriolar recycling endosome, *J. Cell Biol.,* 135: 913–924, 1996.

89. Green, E.G. et al. Rab 11 is associated with transferrin-containing recycling compartments in K562 cells, *Biochem. Biophys. Res. Commun.*, 239: 612–616, 1997.

90. Mamoto, A. et al. Rab11BP/rabiphilin-11, a downstream target of Rab11 small G protein implicated in vesicle recycling, *J. Biol. Chem.*, 274: 25517–25524, 1999.

91. Aniento, F. et al. An endosomal ε-COP is involved in the pH-dependent formation of transport vesicles destined for late endosomes, *J. Cell Biol.*, 133: 29–41, 1996.

92. Whitney, J. et al. Cytoplasmic coat protein involved in endosome function, *Cell*, 83: 703–713, 1995.

93. Daro, E. et al. Inhibition of endosome function in CHO cells bearing a temperature-sensitive defect in the coatomer (COPI) component ε-COP, *J. Cell. Biol.*, 139: 1747–1759, 1997.

94. McMahou, H.T. et al. Cellubrevin is a ubiquitous tetanus-toxin substrate homologous to a putative synoptic vesicle fusion protein, *Nature*, 364: 346–349, 1993.

95. Galli, T. et al. Tetanus toxin-mediated cleavage of cellubrevin impairs exocytosis of transferrin receptor-containing vesicles in CHO cells, *J. Cell Biol.*, 125: 1015–1024, 1994.

96. Annaert, W.G. et al. Export of cellubrevin from the endoplasmic reticulum is controlled by BAP31, *J. Cell. Biol.*, 139: 1397–1410, 1997.

97. Stoorvogel, W., Oorshot, V., and Genze, H.J. A novel class of clathrin-coated vesicles budding from endosomes, *J. Cell. Biol.*, 132: 21–33, 1996.

98. Schonhorn, J.E. and Wessling-Resnick, M. Brefeldin down-regulates the transferrin receptor in K562 cells, *Mol. Cell. Biochem.*, 135: 159–169, 1994.

99. D'Souza-Schorey, C. et al. A regulatory role for ARF6 in receptor-mediated endocytosis, *Science*, 267: 1175–1178, 1995.

100. Peters, P.J. et al. Overexpression of wild-type and mutant ARF1 and ARF6: distinct perturbations of nonoverlapping membrane compartments, *J. Cell. Biol.*, 128: 1003–1017, 1995.

101. D'Souza-Schorey, C. et al. ARF6 targets recycling vesicles to the plasma membrane: insights from an ultrastructural investigation, *J. Cell. Biol.*, 140: 603–616, 1998.

102. Radhakrishna, H. and Donaldson, J.G. ADP-ribosylation factor 6 regulates a novel plasma membrane recycling pathway, *J. Cell. Biol.*, 139: 49–61, 1997.

103. Franco, M. et al. EFA6, a sec7 domain-containing exchange factor for ARF6, coordinates membrane recycling and cytoskeleton organization, *EMBO J.*, 18: 1480–1491, 1999.

104. Lamaze, C. et al. Regulation of receptor-mediated endocytosis by Rho and Rac, *Nature*, 382: 177–179, 1996.

105. Jess, T.J. et al. Phosphatidylinositol 3'-kinase, but not p70 ribosomal S6 kinase, is involved in membrane protein recycling: wortmannin inhibits glucose transport and downregulates cell-surface transferrin receptor numbers independently of any effect on fluid-phase endocytosis in fibroblasts, *Cell Signal*, 8: 297–304, 1996.

106. Spiro, D.J. et al. Wortmannin alters the transferrin receptor endocytic pathway *in vivo* and *in vitro*, *Mol. Biol. Cell.*, 7: 355–367, 1996.

107. Dautry-Varsat, A., Ciechanoven, A., and Lodish, H.F. pH and recycling of transferrin during receptor-mediated endocytosis, *Proc. Natl. Acad. Sci. U.S.A.*, 80: 2258–2262, 1983.

108. Presley, J.F. et al. Bafilomycin A1 treatment retards transferrin receptor recycling more than bulk membrane recycling, *J. Biol. Chem.*, 272: 13929–13936, 1997.

109. Altan, N. et al. Tamoxifen inhibits acidification in cells independent of the estrogen receptor, *Proc. Natl. Acad. Sci. U.S.A.*, 96: 4432–4437, 1999.

110. Henkel, J.R. et al. Selective perturbation of early endosome and/or trans-Golgi network pH but not lysosome pH by dose-dependent expression of influenza M2 protein, *J. Biol. Chem.*, 274: 9854–9860, 1999.

111. Codina, J., Gurich, R., and Dubose, T.D. Peptides derived from the human transferrin receptor stimulate endosomal acidification via a Gi-type protein, *Kidney Int.*, 55: 2376–2382, 1999.

112. Stoorvogel, W., Geuze, H.J., and Strous, G.J. Sorting of endocytosed transferrin and asialoglycoprotein occurs immediately after internalization in HepG2 cells, *J. Cell. Biol.*, 104: 1261–1268, 1987.

113. Callus, B.A. et al. Effects of overexpression of the transferrin receptor on the rates of transferrin recycling and uptake of non-transferrin-bound iron, *J. Biochem.*, 238: 463–469, 1996

114. Warren, R.A., Green, F.A., and Enns, C.A. Saturation of the endocytic pathway for the transferrin receptor does not affect the endocytosis of the epidermal growth factor receptor, *J. Biol. Chem.*, 272: 2116–2121, 1997.

115. Marsh, E.W. et al. Oligomerized transferrin receptors are selectively retained by a lumenal sorting signal in a long-lived endocytic recycling compartment, *J. Cell. Biol.*, 129: 1509–1522, 1995

116. Núñez, M.T. et al. Mobilization of iron from endocytic vesicles. The effects of acidification and reduction, *J. Biol. Chem.*, 265: 6688–6692, 1990.

117. Núñez, M.T. et al. Apotransferrin and holotransferrin undergo different endocytic cycles in intestinal epithelia (Caco-2) cells, *J. Biol. Chem.*, 272: 19425–19428, 1997.

117a. Fleming, M.D. et al. Nramp2 is mutated in the anemic Belgrade rat: evidence of a role for Nramp2 in endosomal iron transport, *Proc. Natl. Acad. Sci. U.S.A.*, 95: 1148–1153, 1998.

117b. Gruenheid, S. et al. The iron transport protein NRAMP2 is an integral membrane glycoprotein that colocalizes with transferrin in recycling endosomes, *J. Exp. Med.*, 189: 831–841, 1999.

118. Raso, V., Brown, M., and McGrath, J. Intracellular targeting with low pH-triggered bispecific antibodies, *J. Biol. Chem.*, 272: 27623–27628, 1997.

119. Pearse, B.M. F. and Robinson, M.S. Clathrin, adaptors and sorting, *Annu. Rev. Cell. Biol.*, 6: 151–171, 1990

120. Rothenberger, S., Iacopetta, B.J., and Kühn, L.C. Endocytosis of the transferrin receptor requires the cytoplasmic domain but not its phosphorylation site, *Cell*, 49: 423–431, 1987.

121. Trowbridge, I.S., Collawn, J.F., and Hopkins, C.R. Signal-dependent membrane protein trafficking in the endocytic pathway, *Annu. Rev. Cell. Biol.*, 9: 129–161, 1993.

122. Davis, C.G. et al. The low density lipoprotein receptor. Identification of amino acids in cytoplasmic domain required for rapid endocytosis, *J. Biol. Chem.*, 262: 4075–4082, 1987.

123. Chen, W.J., Goldstein, J.L., and Brown, M.S. NPXY, a sequence often found in cytoplasmic tails, is required for coated pit-mediated internalization of the low density lipoprotein receptor, *J. Biol. Chem.*, 265: 3116–3123, 1990.

124. Jing, S.Q. et al. Role of the human transferrin receptor cytoplasmic domain in endocytosis: localization of a specific signal sequence for internalization, *J. Cell. Biol.*, 110: 283–294, 1990.

125. McGraw, T.E. and Maxfield, F.R. Human transferrin receptor internalization is partially dependent upon an aromatic amino acid on the cytoplasmic domain, *Cell. Regul.*, 1: 369–377, 1990.

126. Alvarez, E., Girones, N., and Davis, R.J. A point mutation in the cytoplasmic domain of the transferrin receptor inhibits endocytosis, *Biochem. J.,* 267: 31–35, 1990.

127. Collawn, J.F. et al. Transferrin receptor internalization sequence YXRF implicates a tight turn as the structural recognition motif for endocytosis, *Cell,* 63: 1061–1072, 1990.

128. Collawn, J.F. et al. Transplanted LDL and mannose-6-phosphate receptor internalization signals promote high-efficiency endocytosis of the transferrin receptor, *EMBO J.,* 10: 3247–3253, 1991.

129. McGraw, T.E. et al. Mutagenesis of the human transferrin receptor: two cytoplasmic phenylalanines are required for efficient internalization and a second-site mutation is capable of reverting an internalization-defective phenotype, *J. Cell Biol.,* 112: 853–861, 1991.

130. Girones, N. et al. Mutational analysis of the cytoplasmic tail of the human transferrin receptor. Identification of a sub-domain that is required for rapid endocytosis, *J. Biol. Chem.,* 266: 19006–19012, 1991.

131. Subtil, A., Delepierre, M., and Dautry-Varsat, A. An alpha-helical signal in the cytosolic domain of the interleukin 2 receptor beta chain mediates sorting toward degradation after endocytosis, *J. Cell. Biol.,* 136: 583–595, 1997.

132. Collown, J.F. et al. YTRF is the conserved internalization signal of the transferrin receptor; a second YTRF signal at the position 31-34 enhances endocytosis, *J. Biol. Chem.,* 268: 21686–21692, 1993.

133. Pytowski, B., Judge, T.W., and McGraw, T.E. An internalization motif is created in the cytoplasmic domain of the transferrin receptor by substitution of a tyrosine at the first position of a predicted tight turn, *J. Biol. Chem.,* 270: 9067–9073, 1995.

134. Warren, R.A., Green, F.A., and Enns, C.A. Saturation of the endocytic pathway for the transferrin receptor does not affect the endocytosis of the epidermal growth factor, *J. Biol. Chem.,* 272: 2116–2121, 1997.

135. Warren, R.A. et al. Distinct saturable pathways for the endocytosis of different tyrosine motifs, *J. Biol. Chem.,* 273: 17056–17063, 1998.

136. Nordeng, T.W. and Bakke, O. Overexpression of proteins containing tyrosine-or leucine-based sorting signals affects transferrin receptor trafficking, *J. Biol. Chem.,* 274: 21139–21148, 1999.

137. Zerial, M. et al. Phosphorylation of the human transferrin receptor by protein kinase C is not required for endocytosis and recycling in mouse 3T3 cells, *EMBO J.,* 6: 2661–2667, 1987.

138. Dargemont, C. et al. The internalization signal and the phosphorylation site of transferrin receptor are distinct from the main basolateral sorting information, *EMBO J.,* 12: 1713–1721, 1993.

139. Alvarez, A., Girones, N., and Davis, R.J. Inhibition of the receptor-mediated endocytosis of diferric transferrin is associated with the covalent modification of the transferrin receptor with palmitic acid, *J. Biol. Chem.,* 265: 16644–16655, 1990.

140. Jing, S.Q. and Trowbridge, I.S. Nonacylated human transferrin receptors are rapidly internalized and mediate iron uptake, *J. Biol. Chem.,* 265: 11555–11559, 1990.

141. Odorizzi, G. et al. Apical and basolateral endosomes of MDCK cells are interconnected and contain a polarized sorting mechanism, *J. Cell. Biol.,* 135: 139–152, 1996.

142. Odorizzi, G. and Trowbridge, I.S. Structural requirements for basolateral sorting of the human transferrin receptor in the biosynthetic and endocytic pathways of Madin-Darby canine kidney cells, *J. Cell Biol.,* 137: 1255–1264, 1997.

143. Hemar, A. et al. Dendroaxonal transcytosis of transferrin in cultured hippocampal and sympathetic neurons, *J. Neurosci.,* 17: 9026–9034, 1997.

144. West, A.E., Neve, R.L., and Buckley, K.M. Identification of a somatodendritic targeting signal in the cytoplasmic domain of the transferrin receptor, *J. Neurosci.*, 17: 6038–6047, 1997.

145. Schmidt, A., Hannah, M.J., and Huttner, W.B. Synaptic-like microvesicles of neuroendocrine cells originate from a novel compartment that is continuous with the plasma membrane and devoid of transferrin receptor, *J. Cell. Biol.*, 137: 445–458, 1997.

146. West, A.E., Neve, R.L., and Buckley, K.M. Targeting of the synaptic vesicle protein synaptobrevin in the axon of cultured hippocampal neurons: evidence for two distinct sorting steps, *J. Cell Biol.*, 139: 917–927, 1997.

147. Testa, U., Pelicci, P.G., Thomopoulos, P., Titeux, M., Rochant, H. The number of the transferrin receptors on human hematopoietic cell lines is influenced by membrane phospholipids, *Biochem. Int.*, 7: 169–178, 1983.

148. Klausner, R.D., Harford, J., and van Renswoude, J. Rapid internalization of the transferrin receptor in K562 cells is triggered by ligand binding or treatment with a phorbol ester, *Proc. Natl. Acad. Sci. U.S.A.*, 81: 3005–3009, 1984.

149. May, W.S., Jacobs, S., and Cuatrecasas, P. Association of phorbol ester-induced hyperphosphorylation and reversible regulation of transferrin membrane receptors in HL-60 cells, *Proc. Natl. Acad. Sci. U.S.A.*, 81: 2016–2020, 1984.

150. May, W.S. et al. Mechanism of phorbol diester-induced regulation of surface transferrin receptor involves the action of activated protein kinase C and an intact cytoskeleton, *J. Biol. Chem.*, 260: 9419–9426, 1985.

151. Davis, R.J. et al. Identification of serine 24 as the unique site on the transferrin receptor phosphorylated by protein kinase C, *J. Biol. Chem.*, 261: 9034–9041, 1986.

152. Rothenberger, S., Iacopetta, B.J., and Kühn, L.C. Endocytosis of the transferrin receptor requires the cytoplasmic domain but not its phosphorylation site, *Cell*, 49: 423–431, 1987.

153. Zerial, M. et al. Phosphorylation of the human transferrin receptor by protein kinase C is not required for endocytosis and recycling in mouse 3T3 cells, *EMBO J.*, 6: 2661–2667, 1987.

154. Davis, R.J. and Meisner, H. Regulation of transferrin receptor cycling by protein kinase C is independent of receptor phosphorylation at serine 24 in Swiss 3T3 fibroblasts, *J. Biol. Chem.*, 262: 16041–16047, 1987.

155. McGraw, T.E., Dunn, K.W., and Maxfield, F.R. Phorbol ester treatment increases the exocytic rate of the transferrin receptor recycling pathway independent of serine-24 phosphorylation, *J. Cell Biol.*, 106: 1061–1066, 1988.

156. Eichholtz, T. et al. Activation of protein kinase C accelerates internalization of transferrin receptor but not of major histocompatibility complex class I, independent of their phosphorylation status, *J. Biol. Chem.*, 267: 22490–22495, 1992.

157. Schonhorn, J.E., Akompong, T., and Wessling-Resnick, M. Mechanism of transferrin receptor down-regulation in K562 cells in response to protein kinase C activation, *J. Biol. Chem.*, 270: 3698–3705, 1995.

158. Fallon, R.J. et al. Defective asialoglycoprotein receptor endocytosis mediated by tyrosine kinase inhibitors. Requirement for a tyrosine in the receptor internalization signal, *J. Biol. Chem.*, 269: 11011–11017, 1994.

159. Salmerón, A. et al. Transferrin receptor induces tyrosine phosphorylation in T cells and is physically associated with the TCR zeta-chain, *J. Immunol.*, 154: 1675–1683, 1995.

160. Cotlin, L.F. et al. Casein kinase II activity is required for transferrin receptor endocytosis, *J. Biol. Chem.*, 274: 30550–30556, 1999.

161. Beauchamp, J.R. and Woodman, P.G. Regulation of transferrin receptor recycling by protein phosphorylation, *Biochem. J.,* 303: 647–655, 1994.

162. Runnegar, M. et al. Transferrin receptor recycling in rat hepatocytes is regulated by protein phosphatase 2A, possibly through effects on microtubule-dependent transport, *Hepatology,* 26: 176–185, 1997.

163. Lamb, J.E. et al. Internalization and subcellular localization of transferrin and transferrin receptors in HeLa cells, *J. Biol. Chem.,* 258: 8751–8758, 1983

164. Ajioka, R.S. and Kaplan, J. Intracellular pools of transferrin receptors result from constitutive internalization of unoccupied receptors, *Proc. Natl. Acad. Sci. U.S.A.,* 83: 6445–6449, 1986.

165. Hopkins, C.R., Intracellular routing of transferrin and transferrin receptors in epidermoid carcinoma A431 cells, *Cell,* 35: 321–330, 1983.

166. Hopkins, C.R. and Trowbridge, I.S. Internalization and processing of transferrin and the transferrin receptor in human carcinoma A431 cells, *J. Cell. Biol.,* 97: 508–521, 1983.

167. Willingham, M.C. et al. Morphologic characterization of the pathway of transferrin endocytosis and recycling in human KB cells, *Proc. Natl. Acad. Sci. U.S.A.,* 81: 175–179, 1984.

168. Yamashiro, D.J. et al. Segregation of transferrin to a mildly acidic (pH 6.5) para-Golgi compartment in the recycling pathway, *Cell,* 37: 789–800, 1984.

169. Octave, J.N. et al. Transferrin uptake by cultured rat embryo fibroblasts. The influence of temperature and incubation time, subcellular distribution and short-term kinetic studies, *Eur. J. Biochem.,* 115: 611–618, 1981.

170. Snider, M.D. and Rogers, O.C. Intracellular movement of cell surface receptors after endocytosis: resialylation of asialo-transferrin receptor in human erythroleukemia cells, *J. Cell. Biol.,* 100: 826–834, 1985.

171. Snider, M.D. and Rogers, O.C. Membrane traffic in animal cells: cellular glycoproteins return to the site of Golgi mannosidase I, *J. Cell. Biol.,* 103: 265–275, 1986.

172. Woods, J.W., Doriaux, M., and Farquhar, M.G. Transferrin receptors recycle to *cis* and middle as well as *trans*-Golgi cisternae in Ig-secreting myeloma cells, *J. Cell. Biol.,* 103: 277–286, 1986.

173. Stein, B.S. and Sussman, H.H. Demonstration of two distinct transferrin receptor recycling pathways and transferrin-independent receptor internalization in K562 cells, *J. Biol. Chem.,* 261: 10319–10331, 1986.

174. Fishman, J.B. and Fine, R.E. A *trans*-Golgi-derived exocytic coated vesicle can contain both newly synthesized cholinesterase and internalized transferrin, *Cell,* 48: 157–164, 1987.

175. Stoorvogel, W. et al. The pathways of endocytosed transferrin and secretory protein are connected in the trans-Golgi reticulum, *J. Cell. Biol.,* 106: 1821–1829, 1988.

176. Draper, R.K. et al. Antibodies to clathrin inhibit endocytosis but not recycling to the *trans*-Golgi network *in vitro, Science,* 248: 1539–1541, 1990.

177. Robertson, B.J., Park, R.D., and Snider, M.D. Role of vesicular traffic in the transport of surface transferrin receptor to the Golgi complex in cultured human cells, *Arch. Biochem. Biophys.,* 292: 190–198, 1992.

178. Jin, M. and Snider, M.D. Role of microtubules in transferrin receptor transport from the cell surface to endosomes and the Golgi complex, *J. Biol. Chem.,* 268: 18390–18397, 1993.

179. Futter, C.E. et al. Newly synthesized transferrin receptors can be detected in the endosome before they appear on the cell surface, *J. Biol. Chem.,* 270: 10999–11003, 1995.

180. Stinchcombe, J.C. et al. Anterograde and retrograde traffic between the rough endoplasmic reticulum and the Golgi complex, *J. Cell. Biol.*, 131: 1387–1401, 1995.

181. Hunt, R.C. and Marshall-Carlson, L. Internalization and recycling of transferrin and its receptor. Effect of trifluoperazine on recycling in human erythroleukemic cells, *J. Biol. Chem.*, 261: 3681–3686, 1986.

182. Grasso, J.A. et al. Calmodulin dependence of transferrin receptor recycling in rat reticulocytes, *Biochem. J.*, 266: 261–272, 1990.

183. Apodaca, G., Enrich, C., and Mostov, K.E. The calmodulin antagonist, W-13, alters transcytosis, recycling, and the morphology of the endocytic pathway in Madin-Darby canine kidney cells, *J. Biol. Chem.*, 269: 19005–19013, 1994.

184. Morimoto, T. et al. Calcium-dependent transmitter secretion from fibroblasts: modulation by synaptotagmin I, *Neuron*, 15: 689–696, 1995.

185. Chapin, S.J. et al. Calmodulin binds to the basolateral targeting signal of the polymeric immunoglobulin receptor, *J. Biol. Chem.*, 271: 1336–1342, 1996,

186. Morgan, E.H. Calcium chelators induce association with the detergent-insoluble cytoskeleton and functional inactivation of the transferrin receptor in reticulocytes, *Biochim. Biophys. Acta*, 981: 121–129, 1989.

187. Vidal, M. et al. A GTP-binding protein modulates a Ca^{2+} pump present in reticulocyte endocytic vesicles, *Biochem. Mol. Biol. Int.*, 35: 889–898, 1995.

188. Sainte-Marie J. et al. Transferrin receptor functions as a signal-transduction molecule for its own recycling via increases in the internal Ca^{2+} concentration, *Eur. J. Biochem.*, 250: 689–697, 1997.

189. Poss, K.D. and Tonegawa, S. Reduced stress defense in heme oxygenase 1-deficient cells, *Proc. Natl. Acad. Sci. U.S.A.*, 94: 10925–10930, 1997.

190. Pantopoulos, K. and Hentze, M.W. Rapid responses to oxidative stress mediated by iron regulatory protein, *EMBO J.*, 14: 2917–2924, 1995.

191. Malorni, W. et al. Menadione-induced oxidative stress leads to a rapid down-modulation of transferrin receptor recycling, *J. Cell. Sci.*, 106: 309–318, 1993.

192. Malorni, W. et al. Oxidative stress leads to a rapid alteration of transferrin receptor intravesicular trafficking, *Exp. Cell. Res.*, (in press), 1998.

193. Gatter, K.C. et al. Transferrin receptors in human tissues: their distribution and possible clinical relevance, *J. Clin. Pathol.*, 36: 539–545, 1983.

194. Johansson, A. et al. Functional, morphological, and phenotypical differences between rat alveolar and interstitial macrophages, *Am. J. Respir. Cell. Mol. Biol.*, 16: 582–588, 1997.

195. Franklin, W.A. et al. Expansion of bronchial epithelial cell populations by *in vitro* culture of explants from dysplastic and histologically normal sites, *Am. J. Respir. Cell. Mol. Biol.*, 15: 297–304, 1996.

196. Whitney, J.F. et al. Transferrin receptor expression in nonsmall cell lung cancer. Histopathologic and clinical correlates, *Cancer*, 76: 20–25, 1995.

197. Racz, A. et al. Expression analysis of genes at 3q26-q27 involved in frequent amplification in squamous cell carcinoma, *Eur. J. Cancer*, 35: 641–646, 1999.

198. Vogel, W. et al. Heterogeneous distribution of transferrin receptors on parenchymal and nonparenchymal liver cells: biochemical and morphological evidence, *Blood*, 69: 264–270, 1987.

199. Vogel, W. et al. Heterogeneous distribution of transferrin receptors on parenchymal and nonparenchymal liver cells: biochemical and morphological evidence, *Blood*, 69: 264–270, 1987.

200. Young, S.P. and Aise, P. The liver and iron, in *Liver Biology and Pathology*, Arias, I.M. et al., Eds., New York, Raven Press, 535, 1988.

201. Lombard, M. et al. Optimizing the immunohistochemical signal from the transferrin receptor in liver tissue, *Histochem. J.,* 21: 223–227, 1989.

202. Trinder, D., Zak, O., and Aisen, P. Transferrin receptor-independent uptake of differic transferrin by human hepatoma cells with antisense inhibition of receptor expression, *Hepatology,* 23: 1512–1520, 1996.

203. Trinder, D., and Aisen P. Inhibition of uptake of transferrin-bound iron, *Hepatology,* 26: 691–198, 1997.

204. Hirose-Kumagai, A., Sakai, H., and Akamatsu, N. Increase of transferrin receptors in hepatocytes during rat liver regeneration, *Int. J. Biochem.,* 16: 601–605, 1984.

205. Hirose-Kumagai, A. and Akamatsu, N. Change in transferrin receptor distribution in regenerating rat liver, *Biochem. Biophys. Res. Commun.,* 164: 1105–1112, 1989.

206. Cairo, G. and Pietrangelo, A. Transferrin receptor gene expression during rat liver regeneration. Evidence for post-transcriptional regulation by iron regulatory factor B, a second iron-responsive element-binding protein, *J. Biol. Chem.,* 269: 6405–6409, 1994.

207. Cairo, G., Tacchini, L., and Pietrangelo, A. Lack of coordinate control of ferritin and transferrin receptor expression during rat liver regeneration, *Hepatology,* 28: 173–178, 1998.

208. Sciot, R. et al. Lack of hepatic transferrin receptor expression in hemochromatosis, *Hepatology,* 7: 831–837, 1987.

209. Sciot, R. et al. Hepatocellular transferrin receptor expression in secondary siderosis, *Liver,* 9: 52–61, 1989.

210. Sciot, R. et al. Transferrin receptor expression in rat liver: immunohistochemical and biochemical analysis of the effect of age and iron storage, *Hepatology,* 11: 416–427, 1990.

211. Hubert, N. et al. Regulation of ferritin and transferrin receptor expression by iron in human hepatocyte cultures, *J. Hepatol.,* 18: 301–312, 1993.

212. Sciot, R. et al. Transferrin receptor expression in human hepatocellular carcinoma: an immunohistochemical study of 34 cases, *Histopathology,* 12: 53–63, 1988.

213. Sciot, R., Van Eyken, P., and Desmet, V.J. Transferrin receptor expression in benign tumours and in hepatoblastoma of the liver, *Histopathology,* 16: 59–62, 1990.

214. Pascale, R.M. et al. Transferrin and transferrin receptor gene expression and iron uptake in hepatocellular carcinoma in the rat, *Hepatology,* 27: 452–461, 1998.

215. Rogers, J.T. et al. Translational control during the acute phase response. Ferritin synthesis in response to interleukin-1, *J. Biol. Chem.,* 265: 14572–14578, 1990.

216. Hirayama, M. et al. Regulation of iron metabolism in HepG2 cells: a possible role for cytokines in the hepatic deposition of iron, *Hepatology,* 18: 874–880, 1993.

217. Franco, A. et al. Transferrin receptor mediates uptake and presentation of hepatitis B envelope antigen by T lymphocytes, *J. Exp. Med.,* 175: 1195–1205, 1992.

218. Jefferies, W.A. et al. Transferrin receptor on endothelium of brain capillaries, *Nature,* 312: 162–163, 1984.

219. Morris, C.M. et al. Brain iron homeostasis, *J. Inorg. Biochem.,* 47: 257–265, 1992.

220. Morris, C.M. et al. Distribution of transferrin receptors in relation to cytochrome oxidase activity in the human spinal cord, lower brainstem and cerebellum, *J. Neurol. Sci.,* 111: 158–172, 1992.

221. Moos, T. Immunohistochemical localization of intraneuronal transferrin receptor immunoreactivity in the adult mouse central nervous system, *J. Comp. Neurol.,* 375: 675–692, 1996.

222. Hulet, S.W., Powers, S., and Connor, J.R. Distribution of transferrin and ferritin binding in normal and multiple sclerotic human brains, *J. Neurol. Sci.,* 165: 48–55, 1999.

223. Friden, P.M. et al. Characterization, receptor mapping and blood–brain barrier trans-
     cytosis of antibodies to the human transferrin receptor, *J. Pharmacol. Exp. Ther.*, 278:
     1491–1498, 1996.
224. Friden, P.M. Receptor-mediated transport of therapeutics across the blood–brain
     barrier, *Neurosurgery*, 35: 294–298, 1994.
225. Friden, P.M. et al. Blood–brain barrier penetration and *in vivo* activity of an NGF
     conjugate, *Science*, 259: 373–377, 1993.
226. Deschamps, L. et al. Receptor-mediated transcytosis of transferrin through
     blood–brain barrier endothelial cells, *Am. J. Physiol.*, 270: H1149-H1158, 1996.
227. Dehouck, B. et al. A new function for the LDL receptor: transcytosis of LDL across
     the blood–brain barrier, *J. Cell Biol.*, 138: 877–889, 1997.
228. Ueda, F. et al. Rate of ^{59}Fe uptake into brain and cerebrospinal fluid and the influence
     thereon of antibodies against the transferrin receptor, *J. Neurochem.*, 60: 106–113,
     1993.
229. van Gelder, W. et al. Quantification of different transferrin receptor pools in primary
     cultures of porcine blood–brain barrier endothelial cells, *J. Neurochem.*, 64: 2708–2715,
     1995.
230. Rothenberger, S. et al. Coincident expression and distribution of melanotransferrin
     and transferrin receptor in human brain capillary endothelium, *Brain Res.*, 712:
     117–121, 1996.
231. Bradbury, M.W.B. Transport of iron in the blood–brain-cerebrospinal fluid system,
     J. Neurochem., 69: 443–454, 1997.
232. Reifenberger, P.R. and Wechsler, W. Transferrin receptor expression in tumors of the
     human neurons system: relation to tumor type, grading and tumor growth fraction,
     Virchows. Arch. A. Pathol. Anat. Histopathol., 416: 491–496, 1990.
233. Recht, L. et al. Transferrin receptor in normal and neoplastic brain tissue: implications
     for brain-tumor immunotherapy, *J. Neurosurg.*, 72: 941–945, 1990.
234. Laske, D.W. et al. Intraventricular immunotoxin therapy for leptomeningeal neoplasia,
     Neurosurgery, 41: 1039–1049, 1997.
235. Laske, D.W. et al. Tumor regression with regional distribution of the targeted toxin
     TF-CRM107 in patients with malignant brain tumors, *Nat. Med.*, 3: 1362–1368, 1997.
236. Li, J.Y. et al. Genetically engineered brain drug delivery vectors: cloning, expression
     and *in vivo* application of an anti–transferrin receptor single chain antibody–strepta-
     vidin fusion gene and protein, *Protein Eng.*, 12: 787–796, 1999.
237. Penichet, M.L. An antibody–avidin fusion protein specific for the transferrin receptor
     serves as a delivery vehicle for effective brain targeting: initial applications in anti-
     HIV antisense drug delivery to the brain, *J. Immunol.*, 163: 4421–4426, 1999.
238. Hu, J. and Connor, J.R. Demonstration and characterization of the iron regulatory
     protein in human brain, *J. Neurochem.*, 67: 838–844, 1996.
239. Soyer, H.P. et al. Transferrin receptor expression in normal skin and in various
     cutaneous tumors, *J. Cutan. Pathol.*, 14: 1–5, 1987.
240. Zambruno, G. et al. Epidermal growth factor and transferrin receptor expression in
     human embryonic and fetal epidermal cells, *Arch. Dermatol. Res.*, 282: 544–548,
     1990.
241. Taylor, A., Hogan, B.L., and Watt, F.M. Biosynthesis of EGF receptor, transferrin
     receptor and collagen by cultured human keratinocytes and the effect of retinoic acid,
     Exp. Cell. Res., 159: 47–54, 1985.
242. Wiley, H.S. and Kaplan, J. Epidermal growth factor rapidly induces a redistribution
     of transferrin receptor pools in human fibroblasts, *Proc. Natl. Acad. Sci. U.S.A.*, 81:
     7456–7460, 1984.

243. Viac, J. et al. Virus expression. EGF and transferrin receptors in human papillomas, *Virchows. Arch. A. Pathol. Anat. Histopathol.,* 411: 73–77, 1987.
244. Lu, J.P., Hayashi, K., and Awai, M. Transferrin receptor expression in normal, iron-deficient and iron-overloaded rats, *Acta Pathol. Jpn.,* 39: 759–764, 1989.
245. Bradbury, M.W., Raja, K., and Ueda, F. Contrasting uptakes of ^{59}Fe into spleen, liver, kidney and some other soft tissues in normal and hypotransferrinaemic mice. Influence of an antibody against the transferrin receptor, *Biochem. Pharmacol.,* 47: 969–974, 1994.
246. Valentijn, J.A. et al. Rab4 associates with the actin terminal web in developing rat pancreatic acinar cells, *Eur. J. Cell. Biol.,* 72: 1–8, 1997.
247. Fernández-Pol, J.A. and Klos, D.J. Isolation and characterization of normal rat kidney cell membrane proteins with affinity for transferrin, *Biochemistry,* 19: 3904–3912, 1980.
248. Fleming, S. and Jones, D.B. Immunocytochemical evidence for transferrin-dependent proliferation during renal tubulogenesis, *J. Anat.,* 153: 191–201, 1987.
249. Thesleff, I. et al. The role of transferrin receptors and iron delivery in mouse embryonic morphogenesis, *Differentiation,* 30: 152–158, 1985.
250. Gibson, A. et al. Sorting mechanisms regulating protein traffic in the apical transcytotic pathway of polarized MDCK cells, *J. Cell. Biol.,* 143: 81–94, 1998.
251. Fialka, I., Steinlein, P., and Huber, L.A. Identification of syntenin as a protein of the apical early endocytic compartment in Madin-Darby canine kidney cells, *J. Biol. Chem.,* 274: 26233–26239, 1999.
252. Leung, S.M. et al. SNAP-23 requirement for transferrin recycling in streptolysin-O-permeabilized Madin-Darby canine kidney cells, *J. Biol. Chem.,* 273: 17732–17741, 1998.
253. Walker, R.A. and Day, S.J. Transferrin receptor expression in non-malignant and malignant human breast tissue, *J. Pathol.,* 148: 217–224, 1986.
254. Schulman, H.M. et al. Transferrin receptor and ferritin levels during murine mammary gland development, *Biochim. Biophys. Acta,* 1010: 1–6, 1989.
255. Tonik, S.E., Shindelman, J.E., and Sussman, H.H. Transferrin receptor is inversely correlated with estrogen receptor in breast cancer, *Breast Cancer Res. Treat.,* 7: 71–76, 1986.
256. Vandewalle, B., Revillion-Carette, F., and Lefebvre, J. Involvement of cell surface transferrin receptor in the assessment of estradiol stimulating effect on cultured breast cancer cells, *Anticancer Res.,* 8: 495–498, 1988.
257. Inoue, T. et al. Differences in transferrin response and numbers of transferrin receptors in rat and human mammary carcinoma lines of different metastatic potentials, *J. Cell. Physiol.,* 156: 212–217, 1993.
258. Elliott, R.L. et al. Breast carcinoma and the role of iron metabolism. A cytochemical, tissue culture, and ultrastructural study, *Ann. N.Y. Acad. Sci.,* 698: 159–166, 1993.
259. Cavanaugh, P.G. and Nicolson, G.L. Selection of highly metastatic rat MTLn2 mammary adenocarcinoma cell variants using *in vitro* growth response to transferrin, *J. Cell. Physiol.,* 174: 48–57, 1998.
260. Cavanaugh, P.G. et al. Transferrin receptor overexpression enhances transferrin responsiveness and the metastatic growth of rat mammary adenocarcinoma cell line, *Breast Cancer Res. Treat.,* 56: 203–217, 1999.
261. Poola, I. and Kiang, J.G. The estrogen-inducible transferrin receptor-like membrane glycoprotein is related to stress-regulated proteins, *J. Biol. Chem.,* 269: 21762–21769, 1994.
262. Basar, I. et al. Transferrin receptor activity as a marker in transitional cell carcinoma of the bladder, *Br. J. Urol.,* 67: 165–168, 1991.

263. Asamoto, M. et al. Immunohistochemical analysis of c-erbB-2 oncogene product and epidermal growth factor receptor expression in human urinary bladder carcinomas, *Acta Pathol. Jpn.,* 40: 322–326, 1990.

264. Seymour, G.J. et al. Transferrin receptor expression by human bladder transitional cell carcinomas, *Urol. Res.,* 15: 341–344, 1987.

265. Rahman, S.A. et al. Flow cytometric evaluation of transferrin receptor in transitional cell carcinoma, *Urol. Res.,* 25: 325–329, 1997.

266. Smith, N.W. et al. Transferrin receptor expression in primary superficial human bladder tumours identifies patients who develop recurrences, *Br. J. Urol.,* 65: 339–344, 1990.

267. Tanoguchi, H., Tachibana, M., and Murai, M. Autocrine growth induced by transferrin-like substance in bladder carcinoma cells, *Br. J. Cancer,* 76: 1262–1270, 1997.

268. Munns, J. et al. Evaluation of the potential of transferrin-adryamicin conjugates in the treatment of bladder cancer, *Br. J. Urol.,* 82: 284–289, 1998.

269. Keer, H.N. et al. Elevated transferrin receptor content in human prostate cancer cell lines assessed *in vitro* and *in vivo,* *J. Urol.,* 143: 381–385, 1990.

270. Grayhack, J.T. et al. Analysis of specific proteins in prostatic fluid for detecting prostatic malignancy, *J. Urol.,* 121: 295–299, 1979.

271. Wilson, E.M., French, F.S., and Petrusz, P. Transferrin in the rat prostate Dunning tumor, *Cancer Res.,* 42: 243–251, 1982.

272. Kovar, J. et al. Differing sensitivity of non-hematopoietic human tumors to synergistic anti-transferrin receptor monoclonal antibodies and deferoxamine *in vitro,* *Pathobiology,* 63: 65–70, 1995.

273. Bhatti, R.A., Gadarowski, J.J., and Ray, P.S. Metastatic behavior of prostatic tumor as influenced by the hematopoietic and hematogenous factors, *Tumour Biol.,* 18: 1–5, 1997.

274. Rossi, M.C. and Zetter, B.R. Selective stimulation of prostatic carcinoma cell proliferation by transferrin, *Proc. Natl. Acad. Sci. U.S.A.,* 89: 6197–6201, 1992.

275. Donat, S.M. et al. Reversal by transferrin of growth-inhibitory effect of suramin on hormone-refractory human prostate cancer cells, *J. Natl. Cancer Inst.,* 87: 41–46, 1995.

276. Rawlings, N.D. and Barrett, A.J. Structure of membrane glutamate carboxypeptidase, *Biochim. Biophys. Acta,* 1339: 247–252, 1997.

277. Heston, W.D. Characterization and glutamyl-preferring carboxypeptidase function of prostate-specific membrane antigen: a novel folate hydrolase, *Urology,* 49: 104–112, 1997.

278. Holash, J.A. et al. Barrier properties of testis microvessels, *Proc. Natl. Acad. Sci. U.S.A.,* 90: 11069–11073, 1993.

279. Sylvester, S.R. and Griswold, M.D. The testicular iron shuttle: a "nurse" function of the Sertoli cells, *J. Androl.,* 15: 381–385, 1994.

280. Morales, C., Sylvester, S.R., and Griswold, M.D. Transport of iron and transferrin synthesis by the seminiferous epithelium of the rat *in vivo,* *Biol. Reprod.,* 37: 995–1005, 1987.

281. Vannelli, B.G. et al. Immunostaining of transferrin and transferrin receptor in human seminiferous tubules, *Fertil. Steril.,* 45: 536–541, 1986.

282. Petrie, R.G., Jr. and Morales, C.R. Receptor-mediated endocytosis of testicular transferrin by germinal cells of the rat testis, *Cell. Tissue Res.,* 267: 45–55, 1992.

283. Roberts, K.P. and Griswold, M.D. Characterization of rat transferrin receptor cDNA: the regulation of transferrin receptor mRNA in testes and in Sertoli cells in culture, *Mol. Endocrinol.,* 4: 531–542, 1990.

284. Petrylak, D.P. et al. Transferrin receptor expression in testis cancer, *J. Natl. Cancer Inst.*, 86: 636–637, 1994.
285. Wandji, S.A. et al. Initiation *in vitro* of growth of bovine primordial follicles, *Biol. Reprod.*, 55: 942–948, 1996.
286. Fuernkranz, H.A., Schwob, J.E., and Lucas, J.J. Differential tissue localization of oviduct and erythroid transferrin receptors, *Proc. Natl. Acad. Sci. U.S.A.*, 88: 7505–7508, 1991.
287. Wandji, S.A. et al. Initiation of growth of baboon primordial follicles *in vitro*, *Hum. Reprod.*, 12: 1993–2001, 1997.
288. Roy, S.K. and Treacy, B.J. Isolation and long-term culture of human preantral follicles, *Fertil. Steril.*, 59: 783–790, 1993.
289. Woodruff, T.K. et al. Comparison of functional response of rat, macaque, and human ovarian cells in hormonally defined medium, *Biol. Reprod.*, 48: 68–76, 1993.
290. FitzGerald, D.J. et al. Antitumor activity of an immunotoxin in a nude mouse model of human ovarian cancer, *Cancer Res.*, 47: 1407–1410, 1987.
291. FitzGerald, D.J., Willingham, M.C., and Pastan, I. Antitumor effects of an immunotoxin made with *Pseudomonas* exotoxin in a nude mouse model of human ovarian cancer, *Proc. Natl. Acad. Sci. U.S.A.*, 83: 6627–6630, 1986.
292. Ohkawa, K. et al. Clear cell carcinoma of the human ovary synthesizes and secretes a transferrin with microheterogeneity of lectin affinity, *FEBS Lett.*, 270: 19–23, 1990.
293. Lombardi, A. et al. Transferrin in FRTL5 cells: regulation of its receptor by mitogenic agents and its role in growth, *Endocrinology*, 125: 652–658, 1989.
294. Boudier, J.A. et al. Polarized distribution of gamma interferon-stimulated MHC antigens and transferrin receptors in a clonal cell line isolated from Fisher rat thyroid (FRT cells), *Cell. Tissue Res.*, 272: 23–31, 1993.
295. Sirbasku, D.A. et al. Purification of an equine apotransferrin variant (thyromedin) essential for thyroid hormone dependent growth of GH1 rat pituitary tumor cells in chemically defined culture, *Biochemistry*, 30: 295–304, 1991.
296. Eby, J.E., Sato, H., and Sirbasku, D.A. Apotransferrin stimulation of thyroid hormone dependent rat pituitary tumor cell growth in serum-free chemically defined medium: role of FE(III) chelation, *J. Cell. Physiol.*, 156: 588–600, 1993.
297. Tilemans, D. et al. Production of transferrin-like immunoreactivity by rat anterior pituitary and intermediate lobe, *J. Histochem. Cytochem.*, 43: 657–664, 1995.
298. Tampanaru-Sarmesin, A. et al. Transferrin and transferrin receptor in human hypophysis and pituitary adenomas, *Am. J. Pathol.*, 152: 413–422, 1998.
299. Atkin, S.L. et al. Expression of the transferrin receptor in human anterior pituitary adenomas is confined to gonadotrophinomas, *Clin. Endocrinol.*, (Oxford) 44: 467–471, 1996.
300. Sorokin, L.M., Morgan, E.H., and Yeoh, G.C. Transferrin receptor numbers and transferrin and iron uptake in cultured chick muscle cells at different stages of development, *J. Cell. Physiol.*, 131: 342–353, 1987.
301. Jennische, E. Sequential immunohistochemical expression of IGF-I and the transferrin receptor in regenerating rat muscle *in vivo*, *Acta Endocrinol.*, (Copenhagen) 121: 733–738, 1989.
302. Ralston, E., McLaren, R.S., and Horowitz, J.A. Nuclear domains in skeletal myotubes: the localization of transferrin receptor mRNA is independent of its half-life and restricted by binding to ribosomes, *Exp. Cell. Res.*, 236: 453–462, 1997.
303. Aledo, J.C. et al. Identification and characterization of two distinct intracellular GLUT4 pools in rat skeletal muscle: evidence for an endosomal and an insulin-sensitive GLUT4 compartment, *Biochem. J.*, 325: 727–732, 1997.

304. Gentili, C. et al. Ovotransferrin and ovotransferrin receptor expression during chondrogenesis and endochondral bone formation in developing chicken embryo, *J. Cell. Biol.,* 124: 579–588, 1994.

305. Carlevaro, M.F. et al. Transferrin promotes endothelial cell migration and invasion: implication in cartilage neovascularization, *J. Cell. Biol.,* 136: 1375–1384, 1997.

306. Seligman, P.A., Schleicher, R.B., and Allen, R.H. Isolation and characterization of the transferrin receptor from human placenta, *J. Biol. Chem.,* 254: 9943–9946, 1979.

307. Galbraith, G.M. et al. Demonstration of transferrin receptors on human placental trophoblast, *Blood,* 55: 240–242, 1980.

308. Galbraith, G.M., Galbraith, R.M., and Faulk, W.P. Immunological studies of transferrin and transferrin receptors on human placental trophoblast, *Placenta,* 1: 33–46, 1980.

309. Jacobsen, G.K., Jacobsen, M., and Henriksen, O.B. An immunohistochemical study of a series of plasma proteins in the early human conceptus, *Oncodev. Biol. Med.,* 2: 399–410, 1981.

310. McArdle, H.J. and Morgan, E.H. Transferrin and iron movements in the rat conceptus during gestation, *J. Reprod. Fertil.,* 66: 529–536, 1982.

311. Verrijt, C.E. et al. Accumulation and release of iron in polarly and non-polarly cultured trophoblast cells isolated from human term placentas, *Eur. J. Obstet. Gynecol. Reprod. Biol.,* 86: 73–81, 1999.

312. Vanderpuye, O.A., Kelley, L.K., and Smith, C.H. Transferrin receptors in the basal plasma membrane of the human placental syncytiotrophoblast, *Placenta,* 7: 391–403, 1986.

313. Cerneus, D.P. and van der Ende, A. Apical and basolateral transferrin receptors in polarized BeWo cells recycle through separate endosomes, *J. Cell Biol.,* 114: 1149–1158, 1991.

314. Cerneus, D.P., Strous, G.J., and van der Ende, A. Bidirectional transcytosis determines the steady state distribution of the transferrin receptor at opposite plasma membrane domains of BeWo cells, *J. Cell Biol.,* 122: 1223–1230, 1993.

315. Shi, F. et al. Permeability and metabolic properties of a trophoblast cell line (HRP-1) derived from normal rat placenta, *Exp. Cell. Res.,* 234: 147–155, 1997.

316. Verrijt, C.E. et al. Binding of human isotransferrin variants to microvillus and basal membrane vesicles from human term placenta, *Placenta,* 18: 71–77, 1997.

317. Kennedy, M.L., Douglas, G.C., and King, B.F. Expression of transferrin receptors during differentiation of human placental trophoblast cells *in vitro*, *Placenta,* 13: 43–53, 1992.

318. Bierings, M.B. et al. Transferrin receptor expression and the regulation of placental iron uptake, *Mol. Cell. Biochem.,* 100: 31–38, 1991.

319. Georgieff, M.K. et al. Increased placental iron regulatory protein-1 expression in diabetic pregnancies complicated by fetal iron deficiency, *Placenta,* 20: 87–93, 1999.

320. van der Ende, A. et al. Iron metabolism in BeWo chorion carcinoma cells. Transferrin-mediated uptake and release of iron, *J. Biol. Chem.,* 262: 8910–8916, 1987.

321. Bulmer, J.N., Morrison, L., and Johnson, P.M. Expression of the proliferation markers Ki67 and transferrin receptor by human trophoblast populations, *J. Reprod. Immunol.,* 14: 291–302, 1988.

322. Verrijt, C.E. et al. Transferrin in cultured human term cytotrophoblast cells: synthesis and heterogeneity, *Mol. Cell. Biochem.,* 173: 177–181, 1997.

323. Parkkila, S. et al. Association of the transferrin receptor in human placenta with HFE, the protein defective in hereditary hemochromatosis, *Proc. Natl. Acad. Sci. U.S.A.,* 94: 13198–13202, 1997.

324. Testa, U. Transferrin receptors: structure and function. *Curr. Top. Hematol.*, 5: 127–161, 1985.

325. Cazzola, M. et al. Manipulations of cellular iron metabolism for modulating normal and malignant cell proliferation: achievements and prospects, *Blood*, 75: 1903–1919, 1990.

326. Sutherland, R. et al. Ubiquitous cell-surface glycoprotein on tumor cells is proliferation-associated receptor for transferrin, *Proc. Natl. Acad. Sci. U.S.A.*, 78: 4515–4519, 1981.

327. Titeux, M. et al. The role of iron in the growth of human leukemic cell lines, *J. Cell. Physiol.*, 121: 251–256, 1984.

328. Larrick, J.W. and Cresswell, P. Modulation of cell surface iron transferrin receptors by cellular density and state of activation, *J. Supramol. Struct.*, 11: 579–586, 1979.

329. Hamilton, T.A., Wada, H.G., and Sussman, H.H. Identification of transferrin receptors on the surface of human cultured cells, *Proc. Natl. Acad. Sci. U.S.A.*, 76: 6406–6410, 1979.

330. Galbraith, R.M.P. et al. Transferrin binding to peripheral blood lymphocytes activated by phytohemagglutinin involves a specific receptor–ligand interaction, *J. Clin. Invest.*, 66: 1135–1146, 1980.

331. Galbraith, R.M.P. and Galbraith, G.M.P. Expression of transferrin receptors on mitogen-stimulated human peripheral blood lymphocytes: relation to cellular activation and related metabolic events, *Immunology*, 44: 703–710, 1981.

332. Frazier, J.L. et al. Studies of the transferrin receptor on both human reticulocytes and nucleated human cells in culture: comparison of factors regulating receptor density, *J. Clin. Invest.*, 69: 853–865, 1982.

333. Enns, C.A. et al. Radioimmunochemical measurement of the transferrin receptor in human trophoblast and reticulocyte membranes with a specific anti-receptor antibody, *Proc. Natl. Acad. Sci. U.S.A.*, 78: 4222–4225, 1981.

334. Wiley, H.S. and Kaplan, J. Epidermal growth factor rapidly induces a redistribution of transferrin receptor pools in human fibroblasts, *Proc. Natl. Acad. Sci. U.S.A.*, 81: 7456–7460, 1984.

335. Davis, R.J. and Czech, M.P. Regulation of transferrin receptor expression at the cell surface by insulin-like growth factors, epidermal growth factor and platelet-derived growth factor, *EMBO J.*, 5: 653–658, 1986.

336. Ward, D.M. and Kaplan, J. Mitogenic agents induce redistribution of transferrin receptors from internal pools to the cell surface, *Biochem. J.*, 238: 721–728, 1986.

337. Hori, T. et al. Effects of tumor necrosis factor on cell growth and expression of transferrin receptors in human fibroblasts, *Cell. Struct. Funct.*, 13: 425–433, 1988.

338. Damke, H., von Figura, K., and Braulke, T. Simultaneous redistribution of mannose 6-phosphate and transferrin receptors by insulin-like growth factors and phorbol ester, *Biochem. J.*, 281: 225–229, 1992.

339. Chitambar, C.R., Massey, E.J., and Seligman, P.A. Regulation of transferrin receptor expression on human leukemic cells during proliferation and induction of differentiation. Effects of gallium and dimethylsulfoxide, *J. Clin. Invest.*, 72: 1314–1325, 1983.

340. Barker, K.A. and Newburger, P.E. Relationships between the cell cycle and the expression of c-*myc* and transferrin receptor genes during induced myeloid differentiation, *Exp. Cell. Res.*, 186: 1–5, 1990.

341. Hedley, D. et al. Modulation of transferrin receptor expression by inhibitors of nucleic acid synthesis, *J. Cell. Physiol.*, 124: 61–66, 1985.

342. Warren, G., Davoust, J., and Cockcroft, A. Recycling of transferrin receptors in A431 cells is inhibited during mitosis, *EMBO J.*, 3: 2217–2225, 1984.

343. Neckers, L.M. et al. Differential expression of c-*myc* and the transferrin receptor in G1 synchronized M1 myeloid leukemia cells, *J. Cell. Physiol.,* 135: 339–344, 1988.

344. Besancon, F., Bourgeade, M.F., and Testa, U. Inhibition of transferrin receptor expression by interferon-alpha in human lymphoblastoid cells and mitogen-induced lymphocytes, *J. Biol. Chem.,* 260: 13074–13080, 1985.

345. Bourgeade, M.F. et al. Post-transcriptional regulation of transferrin receptor mRNA by IFN gamma, *Nucleic Acid Res.,* 20: 2997–3003, 1992.

346. Enns, C.A. et al. Modulation of the transferrin receptor during DMSO-induced differentiation in HL-60 cells, *Exp. Cell. Res.,* 174: 89–97, 1988.

347. Kanayasu-Toyoda, T. et al. Commitment of neutrophilic differentiation and proliferation of HL-60 cells coincides with expression of transferrin receptor, *J. Biol. Chem.,* 274: 25471–25480, 1999.

348. Ross, S.A., Keller, S.R., and Lienhard, G.E. Increased intracellular sequestration of the insulin-regulated aminopeptidase upon differentiation of 3T3-L1 cells, *Biochem. J.,* 330: 1003–1008, 1998.

349. Fleming, S. and Jones, D.B. Immunocytochemical evidence for transferrin-dependent proliferation during renal tubulogenesis, *J. Anat.,* 153: 191–201, 1987.

350. Rudland, P.S. et al. Iron salts and transferrin are specifically required for cell division of cultured 3T6 cells, *Biochem. Biophys. Res. Commun.,* 75: 556, 1977.

351. Trowbridge, I.S. and López, F. Monoclonal antibody to transferrin receptor blocks transferrin binding and inhibits human tumor cell growth *in vitro, Proc. Natl. Acad. Sci. U.S.A.,* 79: 1175–1179, 1982.

352. Laskey, J. et al. Evidence that transferrin supports cell proliferation by supplying iron for DNA synthesis, *Exp. Cell. Res.,* 176: 87–95, 1988.

353. Cooper, C.E. et al. The relationship of intracellular iron chelation to the inhibition and regeneration of human ribonucleotide reductase, *J. Biol. Chem.,* 271: 20291–20299, 1996.

354. Fernández-Pol, J.A., Bono, V.H. Jr., and Johnson, G.S. Control of growth by picolinic acid: differential response of normal and transformed cells, *Proc. Natl. Acad. Sci. U.S.A.,* 74: 2889–2893, 1977.

355. Gurley, L.R. and Jett, J.H. Cell cycle kinetics of Chinese hamster ovary (CHO) cells treated with the iron-chelating agent picolinic acid, *Cell. Tissue Kinet.,* 14: 269–283, 1981.

356. Fernández-Pol, J.A. Isolation and characterization of a siderophore-like growth factor from mutants of SV40-transformed cells adapted to picolinic acid, *Cell,* 14: 489–499, 1978.

357. Terada, N. et al. Definition of the roles for iron and essential fatty acids in cell cycle progression of normal human T lymphocytes, *Exp. Cell. Res.,* 204: 260–267, 1993.

358. Brodie, C. et al. Neuroblastoma sensitivity to growth inhibition by deferrioxamine: evidence for a block in G1 phase of the cell cycle, *Cancer Res.,* 53: 3968–3975, 1993.

359. Seligman, P.A. et al. Effects of agents that inhibit cellular iron incorporation on bladder cancer cell proliferation, *Blood,* 82: 1608–1617, 1993.

360. Lucas, J.J. et al. Effects of iron-depletion on cell cycle progression in normal human T lymphocytes: selective inhibition of the appearance of the cyclin A-associated component of the p33cdk2 kinase, *Blood,* 86: 2268–2280, 1995.

361. Grazidei, I. et al. The acute-phase protein α1-antitrypsin inhibits transferrin-receptor binding and proliferation of human skin fibroblasts, *Biochim. Biophys. Acta,* 1401: 170–176, 1998.

362. Pelicci, P.G. et al. Hemin regulates the expression of transferrin receptors in human hematopoietic cell lines, *FEBS Lett.,* 145: 350–354, 1982.

363. Testa, U. et al. Transferrin binding to K562 cell line. Effect of heme and sodium butyrate induction, *Exp. Cell. Res.*, 140: 251–260, 1982.

364. Taketani, S. et al. Hemopexin-dependent down-regulation of expression of the human transferrin receptor, *J. Biol. Chem.*, 265: 13981–13985, 1990.

365. Eisenstein, R.S. et al. Regulation of ferritin and heme oxygenase synthesis in rat fibroblasts by different forms of iron, *Proc. Natl. Acad. Sci. U.S.A.*, 88: 688–692, 1991.

366. Poss, K.D. and Tonegawa, S. Heme oxygenase 1 is required for mammalian iron reutilization, *Proc. Natl. Acad. Sci. U.S.A.*, 94: 10919–10924, 1997.

367. Ferris, C.D. et al. Haem oxygenase I prevents cell death by regulating cellular iron, *Nature Cell. Biol.*, 1: 152–157, 1999.

368. McCoubrey, W.K. Jr., Huang, T.J., and Maines, M.D. Heme oxygenase-2 is a hemoprotein and binds heme through heme regulatory motifs that are not involved in heme catalysis, *J. Biol. Chem.*, 272: 12568–12574, 1997.

369. Ward, J.H., Kushner, J.P., and Kaplan, J. Regulation of HeLa cell transferrin receptors, *J. Biol. Chem.*, 257: 10317–10323, 1982.

370. Ward, J.H., Kushner, J.P., and Kaplan, J. Transferrin receptors of human fibroblasts. Analysis of receptor properties and regulation, *Biochem. J.*, 208: 19–26, 1982.

371. Louache, F. et al. Modulation of the expression of transferrin receptors by iron, hemin and protoporphyrin IX, C. R. Seances Acad. Sci. III 297: 291–294, 1983.

372. Louache, F. et al. Regulation of transferrin receptors in human hematopoietic cell lines, *J. Biol. Chem.*, 259: 11576–11582, 1984.

373. Pelosi, E. et al. Expression of transferrin receptors in phytohemagglutinin-stimulated human T-lymphocytes. Evidence for a three-step model, *J. Biol. Chem.*, 261: 3036–3042, 1986.

374. Petrini, M. et al. Constitutive expression and abnormal glycosylation of transferrin receptor in acute T-cell leukemia, *Cancer Res.*, 49: 6989–6996, 1989.

375. Sciot, R. et al. Lack of hepatic transferrin receptor expression in hemochromatosis, *Hepatology*, 7: 831–837, 1987.

376. Testa, E.P. et al. Expression of transferrin receptors in human erythroleukemic lines: regulation in the plateau and exponential phase of growth, *Cancer Res.*, 46: 5330–5334, 1986.

377. Rudolph, N.S. et al. Regulation of K562 cell transferrin receptors by exogenous iron, *J. Cell Physiol.*, 122: 441–450, 1985.

378. Rao, K.K. et al. Effects of alterations in cellular iron on biosynthesis of the transferrin receptor in K562 cells, *Mol. Cell. Biol.*, 5: 595–600, 1985.

379. Mattia, E. et al. Biosynthetic regulation of the human transferrin receptor by desferrioxamine in K562 cells, *J. Biol. Chem.*, 259: 2689–2692, 1984.

380. Testa, U. et al. The iron-chelating agent picolinic acid enhances transferrin receptor expression in human erythroleukaemic cell lines, *Br. J. Haematol.*, 60: 491–502, 1985.

381. Bottomley, S.S., Wolfe, L.C., and Bridges, K.R. Iron metabolism in K562 erythroleukemic cells, *J. Biol. Chem.*, 260: 6811–6815, 1985.

382. Richardson, D.R. and Ponka, P. The molecular mechanisms of the metabolism and transport of iron in normal and neoplastic cells, *Biochim. Biophys. Acta*, 1331: 1–40, 1997.

383. Breuer, W., Epsztejn, S., and Cabantchik, Z.I. Iron acquired from transferrin by K562 cells is delivered into a cytoplasmic pool of chelatable iron(II), *J. Biol. Chem.*, 270: 24209–24215, 1995.

384. Konijn, A.M. et al. The cellular labile iron pool and intracellular ferritin in K562 cells, *Blood*, 94: 2128–2143, 1999.

385. Coccia, E.M. et al. Regulation of expression of ferritin H-chain and transferrin receptor by protoporphyrin IX, *Eur. J. Biochem.*, 250: 764–772, 1997.

386. Mullner, E.W. et al. *In vivo* and *in vitro* modulation of the mRNA-binding activity of iron-regulatory factor. Tissue distribution and effects of cell proliferation, iron levels and redox state, *Eur. J. Biochem.*, 208: 597–605, 1992.

387. Rao, K. et al. Transcriptional regulation by iron of the gene for the transferrin receptor, *Mol. Cell. Biol.*, 6: 236–240, 1986.

388. Testa, U. et al. Differential regulation of transferrin receptor gene expression in human hemopoietic cells: molecular and cellular aspects, *J. Receptor Res.*, 7: 355, 1987.

389. Louache, F. et al. Molecular mechanisms regulating the synthesis of transferrin receptors and ferritin in human erythroleukemic cell lines, *FEBS Lett.*, 183: 223–227, 1985.

390. Seiser, C. et al. Effect of transcription inhibitors on the iron-dependent degradation of transferrin receptor mRNA, *J. Biol. Chem.*, 270: 29400–29406, 1995.

391. Toth, I. et al. Hypoxia alters iron-regulatory protein-1 binding capacity and modulates cellular iron homeostasis in human hepatoma and erythroleukemia cells, *J. Biol. Chem.*, 274: 4467–4473, 1999.

392. Lok, C.N. and Ponka, P. Identification of a hypoxia response element in the transferrin receptor gene, *J. Biol. Chem.*, 274: 24147–24152, 1999.

393. Tacchini, L. et al. Transferrin receptor induction by hypoxia, *J. Biol. Chem.*, 274: 24142–24146, 1999.

394. Salceda, S. and Caro, J. Hypoxia-inducible factor 1α (HIF-1α) protein is rapidly degraded by the ubiquitin–proteasome system under normoxic conditions, *J. Biol. Chem.*, 272: 22642–22647, 1997.

395. Maxwell, P.H. et al. The tumor suppressor protein VHL targets hypoxia-inducible factors for oxygen dependent proteolysis, *Nature*, 399: 271–275, 1999.

396. Bianchi, L., Tacchini, L., and Cairo, G. HIF-1-mediated activation of transferrin receptor gene transcription by iron chelation, *Nucleic Acid Res.*, 27: 4223–4227, 1999.

397. Owen, D. and Kühn, L.C. Noncoding 3′ sequences of the transferrin receptor gene are required for mRNA regulation by iron, *EMBO J.*, 6: 1287–1293, 1987.

398. Kühn, L. Human transferrin receptor expressed from a transfected cDNA is functional in murine cells, in *Proteins of Iron Storage and Transport*, Spik, G. et al., Eds., Amsterdam, Elsevier, 369, 1985.

399. Miskimins, W.K. et al. Cell proliferation and expression of the transferrin receptor gene: promoter sequence homologies and protein interactions, *J. Cell. Biol.*, 103: 1781–1788, 1986.

400. Miskimins, W.K. et al. Use of a protein-blotting procedure and a specific DNA probe to identify nuclear proteins that recognize the promoter region of the transferrin receptor gene, *Proc. Natl. Acad. Sci. U.S.A.*, 82: 6741–6744, 1985.

401. Casey, J.L. et al. Deletional analysis of the promoter region of the human transferrin receptor gene, *Nucleic Acid Res.*, 16: 629–646, 1988.

402. Roberts, M.R., Miskimins, W.K., and Ruddle, F.H. Nuclear proteins TREF1 and TREF2 bind to the transcriptional control element of the transferrin receptor gene and appear to be associated as a heterodimer, *Cell. Regul.*, 1: 151–164, 1989.

403. Beard, P. et al. SV40 activates transcription from the transferrin receptor promoter by inducing a factor which binds to the CRE/AP-1 recognition sequence, *Nucleic Acid Res.*, 19: 7117–7123, 1991.

404. Herrlich, P. et al. *Molecular Biology of Cancer Genes*, Vol. 8, Sloyser, M., Ed., New York, Ellis Harwood Press, 150, 1990.

405. Ouyang, Q., Bommakanti, M., and Miskimins, W.K. A mitogen-responsive promoter region that is synergistically activated through multiple signalling pathways, *Mol. Cell Biol.*, 13: 1796–1804, 1993.

406. Hirsch, S. and Miskimins, W.K. Mitogen induction of nuclear factors that interact with a delayed responsive region of the transferrin receptor gene promoter, *Cell. Growth Differ.,* 6: 719–726, 1995.

407. Roberts, M.R. et al. A DNA-binding activity, TRAC, specific for the TRA element of the transferrin receptor gene copurifies with the Ku autoantigen, *Proc. Natl. Acad. Sci. U.S.A.,* 91: 6354–6358, 1994.

408. Christensen, J., Cotmore, S.F., and Tattersall, P. Two new members of the emerging KCWK family of combinatorial transcriptional modulators bind as a heterodimer to flexibly spaced PuCGpy half-sites, *Mol. Cell. Biol.,* 19: 7741–7750, 1999.

409. Miskimins, W.K., King, F., and Miskimins, R. Phosphatidylinositol 3-kinase inhibitor wortmannin blocks mitogenic activation of the transferrin receptor gene promoter in late G1, *Cell. Growth Differ.,* 8: 565–570, 1997.

410. Lok, C.N., Chan, K.F., and Loh, T.T. Transcriptional regulation of transferrin receptor expression during phorbol ester-induced HL-60 cell differentiation. Evidence for a negative regulatory role of the phorbol-ester-responsive element-like sequence, *Eur. J. Biochem.,* 236: 614–619, 1996.

411. Sieweke, M.H. et al. MafB is an interaction partner and repressor of *Ets-1* that inhibits erythroid differentiation, *Cell,* 85: 49–60, 1996.

412. Marziali, G. et al. Ets-1 and Sp-1 cooperate to induce expression of transferrin receptors during erythroid differentiation, *J. Biol. Chem.,* in press, 2001.

413. Tsien, R.Y., Pozzan. T., and Rink. T.J. T-cell mitogens cause early changes in cytoplasmic free Ca^{2+} and membrane potential in lymphocytes, *Nature,* 295: 68–71, 1982.

414. Mastro, A.M. and Smith, M.C. Calcium-dependent activation of lymphocytes by ionophore, A23187, and a phorbol ester tumor promoter, *J. Cell. Physiol.,* 116: 51–56, 1983.

415. Mills, G.B. et al. Increase in cytosolic free calcium concentration is an intracellular messenger for the production of interleukin 2 but not for expression of the interleukin 2 receptor, *J. Immunol.,* 134: 1640–1643, 1985.

416. Depper, J.M. et al. Activators of protein kinase C and 5-azacytidine induce IL-2 receptor expression on human T lymphocytes, *J. Cell Biochem.,* 27: 267–276, 1985.

417. Depper, J.M. et al. Interleukin 2 (IL-2) augments transcription of the IL-2 receptor gene, *Proc. Natl. Acad. Sci. U.S.A.,* 82: 4230–4234, 1985.

418. Depper, J.M. et al. Regulation of interleukin 2 receptor expression: effects of phorbol diester, phospholipase C, and reexposure to lectin or antigen, *J. Immunol.,* 133: 3054–3061, 1984.

419. Alcantara, O., Javors, M., and Boldt, D.H. Induction of protein kinase C mRNA in cultured lymphoblastoid T cells by iron-transferrin but not by soluble iron, *Blood,* 77: 1290–1297, 1991.

420. Manger, B. et al. A transferrin receptor antibody represents one signal for the induction of IL 2 production by a human T cell line, *J. Immunol.,* 136: 532–538, 1986.

421. Cano, E. et al. Induction of T cell activation by monoclonal antibodies specific for the transferrin receptor, *Eur. J. Immunol.,* 20: 765–770, 1990.

422. Salmeron, A. et al. Transferrin receptor induces tyrosine phosphorylation in T cells and is physically associated with the TCR ζ-chain, *J. Immunol.,* 154: 1675–1683, 1995.

423. Luton, F. et al. Tyrosine and serine protein kinase activities associated with ligand-induced internalized TCR/CD3 complexes, *J. Immunol.,* 158: 3140–3147, 1997.

424. Yamamoto, Y. et al. Interleukin 2 mRNA induction in human lymphocytes: analysis of the synergistic effect of a calcium ionophore A23187 and a phorbol ester, *Eur. J. Immunol.,* 15: 1204–1208, 1985.

425. Robb, R.J. Human interleukin 2, *Methods Enzymol.,* 116: 493–525, 1985.

426. Robb, R.J., Greene, W.C., and Rusk, C.M. Low and high affinity cellular receptors for interleukin 2. Implications for the level of Tac antigen, *J. Exp. Med.,* 160: 1126–1146, 1984.

427. Smith, K.A. T-cell growth factor, *Immunol. Rev.,* 51: 337–357, 1980.

428. Smith, K.A. Regulation of normal and neoplastic T-cell growth, in *Cancer Cells,* Vol. 3, Feramisco, J. et al., Eds., Cold Spring Harbor, New York, Cold Spring Harbor Laboratories, 205, 1985.

429. Waldmann, T.A. The structure, function, and expression of interleukin-2 receptors on normal and malignant lymphocytes, *Science,* 232: 727–732, 1986.

430. Brock, J.H. and Rankin, M.C. Transferrin binding and iron uptake by mouse lymph node cells during transformation in response to concanavalin A, *Immunology,* 43: 393–398, 1981.

431. Galbraith, G.M. et al. Transferrin binding by mitogen-activated human peripheral blood lymphocytes, *Clin. Immunol. Immunopathol.,* 16: 387–395, 1980.

432. Neckers, L.M. and Cossman, J. Transferrin receptor induction in mitogen-stimulated human T lymphocytes is required for DNA synthesis and cell division and is regulated by interleukin 2, *Proc. Natl. Acad. Sci. U.S.A.,* 80: 3494–3498, 1983.

433. Titeux, M. et al. The role of iron in the growth of human leukemic cell lines, *J. Cell Physiol.,* 121: 251–256, 1984.

434. Tormey, D.C., Imrie, R.C., and Mueller, G.C. Identification of transferrin as a lymphocyte growth promoter in human serum, *Exp. Cell. Res.,* 74: 163–169, 1972.

435. Dillner-Centerlind, M.L., Hammarstrom, S., and Perlmann, P. Transferrin can replace serum for *in vitro* growth of mitogen-stimulated T lymphocytes, *Eur. J. Immunol.,* 9: 942–948, 1979.

436. Brock, J.H. The effect of iron and transferrin on the response of serum-free cultures of mouse lymphocytes to concanavalin A and lipopolysaccharide, *Immunology,* 43: 387–392, 1981.

437. Lederman, H.M. et al. Deferoxamine: a reversible S-phase inhibitor of human lymphocyte proliferation, *Blood,* 64: 748–753, 1984.

438. Drobyski, W.R. et al. Modulation of *in vitro* and *in vivo* T-cell responses by transferrin–gallium and gallium nitrate, *Blood,* 88: 3056–3064, 1996.

439. Phillips, J.L. and Azari, P. Effect of iron transferrin on nucleic acid synthesis in phytohemagglutinin-stimulated human lymphocytes, *Cell. Immunol.,* 15: 94–99, 1975.

440. Mendelsohn, J., Trowbridge, I., and Castagnola, J. Inhibition of human lymphocyte proliferation by monoclonal antibody to transferrin receptor, *Blood,* 62: 821–826, 1983.

441. Tanno, Y. and Takishima, T. Enhancing effect of saccharated ferric oxide on human lymphocyte transformation in serum-free medium, *Tohoku J. Exp. Med.,* 136: 463–464, 1982.

442. Brock, J.H. and Stevenson, J. Replacement of transferrin in serum-free cultures of mitogen-stimulated mouse lymphocytes by a lipophilic iron chelator, *Immunol. Lett.,* 15: 23–25, 1987.

443. Lum, J.B. et al. Transferrin synthesis by inducer T lymphocytes, *J. Clin. Invest.,* 77: 841–849, 1986.

444. Keyna, U. et al. The role of the transferrin receptor for the activation of human lymphocytes, *Cell. Immunol.,* 132: 411–422, 1991.

445. Djeha, A. et al. Cytokine-mediated regulation of transferrin synthesis in mouse macrophages and human T lymphocytes, *Blood,* 85: 1036–1042, 1995.

446. Brock, J.H., Mainou-Fowler, T., and Webster, L.M. Evidence that transferrin may function exclusively as an iron donor in promoting lymphocyte proliferation, *Immunology,* 57: 105–110, 1986.

447. Taetle, R., Castagnola, J., and Mendelsohn, J. Mechanisms of growth inhibition by anti-transferrin receptor monoclonal antibodies, *Cancer Res.*, 46: 1759–1763, 1986.

448. Kemp, J.D. et al. Role of the transferrin receptor in lymphocyte growth: a rat IgG monoclonal antibody against the murine transferrin receptor produces highly selective inhibition of T and B cell activation protocols, *J. Immunol.*, 138: 2422–2426, 1987.

449. Sato, A., Bacon, P.E., and Cory, J.G. Studies on the differential mechanisms of inhibition of ribonucleotide reductase by specific inhibitors of the non-heme iron subunit, *Adv. Enzyme Regul.*, 22: 231–241, 1984.

450. Terada, N. et al. Definition of the roles for iron and essential fatty acids in cell cycle progression of normal human T lymphocytes, *Exp. Cell. Res.*, 204: 260–267, 1993.

451. Lucas, J.J. et al. Effects of iron-depletion on cell cycle progression in normal human T lymphocytes: selective inhibition of the appearance of the cyclin A-associated component of the p33cdk2 kinase, *Blood*, 86: 2268–2280, 1995.

452. Thorstensen, K. and Romslo, I. The role of transferrin in the mechanism of cellular iron uptake, *Biochem. J.*, 271: 1–9, 1990.

453. Ellem, K.A. and Kay, G.F. Ferricyanide can replace pyruvate to stimulate growth and attachment of serum restricted human melanoma cells, *Biochem. Biophys. Res. Commun.*, 112: 183–190, 1983.

454. Sun, I.L. et al. Diferric transferrin reduction stimulates the Na$^+$/H$^+$ antiport of HeLa cells, *Biochem. Biophys. Res. Commun.*, 145: 467–473, 1987.

455. Sun, I.L. et al. Reduction of diferric transferrin by SV40 transformed pineal cells stimulates Na$^+$/H$^+$ antiport activity, *Biochim. Biophys. Acta*, 938: 17–23, 1988.

456. Sun, I.L. et al. Retinoic acid inhibition of transplasmalemma diferric transferrin reductase, *Biochem. Biophys. Res. Commun.*, 146: 976–982, 1987.

457. Sun, I.L. et al. Transmembrane redox in control of cell growth. Stimulation of HeLa cell growth by ferricyanide and insulin, *Exp. Cell. Res.*, 156: 528–536, 1985.

458. Diaz-Gil, J.J. et al. Purification of a liver DNA-synthesis promoter from plasma of partially hepatectomized rats, *Biochem. J.*, 235: 49–55, 1986.

459. Suthanthiran, M. et al. Glutathione regulates activation-dependent DNA synthesis in highly purified normal human T lymphocytes stimulated via the CD2 and CD3 antigens, *Proc. Natl. Acad. Sci. U.S.A.*, 87: 3343–3347, 1990.

460. Iwata, S. et al. Thiol-mediated redox regulation of lymphocyte proliferation. Possible involvement of adult T cell leukemia-derived factor and glutathione in transferrin receptor expression, *J. Immunol.*, 152: 5633–5642, 1994.

461. Novogrodsky, A., Suthanthiran, M., and Stenzel, K.H. Ferro-mitogens: iron-containing compounds with lymphocyte-stimulatory properties, *Cell. Immunol.*, 133: 295–305, 1991.

462. Novogrodsky, A., Suthanthiran, M., and Stenzel, K.H. Immune stimulatory properties of metalloporphyrins, *J. Immunol.*, 143: 3981–3987, 1989.

463. Novogrodsky, A. et al. Hydroxyl radical scavengers inhibit lymphocyte mitogenesis, *Proc. Natl. Acad. Sci. U.S.A.*, 79: 1171–1174, 1982.

464. Thorson, J.A. et al. Role of iron in T cell activation: TH1 clones differ from TH2 clones in their sensitivity to inhibition of DNA synthesis caused by IgG mAbs against the transferrin receptor and the iron chelator deferoxamine, *Cell. Immunol.*, 134: 126–137, 1991.

465. Smith, A. et al. Role of heme-hemopexin in human T-lymphocyte proliferation, *Exp. Cell. Res.*, 232: 246–254, 1997.

466. Cotner, T. et al. Simultaneous flow cytometric analysis of human T cell activation antigen expression and DNA content, *J. Exp. Med.*, 157: 461–472, 1983.

467. Pelosi-Testa, E. et al. Mechanisms underlying T-lymphocyte activation: mitogen initiates and IL-2 amplifies the expression of transferrin receptors via intracellular iron level, *Immunology,* 64: 273–279, 1988.

468. Kronke, M. et al. Sequential expression of genes involved in human T lymphocyte growth and differentiation, *J. Exp. Med.,* 161: 1593–1598, 1985.

469. Reed, J.C. et al. Sequential expression of protooncogenes during lectin-stimulated mitogenesis of normal human lymphocytes, *Proc. Natl. Acad. Sci. U.S.A.,* 83: 3982–3986, 1986.

470. Pauza, C.D., Bleil, J.D., and Lennox, E.S. The control of transferrin receptor synthesis in mitogen-stimulated human lymphocytes, *Exp. Cell. Res.,* 154: 510–520, 1984.

471. Salmon, M. et al. Transferrin receptor expression by stimulated cells in mixed lymphocyte culture, *Immunology,* 54: 559–564, 1985.

472. Carotenuto, P. et al. Desferoxamine blocks IL 2 receptor expression on human T lymphocytes, *J. Immunol.,* 136: 2342–2347, 1986.

473. Bierer, B.E. and Nathan, D.G. The effect of desferrithiocin, an oral iron chelator, on T-cell function, *Blood,* 76: 2052–2059, 1990.

474. Testa, U. et al. Differential regulation of iron regulatory element-binding protein(s) in cell extracts of activated lymphocytes versus monocytes–macrophages, *J. Biol. Chem.,* 266: 13925–13930, 1991.

475. Neckers, L.M. et al. Diltiazem inhibits transferrin receptor expression and causes G1 arrest in normal and neoplastic T cells, *Mol. Cell Biol.,* 6: 4244–4250, 1986.

476. Kumagai, N. et al. Comparison of phorbol ester/calcium ionophore and phytohemagglutinin-induced signaling in human T lymphocytes. Demonstration of interleukin 2-independent transferrin receptor gene expression, *J. Immunol.,* 140: 37–43, 1988.

477. Reinherz, E.L. et al. Discrete stages of human intrathymic differentiation: analysis of normal thymocytes and leukemic lymphoblasts of T-cell lineage, *Proc. Natl. Acad. Sci. U.S.A.,* 77: 1588–1592, 1980.

478. Vidal, C. et al. Human T lymphotropic virus I infection deregulates surface expression of the transferrin receptor, *J. Immunol.,* 141: 984–988, 1988.

479. Petrini, M. et al. Constitutive expression and abnormal glycosylation of transferrin receptor in acute T-cell leukemia, *Cancer Res.,* 49: 6989–6996, 1989.

480. Lum, J.B. et al. Transferrin synthesis by inducer T lymphocytes, *J. Clin. Invest.,* 77: 841–849, 1986.

481. Kitada, S. and Hays, E.F. Transferrin-like activity produced by murine malignant T-lymphoma cell lines, *Cancer Res.,* 45: 3537–3540, 1985.

482. Seligman, P.A. et al. Transferrin-independent iron uptake supports B lymphocyte growth, *Blood,* 78: 1526–1531, 1991.

483. Morrone, G. et al. Transferrin-like autocrine growth factor, derived from T-lymphoma cells, that inhibits normal T-cell proliferation, *Cancer Res.,* 48: 3425–3429, 1988.

484. Savarino, A. et al. Modulation of surface transferrin receptors in lymphoid cells *de novo* infected with human immunodeficiency virus type-1, *Cell Biochem. Funct.,* 17: 47–55, 1999.

485. Bayer, A.L., Baliga, P., and Woodward, J.E. Transferrin receptor in T cell activation and transplantation, *J. Leukoc. Biol.,* 64: 19–24, 1999.

486. Teixeira, S. and Kühn, L.C. Post-transcriptional regulation of the transferrin receptor and 4F2 antigen heavy chain mRNA during growth activation of spleen cells, *Eur. J. Biochem.,* 202: 819–826, 1991.

487. Brekelmans, P. et al. Transferrin receptor expression as a marker of immature cycling thymocytes in the mouse, *Cell. Immunol.,* 159: 331–339, 1994.

488. Blue, M.L. et al. Discrete stages of human thymocyte activation and maturation *in vitro*: correlation between phenotype and function, *Eur. J. Immunol.*, 16: 771–777, 1986.

489. Brekelmans, P. et al. Inhibition of proliferation and differentiation during early T cell development by anti-transferrin receptor antibody, *Eur. J. Immunol.*, 24: 2896–2902, 1994.

490. Weber, R.J. and Kinkelman, F.D. Increased expression of the B lymphocyte receptor for transferrin is stimulated by *in vivo* crosslinking of cell surface IgD, *Cell. Immunol.*, 104: 400–408, 1987.

491. Boumpas, D.T. et al. Gene transcription during *in vitro* activation of human B lymphocytes with *Staphylococcus aureus* Cowan I strain, *J. Immunol.*, 8: 2701–2705, 1990.

492. Vora, A. et al. Two mechanisms of transferrin receptor induction by anti-Ig in B cells, *J. Immunol.*, 145: 2099–2104, 1990.

493. Roifman, C.M. and Wang, G. Receptor-mediated activation of human B lymphocytes in a nonphosphotyrosine-dependent manner, *Immunol.*, 149: 1179–1184, 1992.

494. West, M.A., Lucocq, J.M., and Watts, C. Antigen processing and class II MHC peptide-loading compartments in human B-lymphoblastoid cells, *Nature*, 369: 147–151, 1994.

495. Pond, L. and Watts, C. Characterization of transport of newly assembled, T cell-stimulatory MHC class II-peptide complexes from MHC class II compartments to the cell surface, *Immunol.*, 159: 543–553, 1997.

496. Andreesen, R. et al. Expression of transferrin and intracellular ferritin during terminal differentiation of human monocytes, *Blut*, 49: 195–202, 1994.

497. Hirata, T. et al. Expression of the transferrin receptors gene during the process of mononuclear phagocyte maturation, *J. Immunol.*, 136: 1339–1346, 1986.

498. Testa U. et al. Iron up-modulates the expression of transferrin receptors during monocyte–macrophage maturation, *J. Biol. Chem.*, 264: 13181–13190, 1989.

499. Bessis, M. L'ilôt érythroblastique, unité functionelle de la moelle osseuse, *Rev. Fr. Hématol.*, 13: 15–27, 1958.

500. Cline, N.J. and Golde, D.W. Cellular interactions in hemopoiesis, *Nature*, 277: 177–181, 1979.

501. Finch, C.A. and Huebers, H. Perspectives in iron metabolism, *New Engl. J. Med.*, 306: 1520–1530, 1982.

502. Andrews, N.C. Disorders of iron metabolism, *New Engl. J. Med.*, 341: 1986–1995, 1999.

503. Golde, D., Byers, L., and Finley, T. Proliferative capacity of human alveolar macrophage, *Nature*, 247: 373–375, 1974.

504. Badwey, J. and Karmovsky, M. Active oxygen species and the function of phagocytic leukocytes, *Annu. Rev. Biochem.*, 49: 695–727, 1980.

505. Klebanoff, S. and Waltersdorph, A. Prooxidant activity of transferrin and lactferrin, *J. Exp. Med.*, 172: 1293–1305, 1990.

506. Alford, C., King, T., and Campbell, P. Role of transferrin, transferrin receptors, and iron in macrophage listericidal activity, *J. Exp. Med.*, 174: 459–473, 1991.

507. Byrd, T.F. and Horwitz, M.A. Lactoferrin inhibits of promotes *Legionella pneumophila* intracellular multiplication in nonactivated and interferon gamma-activated human monocytes depending upon its degree of iron saturation, *J. Clin. Invest.*, 88: 1103–1112, 1991.

508. Gebran, S.J. et al. Inhibition of *Legionella pneumophila* growth by gamma interferon in permissive A/J mouse macrophages: role of reactive oxygen species, nitric oxide, tryptophan, and iron (III), *Infect. Immun.*, 62: 3197–3205, 1994.

509. Byrd, T.F. and Horwitz, M.A. Chloroquine inhibits the intracellular multiplication of *Legionella pneumophila* by limiting the availability of iron. A potential new mechanism for the therapeutic effect of chloroquine against intracellular pathogens, *J. Clin. Invest.*, 88: 351–357, 1991.

510. Newman, S.L. et al. Chloroquine induces human macrophage killing of *Histoplasma capsulatum* by limiting the availability of intracellular iron and is therapeutic in a murine model of histoplasmosis, *J. Clin. Invest.*, 93: 1422–1429, 1994.

511. Fortier, A.H. et al. Growth of *Francisella tularensis* LVS in macrophages: the acidic intracellular compartment provides essential iron required for growth, *Infect. Immun.*, 63: 1478–1483, 1995.

512. Levitz, S.M. et al. Chloroquine induces human mononuclear phagocytes to inhibit and kill *Cryptococcus neoformans* by a mechanism independent of iron deprivation, *J. Clin. Invest.*, 100: 1640–1646, 1997.

513. Douvas, G.S., May, M.H., and Crowle, A.J. Transferrin, iron, and serum lipids enhance or inhibit *Mycobacterium avium* replication in human macrophages, *J. Infect. Dis.*, 167: 857–864, 1993.

514. Clemens, D.L. and Horwitz, M.A. The *Mycobacterium tuberculosis* phagosome interacts with early endosomes and is accessible to exogenously administered transferrin, *J. Exp. Med.*, 184: 1349–1335, 1996.

515. Byrd, T.F. Tumor necrosis factor α (TNFα) promotes growth of virulent *Mycobacterium tuberculosis* in human monocytes, *J. Clin. Invest.*, 99: 2518–2529, 1997.

516. Kamene, E., Schurr, E., and Gros, P. Infection genomics: Nramp 1 as a major determinant of natural resistance to intracellular infections, *Annu. Rev. Med.*, 49: 275–287, 1998.

517. Gruenheid, S. et al. Natural resistance to infection with intracellular pathogens — the Nramp 1 protein is recruited to the membrane of the phagosome, *J. Exp Med.*, 185: 717–730, 1997.

518. Riwende, B.R. et al. Variations in the Nramp1 gene and susceptiblity to tuberculosis in West Africans, *New Engl. J. Med.*, 338: 640–644, 1998.

519. Barton, C.H. et al. Nramp1: a link between intracellular iron transport and innate resistance to intracellular pathogens, *J. Leuk. Biol.*, 66: 757–762, 1999.

520. Atkinson, P.G. and Barton, C.H. High level expression of Nramp1 G169 in RAW 264.7 cell transfectants: analysis of intracellular iron transport, *Immunology*, 96: 656–662, 1999.

521. Zwilling, B.S. et al. Role of iron in Nramp1-mediated inhibition of mycobacterial growth, *Infect. Immun.*, 67: 1386–1392, 1999.

522. Howe, D. and Mallavia, L.P. *Coxiella burnetii* infection increases transferrin receptors on J774A.1 cells, *Infect. Immun.*, 67: 3236–3241, 1999.

523. Uchiya, K. et al. A *Salmonella* virulence protein that inhibits cellular trafficking, *EMBO J.*, 18: 3924–3933, 1999.

524. Bradley, S.J. et al. Non-transferrin-bound iron induced by myeloablative chemotherapy, *Brit. J. Haematol.*, 99: 337–343, 1997.

525. Olakanmi, O. et al. Iron sequestration by macrophages decreases the potential for extracellular hydroxyl radical formation, *J. Clin. Invest.*, 91: 889–899, 1993.

526. Olakanni, O., Stokes, J.B., and Britigan, B.E. Acquisition of iron bound to low molecular weight chelates by human monocyte-derived macrophages, *J. Immunol.*, 153: 2691–2703, 1994.

527. Rama, R. and Sánchez, J. Transferrin uptake by bone marrow macrophages is dependent of the degree of iron saturation, *Brit. J. Haematol.*, 82: 455–459, 1992.

528. Cheung, D.L. and Hamilton, J.A. Regulation of human monocyte DNA synthesis by colony-stimulating factors, cytokines, and cyclin adenosine monophosphate, *Blood*, 79: 1972–1981, 1992.

529. Lo-Keshwar, B.L. and Lin, H.S. Growth factor-dependent regulation of transferrin receptor in proliferating and quiescent macrophages, *Cell. Immunol.*, 130: 401–415, 1990.

530. Hamilton, T., Weiel, J., and Adams, D. Expression of the transferrin receptor in murine peritoneal macrophages is modulated in the different stages of activation, *J. Immunol.*, 132: 2285–2294, 1984.

531. Hamilton, T., Gray, P., and Adams, D. Expression of the transferrin receptor on murine peritoneal macrophages is modulated by *in vitro* treatment with interferon gamma, *Cell. Immunol.*, 89: 478–488, 1984.

532. Byrd, T. and Horwitz, M. Interferon gamma-activated human monocytes downregulate multiplication of *Legionella pneumophila* by limiting the availability of iron, *J. Clin. Invest.*, 83: 1457–1466, 1989.

533. Loo, V. and Lalonde, R. Role of iron in intracellular growth of *Trypanosoma cruzi*, *Infect. Immun.*, 45: 726–735, 1984.

534. Raventos-Surrez, C., Pollack, S., and Nagel, R. *Plasmodium falciparum*: inhibition of *in vitro* growth by deferoxamine, *Am. J. Trop. Med. Hyg.*, 31: 919–931, 1984.

535. McGowan, S., Murray, J., and Parrish, M. Iron binding, internalization, and fate in human alveolar macrophages, *J. Lab. Clin. Med.*, 108: 587–595, 1986.

536. Álvarez-Hernández, X., Felstein, V., and Brock, J. The relationship between iron release, ferritin synthesis, and intracellular iron distribution in mouse peritoneal macrophages. Evidence for a reduced level of metabolically available iron in elicited macrophages, *Biochim. Biophys. Acta,* 886: 214–226, 1986.

537. Baynes, R. et al. Transferrin receptors and transferrin iron uptake by cultured human blood monocytes, *Eur. J. Cell. Biol.*, 43: 372–380, 1987.

538. Taetle, R. and Honeysett, J.M. γ-Interferon modulates human monocyte/macrophage transferrin receptor expression, *Blood*, 71: 1590–1598, 1988.

539. Gessani, S. et al. Enhanced production of LPS-induced cytokines during differentiation of human monocytes to macrophages. Role of LPS receptors, *J. Immunol.*, 151: 3758–3766, 1993.

540. Testa, U. et al. IFN-beta selectively down-regulates transferrin receptor expression in human peripheral blood macrophages by a post-translational mechanism, *J. Immunol.*, 155: 427–435, 1995.

541. MacMicking, J., Xie, Q., and Nathan, C. Nitric oxide and macrophage function, *Annu. Rev. Immunol.*, 15: 323–350, 1997.

542. Drapier, J.C. et al. Biosynthesis of nitric oxide activates iron regulatory factor in macrophages, *EMBO J.*, 12: 3643–3649, 1993.

543. Recalcati, S. et al. Nitric oxide-mediated induction of ferritin synthesis in J774 macrophages by inflammatory cytokines: role of selective iron regulatory protein-2 downregulation, *Blood*, 91: 1059–1066, 1998.

544. Bouton, C., Oliveira, L., and Drapier, J.C. Converse modulation of IRP1 and IRP2 by immunological stimuli in murine RAW 264.7 macrophages, *J. Biol. Chem.*, 273: 9403–9408, 1998.

545. Mulero, V. and Brock, J.H. Regulation of iron metabolism in murine J774 macrophage: role of nitric oxide-dependent and -independent pathways following activation with gamma interferon and lipopolysaccharide, *Blood,* 94: 2383–2389, 1999.

546. Wein, G. et al. Iron regulates nitric oxide synthase activity by controlling nuclear transcrption, *J. Exp. Med.*, 180: 969–976, 1994.

547. Wein, G., Bydan, C., and Hentze, M.W. Pathways for the regulation of macrophage iron metabolism by the anti-inflammatory cytokines IL-4 and IL-13, *J. Immunol.*, 158: 420–425, 1997.

548. Lin, M. et al. Role of iron in NF-kB activation and cytokine gene expression by rat hepatic macrophages, *Am. J. Physiol.*, 272: G1355-G1364, 1997.

549. Morrison, S.J., Shah, N.M., and Anderson, D.J. Regulatory mechanisms in stem cell biology, *Cell*, 80: 287–298, 1997.

550. Brandt, J. et al. Cytokine-dependent long-term culture of highly enriched precursors of hematopoietic progenitor cells from human bone marrow, *J. Clin. Invest.*, 86: 932–941, 1990.

551. Landsorp, P.M. and Dragowska, W. Long-term erythropoiesis from constant numbers of CD34+ cells in serum-free cultures initiated with highly purified progenitor cells from human bone marrow, *J. Exp. Med.*, 175: 1501–1509, 1992.

552. Mayani, H., Dragowska, W., and Lansdorp, P.M. Cytokine-induced selective expansion and maturation of erythroid versus myeloid progenitors from purified cord blood precursor cells, *Blood*, 81: 3252–3258, 1992.

553. Lansdorp, P.M., Dragowska, W., and Mayani, H. Ontogeny-related changes in proliferative potential of human hematopoietic cells, *J. Exp. Med.*, 178: 787–791, 1993.

554. Mayani, H., Dragowska, W., and Lansdorp, P.M. Characterization of functionally distinct subpopulations of CD34+ cord blood cells in serum-free long-term cultures supplemented with hematopoietic cytokines, *Blood*, 82: 2664–2672, 1993.

555. Mayani, H. and Lansdorp, P.M. Thy-1 expression is linked to functional properties of primitive hematopoietic progenitor cells from human umbilical cord blood, *Blood*, 83: 2410–2417, 1994.

556. Muench, M.O. et al. Expression of CD33, CD38, and HLA-DR on CD34+ human fetal liver progenitors with a high proliferative potential, *Blood*, 83: 3170–3181, 1994.

557. Rusten, L.S. et al. The FLT3 ligand is a direct and potent stimulator of the growth of primitive and committed human CD34+ bone marrow progenitor cells *in vitro*, *Blood*, 87: 1317–1325, 1996.

558. Sutherland, H.J., Blair, A., and Zapf, R.W. Characterization of a hierarchy in human acute myeloid leukemia progenitor cells, *Blood*, 87: 4754–4761, 1996.

559. Sieff, C. et al. Changes in cell surface antigen expression during hemopoietic differentiation, *Blood*, 60: 703–712, 1982.

560. Loken, M. et al. Flow cytometric analysis of human bone marrow cells. I. Normal erythroid development, *Blood*, 69: 255–264, 1982.

561. Gabbianelli, M. et al. Pure human hemopoietic progenitors: permissive action of basic fibroblast growth factor, *Science*, 249: 1561–1564, 1990.

562. Bender, J.G. et al. Identification of CD34-positive cells and their subpopulations from normal peripheral blood and bone marrow using multicolor flow cytometry, *Blood*, 77: 2591–2596, 1991.

563. Taetle, R., Honeysett, J.M., and Trowbridge, I. Effects of antitransferrin receptor antibodies on growth of normal and malignant cells, *Int. J. Cancer*, 32: 343–351, 1983.

564. Shannon, K. et al. Selective inhibition of the growth of human erythroid bursts by monoclonal antibodies against transferrin or the transferrin receptor, *Blood*, 67: 1631–1640, 1986.

565. Larrick, J.W. and Hyman, E.S. Acquired iron-deficiency anemia caused by an antibody against the transferrin receptor, *New Engl. J. Med.*, 311: 214–218, 1984.

566. Lebman, D. et al. A monoclonal antibody that detects expression of transferrin receptors in human erythroid precursor cells, *Blood*, 59: 671–680, 1982.

567. Cotner, T. et al. Characterization of a novel form of transferrin receptor preferentially expressed on normal erythroid progenitors and precursors, *Blood,* 73: 214–222, 1989.

568. Nunez, M.T. et al. Transferrin receptors in developing murine erythroid cells, *Br. J. Haematol.,* 36: 519–526, 1977.

569. Glass, J., Lavidoz, L.M., and Robinson, S.H. Expression of transferrin receptors and ferritin on erythroid cells, *J. Cell. Biol.,* 65: 298–305, 1975.

570. Yamada, H. and Gabudza, T.G. Ferritin content of developing erythroid cells, *J. Lab. Clin. Med.,* 83: 478–485, 1974.

571. Iacopetta, B.J., Morgan, E.H., and Yeoh, G.C.T. Transferrin receptors and iron uptake during erythroid cell development, *Biochim. Biophys. Acta,* 687: 204–212, 1982.

572. Pan, B. and Johnstone, R.M. Fate of transferrin receptor during maturation of sheep reticulocytes *in vitro*: selective externalization of the receptor, *Cell,* 33: 967–978, 1983.

573. Parmeley, R., Haidu, I., and Denys, F. Ultrastructural localization of the transferrin receptor and transferrin on marrow surfaces, *Br. J. Haematol.,* 54: 633–670, 1983.

574. Kausas, G., Muirhead, M., and Dailey, M. Expression of the CD11/CD18 leukocyte adhesion molecule 1 and CD44 adhesion molecules during normal myeloid and erythroid differentiation in humans, *Blood,* 76: 2483–2491, 1990.

575. Sassa, S. Sequential induction of heme pathway enzymes during erythroid differentiation of mouse Friend leukemia virus-infected cells, *J. Exp. Med.,* 143: 305–316, 1976.

576. Hu, H.Y., Gardner, J., and Aisen, P. Inducibility of transferrin receptors on Friend erythroleukemic cells, *Science,* 197: 559–562, 1977.

577. Ross, J. and Sautrer, D. Induction of globin mRNA accumulation by hemin in cultured erythroleukemic cells, *Cell,* 8: 513–522, 1976.

578. Tsiftsoglon, A. et al. Dissociation of iron transport and heme biosynthesis from commitment to terminal maturation of murine erythroleukemia cells, *Proc. Natl. Acad. Sci. U.S.A.,* 80: 7528–7533, 1983.

579. Battistini, A. et al. Positive modulation of hemoglobin, heme, and transferrin receptor synthesis by murine interferon-α and -β in differentiating Friend cells, *J. Biol. Chem.,* 266: 528–536, 1991.

580. Battistini, A. et al. Intracellular heme coordinately modulates globin chain synthesis, transferrin receptor number, and ferritin content in differentaiting Friend erythroleukemia cells, *Blood,* 78: 2098–2109, 1991.

581. Grenick, J.L. and Sassa, S. Hemin control of heme biosynthesis in mouse Friend virus transformed erythroleukemia cells in culture, *J. Biol. Chem.,* 53: 5402–5410, 1978.

582. Hoffman, R. et al. Hemin control of heme biosynthasis and catabolism in a human leukemia cell line, *Blood,* 56: 567–575, 1980.

583. Glass, J. et al. Transferrin receptor, iron transport, and ferritin metabolism in Friend erythroleukemia cells, *Biochim. Biophys. Acta,* 542: 154–162, 1978.

584. Beaumont, C. et al. Transcriptional regulation of ferritin H and L subunits in adult erythroid and liver cells from the mouse, *J. Biol. Chem.,* 264: 7498–7506, 1989.

585. Busfield, S.J. et al. Complex regulation of transferrin receptors during erythropoietin-induced differentiation of J2E erythroid cells, *Eur. J. Biochem.,* 249: 77–84, 1997.

586. Graf, T., Ade, N., and Beng, N. Temperature sensitive mutant of avian erythroblastosis virus suggests a block of differentiation as a mechanism of leukemogenesis, *Nature,* 275: 496–499, 1978.

587. Beng, H. et al. Hormone-dependent terminal differentiation *in vitro* of chicken erythroleukemia cells transformed by TS mutants of avian erythroblastosis virus, *Cell,* 28: 907–919, 1982.

588. Schmidt, J.A. et al. Monoclonal antibodies against novel erythroid differentiation antigens reveal specific effect of oncogenes on the leukemic cell phenotype, *Leuk. Res.,* 10: 257–271, 1986.
589. Schmidt, J.A. et al. Control of erythroid differentiation: possible role of the transferrin cycle, *Cell,* 46: 41–53, 1986.
590. Ponka, P. Tissue-specific regulation of iron metabolism and heme synthesis: distinct control mechanisms in erythroid cells, *Blood,* 89: 1–25, 1997.
591. Ponka, P. Cell biology of heme, *Am. J. Med. Sci.,* 318: 241–256, 1999.
592. Chan, L. and Gerhardt, E.M. Transferrin receptor is hyperexpressed and transcriptionally regulated in differentiating erythroid cells, *J. Biol. Chem.,* 267: 8254–8259, 1992.
593. Chan, R.Y.Y. et al. Regulation of transferrin receptor mRNA expression. Distinct regulatory features in erythroid cells, *Eur. J. Biochem.,* 220: 683–692, 1994.
594. Weiss, G. et al. Regulation of cellular iron metabolism by erythropoietin: activation of iron-regulatory protein and upregulation of transferrin receptor expression in erythroid cells, *Blood,* 89: 680–687, 1997.
595. Nakajima, O. et al. Heme deficiency in erythroid lineage causes differentiation arrest and cytoplasmic iron overload, *EMBO J.,* 18: 6282–6289, 1999.
596. Vaisman, B., Fibach, E., and Konijn, A.M. Utilization of intracellular ferritin iron for hemoglobin synthesis in developing human erythroid precursors, *Blood,* 90: 831–838, 1997.
597. Gelvan, D. et al. Ferritin uptake by human erythroid precursors is a regulated iron uptake pathway, *Blood,* 88: 3200–3207, 1996.
598. Garrick, L.M. et al. Non-transferrin-bound iron uptake in Belgrade and normal rat erythroid cells, *J. Cell Physiol.,* 178: 349–358, 1999.
599. Lebron, J.A. et al. Crystal structure of the hemochromatosis protein HFE and characterization of its interaction with transferrin receptor, *Cell,* 93: 111–123, 1998.
600. Parkkila, S. et al. Association of the transferrin receptor in human placenta with HFE, the protein defective in hereditary hemochromatosis, *Proc. Natl. Acad. Sci. U.S.A.,* 94: 13198–13202, 1997.
601. Feder, J.N. et al. The hemochromatosis gene product complexes with the transferrin receptor and lowers its affinity for ligand binding, *Proc. Natl. Acad. Sci. U.S.A.,* 95: 1472–1477, 1998.
602. Waheed, A. et al. Association of HFE protein with transferrin receptor in crypt enterocytes of human duodenum, *Proc. Natl. Acad. Sci. U.S.A.,* 96: 1579–1584, 1999.
603. Gross, C.N. et al. Co-trafficking of HFE, a nonclassical major histocompatibility complex class I protein, with the transferrin receptor implies a role in intracellular iron regulation, *J. Biol. Chem.,* 273: 22068–22074, 1998.
604. Roy, C.N. et al. The hereditary hemochromatosis protein, HFE, specifically regulates transferrin-mediated iron uptake in HeLa cells, *J. Biol. Chem.,* 274: 9022–9028, 1999.
605. Salter-Cid, L. et al. Transferrin receptor is negatively modulated by the hemochromatosis protein HFE: implications for cellular iron homeostasis, *Proc. Natl. Acad. Sci. U.S.A.,* 96: 5434–5439, 1999.
606. Lebron, J.A. and Bjorkman, P.J. The transferrin receptor binding site on HFE, the class I MHC-related protein mutated in hereditary hemochromatosis, *J. Mol. Biol.,* 289: 1109–1118, 1998.
607. Lebron, J.A., West, A.P., and Bjorkman, P.J. The hemochromatosis protein HFE competes with transferrin for binding to the transferrin receptor, *J. Mol. Biol.,* 294: 239–245, 1999.
608. Bennet, M.J., Lebron, J.A., and Bjorkman, P.J. Crystal structure of the hereditary hemochromatosis protein complexed with transferrin receptor, *Nature,* 403: 46–53, 2000.

609. Riedel, H.D. et al. HFE downregulates iron uptake from transferrin and induces iron-regulatory protein activity in stably transfected cells, *Blood*, 94: 3915–3921, 1999.

609a. Levy, J.E., Montross, L.K., and Andrews, N.C. Genes that modify the hemochromatosis phenotype in mice, *J. Clin. Invest.*, 105: 1209–1216, 2000.

610. Gunshin, M. et al. Cloning and characterization of a mammalian proton-coupled metal ion-transporter, *Nature*, 21: 482–488, 1997.

611. Vulpe, C.D.et al. Hephaestin, a ceruloplasmin homologue implicated in intestinal iron transport, is defective in the sla mouse, *Nature Genet.*, 21: 195–199, 1999.

612. Mukhopadhyvay, C.K., Attich, Z.K., and Fox, P.L. Role of ceruloplasmin in cellular iron uptake, *Science*, 279: 714–717, 1998.

613. Attick, Z.K. et al. Ceruloplasmin ferroxidase activity stimulates cellular iron uptake by a trivalent cation-specific transport mechanism, *J. Biol. Chem.*, 274: 1116–1123, 1999.

614. Harris, Z.L. et al. Targeted gene disruption reveals an essential role for ceruloplasmin in cellular iron efflux, *Proc. Natl. Acad. Sci. U.S.A.*, 96: 10812–10817, 1999.

614a. Mukhopadhyvay, C.K., Mazumder, B., and Fox, P.L. Role of hypoxia inducible factor-1 in transcriptional activation of ceruloplasmin by iron deficiency, *J. Biol. Chem.*, 275: 21048–21054, 2000.

615. Donovan, A. et al. Positional cloning of zebrafish ferroportin 1 identifies a conserved vertebrate iron exporter, *Nature*, 403: 776–781, 2000.

616. McKie, A.T. et al. A novel duodenal iron-regulated transporter, IREG-1, implicated in the basolateral transfer of iron to the circulation, *Molecular Cell*, 5: 299–309, 2000.

616a. Abboud, S. and Haile, D.J. A novel mammalian iron-regulated protein involved in intracellular iron metabolism, *J. Biol. Chem.*, 275: 19906–19912, 2000.

617. Canonne-Hergaux, F., Gruenheid, S., and Gros, P. Cellular and subcellular localization of the Nramp2 iron transporter in the intestinal brush border and regulation by dietary iron, *Blood*, 93: 4406–4417, 1999.

618. Tandy, S. et al. Nramp2 expression is associated with pH-dependent iron uptake across the apical membrane of human intestinal Caco-2 cells, *J. Biol. Chem.*, 275: 1023–1029, 2000.

619. Koutnikova, H. et al. Studies of human, mouse and yeast homologues indicate a mitochondrial function for frataxin, *Nature Genet.*, 16: 345–351, 1997.

620. Rotig, A. et al. Aconitase and mitochondrial iron-sulfur protein deficiency in Friedreich ataxia, *Nature Genet.*, 17: 215–217, 1997.

621. Kispal, G. et al. The mitochondrial proteins Atm1 and Nfs1 are essential for biogenesis of cytosolic Fe/S proteins, *EMBO J.*, 18: 3981–3989, 1999.

6 Soluble Transferrin Receptor

CONTENTS

6.1   Introduction 371
6.2   Generation of Soluble Transferrin Receptor 371
6.3   Circulation in Human Serum..374
6.4   Clinical Relevance..374
References..379

6.1   INTRODUCTION

Recent studies have shown that, in addition to membrane-bound transferrin receptor, a soluble form of transferrin receptor is released from the surfaces of cells. This soluble truncated form of the receptor can be detected in human serum by using sensitive immunoassays, and the initial clinical experience measuring this new parameter indicates that soluble transferrin receptor levels reflect the total body mass of tissue receptor.

The physicochemical nature of the circulating receptor was clarified by biochemical studies involving specific monoclonal antibodies.[1] The circulating soluble receptor was purified from human serum and biochemically characterized. The serum receptor exhibited a molecular weight of 85 kDa on SDS-PAGE analysis under both non-reducing and reducing conditions. Amino terminal sequence analysis showed that residues 1 through 19 of the soluble serum receptor were identical to residues 101 through 119 of the membrane-bound receptor. This finding clearly indicates that the circulating receptor is a truncated form that lacks the cytoplasmic and transmembrane domains of the intact membrane-bound receptor (see Figure 6.1).

This conclusion was further supported by an additional study showing that antisera against the extracellular domain exhibited reactivity against both purified intact receptor and immunopurified circulating receptor, whereas antisera against the intracellular domain reacted only with intact receptor.[2] The same study provided clear and direct evidence that >99% of circulating transferrin receptor is represented by a truncated form lacking the intracellular and transmembrane domains, whereas <1% of the total circulating transferrin receptor is intact, consistent with an exosomal origin.[2]

6.2   GENERATION OF SOLUBLE TRANSFERRIN RECEPTOR

Biochemical studies support the concept that the soluble transferrin receptor represents the extracytoplasmic domain of the cellular transferrin receptor, generated as

TRANSFERRIN RECEPTOR

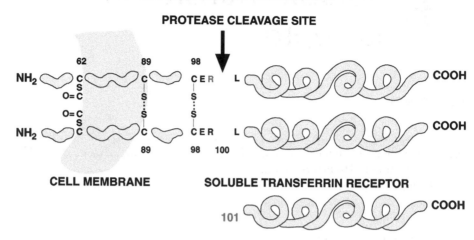

FIGURE 6.1 Schematic representation of the soluble transferrin receptor. The site of protease cleavage is indicated by an arrow.

a consequence of alternate RNA splicing or released from the cell surface through proteolytic cleavage by a membrane-based protease. However, because the truncation site is located adjacent to the cell surface and the amino terminus of the molecule is oriented to the cytoplasm, proteolytic cleavage of intact cellular receptor seems more likely.

In fact, two recent studies have provided direct evidence in favor of the proteolytic cleavage mechanism: (1) cell membranes isolated from HL-60 promyelocytic cells readily released transferrin receptors, and this release may be blocked by a cocktail of protease inhibitors[3]; and (2) the proteolytic activity responsible for the degradation of membrane transferrin receptor, and thus for the release of a soluble truncated form, is specifically inhibited by protease inhibitors PMSF and DIFP, thus suggesting involvement of a serine protease.[4]

Recent studies suggest that the O-linked glycosylation site at position 104 may play a role in mediating the susceptibility of intact transferrin receptor to proteolytic attack.[5] To elucidate the effect of the O-linked carbohydrate on transferrin receptor function, the oligosaccharide was eliminated by replacing Thr 104 with Asp, and the mutated cDNA was expressed in a cell line lacking endogenous transferrin receptor. The elimination of the oligosaccharide at Thr 104 produced a form of the receptor markedly susceptible to proteolytic cleavage, giving rise to a 78-kDa soluble receptor that can bind transferrin and is released into the growth medium.[5]

Treatment of the cells with neuraminidase, which removes sialic acid residues from the membrane transferrin receptor, enhances receptor cleavage.[6] This observation suggests that sialic acid residues are important for protection from proteolytic cleavage.

The blood cells of chickens, sheep, rats, and mice produced little membrane proteolytic activity capable of proteolyzing transferrin receptor.[7] Interestingly, in these species, the serum soluble transferrin receptor is associated with a vesicle (exosome) and is full length.[8]

Studies of an endocytosis-deficient mutant of transferrin receptor lacking an internalization signal localized at the cell surface have shown that this mutant is cleaved less effficiently than the wild-type transferrin receptor.[9] This observation suggests that the protease mainly responsible for transferrin receptor cleavage is localized to an intracellular compartment.

The importance of the endosomal compartment in the cleavage of transferrin receptor is also supported by the observation that a significant proportion of newly synthesized transferrin receptors are detected in the endosome before they appear on the cell surface.[10] Since the soluble transferrin receptor contains all the extracellular part of the receptor, including the transferrin binding domain, and the plasmatic molar concentration of transferrin is much higher than that of soluble transferrin receptor, the the soluble receptor must circulate bound to transferrin. The majority of the soluble receptor can be isolated and thus removed from serum using precipitating antibodies to human transferrin.[11] Thus, it seems reliable to assume that the soluble transferrin receptor circulates in the form of a complex with one molecule of transferrin. However, the small amounts of soluble transferrin receptor in serum compared with the amount in transferrin make the soluble transferrin receptor unlikely to modulate iron uptake detectably.

The cellular mechanisms responsible for the production of soluble transferrin receptor remain to be determined. Several studies focused on the cellular mechanisms responsible for the extrusion of the transferrin receptor from the cell membrane during transition from reticulocytes to young red blood cells. These studies[12-16] demonstrated that small 50-nm particles termed exosomes and containing transferrin receptor are produced when sheep reticulocytes are cultured in vitro. The exosomes arise from internal bleb formation within the endocytic vesicles, and the result is a multivesicular endosome. The exosomes contained in this structure are released by exocytosis. Under this cellular system, transferrin receptor and other reticulocyte membrane proteins no longer needed in mature erythrocytes are lost.[15]

In vitro studies have shown that both reticulocytes[7-14] and in vitro-grown leukemic cells[3,17,18] release intact exosomal and soluble truncated transferrin receptors. These in vitro studies provided evidence that the receptor was cleaved from exosomes in reticulocytes, and thus exosomes represent the major sources of soluble truncated transferrin receptor.

This conclusion is directly supported by the observation that the size of one of the transferrin receptor peptides retained in exosomes corresponds roughly to the expected size of the cytoplasmic domain of the transferrin receptor. It was recently proposed that proteases released from circulating granulocytes may be responsible for proteolysis of transferrin receptors from released exosomes.[19]

The amount of soluble transferrin receptor released from cells and the amount of membrane-bound transferrin receptor are influenced directly by the iron status of the cells. Thus, iron deficiency leads to a marked increase in the level of soluble transferrin receptor released from the cells, whereas an opposite effect is elicited following iron load.

6.3 CIRCULATION IN HUMAN SERUM

The observation of variations based on iron status stimulated the development of reliable methods to quantify the levels of circulating soluble transferrin receptor in human serum. The immunoenzymatic quantification methods using monoclonal[20] or polyclonal[21] anti-transferrin receptor provided the most accurate quantification of soluble transferrin receptor in serum; for normal subjects, values corresponded to 2.2 to 5.6 mg/liter serum. Alternatively, serum transferrin receptor may be evaluated by antibody-coated latex agglutination nephelometry in which the soluble receptor is measured in the form of a complex with transferrin.[22]

The level of soluble transferrin receptor changes during ontogenesis. The level increases during fetal life from 20 to 42 weeks, and at birth reaches values two times higher than levels observed in adults.[23] At one year of age, soluble transferrin receptor levels are slightly higher than those observed in adults.[24]

Recent studies focused on evaluating the variability of serum transferrin receptor levels, and indicate that interindividual and intraindividual variabilities are relatively wide.[25-26] The estimated interindividual coefficient of variation was 20.8% and the intraindividual value was 13.6%.

Particularly important is a recent study in which soluble transferrin receptor levels were evaluated using a quantitative two-site immunoenzymatic technique calibrated from 3 to 80 nmol/L.[27] Hematologically normal adult subjects exhibited individual values ranging from 7.6 to 37.7 nmol/L (mean value: 19.6 nmol/L +5) distributed according to a Gaussian curve. There was no correlation with age (from 19 to 79 years), and no significant differences were noted between men and women or between pre- and post-menopausal women. Interestingly, subjects residing at high altitudes had soluble transferrin receptor concentrations about 9% higher than those living nearer to sea level.[27]

The development of these sensitive methods for the quantification of the plasmatic levels of soluble transferrin receptor led to determination of a possible physiopathological significance of this marker in the diagnosis of different types of anemia.

6.4 CLINICAL RELEVANCE

Several studies focused on evaluating the level of serum transferrin receptor in iron deficiency. An initial study,[28] later confirmed by other studies,[29,30] provided evidence that serum transferrin receptor levels are clearly increased in patients with iron deficiencies. Particularly relevant were the findings from a study in which repeated phlebotomies were performed on normal volunteer subjects to produce varying degrees of iron deficiency.[29]

During the phase of storage iron depletion, a rapid fall in serum ferritin associated with minimal changes of serum ferritin receptor was observed. During tissue iron deficiency, the serum receptor level rose in direct proportion to the induced deficit of functional iron. Relatively little change in serum ferritin occurred after iron stores were fully depleted.[29]

This study clearly indicated that serum ferritin is the most sensitive index of iron status in the presence of residual iron stores, whereas the serum receptor is more sensitive in functional iron deficiency.

This study also showed that the evaluation of the serum transferrin receptor/ferritin ratio represents the most sensitive way to evaluate the development of iron deficiency. This ratio increases from about 100 when iron stores are complete to more than 2000 when a marked deficiency of functional iron is developed.[29]

The serum transferrin receptor/ferritin ratio also represents a valuable tool to distinguish anemia of iron deficiency from anemia of chronic disease.[31] Interestingly, iron-deficient patients do not show the extreme values observed in patients with chronic diseases.[31]

When a patient with iron deficiency anemia was treated with intravenous iron, serum transferrin receptor levels showed a transient rise just after the initiation of iron supplementation, and then returned to normal levels with the improvement of anemia. Reticulocyte counts also increased after iron supplementation and thereafter returned to normal levels. However, the peak reticulocyte counts were observed two days after the peak of serum transferrin receptor level. This discrepancy in the timing of the rises of these two parameters after iron supplementation can be explained by considering that polychromatic erythroblasts, in addition to reticulocytes, represent important sources of serum transferrin receptor.

The measurement of serum transferrin receptor levels in pregnant women is also of some utility to differentiate the dilutional anemia due to an increase in plasma volume during the second trimester of pregnancy from a condition of iron deficiency. The identification of subjects with low ferritin levels associated with high serum transferrin receptor levels greatly helps to identify pregnant women with iron deficiencies.[30]

The concepts that serum ferritin levels reflect iron storage levels and serum transferrin receptor levels are related to the levels of functional iron are further supported by two additional observations:

1. Serum receptors are nearly identical in men and women,[21,27] whereas serum ferritin levels show a striking sex difference, related to lower iron stores present in females than in males due to iron loss from menstruation.
2. Iron absorption in iron-replete normal subjects inversely correlates with serum ferritin levels, but not with serum receptor levels.[32]

These two observations also indicate that serum transferrin receptor levels are not related to the levels of the iron storage compartment.

In vitro studies further indicate that iron supply modulates the level of soluble transferrin receptors released from the cells. They show that normal human erythroblasts[33] and erythroleukemic cell lines[34] released relatively large amounts of soluble transferrin receptors, whose level was affected by the amount of iron supplied to the cells. The addition of diferric transferrin to these cells induced a dose-dependent decrease in cellular and soluble transferrin receptor expression.

Contemporaneous analysis of both soluble transferrin receptor levels and extent of transferrin receptor expression on the membranes of erythroblasts derived from patients with different forms of anemia provided some interesting data:[35]

1. In iron deficiency, a high concentration of soluble transferrin receptor in serum is largely related to an increased expression of transferrin receptors on the membranes of erythroid cells.
2. In anemia of chronic disease and myelodysplasia, a discrepancy between low expression of transferrin receptors on the membranes of erythroblasts and normal or elevated serum values of soluble receptors was seen.

In patients with laboratory evidence of iron-deficient erythropoiesis, the identification of true iron deficiency as the underlying cause is very important from a clinical view because of the prompt response of the condition to therapy and the clue it may provide for detecting potentially serious gastrointestinal disorders. Distinguishing between iron deficiency anemia and the anemia that accompanies infection, inflammation, or malignancy, commonly termed anemia of chronic disease, is often difficult.

A multiparametric study[36] of patients affected either by iron deficiency anemia or by anemia of chronic disease showed elevated levels in the former and normal levels in the latter. This finding indicates that the evaluation of serum transferrin receptor levels is a more reliable way of distinguishing iron deficiency anemia from anemia of chronic disease than the evaluation of serum ferritin levels. In fact, serum ferritin levels are disproportionately elevated in relation to iron stores in patients with inflammation or chronic disease, whereas serum transferrin receptor levels are not affected by these disorders.

Additional studies were conducted to evaluate the levels of serum transferrin receptors in anemias associated with reduced or enhanced erythropoiesis. A series of studies provided evidence of reduced levels of serum transferrin receptors in patients with aplastic anemia[11] and those undergoing high-dose myeloablative chemotherapy before autologous stem cell transplantation.[37]

The decrease of serum transferrin receptors observed corresponds to 50 to 60%, which represents the estimated percentage of total serum transferrin receptor related to developing red cells, the remainder attributable to other tissues.

In patients undergoing autologous stem cell transplantation, the initial decline of soluble transferrin receptor levels is followed by a rescue that parallels the kinetics of recovery of highly fluorescent reticulocytes (immature reticulocytes) and precedes the hemoglobin rescue.[38]

On the other hand, increased serum transferrin receptor levels in patients with hemolytic anemia have been observed. Markedly increased levels of serum transferrin receptor were observed in patients with autoimmune hemolytic anemia,[11,28] hereditary spherocytosis,[11] sickle cell anemia,[2,20] and β-thalassemia.[11]

Serum receptor elevations are significantly higher in patients with ineffective erythropoiesis due to thalassemia than in patients with hemolytic anemia.[11] One interesting finding was that the elevation in circulating receptor was directly proportional to the increase in erythroid precursor mass by comparison with the erythron transferrin uptake. It was then suggested that serum receptor can be used as a quantitative measure of total erythropoiesis.

Soluble transferrin receptor was also evaluated in 230 thalassemics after bone marrow transplantation. Patients who received bone marrow from HLA-identical sibling donors heterozygous for β-thalassemia displayed significantly higher levels of soluble transferrin receptors than patients transplanted with marrow from normal sibling donors.[39] This indicates that the increased erythropoiesis observed in thalassemic patients tranplanted with heterozygous donors is responsible for increased soluble transferrin receptor levels.

Studies of hereditary and acquired hemochromatosis provided evidence that the majority of these patients displayed virtually normal levels of serum transferrin receptor.[40]

The nature of relationship of transferrin receptor and erythropoiesis was further examined by studying megaloblastic anemia, a condition in which total erythropoiesis is markedly increased, and viable erythrocytes are not effectively released into the circulation because the majority of erythroid precursors mature abnormally and are destroyed in bone marrow (dyserythropoiesis).

High serum transferrin receptor levels were observed in patients who were clearly anemic and had high lactate dehydrogenase levels.[41] The elevated receptor levels rose further with cobalamin therapy, as effective erythropoiesis replaced ineffective erythropoiesis. These high levels persisted until the increased erythropoiesis returned to normal.

Initial studies suggest that soluble transferrin receptors are not modified by changes in the level of iron storage. However, a recent study performed on 150 adult subjects from rural Zimbabwe showed that iron overload, as indicated by increased ferritin levels, is associated with a decrease of the level of soluble transferrin receptor.[42]

It is well known that membrane transferrin receptor is highly expressed in tissues undergoing high rates of cell proliferation. On the basis of this observation, the possibility that serum transferrin receptor levels are increased in hematological malignancies independently of erythropoiesis was examined.[37] A significant two- to three-fold elevation was observed in patients with polycythemia or myelofibrosis, whereas no increases were found in patients with chronic granulocytic leukemia, acute myeloid leukemia, and myelodysplasia.[37]

Among patients with lymphoproliferative disorders, serum receptor levels were normal in lymphoma, multiple myeloma, and hairy cell leukemia, but they were significantly elevated in chronic lymphocytic leukemia, where the levels reflected the clinical stage of the disease. While the mechanisms responsible for the inreased levels of serum receptor in chronic lymphocytic leukemia remain to be determined, two possible mechanisms can be ruled out:

1. Circulating leukemic lymphocytes are the source of the increased serum receptor since leukemic lymphocytes possess only low levels of membrane transferrin receptors.
2. Shedding of transferrin receptors from expanded erythropoiesis as the source of increased transferrin receptors is unlikely because erythroid marrow activity was not increased in the majority of patients with chronic lymphocytic leukemia.

Particularly relevant was the addition of serum transferrin receptor evaluation to the clinical assessment and definition of some anemic conditions. On the basis of several criteria mainly based on pathophysiology, standard clinical laboratory tests, and ferrokinetic studies, anemic conditions are usually classified as: (1) anemias due to hyperdestruction with variants of hemolysis or ineffective erythropoiesis; (2) anemias due to intrinsic marrow hypoproliferation; and (3) anemias related to defective erythropoietin production. The ferrokinetic studies may not become part of the standard clinical practice due to their cost and time required for each test.

Thus, recent studies have shown that the measurements of some parameters, including reticulocyte index, erythroid cellularity, serum transferrin and ferritin levels, erythropoietin plasma concentration, and serum transferrin receptor evaluation allow careful determination of the majority of anemic conditions, including anemias due to mixed mechanisms (i.e., hypoproliferation associated with ineffective erythropoiesis).[43]

The analysis of these different biochemical parameters in different anemias showed also that:

1. An excellent inverse correlation was observed between hematocrit and erythropoietin levels on the one hand, and between hematocrit and serum transferrin receptor on the other.
2. The slope of the regression of serum transferrin receptor versus hematocrit was very similar to the slope of the regression of erythropoietin versus hematocrit.
3. A very good direct correlation was observed by plotting serum transferrin receptor levels versus erythron transferrin uptake values derived from ferrokinetic studies.[43]

The changes in serum ferritin, serum soluble transferrin receptor, and reticulocyte index provide clear distinctions among the four major categories of anemia. Iron deficiency anemia is recognized by low serum ferritin, elevated serum transferrin receptor levels, and a normal reticulocyte index.

Hypoproliferative anemia is identified by reciprocal changes in serum ferritin and serum transferrin receptor levels.

Maturation abnormalities are recognized by high serum ferritin and serum receptor and a normal reticulocyte index, while hemolytic disorders are associated with elevated levels of all three laboratory indices.

Anemia is a typical feature of chronic renal failure and the majority of patients on hemodialysis require red blood transfusions. The main mechanism involved in the development of this anemic condition is inadequate erythropoietin production by the kidneys. Clinical trials involving renal failure patients on hemodialysis showed that erythropoietin was effective in correcting anemia, but the level of response to this therapy was highly variable among patients, both in terms of the dose of erythropoietin and the kinetics of hemoglobin recovery. Studies of serum transferrin receptor levels were of value in predicting the responses to erythropoietin in these patients.[44]

Low baseline serum transferrin receptor levels and increments greater than 20% within two weeks of treatment are predictive of a favorable response to erythropoietin therapy.

The measurement of serum transferrin receptor levels represents a valuable predictor of the outcomes of erythropoietin therapy in patients with oncologic diseases. Among patients with hematologic malignancies treated with erythropoietin, the most likely to respond were those with consistent increases (>50%) of serum transferrin receptor levels after 2 weeks of treatment.[45]

REFERENCES

1. Shih, Y.J. et al. Serum transferrin is a truncated form of tissue receptor, *J. Biol. Chem.*, 265: 19077–19081, 1990.
2. Shih, Y.J. et al. Characterization and quantitation of the circulating forms of serum transferrin receptor using domain-specific antibodies, *Blood*, 81: 234–238, 1993.
3. Chitambar, C.R. and Zivkovic, Z. Release of soluble transferrin receptor from the surface of human leukemic HL-60 cells, *Blood*, 74: 602–608, 1990.
4. Baynes, R.D. et al. Production of the serum form of the transferrin receptor by a cell membrane-associated serine protease, *Proc. Soc. Exp. Biol. Med.*, 204: 65–69, 1993.
5. Rutledge, E.A. et al. Elimination of the O-linked glycosylation site at Thr 104 results in the generation of a soluble human transferrin receptor, *Blood*, 83: 580–586, 1994.
6. Rutledge, E.A. and Enns, C.A. Cleavage of the transferrin receptor is influenced by the composition of the O-linked carbohydrate at position 104, *J. Cell. Physiol.*, 168: 284–293, 1996.
7. Ahn, J. and Johnstone, R.M. Origin of soluble truncated transferrin receptor, *Blood*, 81: 2442–2451, 1993.
8. Johnstone, R.M. et al. Exosome formation during maturation of mammalian and avian reticulocytes: evidence that exosome release is a major route for externalization of obsolete membrane proteins, *J. Cell. Physiol.*, 147: 27–37, 1991.
9. Rutledge, E.A., Green, F.A., and Enns, C.A. Generation of the soluble transferrin receptor requires cycling through an endosomal compartment, *J. Biol. Chem.*, 269: 31864–31868, 1994.
10. Futter, C.A. et al. Newly synthesized transferrin receptors can be detected in the endosome before they appear on the cell surface, *J. Biol. Chem.*, 270: 10999–11003, 1995.
11. Huebers, H.A. et al. Intact transferrin receptors in human plasma and their relation to erythropoiesis, *Blood*, 75: 102–107, 1990.
12. Paui, B.T. and Johnstone, R. Fate of the transferrin receptor during maturation of sheep reticulocytes *in vitro*: selective externalization of the receptor, *Cell*, 33: 967–975, 1983.
13. Johnstone, R.M., Bianchini, A., and Teng, K. Reticulocyte maturation and exosome release: transferrin receptors containing exosomes show multiple plasma membrane functions, *Blood*, 74: 1844–1852, 1989.
14. Chitambar, C.R., Loebel, A.R., and Noble, N.A. Shedding of transferrin receptors from rat reticulocytes during maturation *in vitro*: soluble receptor is derived from receptor shed in vesicles, *Blood*, 78: 2444–2452, 1991.
15. Johnstone, R.M. Maturation of reticulocytes: formation of exosomes as a mechanism for shedding membrane proteins, *Biochem. Cell. Biol.*, 70: 179–188, 1992.
16. Shintani, N. et al. Expression and extracellular release of transferrin receptors during peripheral erythroid progenitor cell differentiation in liquid culture, *Blood*, 83: 1209–1215, 1994.
17. Baynes, R.D., Shih, Y.J., and Cook, J.D. Production of soluble transferrin receptor by K562 erythroleukemia cells, *Br. J. Haematol.*, 78: 450–455, 1991.

18. Baynes, R.D. et al. Characterization of transferrin receptors released by K562 erythroleukemic cells, *Proc. Soc. Exp. Biol. Med.*, 197: 416–423, 1991.

19. Johnstone, R.M. Cleavage of the transferrin receptor by human granulocytes: preferential proteolysis of the exosome-bound TfR, *J. Cell. Physiol.*, 168: 333–345, 1996.

20. Flowers, C.H. et al. The clinical measurement of serum transferrin receptor, *J. Lab. Clin. Med.*, 114: 368–377, 1989.

21. Cazzola, M. and Begiun, Y. New tools for clinical evaluation of erythron function in man, *Br. J. Haematol.*, 80: 278–284, 1992.

22. Hikawa, A. et al. Soluble transferrin receptor–transferrin complex in serum: measurement by latex agglutination nephelometric immunoassay, *Clin. Chim. Acta*, 254: 159–172, 1996.

23. Carpani, G. et al. Soluble transferrin receptor in the study of fetal erythropoietic activity, *Am. J. Hematol.*, 52: 192–196, 1996.

24. Young, G.S. and Zlotkin, H.S. Percentile estimates for transferrin receptor in normal infants 9–15 mo. of age, *Am. J. Clin. Nutr.*, 66: 342–346, 1997.

25. Cooper, M.J. and Zlotkin, S.H. Day-to-day variation of transferrin receptor and ferritin in healthy men and women, *Am. J. Clin. Nutr.*, 64: 738–742, 1996.

26. Maes, M. et al. Components of biological variation in serum soluble transferrin receptor: relationship to serum iron, transferrin and ferritin concentrations, and immune and haematological variables, *Scand. J. Clin. Lab. Invest.*, 57: 31–41, 1997.

27. Allen, J. et al. Measurement of soluble transferrin receptor in serum of healthy adults, *Clin. Chem.*, 44: 35–39, 1998.

28. Kohgo, Y. et al. Serum transferrin receptor as a new index of erythropoiesis, *Blood*, 70: 1955–1958, 1987.

29. Skikne, B.S., Flowers, C.H., and Cook, J.D. Serum transferrin receptor: a quantitative measure of tissue iron deficiency, *Blood*, 75: 1870–1876, 1990.

30. Carriaga, M.T. et al. Serum transferrin receptor for the detection of iron deficiency in pregnancy, *Am. J. Clin. Nutr.*, 54: 1077–1081, 1991.

31. Punnonen, K. and Rajamaki, A. Serum transferrin receptor and its ratio to serum ferritin in the diagnosis of iron deficiency, *Blood*, 89: 1052–1057, 1997.

32. Cook, J.D., Dassenko, S., and Skikne, B.S. Serum transferrin receptor as an index of iron absorption, *Br. J. Haematol.*, 75: 603–609, 1990.

33. Shintani, N. et al. Expression and extracellular release of transferrin receptors during peripheral erythroid progenitor cell differentiation in liquid culture, *Blood*, 83: 1209–1215, 1994.

34. Baynes, R.D. Transferrin reduces the production of soluble transferrin receptor, *Proc. Soc. Exp. Biol. Med.*, 209: 286–294, 1995.

35. Kuiper-Kramer, E.P. et al. Relationship between soluble transferrin receptors in serum and membrane-bound transferrin receptors, *Acta Haematol.*, 99: 8–11, 1998.

36. Ferguson, B.J. et al. Serum transferrin receptor distinguishes the anemia of chronic disease from iron deficiency anemia, *J. Lab. Clin. Med.*, 119: 385–390, 1992.

37. Klemov, D. et al. Serum transferrin receptor measurements in hematologic malignancies, *Am. J. Hematol.*, 34: 193–198, 1990.

38. Testa, U. et al. Autologous stem cell transplantation: evaluation of erythropoietic reconstitution by highly fluorescent reticulocyte counts, erythropoietin, soluble transferrin receptors, ferritin, TIBC and iron dosages, *Br. J. Haematol.*, 96: 762–775, 1997.

39. Centis, F. et al. Correlation between soluble transferrin receptor and serum ferritin levels following bone marrow transplantation for thalassemia, *Eur. J. Haematol.*, 54: 329–333, 1995.

40. Baynes, R.D. et al. Serum transferrin receptor in hereditary hemochromatosis and African siderosis, *Am. J. Hematol.,* 45: 288–292, 1994.

41. Carmel, R. and Skikne, B.S. Serum transferrin receptor in the megaloblastic anemia of cobalamin deficiency, *Eur. J. Haematol.,* 49: 246–250, 1992.

42. Khumalo, H. et al. Serum transferrin receptors are decreased in the presence of iron overload, *Clin. Chem.,* 44: 40–44, 1998.

43. Beguin, Y. et al. Quantitative assessment of erythropoiesis and functional classification of anemia based on measurements of serum transferrin receptor and erythropoiesis, *Blood,* 81: 1067–1076, 1993.

44. Beguin, Y. et al. Early prediction of response to recombinant human erythropoietin in patients with the anemia of renal failure by serum transferrin receptor and fibrinogen, *Blood,* 82: 2010–2016, 1993.

45. Cazzola, M. et al. Subcutaneous erythropoietin for treatment of refractory anemia in hematologic disorders. Results of a phase I/II clinical trial, *Blood,* 79: 29–37, 1992.

7 Alternative Iron Uptake Systems

CONTENTS

7.1 Transferrin-Independent Iron Uptake 383
7.2 Proton-Coupled Metal Ion Transporter 386
7.3 Mechanisms of Iron Transport In and Out of Cells: Role of
Ceruloplasmin 390
7.4 HFE 395
7.5 Melanotransferrin 399
References...401

7.1 TRANSFERRIN-INDEPENDENT IRON UPTAKE

The identification of membrane transport systems involved in translocation of iron across the cell membrane using biochemical reconstitution approaches has been greatly hampered by the physicochemical properties of this metal. Although biologically active Fe^{2+} and Fe^{3+} are relatively soluble under acidic pH conditions, at neutral pH under atmospheric conditions, spontaneous oxidation gives rise to the formation of insoluble polymeric aggregates of $Fe(OH)_3$.

Furthermore, Fe^{2+}, through Fenton chemistry, elicits the formation of hydroxyl radicals endowed with a high oxidative activity and therefore highly dangerous for biological systems. To counter these problems, iron rarely exists in a free form in biological systems. Under physiological conditions, iron is complexed with iron-binding proteins such as transferrin (Tf) or lactoferrin, or bound to low molecular weight chelators such as citrate or ATP. Iron biovailability is finely regulated in mammalian cells through translation control of transferrin receptor and ferritin synthesis.

However, studies carried out on HeLa cells, fibroblasts, and K-562 erythroleukemia cells provided the first lines of evidence in favor of a Tf-independent Fe transport system.[1,2] The main feature of this new pathway of iron uptake is its complete independence of transferrin. Initial studies showed that this pathway is protease-sensitive (and is therefore mediated by a membrane protein), mediates the uptake of iron available for cellular metabolism, and is not inhibited by agents such as $NHCl_4$ or chloroquine, which are known to block iron acquisition from transferrin. These studies also showed that the Tf-independent iron uptake is inhibited by other divalent cations such as Cu^{2+} or Cd^{2+} and to a lower extent by Zn^{2+} and Mn^{2+}.

The earliest evidence for a mammalian cell-surface ferrireductase is derived from studies which showed that cells possess the ability to convert ferricyanide to ferrocyanide. Since the iron cyanide derivatives are membrane impermeable, it was argued that reduction of ferrocyanide must occur at the cell surface.

Subsequent studies suggest that Tf-independent iron uptake may be mediated through ferrireductase activity. An initial study was based on the comparative analysis of iron uptake mediated by ^{55}Fe–Citrate or ^{55}Fe–transferrin, and ^{55}Fe–Tf–Sepharose beads. The analysis of iron uptake mediated by Sepharose beads bound to ^{55}Fe–transferrin showed some interesting findings. Tf-dependent iron uptake occurs without employing receptor-mediated endocytosis.[3] The iron of Fe–Tf–Sepharose was reduced and taken up by human fibroblasts in a time- and concentration-dependent fashion. Four subsequent studies confirmed in different cellular systems the existence of this membrane oxidoreductase activity associated with Tf-independent iron uptake.[4-7]

A study carried out in K-562 erythroleukemic cells showed the existence of this oxidoreductase activity by investigating the extracellular reduction of ferricyanide to ferrocyanide catalyzed by K-562 cells. The observations that membrane-impermeable ferricyanide competitively inhibits Tf-independent iron uptake from ^{55}Fe–nitriloacetic acid strongly suggest that this ferrireductase activity is coupled to the iron transport mechanism.[4] Furthermore, in this cellular system, the membrane ferrireductase activity was not modulated by extracellular or intracellular iron level.[4]

Experiments on the hepatoma HepG2 cell line also supported the existence of a ferrireductase involved in the uptake of non-transferrin-bound iron.[5] This conclusion was based on the observation that Fe^{2+}-specific chelators such as bathophenanthroline disulfonate markedly inhibited iron uptake, whether ascorbate was present or not, thus indicating that Fe^{3+} uptake is dependent upon reduction to the ferrous state.[5]

Iron loading in these cells stimulates transferrin-independent iron uptake. Experiments on HeLa and human skin fibroblasts using analysis of ferric citrate uptake (either labeled at the level of iron or at the level of citrate) were compatible with the existence of a membrane ferrireductase activity involved in the mechanism of non-transferrin iron uptake.[6] Finally, studies carried out on a human cell line derived from human duodenal microvillus membranes further confirmed that a ferrireductase activity is involved in the mechanism of non-transferrin iron uptake.[7] The ferrireductase activity was also observed on isolated cell membranes and was clearly sensitive to trypsin treatment, thus indicating that it is ascribable to a membrane protein.

Iron may be acquired through a non-receptor mediated mechanism from two different sources: transferrin-bound iron and non-transferrin-bound iron.[8] Experiments on isolated hepatocytes indicate that non-transferrin-bound Fe and transferrin–Fe may represent sources of iron for transferrin receptor-independent iron uptake; both use the same cellular pathway and a common iron carrier.[8]

The mechanism of non-transferrin-bound iron was characterized in detail in primary cultures of normal rat hepatocytes using ^{59}Fe–citrate. It was shown that:

1. Optimal iron internalization occurred at an Fe:citrate ratio of 1:100 at neutral pH (7.40) and at an extracellular Ca^{2+} concentration of 1 mM.
2. The transition metals Zn^{2+}, Co^{2+}, and Ni^{2+} inhibited Fe uptake when added in excess with respect to Fe.

3. The rate of Fe internalization was similar both in hepatocytes derived from normally-fed rats and from iron-loaded animals.
4. The rate of Fe internalization from Fe–citrate was about 20 times as great as that from Fe–transferrin.[9] These observations suggest that non-transferrin-bound iron uptake by hepatocytes is not regulated by cellular Fe levels.

Exposure of leukemic HL-60 cells to polyvalent cationic metals, such as gallium and aluminum, elicited a marked increase in the rate of iron uptake from low molecular weight chelating agents.[10] This enhancement of transferrin-independent iron uptake by trivalent metals does not require protein synthesis, does not depend on reductase activity, and is seemingly due to a change in membrane fluidity.[10]

Among normal cells, the mechanism of transferrin-independent iron uptake seems to be particularly important in monocytes–macrophages. Macrophages have a pronounced capacity to take up iron through a variety of low molecular weight complexes.[11,12] This process involves the internalization of extracellular iron and its subsequent transfer to cytoplasmic ferritin,[11,12] thus limiting the potential for iron to induce the formation of dangerous hydroxyl radicals. The physiological role of this system may be particularly relevant under conditions of iron overload when transferrin is fully saturated and the concentrations of low molecular weight chelates increase. In these conditions, an increased uptake of non-transferrin-bound iron by macrophages reduces the concentration of iron that may induce tissue damage.

These observations suggest that the mammalian transferrin-independent iron transport system is apparently similar to iron transport systems present in both bacteria and plants, which require the activities of both a surface reductase and a ferrous metal transporter.

A membrane protein involved in the mechanism of transferrin-independent iron uptake was recently identified[13] using an approach based on functional assay through microinjection of Xenopus oocytes. The strategy takes advantage of the observation that phorbol esters greatly stimulate non-transferrin iron uptake by K-562.[14] This phenomenon is related to the induction of megakaryocytic differentation and involves a transcriptional mechanism since it was inhibited by actinomycin D.[13]

Injecting mRNA extracted from phorbol ester-treated K-562 cells into Xenopus oocytes and analyzing the level of non-transferrin iron uptake made it possible to isolate an mRNA species and the corresponding cDNA encoding for a membrane protein called SFT (stimulator of Fe transport).[13] Transport properties associated with SFT-mediated uptake resemble those defined for non-Tf-bound Fe uptake by K-562 cells. The SFT gene encodes for a transmembrane protein composed of 338 amino acids. Kyte–Doolittle analysis of the amino acid sequence revealed six transmembrane-spanning domains. At the level of the first cytoplasmic loop, a 5-amino acid REIHE motif was observed. This motif is also present in the yeast iron transporter Ftr1 and is required for the protein activity since its mutation leads to a complete loss of the capacity to mediate iron uptake.[13]

The SFT protein synthesized in vitro from the corresponding cDNA forms a homodimer of about 87 kDa. Transfection experiments with SFT cDNA provided clear evidence that SFT mediates transport functions highly reminiscent of those observed in K-562 cells for Tf-independent Fe uptake. Iron uptake mediated by SFT is an active process requiring energy.

In addition to its function related to Tf-independent iron uptake, SFT is also able to stimulate Fe uptake from transferrin. This conclusion is supported by investigation of ^{55}Fe-transferrin uptake into HeLa cells transfected with SFT cDNA. Furthermore, experiments with intracellular SFT localization using a SFT/GFP chimeric protein showed that SFT is localized at the plasmalemma (suggesting that SFT may traffic between cell surfaces and endocytic compartments) and at a juxtanuclear region corresponding to a recycling compartment which also contains recycling transferrin receptors. These observations suggest that SFT, in addition to its role in mediating transferrin-independent iron uptake, may play an additional role in transferrin-mediated iron uptake by transporting Fe released in endosomal compartments.

SFT does not display any significant homology with yeast, plant, and bacterial Fe transporters. The only similarity of these proteins is a similar molecular weight and a secondary structure with 6 to 8 predicted membranous domains (Figure 7.1). However, the first cytoplasmic loop of the 5-amino acid REIHE motif is identical to that observed in the yeast iron transporter Ftr1.

In both compounds, the integrity of this motif is crucial for the functioning of the transporter. Furthermore, SFT displays no structural or functional homology with the Nramp family recently identified as H^+-coupled metal ion transporters mediating the uptake of multiple divalent cations including Fe^{2+}. However, the relationship between Nramp transporters and SFT, and their respective roles in iron absorption, and particularly in dietary iron absorption, remain to be fully clarified.

SFT functionally stimulates iron uptake regardless of whether cells are presented with ferric or ferrous forms. Transfection of the SFT gene into HeLa cells resulted in a consistent increase in iron uptake. The K_m for Fe uptake was not modified by SFT expression, while the V_{max} of iron transport doubled.[14] However, SFT was unable to stimulate iron uptake when the cells were grown in the absence of Cu, thus suggesting an essential role of Cu in SFT function.[15] Copper deficiency markedly reduced ferrireductase activity required for SFT function in Fe^{3+} transport. It was proposed that the ferrireductase acts upstream with respect to SFT by reducing Fe^{3+} to Fe^{2+}, which is then transported by SFT.

7.2 PROTON-COUPLED METAL ION TRANSPORTER

A recently isolated proton-coupled metal ion transporter, called DCT1, has an unusually broad substrate range that includes Fe^{2+}, Zn^{2+}, Mn^{2+}, Co^{2+}, Cd^{2+}, Cu^{2+}, Ni^{2+} and Pb^{2+}.[16] It was identified by an approach similar to that used for the isolation of SFT: by screening for iron uptake in oocytes. The DCT1 complementary DNA was isolated from a cDNA library prepared using duodenal mRNA from rats fed low-iron diets. The isolated cDNA encodes a 561-amino acid protein with 12 putative transmembrane domains, and predicted glycosylation sites at the fourth extracellular loop; a consensus transport motif was also present at the level of the fourth cytoplasmic loop[16] (Figure 7.1).

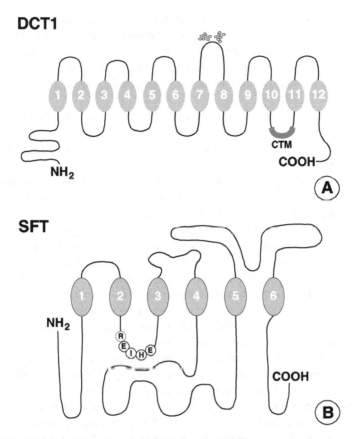

FIGURE 7.1 Hypothetical membrane orientations of DCT1 (divalent cation transporter 1) and SFT (stimulator of Fe transport). The numbered ovals indicate the transmembrane domains of the two membrane proteins. (Redrawn with modifications from Gutierrez, J.A. et al. *J. Cell Biol.,* 139: 895–905, 1997 (SFT) and Gunshlin, H. et al., *Nature,* 388: 482–488, 1997 (DCT1).)

Comparison of the sequence of DCT1 gene with the sequences of other known genes indicates that DCT1 belongs to the very ancient family of mammalian Nramp proteins found in invertebrates, plants, fungi, and bacteria.[17] Members of this family are defined by highly conserved hydrophobic cores composed of 10 transmembrane domains (with several invariant charged residues in these domains), and helical periodicity of sequence conservation.[17] The Nramp family includes two members found in mammalian cells and known as Nramp1 and Nramp2.

The predicted amino acid sequence of human Nramp1 shows an integral membrane protein of 12 transmembrane domains, a glycosylated extracytoplasmic loop, and several putative phosphorylation sites in the predicted intracellular loop.[18] The Nramp1 gene controls the innate resistance or susceptibility of mice to infection by a group of unrelated intracellular bacteria or parasites and is selectively expressed in professional phagocytes.[19]

The Nramp1 gene encodes an integral membrane protein that shares structural characteristics with ion channel and transporters.[17] Nramp1 is located on internal organelle membranes of activated macrophages, and functionally associated with the phagocytic apparatus.[19] The synthesis of Nramp1 is stimulated in macrophages by interferon-γ and by iron deprivation.[20] Its exact function is not known, but its structure strongly suggests that it acts as metal transporter. Gene transfer experiments in oocytes indicate that Nramp1 stimulates iron uptake, but to a lesser extent than Nramp2.[16]

Recent studies of gene transfer in COS-1 cells showed that Nramp1 does not enhance iron uptake, and its expression is associated with reduced iron uptake.[21] These observations suggest that Nramp1 may play a role in a salvage pathway of iron recycling.

Nramp 1 may transport Mn^{2+}, Fe^{2+}, and Zn^{2+} into the cytoplasm of macrophages and, after the generation of phagosomes, removes these metals from the organelles, thus preventing the growth of engulfed bacteria.[22]

A second mammalian gene, Nramp2, that encodes a highly similar protein to Nramp1 was recently isolated.[23] The full-length cDNA for human Nramp2 is 4142 bp in length and encodes for a protein of 561 amino acid residues with a molecular weight of 61 kDa.[24] The predicted amino acid sequence of Nramp2 indicates that it shares 64% homology with Nramp1; however, the NH_2-terminal cytoplasmic domain of Nramp2 exhibits only 21% sequence homology with the corresponding NH_2-terminal region of Nramp1.[24] Unlike its phagocyte-specific Nramp1 counterpart, Nramp2 is expressed uniformly in most tissues tested.

The comparative analysis of the predicted amino acid sequences of rat DCT1 and human Nramp2 proteins showed that they display 92% identity.[16] This suggests that human DCT1 and human Nramp2 should display 100% identity.

Analysis of the pattern of expression of DCT1 (Nramp2) revealed an almost uniformly pattern of expression. DCT1 mRNA was observed in proximal intestine, kidney, thymus and brain at higher levels and in testes, liver, colon, heart, spleen, skeletal muscle, lung, bone marrow, and stomach at lower levels.[12] DCT1 mRNA expression was highest in the duodenum and decreased toward the colon. Following diet-induced iron deficiency, DCT1 mRNA levels markedly increased in the duodenum and, to a lesser extent, in other tissues.

In situ hybridization analysis on tissue sections showed that DCT1 in the small intestine is markedly expressed in enterocytes lining the villi, particularly in the crypt cells. DCT1 labelling was most prominent in proximal tubule segments of the kidney and was also present on collecting ducts. DCT1 mRNA was expressed in Sertoli cells of the seminiferous tubules of the testes. DCT1 expression was observed in cortical, but not in medullary thymocytes; and was observed in the neurons, but not in the glial or ependymal cells of the brain.[16]

Functional studies of oocytes transfected with DCT1 cDNA showed that DCT1 is involved in the transport of Fe^{2+}, Zn^{2+}, Cd^{2+}, Mn^{2+}, Cu^{2+}, and Co^{2+} and, to a lesser extent, of Pb^{2+} and Ni^{2+}. These observations suggest that DCT1 may represent a key mediator of intestinal iron obsorption and its expression may be disturbed in hereditary hemochromatosis.[16]

This interpretation is directly supported by a recent study showing the involvement of DCT1/Nramp 2 in the origin of hereditary microcytic anemia in mice.[25] A missense mutation of Nramp2 gene was observed. This mutation introduced a charged amino acid into transmembrane domain. It was proposed that the introduction of a charged amino acid for Gly 185 may interfere with the function of Asp 192.

A mutation of the Nramp2 gene was also responsible for hereditary, microcytic, hypochromic anemia characterized by reduced reticulocyte iron uptake and decreased gastrointestinal iron absorption in Belgrade rats. A glycine-to-arginine missense mutation (G185R) was present in the Belgrade rat Nramp2 gene; the mutated Nramp2 failed to transport iron.[26] It is of interest that a similar missense mutation at transmembrane domain 4 (G169D) was shown to inactivate Nramp1 and, as a consequence, allow susceptibility to infection by intracellular parasites.[27]

The study on mice with microcytic anemia indicates that, although DCT1/Nramp2 is expressed in all tissues, its more relevant function is in the intestine, where it contributes to dietary iron absorption required to adequately sustain hemoglobin synthesis.

The studies in the anemic Belgrade rats suggest a possible role of Nramp2 in endosomal iron release at the level of erythroid precursors. Investigation of the transferrin cycle in reticulocytes of these animals showed that a large part of the endocytosed iron is returned to the extracellular space with transferrin, indicating a defect of the transfer of iron from the endosomes. These observations suggest that Nramp2 plays an important role in iron uptake in intestinal cells and endosomal iron release during the transferrin cycle in erythroid cells[26] (Figure 7.2).

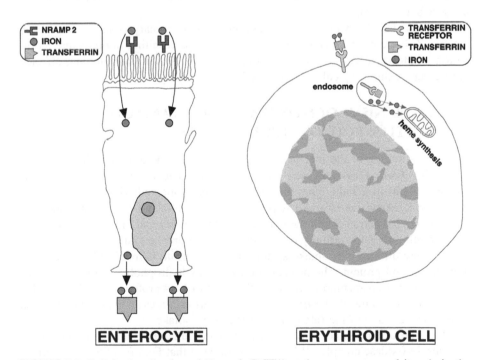

FIGURE 7.2 Cellular localization of Nramp2 (DCT1) at the enterocytes and its role in the mechanism of intestinal iron absorption. By comparison, erythroblasts acquire iron mainly through the transferrin–transferrin receptor pathway.

The function of Nramp2 is further supported by studies of yeast.[28] Nramp2 is homologous to two yeast Smf1 and Smf2 transport divalent cations. The question was whether it can functionally complement Smf1 and Smf2-deficient yeast cells.[28] Experiments have shown that Nramp2, but not Nramp1, can complement Smf1/Smf2-deficient cells, thus indicating that Nramp2 can transport divalent cations, particularly Mn^{2+}, in yeast.[28]

A series of proteins involved in zinc transport in mammalian cells have been cloned recently. All these Zn^{2+} transporters are membrane proteins with multiple membrane spanning domains; they usually exhibit in their intracytoplasmic domains a histidine-rich region.[29] ZnT1 was the first to be cloned and its function relates to the control of Zn^{2+} efflux. ZnT1, whose expression is observed in virtually all tissues, is located in enterocytes of the small intestine and in renal tubular cells at the basolateral membranes. ZnT1 is not expressed in the ileum and colon. ZnT1 expression is regulated by the level of zinc intake. Dietary zinc supplementation is accompanied by a significant rise of the level of ZnT1.

Similar arguments may be made also for ZnT2, whose function may be relevant in the control of Zn^{2+} uptake/efflux in the intestines, kidneys, and testes. Unlike ZnT1, ZnT2 is located intracellularly in the cytoplasmic vesicular compartments.

ZnT3 seems to be specifically involved in the control of Zn^{2+} uptake in the neurons and testes. ZnT3 expression in the brain is confined to vesicles in the synaptic termini of glutamatergic neurons, thus suggesting an important role for zinc in these brain regions.

ZnT4 is also highly expressed in neurons and in mammary glands, where it may act as a Zn^{2+} exporter. Interestingly, a point mutation resulting in premature termination is known to cause a lethal mouse syndrome. Before succumbing, affected mouse pups exhibit typical syndromes of zinc deficiency.

7.3 MECHANISMS OF IRON TRANSPORT IN AND OUT OF CELLS: ROLE OF CERULOPLASMIN

The understanding of the mechanisms responsible for the uptake of iron into and out of *Saccharomyces cerevisiae* greatly contributed to our understanding of the mechanisms of iron and copper transport into eukaryotic cells (Figure 7.3). Uptake of iron in *S. cerevisiae* is a two-step process. Extracellular Fe^{3+} is first reduced to Fe^{2+} through the action of cell membrane reductases (Fre1 and Fre2) and transported into the cells via a low-affinity or high-affinity iron uptake system.[30,31]

The low-affinity iron uptake system of *S. cerevisiae* is mediated by the transmembrane Fet4 protein. The protein contains many hydrophobic residues arranged in six putative transmembrane regions. The level of Fet4 protein is regulated by the concentration of intracellular iron, i.e., iron starvation causes an increase of the level of Fet4 protein and of the rate of low-affinity iron uptake.[32]

The high-affinity system involves two proteins called Fet3 and Ftr1 which form a permease-oxidase complex.[33] It was hypothesized that Fet3 functions as a ferroxidase and Ftr1 as a membrane iron transporter. Fet3 is a membrane protein showing a high level of homology with the multicopper oxidase family of proteins that includes ceruloplasmin. However, unlike other multicopper oxidases which are

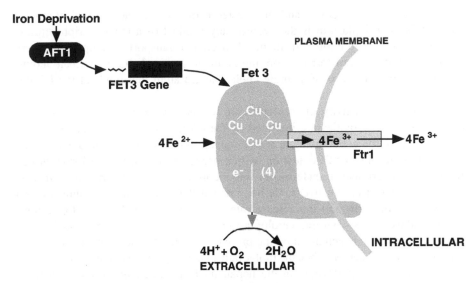

FIGURE 7.3 Mechanism of iron uptake by the Fet3/Ftr1 oxidase-permease complex in *Saccharomyces cervisiae*. Fet3 expression is stimulated by iron deprivation through a transcriptional mechanism involving interaction of the transcription factor AFT1 with the 5′ flanking region of Fet3 gene. Fet3 and Ftr1 form a complex at the cell membrane. Fe^{2+} is oxidized to Fe^{3+} through the oxidase activity of Fet3; Fe^{3+} then interacts with the Ftr1 transporter which mediates its translocation into the cell. (Reprinted with modifications from Eide, D. *Curr. Opin. Cell Biol.*, 9: 573–577, 1997. With permission.)

secreted proteins, Fet3 is an integral membrane protein with a single transmembrane domain. Fet3, whose synthesis is induced by iron starvation, possesses a multioxidase activity and, particularly, a ferroxidase activity that potentiates the spontaneous oxidation of Fe^{2+} to Fe^{3+}. Recent studies have shown that Fet3 possesses an intrinsic ferroxidase activity even in the absence of a specific acceptor and is capable of iron loading apotransferrin. Because of these properties, Fet3 is regarded as homologous to the human plasma protein ceruloplasmin.[34] Fe oxidized by Fet3 is transported into the cells by Ftr1, an integral membrane protein with six potential transmembrane domains and several metal-binding sites.[32]

The important role played by the multicopper oxidase Fet3 suggests the existence of a connection between iron and copper metabolism. In yeast cells, a deficiency of copper resulted in decreased iron uptake.

Recent studies indicate that such a link between iron and copper uptake exists also in mammalian cells. Particularly relevant were the studies of patients with deficiencies of ceruloplasmin.

Ceruloplasmin is a 132-kDa blue copper-containing α-2 glycoprotein. It is synthesized mainly in the liver and binds 90 to 95% of plasma copper in vertebrates. This protein is a multicopper oxidase synthesized in hepatocytes as a single-chain polypeptide and secreted as a holoprotein with six atoms of copper per molecule.[31] Ceruloplasmin is composed of a single polypeptide chain of 1046 amino acid residues with a molecular weight of 132,000.

Its amino acid sequence and the existence of a three-fold internal homology suggest that ceruloplasmin is the evolutionary product of a tandem triplication of ancestral genes. In addition to its main role in copper transport, ceruloplasmin exerts oxidase activity toward ferrous iron and aromatic amines.[31] Its prooxidant activity depends on the presence of a single chelatable Cu atom that binds at the His 426 residue.[35]

The oxidative activity of ceruloplasmin seems to be specific for iron since its other prooxidant activities are inhibited by the addition of an iron chelator and are thus related to iron contamination. It was suggested that the main role of ceruloplasmin is the control of Fe 2^+ oxidation, with subsequent incorporation of iron into the transport compartment (transferrin) or into the iron storage compartment (ferritin).

Studies performed in liver cells have shown that apotransferrin stimulates iron release from the cells and this release rate is greatly enhanced by ceruloplasmin.[36] Other studies have shown that ceruloplasmin can load iron into ferritin and that its ferroxidase activity is enhanced by the presence of ferritin. The optimal iron loading into ferritin was observed when the molar ratio of ceruloplasmin and ferritin was 1.[37,38]

A recent study performed in the hepatoma HepG2 cell line offers a different view of the role of cerulopasmin in iron metabolism.[39] In a first set of experiments, ceruloplasmin did not modify iron release from cells, whether the cells were iron-deficient or iron-repleted. Ceruloplasmin did not modify iron uptake in untreated or iron-loaded cells, but significantly stimulated uptake by iron-deficient cells.[39]

This stimulatory effect on iron uptake was observed at concentrations (10 to 30 µg/ml) well below the normal plasmatic concentration of this protein (210 to 450 µg/ml). The stimulatory effect was due to a decrease in K_m (from 1.2 µM to 0.4 µM) without any change in V_{max}.

The observation that the stimulatory effect of ceruloplasmin on iron uptake was observed only in iron-deficient cells indicates that a cellular component induced by iron deficiency is required for this effect; this component is neosynthesized since cycloheximide addition blocks the effect of iron deprivation. Consistent with these observations and with the role of ceruloplasmin in iron uptake, it was observed that iron depletion induces a four- to five-fold stimulation of ceruloplasmin synthesis and of ceruloplasmin mRNA level. This phenomenon was due to an increase in ceruloplasmin gene transcription.[39]

These observations are consistent with a model in which iron depletion in the liver causes increased synthesis and the release of ceruloplasmin that interacts with an iron transporter, whose synthesis is also increased by iron deprivation, resulting in increased iron uptake.

Recent studies on patients with aceruloplasminemia indicate that ceruloplasmin plays an important role in iron mobilization, but not in copper distribution to tissues. These patients have normal copper tissue levels and increased iron accumulation in tissues.

Aceruloplasminemia is an autosomic recessive disease characterized by mutations in the ceruloplasmin gene, absence of serum ceruloplasmin, iron accumulation in the brain, liver, pancreas, and other tissues, and pigment degeneration of the retina. Studies on three different families of aceruloplasminemic patients have shown different types of genetic mutations in the ceruloplasmin gene: insertion of adenine in exon 3 resulting in a premature stop codon[40]; a G- to A-substitution in exon 15 resulting in a nonsense mutation at amino acid Trp 858[41]; a 5-bp insertion at amino acid 410 (exon 7) resulting in a frame-shift mutation and a truncated open reading frame[42]; a G- to A-transition at the splice acceptor site that introduced a premature termination codon at amino acid position 991 by defective splicing[43]; an A- to G-transition at the splice acceptor site of the intron 6 causing an 8-bp deletion in ceruloplasmin mRNA by defective splicing, resulting in a premature termination codon at amino acid position 388.[44]

From a biochemical view, these patients exhibit undetectable levels of ceruloplasmin, very low cupremia, decreased sideremia, and significantly increased ferritinemia.[43] From a clinical view, they show neurologic symptoms (usually chorioathetotic involuntary movements and cerebellar ataxia), diabetes mellitus, and retinal degeneration. Postmortem anatomic pathology studies of some patients showed markedly increased intracytoplasmatic deposition of iron, but not copper, in the brain, liver, pancreas, heart, kidney, spleen, and thyroid.

Studies of the expression of the ceruloplasmin gene in the central nervous system showed that mRNA encoding ceruloplasmin is expressed in specific populations of glial cells associated with the brain microvasculature, surrounding dopaminergic melanized neurons in the substantia nigra and within the internuclear layer of the retina.[45] Magnetic resonance imaging of the brain showed increased basal ganglial iron content. In line with this pathogenetic interpretation, a recent report indicates that treatment of aceruloplasminemic patients who have neurological symptoms with the iron chelator desferrioxamine decreased brain iron stores and prevented progression of neurological symptoms.[46]

It is now clear that aceruloplasminemia is a disease of iron metabolism which causes hemosiderosis. It was also proposed that ceruloplasmin is a main regulator of iron export from the cells. This conclusion is supported by a recent study showing that ceruloplasmin greatly accelerates iron release from liver cells.[36] This conclusion is questioned by Mukhopadhyay et al. who showed a role of ceruloplasmin in iron uptake and not in iron export from the cells.[39] According to this view, the iron overload observed in aceruloplasminemia may be related to defective ceruloplasmin-mediated iron uptake into the liver resulting in accumulation into the plasma of iron in the form of low-molecular weight chelates at a level sufficiently high to permit their uptake by most tissues.

The widespread effects of ceruloplasmin on iron accumulation in different tissues implies that this protein must be synthesized in different tissues. Although ceruloplasmin was originally considered a plasma protein synthesized and secreted by liver cells, several studies have shown its synthesis in extrahepatic tissues, lymphocytes, and macrophages. Particularly relevant are the studies performed in monocytes–macrophages and in the brain.

Recent studies have shown that the promonocytic human cell line U937 synthesized ceruloplasmin following stimulation with zymosan.[47] A subsequent study performed on U937 cells and on normal monocytes has shown that IFN-γ greatly stimulated ceruloplasmin synthesis by monocytic cells.[48] This activity was specific for IFN-γ since several other cytokines were unable to stimulate ceruloplasmin synthesis. These observations suggest that monocytic cell-derived ceruloplasmin may partially contribute to the defense mechanisms of phagocytic cells by its fer-roxidase activity and its stimulatory effects on iron export out of the cells which exerted inhibitory effects on the growth of invasive microrganisms.[48]

The unique involvement of the central nervous system allows us to distinguish aceruloplasminemia from other inherited and aquired iron storage disorders, and suggests that ceruloplasmin is important in iron homeostasis in the normal brain. The most striking clinical feature of aceruloplasminemia is the progressive neuro-degeneration which produces important clinical symptoms including dementia. Under physiological conditions, ceruloplasmin is unable to cross the blood–brain barrier, so it must be argued that the protein must be synthesized in the brain.

Recent studies have shown that ceruloplasmin is synthesized in some regions of the brain by a few specialized cell types. One study carried out in mice showed ceruloplasmin expression at the brain–CSF barrier. In situ hybridization studies showed that this protein is synthesized by astrocytes within the retina and throughout the cerebral microvasculature. The highest levels of ceruloplasmin mRNA were observed in the eye.[49] It is noteworthy that ceruloplasmin mRNA was not observed in all astrocytes, but only in astrocytes predominantly surrounding the microvascu-lature. It can be regarded as a molecular marker for this astrocyte subpopulation.

Biosynthetic studies on primary cultures confirmed that astrocytes, but not neu-rons, are capable of synthesizing ceruloplasmin.[49] Similar studies of the human brain showed abundant ceruloplasmin gene expression in specific populations of glial cells associated with the brain microvasculature, surrounding dopaminergic melanized neurons in the substantia nigra, and within the inner nuclear layer of the retina.[45]

It is of interest that increased regional brain concentrations of ceruloplasmin were observed in several neurodegenerative disorders, including Alzheimer's disease, Parkinson's disease, Huntington's disease, and progressive supranuclear palsy. These increases in ceruloplasmin content related to an acute-phase response and/or to a compensatory increase in oxidative stress.[50]

The pattern of expression of the ceruloplasmin gene in the brain and other tissues has important implications for the understanding of the role played by this protein in iron metabolism.

The patterns of ceruloplasmin expression in different tissues revealed some interesting features. Ceruloplasmin synthesis was detected in the spleen at the level of the reticuloendothelial cells, localized near the periarteriolar lymphoid foci. It is tempting to suggest that the ferroxidase activity present in these reticuloendothelial cells facilitates the export of iron out of these cells after they ingested and digested senescent red blood cells.

Ceruloplasmin is actively synthesized and secreted by hepatic cells; the protein released in the extracellular medium facilitates iron export out of the hepatocytes, thus preventing potentially dangerous iron accumulation. Ceruloplasmin synthesis was also detected in the bronchiolar epithelia of the lung, but not in the lung parenchyma. It was suggested that the ceruloplasmin synthesized in ciliated epithelia may play important roles by acting as an antioxidant and preventing the growth of pathogens through limitation of the available iron.

Based on observations made in other tissues, it is reasonable to assume that ceruloplasmin plays an important role in the control of iron influx/efflux of cells in the central nervous system. A recent report shows that ceruloplasmin is present in astrocytes in a membrane-bound form. The protein is directly anchored to the cell surface via a glycosylphospatidylinositol anchor.

Membrane-bound ceruloplasmin has a structure identical to that of the secreted form of the protein and also possesses oxidase activity.[51] The cell surface localization of ceruloplasmin is not seen on hepatocytes and cells of the choroid plexus, both of which are known to synthesize and secrete ceruloplasmin. The GPI-anchored form of ceruloplasmin present on astrocytes seemingly has an important role in the transport of iron in neurons and glial cells in the central nervous system.

Another genetic syndrome, Hallervorden–Spatz syndrome, is associated with massive iron deposits in the basal ganglia. The genetic defect is unknown, but it maps to chromosome 20p12.3-p13 and homozygotes affected by this syndrome exhibit normal levels of ferritin, transferrin and ceruloplasmin.[52] It is tempting to suggest that this syndrome may derive from the mutation of a gene involved in iron metabolism in the brain.

7.4 HFE

Hereditary hemochromatosis is a common autosomal recessive disorder of iron metabolism characterized by progressive iron overload of the parenchymal organs. It is one of the most common diseases among individuals of Northern European descent (1/200 to 1/400 individuals have hereditary hemochromatosis, with estimated carrier frequency ranging from 1/8 to 1/10).

The disease is characterized by excessive iron accumulations in a series of organs that lead to severe functional damage. The biochemical mechanisms responsible for the increased iron absorption are unknown.

Recent studies led to the identification of the putative gene responsible for hereditary hemochromatosis. The first indication suggesting the location of this gene near the major histocompatibility complex (MHC) on chromosome 6 was the important observation of significantly increased frequency of the HLA-A3 in hemochromatosis patients.[53] Subsequent studies confirmed the linkage of the hemochromatosis gene to the HLA locus and located this gene approximately 1 to 2 cM from the HLA-A gene.

Twenty years after this initial observation, a candidate gene for this common genetic disorder has finally been identified.[54] This candidate gene is located on the short arm of chromosome 6, about 4 mb telomeric to the MHC complex.[54] It encodes a 343-amino acid protein that exhibits significant similarity to the HLA class I molecules and is widely expressed.[54] This gene was designed HLA-H or HFE.

It is important to note that before the discovery of the HLA-H gene, two studies on mice made deficient in β-2 microglobulin expression revealed a phenotype resembling hereditary hemochromatosis and characterized by parenchymal iron overload.[55,56]

These studies suggest a major role of β-2 microglobulin, and possibly of some class I gene products, in the control of iron absorption. The HFE gene, which is similar to MHC class-I-like genes, is located over 4 mb telomeric of the most distal cluster of previously known class I genes within MHC. The HFE gene shares ~58%

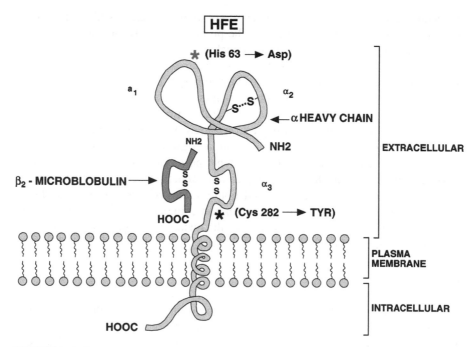

FIGURE 7.4 Structure of HFE, the candidate protein for hereditary hemochromatosis. HFE is a major histocompatibility complex class I-like protein (HLA-H) that associates with β-2 microglobulin. The positions of the two mutations of this protein observed in hereditary hemochromatosis are shown. (Redrawn with modifications from Feder J.N. et al. *Nature Genet.*, 13: 399–408, 1996.)

homology with HLA-A2 on an amino acid level and possesses a structure very similar to those of MHC class I molecules: α-1 and α-2 domains for peptide binding, an Ig-like α-3 domain, a transmembrane region, and a cytoplasmic region.[54] As with other HLA class I molecules, the proper conformation of the α-3 domain of HFE is probably necessary for non-covalent interaction with β-2 microglobulin and correct cell surface presentation (Figure 7.4).

Despite its MHC class I similarities, HLA-H differs in some structural features from the other HLA members of this family involved in antigen presentation:

1. Only two of four Tyr residues in the peptide-binding region are conserved.
2. The presence of a proline residue at position 188, in analogy with that found in the Fc receptor and other non-peptide binding MHC molecules, indicates that HLA-H is not involved in antigen presentation but rather in internalization/recycling of ligand via a receptor-mediated pathway.

A role of this gene in hemochromatosis is supported by the high frequency of a G- to A-transition at nucleotide 845 of the open reading frame in hemochromatosis patients. Eighty-three percent of the patients genotyped by Feder and co-workers were indeed homozygous for this mutation which is predicted to replace the cysteine residue 282 with a tyrosine in the α-3 domain of the molecule. This may prevent the HLA-H molecule from folding properly, and thus from interacting with β-2 microglobulin and appearing on cell surfaces.

Co-immunoprecipitation experiments performed on cells transfected with wild-type HLA-H or the C282Y mutated gene showed that wild-type HLA-H binds β-2 microglobulin, while the C282Y mutation completely abrogates this interaction.[57] Immunofluorescence experiments showed that wild-type HLA-H is expressed on the cell surface, while the C282Y mutant protein is only localized intracellularly.[57] The C282Y mutant HLA-H protein is retained in the endoplasmic reticulum and middle Golgi compartment where it is subjected to accelerated degradation.[58]

The significance of the alteration of this cysteine residue is strengthened by the fact that mice deficient in β-2 microglobulin exhibit signs of iron overload.[55,56] An additional C-G transition at nucleotide 187, responsible for the substitution of a histidine by an aspartic acid at residue 63, was found in excess in heterozygotes for the first mutation, but the evidence for causality of this second variant is less certain.[54] This mutation does not impair the interaction between HLA-H and β-2 microglobulin.

Subsequent studies confirmed the high frequency of C282Y mutation in hemochromatosis, and reported 72 to 92% of French patients,[59,60] 100% of Australian patients,[61] 64 to 66% of Italian patients,[62,63] and 51% of German patients[64] were homozygous for this mutation, thus providing further evidence that this gene is the hemochromatosis gene.

The C282Y mutation exhibits a low frequency where the hemochromatosis disease is rare or absent. In the Ashkenakic Jewish population, the frequency for the C282Y mutation is 0.013, compared with a frequency of 0.070 observed among populations of North European origin.[65] The C282Y mutation was virtually absent among ethnic groups from Algeria, Ethiopia, and Senegal.[66] In conclusion, these studies have shown that the distribution of C282Y mutation coincides with distribution of populations in which hemochromatosis has been reported. The results are fully consistent with the theory of a North European origin for this mutation. In contrast, the H63D polymorphism is more widely distributed and its connection with hemochromatosis remains unclear.

Interestingly, a recent study evaluated the concordance between the genetic origin and previous clinical diagnosis in families with hereditary hemochromatosis.[67] The major findings of this study were: (1) all subjects previously diagnosed as homozygous or heterozygous for hereditary hemochromatosis carry at least one C282Y mutation; (2) of 127 subjects homozygous for the C282Y mutation, 105 met criteria for clinical diagnosis of hereditary hemochromatosis; and (3) in subjects heterozygous for the C282Y mutation, 4.8% have iron overload in the range previously diagnosed as homozygous according to clinical criteria.

Northern blot analysis showed that HFE is present in two different forms of mRNA, the major transcript being a 4-Kb mRNA weakly expressed in all tissues except the brain, with some suggestion of higher levels in liver and intestine.[54]

In a subsequent study, the expression of HFE in the gastrointestinal tract was investigated in detail by immunohistochemistry.[68] In the esophagus, the HFE protein shows a weak positivity, with the reaction distributed around the entire plasma membranes of the stratified squamous epithelial cells. The expression in the stomach was more evident than in the esophagus, with reactivity restricted to the basoloteral plasma membranes of epithelial cells. In the small intestine, the staining was primarily intracellular and perinuclear; apical and basolateral surfaces showed no significant reactivity. The staining was most intense in the epithelial cells in crypts, while the surface epithelial cells in upper portions and tips of the villi did not exhibit reactivity. Finally, a weak staining was observed in the large intestine. It was hypothesized that HFE may act as a barrier to iron transport in these tissues and its loss of function may contribute to the development of hemochromatosis.

In the placenta, the HFE protein is expressed in the plasma membranes of the syncytiotrophoblast cells, forming a complex with both β-2 microglobulin and transferrin receptor.[69] The possible physiological relevance of this finding is unknown, but may represent a tentative and explanation of the polymorphic frequency of the C282Y mutation. It was suggested that heterozygosity of the C282Y mutation must confer some positive selective advantage.[70] Thus, the C282Y mutation may have improved survival during infancy under conditions of iron deficiency and might also confer an advantage to the fetus under conditions of maternal iron deprivation.[69]

In spite of rapid progress in our understanding of the genetic basis of hereditary hemochromatosis, the function of HFE/HLA-H remains largely unknown. Three hypotheses have been proposed to explain the physiological role of this protein in the control of iron metabolism:[54]

1. HFE is a receptor for an iron-binding protein; since the C282Y mutation disrupts the capacity of HFE to interact with β-2 microglobulin to be expressed at the level of cell surface and seemingly inactivates the protein, it must be assumed that it plays a role as a negative physiological regulator of iron absorption. A variant of this hypothesis is that the HFE protein alone is not a receptor for an iron-binding protein, but is a co-receptor that interacts with and modulates the activities of other iron–ligand receptors, such as the transferrin receptor.
2. HLA-H/HFE may be involved in the signal transduction machinery in a step involved in iron sensing and consequent modulation of iron regulatory proteins.
3. HLA-H/HFE may act through association with components of the immune system involved in iron metabolism.

A recent study offers evidence of the mechanisms through which HFE may modulate iron metabolism. Wild-type HFE protein forms stable complexes with the transferrin receptor and induces inhibition of transferrin binding to its receptor by lowering the affinity of the ligand for the receptor. The mutant H63D HFE protein also exhibits the capacity to complex with transferrin receptor, but does not modify the affinity of this receptor for its ligand. Finally, the C282Y mutant protein completely failed to interact with the transferrin receptor.[71]

This observation was supported by the analysis of the crystal structure of the hemochromatosis protein HFE. This analysis showed that hemochromatosis mutations concern a region involved in pH-dependent interaction with transferrin receptor.[72] Soluble HFE and transferrin receptor were able to interact at the basic pH of the cell surface, but not at the acidic pH levels of intracellular vesicles.[72] The pH dependence of the HFE–transferrin receptor interaction implies that HFE enters the cells with the transferrin–transferrin receptor complex and subsequently dissociates from transferrin receptor at the acidic vesicles. Finally, it was shown that the stoichiometry of transferrin receptor–HFE (2:1) differed from transferrin receptor–transferrin stoichiometry (2:2). These findings suggest that HFE, transferrin and transferrin receptor form a ternary complex, and clearly support the existence of a molecular link between HFE and a key regulator of iron metabolism, the transferrin receptor.

Despite the limitations in our understanding of the mechanisms through which HLA-H affects iron uptake, recent studies provide clear evidence that mutations of this gene are responsible for hereditary hemochromatosis. In fact, mice made deficient in HLA-H expression through a targeted disruption of the murine homologue of the HFE gene displayed a hemochromatosis phenotype characterized by a progressive increase in transferrin saturation and hepatic iron accumulation.[73] The HFE knockout model resembles the β-2 microglobulin knockout mouse, which also has excessive iron storage. The knockout mouse model of hereditary hemochromatosis not only provides definitive evidence about the implication of the HFE gene in the pathogenesis of hereditary hemochromatosis, but also provides a useful model to investigate the biochemical and cellular mechanisms responsible for increased iron accumulation in hemochromatosis. It will serve as a valuable tool for evaluating novel therapeutic approaches for the treatment of iron accumulation.

Interestingly, recent studies have shown a high frequency of HFE mutations among patients affected by porphyria cutanea tarda (PCT). Sporadic PCT is a skin disease associated with hepatic siderosis caused by a reduced activity of uroporphyrinogen decarboxylase in the liver. Mild to moderate iron overload is common in PCT, as iron is one of the factors triggering the clinical manifestation of the disease through inactivation of the URO-D enzyme. In 41 PCT patients in the United Kingdom, high frequencies (44%) of the Cys 282 Tyr HFE mutation were observed.[74] In 68 Italian PCT patients, high frequencies (~50%) of the HFE His 63 Asp mutation were found.[75] These observations suggest the inheritance of one or more hemochromatosis genes is an important factor in determining susceptibility to sporadic PCT.

The idea that HLA class I may be involved in iron metabolism is further reinforced by the observation that a link between the HFE H63D mutation and HLA-A29 phenotype was also observed in non-classical forms of hereditary hemochromatosis.[76] These observations support a model in which the immunological system through HLA-class I plays a major physiological role in the regulation of iron load.

7.5 MELANOTRANSFERRIN

Melanotransferrin, initially called p97, is a cell surface glycoprotein present in most human melanomas, but only in trace amounts in normal adult tissues.[77] The cloning of melanotransferrin cDNA showed that the corresponding protein is composed of a 719-residue protein comprising extracellular domains of 342 and 352 residues and a C terminal 25-amino acid residue of predominantly hydrophobic amino acids that act as a membrane anchor.[77] Each extracellar domain contains 14 cysteine residues that form seven intradomain disulfide bridges. The two extracellular domains are highly homologous to each other (46% amino acid sequence homology) and to corresponding domains of lactoferrin (40% homology) and transferrin (39% homology).

Studies of the iron-binding properties of melanotransferrin, based on spectroscopy, amino acid sequence comparison, and modeling, have shown that this protein, at variance with other transferrins, binds only one Fe^{3+} atom per molecule.[78] This iron atom is bound at its N terminal site, while its C terminal site does not bind iron at all.

Melanotransferrin was first identified as a melanoma-associated surface glycoprotein,[77] but subsequent studies have shown that it is expressed in many other tumors including lymphomas and brain tumors.[79] It is also expressed in a wide range of cultured cell types, including renal and intestinal epithelial cells.[80]

Melanomas express high levels of melanotransferrin, whereas normal melanocytes have virtually undetectable melanotransferrin. In contrast to its wide expression pattern in proliferating cells, melanotransferrin shows a restricted pattern of expression in normal human tissues. Its expression is detectable in fetal intestine, cord blood, sweat gland ducts, and living liver sinusoidal cells. Recent studies have shown that melanotronsferrin is localized at brain, with selective localization to capillary brain endothelium.[81] This peculiar localization of melanotransferrin, similar to that observed for transferrin receptors, suggests that melanotranferrin may play a role in iron transport within the human brain.

One unique property of melanotransferrin is its anchor to the cell membrane through a glycosyl-phosphatidylinositol (GPI) structure. This phenomenon was initially observed in intestinal epithelial cells[82] and subsequently confirmed in melanoma cells.[83]

Treatment of cells with phosphatidylinositol-specific phospholipase C resulted in the release from the cell membranes of soluble melanotransferrin.[82,83] Studies of fetal porcine small intestine showed that melanotransferrin was localized to the apical surfaces of the enterocytes at the level of patches of flat or invaginated apical membrane domains rather than at the surfaces of microvilli. Caveolae were not found in association with these labeled microdomains.[80] Furthermore, the analysis of the different biosynthetic steps of melanotransferrin suggests a model in which the protein is exported at the cell surface through an exocytic process by way of detergent-insoluble glycolipid "rafts" that fuse with the plasma membranes at restricted sites devoid of microvilli.[84] The function of melanotransferrin in mediating iron uptake remains difficult to establish because cells expressing melanotransferrin also synthesize transferrin receptors.

To bypass this limitation, the functional role of melanotransferrin was studied using CHO cell lines defective in both endogenous melanotransferrin and transferrin receptor.[85] This experimental approach showed that melanotransferrin is capable of binding and internalizing Fe into cells from Fe–citrate, but not from Tf .[85] CHO cells transfected with melanotransferrin doubled their capacity to internalize iron via non-transferrin-bound iron; this iron transport mechanism was temperature-sensitive and saturated at a media iron concentration of 25 µg/ml.[85] It was proposed that melanotransferrin binds iron and is then internalized.

Under normal conditions, only trace amounts of free iron are observed, so it is reasonable to assume that melanotransferrin does not function in the normal recirculation of iron within the body. However, it may play a role in certain physiopathological conditions, such as iron overload, by acting as an iron scavenger to reduce the levels of potentially toxic free iron for the cells.

Furthermore, melanotransferrin may play a role as a scavenger of the iron released during normal cell death.[85] Finally, it is possible that melanotransferrin is simply a member of an ancestral metal transport system, whose function was replaced during evolution by a more evolved iron transport system, mainly represented by the transferrin receptor.[85]

REFERENCES

1. Sturrok, A. et al. Characterization of a transferrin-independent uptake system for iron in HeLa cells, *J. Biol. Chem.,* 265: 3139–3145, 1990.
2. Innan, R.S. and Wessling-Resnick, M. Characterization of transferrin-independent iron transport in K562 cells: unique properties provide evidence for multiple pathways of iron uptake, *J. Biol. Chem.,* 268: 8521–8528, 1993.
3. Oshiro, S. et al. Redox, transferrin-independent, and receptor-mediated endocytosis iron uptake systems in cultured human fibroblasts, *J. Biol. Chem.,* 268: 21586–26591, 1993.
4. Inman, R.S., Coughlan, M.M., and Wessling-Resnick, M. Extracellular ferrireductase activity in K562 cells is coupled to transferrin-independent transport, *Biochemistry,* 33: 11850–11857, 1994.
5. Jordan, I. and Kaplan, J. The mammalian transferrin-independent iron transport system may involve a surface ferrireductase activity, *Biochem. J.,* 302: 875–879, 1994.
6. Randell, E.W. et al. Uptake of non-transferrin-bound iron by both reductive and nonreductive processes is modulated by intracellular iron, *J. Biol. Chem.,* 269: 16046–16053, 1994.
7. Riedel, H.-D. et al. Characterization and partial purification of a ferrireductase from human duodenal microvillus membranes, *Biochem. J.,* 309: 745–748, 1995.
8. Trinder, D. and Morgan, E. Inhibition of uptake of transferrin-bound iron by human hepatoma cells by non-transferrin-bound iron, *Hepatology,* 26: 691–698, 1997.
9. Baker, E., Baker, S.M., and Morgan, E.H. Characterization of non-transferrin bound iron (ferric citrate) uptake by rat hepatocytes in culture, *Biochim. Biophys. Acta,* 1380: 21–30, 1998.
10. Olakami, O. et al. Polyvalent cationic metals induce the rate of transferrin-independent iron acquisition by HL-60 cells, *J. Biol. Chem.,* 272: 2599–2606, 1997.
11. Olakami, O. et al. Iron sequestration by macrophages decreases the potential for extracellular hydroxyl radical formation, *J. Clin. Invest.,* 91: 889–899, 1993.
12. Olakami, O., Stokes, J.B., and Britigan, B.E. Acquisition of iron bound to low molecular weight chelates by human monocyte-derived macrophages, *J. Immunol.,* 153: 2961–2703, 1994.
13. Guitierrez, J.A. et al. Functional expression, cloning and characterization of SFT, a stimulator of Fe transport, *J. Cell. Biol.,* 139: 895–905, 1997.
14. Akompong, T., Inman, R.S., and Wessling-Resnick, M. Phorbol esters stimulate non-transferrin iron uptake by K562 cells, *J. Biol. Chem.,* 270: 20937–20941, 1995.

15. Yu, J. and Wessling-Resnick, M. Influence of copper depletion on iron uptake mediated by SFT, a stimulator of Fe transport, *J. Biol. Chem.,* 273: 6909–6915, 1998.

16. Gunshlin, H. et al. Cloning and characterization of a mammalian proton-coupled metal-ion transporter, *Nature,* 388: 482–488, 1997.

17. Cellier, M. et al. Nramp defines a family of membrane proteins, *Proc. Natl. Acad. Sci. U.S.A.,* 42: 10089–10093, 1995;.

18. Cellier, M. et al. Human natural resistance-associated macrophage protein: cDNA cloning, chromosomal mapping, genomic organization, and tissue-specific expression, *J. Exp. Med.,* 180: 1741–1752, 1994.

19. Vidal, S., Gros, P., and Skamene, E. Natural resistance to infection with intracellular parasites: molecular genetics identifies Nramp1 as the Bcg/Ity/Lsh locus, *J. Leukoc. Biol.,* 58: 382–390, 1995.

20. Atkinson, G.P., Blackwell, J.M., and Borton, C.H. Nramp 1 encodes a 65-kDa interferon-gamma-inducible protein in murine macrophages, *Biochem. J.,* 325: 779–786, 1997.

21. Atkinson, P.G. and Barton, C.H. Ectopic expression of Nramp1 in COS-1 cells modulates iron accumulation, *FEBS Lett.,* 425: 239–242, 1998.

22. Supek, F., Supekova, L., and Nelson, N. Function of metal-ion homeostasis in the cell division cycle, mitochondrial protein processing, sensitivity to mycobacterial infection and brain function, *J. Exp. Biol.,* 200: 321–330, 1997.

23. Vidal, S. et al. Cloning and characterization of a second NRAMP gene on chromosome 12 q13, *Mamm. Genome,* 6: 224–230, 1995.

24. Kishi, F. and Tabuchi, M. Complete nucleotide sequence of human NRMP2 cDNA, *Mol. Immunol.,* 34: 839–842, 1997.

25. Fleming, M.D. et al. Microcytic anemia mice have a mutation in Nramp 2, a candidate iron transporter, *Nat. Genet.,* 16: 383–386, 1997.

26. Fleming, M.D. et al. Nramp2 is mutated in the anemic Belgrade (b) rat: evidence of a role for Nramp2 in endosomal iron transport, *Proc. Natl. Acad. Sci. U.S.A.,* 95: 1148–1153, 1998.

27. Vidal, SM. et al. Natural resistance to infection with intracellular parasites: isolation of a candidate for Bcg, *Cell,* 73: 469–485, 1993.

28. Pinner, E. et al. Functional complementation of the yeast divalent cation transporter family SMF by NRAMP2, a member of the mammalian natural resistance-associated macrophage protein family, *J. Biol. Chem.,* 272: 28933–28938, 1997.

29. McMahon, R.J. and Cousins, R.J. Mammalian zinc transporters 1, 2, *J. Nutr.,* 128: 667–670, 1998.

30. Eide, D. Molecular biology of iron and zinc uptake in eukaryotes, *Curr. Opinion Cell Biol.,* 9: 573–577, 1997.

31. Winzerling, J.J. and Haw, J.H. Comparative nutrition of iron and copper, *Annu. Rev. Nutr.,* 17: 501–526, 1997.

32. Dix, D. et al. Characterization of the FET4 protein of yeast, *J. Biol. Chem.,* 272: 11770–11777, 1997.

33. Stearman, R. et al. A permease–oxidase complex involved in high-affinity iron uptake in yeast, *Science,* 271: 1552–1557, 1996.

34. De Silva, D. et al. Purification and characterization of Fet 3 protein, a yeast homologue of ceruloplasmin, *J. Biol. Chem.,* 272: 14208–14213, 1997.

35. Mukhopadhyay, C.K. et al. Identification of the prooxidant site of human ceruloplasmin: a model for oxidative damage by copper bound to protein surfaces, *Proc. Natl. Acad. Sci. U.S.A.,* 94: 11546–11551, 1997.

36. Young, S.P., Fahmy, M., and Golding, S. Ceruloplasmin, transferrin and apotransferrin facilitate iron release from human liver cells, *FEBS Lett.,* 411: 93–97, 1997.

37. Juan, S.H., Guo, J.H., and Aust, S.D. Loading of iron into recombinant rat liver ferritin heteropolymers by ceruloplasmin, *Arch. Biochem. Biophys.,* 341: 280–286, 1997.

38. Reilly, C.A. and Aust, S.D. Stimulation of the ferroxidase activity of ceruloplasmin during iron loading ferritin, *Arch. Biochem. Biophys.,* 347: 242–248, 1997.

39. Mukhopadhyay, C.K., Attieh, Z.K., and Fox, P.L. Role of ceruloplasmin in cellular iron uptake, *Science,* 279: 714–717, 1998.

40. Ohamoto, N. et al. Hereditary ceruloplasmin deficiency with hemosiderosis, *Hum. Genet.,* 97: 755–758, 1996.

41. Takahashi, Y. et al. Characterization of a nonsense mutation in the ceruloplasmin gene resulting in diabetes and neurodegenerative disease, *Hum. Mol. Genet.,* 5: 81–84, 1996.

42. Harris, Z.L. et al. Aceruloplasminemia: molecular characterization of this disorder of iron metabolism, *Proc. Natl. Acad. Sci. U.S.A.,* 92: 2539–2543, 1995.

43. Yoshida, K. et al. Mutation in the ceruloplasmin gene is associated with systematic hemosiderosis in humans, *Nature Genet.,* 9: 267–272, 1995.

44. Yakaki, M. et al. A novel splicing mutation in the ceruloplasmin gene responsible for hereditary ceruloplasmin deficiency with hemosiderosis, *J. Neurol. Sci.,* 156: 30–34, 1998.

45. Klomp, L.W. and Gitlin, J.D. Expression of the ceruloplasmin gene in the human retina and brain: implications for a pathogenetic model in aceruloplasminemia, *Hum. Mol. Genet.,* 5: 1989–1996, 1996.

46. Miyajima, H. et al. Use of desferrioxamine in the treatment of aceruloplasminemia, *Ann. Neurol.,* 41: 404–407, 1997.

47. Ehrenwald, E. and Fox, P.L. Role of endogenous ceruloplasmin in low density lipoprotein oxidation by human U937 monocytic cells, *J. Clin. Invest.,* 97: 884–890, 1996.

48. Mazumder, B. et al. Induction of ceruloplasmin synthesis by IFN-gamma in human monocytic cells, *J. Immunol.,* 159: 1938–1944, 1997.

49. Klomp, L.W.J. et al. Ceruloplasmin gene expression in the murine central nervous system, *J. Clin. Invest.,* 98: 207–215, 1996.

50. Loeffler, D.A. et al. Increased regional brain concentration of ceruloplasmin in neurodegenerative disorders, *Brain Res.,* 738: 265–274, 1996.

51. Patel, B.N. and David, S. A novel glycosylphosphatidylinositol-anchored form of ceruloplasmin is expressed by mammalian astrocytes, *J. Biol. Chem.,* 272: 20185–20190, 1997.

52. Taylor, T.D. et al. Homozygosity mapping of Hallervorden–Spatz syndrome to chromosome 20p 12.3-p13, *Nature Genet.,* 14: 479–481, 1996.

53. Simon, M. et al. Association of HLA-A3 and HLA-B14 antigenes with idiopathic hemochromatosis, *Gut,* 17: 332–334, 1976.

54. Feder, J.N. et al. A novel MHC class I-like gene is mutated in patients with hereditary hemochromatosis, *Nature Genet.,* 13: 399–408, 1996.

55. De Sousa, M. et al. Iron overload in beta-2microglobulin-deficient mice, *Immunology Lett.,* 39: 105–111, 1994.

56. Rothenberg, B.E. et al. Beta 2 knockout mice develop parenchymal iron overload: a putative role for class I genes of the major histocompatibility complex in iron metabolism, *Proc. Natl. Acad. Sci. U.S.A.,* 93: 1529–1534, 1996.

57. Feder, J.N. et al. The hemochromatosis founder mutation in HLA-H disrupts beta-2 microglobulin interaction and cell surface expression, *J. Biol. Chem.*, 272: 14025–14028, 1997.

58. Waheed, A. et al. Hereditary hemochromatosis: effects of C282Y and H63D mutations on association with beta-2 microglobulin, intracellular processing, and cell surface expression of the HFE protein in COS-7 cells, *Proc. Natl. Acad. Sci. U.S.A.*, 94: 12384–12389, 1997.

59. Borot, N. et al. Mutations in the MHC class I-like candidate gene for hemochromatosis in French patients, *Immunogenetics*, 45: 320–324, 1997.

60. Jouanolle, A.M. et al. A candidate gene for hemochromatosis: frequency of the C282Y and H63D mutations, *Hum. Genet.*, 100: 544–547, 1997.

61. Jazwinska, E.C. et al. Haemachromatosis and HLA-H, *Nat. Genet.*, 14, 249–251, 1996.

62. Carella, M. et al. Mutation analysis of the HLA-H gene in Italian hemochromatosis patients, *Am. J. Hum. Genet.*, 60: 828–832, 1997.

63. Piperno, A. et al. Heterogeneity of hemochromatosis in Italy, *Gastroenterology*, 114: 996–1002, 1998.

64. Gottschalk, R. et al. HFE codon 63/282 (H63D/C282Y) dimorphism in German patients with genetic hemochromatosis, *Tissue Antigens*, 51: 270–275, 1998.

65. Beutler, E. and Gelbart, T. HLA-H mutation in Askenazi Jewish population, *Blood Cell Mol. Dis.*, 23: 95–98, 1997;.

66. Roth, M. et al. Absence of the hemochromatosis gene Cys 282 Tyr mutation in three ethnic groups from Algeria, Ethiopia and Senegal, *Immunogenetics*, 46: 222–225, 1997.

67. Crawford, D.H.G. et al. Expression of HLA-linked hemochromatosis in subjects homozygous or heterozygous for the C282Y mutation, *Gastroenterology*, 114: 1003–1008, 1998.

68. Parkkila, S. et al. Immunohistochemistry of HLA-H, the protein defective in patients with hereditary hemochromatosis, reveals unique pattern of expression in gastrointestinal tract, *Proc. Natl. Acad. Sci. U.S.A.*, 94: 2534–2539, 1997.

69. Parkkila, S. et al. Association of the transferrin receptor in human placenta with HFE, the protein defective in hereditary hemochromatosis, *Proc. Natl. Acad. Sci. U.S.A.*, 94: 13198–13202, 1997.

70. Bothwell, T.H. et al. in *Hereditary Hemochromatosis, The Metabolic and Molecular Bases of Inherited Disease*, Scriver, C.R. et al., Eds., New York, McGraw-Hill, 2246–2247, 1995.

71. Feder, J.N. et al. The hemochromatosis gene product complexes with the transferrin receptor and lowers its affinity for ligand binding, *Proc. Natl. Acad. Sci. U.S.A.*, 95: 1472–1477, 1998.

72. Lebrou, J.A. et al. Crystal structure of the hemochromatosis protein HFE and characterisation of its interaction with transferrin receptor, *Cell*, 93: 111–123, 1998.

73. Zhou, X.Y. et al. HFE gene knockout produces mouse model of hereditary hemochromatosis, *Proc. Natl. Acad. Sci. U.S.A.*, 95: 2492–2497, 1998.

74. Roberts, A.G. et al. Increased frequency of the haemochromatosis Cys 282 Tyr mutation in sporadic porphyria cutanea tarda, *Lancet*, 349: 321–323, 1997.

75. Sampietro, M. et al. High prevalance of the His 63 Asp HFE mutation in Italian patients with porphyria cutanea tarda, *Hepatology*, 27: 181–184, 1998.

76. Porto, G. et al. Major histocompatibility complex class I association in iron overload: evidence for a new link between the HFE H63D mutation, HLA-A29, and non-classical forms of hemochromatosis, *Immunogenetics*, 47: 404–410, 1998.

77. Rose, T.M. et al. Primary structure of the human melanoma associated antigen p97 (melanotransferrin) deduced from the mRNA sequence, *Proc. Natl. Acad. Sci. U.S.A.,* 83, 1261–1265, 1986.
78. Baker, E.N. et al. Human melonatransferrin (p97) has only one functional iron-binding site, *FEBS Lett.,* 298, 215–218, 1992.
79. Dorcas, D. et al. Molecular detection of tumor-associated antigens shared by human cutaneous melanomas and gliomas, *Am. J. Pathol.,* 150: 2143–2152, 1997.
80. Real, F.X. et al. Class 1 (unique) tumor antigenes of human melanoma: identification of unique and common epitopes on a 90-kDa glycoprotein, *Proc. Natl. Acad. Sci. U.S.A.,* 85: 3965–3969, 1988.
81. Rothenberger, S. et al. Coincident expression and distribution of melanotransferrin and transferrin receptor in human brain capillary endothelium, *Brain. Res.,* 712: 117–121, 1996.
82. Alemany, R. et al. Glycosyl phosphatidylinositol membrane anchoring of melanotransferrin (p97): apical compartmentalization in intestinal epithelial cells, *J. Cell Sci.,* 104, 1155–1162, 1993.
83. Food, M.R. et al. Transport and expression in human melanomas of a transferrin-like glycosylphosphatidylinositol-anchored protein, *J. Biol. Chem.,* 269, 3034–3040, 1994.
84. Darrielsen, E.M. and Van Deurs, B. A transferrin-like GPI-linked iron-binding protein in detergent-insoluble noncaveolar microdomains at the apical surface of fetal intestinal cells, *J. Cell. Biol.,* 131: 939–950, 1995.
85. Kennard, M.L. et al. A novel iron uptake mechanism mediated by GPI-anchored human p97, *EMBO J.,* 14: 4178–4186, 1995.

8 Iron-Responsive Elements and Iron Regulatory Proteins

CONTENTS

8.1 Introduction: The Discovery of Iron-Responsive Elements (IREs) and
Iron Regulatory Proteins (IRPs) 407
8.2 Secondary and Tertiary Structures of Iron Response Elements (IREs).......410
8.3 Iron Regulatory Proteins...415
8.4 Role of IRP in Mediating Transferrin Receptor mRNA Stability 422
8.5 Role of IRP in the Control of Ferritin Translation 425
8.6 Evolution of IRPs...426
8.7 Regulation of IRP Activity by Nitric Oxide 429
8.8 Regulation of IRP by Oxidative Stress 432
8.9 Role of Phosphorylation in the Control of IRP Activity 435
8.10 Role of Heme in the Regulation of IRP Activity.....................................436
8.11 IRPs Coordinate In Vivo Iron Metabolism..436
8.12 Tissue-Dependent Regulation of IRP Activity 437
8.13 Conclusions 439
References...440

8.1 INTRODUCTION: THE DISCOVERY OF IRON-RESPONSIVE ELEMENTS (IRES) AND IRON REGULATORY PROTEINS (IRPS)

The coordinated and controlled expression of gene information represents a key problem in all forms of life. This problem is particularly important when a set of genes involved in a metabolic pathway (i.e., iron metabolism) must be finely and sequentially regulated.

The pattern of gene regulation involves early (gene transcription), intermediate (mRNA processing, maturation, and stability), and late events (mRNA translation and post-translational modifications of proteins). Most gene regulatory mechanisms act at the level of gene transcription. However, increasing evidence indicates that several important regulatory mechanisms act at the mRNA processing and mRNA stability stages. mRNA processing in eukaryotic cells can be divided into four steps: 5′ capping, 3′ cleavage/polyadenylation, RNA splicing, and transport of mRNA through nuclear pore complexes into the cytoplasm.

An important mechanism of control involves the regulation of mRNA stability. The concentration of a given mRNA depends on both its rate of synthesis and its rate of degradation. An increase in mRNA stability produces an increase in the level of synthesis of the corresponding protein, and thus a net increase in the level of gene expression. The stability of mRNA also determines how rapidly the synthesis of the encoded protein can be maintained or shut down.

The stability of cytoplasmic mRNAs varies widely, even though the majority of eukaryotic mRNAs have half-lives of many hours. In some cases, the degradation rates of specific mRNAs change in response to extracellular signals. Examples of such regulation are the genes encoding the milk protein casein and the transferrin receptor.

The level of mRNA encoding casein is dramatically increased following exposure of mammary gland cells to prolactin, a phenomenon largely related to a marked improvement of casein mRNA stability induced by the hormone.

In the second example, a decrease in iron levels produces an increase in the rate of transferrin receptor synthesis that is dependent upon augmentation of transferrin receptor mRNA levels due to an improvement of mRNA stability. Thus, when iron stores in the cell are sufficient to adequately sustain cellular metabolism, the import of the transferrin–iron complex is reduced by increasing the degradation rate of transferrin receptor mRNA, quickly resulting in a decrease in the level of transferrin receptor mRNA, and a consequent decrease in the level of transferrin receptor.

An opposite phenomenon is observed when cellular iron levels fall: transferrin receptor mRNA is stabilized and increased synthesis of the transferrin receptor protein occurs. The regulation of transferrin receptor RNA stability depends on the presence of iron-responsive elements (IREs) present in the 3′ untranslated region of the receptor mRNA and on cytoplasmic proteins initially called IRE-binding proteins (IRE-BPs) and now known as iron response proteins (IRPs). Their conformations and activities as RNA-binding proteins change at high and low iron levels, respectively.

An additional mechanism controlling gene regulation involves the control of mRNA translation. One example of such a mechanism is the iron-dependent regulation of ferritin synthesis, a process controlled by the same IRP involved in the control of transferrin receptor mRNA stability. When intracellular stores of iron are low, translation of ferritin mRNA is repressed, thus rendering the endocytosed iron available for iron-requiring enzymes. When iron is in excess, ferritin mRNA translation is repressed and ferritin proteins are synthesized and bind the iron endocytosed within the cell. This translation regulation of ferritin synthesis is mediated through the binding of IRP to an IRE sequence present at the level of the 5′ untranslated region of ferritin mRNA.

The discovery of IREs and IRPs mainly resulted from:

1. Studies revealing a coordinated mechanism regulating the level of transferrin receptor and ferritin synthesis inversely according to iron levels. Iron loading produces an increase in ferritin synthesis accompanied by a decrease of transferrin receptor expression, while an opposite phenomenon is elicited by iron deprivation.
2. Isolation of ferritin and transferrin receptor cDNA clones.

Comparison of the sequences and the analysis of some deletion mutants led to the identification of a *cis* regulatory element in the 5′ untranslated region of ferritin H and L mRNAs.[1,2] Deletion of the 5′ untranslated region of ferritin mRNA caused a loss of its iron-dependent regulation.

Further analysis of this 5′ untranslated region led to the identification of a 35-bp region that is necessary and sufficient for translational regulation by iron.[3,4] Examination of all the cloned ferritin genes confirmed the presence in the 5′ untranslated region of the regulatory element required for iron-dependent regulation. The element is known as the iron-responsive element (IRE).[5] The sequences present at the IRE form stem–loop structures with loop-conserved sequences of CAGUGN. Due to the presence of the IRE sequence, ferritin mRNA is poorly translated in the presence of low iron concentrations; when iron concentration is high and when the IRE is deleted, ferritin translation is repressed.

The repression of ferritin mRNA in iron-deprived cells is attributed to a protein that may be capable of interacting with the IRE sequence. Studies of rat liver and *in vitro* rabbit reticulocyte translation provided the first evidence that the repression of ferritin translation may be due to the binding of a protein to the IRE sequence.[6-9] This protein was initially termed IRE-binding protein (IRE-BP), and is now known as iron response protein (IRP). IRP is constantly expressed in cells, but its RNA-binding activity changes as a function of iron concentration. When iron concentration is low, the IRP RNA-binding activity is high, with consequent repression of ferritin mRNA translaton, while when iron concentration is high the RNA-binding activity of IRP is inhibited and, consequently, ferritin mRNA synthesis is repressed.[10-12]

These initial studies led to development of an electrophoretic mobility shift assay to study the RNA-binding activity of IRP. The technique is based on incubation of labelled IRE-containing RNA in the presence of a cytoplasmic extract that contains IRP. The presence of the RNA-protein complex resulted in the formation of a typical band; its migration was consistently retarded, compared to migration of the free probe.[6,7]

In contrast to ferritin mRNA, whose level remains virtually unmodified when iron levels are changed, transferrin receptor mRNA levels markedly changed in response to modifications of iron concentration, without concomitant changes in mRNA translation efficiency. The analysis of several cDNA clones containing various deletions at the 3′ untranslated region and their expression after transfection in host cells showed the existence of two *cis* regulatory elements present at two areas of about 150 bases within the 2.5 Kb 3′ untranslated region of the transferrin receptor mRNA.[5,12,13] Transferrin receptor expression was found to be modulated by iron through post-transcriptional mechanisms modulating transferrin receptor mRNA stability.[14] Thus, iron loading leads to a rapid degradation of transferrin receptor mRNA, while iron deprivation produces stabilization of this mRNA species.

It is of interest that mutagenesis studies of the 3′ untranslated regulatory region of the transferrin receptor mRNA revealed two types of mutations. Some mutations abrogate iron-dependent regulation of transferrin receptor, while other mutations produce an elevated and unregulated expression of transferrin receptor mRNA.[5,12,13] The analysis of the sequence of this regulatory region showed the existence of five IREs[5,12-14] that exhibit clear homologies to the IRE present in the untranslated region of the ferritin mRNA.

These studies provided clear evidence of the existence of a common regulatory element located in the 5' untranslated region of the ferritin mRNA and the 3' untranslated region of the transferrin receptor mRNA. The differential location of the IRE common sequences in different regions of these two mRNAs is related to the opposite effects of iron on ferritin and transferrin receptor RNAs: increased translaton of ferritin mRNA and destabilization of transferrin receptor mRNA.

Other studies led to the purification and preliminary characterization of the iron regulatory protein (IRP). When purified at homogeneity from rabbit red cells and liver, IRP exhibits a 95-kDa molecular weight and the ability to bind IRE sequences.[8] The activated IRP binds the IRE with a 1:1 stoichiometry with a high affinity ($K_d = 10^{-11}$ M).[11,15]

IRP is localized in the cytosol where it is present as free cytosolic protein or RNA-bound protein.[15] A significant proportion of IRP present in the cytoplasmic extracts is unable to bind IRE. It becomes able to do so after treatment of the cytoplasmic extracts with β-mercaptoethanol.[10] Addition of an iron chelator to the cytoplasmic extract did not activate IRP, thus indicating that the stimulatory effect of iron deprivation on IRP activity cannot be mediated by a direct effect of a chelator.

The analysis of the sequence of a cDNA clone encoding human IRP showed that this protein resembles mitochondrial aconitase, an enzyme containing a 4Fe–4S cluster.[16-18]

It is now apparent that two IRPs (IRP1 and IRP2) are present in mammalian cells. Their characterization and properties will be outlined below.

8.2 SECONDARY AND TERTIARY STRUCTURES OF IRON RESPONSE ELEMENTS (IREs)

Based on secondary structure predictions, IREs can be folded in a stem–loop structure (Figure 8.1) with a double-stranded, upper, 5-bp helix of variable sequence (upper stem) and a 6-nucleotide loop with the consensus sequence 5'-CAGUGN-3' (loop). Below the paired upper stem is a small asymmetrical bulge with an unpaired cytosine residue at the first nucleotide 5' of the stem. The structure of this bulge is variable. In some cases the bulge is composed of a single unpaired C nucleotide residue. In other IREs, the C nucleotide and two additional 5' nucleotides are apposed to one single free 3' nucleotide. Below the bulge, a second stem (lower stem) is observed. The base pair region shows considerable sequence variability from one IRE to another.

The basic structure of the IRE is common to all the IREs observed in various mRNAs analyzed. However, some IREs showed bulge domains that were atypical. Analysis of the structure of IRE present in crayfish ferritin mRNA showed a guanine instead of a cytosine at the expected position of the bulge.[19] Furthermore, the predicted structure of the IRE of the *qoxD* gene of *Bacillus subtilis* showed complete absence of the bulge.[20] Finally, in other IREs, the upper stem exhibits an atypical structure. The IRE of the mouse glycolate oxidase mRNA displayed in the upper stem an unusual A*A mismatch[21] (Figure 8.2), and did not confer iron-dependent regulation.

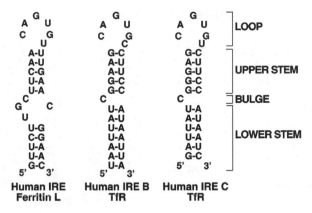

FIGURE 8.1 The basic structures of IREs are similar for transferrin receptor mRNA and ferritin mRNA. The IRE is composed of four regions: loop, upper stem, bulge, and lower stem. The loop exhibits the typical consensus sequence 5'-CAGUGN-3' required for the binding of IRE to IRP. Below the paired upper stem is a small asymmetrical bulge with an unpaired cytosine (C) residue as the first nucleotide 5' of the stem. The structure of this bulge is variable. Below the bulge is the lower stem, whose structure is a base-paired region that varies considerably from one IRE to another.

The conformity of the stuctures of different IREs suggests the presence of a specific three-dimensional structure that aids the formation of a specific conformation that enables these structures to interact with IRP. Footprinting experiments have shown that the extent of interaction of IRP with IRE is limited to the IRE itself.[22] RNA sequences near the IRE, although not involved in the binding with IRP, undergo structural changes when IRP binds to the IRE.[22]

Other studies have shown that two types of mutations markedly reduce the affinity with which IRP binds the IRE: (1) mutations disrupting the base pairing in the upper stem of IRE; and (2) mutations increasing the size of the loop or increasing the distance between the UC bulge and the loop.[23,24] It is clear that the primary nucleotide sequences of the upper and lower stems are less important than the ability to form base pairing within the paired nucleotides of the stems.[23]

These observations suggest that the seven highly conserved nucleotides at the loop and the bulge are the primary determinants of binding specificity and represent the potential sites involved in interaction with IRP.

To better define the structures of the IRE regions involved in IRP binding, Henderson and coworkers prepared a pool of 16,384 IRE molecules randomized at the nucleotide corresponding to the loop and the bulge. Mutants were screened according to their capacity to bind IRP. This screening showed two major clones of high-affinity RNA ligands: the optimal loop sequences were 5'-CAGUGN-3' (wild-type loop) and 5'-UAGUAN-3'.[25]

This observation is very important in that it raises the possibility that base pairing within the IRE loop at nucleotide positions 1 and 5 is required for the formation of the loop that frees nucleotides 2 to 4 to interact with IRP. A change from a C to a G in the first nucleotide of the loop was tolerated as long as a complementary change

Different types of IREs

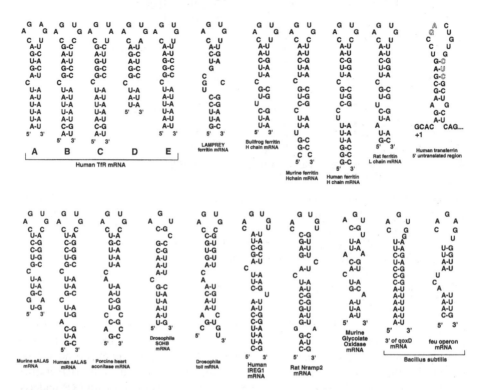

FIGURE 8.2 Structures of IREs observed in different mRNA species. IREs can be subdivided into two groups according to position in the mRNA molecule with respect to the coding sequence. Ferritin-like IREs exhibit similar structures and are located near the 5' ends of mRNAs encoding ferritin subunits H and L, mitochondrial aconitase, succinate dehydrogenase, IREG-1 (a duodenal iron transporter at the basolateral membrane), and aminolevulinate synthase expressed in erythroid cells (eALAS). Transferrin receptor-like IREs are observed only at the untranslated 3' region of transferrin receptor mRNA. Examples of IREs present in genes of a number of species are shown.

from a G to a C occurred at the level of the fifth nucleotide.[25] The interaction between nucleotide G at position 5 of the loop and nucleotide C at position 1 of the loop is suggested by several additional observations:

1. Ribonuclease T1 was unable to cut at position G5 of the loop; it cuts at position G3.[22]
2. The nucleotide C at position 1 of the loop was not methylated.[26]
3. Loss of an imino resonance signal in the region characteristic of G–C paired residues in the ^{1}H-NMR spectrum of IRE mutants with G5 to A mutation.[27]

Additional spectroscopic and structural studies confirmed and extended these observations. A thermodynamic and spectroscopic study analyzed the ferritin IRE structure. In predictions of RNA secondary structures based on thermodynamic parameters, loops are considered to be destabilizing, so some factors within IRE loops must contribute to their stability. In fact, stability measurements showed that the loop has 2.9 Kcal/mol more free energy than predicted. This thermodynamic stability is favored by hydrogen bonding between the C1 and G5 nucleotides of the loop.[28]

A second study used multi-dimensional nuclear magnetic resonance and computational approaches to investigate the structure and dynamics of IRE in solution. This technique provided clear evidence that C1 and G5 residues of the loop are base paired to each other and, through their binding, they contribute to the stability of the IRE loop.[29] Furthermore, the A2 nucleotide of the loop stacks directly above the G5 base of the loop. This study also suggested that loop residues C1 and G5 do not interact directly with IRP.[29]

Analysis of several mutants defined IREs specific for IRP1 and IRP2. The specific IRP1 and IRP2 ligands were originated by the systematic evolution of ligands by exponential enrichment (SELEX) procedure.[30,31] This procedure defined IRE mutants with mutated loop sequences capable of selectively binding IRP1 (UAGUAC sequence of the loop) or IRP2 (GGGAGU sequence of the loop). These findings indicate that IRP2 binding requires the conserved G–C base pair in the terminal loop, whereas IRP1 binding requires the G–C or the engineered U–A base pair. Competition experiments showed that these selective IRP1 and IRP2 ligands competed with the canonical IRE binding sequence for binding to purified IRP1 and IRP2 proteins, respectively.[31] These observations suggest possible roles for IRP1 and IRP2 in the regulation of unique RNA targets *in vivo*.[30]

Transfection experiments of IRE mutants specific for either IRP1 or IRP2 showed that these IREs and wild-type IRE maintain their IRP specificity *in vivo* and are able to mediate translational control.[32]

Additional understanding of a possible preferential binding of IRP1 and IRP2 to specific IRE structures derived from the study of IRE mutants containing mutations at their bulges. The two types of bulges present in natural IREs were mutated and the effects of such mutations on IRP1 and IRP2 binding were analyzed. The conversion of the internal loop/bulge in the ferritin IRE to a C bulge by deletion of the U base in the bulge resulted in a marked decline of IRP2 binding (>95%), compared to a mild decline of IRP1 binding (about 13%).[33] This observation suggests that IREs containing internal loops/bulges (such as ferritin IRE) bind IRP2 much more than IREs containing C bulges (such as transferrin receptor IRE). This assumption was proven by showing that all natural C-bulge IREs bind IRP2 much more poorly than IRP1, when compared with ferritin IRE.[33]

The role of IRE in the *in vivo* regulation of ferritin synthesis is directly supported by studies of a recently identified autosomal dominant disorder called hereditary hyperferritinemia–cataract syndrome.[34] This disorder is characterized by elevated levels of serum ferritin and congenital nuclear cataracts in the absence of iron overload.[34] That phenomenon suggests that the increase in ferritin levels (mostly L-ferritin) observed in this syndrome may be due to a misregulation of ferritin L

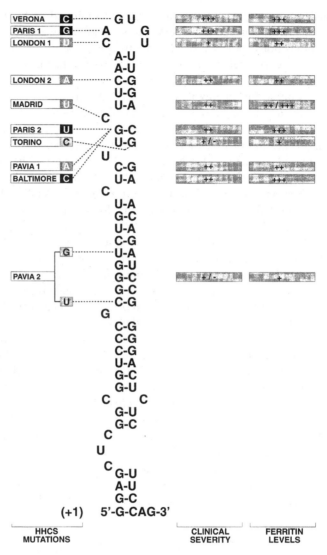

FIGURE 8.3 Mutations of the L-ferritin IRE observed in patients with hereditary hyperferritinemia–cataract syndrome. The left column illustrates amino acid substitutions of each mutation. The boxes on the right show clinical severity and serum ferritin levels: +/–, slight; +, mild; ++, moderate; and +++, high. The effects of these mutations on ferritin synthesis relate mainly to their affinity for binding IRP.

gene expression. This hypothesis was directly confirmed when analysis of the ferritin L gene in these patients, showed single mutations,[34-38] double mutations,[36] or a 29-bp deletion[39] (Figure 8.3). The majority of these mutations disrupt IRE–IRP binding *in vitro* and through this mechanism lead to an increased ferritin synthesis.

The clinical severity of hereditary hyperferritinemia–cataract syndrome (HHCS) is highly variable, based on the level of loss of affinity for IRP by mutated IREs[40]

(Figure 8.3). The biochemical impacts on IRE function are variable in that some HHCS mutations produced a complete loss of IRE–IRP recognition with minimal effects on IRE stability; other mutations changed the stability and secondary structure of the IRE.[40]

IRE structures can be subdivided into two categories: ferritin-like IRE or IRE translation regulators; and transferrin receptor-like IRE or IRE stability/turnover regulators. The ferritin-like IREs exhibit similar structures and are located near the 5′ ends of mRNAs encoding ferritin subunits H and L, mitochondrial aconitase succinate dehydrogenase, and aminolevulinate synthase expressed in erythroid cells (eALAS). Unlike transferrin receptor-like IREs, the ferritin-like IREs have high G/C content.

The regulatory role of ferritin-like IREs on translation is position-dependent. Ferritn and eALAS IREs are located 16 to 42 nucleotides downstream of the cap structure of the mRNA. (The cap structure consists of a residue of 7-methylguanosine added by a dimering capping enzyme at the 5′ end of the nascent mRNA when it reaches a length of 25 to 30 nucleotides.)

To evaluate a possible position-dependent effect of IRE on its capacity to modulate translation, specific constructs were made to vary the distance between the IRE and the 5′ end of the mRNA. These studies showed that the length of the RNA spacer, rather than its nucleotide sequence or predicted secondary structure represented the major determinant of IRE function.[41,42] When the position of the IRE was preserved, sequences flanking the IRE did not seem to play a relevant role. Thus, these sequences can be changed without significantly affecting the regulatory function of IRE.

A transferrin receptor-like IRE element was observed only in the 3′ untranslated region of transferrin receptor mRNA. However, a functional IRE was found at the 3′ untranslated region of the mRNA encoding IREG-1, an iron transporter found on the basolateral sides of villus duodenal cells. The five IREs observed in the untranslated 3′ region of the transferrin receptor act as modulators of receptor mRNA stability.

Transferrin receptor mRNA regulation requires both IREs and an untranslated mRNA non-IRE base-paired region present between IREb and IREc.[13,14] The transferrin receptor mRNA IREs display some characteristics similar to ferritin-like IREs. Instead of being rich in G/C base pairs, sequences in the lower stem of the IRE are rich in A/U base pairs (see Figure 8.2). This finding is particularly interesting in that AU-rich sequences turn over rapidly in several mRNA species.

Based on these differences in the sequences of ferritin-like IRE and transferrin receptor IREs, it is not surprising that these two IREs display only limited sequence identity, with the major sequence divergence at the lower stem.

8.3 IRON REGULATORY PROTEINS

IRPs are functionally detected *in vitro* by incubating a cytoplasmic protein extract in the presence of a radiolabeled IRE-containing RNA. A modification of this technique including cross-linking of RNA–protein complexes with ultraviolet light provided the first estimates of the molecular weights of IRPs. No distinction was made between IRP1 and IRP2. Molecular weights of about 100 kDa with rat liver

```
                          ┌─────────────┐
                          │ HUMAN IRP 1 │
                          └─────────────┘

  1 cprktrtqnl ppwlsnklyk nievpfkpar vilqdftgvp avvdfaamrd avkklggdpe
 61 kinpvcpadl vidhsiqvdf nrradslqkn qdlefernre rfeflkwgsq afhnmriipp
121 gsgiihqvnl eylarvvfdg dgyyypdslv gtdshttmid glgilgwgvg gieaeavmlg
181 qpismvlpqv igyrlmgkph plvtstdivl titkhlrqvg vvgkfveffg pgvaqlsiad
241 ratianmcpe ygataaffpv devsitylvq tgrdeeklky ikkylqavgm frdfndpsqd
301 pdftqvveld lktvvpccsg pkrpqdkvav sdmkkdfesc lgakqgfkgf gvapehhndh
361 ktfiydntef tlahgsvvia aitsctntsn psvmlgagll akkavdagln vmpyiktsls
421 pgsgvvtyyl qesgvmpyls qlgfdvvgyg cmtcignsgp lpepvveait qgdlvavgvl
481 sgnrnfegrv hpntranyla spplviayai agtiridfek eplgvnakgq qvflkdiwpt
541 rdeiqaverq yvipgmfkev yqkietvnes wnalatpsdk lffwnsksty iksppffenl
601 tldlqppksi vdayvllnlg dsvttdhisp agniarnspa aryltnrglt prefnsygsr
661 rgndavmarg tfanirllnr flnkqapqti hlpsgeildv fdaaeryqqa glplivlagk
721 eygagssrdw aakgpfllgi kavlaesyer ihrsnlvgmg vipleylpge nadalgltgq
781 erytiiipen lkpqmkvqvk ldtgktfqav mrfdtdvelt yflnggilny mirkmak
```

FIGURE 8.4 Amino acid sequence of human IRP1.

extracts[6] and a doublet of about 97 and 103 kDa with extracts from human placenta have been observed for IRE/IRP complexes.

The initial assays provided only approximate molecular weights of IRP. Subsequent studies were based on the purification of IRP from various tissues by affinity chromatography using an IRE-containing RNA column[43,44] or by standard biochemical purification procedures.[8] IRP purified from either rabbit or human liver exhibited a molecular weight of about 90 kDa and migrated as a single molecular species.[8,43] IRP purified from human placenta migrated as molecular doublet with two bands of 95 and 100 kDa, respectively.

A further step toward understanding the structure and function of IRPs was achieved by the cloning of the IRP1 and IRP2 genes. A degenerate oligonucleotide probe derived from a single peptide sequence of purified human IRP was used to isolate a cDNA clone that encodes human IRP[16] (Figure 8.4). This cDNA encodes a protein of 87 kDa with a slightly acidic isoelectric point and an mRNA species of about 3.6 Kb widely expressed in various tissues.[16] The initial analysis showed a nucleotide-binding consensus sequence and regions of histidine and cysteine clusters. The gene encoding IRP1 was located on chromosome 9.[16]

Using a polymerase chain approach based on the sequences from that study, a full length cDNA clone encoding human IRP1 was isolated.[45] Hybridization to genomic DNA and mRNA and sequencing data indicated a 40-Kb IRP1 gene encoding a 4.0-Kb mRNA that translated into a protein of 98,400 Da.[45] Recombinant IRP1 produced *in vitro* from the full length cDNA binds to ferritin IRE with high affinity; this property was lost in the IRP1 mutant through a deletion of 132 amino acids at the COOH terminus.

Rabbit IRP1 was also cloned and its sequence exhibits a greater than 90% identity with human IRP1.[46]

The cloning of IRP1 allowed us to compare its deduced amino acid sequence with those of other proteins in data banks. Computer-based analysis of data banks showed a significant level of sequence conservation between the human IRP and the Fe–S protein, aconitase, a mitochondrial enzyme of the citric acid cycle catalyzing the stereospecific dehydration and rehydration reactions involved in the conversion of citrate to isocitrate through the intermediate metabolite, *cis*-aconitate.[17]

IRP exhibited about 30% sequence identity with aconitase. A significant level of identity was observed also between IRP and isopropylmalate isomerase.[18]

The similarity of IRP and aconitase is even more relevant if the comparison is restricted to the amino acid residues that contribute to the formation of the catalytic site of aconitase, including the three cysteine residues known to coordinate the Fe–S cluster. All 18 active site residues in aconitase are identical in primary sequence alignment to human IRP.

The crystal structure of mitochondrial aconitase[47] and the computer-simulated three-dimensional structure of IRP1[17] allowed us to propose a mode of IRP1 function based on the presence or absence of the iron–sulfur cluster. The presence of the Fe–S cluster in the IRP1 protein leads to an inhibition of its capacity to bind IRE and concomitantly favors its enzymatic activity as aconitase. The crystal structure of mitochondrial aconitase provided evidence that the protein is composed of globular domains. Domains 1 to 3 are separated from the fourth domain by a single hinge, resulting in the formation of a cleft between domains 1 through 3 and domain 4. The Fe–S cluster is located at this cleft and anchored to the protein by ligation of three of the four iron molecules to cysteines at domains 1 through 3.[48] The fourth domain is involved in the binding of enzyme substrates such as citrate[47,49] (Figure 8.5).

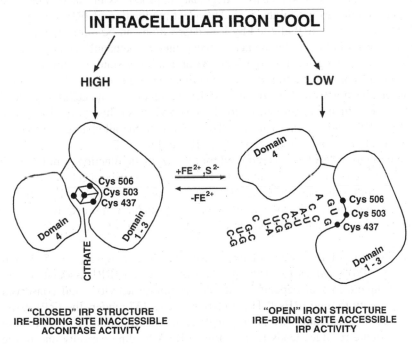

FIGURE 8.5 Three-dimensional changes of IRP1 conformation under conditions of low and high iron loading. In iron-deprived cells, the IRP1 iron–sulfur cluster is disrupted and the RNA binding site is exposed, thus allowing the switch from the aconitase to the IRE-binding activity. In iron-loaded cells, the iron–sulfur cluster is present at the active site of the IRP1 molecule, thus allowing the closure of the cleft and the masking of the IRE-binding site located in the cleft.

These observations led to the hypothesis that IRP may, based on the iron status of the cell, switch from an enzymatic aconitase function to an RNA-binding function, IRP may correspond to cytosolic aconitase. This hypothesis was directly supported by the observation that both purified IRP and *in vitro*-synthesized recombinant IRPs possess aconitase activity.[50]

The demonstration of an Fe–S cluster within IRP is of fundamental importance for understanding the iron-dependent modulation of IRP activity. Several studies have shown an inverse correlation between IRP activity in cellular extracts and the level of the intracellular chelatable iron pool. Thus, iron deprivation, through a post-translational mechanism, shifts IRP to a high-affinity state, while iron loading through addition of iron salts or hemin to cells decreases IRP affinity about 100-fold[11,15] (Figure 8.5). These observations suggest that IRP is present in cells grown under physiologic iron conditions in three different forms: (1) a pool of free inactive low-affinity IRP; (2) a pool of mRNA-bound high-affinity IRP; and (3) a pool of free high-affinity IRP.[15]

The iron-dependent modulation of IRP1 activity is mediated by post-translational mechanisms. Iron loading down-modulates IRP1 activity, but this decrease is not accompanied by any significant change in the level of the protein. Iron loading does not affect the rate of IRP synthesis or degradation; protein synthesis inhibitors, such as cycloheximide, do not inhibit iron-dependent modulation of IRP activity.[51]

The activation of IRP elicited by iron deprivation can be mimicked *in vitro* by reducing agents. In fact, reducing agents like β-mercaptoethanol[52] and cysteine[53] are capable of converting inactive IRP into the high-affinity state to a level similar to that induced by iron chelation *in vivo*. *In vitro* activation is observed only with IRP that has not been activated *in vivo* before the preparation of the cellular lysate. The effect of the reducing agent can be totally reversed by sulfhydryl oxidizing agents only if the IRP is not bound to an IRE. Treatment with SH-alkylating agents elicited an irreversible blockade of the capacity of IRP to bind IRE.

An intact 4Fe–4S cluster is required for the enzymatic function of mitochondrial aconitase in the tricarboxylic acid cycle. By analogy, a similar mechanism may also apply to cytoplasmic aconitase (IRP1). The 4Fe–4S cluster within the cleft of the IRP1 molecule triggers the switch of the protein from an iron sensor to an enzyme. A number of studies support such an assumption:

1. Mutation of the three putative cluster cysteine (Cys 437, Cys 503, and Cys 506) residues to serine residues produces an IRP1 associated with complete loss of capacity to display aconitase activity[54] and conserved capacity to bind IREs.[54] The mutation of Cys 437 renders the IRP1 resistant to inactivation by N-ethylmaleimide.[54,55] Mutation of any of these cysteine residues leads to constitutive RNA-binding capacity and to the loss of the iron-dependent regulation.[55] These observations strongly suggest that cysteines 437, 503, and 506 are involved in the binding of the 4Fe–4S cluster to IRP1.
2. Mutation of arginine residues present at the aconitase active site in the fourth domain of IRP1 facing the cleft produces losses of both aconitase activity and IRE binding activity.

3. Mutation of Ser 778 produces a loss of aconitase enzymatic activity, but no modification of RNA-binding capacity.[56]

4. Citrate and isocitrate substrates prevent the activation of IRP1 mediated by β-mercaptoethanol, a phenomenon seemingly due to the interaction of the substrate near the cleft, thus inhibiting the interaction of IRP1 with IRE.[57]

5. Ultraviolet cross-linking experiments identified the regions of IRP1 involved in IRE binding and most of them are predicted to be located in the cleft of IRP1.[58,59] In these experiments, recombinant wild-type and mutant IRP1 proteins are incubated with a radiolabeled target IRE sequence and subjected to ultraviolet irradiation. Covalently linked RNA–protein complexes are digested with hydroxylamine, and labeled peptides are then separated by electrophoresis and sequenced. This procedure showed that IRP1 interacts with IREs through the region between amino acid residues 116 and 151.[58,59] The RNA–protein cross-link site is represented by Ser 127, which resides in the cleft near the 4Fe–4S cluster.[58]

6. Mutational studies led to the identification of several IRP1 mutants, with changes at the two regions of domain 4. They exhibited marked reductions in the affinity for wild-type IRE and normal aconitase activity.[60] Biochemical analyses of these mutants showed that arginines 728 and 732 contact the IRE bulge, whereas region 685 through 689 is the location required for recognition of the IRE loop.[60] These two mutants affect a local area of domain 4 assumed to be the surface at the edge of the cleft.

7. The presence or lack of 4Fe–4S within IRP1 modifies its susceptibility to phosphorylation and degradation by proteases.[61]

These observations are compatible with a model of IRP1 structure and function in which the IRE-binding domain is located within the cleft. When the iron–sulfur cluster is bound to IRP1, the cleft is closed and IRP1 cannot interact with IRE; when the iron–sulfur cluster is not bound to the IRP1, the cleft is open and the IRE binding site may interact with IRE.

Recent protein footprinting studies support this model. These studies showed structural changes in IRP1 as a function of the presence or absence of the iron–sulfur cluster.[62] Binding of the ferritin IRE or transferrin receptor IRE induced protection against proteolysis in the IRP1 region spanning amino acids 80 to 187 at the level of the cleft. The binding of IRE to IRP1 also induced protection at Arg 721 and Arg 728.[62] In contrast, the aconitase form of IRP1 exhibited a more compact structure and strong reductions of proteolytic cleavage at the two domains corresponding to amino acid residues 149 to 187 and 721 to 735.

IRP1 displays a double role; it is involved in the post-transcriptional regulation of genes involved in iron metabolism and acts as an enzyme with aconitase activity. Most studies have stressed its regulation of iron metabolism. Only a few concerned its function as a cytoplasmic aconitase.

A recent study performed with an elegant experimental approach suggests that the function of IRP1 as a cytoplasmic aconitase may have physiologic significance.[63] Rabbit IRP1 was expressed in aconitase-deficient *Saccharomyces cerevisiae*, and conditions affecting the aconitase function of IRP1 were investigated. The IRP1 expression may complement the aconitase mutation in *Saccharomyces cerevisiae*.[63] Interestingly, hyperexpression of cytosolic $NADP^+$-dependent isocitrate dehydrogenase in this yeast improved the capacity of IRP1 to display aconitase activity and favored the conversion of IRP1 from IRE-binding activity to the cytoplasmic aconitase function.

Recent studies also suggest a possible role for metabolites of the tricarboxylic acid pathway in the control of IRP1 activity.[64] Oxalamate, a tricarboxylic acid generated from the condensation of oxaloacetate with glyoxylate, acts as an inhibitor of aconitase and RNA-binding activity of IRP1. This inhibitory activity was observed in both iron-loaded and iron-depleted cells. Furthermore, the inhibition of IRP1 activity by oxalamate cannot be halted by addition of a reducing agent. The inhibitory effect of oxalamate on IRP1 activity may be related to two main mechanisms: (1) interaction of oxalamate with the 4Fe–4S cluster of IRP1; and (2) modification of IRP1 amino acid residues, such as cysteine and arginine, involved in its interaction with IREs.

A second IRE-binding protein, called IRP2, has been identified and characterized. The discovery of this second molecular species arose from the analysis of the IRE-binding protein observed in RNA band-shift assays performed on cellular extracts derived from murine cells. *In vitro* assays revealed the formation of two IRE–protein complexes in rat and mouse extracts. In addition to the canonical band, a second, faster migrating complex was observed and initially called complex B. Initial biochemical studies based on antibody reactivity and UV cross-linking experiments followed by partial sequence of cross-linked proteins provided preliminary evidence that IRP responsible for complex B formation is different from IRP1.[65]

These studies also showed that the protein responsible for B complex formation has an apparent molecular weight of about 105 kDa.[65] Subsequent biochemical and functional characterization studies of the IRP protein responsible for complex B formation based on purification of IRP2 from rat liver showed that:

1. The activities of both IRP1 and IRP2 markedly decreased in iron-saturated cells; however, unlike IRP1 protein levels which do not change following iron loading, IRP2 protein levels decreased to undetectable levels in cells loaded with iron.
2. Unlike IRP1, IRP2 does not have aconitase activity.[66]

This finding suggests major structural differences between IRP1 and IRP2.

A full understanding of the structural differences between IRP2 and IRP1 was based on the comparison of their amino acid sequences (Figure 8.6). Cloning of human IRP2 cDNA showed several interesting findings.[67] The protein has a predicted molecular mass of 105 kDa and contains 963 amino acids in contrast to the 98-kDa mass and 889 amino acids of IRP1. IRP2 can be arranged in four domains and exhibits 57% identity and 79% similarity in amino acid sequence with IRP1. IRP2 contains a 79-amino acid insertion at the level of residue 136 as compared to IRP1; this difference accounts for the different molecular weights of the two proteins.

HUMAN IRP 2

```
  1 mdapkagyaf eylietlnds shkkffdvsk lgtkydvlpy sirvlleaav rncdgflmkk
 61 edvmnildwk tkqsnvevpf fparvllqdf tgipamvdfa amreavktlg gdpekvhpac
121 ptdltvdhsl qidfskcaiq napnpgggdl gkagklsplk vqpkklpcrg qttcrgscds
181 gelgrnsgtf ssqientpil cpfhlqpvpe petvlknqev efgrnrerlq ffkwssrvlk
241 nvavippgtg mahqinleyl srvvfeekdl ifpdsvvgtd shitmvnglg ilgwgvggie
301 teavmlglpv sltlpevvgc eltgssnpfv tsidvvlgit khlrqvgvag kfveffgsgv
361 sqlsivdrtt ianmcpeyga ilsffpvdnv tlkhlehtgf skaklesmet ylkavklfrn
421 dqnssgepey sqviqinlns ivpsvsgpkr pqdrvavtdm ksdfqaclne kvgfkgfqia
481 aekqkdivsi hyegseykls hgsvviaavi sctnncnpsv mlaagllakk aveaglrvkp
541 yirtslspgs gmvthylsss gvlpylsklg feivgygcst cvgntaplsd avlnavkqgd
601 lvtcgilsgn knfegrlcdc vranylaspp lvvayaiagt vnidfqtepl gtdptgkniy
661 lhdiwpsree vhrveeehvi lsmfkalkdk iemgnkrwns leapdsvlfp wdlkstyirc
721 psffdkltke pialqaiena hvllylgdsv ttdhispags iarnsaaaky ltnrgltpre
781 fnsygarrgn davmtrgtfa niklfnkfig kpapktihfp sgqtldvfea aelyqkegip
841 liilagkkyg sgnsrdwaak gpyllgvkav laesyekihk dhligigiap lqflpgenad
901 slglsgretf sltfpeelsp gitlniqtst gkvfsviasf eddveitlyk hggllnfvar
961 kfs
```

FIGURE 8.6 Amino acid structure of human IRP2.

IRP2 was cloned also from rat liver[68] and confirmed the findings obtained from human IRP2. Rat liver IRP2 shares 61% amino acid identity and 79% amino acid similarity with IRP1. An important difference between IRP1 and IRP2 is the insertion into IRP2 of a 73-amino acid domain located 139 amino acids from the initiation translation codon. The analysis of rat liver IRP2 sequences also offered a way to understand the structural basis responsible for the absent aconitase activity of IRP2. In IRP1, all the 18 active site residues of mitochondrial aconitase are present. Sixteen of the 18 residues are present in IRP2, but two residues essential for enzymatic activity in mitochondrial aconitase, Arg 474 and Ser 669, are substituted with other amino acid residues.[68]

Interestingly, IRP2 is encoded by multiple transcripts of 6.4, 4.0, and 3.7 Kb, respectively. The 3' untranslated region of IRP2 contains multiple polyadenylation signals which could account for the 4.0- and 3.7-Kb mRNAs.

IRP2 activity is modulated according to iron levels in a manner similar to that observed for IRP1. Iron deprivation increases activities of both IRP1 and IRP2, while an opposite effect is elicited by iron loading. In spite these similarities, the mechanisms responsible for the changes in IRE-binding activity are distinct for both proteins. Thus, treatment of cells with iron salts results in a marked decline of IRP2 binding activity associated with a marked decline of IRP2 protein levels.[67] This observation suggests that a degradation or blockade of the synthesis of IRP2 is responsible for the reduction in IRP2 binding activity observed following iron loading. Furthermore, unlike IRP1, protein synthesis inhibitors prevented increases in IRP2 IRE-binding activity and in IRP2 protein levels in response to iron deprivation, thus suggesting that *de novo* protein synthesis is required for the IRP2 induction observed under these conditions.[69,70]

Similarly, down-modulation of IRP2 IRE-binding activity by iron loading requires a translation event.[68] Subsequent studies have further clarified the mechanisms responsible for IRP2 activity modulation by iron. An iron-dependent decrease in IRP2 levels was not due to a decrease in IRP2 synthesis, but to an increased rate of IRP2 degradation, as shown by pulse-chase experiments.[70] This degradation event requires active protein synthesis and is inhibited by proteasome inhibitors, but not by lysosomal or calpain II inhibitors.[71]

The increased sensitivity of IRP2 to proteasome degradation is due to the additive 73-amino acid sequence present in this proten. The deletion of this sequence abrogates the rapid iron-dependent degradation of IRP2, whereas the transfer of this 73-amino acid sequence into the IRP1 protein at the corresponding position confers to the mutated IRP1 protein the capacity for iron-dependent degradation.[72] Point mutation studies have shown the important roles of some cysteine residues in determining the sensitivity of IRP2 to proteolytic degradation. Inhibition of IRP2 degradation abrogated the iron-dependent regulation of IRP2 IRE activity.[72]

Recent studies have in part clarified the proteolytic system involved in iron-dependent degradation of IRP2 which includes ubiquitination of the protein. In ubiquitin-mediated proteolysis, proteins to be degraded are tagged by linkage to a small, highly conserved protein known as ubiquitin. Covalent attachement of ubiquitin to proteins signals their destruction by the 26S proteasome, a large complex with multiple proteolytic activities. Studies of IRP2 involving the use of proteasome inhibitors led to the identification of two biochemical events involved in iron-dependent degradation of IRP2: (1) iron-dependent oxidation at the degradation domain of IRP2, a process requiring the integrity of cysteine residues present at this domain; and (b) ubiquitination of IRP2 in iron-loaded cells.[73] The oxidation event precedes the ubiquitination, thus indicating that oxidatively modified IRP2 is a substrate for ubiquitination.[73]

8.4 ROLE OF IRP IN MEDIATING TRANSFERRIN RECEPTOR mRNA STABILITY

Iron-dependent regulation of transferrin receptor mRNA levels may be ascribed to the 3′ untranslated region of this mRNA.[13] Deletion of this 3′ untranslated region produces a high level of transferrin receptor expression (not modulated by iron) in transfectant cell lines. Transfer of this 3′ untranslated region of the transferrin receptor mRNA to the 3′ untranslated region of reporter genes, such as HLA-A2[5] or human growth hormone[14] led to iron-dependent regulation by the reporter genes. Deletion experiments of different portions of this regulatory region showed two required areas composed of 200 nucleotides and separated by about 250 nucleotides.

The regulatory region of transferrin receptor RNA contains 5 IREs (A, B, C, D, and E) and is large, thus making the prediction of its secondary structure difficult. Two models have been proposed to describe the structure of this regulatory region,[5,14] but only the model predicting that all the IREs present in the 3′ untranslated region of the transferrin receptor mRNA are in a hairpin conformation received experimental support.[14]

In vivo RNA footprinting experiments have directly supported this model.[74,75] These experiments were based on the permeabilization of the cells, followed by treatment with ribonuclease T1, and extraction of RNA. The most 5′ IRE of the 3′ untranslated region of transferrin receptor mRNA was analysed by the ligation-mediated polymerase chain reaction.[75] The pattern of RNAase T1 cleavage was compatible with a structure involving stem loops. Nucleotides involved in stem formation are not cleaved by RNAase T1 since they are involved in base pair formation.

Studies of binding to IRP of different segments of the 3′ untranslated region of the transferrin receptor showed that a 125-bp fragment of this region of the transferrin receptor mRNA containing three of the five IREs was sufficient for the iron-dependent regulation of the transferrin receptor.[5,14] Two of the five IREs of the transferrin receptor mRNA replaced the ferritin IRE in a translation assay in which these elements were placed in the 5′ untranslated region of a reporter gene.[76]

The key role of IREs present at the 3′ untranslated region in the control of transferrin receptor mRNA stability was directly supported by experiments based on deletion or mutation of the IREs present in this mRNA and on correlation of IRE-binding activity and transferrin receptor mRNA stability. Studies showed a direct correlation between the level of IRE-binding activity and the stability of transferrin receptor mRNA, thus suggesting that IRP, through interaction with the IREs present in the 3′ untranslated region of transferrin receptor mRNA, stabilizes this mRNA.[12]

A second line of evidence was based on mutational studies in which the transferrin receptor IREs were mutated to inactivate their capacity to bind IRP; these mutations produced uncontrolled synthesis of an unstable mRNA.[77]

The nucleases involved in the degradation of transferrin receptor mRNA, as well as the endonucleolytic cleavage and degradation pathways, remain to be determined. Studies of the human plasmocytoma cell line ARH-77 and of murine fibroblasts led to the identification in iron-treated cells of a shorter transferrin receptor mRNA that represents a degradation product of the full size mRNA.[78]

These shorter mRNAs result from endonuclease cleavage of transferrin receptor mRNA occurring at a single-stranded region near one of the iron-responsive elements contained in the 3′ untranslated region.[78] This 3′ endonuclease cleavage product is polyadenylated at the same extent as full-length mRNA, thus indicating that transferrin receptor mRNA does not require deadenylation prior to degradation.[78]

These observations suggest that IRP increases transferrin receptor mRNA stability through interaction with the IRE sequences located in its 3′ untranslated region; this binding masks the endonuclease cleavage sites of this mRNA and thus improves its stability.

Additional studies focused on evaluating the main metabolic steps required for transferrin receptor mRNA degradation. The need for transcriptional and translational events for iron-dependent degradation of transferrin receptor mRNA was investigated. In iron-deficient cells, about 95% of IRP co-localizes with transferrin receptor mRNA at the rough endoplasmic reticulum. In iron-loaded cells, the majority of IRP dissociates from transferrin receptor mRNA, and this event precedes the degradation of this mRNA.

Transcription and translation inhibitors completely suppressed iron-dependent degradation of transferrin receptor mRNA, but did not interfere with IRP dissociation from transferrin receptor mRNA.[79] These observations suggest the role of a *trans*-acting element involved in the regulation of transferrin receptor mRNA degradation. They also suggest that this *trans*-acting factor is encoded by a very unstable mRNA. Experiments with specific inhibitors of RNA polymerases II and III suggest that the latter is required for transcription of the *trans*-acting regulator of transferrin receptor mRNA stability.[79]

Some components of the biochemical machinery involved in transferrin receptor mRNA degradation pathway must be synthesized *de novo*. In this context, transfection experiments using a construct containing the 3′ untranslated region of the transferrin receptor mRNA without functional IREs and with a functional IRE introduced into the 5′ untranslated region showed that a marked translational repression of this chimeric RNA in iron-deprived cells did not cause significant change in mRNA levels.[80] This observation indicates that transferrin receptor mRNA decay does not require the association of this mRNA species with ribosomes.

More recently, a novel assay characterized the translation-dependent step during the iron-dependent decay of transferrin receptor mRNA. The 3′ untranslated region of the human transferrin receptor gene containing all five IREs and the rapid turnover determinant was linked to the coding sequence of mouse thymidine kinase cDNA.[81] In transfected cells, the expression of this reporter gene was modulated by iron and endogenous transferrin receptor. Experiments with selected mutants and with translation inhibitors provided evidence that the iron-dependent degradation of the reporter mRNA is uncoupled from ongoing translation.[81] These observations suggest that the inhibition of translation intereferes with a rate-limiting step of iron-induced degradation of transferrin receptor mRNA in a *trans*-acting mechanism by blocking IRP inactivation.

The mechanism of control of transferrin receptor mRNA stability by IRP must be considered in a more general context.[82] Recent studies have led to a better understanding of the mechanisms modulating mRNA degradation. The steady state of cytoplasmic mRNA is dependent on the rates of both mRNA synthesis and degradation. After transcription, the mRNA undergoes a series of processing events and is transported from the nucleus to the cytoplasm for its final translation. A unique property of the mRNA is its peculiar rate of turnover at each of these steps. Eukaryotic mRNA exhibits a wide range of half-lives and the control of their degradation rate is an important mechanism in the control of gene expression. Short-lived cytokine and cellular oncogene mRNAs exhibit half-lives of about 5 to 30 minutes. Other mRNAs display half-lives of several days. The half-life of transferrin receptor is less than 1 hour in iron-repleted cells and more than 24 hours in iron-depleted cells.

In general, stabilizing elements of mRNAs are the 5′ cap and 3′ polyadenylated tail, and they provide basal non-specific protection of mRNA from degradation mediated by exonucleases. The differential stability of mRNA is largely determined by some specific sequences at the 3′ untranslated region and by the interaction of RNA-binding protein with these sequences.

The 3′ untranslated regions of rapidly decaying mRNAs, such as those encoding many cytokines or proto-oncogenes like c-*fos* and c-*myc*, contain adenosine- or uridine-rich elements (AREs) characterized by multiple copies of the pentanucleotide AUUUA. These AREs direct asynchronous deadenylation, suggesting a progressive poly(A) digestion followed by mRNA decay. The mechanism by which AREs exert post-transcriptional control of gene expression appears to be mediated by their interactions with cytoplasmic and nuclease RNA-binding proteins, such as AUF1 and tris-tetraprolin.

Transferrin receptor mRNA does not contain AREs and its stability is controlled by a unique mechanism outlined above, and different from mechanisms observed for the majority of unstable mRNAs.

8.5 ROLE OF IRP IN THE CONTROL OF FERRITIN TRANSLATION

Recent studies have in part clarified the mechanism of mRNA translation in eukaryotic cells. Efforts have been made to understand translation initiation in eukaryotic cells because most of regulatory mechanisms, including the control of ferritin mRNA translation act at this level.[83,84]

The mechanism of initiation of translation in eukaryotic cells is unique in that the small 40S ribosomal subunit interacts not with the AUG codon, but with the 5' end of the mRNA. The 40S ribosomal subunit, carrying Met-tRNA-eIF2, GTP, and other factors, migrates through the 5' untranslated region until it encounters the AUG codon, which is recognized by base pairing with the anticodon in Met-tRNA. When a 60S large ribosomal subunit joins the 40S subunit, the selection of the start codon is then fixed.

Some studies have clarified the mechanisms responsible for IRE-dependent control of ferritin synthesis. Transfection experiments with different constructs provided evidence that the IRP-mediated block of ferritin is translation-dependent.[41,42] In fact, an IRE element placed more than 60 nucleotides downstream from the cap structure was not very efficient in mediating repression of ferritin translation. In contrast, an IRE element placed less than 60 nucleotides from the cap structure of the mRNA mediated efficient repression of ferritin tanslation.

To explore the translational regulation of ferritin synthesis by IRP–IRE interaction, a rabbit reticulocyte cell-free translation system was developed using *in vitro* transcribed, capped mRNA and recombinant IRP1.[85] It was possible to reconstitute the translation regulation of ferritin mRNA.[85] Subsequent studies based on sucrose gradient separation showed that the effect of IRP binding to ferritin IRE mRNA prevents the recruitment of the small 40S ribosome subunit to the ferritin mRNA.[86]

To better clarify the molecular mechanisms responsible for the translational repression of ferritin mRNA mediated by activated IRP, additional studies focused on evaluating the effect of IRP binding to ferritin IRE mRNA on the sequential binding of translation initiation factors to the mRNA. IRP1 did not block the assembly of the cap-binding complex eIF4F which includes factors eIF4E, eIF4G, and eIF4A. However, this assembly was not functional in that eIF4F complex assembled in the presence of a IRP–IRE complex was unable to interact with the small ribosomal subunit-associated eIF3.[87]

The capacity to prevent recruitment of the 43S translation preinitiation complex was analyzed for different IREs located at different positions downstream of the cap structure. IREs 34 or 64 nucleotides downstream of the cap can prevent the recruitment of the 43S translation reinitiation complex, while 43S recruitment to IRE was not impeded.[88]

Some effects related to the IRP–IRE interaction are observed also at the IRE.100 mRNA. After complex recruitment, the small ribosomal subunit appeared to be stalled as a consequence of the IRE–IRP complex on the IRE.100 mRNA. This finding indicates that IRE.100 does not prevent 43S initiation recruitment; it reduces translational efficiency in the presence of IRP.

8.6 EVOLUTION OF IRPs

Electromobility shift assays revealed specific IRP-IRE complexes in extracts prepared from mammals, fishes, flies, and frogs, but the complexes have not been reported in yeasts, bacteria, and plants.[89] However, subsequent studies have shown that plants, yeasts, and bacteria possess aconitase enzymes homologous to IRP.

Studies carried out in bacteria have shown the presence of aconitases that may play a role in iron metabolism and virulence. Studies of *Escherichia coli* found two aconitases: a stationary-phase aconitase (AcnA) induced by iron and oxidative stress, and an unstable aconitase (AcnB) synthesized during the exponential growth phase.[90] The existence of a third aconitase (AcnC) has been suggested.

AcnA and AcnB resemble the bifunctional iron regulatory proteins of vertebrates in that they exhibit alternative mRNA-binding and enzymatic activities. Sequence comparisons showed that AcnA is 53% identical to human IRP1 and 2 to 29% identical to mitochondrial aconitases, while AcnB is only 15 to 17% identical to AcnA, the IRPs, and mitochondria aconitase.[91]

Interestingly, a gel retardation assay showed that AcnA and AcnB are able to interact with the 3′ untranslated regions of AcnA and AcnB mRNA. No sequences exhibiting features of IREs for their primary or secondary structures were found in this mRNA region. However, three putative stem–loop structures were found in the 3′ untranslated regions of AcnA and AcnB mRNAs. Oxidative stress decreased the aconitase activity of AcnA and AcnB, but increased their RNA-binding activity, which led to an increase of AcnA and AcnB mRNA stability.[90] These observations suggest that the RNA-binding activity of *E. coli* aconitases is modulated by the absence or the presence of an iron–sulfur cluster, thus explaining the reciprocal relationship between the enzyme and RNA-binding activity.

Aconitase structure and function were also explored in *Bacillus subtilis*, which requires the enzyme for sporulation, an iron-dependent process. *B. subtilis* aconitase is an RNA-binding protein able to interact with the IRE sequences at the ferritin mRNA. Its aconitase activity is greatly potentiated when the bacteria are grown in the presence of iron salts. RNA-binding activity increases considerably in iron-deprived bacteria.[92]

IRE-like sequences were observed in the untranslated regions of two bacterial mRNAs: the 3′ *quoxD* gene and the operon of the *feu* genes[92] (Figure 8.2). The *quoxD* gene encodes a major iron-containing protein involved in electron transport, and *feuA* and *feuB* genes are involved in iron uptake.

These observations were confirmed in other bacteria, such as *Legionella pneumophila*. Aconitase is the major iron-containing protein of these bacteria. Aconitase is composed of 891 amino acids and exhibits great homology with human IRP1.[93] These observations suggest that prokaryotic aconitase is the ancestor of eukaryotic IRP1.

Aconitases are also found in plants that also contain cytoplasmic and mitochondrial enzymes; 50 to 90% of total aconitase activity resides in the cytosol and functions in the glyoxalate cycle. Plant aconitase was purified from melon seeds, its NH_2 amino acid sequence was determined, and this peptide sequence was used to isolate cDNA encoding the protein. Sequence analysis and comparison of this sequence with the protein sequences of mammalian aconitases produced interesting findings:

1. Plant aconitase is more closely related to IRP (sequence homology greater than 70%) than it is to mammalian mitochondrial aconitase (similarity of 43%).
2. The 23 amino acids forming the active site of IRP are conserved in plant aconitase.[94]

Similar observations were also made in a cytoplasmic aconitase of tobacco that was recently cloned. Its deduced amino acid sequence shared 61% identity with human IRP1.[95] Interestingly, the RNA-binding activity of tobacco aconitase is stimulated by nitric oxide.

Aconitase is present in parasites, such as *Trypanosoma brucei*, and its structure was recently determined. The gene of *Trypanosoma brucei* aconitase encodes a 98-kDa protein, localized at the cytosol and the mitochondria. This protein exhibits 74% sequence similarity to human IRP1. It functions in mitochondria in the citric cycle; in the cytosol, it acts as an RNA-binding protein.[96]

The capacity of *Trypanosoma brucei* to function in the mitochondria as an enzyme of the Krebs cycle is surprising in that this aconitase lacks an arginine of fundamental importance for the formation of the catalytic site of the enzyme. The arginine is replaced by a leucine residue.

These observations suggest that animal IRPs evolved from a multi-compartmentalized ancestral aconitase. It is of interest that *Trypanosoma brucei* aconitase does not act as a modulator of the transferrin receptors expressed in these protozoa.[97] This finding is probably related to the absence of IRE elements in the mRNAs encoding for the two subunits of *Trypanosoma brucei* transferrin receptor. However, it may also relate to the lack of similarity of trypanosomal and mammalian transferrin receptors.

An IRE-binding protein was observed also in extracts of *Leishmania terentolae*. It specifically interacts with mammalian IRE sequences but was unable to bind to IRE mutants that lost the capacity to interact with mammalian IRPs.[98] Interestingly, the IRE-binding activity of this protein was not increased following growth of the parasite in iron-deficient medium.

Particularly relevant to understanding the evolution of this protein were the studies of *Drosophila* iron regulatory protein. Electromobility shift assays provided evidence of specific complexes with mammalian IREs in extracts prepared from a *Drosophila melanogaster* cell line[89] and *Drosophila* embryos.[99]

Functional IRE-binding activities have been identified in the untranslated regions of the mRNA encoding *Drosophila* ferritin mRNA[100] and *Drosophila* succinate dehydrogenase, an enzyme of the citric acid cycle.[99,101]

IRPs have also been identified in *Drosophila*. A degenerate PCR-primer strategy revealed homologues of IRP1 in cells of *Drosophila melanogaster*. In contrast to vertebrates, *Drosophila* expresses two IRP homologues (*Drosophila* IRP1A and *Drosophila* IRP1B) in cells. These two proteins are homologous to IRP1, but are distinct from vertebrate IRP2.[102] IRP1A and IRP1B are co-expressed in all tissues.

Interestingly, antibodies raised against purified human IRP1 are able to interact with *Drosophila* IRPs. *Drosophila* apparently does not contain an IRP2 protein; it contains two IRP1 homologues. The reason for such a pattern of expression is unknown, but one suggestion is that the two *Drosophila* IRP1 homologues may bind to different RNA targets.

This observation led to some interesting speculation on the evolutionary changes of IRPs. It was suggested that duplication of the IRP gene occurred prior to the divergence of invertebrates and vertebrates. Gene duplication in invertebrates generates two IRP genes maintaining similar structures. The IRP gene in vertebrates evolved into IRP1 and IRP2 by insertion of a 73-amino acid exon.

Other interesting observations were made for *Drosophila* ferritin mRNA. Several mRNAs encode the same ferritin subunit of *Drosophila*, but only one possesses an IRE sequence in its 5' untranslated region, that can bind mammalian IRP1.[103]

Particularly interesting are studies of lampreys, vertebrates of ancient evolutionary origin. Two genetically distinct IRPs have been cloned; IRP1 and IRP2.[104] These two genes are 72% identical to each other and exhibit 74% sequence identity to their presumed homologues in mammals.[104]

Since lamprey IRP1 shares more sequence identity to human IRP1 than to human IRP2 and exhibits the conservation of amino acid residues required for aconitase activity, it may represent the lamprey homologue of mammalian IRP1.

On the other hand, lamprey IRP2 shares more identity with mammalian IRP2 than to mammalian IRP1, exhibits some substitutions at the amino acid residues required for aconitase activity, and may represent the lamprey homologue of mammalian IRP2. These propose that the control of iron metabolism based on two IRPs (IRP1 and IRP2) appeared in the early stages of vertebrate evolution, since lampreys are the only extant representatives of the early agnathan stage of vertebrate evolution.

These observations suggest a pattern of evolution of IRPs and aconitase. Two different types of genes must be considered: (1) the gene encoding mitochondrial aconitase; and (2) the two genes encoding cytoplasmic aconitases IRP1 and IRP2. Members of both families have been observed in invertebrates, such as *Drosophila melanogaster*, thus suggesting that the specialization of mitochondrial and cytoplasmic aconitases occurred early during evolution. Plants and some protozoa have a single, IRP-related aconitase, localized in both mitochondria and cytosol. In addition, bacterial aconitases are more related to mammalian IRPs than to mitochondrial aconitases. These observations suggest a two-branch evolutionary pattern for the aconitase gene family: one branch represented by animal IRPs, plant aconitases, bacterial aconitases, and protozoan aconitases; and the other represented by mitochondrial aconitases of yeasts and animals.[96] These findings suggest that the ancestor of IRP was an enzyme of mitochondrial energy metabolism acquired from a protobacterial endosymbiont.[96]

8.7 REGULATION OF IRP ACTIVITY BY NITRIC OXIDE

It is now clear that signals other than iron are able to modulate IRP1 and IRP2 activity and, through this mechanism, modulate iron metabolism.

Nitric oxide (NO) can interact with most transition metals and particularly both ferrous and ferric iron.[105] The capacity of NO to interact with iron dictates many of its main functions, such as the activation of guanylate cyclase which proceeds through the direct binding of NO to heme iron.[105] NO is also able to interact with iron–sulfur proteins.

Initial studies have shown that mitochondrial aconitase may represent a target of NO; in fact, NO synthesized by activated macrophages forms EPR-detectable nitrosyl–iron complexes, whose appearance is concomitant with the loss of mito-chondrial aconitase activity.[106] These observations suggest that NO targets the iron–sulfur center of mitochondrial aconitase, thus inhibiting enzyme activity. These observations prompted studies of the effects of NO on IRP activity and iron metab-olism. To demonstrate such effects in initial studies, murine macrophages were stimulated to produce NO through activation of NO synthase by treatment with interferon-γ or bacterial lipopolysaccharide.[107,108]

The induction of NO production in these cells stimulated IRP activity, with consequent repression of ferritin synthesis and increased transferrin receptor mRNA expression.[107,108] The effect of NO on IRP activity was not cell-specific in that it was observed in other cell types. Stimulation of NO synthesis in erythroleukemia K-562 cells elicited a significant rise of IRP activity, while inhibition of NO synthesis elicited an opposite effect.[108]

The relationship between NO and iron metabolism is more restricted in that iron levels modulate nitric oxide synthase activity. Studies of murine macrophage cell lines showed that iron deprivation augments the level of NO synthase activity induced by interferon-γ, while an opposite effect is elicited by iron loading.[109]

The modulation of IRP activity by NO under physiological conditions requires active NO synthase in the cells. This was clearly demonstrated by transfection of fibroblasts lacking endogenous NO synthase activity with the gene encoding murine-inducible NO synthase.[110] Induction of NO synthesis in these cells via activation of NO synthase activity stimulated IRE-binding activity.

Subsequent studies confirmed these initial observations.[111,112] They were based on the analysis of the effects of NO-releasing drugs on IRP activity in K-562 cells. These studies showed that among different molecular species, only nitrogen mon-oxide (and not its nitrosium ion) was able to stimulate IRP activity.[112]

Similar observations of neurons showed that both NO-releasing drugs and phys-iologic stimuli increased NO release. For example, glutamatergic stimulation with N-methyl-D-aspartate produced an increase in IRP binding to IRE sequences.[113]

NO-mediated activation of IRP activity controls hepatic iron metabolism during acute inflammation.[114] Liver cells exposed to inflammatory stimuli showed an increase in IRP activity that could be inhibited by a selective inhibitor of NO synthase.[114]

Two hypotheses have been proposed to explain the mechanism by which NO reacts with and modulates IRP activity. The first hypothesis suggests that IRP modulation by NO results from a reduced intracellular iron pool due to a direct interaction of NO with iron, thus leading to cellular iron deficiency. The second hypothesis suggests that NO directly interacts with IRP1, particularly with its iron–sulfur center.

The first hypothesis was initially supported only by indirect evidences: the observation that the NO-mediated increase in IRP1 RNA-binding activity occurred followed kinetics strictly resembling the kinetics of iron chelators.[115] However, a recent study provided direct evidence that NO may lower intracellular iron concentration. LMTK⁻ cells incubated with a NO-releasing drug showed a marked potentiation of iron release.[116]

The second hypothesis is supported two observations:[116] (1) the protection exerted against the stimulatory effect of NO on IRP activity by aconitase substrates able to bind to the Fe–S center; and (2) the rapidity of the effect of NO on IRP activity in cell-free systems and in whole cells under appropriate experimental conditions.

The experiments in whole cells were based on co-culture experiments. Adherent monolayers of a macrophage cell line were first induced to produce NO, and then covered with a non-adherent T cell line. Under these conditions, IRP1 activity was induced as early as 1 hour after the start of co-culture.[117]

A recent study suggests a possible mechanism through which NO activates IRP1 activity.[116] The effects of NO generators, such as S-nitroso-N-acetylpenicillamine and spermine-NONOate, on lysates derived from iron-loaded or iron-deprived cells were investigated. NO-releasing drugs increased IRP1 activity only in lysates derived from iron-loaded cells, while no stimulatory effect was observed in lysates derived from iron-deprived cells.[116] This finding suggests that NO directly interacts with 4Fe–S.

Several studies focused on possible differential effects of NO on IRP1 and IRP2 activity.[118,119] The investigations involved different cellular systems and different stimulators of NO release. Results were variable and, in some cases, contradictory. These studies have largely concerned murine cells because gel shift assays with cell extracts derived from murine species can distinguish the two different IRP forms. Studies in rat hepatoma cells showed that stimulation of NO production either by NO-releasing drugs or by IFN-γ elicited a significant rise of IRP–IRE binding activity, but no changes in IRP2 activity.[119]

The modulation of IRP1 and IRP2 activities in cells of macrophage lineage, particularly J774 and RAW 264.7 cell lines, was investigated. A study carried out in J774 cells showed that treatment of these cells with IFN-γ resulted in stimulation of NO production associated with a moderate increase in IRP1 activity and marked down-modulation of IRP2 activity.[120]

However, a subsequent study showed that the effect of NO on IRP1 and IRP2 activity is more complex than proposed in the previous study. NO increased both IRP1 and IRP2 activities, while IFN-γ decreased IRP2 activity through a NO-independent mechanism.[121]

Studies of J774 cells suggest that IRP1 and IRP2 are differentially modulated by NO, and that the effects of IFN-γ on macrophage iron metabolism are complex and mediated through NO-dependent and NO-independent mechanisms. Three studies focused on evaluating the modulation of IRP1 and IRP2 in the macrophage cell line 264.7.

In one study, lysates derived from this cell line were incubated with NO donors and subsequently exposed to glutathione or thioredoxin, a 12-kDa protein present in many species of plants and mammals that functions in cells as a major protein disulfide reductase. Both reducing species enhanced the RNA-binding activity of IRP1 exposed to NO donors. In contrast, an inhibition of IRP2 activity after exposure to NO *in vitro* was restored by the thioredoxin system, but not by GSH.[122]

In a second study, the effect of generators of different forms of NO (NO or NO+) on IRP1 and IRP2 activities was investigated. These studies showed that while NO generators elicited an increase in IRP1 activity, with no change in IRP2 activity, NO+ donors, such as sodium nitroprusside, do not modify IRP1 activity and elicited a marked decline of IRP2 activity, a phenomenon dependent on proteosome-dependent protein degradation.[123]

The NO+-mediated decrease of IRP2 activity was followed by a rapid decline of transferrin receptor mRNA levels and an increase in ferritin synthesis; both effects were inhibited by proteosome inhibitors.[123]

The third study analyzed the effects of IFN-γ and LPS on IRP1 and IRP2 expression in detail. The stimulation of RAW 264.7 cells with these agents induced a moderate increase of IRP1 activity associated with a marked decline of IRP2 activity due to proteolytic degradation.[124] Interestingly, these modulatory effects on IRP activity were associated with a pronounced decline of transferrin receptor mRNA levels. This phenomenon is prevented by inhibitors of inducible nitric oxide synthase.[124]

The effects of NO on IRP1 gene expression were recently investigated in detail. Surprisingly, studies of RAW 264.7 macrophage cell lines and primary murine macrophages showed that endogenous and exogenous NO exhibited differential effects on IRP1 activity and IRP1 synthesis.[125] In line with previous reports, an increase of NO production in these cells elicited a clear increase of IRP1 activity associated with a decrease of IRP1 content — a phenomenon seemingly related to decreased IRP1 synthesis due to a decrease of IRP1 mRNA levels.[125]

The reasons for this double effect of NO on IRP1 remain to be determined. It seems logical to assume the existence of a protective mechanism in that the activation of IRP1 activity by NO may last for several hours and may represent a harmful situation for cells. Prolonged IRP1 activation may lead to a complete inhibition of ferritin synthesis, thus abolishing the capacity of cells to chelate free iron and exposing them to the potential toxic effects of free iron. This danger is tempered by a concomitant decrease in IRP1 expression.

Additional studies have investigated the effect on IRP activity of peroxynitrite, a compound resulting from the reaction of NO with O_2^-. Peroxynitrite is a strong oxidant and its formation is observed in some cells, notably macrophages, under certain physiopathological conditions. One important biological effect of peroxynitrite is inhibition of mitochondrial respiration, a phenomenon mediated through inactivation of Fe–S cluster-containing enzymes. Given these properties, it is not surprising that peroxynitrite inactivates the enzymatic activities of both mitochondrial aconitase and IRP1.[126,127] However, despite its capacity to inactivate the aconitase activity of IRP1, peroxynitrite was unable to activate the IRE-binding activity of IRP1.[128] Peroxynitrite-treated IRP1 exhibits increased sensitivity to activation by suboptimal β-mercaptoethanol. This effect is prevented by aconitase substrate, thus suggesting that the Fe–S cluster is the target of peroxynitrite.[129] Furthermore, a single mutation of redox-active Cys 437 rendered the IRP1 mutant insensitive to oxidation by peroxynitrite. The effect of peroxynitrite extends also to IRP2, whose activity is inhibited by peroxynitrite.

8.8   REGULATION OF IRP BY OXIDATIVE STRESS

Several lines of evidence indicate an important biological link exists between iron metabolism and oxidative stress. Recent studies indicated that this link may be mediated in large part by a modulation of IRP activity exerted by oxidative radicals.

Initial studies showed that hydrogen peroxide (H_2O_2) activates IRP1 activity in living cells but not when added *in vitro*.[130,131] Incubation of fibroblasts in the presence of hydrogen peroxide elicited a repression of ferritin synthesis associated with an increase of transferrin receptor mRNA expression.[131]

These effects on ferritin and transferrin receptor expression prompted evaluation of the effect of hydrogen peroxide on IRP activity. IRP activity markedly increased after incubation of the cells for 90 minutes with hydrogen peroxide. This activation was accompanied by a concomitant loss of aconitase activity, thus suggesting that the effect may be due to disassembly of the Fe–S cluster. This conclusion was confirmed by the observation that citrate was unable to lock IRE-binding activity in cell extracts from hydrogen peroxide-treated cells.[131]

These initial studies showed that the effect of hydrogen peroxide on IRP cannot be direct, as suggested by two lines of evidence: (1) *in vitro* addition of hydrogen peroxide to cell extracts failed to activate IRP; and (2) the stimulatory effect of hydrogen peroxide on IRP activity was inhibited by okadaic acid, an inhibitor of type I and II phosphatases. This implies that phosphorylation is involved in this process.

Different results were obtained in liver cells exposed to phorone, a glutathione-depleting drug that alters the ratio of anti-oxidant and pro-oxidant molecules and enhances the biological effects induced by reactive oxygen species released during cellular metabolism.[132] Treatment of liver cells with this drug induced increased ferritin synthesis due to a moderate decrease of IRP activity and an increase of ferritin H and L gene transcription.[132]

It was hypothesized that the effect of phorone may be related to the oxidative stress it induces, as suggested also by its capacity to induce heme oxygenase, an oxidative stress-responsive gene. Finally, phorone induced an increase in the intracellular pool of free iron.[132] Because of the complexities of the effects of phorone, it is difficult to ascribe its effects on IRP directly to the generation of oxidative radicals.

Subsequent studies on liver cells confirmed these observations. Treatment of rat liver lysates with xanthine oxidase, an enzyme that generates the formation of both superoxide and hydrogen peroxide, caused a marked inhibition of IRP activity, reversed by β-mercaptoethanol addition.[133] This IRP inhibition was prevented by the addition of either superoxide dismutase or catalase, thus indicating that both super-oxide and hydrogen peroxide are involved in this phenomenon.[133]

Iron chelators and hydroxyl radical scavengers also prevented IRP inhibition by xanthine oxidase treatment.[133] These observations suggest that IRP inhibition by superoxide in liver cells may exert a protective effect against oxidative injury.

Additional studies provided better understanding of the mechanisms by which hydrogen peroxide modulates IRP activity. Hydrogen peroxide activates IRP through a mechanism different from that used by NO.[115] The stimulatory effect of hydrogen peroxide on IRP activity is rapid, requiring only 15 minutes of exposure of the cells to hydrogen peroxide. The effect is limited to IRP1 since IRP2 activity is apparently not affected after incubation of the cells with hydrogen peroxide.[115]

Another study addressed the important issue of the effects of extracellular and intracellular hydrogen peroxide on IRP activity. Concentrations of extracellular hydrogen peroxide as low as 10 μM are sufficient to induce IRP activity. IRP1 also responds differently to oxidative stress from extracellular addition of hydrogen peroxide, as compared with the intracellular accumulation of hydrogen peroxide in response to pharmacological stimulation.[134] These observations suggest that the stimulatory effects of hydrogen peroxide on IRP1 activity are not related to a direct chemical attack of hydrogen peroxide at the 4Fe–4S cluster, but are seemingly related to a signal induced by stimulation of the cells with extracellular hydrogen peroxide.

To better explain the cellular and biochemical mechanisms responsible for hydrogen peroxide-mediated IRP1 activation, a unique *in vitro* system was developed to reconstitute the response of IRP1 to extracellular hydrogen peroxide.[135] IRP1 in the cytosolic fraction failed to be directly activated upon hydrogen peroxide addition. The IRP1 activation also required a membrane-associated component.[135] The evaluation of different experimental conditions indicates that IRP1 activation by hydrogen peroxide requires energy supplied by ATP–GTP and involves at least one phosphorylation–dephosphorylation step.[135]

The *in vitro* effects of hydrogen peroxide on purified recombinant IRP1 were carefully analyzed. Addition of 100 μM hydrogen peroxide to purified IRP1 elicited a loss of aconitase activity. The loss was prevented by ferrous, but not by ferric ion addition.[136] However, *in vitro* hydrogen peroxide addition to purified IRP1 failed to activate its IRE-binding activity.

The aconitase activity of IRP1 is rapidly lost after incubation with hydrogen peroxide because the 4Fe–4S cluster is converted to the 3Fe–4S form by release of a single ferrous ion per molecule.[136] The protective effect of ferrous sulfate on IRP1 inactivation by hydrogen peroxide may be explained by the capacity to replenish the iron lost at the active site.

Other studies focused on the effects of other oxidative stress-inducing agents on IRP activity, particularly the effects of quinines. These compounds are pro-oxidants that are reduced to semiquinones and hydroquinones within the cells. These intermediary compounds react with O_2 and O_2^- to generate quinone, superoxide, and hydrogen peroxide.

Incubation of cells with menadione elicited a biphasic effect on IRP1 activity. An initial modest activation at 15 to 30 minutes was followed by a progressive inactivation occurring at 60 to 120 minutes involving both IRE binding and aconitase activity.[137] The loss of IRE binding and aconitase activity induced by menadione occurs through a mechanism independent of the Fe–S cluster switch, and does not involve activation of proteolytic degradation of IRP1.[137]

Recent studies evaluated the effect of hypoxia on IRP activity. Initial experiments in murine hepatoma cell lines showed that hypoxia elicited a three-fold decrease of IRP1 activity and no change in IRP2 activity. The IRP1 hypoxic inactivation was blocked by iron chelators, thus suggesting a role for iron in this phenomenon.[138] The inhibitory effect of hypoxia on IRP1 activity was fully reversed by reoxygenation. The inhibitory effect does not involve protein degradation, nor protein synthesis, thus indicating it is due to a change in the IRE-binding activity of IRP1.

These findings were confirmed in a study evaluating the effects of hypoxia on ferritin synthesis and IRP activity in mouse macrophage cells. Hypoxia elicited a decline of IRP1 activity associated with an increase of ferritin synthesis.[139]

Experiments on human hepatoma and erythroleukemia cell lines provided different results from those observed in murine cell lines. Exposure of human hepatoma Hep3B cells and human erythroleukemia K-562 cells elicited marked and moderate increases of IRP1–IRE activity, respectively.[140] This phenomenon was associated with a concomitant increase in transferrin receptor expression. However, the activation of IRP1 and stimulation of the transcription of transferrin receptor gene by HIF-1α both contribute to the enhanced transferrin receptor gene expression induced by hypoxia.[141]

The different effects of hypoxia on IRP–IRE binding activity (i.e., inhibition in murine cell lines and stimulation in human cell lines) may be related to the different animal origins of the cells used in these studies and to the different ratios of IRP1/IRP2 expression in these cells.

Additional experiments focused on the differential effects of hypoxia on both IRP1 and IRP2 activities. Hypoxia increased IRP2 RNA-binding activity by increasing IRP2 protein levels concomitant with a decrease in IRP1 RNA-binding activity.[142] Activation of IRP2 by hypoxia was not dependent on HIF or related to impaired proteasomal function.[142] These findings suggest that the differential regulation of IRP1 and IRP2 may contribute to the regulation of a specific RNA target whose function is required for response and/or adaption to hypoxia.[142]

In addition to nitric oxide and reactive oxygen species, iron may also affect IRP activity. Previous studies have shown that reactive iron species in biological fluids activate the iron–sulfur cluster of aconitase.[143] Recent studies suggest that some agents that modulate the iron–sulfur cluster also affect IRP activity. The exposure of skin fibroblasts to ultraviolet A radiation components of sunlight elicited the immediate release of free iron into the cells through proteolysis of ferritin.[144] Few minutes later, dose-dependent decreases of both IRP1 activity and ferritin synthesis were observed.[144]

A unique situation is observed with doxorubicin and its metabolites. A secondary alcohol metabolite called doxorubicinol can release low molecular weight iron from the 4Fe–4S cluster of IRP1, followed by the disassembly of cluster.[145] Surprisingly, doxorubicinol-dependent iron release and cluster disassembly eliminate aconitase activity and irreversibly abolish the capacity of IRP1 to function as an IRE-binding protein.[145] This effect of doxorubicin may play a role in the induction of cardiotoxicity induced by this drug.

8.9 ROLE OF PHOSPHORYLATION IN THE CONTROL OF IRP ACTIVITY

IRPs are targets of protein kinase C, and these phosphorylation events produce a modulation of IRE-dependent RNA binding activity.[146-148] Protein kinase C activators, such as phorbol esters, stimulated phosphate incorporation into both IRP1 and IRP2 in HL-60 cells. As a consequence of these phosphorylation events, both IRP proteins exhibited significant increases of IRE-binding capacity, associated with a concomitant increase of transferrin receptor mRNA expression.[147] IRP2 is more abundantly phosphorylated than IRP1, thus suggesting that the basal function requires phophorylation.[147] The stimulatory effect of phosphorylation on IRP2 activity seems to be related to a redox mechanism, with potential sites for this redox mechanism at some of the cysteine residues present in the 73-amino acid degradation motif of IRP2.

In vitro biochemical studies of the phosphorylation pattern of the entire IRP1 molecule and synthetic peptide fragments corresponding to some areas of the molecule and containing putative phosporylation sites showed that Ser 138 and Ser 711 sites are phosphorylated by protein kinase C.[146] These two sites are located at a strategic position of the IRP molecule: they are near the cleft involved in both RNA binding and aconitase activity.

Subsequent studies showed that purified IRP1 apoprotein (the form lacking the Fe–S cluster and exhibiting high affinity for binding to IRE) is a preferred substrate for protein kinase phosphorylation, as compared to IRP1 holoprotein (the form containing the 3Fe–4S or the 4Fe–4S cluster).

The presence of the Fe–S cluster determines structural alterations in IRP1 that reduce the accessibility of the PKC phosphorylation sites. Following phosphorylation, the affinity of IRP1 for binding to IRE sequences increases about 20-fold.[148] Interestingly, IRP1 phosphorylation by PKC was specifically inhibited by RNA species containing functional IREs. Finally, studies of limited proteolysis provided evidence that transition of IRP1 from the c-aconitase form to the IRP active form is associated with a conformational change favored by phosphorylation.[148]

To better understand the mechanism responsible for the increase of IRE binding activity by IRP1 elicited by phosphorylation, experiments on IRP1 mutants were performed. Mutations of the Ser 138 residues to either phosphomimetic glutamate or aspartate or to non-phosphorylable alanine produced IRP1 mutants endowed with normal IRE binding activity, reduced aconitase activity, and increased sensitivity to oxygen-dependent degradation of the 4Fe–4S cluster.[149]

All these observations suggest that IRP1 phosphorylation may represent a mechanism, independent of iron status, that favors IRP1 in its IRE-binding active form. This mechanism may be particularly active in cells, such as erythroid cells and macrophages, that, despite the high levels of iron to which they are exposed, maintain high levels of IRP1 activity.

8.10 ROLE OF HEME IN THE REGULATION OF IRP ACTIVITY

Initial studies suggest that heme may be responsible for modulation of IRP activity. Incubation of several types of cell lines in the presence of hemin resulted in inhibition of IRP binding to IRE comparable to that observed when the cells are grown in the presence of iron salts.[150] However subsequent studies have shown that the inhibitory effect of heme on IRP activity is due in large part to the iron released by enzymatic cleavage of the hemin entering the cells through the action of heme oxygenase.[151]

Heme *in vitro* can interact with IRP and, following this interaction, inhibits the binding of IRP to IRE.[152] However, the specificity of the interaction of hemin with IRP was questioned.[153]

While it seems unlikely that heme plays a significant role in the modulation of IRP activity under physiological conditions, recent observations suggest that heme, together with iron, may be involved in the degradation of IRP2.[154]

Finally, preliminary evidence indicates that δ-aminolevulinic acid activates IRP1 activity through a process involving oxidative stress.[155] These studies involved supplying exogenous δ-aminolevulinic acid to the cells or increasing the intracellular concentration of this heme precursor by incubating the cells in the presence of an inhibitor of aminolevulinic acid dehydradatase. The physiological role of δ-aminolevulinic acid remains to be determined.

8.11 IRPs COORDINATE *IN VIVO* IRON METABOLISM

As mentioned, several *in vitro* studies have shown that IRPs, by binding to the IRE sequences present in the different mRNA species, control virtually all the steps of iron metabolism, from intestinal iron absorption to iron uptake and storage. Thus, IRPs play a key role in iron homeostasis. A limited number of studies investigated whether the mechanisms responsible for *in vitro* the modulation of IRP activity also operate *in vivo*.

As discussed in the previous sections, multiple factors modulate the RNA-binding activity of IRP1 and IRP2: iron, hypoxia, oxidative stress, nitric oxide, and phosphorylation by protein kinase C. The most important factor from a physiological view is iron, so it was logical to evaluate the *in vivo* effects of perturbations of iron status on IRP activity and some of its target genes.

The effect of dietary iron intake on IRP activity was investigated in rats fed diets with different levels of iron content to determine conditions of normal iron intake and iron deficiency. Animals fed low iron levels showed significant increases in both IRP1 and IRP2 levels and marked declines of liver ferritin levels and mitochondrial aconitase activity.[156] Interestingly, an inverse relationship between blood hemoglobin levels and IRP activity was observed in these animals.[156] A second study involving kinetic analysis of IRP levels following dietary iron deprivation provided additional interesting information.[157] Liver IRP activity rapidly increased on day 1 in rats fed iron-deficient diets. The increases peaked at day 4 for IRP1 and at day 7 for IRP2. A marked decline of liver ferritin content and a significant decrease of mitochondrial aconitase were observed.[157] The effect of iron deficiency on mitochondrial aconitase activity is not surprising in that iron regulates mitochondrial aconitase synthesis in cultured cells through a translational mechanism dependent upon an IRE present in the 5' untranslated region of mitochondrial aconitase mRNA.[158]

Other evidence indicates that the IRP–IRE system controls expression of the major proteins involved in iron metabolism in intact cells. Two different experimental approaches were used. One study investigated the effect of the expression of a constitutively activated IRP1 mutant.[159] This mutant elicited a marked enhancement of transferrin receptor synthesis associated with an inhibition of ferritin expression.[159] Importantly, the cells expressing this constitutively activated IRP are unable to change the rates of synthesis of transferrin receptor and ferritin in response to changes in iron loading.

In a second study, an IRE was overexpressed into Caco-2 intestinal cells. The expression triggered a competition with cellular mRNAs possessing this structural element in their untranslated regions, leading to inhibition of transferrin receptor synthesis and stimulation of ferritin synthesis.[160]

8.12 TISSUE-DEPENDENT REGULATION OF IRP ACTIVITY

In addition to the mechanisms involved in modulation of IRP activity in virtually all cell types, some tissues exhibit unique mechanisms of control of IRP activity. IRP activity in these tissues is mainly modulated via extracellular signals dependent upon interaction of hormonal regulators with their membrane recptors. Thus, in liver cells IRP activity is modulated by thyroid hormones and that leads to a decrease of IRP activity. Thyroid hormone also potentiated the inhibitory effect of iron loading on IRP activity.[161] As a consequence of this reduced IRP activity, thyroid hormone increased ferritin synthesis in liver cells.

Interesting observations were also made about the modulation of IRP activity in different types of pituitary cells by different types of hormones and growth factors. The pituitary gland is an active secretory organ requiring iron to sustain its metabolism. The anterior pituitary gland is involved in the secretion of a series of hormones, including thyroid-stimulating hormone (TSH), whose secretion is stimulated by thyrotropin (TRH), and epidermal growth factor (EGF). TRH and EGF also exert a stimulatory effect on prolactin release by anterior pituitary cells and stimulate pituitary cell proliferation.

The effects of TRH and EGF on IRP activation in pituitary cells involved in TSH (thyrotrophs) and prolactin (lactotrophs) release were investigated. Interestingly, pituitary cells predominantly express IRP2. TRH and EGF stimulated IRP1 and IRP2 activity in thyrotrophs, and elicited inhibitory effects on IRP1 and IRP2 activity in lactotrophs.[162] The stimulatory effect in thyrotrophs involves PKC activation, while the inhibitory effect in lactotrophs implicates both mitogen-activated protein kinase and phosphatase activity.[162]

The effect of growth factor in mediating IRP modulation was demonstrated in several types of hemopoietic cells. In quiescent T lymphocyes, IRP activity was very low and was induced following proliferative activation with a mitogenic lectin. This phenomenon involves induction of the expression and activity of both IRP1 and IRP2.[163] The mechanisms responsible for IRP activation in these cells are unknown, but they are certainly complex, involving effects on mRNA and proteins.[164] Induction of IRP activity was also observed in T cell lines triggered to proliferate by interleukin-2.[165]

Similarly, monocytes freshly isolated from peripheral blood express low levels of IRP; expression is greatly stimulated when these cells are induced to maturate to macrophages by macrophage colony-stimulating factor (M-CSF).[163] The activation of IRP during the differentiation of monocytes to macrophages is accompanied by a marked increase of transferrin receptor and ferrtin expression.[163] The induction of transferrin receptor expression is an expected finding on the basis of the concomitant IRP activation. However, the marked increase in feritin content is an unexpected finding in view of the parallel increase in IRP activity. This last finding suggests the mechanisms of control of ferritin gene expression are peculiar to monocytes–macrophages, and implies a main effect at the level of transcription.

Evidence indicates that erythropoietin controls IRP activity in erythroid cells. Preliminary studies of erythroleukemia cell lines have shown that erythropoietin addition stimulates IRP activity.[166] Additional studies explored IRP1 and IRP2 expression in cultures of primary erythroblasts. IRP1 expression is predominant over IRP2 expression in these cells and IRP1 is to a large extent constitutively active in these cells.[164]

A cell culture system allowing the selective growth of cells pertaining to the different hemopoietic lineages starting from undifferentiated CD34+ hemopoietic progenitors demonstrated that IRP1 is preferentially expressed at mRNA and protein levels in cells of erythroid lineage, as compared to its expression in other hemopoietic lineages, including granulocytic, monocytic, and megakaryocytic cells.

This finding implies a unique mechanism of IRP1 gene regulation involving its preferential expression in erythroid cells through mechanisms dependent on the control of both mRNA expression and IRE-binding activity.[164] This elevated IRP1 activity observed in erythroid cells greatly contributes to maintaining the elevated levels of transferrin receptor expression observed in these cells.

On the other hand, a recent study suggests that IRP2 expression in many cell types may be transcriptionally controlled through the nuclear protein c-*myc*. The c-*myc* proto-oncogene is a nuclear protein that functions as a transcriptional factor forming a heterodimer with *MAX*, another nuclear protein. The *myc/MAX* heterodimer regulates the transcriptional activities of target genes, mainly genes involved in the control of cell proliferation. In target cells, the *myc/MAX* complex may act as a stimulator or a repressor of transcription. Overexpression of c-*myc* induces a decrease of ferritin H transcription, associated with an increased expression of IRP2 mRNA. No changes in transferrin receptor and ferritin chain L mRNA levels were induced.[167] The effect of c-*myc* on ferritin H and IRP2 expression is related to an effect on gene transcription. The stimulation of IRP2 gene transcription by c-*myc* may be responsible for the induction of the expression of this gene observed during the induction of T lymphocyte proliferation and monocyte-to-macrophage maturation.

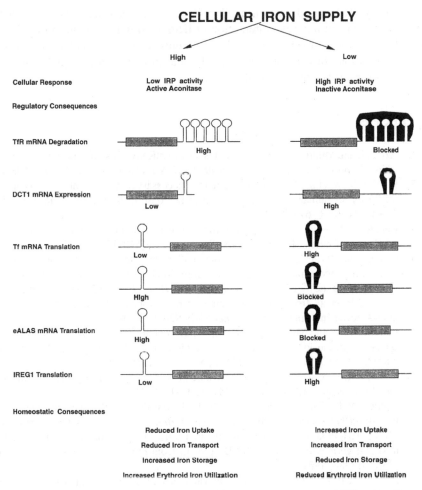

FIGURE 8.7 Coordinated control of iron metabolism by IRPs. The effects on the expression of proteins involved in iron absorption, uptake, and storage are shown. The activity of IRPs is modulated according to the level of iron loading.

8.13 CONCLUSIONS

IRPs play a key role in the control of the expression of the proteins involved in iron metabolism. Particularly, they orchestrate a coordinated pattern of expression of these proteins. As outlined in Figure 8.7, IRPs modulate the expression of the proteins involved in all steps of iron metabolism: proteins involved in iron absorption in the duodenal cells, e.g., the basolateral membrane transporter IREG-1; proteins implicated in cellular iron processing, such as Nramp2; proteins involved in plasmatic iron transport (transferrin) and membrane transport (transferrin receptor); and proteins involved in iron storage (ferritin). The expression of all these proteins is modulated by IRPs based on iron availability through post-transcriptional mechanisms and dependent upon the interaction of IRPs with IREs present in the 3′ or 5′ untranslated regions of the mRNAs of these proteins. These coordinated mechanisms of gene regulation allow organisms to finely tune iron absorption, uptake, storage, and excretion rates to physiological needs.

REFERENCES

1. Aziz, N. and Munro, H.N. Iron regulates ferritin mRNA translation through a segment of its 5′ untranslated region, *Proc. Natl. Acad. Sci. U.S.A.,* 84: 8478–8482, 1987.
2. Hentze, M.W. et al. A *cis*-acting element is necessary and sufficient for translational regulation of human ferritin expression in response to iron, *Proc. Natl. Acad. Sci. U.S.A.,* 84: 6730–6734, 1987.
3. Hentze, M.W. et al. Identification of the iron-responsive element for the translational regulation of human ferritin mRNA, *Science,* 238: 1570–1572, 1987.
4. Caughman, S.W. et al. The iron-responsive element is the single element responsible for iron-dependent translation regulation of ferritin biosynthesis. Evidence for function as the binding site for a translational repressor, *J. Biol. Chem.,* 263: 19048–19053, 1988.
5. Casey, J.L. et al. Iron-responive elements: regulatory RNA sequences that control mRNA levels and translation, *Science,* 240: 924–928, 1988.
6. Leibold, E.A. and Munro, H.N. Cytoplasmic protein binds *in vitro* to a highly conserved sequence in the 5′ untranslated region of ferritin heavy- and light-subunit mRNAs, *Proc. Natl. Acad. Sci. U.S.A.,* 85: 2171–2175, 1988.
7. Rouault, T.A. et al. Binding of a cytosolic protein to the iron-responsive element of ferritin messenger RNA, *Science,* 241: 1207–1210, 1988.
8. Walden, W.E., Patino, M.M., and Gaffield, L. Purification of a specific repressor of ferritin mRNA translation from rabbit liver, *J. Biol. Chem.,* 264: 13765–13769, 1989.
9. Brown, P.H. et al. Requirements for the translational repression of ferritin transcripts in wheat germ extracts by a 90-kDa protein from rabbit liver, *J. Biol. Chem.,* 264: 13383–13386, 1989.
10. Hentze, M.W. et al. Oxidation–reduction and the molecular mechanism of a regulatory RNA–protein interaction, *Science,* 244: 357–359, 1989.
11. Haile, D.J. et al. Regulation of interaction of the iron-responsive element binding protein with iron-responsive elements, *Mol. Cell. Biol.,* 9: 5055–5061, 1989.
12. Mullner, E.W., Neupert, B., and Kuhn, L.C. A specific mRNA binding factor regulates the iron-dependent stability of cytoplasmic transferrin receptor mRNA, *Cell* 58: 373–382, 1989.
13. Owen, D. and Kuhn, L.C. Non-coding sequences of the transferrin receptor gene are required for mRNA regulation by iron, *EMBO J.,* 6: 1287–1293, 1987.
14. Mullner, E.W. and Kuh, L.C. A stem-loop in the 3′ untranslated region mediates iron-dependent regulation of transferrin receptor mRNA stability in the cytoplasm, *Cell,* 58: 815–825, 1988.
15. Barton, H.A. et al. Determinant of the interaction between the iron-responsive element-binding protein and its binding site in rat L-ferritin mRNA, *J. Biol. Chem.,* 265: 7000–7008, 1990.
16. Rouault, T.A. et al. Cloning of the cDNA encoding an RNA regulatory protein — the human iron-responsive element-binding protein, *Proc. Natl. Acad. Sci. U.S.A.,* 87: 7958–7962, 1990.

17. Rouault, T.A. et al. Structural relationhip between an iron-regulated RNA-binding protein (IRE-BP) and aconitase. Functional implications, *Cell*, 64: 881–883, 1991.
18. Hentze, M.W. and Argos, P. Homology between IRE-BP, a regulatory RNA-binding protein, aconitase and isopropylmalate isomerase, *Nucleic Acid Res.*, 19: 1739–1740, 1991.
19. Huang, T.S. et al. An atypical iron responsive element (IRE) within crayfish ferritin mRNA and an iron regulatory protein 1 (IRP1)-like protein from crayfish hepatopancreas, *Insect Biochem. Mol. Biol.*, 29: 1–9, 1999.
20. Alén, C. and Soneshein, A.L. *Bacillus subtilis* aconitase is an RNA-binding protein, *Proc. Natl. Acad. Sci. U.S.A.*, 96: 10412–10417, 1999.
21. Kohler, S.A., Menotti, E., and Kuhn, L.C. Molecular cloning of mouse glycolate oxidase, *J. Biol. Chem.*, 274: 2401–2407, 1999.
22. Harrel, M.C. et al. Ferritin mRNA: interactions of iron regulatory element with translational regulator protein P-90 and the effect on base-paired flanking regions, *Proc. Natl. Acad. Sci. U.S.A.*, 88: 4166–4170, 1991.
23. Bettany, A.J., Eisenstein, R.S., and Munro, H.N. Mutagenesis of the iron regulatory element further defines a role for RNA secondary structure in the regulation of ferritin and transferrin receptor expression, *J. Biol. Chem.*, 267: 16531–16537, 1992.
24. Jaffrey, S.R. et al. The interaction between the iron-responsive element binding protein and its cognate RNA is highly dependent upon both RNA sequence and structure, *Nucleic Acid Res.*, 21: 4627–4631, 1993.
25. Henderson, B.R. et al. Optimal sequence and structure of iron-responsive elements, *J. Biol. Chem.*, 269: 17481–17489, 1994.
26. Wang, J.H., Sezekan, J.R., and Theil, E.C. Structure of the 5′ untranslated regulatory region of ferritin mRNA studied in solution, *Nucleic Acid Res.*, 18: 4463–4470, 1990.
27. Sierrputowska-Gracz, H., McKenzie, R.A., and Theil, E.C. The importance of a single G in the hairpin loop of the iron responsive element (IRE) in ferritin mRNA structure: an NMR spectroscopy study, *Nucleic Acid Res.*, 23: 146–153, 1995.
28. Laing, L.G. and Hall, K.B. A model of the iron responsive element RNA hairpin loop structure determined from NMR and thermodynamic data, *Biochemistry*, 35: 13586–13596, 1996.
29. Addess, K.J. et al. Structure and dynamics of the iron responsive element RNA: implications for binding of the RNA by iron regulatory binding proteins, *J. Mol. Biol.*, 274: 72–83, 1997.
30. Henderson, B.R., Menotti, E., and Kuhn, L.C. Iron regulatory proteins 1 and 2 bind distinct sets of RNA target sequences, *J. Biol. Chem.*, 271: 4900–4908, 1996.
31. Butt, J. et al. Differences in the RNA binding sites of iron regulatory proteins and potential target diversity, *Proc. Natl. Acad. Sci. U.S.A.*, 93: 4345–4349, 1996.
32. Menotti, E., Henderson, B.R., and Kuhn, L.C. Translational regulation of mRNAs with distinct IRE sequences by iron regulatory proteins 1 and 2, *J. Biol. Chem.*, 273: 1821–1824, 1998.
33. Ke, Y. et al. Loops and bulge/loops in iron-responsive element isoforms influence iron regulatory protein binding, *J. Biol. Chem.*, 273: 23637–23640, 1998.
34. Beaumont, C. et al. Mutation in the iron responsive element of the L ferritin mRNA in a family with dominant hyperferritinemia and cataract, *Nature Genet.*, 11: 444–446, 1995.
35. Girelli, D. et al. Molecular basis for the recently described hereditary hyperferritinemia–cataract syndrome: a mutation in the iron-responsive element of ferritin L-subunit gene, *Blood*, 90: 4050–4053, 1995.

36. Cazzola, M. et al. Hereditary hyperferritinemia–cataract syndrome: relationship between phenotypes and specific mutations in the iron responsive element of ferritin light-chain mRNA, *Blood,* 90: 814–821, 1997.

37. Arnold, J.D. et al. Hyperferritinaemia in the absence of iron overload, *Gut,* 41: 408–410, 1997.

38. Mumford, A.D. et al. Hereditary hyperferritinemia–cataract syndrome: two novel mutations in the L-ferritin iron-responsive element, *Blood,* 91: 367–368, 1998.

39. Girelli, D. et al. Hereditary hyperferritenemia–cataract syndrome caused by a 29-base pair deletion in the iron responsive element of ferritin L-subunit gene, *Blood,* 10: 2084–2088, 1997.

40. Allenson, C.R., Cazzola, M., and Rouault, T.A. Clinical severity and thermodynamic effects of iron-responsive element mutations in hereditary hyperferritinemia–cataract syndrome, *J. Biol. Chem.,* 274: 26439–26447, 1999.

41. Goossens, B. et al. Translational repression by a complex between the iron-responsive element of ferritin mRNA and its specific cytoplasmic binding protein is position-dependent *in vivo, EMBO J.,* 12: 4127–4133, 1990.

42. Goossens, B. and Hentze, M.W. Position is the critical determinant for function of iron-responsive elements as translational regulators, *Mol. Cell. Biol.,* 12: 1959–1966, 1992.

43. Rouault, T.A. et al. The iron-responsive element binding protein: a method for the affinity purification of a regulatory RNA-binding protein, *Proc. Natl. Acad. Sci. U.S.A.,* 86: 5768–5772, 1989.

44. Neupert, B. et al. A high yeld affinity purification method for specific RNA-binding proteins: isolation of the iron regulatory factor from human placenta, *Nucleic Acid Res.,* 18: 51–64, 1990.

45. Hirling, H. et al. Expression of active iron regulatory factor from a full-length human cDNA by *in vitro* transcription/translation, *Nucleic Acid Res.,* 20: 33–39, 1992.

46. Patino, M.M. and Walden, W.E. Cloning of a functional cDNA for the rabbit ferritin mRNA repressor protein. Demonstration of a tissue-specific pattern of expression, *J. Biol. Chem.,* 267: 19011–19016, 1992.

47. Robbins, A.H. and Stout, C.D. The structure of aconitase, *Proteins,* 5: 289–312, 1989.

48. Lauble, H., Kennedy, M.C., and Beinert, H. Crystal structure of aconitase with isocitrate and nitroisocitrate bound, *Biochemistry,* 31: 2735–2748, 1992.

49. Goosdell, D.S. et al. Automated docking in crystallography: analysis of the substrates of aconitase, *Proteins,* 17: 1–10, 1993.

50. Kaptain, S. et al. A regulated RNA binding protein also possesses aconitase activity, *Proc. Natl. Acad. Sci. U.S.A.,* 88: 10109–10113, 1991.

51. Tang, C.K. et al. Iron regulates the activity of the iron-responsive element binding protein without changing its rate of synthesis or degradation, *J. Biol. Chem.,* 267: 24466–24470, 1992.

52. Haile, D.J. Regulation of genes of iron metabolism by the iron-response proteins, *Am. J. Med. Sci.,* 318: 230–240, 1999.

53. Constable, A. et al. Modulation of the RNA-binding activity of a regulatory protein by iron *in vitro*: switching between enzymatic and genetic function? *Proc. Natl. Acad. Sci. U.S.A.,* 89: 4554–4558, 1992.

54. Philippot, C.C. et al. Modification of a free Fe–S cluster cysteine residue in the active iron-responsive element-binding protein prevents RNA binding, *J. Biol. Chem.,* 268: 17655–17658, 1993.

55. Hirling, H., Hnderson, B.R., and Kuhn, L.C. Mutational analysis of the [4Fe-4S]-cluster converting iron regulatory protein from its RNA-binding form to cytoplasmic aconitase, *EMBO J.,* 13: 453–461, 1994.

56. Philippot, C.C., Klausner, R.D., and Rouault, T.A. The bifunctional iron-responsive element binding protein/cytosolic aconitase: the role of active-site residues in ligand binding and regulation, *Proc. Natl. Acad. Sci. U.S.A.,* 91: 7321–7325, 1994.

57. Haile, D.J. et al. Cellular regulation of the iron-responsive element binding protein: disassembly of the iron–sulfur cluster results in high-affinity RNA binding, *Proc. Natl. Acad. Sci. U.S.A.,* 89: 11735–11739, 1992.

58. Basilon, J.P. et al. The iron-responsive element-binding protein: localization of the RNA-binding site to the aconitase active-site cleft, *Proc. Natl. Acad. Sci. U.S.A.,* 91: 574–578, 1994.

59. Neupert, B., Menotti, E., and Kuhn, L.C. A novel method to identify nucleic acid binding site in proteins by scanning mutagenesis: application to iron regulatory protein, *Nucleic Acid Res.,* 23: 2579–2583, 1995.

60. Kaldy, P. et al. Identification of the RNA-binding surfaces in iron regulatory protein-1, *EMBO J.,* 8: 6073–6083, 1999.

61. Schalinske, L. et al. The iron–sulfur cluster of the iron-regulatory protein 1 modulates the accessibility of RNA binding and phosphorylation sites, *Biochemistry,* 36: 39503958, 1997.

62. Geogout, V. et al. Ligand-induced structural aterations in human iron regulatory protein-1 revealed by protein footprinting, *J. Biol. Chem.,* 274: 15052–15058, 1999.

63. Narahari, J. et al. The aconitase function of iron regulatory protein-1, *J. Biol. Chem.,* 275: 16227–16234, 2000.

64. Festa, M. et al. Oxalamate, a competitive inhibitor of aconitase, modulates the RNA binding activity of iron-regulatory proteins, *Biochem. J.,* 348: 315–320, 2000.

65. Henderson, B.R., Seiser, C., and Kuhn, L.C. Characterization of a second RNA-binding protein in rodents with specificity for iron-responsive elements, *J. Biol. Chem.,* 268: 27327–27334, 1993.

66. Guo, B., Yu, Y., and Leibold, E.A. Iron regulates cytoplasmic levels of a novel iron-responsive element-binding protein without aconitase activity, *J. Biol. Chem.,* 269: 24252–24260, 1994.

67. Samaniego, F. et al. Molecular characterization of a second iron-responsive element binding protein, iron regulatory protein 2, *J. Biol. Chem.,* 269: 30904–30910, 1994.

68. Guo, B. et al. Characterization and expression of iron regulatory protein 2 (IRP2), *J. Biol. Chem.,* 270: 16529–16535, 1995.

69. Henderson, B.R. and Kuhn, L.C. Differential modulation of the RNA-binding proteins IRP-1 and IRP-2 in response to iron, *J. Biol. Chem.,* 270: 20509–20515, 1995.

70. Pantopoulos, K., Gray, N.K., and Hentze, M.W. Differential regulation of two related RNA-binding proteins, iron regulatory protein (IRP) and IRPb, *RNA,* 1: 155–163, 1995.

71. Guo, B. et al. Iron regulates the intracellular degradation of iron regulatory protein 2 by the proteasome, *J. Biol. Chem.,* 270: 21645–21651, 1995.

72. Iwai, K., Klausner, R.D., and Rouault, T.A. Requirements for iron-regulated degradation of the RNA binding protein, iron regulatory protein 2, *EMBO J.,* 14: 5350–5357, 1995.

73. Iwai, K. et al. Iron-dependent oxidation, ubiquitination, and degradation of iron regulatory protein 2: implications for degradation of oxided proteins, *Proc. Natl. Acad. Sci. U.S.A.,* 95: 4924–4928, 1998.

74. Horowitz, J.A. and Harford, J.B. The secondary structure of the iron regulatory region of the transferrin receptor mRNA deduced by enzymatic cleavage, *New Biol.,* 4: 330–338, 1992.

75. Bertrand, E. et al. Visualization of the interaction of Fe regulatory protein with RNA *in vivo, Proc. Natl. Acad. Sci. U.S.A.,* 90: 3496–3500, 1993.

76. Koeller, D.M., Casey, J.L., and Hentze, M.W. A cytosolic protein binds to structural elements within the iron regulatory region of the transferrin receptor mRNA, *Proc. Natl. Acad. Sci. U.S.A.,* 86: 3574–3578, 1989.

77. Casey, J.L. et al. Iron regulation of transferrin receptor mRNA levels requires iron-responsive elements and a rapid turnover determinant in the 3′ untranslated region of the mRNA, *EMBO J.,* 8: 3693–3699, 1989.

78. Binde, R. et al. Evidence that the pathway of transferrin receptor mRNA degradation involves an endonucleolytic cleavage within the 3′ UTR and does not involve poly(A) tail shortening, *EMBO J.,* 13: 1969–1980, 1994.

79. Seiser, C. et al. Effect of transcription inhibitors on the iron-dependent degradation of transferrin receptor mRNA, *J. Biol. Chem.,* 270: 29400–29406, 1995.

80. Koeller, D.M., Horowitz, J.A., and Casey, J.L. Translation and the stability of mRNAs encoding the transferrin receptor and c-*fos,* *Proc. Natl. Acad. Sci. U.S.A.,* 88: 7778–7782, 1991.

81. Posch, M. et al. Characterization of the translation-dependent step during iron-regulated decay of transferrin receptor mRNA, *J. Biol. Chem.,* 274: 16611–16618, 1999.

82. Sachs, A.B., Sarnow, P., and Hentze, M.W. Starting at the beginning, middle and end: translation initiation in eukaryotes, *Cell,* 89: 831–838, 1997.

83. Kozak, M. Initiation of translation in prokaryotes and eukaryotes, *Gene,* 234: 187–208, 1999.

84. Preiss, T. and Hentze, M.W. From factors to mechanisms: translation and translational control in eukaryotes, *Curr. Opin. Genet. Devel.,* 9: 515–521, 1999.

85. Gray, N.K. et al. Recombinant iron regulatory factor functions as an iron-responsive element-binding protein, a translation repressor and an aconitase. A functional assay for translational repression and direct demonstration of the iron switch, *Eur. J. Biochem.,* 218: 657–667, 1993.

86. Gray, N.K. and Hentze, M.W. Iron regulatory protein prevents binding of the 43S translation pre-initiation complex to ferritin and eALAS mRNAs, *EMBO J.,* 13: 3882–3891, 1994.

87. Muckenthaler, M., Gray, N.K., and Hentze, M.W. IRP-1 binding to ferritin mRNA prevents the recruitment of the small ribosomal subunit by the cap-binding complex eIF4, *Molecular Cell,* 2: 383–388, 1998.

88. Paraskeva, E. et al. Ribosomal pausing and scanning arrest as mechanisms of translational regulation from cap-distal iron-responsive element, *Mol. Cell. Biol.,* 19: 807–816, 1999.

89. Rothenbeyer, S., Mullner, E.W., and Kuhn, L.C. The mRNA binding protein which controls ferritin and transferrin receptor expression is conserved during evolution, *Nucleic Acid Res.,* 18: 1175–1179, 1990.

90. Tang, Y. and Guest, J.R. Direct evidence for mRNA binding and post-transcriptional regulation by *Escherichia coli* aconitases, *Microbiology,* 145: 3069–3079, 1999.

91. Prodromou, C., Artymiuk, P.J., and Guest, J.R. The aconitase of *Escherichia coli.* Nucleotide sequence of the aconitase gene and amino acid sequence similarity with mitochondrial aconitases, the iron-responsive-element-binding protein isopropyl-malate isomerases, *Eur. J. Biochem.,* 204: 599–609, 1992.

92. Alén, C. and Sonenshein, A.L. *Bacillus subtilis* aconitase is an RNA-binding protein, *Proc. Natl. Acad. Sci. U.S.A.,* 16: 10412–10417, 1999.

93. Megnaud, J.M. and Horwitz, M.A. The major iron-containing protein of *Legionella pneumophila* is an aconitase homologous with the human iron-responsive element-binding protein, *J. Bacteriol.,* 175: 5666–5676, 1993.

94. Peyret, P., Perez, P., and Alric, M. Structure, genomic organization, and expression of the *Arabidopsis thaliana* aconitase gene, *J. Biol. Chem.,* 270: 8131–8137, 1995.

95. Navarre, D.A. et al. Nitric oxide modulates the activity of tobacco aconitase, *Plant Physiol.,* 122: 573–582, 2000.

96. Saas, J. et al. A developmentally regulated aconitase related to iron-regulatory protein-1 is localized in the cytoplasm and in the mitochondrion of *Trypanosoma brucei,* *J. Biol. Chem.,* 275: 2745–2755, 2000.

97. Fast, B. et al. Iron-dependent regulation of transferrin receptor expression in *Trypanosoma brucei, Biochem. J.,* 342: 691–696, 1999.

98. Meehm, H.A., Lundberg, R.A., and Connel, C.J. A trypanosomatid protein specifically interacts with a mammalian iron-responsive element, *Parasitol. Res.,* 86: 109–114, 2000.

99. Gray, N.K. et al. Translational regulation of mammalian and *Drosophila* citric acid cycle enzymes via iron-responsive elements, *Proc. Natl. Acad. Sci. U.S.A.,* 93: 4925–4930, 1996.

100. Charlesworth, A. et al. Isolation and properties of *Drosophila melanogaster* ferritin, *Eur. J. Biochem.,* 247: 470–475, 1997.

101. Kohler, S.A., Henderson, B.R., and Kuhn, L.C. Succinate dehydrogenase B mRNA of *Drosophila melanogaster* has a functional iron-responsive element in its 5′ untranslated region, *J. Biol. Chem.,* 270: 30781–30786, 1995.

102. Muckenthaler, M. et al. Iron-regulatory protein-1 (IRP-1) is highly conserved in two invertebrate species, *Eur. J. Biochem.,* 254: 230–237, 1998.

103. Lind, M.I. et al. *Drosophila* ferritin mRNA: alternative RNA splicing regulates the presence of the iron-responsive element, *FEBS Lett.,* 436: 476–482, 1998.

104. Andersen, O. et al. Regulation of iron metabolism in the sanguivore lamprey *Lampetra fluviatilis*: molecular cloning of two ferritin subunits and two iron regulatory proteins (IRPs) reveals evolutionary conservation of the iron regulatory element (IRE)/IRP regulatory system, *Eur. J. Biochem.,* 254: 223–229, 1998.

105. Cooper, C.E. Nitric oxide and iron proteins, *Biochim. Biophys. Acta,* 1411: 290–309, 1999.

106. Drapier, J., Pellat, C., and Henry, Y. Generation of EPR-detectable nitrosyl iron complexes in tumor target cells co-cultured with activated macrophages, *J. Biol. Chem.,* 266: 10162–10167, 1991.

107. Drapier, J.C. et al. Biosynthesis of nitric oxide activates iron regulatory factor in macrophages, *EMBO J.,* 12: 3643–3649, 1993.

108. Weiss, G. et al. Translational regulation via iron-responsive elements by the nitric oxide/NO-synthase pathway, *EMBO J.,* 12: 3651–3657, 1993.

109. Weiss, G. et al. Iron reduces nitric oxide synthase activity by controlling nuclear transcription, *J. Exp. Med.,* 180: 969–976, 1994.

110. Pantopoulos, K. and Hentze, M.W. Nitric oxide signaling to iron-regulatory protein: direct control of ferritin mRNA translation and transferrin receptor mRNA stability in transfected fibroblasts, *Proc. Natl. Acad. Sci. U.S.A.,* 92: 1267–1271, 1995.

111. Oria, R. et al. Effect of nitric oxide on expression of transferrin receptor and ferritin and on cellular iron metabolism in K562 human erythroleukemia cells, *Blood,* 85: 2962–2966, 1995.

112. Richadson, D.R. et al. The effect of redox-related species of nitrogen monoxide on transferrin and iron uptake and cellular proliferation of erythroleukemia (K562) cells, *Blood,* 86: 321–3219, 1995.

113. Jaffrey, S.R. et al. The iron-responsive element binding protein: a target for synaptic actions of nitric oxide, *Proc. Natl. Acad. Sci. U.S.A.,* 91: 12994–12998, 1994.

114. Cairo, G. and Pietrangelo, A. Nitric-oxide-mediated activation of iron-regulatory protein controls hepatic iron metabolism during acute inflammation, *Eur. J. Biochem.,* 232: 358–363, 1995.

115. Pantopoulos, K., Weiss, G., and Hentze, M.W. Nitric oxide and oxidative stress control mammalian iron metabolism by different pathways, *Mol. Cell. Biol.,* 16: 3781–3788, 1996.

116. Wandrop, S.L., Watts, R.N., and Richardson, D.R. Nitrogen monoxide activates iron regulatory protein 1 RNA-binding activity by two possible mechanisms: effect on the [4Fe–4S] cluster and iron mobilization from cells, *Biochemistry,* 39: 2748–2758, 2000.

117. Bouton, C. Nitrosative and oxidative modulation of iron regulatory proteins, *Cell Mol. Life Sci.,* 55: 1043–1053, 1999.

118. Bouton, C., Oliveira, L., and Drapier, J.C. Converse modulation of IRP1 and IRP2 by immunological stimuli in murine RAW 264.7 macrophages, *J. Biol. Chem.,* 273: 9403–9408, 1998.

119. Philips, J.D. et al. Differential regulation of IRP1 and IRP2 by nitric oxide in rat hepatoma cells, *Blood,* 87: 2983–2992, 1996.

120. Recalcati, S. et al. Nitric oxide-mediated induction of ferritin synthesis in J774 macrophages by inflammatory cytokines: role of selective iron regulatory protein-2 downregulation, *Blood,* 91: 1059–1066, 1998.

121. Mulero, V. and Brock, J. Regulation of iron metabolism in murine J774 macrophages: role of nitric oxide-dependent and independent pathways following activation with gamma interferon and lipopolysaccharide, *Blood,* 94: 2383–238, 1999.

122. Oliveira, L., Bouton, C., and Drapier, J.C. Thioredoxin activation of iron regulatory proteins, *J. Biol. Chem.,* 274: 516–521, 1999.

123. Kim, S. and Ponka, P. Control of transferrin receptor expression via nitric oxide-mediated modulation of iron-regulatory protein 2, *J. Biol. Chem.,* 274: 33035–33042, 1999.

124. Kim, S. and Ponka, P. Effects of interferon-γ and lipopolysaccharide on macrophage iron metabolism are mediated by nitric oxide-induced degradation of iron regulatory protein 2, *J. Biol. Chem.,* 275: 6220–6226, 2000.

125. Oliveira, L. and Drapier, J.C. Down-regulation of iron regulatory protein 1 gene expression by nitric oxide, *Proc. Natl. Acad. Sci. U.S.A.,* 97: 6550–6555, 2000.

126. Hausladen, A. and Fridovich, I. Superoxide and peroxynitrite inactivate aconitases, but nitric oxide does not, *J. Biol. Chem.,* 269: 29405–29408, 1994.

127. Castro, L., Rodriguez, M., and Radi, R. Aconitase is readily inactivated by peroxynitrite, but not by its precursor, nitric oxide, *J. Biol. Chem.,* 269: 29409–29415, 1994.

128. Bouton, C., Riveau, M., and Drapier, J.C. Modulation of iron regulatory protein functions. Further insights into the role of nitrogen- and oxygen-derived reactive species, *J. Biol. Chem.,* 271: 2300–2306, 1996.

129. Bouton, C., Hirling, H., and Drapier, J.C. Redox modulation of iron regulatory proteins by peroxynitrite, *J. Biol. Chem.,* 272: 19969–19975, 1997.

130. Martins, E.A., Robalinho, R.L., and Meneghini, R. Oxidative stress induces activation of a cytosolic protein responsible for control of iron uptake, *Arch. Biochem. Biophys.,* 316: 128–134, 1995.

131. Pantopoulos, K. and Hentze, M.W. Rapid responses to oxidative stress mediated by iron regulatory protein, *EMBO J.,* 14: 2917–2924, 1995.

132. Cairo, G. et al. Induction of ferritin synthesis by oxidative stress, *J. Biol. Chem.,* 270: 700–703, 1995.

133. Cairo, G. et al. Superoxide and hydrogen peroxide-dependent inhibition of iron regulatory protein activity: a protective stratagem against oxidative injury, *FASEB J.,* 10: 1326–1335, 1996.

134. Pantopoulos, K. et al. Differences in the regulation of iron regulatory protein-1 (IRP-1) by extra- and intracellular oxidative stress, *J. Biol. Chem.,* 272: 9802–9808, 1997.

135. Pantopoulos, K. and Hentze, M.W. Activation of iron regulatory protein-1 by oxidative stress *in vitro, Proc. Natl. Acad. Sci. U.S.A.,* 95: 10559–10563, 1998.

136. Brazzolotto, X. et al. Human cytoplasmic aconitase (iron regulatory protein 1) is converted into its [3Fe–4S] form by hydrogen peroxide *in vitro,* but is not activated for iron-responsive element binding, *J. Biol. Chem.,* 274: 21625–21630, 1999.

137. Gehring, N.H., Hentze, M.W., and Pantopoulos, K. Inactivation of both RNA binding and aconitase activities of iron regulatory protein-1 by quinone-induced oxidative stress, *J. Biol. Chem.,* 274: 6219–6225, 1999.

138. Hanson, E.S. and Leibold, E.A. Regulation of iron regulatory protein 1 during hypoxia and hypoxia/reoxygenation, *J. Biol. Chem.,* 273: 7588–7593, 1998.

139. Kuriyama-Matsumura, K. et al. Regulation of ferritin synthesis and iron regulatory protein 1 by oxygen in mouse peritoneal macrophages, *Biochem. Biophys. Res. Commun.,* 249: 241–246, 1998.

140. Toth, I. et al. Hypoxia alters iron-regulatory protein 1 binding capacity and modulates cellular iron homeostasis in human hepatoma and erythroleukemia cells, *J. Biol. Chem.,* 274: 4467–4473, 1999.

141. Tacchini, L. et al. Transferrin receptor induction by hypoxia, *J. Biol. Chem.,* 274: 24143–14149, 1999.

142. Hanson, E.S., Foot, L.M., and Leibold, E.A. Hypoxia post-transcriptionally activates iron-regulatory protein 2, *J. Biol. Chem.,* 274: 5047–5052, 1999.

143. Mumby, S. et al. Reactive iron species in biological fluids activate the iron–sulphur cluster of aconitase, *Biochim. Biophys. Acta,* 1380: 102–108, 1998.

144. Pourzand, C. et al. Ultraviolet A radiation induces immediate release of iron in human primary skin fibroblasts: the role of ferritin, *Proc. Natl. Acad. Sci. U.S.A.,* 96: 6751–6756, 1999.

145. Minotti, G. et al. The secondary alcohol metabolite of doxorubicin irreversibly inactivates aconitase/iron regulatory protein-1 in cytosolic fractions from human myocardium, *FASEB J.,* 12: 541–552, 1998.

146. Eisensein, R.S. et al. Iron responsive element binding protein. Phosphorylation by protein kinase C, *J. Biol. Chem.,* 268: 27363–27370, 1993.

147. Schalinske, K.L. and Eisenstein, R.S. Phosphorylation and activation of both iron regulatory proteins 1 and 2 in HL-60 cells, *J. Biol. Chem.,* 271: 7168–7176, 1996.

148. Schalinske, K.L. et al. The iron–sulfur cluster of iron regulatory protein 1 modulates the accessibility of RNA binding and phosphorylation sites, *Biochemistry,* 36: 3950–3958, 1997.

149. Brown, N.M. et al. Novel role of phosphorylation in Fe–S cluster stability revealed by phosphomimetic mutations at Ser-138 of iron regulatory protein 1, *Proc. Natl. Acad. Sci. U.S.A.,* 95: 15235–15240, 1998.

150. Lin, J.J. et al. Derepression of ferritin messenger RNA translation by hemin *in vitro, Science,* 274: 74–77, 1990.

151. Eisenstein, R.S. et al. Regulation of ferritin and heme oxygenase synthesis in rat fibroblasts by different forms of iron, *Proc. Natl. Acad. Sci. U.S.A.,* 88: 688–692, 1991.

152. Goessling, L.S. et al. Enhanced degradation of the ferritin repressor during induction of ferritin messenger RNA translation, *Science,* 256: 670–673, 1992.

153. Haile, D.J. et al. The inhibition of the iron responsive element RNA–protein interaction by heme does not mimic *in vivo* iron regulation, *J. Biol. Chem.,* 265: 12786–12789, 1990.

154. Goessling, L.S., Mascotti, D.P., and Thach, R.E. Involvement of heme in the degradation of iron-regulatory protein 2, *J. Biol. Chem.,* 273: 1255–1257, 1998.

155. Carvalho, H. et al. Human precursor δ-aminolevulinic acid induces activation of the cytosolic iron regulatory protein 1, *Biochem. J.,* 328: 827832, 1997.

156. Chen, O.S., Schalinske, E., and Eisenstein, R.S. Dietary iron intake modulates the activity of iron regulatory proteins and the abundance of ferritin and mitochondrial aconitase in rat liver, *J. Nutr.,* 127: 238–248, 1997.

157. Chen, O.S. et al. Dietary iron intake rapidly influences iron regulatory proteins, ferritin subunits and mitochondrial aconitase in rat liver, *J. Nutr.,* 128: 525–535, 1998.

158. Schalinske, K.L., Chen, O.S., and Eisenstein, R.L. Iron differentially stimulates translation of mitochondrial aconitase and ferritin mRNAs in mammalian cells, *J. Biol. Chem.,* 273: 3740–3746, 1998.

159. De Russo, P.A. et al. Expession of iron regulatory protein 1 abolishes iron homeostasis in mammalian cells, *J. Biol. Chem.,* 270: 15451–15454, 1995.

160. Garate, M.A. and Nunez, M.T. Overexpression of the ferritin iron-responsive element decreases the labile iron pool and abolishes the regulation of iron absorption by intestinal epithelial (Caco-2) cells, *J. Biol. Chem.,* 275: 1651–1655, 2000.

161. Leedman, P.J. et al. Thyroid hormone modulates the interaction between iron regulatory proteins and the ferritin mRNA iron-responsive element, *J. Biol. Chem.,* 271: 12017–12023, 1996.

162. Thomson, A.N., Rogers, J.T., and Leedman, P.J. Thyrotropin-releasing hormone and epidermal growth factor regulate iron regulatory protein binding in pituitary cells via protein kinase C-dependent and -independent signalling pathways, *J. Biol. Chem.,* in press 2000.

163. Testa, U. et al. Differential regulation of iron regulatory element-binding protein(s) in cell extracts of activated lymphocytes versus monocytes–macrophages, *J. Biol. Chem.,* 266: 13925–13930, 1993.

164. Sposi, N.M. et al. Mechanisms of differential transferrin receptor expression in normal hematopoiesis, *Eur. J. Biochem.,* 267: 6762–6774, 2000.

165. Seiser, C., Teixeira, S., and Kuhn, L.C. Interleukin-2-dependent transcriptional and post-transcriptional regulation of transferrin receptor mRNA, *Eur. J. Biochem.,* 268: 13074–13080, 1993.

166. Weiss, G. et al. Regulation of cellular iron metabolism by erythropoietin: activation of iron-regulatory protein and upregulation of transferrin receptor expression in erythroid cells, *Blood,* 89: 680–687, 1997.

167. Wu, K.J., Pollack, A., and Dalla-Favera, R. Coordinated regulation of iron-controlling genes, H-ferritin and IRP2, by c-*myc*, *Science,* 283 676–679, 1999.

9 Ferritin

CONTENTS

9.1 Introduction 449
9.2 Evolution of Ferritins 450
9.3 Three-Dimensional Structure of Ferritin 457
9.4 Ferroxidase Activity of Ferritin 459
9.5 Role of Phosphate In Iron Oxidation and Iron-Core Formation 461
9.6 Iron Release From Ferritin 462
    9.6.1 Ferritin as an Iron Donor in Erythroid Cells 463
    9.6.2 Ferritin as an Iron Donor in Hepatic Cells 464
9.7 Hemosiderin 466
9.8 Serum Ferritin 467
9.9 Ferritin Genes 474
9.10 Transcriptional Control of Ferritin Gene Expression 477
9.11 Ferritin and Endothelium 484
9.12 Ferritin in Tumor Cells 487
9.13 Ferritin Expression and Function in Selected Tissues 492
    9.13.1 Ferritin Expression and Function in the Eye 492
    9.13.2 Expression and Function of Ferritin in the Skin 494
    9.13.3 Ferritin Expression in the Brain 495
    9.13.4 Ferritin Expression and Function in Macrophages 504
9.14 Unexpected Functions of Ferritin 513
References 515

9.1 INTRODUCTION

All living forms require iron for their metabolism. The large majority of iron is in molecular forms bound to proteins. However, a very small amount is present as a free iron pool where it is complexed with low molecular weight compounds. Although free iron concentration within the cells is very low, it is potentially harmful, due to the possible generation of highly reactive hydroxyl radicals. In vertebrates, the defenses against the toxic effects of Fe are mediated in different compartments: in the plasma, in extracellular spaces, mediated by transferrin, and within the cells, mediated by ferritin.

The importance of the iron storage function of ferritin is supported by the observation that this protein is ubiquitously maintained in all living forms from bacteria to eukaryotic cells. Furthermore, ferritin possesses the capacity to store iron under the Fe^{3+} form, a safe form that is unable to catalyze the production of free radicals until it is immobilized.

Mammalian ferritins are heteropolymers composed of two different chains, known as the ferritin H and ferritin L subunits. The H subunit displays an intrinsic ferroxidase activity that seems necessary for iron incorporation into the ferritin molecule. The L subunit is required for iron mineralization inside the ferritin cavity. In prokaryotes and plants, ferritins are made of 24 identical subunits that all exert ferroxidase activity.

The unique role of ferritin in intracellular iron storage is related to its capacity to store very large amounts of iron (up to 4000 Fe atoms per ferritin molecule). The biological relevance of ferritin in iron storage is supported by phenotypical analysis of animals rendered genetically deficient in the synthesis of ferritin. The deletion of both FerH alleles (FerH- and FerH-) was lethal. Embryos of animals with double knockouts of FerH genes died between 3.5 and 9.5 days of development.[1] This observation suggests that the function of FerH cannot be replaced by the other ferritin subunit, FerL.

The pattern of expression of FerH in mice embryos at 7 to 9 days of development shows the selective presence of this protein in the heart and brain, thus suggesting a role for this protein in these tissues.[1]

The possible function of ferritin was also investigated through overexpression of the FerH subunit in some cell types. One study analyzed the effects of FerH overexpression in erythroid cells.[2] FerH overexpression in Friend erythroleukemia cells elicited three important biological effects: (1) reduction of the size of the intracellular iron pool; (2) reduction in the production of reactive oxygen species following oxidative stress; (3) an increase in the membrane glycoprotein responsible for multidrug resistance.[2] This observation further reinforces the idea that ferritin H expression is required for protection against the potential toxic effects of free iron related to the generation of reactive oxygen species.

9.2 EVOLUTION OF FERRITINS

Plant and animal ferritins arose from a common ancestral precursor. They were highly conserved during evolution, and are present in plants, bacteria, and all types of animals.[3] However, plant, bacterial, and insect ferritins display some structural and functional features different from those observed for vertebrate ferritins. Furthermore, the regulatory mechanisms of control of ferritin gene expression changed during evolution in that an iron-response element (IRE) is found in the 5' untranslated regions of ferritin mRNAs of higher animals. In contrast, ferritins of plants, yeasts, and bacteria do not possess IREs.

Plants, because of their immobility, must tightly regulate iron homeostasis to prevent iron toxicity and deficiency. Plant ferritins are composed of only one subunit chain that displays features of both H and L type animal ferritin subunits. The coding region of the plant ferritin subunit is longer than the coding regions encoding for animal ferritin subunits. An extra sequence present at the N terminus of a plant ferritin encodes a transit peptide responsible for the localization of the ferritin at a peculiar organelle, the plastid, and also an extension peptide of about 24 amino acid residues specific for plant ferritins.[4] It is of interest that amino acid residues required for oxidation properties of H-type animal ferritin are conserved in plant ferritins.

Plant ferritin genes have been cloned and sequenced from different plant species including maize,[5] pea,[6] soybean,[7] and *Arabidopsis thaliana*.[8] Their structural organization shows some divergence as compared to the organization of animal ferritin genes. Seven introns are found in plant ferritin genes and three in animal ferritin genes. That suggests that the ferritins observed in animals originated through an evolutionary process of exon shuffling.[9]

Plant ferritin genes differ from animal ferritin genes not only in structural organization, but also in regulation. A key regulator of ferritin synthesis in both plants and animals is iron. Iron supplies induce ferritin synthesis in both plant and animal cells. However, in animal cells, the stimulation of ferritin synthesis by iron is mediated through a translational mechanism, while in plant cells, this phenomenon is mediated through a transcriptional mechanism.

Addition of soluble iron salts to plant cell cultures elicited a marked stimulation of ferritin mRNA that preceded the subsequent increase of ferritin content.[10,11] The molecular mechanisms responsible for the induction of plant ferritin gene expression by iron were investigated in soybeans. An upstream promoter element of the ferritin gene contained in a segment of about 80 nucleotides was involved in this transcriptional regulation. This DNA sequence did not contain any IRE-like element and did not display any other sequence similarity.[7] This regulatory element acted as a derepressional regulator; iron supply inhibited the binding of a repressor to this target sequence.[7]

Additional studies in maize and *Arabidopsis thaliana* provided evidence that a sequence present at position −100 in the promoter region (called the "iron-dependent-regulatory sequence") is required for the iron-dependent regulation of ferritin genes. Thus, in plants, the DNA target mechanisms of gene regulation by iron diverged from the correspondent animal iron gene regulation, even though the signal (iron) and the gene target (ferritin) were conserved.

In addition to iron, other stimuli including development, abscisic acid and reactive oxidative species influence plant ferritin gene expression. Two maize ferritin genes have been characterized and named Fer1 and Fer2. The Fer2 gene is regulated by a cellular pathway involving the plant hormone, abscisic acid, while the Fer1 gene is regulated by an abscisic acid-independent pathway that involves iron.

Particularly interesting are the observations of ferritin expression during plant development. Ferritin accumulates in non-green plastids; only low levels of this protein are observed in the chloroplasts, where the photosynthetic process is active. Ferritin is expressed only in roots and leaves of young plantlets, and remains undetectable in the corresponding organs of adult plants.[4] It was proposed that plant ferritin may represent an important iron source during the initial stages of development, providing the iron required for the synthesis of iron proteins implicated in photosynthesis.[4]

Similar observations have been made also on ferritin expression during development in plants of the legume family. At the nodules utilized for a symbiotic process between plant and soil bacteria, ferritin synthesis is observed only during initial stages of nodule development and shuts down when the nodules become mature, as evidenced by their capacity to fix nitrogen.

The analysis of the structure of pea seed ferritin provided evidence that it is composed of 24 subunits assembled into a hollow shell. An N terminal extension of

71 amino acid residues contains a transit peptide consisting of the first 47 residues; it is responsible for plastid targeting. The second part of the extension (extension peptide) contains a specific sequence of 24 amino acids. This extension peptide is a specific feature of plant ferritins and is required for optimal ferroxidase activity.

Another specific feature of pea seed ferritin is its high phosphate content, a phenomenon seemingly related to the high phosphate concentrations found within plastids. The presence of ferritin in peas and *Phaseolus vulgaris* plants, as well as in other Leguminaceae, is related also to a defense mechanism operating in root nodules via sequestration of catalytic iron. In plant nodules, ferritin increases early in the nodulation process and declines concomitantly with increases in nitrogenase activity and heme and non-heme iron. Ferritin synthesis may be reactivated in senescent nodules after nitrate loading through an anti-oxidative response mechanism. Finally, ferritin is also involved in early stages of germination. Ferritin stored in seeds is degraded, thus providing the iron required for early phases of germination.

These studies suggest an important role for ferritin as a transient iron buffer system used to sustain iron-dependent processes related to plant development such as photosynthesis, nitrogen fixation and germination.[4] Ferritin synthesis in these processes seems to be modulated through post-transcriptional mechanisms.

The observations of ferritin's role in plant physiology promoted studies of the development of transgenic plants that overexpress this protein. Tobacco plants over-producing ferritin did not show any alteration in their morphology, growth rate, or photosynthetic function, but exhibited increased resistance to infections by necro-pathic pathogens.[12] Ferritin was overexpressed in rice seeds, and the transgenic rice plants overexpressing this protein did not display modifications of their growth or develop rice grains of normal morphology and size.[13] These types of transgenic plants may be useful as food sources to counteract the problem of iron deficiency in some underdeveloped areas of the world.

Ferritins are present in the majority of bacteria. Bacterial ferritins can be classified, according to their structural characteristics, into two categories: (1) bacterioferritins, that contain heme and are found in *Escherichia coli*, *Neisseria gonorrhoeae*, *Brucella melitensis*, and *Pseudomonas*; and (2) non-heme-containing bacterial ferritins such as those found in *Helicobacter pylori*, *Porphyromonas gengivalis*, *Campylobacter jejuni*, and *Haemophilus influenzae*.

Bacterioferritins are formed by a single subunit, with low amino acid sequence homology (about 20%) with eukaryotic ferritins, and with a three-dimensional structure similar to those observed for vertebrate ferritins consisting of a four-helix bundle conformation.[14] Bacterioferritins exhibit seven amino acid residues that constitute the ferroxidase centers of FerH chains. However, the ferroxidase centers exhibit some differences with respect to the corresponding centers of mammalian FerH chains.[14] The two iron atoms of bacterioferritins have histidine coordination and are bridged by two carboxylate residues (Glu 62) (Figure 9.1). As a consequence of these differences, one site at each ferroxidase center is similar in bacterioferritins and mammalian FerH; the other site is different.[14] Bacterioferritin contains up to 12 heme molecules of unknown function, localized at the two bacterioferritin subunits through binding to symmetry-related methionine residues (Met 52 and Met 52′).

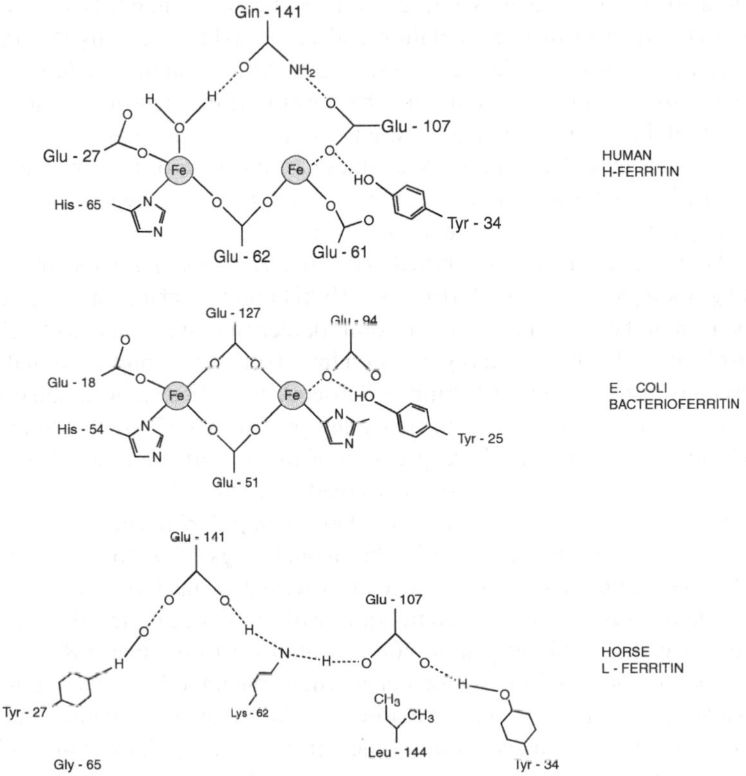

FIGURE 9.1 Structures of the iron-binding sites of ferritin chains: human ferritin H chain (top), *E. coli* bacterioferritin (middle), and horse ferritin L chain (bottom).

A particular type of bacterioferritin observed in *Desulfovibrio desulfuricans* contains a unique type of non-covalently bound heme molecule (iron uroporphyrin) with peculiar biophysical properties.[15] This bacterioferritin is considered the most divergent of the group so far characterized. The purified protein is a homodimer with a mass of 52 kDa. The monomers are linked by an iron-coproporphyrin group and each monomer contains a diferric center. The function of ferritin in this anaerobic bacterium is unknown. However, the association of the ferritin gene within an operon with genes involved in oxygen detoxification led to speculation about a possible role of this protein in buffering the deleterious effects of non-controlled oxidation of ferrous iron.

The kinetics of binding of iron to bacterioferritin was analyzed in detail, leading to the identification of three steps: (1) an initial rapid phase, corresponding to the binding of two Fe^{2+} ions to the binuclear ferroxidase site; (2) a slower second phase corresponding to the oxidation of the Fe^{2+}–bacterioferritin complex; and (3) a third phase corresponding to the mineral core formation.[14]

The non-heme containing ferritin-like proteins expressed in some bacteria have been partially characterized. Non-heme bacterial ferritins exhibit the conservation of amino acid residues, such as Glu 27, Tyr 34, Glu 61, Glu 62, His 65, Glu 107, and Gln 141, involved in iron chelation and in ferroxidase activity.[16] These non-heme bacterial ferritins display the same structures as mammalian ferritins under electron microscopy analysis. Non-heme bacterial ferritin seems to be important for the growth of bacteria under iron-restricted conditions, as suggested by studies performed on bacteria deficient in the synthesis of this type of ferritin. The mutants ceased growth under iron starvation more rapidly than wild-type bacteria, which have the capacity to synthesize bacterial ferritin.[17]

Non-heme bacterial ferritins studied in detail in *Helicobacter pylori* show several interesting findings. Non-heme ferritin is a 19-kDa protein abundantly expressed in the cytoplasm of the bacterium where it forms paracrystalline inclusions.[18] All amino acids involved in chelation of inorganic iron by ferritins from higher animal species are conserved in the *H. pylori* ferritin.[18] Subsequent studies showed some identity of *H. pylori* ferritin with neutrophil activating protein A (HPNAP).[19] Studies with bacterial mutants deficient in the expression of this ferritin suggest it has a role in iron starvation and survival in response to oxidative stress.[20]

Characterization of the molecular structure of *H. pylori* ferritin showed a three-dimensional structure characterized by hexagonal rings of 9- to 10-nm diameter, with a hollow central core similar to that observed in bacterial DNA-protecting proteins (Dps). This structure is compatible with a dodecameric oligomer with a central hole capable of binding up to 500 iron atoms per oligomer.[21] Recent studies have somewhat elucidated the molecular mechanisms underlying the regulation of *Helicobacter pylori* ferritin. This ferritin is induced by iron loading through a mechanism involving the inactivation of the ferric uptake regulator (Fur) which acts as a repressor when active (i.e., under iron-restricted conditions).

Particularly interesting are the studies of a ferritin-like molecule containing non-heme iron observed in *Listeria innocua*, a Gram-positive bacterium.[22-24] The protein purified from these bacteria exhibits a molecular weight of about 240 kDa and is composed of a single type of subunit of about 18 to 20 kDa.[22] This observation suggests that the *Listeria innocua* ferritin molecule was composed of 12 subunits, not 24 subunits, as observed in mammalian ferritins and in bacterioferritins.[22]

The analysis of the amino acid sequence showed a high similarity to the family of DNA-binding proteins known as Dps. Among the proteins of the ferritin family, the highest similarity was found with mammalian FerL.[22] The three-dimensional structure of ferritin isolated from *Listeria innocua* was recently determined.[23] The crystallized dodecamer displayed typical features of mammalian ferritins, such as the negatively charged channels located along the three-fold symmetry axes which are required for iron entry into the cavity and iron deposit.[23]

The ligands involved in iron binding of *Listeria* ferritin are similar to those observed in other ferritins; however, the iron-binding site of *Listeria* ferritin displayed a unique feature in that it has ligands belonging to two different subunits, not contained within a four-helix bundle.[23] The kinetics and the three phases of iron binding and incorporation have been determined:[24,25] (1) an initial rapid phase in which Fe^{2+} binds to apoferritin; (2) a slower second phase corresponding to the oxidation of the iron complex; and (3) a third phase of iron mineralization. This process showed some differences with mammalian ferritins. In *Listeria innocua* ferritin, the final product of dioxygen reduction is H_2O, while in the case of 24-subunit ferritins, such as mammalian ferritins and bacterioferritins, the product is H_2O_2.[24,25]

Ferritins have been characterized also in parasites. Two different ferritins of different primary amino acid sequences have been found in the parasite *Schistosoma mansoni*, one of which is more prevalent in females and the other more prevalent in males. Ferritin 1, also called yolk ferritin, is predominantly found in mature egg-laying females, where it is localized in the vitellarium and the ovary and functions as an iron storage protein used for embryonic development.[26] Ferritin 2, also called soma ferritin, is present in both sexes, and is expressed in virtually all tissues.[26] The mRNAs encoding both parasite ferritins do not possess IREs in their 5′ untranslated regions.[27]

The similarity between *S. mansoni* ferritins and vertebrate H-chain ferritins exceeds 50% identity, including the ferroxidase centers. A ferritin was isolated also from *Echinococcus granulosus*, with a deduced amino acid sequence showing 56% homology to the heavy chains of human ferritin.[28]

Particular attention was focused on insect ferritins because they have unique properties. Most insect ferritins have molecular larger weights (>600 kDa) than vertebrate ferritins, as the result of containing larger subunits (24-, 26-, and 32-kDa).[29,30] These subunits are glycosylated and found in the endoplasmic reticulum of insect cells and in the blood (hemolymph).

These ferritin subunits, at variance with vertebrate ferritins, are secreted as a consequence of signal peptides that drive these molecules to the secretory pathway. Ferritin is present in many tissues of insects, including midgut, hemolymph, malpighian tubules, and body fat, and is released into the gut lumen as part of its role in iron transport. The analysis of the amino acid sequences indicates that the smaller subunits of insect ferritins are homologues of the vertebrate ferritin heavy chains, whereas the larger ferritin subunits are homologues of the vertebrate light chains. However, the marked differences in the sequences of insect ferritins led to the suggestion that they may play a particular role in metabolism. In insects, ferritin is visible in the vacuolar systems of tissues that filter the hemolymph; in Lepidoptera, ferritin is abundant in the hemolymph. Sequences reported for insect-secreted ferritins from Lepidoptera and Diptera have high sequence diversity.

Insect ferritins have been characterized in *Drosophila melanogaster* and in the yellow fever mosquito, *Aedes aegypti*. Ferritins present in *Aedes aegypti* are heterodimers, with molecular weights of about 660 kDa composed of ferritin subunits of different molecular weights (24-, 26-, 28-, and a minor component of 30-kDa). The 24- and 26-kDa subunits have the same N terminus sequences and therefore the products of the same gene, whereas the 28-kDa subunit is the product of a different gene.[30] The nucleotide sequence of the gene encoding the 24- to 26-kDa ferritin subunits was cloned and the analysis of this sequence showed some interesting findings: (1) the presence of an IRE in the 5′ untranslated region of the corresponding mRNA; (2) the presence of residues required for the ferroxidase activity, as observed in vertebrate H ferritins; and (3) comparison of sequences showed a higher similarity with vertebrate H ferritin than with L ferritin.

The structure of the ferritin gene encoding the 24- and 26-kDa subunits was recently determined and showed the presence of an additional intron (the gene contains four introns, not three introns as found in vertebrate ferritin gene) located between the cap site and the start codon and an unusually long second intron.[31] Quantitative Southern blotting analysis showed that this ferritin gene is present in the genome of the mosquito as a single gene copy, and no pseudogenes are observed, a finding consistently observed in vertebrates.[31]

Biochemical studies performed on ferritins isolated from *Drosophila melanogaster* revealed heterodimers composed of subunits of different molecular weights (24-, 26-, and 28-kDa).[32] The 24-, 26-, and 28-kDa subunits are encoded by two closely linked genes, Fer1 and Fer2, which possess 3 and 2 introns, respectively, and represent the homologues of vertebrate FerH and FerL genes, respectively.[33] Adams[34] suggested the existence also of a third ferritin that encodes a cytoplasmic ferritin subunit. Northern blot analysis of *D. melanogaster* ferritin transcripts showed three bands of about 1.1, 1.3, and 1.5 Kb.[35] These three bands correspond to four mRNAs, two of which contain IREs and two of which lack IREs. Interestingly, the IRE is located at a region of the gene corresponding to an intron and the alternative splicing of this intron generates mRNAs either containing or lacking an IRE element.[35] Iron loading determines a shift of these mRNA species toward those lacking IREs.

Particularly interesting are the studies of the brown plant hopper, *Nilaparvata lugens*, one of the five major pests of rice plants. The ferritin of this insect was discovered in studies related to the identification of protein which bind to the lectin GNA (*Galantus nivalis* agglutinin). The major binding protein in the *N. lugens* midgut is ferritin. The purified protein exhibits a molecular weight of more than 400 kDa and is composed of three subunits of 20-, 26-, and 27-kDa. Two cDNAs encoding two of these three subunits have been isolated. One, the Fersub1 cDNA carries a stem–loop IRE structure upstream from the start codon.

Among invertebrates, particularly interesting are the observations made in the snail *Lymnae stagnalis*. Two peculiar ferritin isoforms have been observed: one is expressed in almost every tissue and has been termed soma ferritin, whereas the other is accumulated in the yolk platelets of the growing oocytes and has been termed yolk ferritin.[36,37] Comparison of the primary sequences showed that soma ferritin exhibits 50 to 70% sequence identity with subunits of vertebrate ferritins, whereas yolk ferritin exhibits lower level (31 to 42%) of identity with vertebrate ferritin subunits. Interestingly, in soma ferritin mRNA, an IRE was observed, while IRE was absent in yolk ferritin mRNA.[36,37]

Ferritins have been described also in other invertebrates. In the crustacean *Artemia salina*, ferritin was found to be the major storage protein of the cysts.[38] However, it is unknown whether this protein is able to store iron as a true ferritin, and its relationship to the ferritin family is based on sequence similarity. Ferritin was isolated and characterized also from the freshwater crayfish *Pacifastacus lenisculus*.[39] The deduced amino acid sequence of this ferritin exhibits a closer similarity to vertebrate ferritins (heavy subunits) than to insect or plant ferritins, and contains the conserved H-specific residues of the ferroxidase centers of the vertebrate ferritin.[39] Interestingly, an IRE was observed in the 5′ untranslated region of crayfish ferritin mRNA.

Two ferritin subunits were detected n the Atlantic salmon and the corresponding cDNAs have been cloned: (1) the H (heavy) subunit was composed of 177 amino acid residues and was expressed in various tissues, including the liver, gonads, kidney, heart, and spleen; and (2) the M (middle) subunit was composed of 176 amino acid residues and was found almost exclusively in the gonads.[40] Fish ferritins are distantly related to mammalian ferritins, as suggested by the absence of cross-reactivity between fish ferritins and anti-mammalian ferritin.

In the octopus, a ferritin-like molecule was isolated; its pattern of expression extended to virtually all tissues and had a consistent degree of homology (>60%) to soma ferritin of the gastropod mollusc *Limnaea stagnalis* and vertebrate H ferritin.[44] All residues involved in metal binding and ferroxidase activity are conserved in these proteins.

The observations on the structures of crustaceous ferritins and comparison to the structures of ferritins of other animal species led to some interesting speculations. It was suggested that two different types of ferritin exist in animal species: insect-type or secretory ferritins which are predominant in insects and snail oocytes, and vertebrate-type ferritins localized in the cytoplasm and predominant in vertebrates and crustacea.

In more developed animal species, such as primates, rodents, marsupials, birds, and amphibians, ferritin amino acid sequences are highly conserved. The description in detail of the biochemical and molecular properties of mammalian ferritins will be discussed in the next sections. Even in animals distantly related to humans, such as the amphibians, two ferritin chains (H and L) are observed and they form heteropolymers. The amino acid sequences of these ferritin chains exhibit >70% identity with the corresponding human ferritin chains.[42] Furthermore, the three-dimensional structures of amphibian H and L ferritin chains are highly comparable to those observed for the corresponding human ferritin chains.[43,44]

9.3 THREE-DIMENSIONAL STRUCTURE OF FERRITIN

The analysis of the three-dimensional structure of the ferritin molecule was based on the X-ray diffraction analyses of ferritin crystals.

The apoferritin molecule is quasi-spherical and is composed of 24 subunits arranged in pairs along the 12 walls of a quasi-rhombododecahedron.[43,45,46] Apoferritin crystals have face-centered cubic lattices, faceted by hexagonal planes.[47,48] Direct atomic force microscopy revealed that the so-called nuclear critical-size clusters formed during the crystallization of ferritin consist of planar arrays of one or two mononuclear layers containing five to ten rods of up to seven molecules each.[48] Each subunit exhibits an identical structure folded into four α-helix bundles over 5 nm in length. The A–B and C–D bundles are bridged by short A–B and C–D loops, respectively; a long loop (B–C loop) connects the A–B and C–D bundles. A fifth shorter helix (E) is located about 60° and terminates at the C terminal end.

The subunits are arranged to resemble a hollow, symmetrical shell with outside and inside diameters corresponding to 125 and 80 Å, respectively. The internal cavity of ferritin, with an 80-Å diameter, is capable of storing up to 4500 Fe^{3+} atoms as a polynuclear ferric oxhydroxide–phosphate complex. Each apoferritin molecule is assembled from 24 structurally equivalent subunits. The interaction among these subunits results in the assembly of a molecule notably stable to heat treatment and incubation with urea or guanidium chloride. This stability is dependent upon the presence of a large number of intra- and inter-subunit salt bridges and hydrogen bonds.

In the assembled 24-subunit molecule, the different subunits interact following 2, 3, and 4 symmetry. The formation of anti-parallel pairs interacting along their long apolar interfaces around two-fold symmetry axes seemingly represent the first assembly intermediates. Through the interaction of three subunits, eight hydrophilic channels are formed in the ferritin shell; these channels are responsible for iron entry and exit. Through the interaction of four other subunits, six hydrophobic channels are formed; these channels are used by iron chelators and iron chelator complexes to enter and exit the ferritin molecule.

The comparison of the amino acid sequences of human H chain ferritin and human L chain ferritin showed a high degree of conservation of the amino acid residues involved in subunit interaction, and this finding represents the structural basis to explain the interaction between H and L chains to form heteropolymeric ferritin molecules.[46]

A key feature of the internal structures of ferritin subunits is a central hydrophilic region surrounded by hydrophobic residues. However, this region exhibits differences between the H and L subunits. Each H subunit has a narrow 1-Å channel through the four α-helix bundle that forms the dinuclear metal binding site, whereas this channel is blocked by a salt bridge (Lys 62 vs. Glu 107) inside the four α-helix bundle of the ferritin L chain.[49]

The studies on the three-dimensional structure of ferritin have elucidated the various biochemical processes involved in iron storage into the ferritin molecule: iron (Fe^{2+}) uptake into the protein shell, conversion of ferrous iron to ferric ion at the ferroxidase center, and migration of Fe^{3+} in the ferritin cavity, where its deposition as ferrihydrite occurs.

The structure of the ferritin molecule responsible for iron uptake into the shell is represented by the three-fold channel. This channel is hydrophilic, and substitution of the carboxylate residues (glutamates and aspartates) with other amino acids (such as alanine or leucine) gives rise to a marked inhibition of iron uptake into the ferritin shell.[50] The movement of iron along the three-fold channel implies the passage of iron through this channel and its movement from this structure to the iron oxidation center along a hydrophilic pathway. Iron is seemingly driven during its passage through this pathway by an electrostatic gradient.[51]

Iron oxidation occurs at the di-iron center located within the four bundles of each ferritin H subunit (Figure 9.1). This center can bind and oxidize two iron atoms ($Fe^{2+} \rightarrow Fe^{3+}$) through a process detailed in Section 9.5. In the di-iron center of the human ferritin H subunit one iron atom has histidine coordination (His 65); the other atom has Glu 61 at the equivalent position and the two sites are bridged by only one carboxylate residue (Glu 62) (Figure 9.1). Substitutions of these amino acids or other amino acids involved in the formation of this center (examples of these mutations are E62K, H65G, and Y34F) resulted in a decrease of ferroxidase activity.[52-54]

   Putative ferrihydrite nucleation sites have been located on the inner surfaces of the ferritin shells. These sites may function as iron binding sites and are more represented in the L chains (Glu 53, 56, 57, 60, and 63) than in the H chains (Glu 61, 64, and 67). One iron atom is ligated by two Glu residues. The presence of a higher number of glutamate sites in L ferritin chains than in H ferritin chains explains why L chains are more efficient than H chains in promoting ferrihydrite nucleation. In fact, L ferritin chains incorporate iron *in vitro* with specificity and efficiency under some conditions. When ferritins are subjected *in vitro* to high iron loading, near the saturation limit of 4000 Fe atoms per molecule, the homopolymers of human H chains tend to form insoluble aggregates, caused by non-specific iron hydrolysis, whereas the homopolymers of human L chains remain soluble and incorporate most of the available iron.[55]

   Ferritin H mutants which have lost ferroxidase activity are unable to display the iron saturation properties of L ferritin homopolymers under conditions of high iron loading.[55] The substitution for Ala of Glu residues present in the inner cores (Glu 61, Glu 64, and Glu 67) of the ferritin H mutant with inactivated ferroxidase centres abrogated the iron uptake capacity, thus suggesting that carboxyl groups facing the central cavity are required for the binding of Fe^{3+} and to promote ferrihydrite nucleation.[56,57] The substitution of His for Glu 57 and Glu 60 abolished the efficiency of iron incorporation typical of L ferritin.[58] In conclusion, these observations suggest a main role for carboxylic residues exposed into the core cavity in the process of iron mineralization by ferritin.

9.4 FERROXIDASE ACTIVITY OF FERRITIN

Iron represents the substrate of the ferroxidase activity of ferritin and, after oxidation at the ferroxidase site, it is translocated to the inner protein cavity to form the mineral core of ferritin.

   The process and the mechanism of iron oxidation by ferritin have been partially elucidated. Studies using recombinant preparations of vertebrate ferritin subunits, including human ferritins, have shown that the ferroxidase activity is associated only with H-type subunits, while L-type subunits lack this activity. The iron oxidation occurs at the ferritin ferroxidase site and different stages of this process can be distinguished. The initial reaction involves the transient reaction of two Fe^{2+} atoms with O_2 at the ferroxidase site, with subsequent release of the expected hydrogen peroxide and ferric–oxo species.[59] This process can be described according to the following equations, as proposed by Yang et al.[60]:

$$2Fe^{2+} + P^2 \rightarrow \left[Fe_2 - P\right]^{2+4} \tag{9.1}$$

$$\left[Fe_2 - P\right]^{2+4} + O_2 + 3H_2O \rightarrow \left[Fe_2O\left(OH\right)_2 - P\right]^{2+2} + H_2O_2 + 2H^+ \tag{9.2}$$

where P^2 is the apoferritin molecule with a net surface charge z and $[Fe_2 - P]^{2+4}$ and $[Fe_2O(OH)_2 - P]^{2+2}$ represent a dinuclear Fe^{2+} complex at the ferroxidase site and a hydrolyzed µ-oxo-bridged dinuclear (III) complex at the same site, respectively.[60]

During the ferroxidase reaction, hydrogen peroxide is produced as a consequence of the two-electron reduction of dioxygen at the di-iron site. Recent studies suggest the transient production of a peroxo intermediate during the ferroxidation reaction. Stopped-flow absorption and rapid freeze–quench Mossbauer studies showed the release during Fe oxidation of an intermediate compound identified according its properties as a peroxodiferric intermediate.[61] Raman spectroscopy studies were used to characterize the transient species trapped by freeze–quenching, and provided definitive evidence that it corresponds to a peroxodiferric μ-1, 2-bridged diferric peroxide intermediate.[62] X-ray absorption spectroscopy and Mossbauer analyses, showed that this peroxodiferric species exhibits an unusually short Fe–Fe distance corresponding to 2.53 Å.[63] This small Fe–Fe distance implies a small Fe–O–O angle, a condition that favors the decay of the peroxodiferric complex by the release of hydrogen peroxide and μ-oxo or μ-hydroxo diferric precursors.[63]

In subsequent steps, the μ-oxo bridged di-iron intermediates split and the mono-nuclear Fe^{3+} complexes move to the cavity where they hydrolyze. The transfer of iron from the ferroxidase centers to the mineral surfaces of ferritins occurs in a range of 5 to 10 minutes, leaving the ferroxidase center available for new cycles of iron oxidation. During the migration of the iron from the ferroxidase site to the mineral core, further iron hydrolysis occurs following the reaction: $2Fe^{2+} + O_2 + 4H_2O \rightarrow 2FeO(OH) + H_2O_2 + 4H^+$. Once sufficient core has formed, the oxidation/hydrolysis reaction shifts to a mineral surface mechanism, becomes autocatalytic, and follows the following global reaction:

$$4Fe^{2+} + O_2 + 6H_2O \rightarrow 4FeO(OH)_{(core)} + 8H^+ \qquad (9.3)$$

The movement of iron from ferroxidase centers to the central core should allow the regeneration of these ferroxidase centers, making them available for other cycles of iron oxidation. However, the rate of ferroxidase regeneration is greatly influenced by the presence of ferritin L chains within the ferritin molecules. In fact, at 10 to 40 Fe atoms/ferritin molecule μ-oxobridged dimers persisted for several hours in the ferroxidase centers of recombinant human H subunit ferritin, while at these iron loading ratios, all the iron was present in clusters in preparations of natural ferritin purified from horse spleen.[64] Other experimental approaches, such as the addition of one Fe atom/ferritin molecule every 10 minutes, showed that ferritins containing L chains exhibited faster kinetics of reutilization of ferroxidase centers.[60] These findings suggest that L subunits not only promote iron mineralization, but also promote the reutilization of ferritin H ferroxidase centers.[65]

In vitro studies suggest that, in addition to the catalytic reaction, spontaneous iron oxidation reactions and the formation of water occur when ferritin is exposed to high iron concentrations, with a stoichiometry of four Fe^{2+} ions per O_2 molecule.[66] This type of reaction requires high iron concentrations and high pH values and its physiological significance is very limited. On the other hand, catalytic-mediated iron oxidation by ferritin seems to have physiological relevance as supported by the observation that ferritin can still incorporate iron, although with lower efficiency, when iron is supplied as a chelate with citrate.[58]

The physiological relevance of ferroxidase for the iron-binding activity of ferritin is directly supported by recent studies based on the overexpression in HeLa cells of both wild-type ferritin H chain and ferritin H chain mutants devoid of ferroxidase activity.[67] The mutant ferritin had no effect on cell growth or iron phenotype, while the wild-type ferritin H chain elicited a significant decrease of cell growth associated with development of an iron-deficient phenotype (i.e., a decrease of the free iron pool and increases in IRP activity and transferrin receptor expression) and increased resistance to H_2O_2-mediated toxicity.[67] These observations clearly indicate that the integrity of the ferroxidase activity is strictly required to maintain the iron-binding capacity of ferritin.

Aust and coworkers challenged the widely accepted view that ferritin possesses intrinsic ferroxidase activity. They suggested that the ferroxidase catalytic activity attributed to ferritin may be due to Fe^{2+} autoxidation facilitated by the buffer used for the *in vitro* reaction.[68] They showed that, in the absence of a buffer, no ferroxidase activity was observed with either horse spleen ferritin or the recombinant H chain homopolymer of rat liver ferritin.[69] They suggested that ceruloplasmin may be responsible for Fe^{2+} to Fe^{3+} oxidation and loading into the ferritin H chain.[70]

However, recent studies have shown that the experiments questioning the ferroxidase activity of ferritin were flawed by inadequate pH controls. During the incubation of ferritin with iron ($FeSO_4$) in the absence of a buffer, a drop in pH values was observed and this phenomenon was responsible for the low ferroxidase activity.[66] The buffers commonly used in this type of reaction usually retard rather than accelerate Fe^{2+} oxidation.[66] More importantly, animals made genetically deficient in ceruloplasmin production by gene targeting exhibited increased iron accumulation at the tissue iron storage compartments, such as the liver and the reticuloendothelial system, thus ruling out a role for ceruloplasmin as an iron donor to ferritin.[71]

9.5 ROLE OF PHOSPHATE IN IRON OXIDATION AND IRON-CORE FORMATION

Iron is stored in ferritins and bacterioferritins in the form of an inorganic complex, whose composition and properties are highly variable, particularly their iron and phosphate content.[65] Phosphate content was highest in bacterioferritins, intermediate in mammalian ferritins, and lowest in mollusc ferritins. The phosphate content of ferritins influenced the structures of the cores which tended to be more amorphous when the ferritin content was high and more crystalline when the phosphate content was low.[72]

Experiments involving removal of phosphate from horse spleen ferritin by changing the pH or by redox agents suggested that phosphates reside primarily on the surfaces of the cores.[73] In contrast, the phosphate present in plant ferritins and in bacterioferritins is uniformily distributed throughout the mineral cores.[72] The structure of the iron core does not represent an intrinsic property of the protein shell, but depends on the milieu in which iron loading occurred. Bacterial and plant apoferritin molecules loaded with iron in the absence of phosphate develop crystalline iron cores resembling those observed in mammalian ferritins.[74] On the other hand, phosphate content of bacterioferritins seems to be high since they develop into the plastids in which phosphate concentration is particularly elevated.

A physiological role for phosphate in ferritin function was suggested in 1991 on the basis of a study of the effect of phosphate on iron incorporation into apoferritin.[75] These authors reported that phosphate accelerated the rate of oxidation of Fe^{2+} to Fe^{3+} and improved the rate of transfer of Fe^{3+} to the mineral core.[74] The acceleration of Fe^{2+} oxidation by phosphate was interpreted as an effect of anion binding on the redox potential of the ferroxidase center.

Phosphate increased also the Fe^{2+} binding to apoferritin under anaerobic conditions, promoting its oxidation to Fe^{3+}.[76] Additional studies showed that phosphate was not only responsible for promoting Fe^{2+} binding to native ferritin cores, but also that phosphate is in part released from ferritin following reduction of the core.[73] All these observations suggest that a redox surface involving phosphate and iron is formed on the ferritin core. This surface may play an important role in catalyzing iron redox reactions associated with iron deposition and release.

These studies used natural preparations of ferritin already containing mineral cores. More recent studies of the role of phosphate used horse ferritin preparations with *in vitro* reconstituted iron cores of various size.[77] Incubation of Fe^{2+} and phosphate (in a 1:4 ratio) under anaerobic conditions allowed the formation of a layer with an iron/phosphate ratio of about 1:3. The amounts of iron and phosphate increased following the increase of the mineral core up to 2500 iron atoms. Interestingly, Mossbauer spectroscopic analyses showed that most iron incorporated at the iron/phosphate surface is present as Fe^{3+}, indicating that, under anaerobic conditions, incoming Fe^{2+} undergoes a redox reaction during the formation of the iron/phosphate layer.[77]

9.6 IRON RELEASE FROM FERRITIN

The *in vitro* release of iron from ferritin involves the reduction of Fe^{3+} present in the iron core and formation of a complex with a chelator molecule of the released Fe^{2+}. This process is slow and its mechanism is not well known. The *in vivo* mechanisms of iron release are largely unknown.

A large number of studies reviewed by Harrison and Arosio[3] have shown that many small reductants, such as dithionite, thioglycollate, reduced flavins, and ascorbate can mediate *in vitro* iron release from transferrin. The H subunit of ferritin also possesses intrinsic redox activity.[78] It was suggested that this activity may play a role in the mechanism of iron release. This hypothesis, however, remains to be directly supported by experimental evidence. Nevertheless, preliminary observations indicate that recombinant human ferritin H releases its iron twice as rapidly as ferritin L.[78]

The junction of three ferritin subunits seems to be the structure of the ferritin molecule mainly involved in iron release. This conclusion was related to the analysis of the kinetics of iron release and the study of the three-dimensional structures of selected ferritin mutants.[79] The kinetics and extent of iron release from the L134P variant of amphibian H chain ferritin were greatly potentiated as compared to those observed in the wild-type protein.[79] X-ray crystallography studies showed that the ferritin mutant, in which the conserved leucine 134 was replaced by proline, showed localized unfolding at the three-fold axis.[79]

These findings strongly suggest that the three-fold channels of the ferritin molecule are used for iron entry and exit. Studies of the entry and exit from the ferritin molecule of small molecular species of 7 to 9 Å provided evidence that the limitation factors for the transit along the three-fold channel are molecular charge and polarity. A negatively charged nitroxide was completely excluded from the interior of the protein, while positively charged and polar nitroxide radicals penetrated the protein shell.[80,81] The capacity of ferritin to function as a donor of iron for cellular metabolism was investigated in erythroid cells and hepatocytes.

9.6.1 Ferritin as an Iron Donor in Erythroid Cells

Erythroid cells possess unique mechanisms that control iron metabolism. Iron metabolism is entirely dedicated to sustaining the synthesis of elevated amounts of heme required for hemoglobin synthesis.[82] Iron incorporation in erythroid cells was studied in detail and showed that:

1. The rate of synthesis of heme in erythroid cells is tuned to the levels of iron uptake.[83] The rate of synthesis of transferrin receptors is particularly high with consequent high rates of iron uptake and heme synthesis.
2. Iron incorporated into erythroid cells via internalization of transferrin bound to membrane receptors is rapidly targeted to mitochondria, where it is used for heme synthesis. Iron continues to be targeted to mitochondria even when heme synthesis is inhibited by succinylacetone.[84]

These initial studies have not considered ferritin capable of donating iron for heme synthesis. However, recent studies provided evidence of a role of ferritin as an iron donor in erythroid cells. They suggest that ferritin endogenously synthesized by erythroid cells or exogenously endocytosed through membrane receptors may act as an iron donor. Exogenous ferritin is bound and internalized by developing human erythroid cells through a specific and saturable process involving a membrane receptor.[85] This receptor displayed the preferential binding of acidic isoferritin.[85] Its expression was regulated according to the iron status of the cell. Iron deprivation stimulated receptor expression, while an opposite effect was induced by iron loading.[86]

The iron internalized by the uptake of exogenous ferritin is utilized for cellular metabolism. Part of this iron is incorporated into heme; ferritin uptake induced iron-mediated down-regulation of transferrin receptor expression.[86] These findings suggest that ferritin uptake by erythroid cells may constitute an additional iron uptake system that cooperates with the classical transferrin–transferrin receptor iron uptake pathway.[86]

Subsequent studies provided evidence that the iron-donating activity of ferritin requires the proteolytic degradation of the endocytosed ferritin.[87] This conclusion was based on two findings: (1) ferritin is endocytosed by erythroid cells and degraded at the cellular acidic compartment; and (2) trypsin inhibitors, such as leupeptin, and agents that alkalize the lysosomes, such as chloroquine, are able to largely inhibit the degradation of ferritin and the transfer of its iron to hemoglobin.[87]

The capacity of exogenously internalized ferritin to donate iron for cellular metabolism is also supported by analysis of the levels of the cellular labile iron pool. The level of the pool is increased following internalization of holoferritin, and is decreased after endocytosis of apoferritin.[87]

Other studies suggest a potential role for intracellular ferritin (i.e., ferritin endogenously synthesized by erythroblasts) as an iron donor for erythroblasts. Erythroid cell maturation is associated with declines of ferritin content and the level of iron incorporated into ferritin and a progressive rise of iron incorporated into hemoglobin.[88] Kinetics studies provided evidence that the iron incorporated into ferritin is used for heme synthesis. The process is hampered by protease inhibitors, thus suggesting that ferritin degradation is required for utilization of its iron for heme synthesis. Ferritin degradation was active in immature (but not in mature) erythroblasts.[89] These observations indicate that ferritin can mediate cellular iron transport and donate iron for heme synthesis, mainly during early phases of erythroid maturation, when hemoglobin synthesis starts.

Studies in erythroleukemia cell lines confirmed the role of ferritin as an iron source for cellular metabolism. Ferrokinetic studies have shown that a significant amount of iron incorporated into K-562 erythroleukemic cells is bound by ferritin and mobilized to other cellular compartments when necessary.[90] The levels of ferritin in these cells, particularly levels of ferritin H chains, are major determinants of the level of the labile iron pool, as shown by overexpression of ferritin H chains into Friend erythroleukemia cells.[91] Finally, studies of the kinetics of iron mobilization in various subcellular compartments following depletion of the labile iron pool with a fast-acting chelator showed that iron is mobilized into the labile iron pool from intracellular sources with fast release from rapidly accessible cellular sources and slow release from cytosolic ferritin through proteolysis.[92]

Some conditions can modulate the sensitivity of ferritin to proteolytic degradation: (1) ascorbic acid retards the degradation of ferritin in K-562 cells by reducing its autophagocytic degradation[93]; and (2) oxidized ferritin is degraded more rapidly than non-oxidized ferritin.[94] However, the physiological roles of these two mechanisms of control of the rate of ferritin degradation remain to be proven.

Preliminary observations, based on the study of primary avian erythroblastosis, suggest that ferritin H expression is regulated in these cells in a peculiar way in that iron supplementation does not increase the mobilization of ferritin H into polysomes with consequent increase of ferritin H chain synthesis.[95] This finding further suggests that erythroid cells are programmed to express ferritins only in the initial stages of maturation and that this expression is progressively lost during advanced stages of erythroid maturation.

9.6.2 FERRITIN AS AN IRON DONOR IN HEPATIC CELLS

The liver is a major iron storage organ, containing about one-third of total body iron. Under physiological conditions, most liver iron is located within the hepatocytes, with minority in the cells of the reticuloendothelial system within the liver.[96]

The liver is involved in the uptake of iron and in its release when required. Hepatocyte iron within liver cells is bound to several proteins, including ferritin, transferrin, and heme; the majority of this iron is bound to ferritin.

Liver cells can incorporate iron through different pathways: iron bound to transferrin; iron bound to heme and vehiculated by hemopexin; iron bound to hemoglobin complexed with haptoglobin; iron bound to low molecular weight chelators; and iron bound to ferritin. Concerning the pathway of uptake of iron bound to transferrin, it must be understood that hepatocytes preferentially express on their surfaces transferrin receptor 2, a form whose expression is not modulated by iron levels due to the lack of IREs in the 3′ untranslated region of its mRNA.[97] This finding explains the particular tendency of hepatic cells to load iron in hereditary hemochromatosis. In fact, hepatic cells are unable to down-modulate their main membrane receptors involved in transferrin binding, and thus continue to incorporate and accumulate iron.

Several studies have shown that hepatocytes bind and internalize ferritin through a specific receptor-mediated process. Mack et al.[98] used ^{125}I-ferritin to demonstrate that hepatocytes bind this protein with specificity and relatively high affinity; each hepatocyte possesses about 3×10^4 binding sites. The hepatocyte ferritin receptor is able to bind ferritins isolated from different tissues with different ferritin H/L chain ratios.[99] Furthermore, the removal of carbohydrate from ferritin by glycosidase treatment did not change the level or affinity of binding of ferritin to hepatocyte membranes.[99]

Receptors for ferritin were also observed on human hepatocytes.[100] Ferritin is first bound to hepatocytes and then internalized; the uptake of iron mediated through endocytosis of ferritin is inhibited by iron chelators and enhanced by ascorbate.[101] Since chloroquine inhibits ferritin-mediated iron uptake by hepatocytes, it was suggested that this process requires lysosomal degradation and recycling of ferritin receptors.[101]

A study of the binding of recombinant human ferritin homopolymers to isolated hepatocytes provided evidence that these cells can equally bind both ferritin chains, at variance with erythroid cells which preferentially bind ferritin H chains.[102]

The receptor-mediated endocytosis of ferritin into hepatocytes is inhibited by the microtubule-inhibiting drug colchicine, thus showing that the clearance of ferritin by the hepatic ferritin receptor is a microtubule-dependent process.[103] During internalization, ferritin bound to its receptor is endocytosed at the membrane invaginations which form the endocytic vesicles through a process similar to that described for the transferrin–transferrin receptor complex. About 5 minutes after the start of internalization, internalized ferritin is found in the multivesicular endosomes and remains in these subcellular structures up to 60 minutes after the initial binding. After that, internalized ferritin, detached from its receptor, enters a lysosomal compartment where it is in large part degraded via proteolysis.

The rate of ferritin degradation within hepatocytes is controlled by certain conditions. Thus, nutrient deprivation induced an increased rate of ferritin degradation. Ascorbic acid significantly reduced the turnover of ferritin and prevented an increase of the chelatable iron pool.[104] The inhibitory effect of ascorbate on ferritin degradation was observed also in other cell types, such as fibroblasts.[105]

Ferritin degradation was transiently induced in ischemic reperfused liver. The degradation of liver ferritin causes an intracellular iron concentration that allows the translation of ferritin, thus re-establishing normal amounts of ferritin required to limit reperfusion damages.[106]

In addition to hepatocytes, ferritin receptors are present also on other liver cell types, such as reticuloendothelial cells and lipocytes.[107] Ferritin receptors are present only on activated lipocytes and bind preferentially ferritin H chains.[107] As observed in hepatocytes, lipocytes internalize ferritin after its initial binding at the cell membrane. The presence of ferritin receptors on the different cell types in the liver suggests a possible interchange of iron and ferritin between these cell types.

Macrophages release about 50% of the iron ingested through the phagocytosis of red blood cells in the form of ferritin[108,109]; hepatocytes are able to rapidly bind the ferritin released by Kupffer cells.[109] These findings suggest that ferritin, in addition to its activity in the transferrin system, may play a relevant role in the supply of iron to hepatocytes.[109]

Small amounts of ferritin, whose levels are increased in patients with iron overload, were observed in bile and believed to be the products of excretion from hepatic cells. Two theories have been proposed to explain the excretion of ferritin–iron into bile: (1) excretion of iron through transmembrane passage of ferritin molecules; and (2) excretion of ferritin through the fusion of secondary lysosomes with the biliary canalicular membrane.[110]

The effects of the lysosomatropic agent chloroquine and the microtubular inhibitors cytochalasin D and vinblastine on biliary excretion of exogenous and endogenous ferritin were investigated.[111,112] Both types of drugs significantly decreased biliary ferritin excretion, thus indicating that the release of ferritin into the bile is a chloroquine-sensitive, microfilament-dependent process.[111,112]

9.7 HEMOSIDERIN

Hemosiderin is a poorly defined iron–protein complex that forms an insoluble iron storage system thought to be derived from the lysosomal degradation of the ferritin protein shell.[3] The hemosiderin name originally given to a peculiar protein complex should reflect its presumed origin; however, the iron in hemosiderin is non-heme. A peculiar property of hemosiderin is its insolubility, as compared to ferritin which is soluble; however, hemosiderin can be solubilized after purification by incubation with alkaline solutions or solutions containing appropriate detergents.[3] The origin of hemosiderin from ferritin is supported by studies showing its reactivity with anti-ferritin antibodies. Analysis of the proteic component of hemosiderin suggests that it contains degradation products of ferritin chains.

Under normal iron homeostatic conditions, the body tissues contain only low amounts of hemosiderin; however, under conditions of primary or secondary iron overload, the tissue hemosiderin content dramatically increases, particularly in organs such as the liver, pancreas, and heart.

Some biochemical properties of hemosiderin are different from those observed for ferritins. The structures of the iron cores in hemosiderins are highly variable and in part relate to the origin of these protein–iron complexes.[113-116] Three different particle structures have been identified:

1. A structure based on that of ferrihydrite ($5Fe_2O_3 \cdot 9H_2O$), preferentially observed in hemosiderins isolated from tissues of normal subjects and patients with mild iron overload.
2. A highly defective structure based on that of the mineral goethite (α-FeOOH), preferentially observed in hemosiderins isolated from tissues of patients with the iron overload disease, secondary hemochromatosis (thalassemia).
3. A non-crystalline disordered ferrihydrite-like structure, preferentially observed in hemosiderins isolated from patients with primary hemochromatosis.

The methodology usually adopted to determine the structures of the mineral cores of hemosiderins involves analysis of Mossbauer spectroscopy and electron diffraction on samples of purified hemosiderins. This analysis is important because the three different structures have different degrees of tissue toxicity and exhibit different reactivities with chelating drugs.[117] It is of interest that the compositions of the mineral cores of hemosiderins may be related to the disease responsible for the iron overload and also to the medical treatment. Thalassemic patients transfused regularly have higher fractions of their non-heme spleen iron in a goethite-like structure than never-transfused patients.[116]

The pathologic conditions most frequently associated with the accumulation of hemosiderin in the liver are hereditary hemochromatosis[118] and thalassemias.[119] The monitoring of iron accumulation in the liver is a tool for evaluating the severity of iron overload and the response to iron chelation therapy.[120]

9.8 SERUM FERRITIN

Ferritin is detected at significant levels in normal human serum. A large number of studies and a few decades of clinical experience have clearly shown that the level of serum ferritin reflects the concentration of stored iron. Immunoassay of serum ferritin has therefore become widely used in diagnosing iron-related disorders and in surveying of iron status. Furthermore, the evaluation of serum ferritin concentration along with the determination of other serum markers, including transferrin concentration and saturation and soluble transferrin receptor levels, led to differential diagnosis of anemic conditions of difficult classification.

Finally, the study of serum ferritin levels in several neoplasias have shown that ferritin levels are abnormally increased in some tumors and may represent non-specific markers for prognostic evaluation or assessment of the response to therapy. A large number of studies focused on establishing the diagnostic role of serum ferritin levels in the evaluation of iron-deficiency conditions.

In normal adult males, the serum ferritin concentration is around 120 µg/L, while the value in menstruating women is around 45 µg/L. Serum ferritin values increase rapidly in men between 18 and 30 years of age, a phenomenon related to the establishment of iron stores during those years. Serum ferritin values in women remain virtually constant between 18 and 30 years of age. At birth, serum ferritin levels are comparable to those found in normal male adults. These values are higher in female infants than in male infants.[121] There is no obvious explanation for this difference. Analysis of serum ferritin levels during fetal life showed a progressive increase up to the levels observed in newborns.[122]

Several studies have shown the role of serum ferritin evaluation in the diagnosis of iron deficiency. Measurement of serum ferritin is currently the accepted laboratory test for diagnosing iron deficiency, and a ferritin value below 12 µg/L is a specific indicator of iron deficiency.[123] Studies of normal volunteers undergoing repeated phlebotomies[124] and in normal subjects undergoing blood donation[125] clearly showed that levels of serum ferritin clearly reflect the size of body iron stores.

The evaluation of serum ferritin is not of value for assessing the responses of patients with iron deficiency anemia to iron therapy. Patients treated with standard therapy (i.e., oral ferrous sulfate) showed no rise of serum ferritin levels during the first 3 weeks of treatment, although hematologic response occurred.[126]

Other laboratory tests commonly used for the medical diagnosis of anemia, such as the determinations of serum iron, total iron-binding capacity, mean corpuscular volume, and transferrin saturation do not provide additional diagnostic advantage over ferritin.[126] However, the diagnostic impact of serum ferritin evaluation is limited by the fact that ferritin is an acute phase reactant. Thus, the diagnosis of iron deficiency anemia in ill or hospitalized patients can be difficult, since these patients may have normal or increased ferritin values even when iron deficiency exists. To overcome this difficulty, a combined evaluation of serum ferritin and soluble transferrin receptor levels was proposed.

In patients with selective iron deficiencies, the levels of soluble transferrin receptor are increased, while they are low in anemia associated with chronic disease.[127] Patients with a combination of these two conditions (i.e., iron deficiency and chronic disease) have high soluble transferrin receptor levels associated with increased serum ferritin levels.[128] Dividing the ratio of soluble transferrin receptor values by the log of ferritin values offers a simple and clear tool to distinguish the three groups of patients, i.e., those with iron deficiency, those with anemia associated with chronic disease, and those with a combination of both conditions.[128]

The combined evaluation of these two parameters is also of value to identify healthy subjects with subclinical iron deficits. In a study of 65 healthy adults, subclinical iron deficits were observed only among women. These subjects exhibited high soluble transferrin receptor/ferritin ratios that returned to normal after oral iron supplementation.[129] A second study based on the measurement of these two parameters on dried plasma spots confirmed that the evaluation of the soluble transferrin receptor/ferritin ratio allows to identification of iron-depleted individuals.[130]

Serum ferritin levels are markedly increased in primary and secondary iron overload. The measure of serum ferritin levels in such patients provides an approximate estimation of tissue iron loading. However an accurate evaluation of tissue iron stores can be provided only through evaluation of liver iron concentration by biomagnetometry. In patients with consistent iron overload, the correlation between serum ferritin and iron stores is poor.[131] It was recently proposed to evaluate the level of iron contained in ferritin in addition to total serum ferritin. Total serum ferritin closely correlates with total ferritin levels; however, the correlation between ferritin iron concentration and liver iron concentration was poor in both β-thalassemia and hemochromatosis patients.[132] The degree of ferritin iron saturation was about 5% in non-iron-loaded patients; in contrast, in patients with liver cell damage, the ferritin iron saturation in serum was significantly higher than levels found in groups with iron overload disease.[132]

These observations suggest that the ferritin iron saturation reflects the tissue origin of serum ferritin. Iron overload patients show an inverse relationship between serum ferritin levels and soluble transferrin receptor levels.[133] This finding is related to the mechanisms involved in the control of the levels of soluble transferrin receptors, i.e., the rate of erythropoiesis and iron status. Soluble transferrin receptor levels are increased in iron deficiency. The opposite phenomenon is observed in iron overload conditions.

It is of interest that a peculiar form of non-transferrin-bound iron was found in the circulation of iron-overloaded patients. This form is seemingly a molecular aggregate of iron released into the cells in a chemical structure not able to bind to transferrin. Although this iron form is not found in normal serum, its release in serum is not related quantitatively to the levels of tissue iron stores.[134]

Ferritin levels are greatly increased in pediatric patients exhibiting perinatal hemochromatosis[135] and in newborns who have a peculiar syndrome characterized by fetal growth retardation, lactic acidosis, liver hemosiderosis, and iron overload.[136] However, even in these pediatric patients, no evidence indicates that serum ferritin levels directly reflect the levels of tissue iron stores.

Serum ferritin levels are increased in many acute and chronic pathological conditions associated with inflammatory processes. Iron metabolism parameters in inflammatory disease include blockage of tissue iron release, decreased serum iron and total iron binding capacity, and elevated serum ferritin level reflecting augmented ferritin synthesis as part of the acute-phase response. The significance and mechanisms responsible for the elevation in serum ferritin during inflammatory processes remained obscure for a long time and whether it resulted from secretion of ferritin from specific tissues or from leakage of intracellular ferritin due to cell death remained unclear.

Initial studies have shown that induction of fever in normal subjects through injection of low doses of bacterial endotoxin elicited a significant rise in serum ferritin concentration associated with a decline in serum iron levels.[137] Several other *in vitro* and *in vivo* studies confirmed that different inflammatory stimuli elicited increases of ferritin levels; this phenomenon was observed for serum ferritin levels and capacity to synthesize ferritin by many cell types.[138,139]

Some cytokines, such as tumor necrosis factor α, interleukin-6, interleukin-1, and interleukin-8 are key mediators of the inflammatory process. Cytokines are able to stimulate ferritin synthesis by several cell types, including fibroblasts, hepatocytes, endothelial cells, muscle cells, and macrophages.[140-147] The stimulatory effects exerted by the cytokines in most cell types resulted in a preferential, and in many cases, selective stimulation of ferritin H synthesis. One of these studies showed that stimulation of hepatocytes with tumor necrosis factor α resulted in a marked stimulation of both synthesis and secretion of ferritin.[146] This study suggested that the increased ferritin levels observed in inflammatory conditions may be related to increased secretion of ferritin in circulation by tissues stimulated by inflammatory cytokines.

Increases in serum ferritin levels were observed in several clinical conditions associated with acute inflammatory responses. Examples of these conditions are acute myocardial infarction where the increase in serum ferritin levels was moderate and started on the second day post-infarction, reaching a peak corresponding to four times the initial level on the sixth day post-infarction[148]; acute hepatocellular damage, where very high increases of serum ferritin levels, up to >45,000 µg/L, are observed[149]; acute malaria, where the levels of serum ferritin are always increased and correlated with other serum markers of inflammation, such as C-reactive protein[150]; and arthritis and osteoarthritis, where the rise of serum ferritin reflects the degree of joint inflammation.[151]

In some inflammatory conditions associated with infectious processes, such as meningitis, the rises in ferritin levels are better observed in cerebrospinal fluid, where the ferritin levels provide additional guidance in evaluation of the response to therapy.[152]

The study of serum ferritin levels is particularly relevant from diagnostic and prognostic points of view in some chronic conditions associated with inflammatory responses. Serum ferritin concentrations are markedly elevated in both juvenile and adult chronic polyarthritis (Still's disease),[153] and serum ferritin is a sensitive indicator of disease activity.[154] Analysis of the ferritin composition present in the serum of Still's disease patients showed that it is composed almost exclusively of L ferritin chains[155] and is only scarcely (<15%) glycosylated.[156] Although serum ferritin was always elevated in adult and juvenile forms of the disease, adults had significantly higher serum ferritin concentrations.[157,158]

Serum ferritin levels have been explored in detail in acute and chronic liver diseases associated with inflammatory responses. The most common liver diseases are caused by alcohol abuse. Alcohol consumption causes increased levels of serum ferritin and the levels correlate with the amounts of alcohol consumed. Many alcoholic patients have increased liver iron stores. However, the hepatic iron observed in alcoholic patients is deposited in the Kupffer cells, while iron accumulates in hepatocytes in hereditary hemochromatosis.

Serum ferritin is more frequently elevated in patients with alcoholic liver disease (about 60%) than in patients with other chronic diseases such as autoimmune liver disease and hepatitis C.[159] Interestingly, serum ferritin levels decrease rapidly during alcohol abstinence. However, the increased serum ferritin levels observed in alcoholic patients do not necessarily correlate with hepatic iron stores and may therefore reflect inflammatory process rather than increased iron stores.[160] The serum ferritin levels observed in alcoholic patients are largely glycosylated, thus suggesting that the ferritin derives from active cellular secretion and not from damaged cells.[160]

The evaluation of serum ferritin levels in other chronic liver diseases showed interesting results. Non-alcoholic steatohepatitis is a necro-inflammatory disorder of the liver associated with fatty infiltration of hepatocytes. Its cause is unknown, but evidence suggests that its pathogenesis involves increased oxidative stress and mitochondrial dysfunction. Serum ferritin levels are frequently increased, along with moderate increases of hepatic iron stores.[161,162] Patients carrying HFE alleles exhibited higher ferritin levels as compared to patients with normal hemochromatosis genes.[162]

Other studies focused on the possible diagnostic and prognostic significance of serum ferritin levels in viral hepatitis. Serum ferritin levels are increased in a significant proportion of patients chronically infected with hepatitis B virus.[163] Interestingly, patients who had high levels of circulating ferritin had significantly better chances of developing primary hepatocellular carcinomas than patients with lower ferritin levels.[163] High serum ferritin level correlated with poor response to anti-viral therapy with interferon.[164] The administration of an iron chelator improved the response to interferon therapy in the high ferritin group.[164]

Other studies focused on modifications of serum ferritin levels in chronic C hepatitis. Liver iron accumulation is observed in 40 to 50% of patients with active chronic C hepatitis, but it is usually mild.[165] It has been suggested that liver iron accumulation may play a role in the course of chronic C hepatitis and negatively influence the efficiency of interferon-α treatment.[166] Serum ferritin levels are often increased in chronic C hepatitis and these increased values correlate with the activity of the disease, i.e., with alanine aminotransferase levels.[167]

The hepatic iron content and serum ferritin levels of patients with chronic viral hepatitis were significantly lower in responders to interferon therapy as compared to those who failed to respond to this therapy.[168] Serum ferritin levels are correlated with hepatic C virus genotype; the highest values were associated with genotype 16 and the lowest levels with genotype 3a.[169] Hepatitis C patients with high serum ferritin levels exhibited low concentrations of glutathione within hepatic cells.[169] Serum ferritin levels exhibited a very good correlation with hepatic liver content among patients with chronic hepatitis C; the highest serum ferritin levels were observed in patients with cirrhotic livers.[170]

Serum ferritin levels are increased in porphyria cutanea tarda, a disorder of porphyrin metabolism associated with decreased uroporphyrogen decarboxylase in the liver and characterized by photosensitive dermatitis and hepatic siderosis. The key role of iron in pathogenesis is well established since iron overload is one of the main factors triggering the clinical manifestations, and iron depletion represents the cornerstone of therapy.[171] The disease is frequently associated with conditions that favor hepatic iron accumulation, such as the hemochromatosis gene and hepatitis C virus. Serum ferritin levels are increased in all patients with porphyria cutanea tarda and particularly in those homozygous for hereditary hemochromatosis mutant genes.[172]

Serum ferritin levels decline during pregnancy, particularly during the third trimester, as iron stores are depleted by fetoplacental demand and expansion of maternal red cell mass. The values of ferritin observed in maternal serum serve as a prognostic index of pregnancy outcome.[173] Several studies support this conclusion. High values of ferritin (i.e., >40 ng/ml) during the third trimester of pregnancy are associated with increased risk of preterm delivery.[174,175] High serum ferritin level during the third trimester increases the risk of preterm delivery more than eight-fold.[176]

An increased level of ferritin during pregnancy may be related also to fetus infection, such as infant sepsis or chorioamnionitis.[177] On the other hand, decreased serum ferritin levels during pregnancy are related to iron deficiency. Low ferritin concentrations in early pregnancy are associated with low birth weight and increased placental vascularization at term.

Increased serum ferritin levels in newborns are associated with disease leading to iron accumulation, such as neonatal hemosiderosis and a recently described syndrome characterized by fetal growth retardation, lactic acidosis, liver hemosiderosis, hypotransferrinemia, and hyperferritinemia.[178]

The modifications of serum ferritin levels observed in neurological diseases and neoplasias will be discussed in the sections related to ferritin in the central nervous system (Section 9.13.3) and in tumors (Section 9.12), respectively.

The prognostic significance of serum ferritin values in arteriosclerotic disease was evaluated in a series of studies. Several supported the idea that increased serum ferritin may represent a risk factor for the development of ischemic heart disease; high serum ferritin concentrations were positively correlated with the incidence of myocardial infarction.[179] Serum ferritin levels can indicate the presence and progression of carotid artery disease.[180,181] However, other studies failed to demonstrate a clear association between serum ferritin levels and the risk of myocardial infarction.[182] The subject was recently re-evaluated, and only subjects with serum ferritin levels clearly increased over the normal range exhibited increased risks of myocardial infarction.[183] However, this increased risk was most evident in smokers and subjects with diabetes, thus suggesting that ferritin may be a risk factor for ischemic heart disease in the presence of other risk factors.[183]

In conclusion, although some epidemiological studies suggest increased iron stores, reflected by increased serum ferritin levels, represent a risk factor for the development of ischemic heart disease, the issue remains controversial and must be assessed by specific clinical trials.[184] Thus, it is difficult to accept as experimentally demonstrated the hypothesis that female cardiovascular protection can be attributed to the iron loss caused by menstruation and that blood donations are recommended for prevention of ischemic heart disease.[185]

Few studies suggest that increased serum ferritin levels are also associated with increased risk of diabetes. In fact, studies of patients with newly diagnosed diabetes and a group of normal controls showed that the risk of newly diagnosed diabetes is significantly higher among male subjects exhibiting serum ferritin levels higher than 300 ng/ml and female subjects with values higher than 150 ng/ml.[186] However, this study did not clarify whether the increased serum ferritin levels could have been related to an inflammatory condition or to increased iron stores. Among diabetic patients, serum ferritin levels are clearly increased in patients with retinopathy, and particularly in those undergoing photocoagulation.[187]

Serum ferritin levels are also affected in normal subjects by genetic polymorphisms. A recent study showed that serum ferritin levels are significantly higher in individuals homozygous for the haptoglobin 2 allele than in individuals homozygous for the haptoglobin 1 allele or heterozygous for both alleles.[188] The higher serum ferritin concentrations observed in 2/2 individuals were associated with higher L ferritin concentrations in monocytes–macrophages, possibly as a consequence of a delocalization pathway leading to increased iron accumulation in phagocytic cells.[188]

There was no correlation between haptoglobin polymorphism and serum ferritin levels in females and this was seemingly due to menstrual bleeding. Interestingly, a very good correlation was found between serum ferritin levels and monocyte L ferritin levels.[188]

Few studies have focused on the structure of ferritin in serum. They have been hampered by the low levels of ferritin present in normal serum. Initial studies showed that ferritin isolated from serum of hemochromatosis patients exhibited a peculiar biochemical composition in that it was composed of H, L, and G subunits.[189] The G subunit was of a larger molecular weight than the H and L subunits. It stained for carbohydrates, thus indicating that it is glycosylated.[188]

The analysis of ferritin purified from normal horse serum showed several interesting findings: (1) serum ferritin contained lower iron levels than tissue ferritin; (2) the number of subunits composing each serum ferritin molecule was highly variable; (3) subunits with molecular weights higher than those of H and L subunits are present in serum ferritin (weights of about 26, 50, and 58 kDa, respectively); (4) a significant proportion (>50%) of serum ferritin subunits are glycosylated; the glycosylated subunits exhibit high molecular weights that markedly decrease after endoglycosidase treatment; and (5) preliminary amino acid sequences of N terminal peptides derived from 50- and 58-kDa subunits showed that their sequences are different from those of H and L ferritin subunits.[190]

According to these findings, serum ferritin may represent the product of a separate ferritin gene, encoding a secretory ferritin.[189] However, serum ferritin is certainly related to tissue ferritin in that serum ferritin clearly reacts with antibodies raised against tissue ferritins. It was suggested, but not experimentally proven, that the glycosylated form of ferritin in serum may derive from a secretory process, while the unglycosylated form of serum ferritin may be related to cell lysis.[191]

Ferritin circulates bound to proteins in the β-2 region of human serum. More recently, it was shown that both H and L ferritin subunits are able to bind H kininogen. This protein exerts several biological activities and generates, through proteolytic degradation, bradykinin, a potent vasoactive and pro-inflammatory nanopepetide. Since both ferritin and H-kininogen are involved in the inflammatory response, it was suggested that the ferritin–H-kininogen complex may also play a role in the inflammatory response and, particularly, ferritin may control the release of bradykinin from H-kininogen.[192]

In canine serum, ferritin forms a complex with the immunoglobulins in the IgM and IgA groups, but not with the IgG subclass.[193] It is of interest that the level of serum ferritin is a predictor of the rate of intestinal iron absorption. Heme- and non-heme iron absorption is inversely associated with serum ferritin concentration.[194] This finding further reinforces the concept that serum ferritin levels, at least in healthy subjects, represent a valuable measure of the size of body iron stores.

Ferritin was observed in other extracellular fluids, such as milk. In fact, human milk (both colostrum and milk) contains significant levels of ferritin, whose immunological reactivity and pattern of glycosylation were similar to those of serum ferritin.[195]

9.9 FERRITIN GENES

The structures of ferritin genes and of ferritin mRNA encoding H and L subunits were explored in detail in recent years.

Human ferritin H chain cDNA was initially cloned from a human adult liver cDNA library.[196] The analysis of this cDNA showed an open reading frame starting with an ATG codon at position 92 of the cDNA sequence and ending at position 662 with a termination codon (Figure 9.2). The open reading frame is preceded by 91 residues and followed by 137 residues as 5′ and 3′ untranslated sequences. The analysis of human genomic DNA by Southern blot provided preliminary evidence of the existence of multiple genes encoding the H ferritin subunit; however, subsequent studies showed that a single gene encodes the functional ferritin H chain, while the other genes are non-functional pseudogenes.

In subsequent studies based on screening of human liver genomic libraries, the gene encoding human heavy-chain gene was cloned and its structure was determined.[197,198] The ferritin H chain gene is arranged as four exons and three introns (Figure 9.2). The major transcription start site is located in the first exon 24 base pairs downstream of the first T of the TATA box. Intron 1 is about 1.65 Kb long and is placed between mRNA bases 331 and 332; intron 2 is 256 bp long and is placed between bases 478 and 479; finally, intron 3 is 96 bp long and is placed between bases 604 and 605.

The gene encoding the human functional H ferritin gene is located on chromosome 11, at position 11q13.[199] The cDNA encoding human L ferritin was isolated from a cDNA library of the human promonocytic U937 cell line[200] and the human liver.[201] This cDNA showed an open reading frame coding for a 174-amino acid protein (Figure 9.3). The L ferritin gene has a structural organization similar to that observed for the H ferritin gene. The gene is composed of four exons separated by three introns which split the coding sequence at codons 33/34, 82/83, and 124/125, respectively[201] (Figure 9.3). The start of transcription is located 198 bp upstream of the AUG codon; in the 5′ flanking sequence of the gene a TATAA-like box is observed at −30 base pairs from the start of transcription. The gene encoding the L ferritin subunit was localized on chromosome 19.

In addition to the two functional genes coding for L and H ferritin subunits, several pseudogenes have been observed. The pseudogenes are non-functional sequences characterized by close similarities to one or more paralogous genes.[202] Their lack of function is the result of a failure of transcription or translation or production of a protein that does not possess the same functional properties as the protein encoded by the normal paralogue gene. The nucleotide sequences of the pseudogenes differ from those of the paralogous functional genes at crucial points in the gene sequence. The pseudogenes may arise as a consequence of gene duplication by two possible mechanisms: retro-transposition and duplication of genomic DNA. Importantly, pseudogenes arising from retro-transposition (probably generated from single-stranded RNA by RNA polymerase II) represent the so-called "processed pseudogenes", characterized by the absence of both 5′ promoter region and introns.[202]

More than twenty H ferritin pseudogenes have been reported, scattered along the majority of human chromosomes, the majority representing non-functional

H - FERRITIN GENE

```
mRNA

  1  caccgcaccc  tcggactgcc  ccaaggcccc  cgccgccgct  ccagcgccgc  gcagccaccg
 61  cgccgccgc   cgcctctcct  tagtcgccgc  catgacgacc  gcgtccaccc  cgcaggtgcg
121  ccagaactac  caccaggact  cagaggccgc  catcaaccgc  cagatcaacc  tggagctcta
181  cgcctcctac  gttacctgt   ccatgtctta  ctactttgac  cgcgatgatg  tggcttttgaa
241  gaactttgcc  aaatactttc  ttcaccaatc  tcatgaggag  agggaacatg  ctgagaaact
301  gatgaagctg  cagaaccaac  gaggtggccg  aatcttcctt  caggatatca  agaaaccaga
361  ctgtgatgac  tgggagagcg  ggctgaatgc  aatggagtgt  gcattacatt  tggaaaaaaa
421  tgtgaatcag  tcactactgg  aactgcacaa  actggccact  gacaaaaatg  accccattt
481  gtgtgacttc  attgagacac  attacctgaa  tgagcaggtg  aaagccatca  aagaattggg
541  tgaccacgtg  accaacttgc  gcaagatggg  agcgcccgaa  tctggcttgg  cggaatatct
601  ctttgacaag  cacacctggg  agacagtgat  aatgaaagct  aagcctcggg  ctaatttccc
661  atagccgtgg  ggtgacttcc  tggtcaccaa  ggcagtgcat  gcatgttggg  gtttcctta
721  ccttttctat  aagttgtacc  aaaacatcca  cttaagttct  ttgatttgta  ccattccttc
781  aaataaagaa  atttggtacc  c
```

"MTTASTSQVRQNYHQDSEAAINRQINLELYASYYYLSMSYYFDR

PROTEIN

DDVALKNFAKYFLHQSHEEREHAEKLMKLQNQRGGRIFLQDIKKPDCDDWESGLNAME
CALHLEKNVNQSLLELHKLATDKNDPHLCDFIETHYLNEQVKAIKELGDHVTNLRKMG
APESGLAEYLFDKHTWETVIMKAKPRANFP"

FIGURE 9.2 Organization of the human ferritin H gene composed of four exons (boxes) and three introns (lines between boxes) (top). Coding sequence of the human ferritin H gene (middle). The deduced amino acid sequence of the corresponding protein (bottom).

L - FERRITIN GENE

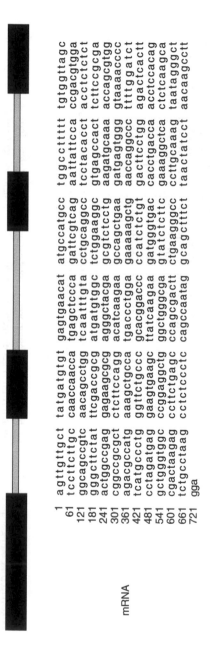

```
         1  agttgttgct  tatgatgtgt  gagtgaacat  atgccatgcc  tggccttttt  tgtggttagc
        61  tccttcttgc  caaccaacca  tgagctccca  gattcgtcag  aattattcca  ccgacgtgga
       121  ggcagccgtc  aacagcctgg  tcaatttgta  cctgcaggcc  tcctacacct  acctctctct
       181  ggcttctat   ttcgaccgcg  atgatgttggc tctgaaggc   gtgagccact  tcttccgcga
       241  actggccgag  gagaagcgcg  aggctacga   gcgtctcctg  aagatgcaaa  accagcgtgg
       301  cggccgcgct  ctcttccagg  acatcaagaa  gccagctgaa  gatgagtggg  gtaaaacccc
       361  agacgccatg  aaagctgcca  tgacctgga   gaaaaagctg  aaccaggccc  ttttggatct
       421  tcatgccctg  ggttctgccc  gcacggaccc  ccatctctgt  gacttcctgg  agactcactt
       481  cctagatgag  gaagtgaagc  ttatcaagaa  gatgggtgac  cacctgacca  acctccacag
       541  gctgggtggc  ccggaggctg  ggctgggcga  gtatctcttc  gaaaggctca  ctctcaagca
       601  cgactaagag  ccttctgagc  ccagcgactt  ctgaagggcc  ccttgcaaag  taatagggct
       661  tctgcctaag  cctctccctc  cagccaatag  gcagcttct   taactatcct  aacaagcctt
       721  gga
```

mRNA

PROTEIN   "MSSQIRQNYSTDVEAAVNSLVNLYLQASYTYLSLGFYFDRDDVA
                   LEGVSHFFRELAEEKREGYERLLKMQNQRGGRALFQDIKKPAEDEWGKTPDAMKAAMT
                   LEKKLNQALLDLHALGSARTDPHLCDFLETHFLDEEVKLIKKMGDHLTNLHRLGGPEA
                   GLGEYLFERLTLKHD"

FIGURE 9.3 Organization of the human ferritin L gene composed of four exons (boxes) and three introns (lines between boxes) (top). Coding sequence of the human ferritin L gene (middle). The deduced amino acid sequence of the corresponding protein (bottom).

processed pseudogenes.[203-205] The L ferritin pseudogenes have been identified. The analysis of the sequences of H ferritin pseudogenes led to some interesting speculations. Although most H ferritin pseudogenes seem to be non-functional ferritin sequences, the majority of them share several conserved mutations. This observation was tentatively explained by suggesting that these sequences may have arisen from a second hypothetical functional H gene.[205]

9.10 TRANSCRIPTIONAL CONTROL OF FERRITIN GENE EXPRESSION

A large number of studies have provided evidence that translational mechanisms play a major role in the control of ferritin expression. The synthesis of ferritin is regulated at the translational level by the cytosolic content of chelatable iron. This response to iron is regulated by the iron-modulated binding to ferritin mRNA of a protein acting as a repressor, the iron regulatory protein. This mechanism is discussed in detail in the chapter on iron regulatory proteins.

However, in addition to the translational mechanisms, transcriptional mechanisms play a key role in the control of ferritin gene expression in some conditions. One condition associated with a transcriptional regulation of ferritin gene expression is stimulation of thyroid cells by thyroid-stimulating hormone (TSH) or cyclic-AMP (cAMP). Under these conditions, TSH and cAMP increase the level of ferritin mRNA,[206] mainly through a transcriptional mechanism.[207] A second condition associated with transcriptional regulation of ferritin gene expression is stimulation of ferritin H mRNA by tumor necrosis factor and other inflammatory cytokines.[208] The stimulation induced by tumor necrosis factor was completely prevented by actinomycin D, thus indicating a transcriptional mechanism.[207] A third condition is stimulation of ferritin mRNA levels by heme[208]; the stimulation of ferritin gene expression was related to both transcriptional and translational mechanisms.[209] The fourth condition is repression of ferritin H transcription induced by the E1A protein of the adenovirus that acts as a potent oncogene capable of modifying the transcriptional activities of selected genes.[210] The elucidation of the molecular mechanisms underlying these conditions of transcriptional regulation greatly contributed to the understanding of the regions of the ferritin H gene promoter involved in the transcriptional regulation of this gene and of the transcriptional factors interacting with these regions.

These observations promoted a series of studies evaluating the functional roles of different regions of the ferritin H gene promoter in the control of the transcription of this gene. The initial studies based on the analysis of the sequences of mouse and rat ferritin gene promoters initially showed the presence of a DNA motif corresponding to a GC-rich box and a CCAAT box, both frequently associated with eukaryotic gene promoters and able to bind ubiquitous transcription factors.[211,212]

The regions of the ferritin H gene promoter required for the basal transcriptional activity of this gene were determined by classic transcriptional assays based on the evaluation of transcriptional activity of promoter fragments ligated upstream of the reporter gene luciferase. The basal transcriptional activity in both NIH 3T3 cells and in thyroid cells increased for promoter fragments of increasing size from 219 to 351, 666, and 1046 bp.[212] These initial studies showed that the first 1000 nucleotides in the 5′ flanking region of the ferritin H gene contain the nucleotide sequences required to sustain the basal transcriptional activity of this gene.

These findings were confirmed in a subsequent study that also showed a 140-bp region, from 0 to −140, containing one CCAAT box and two Sp1 binding sites, is required for minimal promoter activity; however, the basal promoter activity was significantly higher for a 940-bp promoter fragment than for the 140-bp minimal promoter[213] (Figure 9.4). This study showed also that an enhancer regulatory element, lying 4.5 Kb upstream of the ferritin transcription start site, functions like an enhancer-inducible element during chemical differentiation of mouse erythroleukemia cells[212] (Figure 9.4).

Considerable efforts have focused on the analysis of the proximal region of the ferritin H gene promoter contained within the first 200 bp 5′ upstream of the start transcription site. Initial footprinting analysis of this DNA region provided evidence of the existence of two sites named A and B: site A spanning from −132 to −109, and site B spanning from −62 to −45.[214] The B site appeared to be of critical importance for the control of ferritin H gene transcription. Deletion analysis of this region showed that a transcription factor interacting with this site is required for transcription of the ferritin gene.[215] Cells undergoing differentiation or treated with translation inhibitors exhibited an increase of the binding of transcription factors to the B box of the ferritin promoter, a phenomenon associated with an increased rate of ferritin gene transcription.[215,216]

Subsequent studies have shown that the B box binding factors are required for cAMP induction of the ferritin promoter.[216] In fact, the integrity of the B box of the ferritin H promoter is required for the transcriptional response to cAMP. The binding of the B box factor to the ferritin promoter does not seem to involve interactions with typical cAMP response elements.[217] The B box binding factor forms a complex with the transcriptional adaptors CBP and p300, and the formation of this complex is completed by the product of the adenovirus E1A gene.[216]

Since the E1A protein is inhibitory for both ferritin H gene transcription and the formation of the complex between the B box binding factor and CBP/p300, it was proposed that the formation of this complex is required for the transcriptional activity of the ferritin H gene.[217] This conclusion is strengthened by the observation that okadaic acid, a tumor promoter that acts also as a potent phosphatase inhibitor, stimulates the formation of the BBF/p300 complex and enhances ferritin H gene transcription.[218]

The role of the p300 component of this complex is largely related to its capacity to bind the nuclear P/CAF protein, which possesses intrinsic histone acetylase activity.[219] The P/CAF protein, once bound to specific promoters via interaction with p300, may locally induce an open chromatin conformation compatible with transcriptional activity.[220,221] P/CAF seems to play a relevant role in the tissue-specific control of ferritin H gene expression, as suggested by the observation that tissues exhibiting high levels of ferritin H expression also possess high levels of P/CAF.[219] Furthermore, P/CAF overexpression counteracts the inhibitory effects of E1A adenovirus protein on ferritin H gene transcription.[219]

MOUSE H - FERRITIN GENE - PROMOTER REGION

FER - 1   CCATGACAAAGCACTTTTGGGAGCCCAACCCCTCCAAAGGAGCAGAATGCTGAGTCACGGTG
           AP - 1   DVAD   AP - 1/NF - E2

FER - 2   GGTACACTTGCAAATATCAGAATTTCCAGCACACTTCTCGGGGAATCCCATCCTTTTGCAACACT
                                                    NFkB

CCAAT   CTACAGGAAGAGAGGCGGGGGCTGGGCGGCCCACCGCGCTGATTGGCCGGAGCGCGCCTGACGCGCAGGATCCCGCTATAAAGTGC
                                  Sp1   NF - Y / p300   AP - 1

FIGURE 9.4 Organization of the murine ferritin H gene promoter. The region upstream of the transcription start site contains three boxes, Fer-1, Fer-2, and CCAAT which were characterized from a functional and structural point of view. The transcription factors interacting with these boxes are reported.

Independent studies focused on determining the transcriptional mechanisms responsible for the induction of ferritin H expression associated with cell differentiation led to the identification of a transcriptional factor able to interact with the ferritin promoter B box. These studies were based on two cellular systems: (1) Friend erythroleukemia cells stimulated to differentiate by hemin treatment; and (2) normal human monocytes spontaneously differentiating *in vitro* to macrophages.

Cell differentiation in both these systems was associated with increased binding of a transcription factor at the level of a CCAAT element present in reverse orientation at the level of the B box of the ferritin H gene promoter.[212] This element is responsible for binding and for gene activity and also binds the ubiquitous transcription factor NFY[221] (Figures 9.4 and 9.5). This transcription factor is formed by three subunits, A, B, and C. Interestingly, freshly isolated monocytes do not possess the NFYA subunit, whose expression is induced following maturation of these cells to macrophages.[221] The lack of NFYA synthesis is due to post-transcriptional regulation, since NFYA mRNA was present during all stages of maturation of monocytes to macrophages, while NFYA was present only in differentiating macrophages.[223]

In another study, the B subunit of NFY transcription factor was able to bind p300/pCAF nuclear proteins. The binding of this large complex formed by p300/pCAF and NFY allows the recruitment of RNA polymerase II at the ferritin promoter and thus the transcriptional activity of this gene.[224] The formation of this transcriptional complex is potentiated by cAMP and this finding explains the molecular mechanism through which cAMP stimulates the transcription of ferritin H gene promoter.

Studies initially aimed to evaluate the molecular mechanisms through which E1A adenovirus product represses ferritin H gene transcription led to mapping a region of the ferritin promoter located approximately 4.1 Kb 5′ to the transcription initiation site that acts as an enhancer of the ferritin H gene.[225]

This region is called FER-1 and includes an AP-1-like sequence, 22-bp dyad symmetry, and AP-1/NF-E2 binding sites (Figure 9.4). The AP-1 site was shown to bind c-*fos* and c-*jun* and its integrity is required for the repressive action of E1A on ferritin H gene transcription.[225] The dyad symmetry element along did not display any significant enhancer activity, but increased the enhancer activity of the AP-1-like sequence.

A subsequent study better defined the transcription factors able to bind to the AP-1-like and dyad symmetry sites. The AP-1-like site was shown to bind d-*jun*, b-*fos* and ATF1, while the GC-rich elements of the dyad symmetry element bind Sp1 and Sp3 transcription factors.[226]

The activity of the FER-1 enhancer is influenced by the p300/CBP transcriptional co-activators with intrinsic histone acetylase activity. In fact, the capacity of the E1A adenovirus product to inhibit FER-1 enhancer activity depends on its capacity to interact with the p300/CBP co-activators, as suggested by four lines of evidence: (1) mutants of the E1A protein unable to bind the p300/CBP proteins fail to repress ferritin H gene transcription; (2) p300/CBP is able to restore the FER-1 transcriptional activity repressed by the E1A protein; (3) p300/CBP is able to increase FER-1 enhancer activity; and (4) sodium butyrate and other histone deacetylase inhibitors are able to mimic the effect of p300/CBP on ferritin H gene transcription.[227]

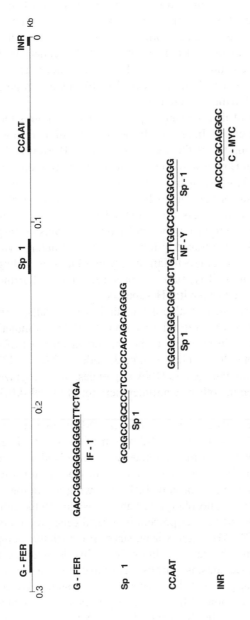

FIGURE 9.5 Organization of the proximal region of the human ferritin H gene promoter. The region 0.3 kb upstream of the transcription start site contains four boxes, G-Fer, Sp1, CCAAT, and INR which were characterized from a functional and structural point of view. The transcription factors interacting with these boxes are reported.

Recent studies provided evidence that the FER-1 enhancer is also involved in mediating the transcriptional response of the ferritin H gene to oxidative stress. In fact, a series of recent studies have shown that oxidative stress can modulate ferritin synthesis. The effect of oxidative stress is biphasic, with initial inhibition of ferritin synthesis due to IRP1 activation, followed by an increase of the level of ferritin synthesis due to stimulation of ferritin gene transcription.[228] The stimulatory effect on transcription is due to an electrophile response element, typically observed in anti-oxidant response genes and present at the FER-1 enhancer. Two electrophile response elements were observed in the FER-1 enhancer. They are arranged in opposing orientations and include the two AP-1-like elements present in FER-1 at both sites of the dyad symmetry element.

The functionality of this oxidant response element requires the integrity of the dyad symmetry element.[228] Interestingly, an electrophilic response element was found also at the promoter of the ferritin L gene.[229] However, the transcriptional response of the ferritin H gene to oxidative stress was more pronounced as compared to that of the ferritin L gene. In line with the other observations on the inhibitory effect of the E1A viral protein on FER-1 activity, the transcriptional response of ferritin H gene to oxidative stress was inhibited by the E1A protein.[230]

Few studies have focused on the molecular mechanisms responsible for the activation of ferritin H gene transcription by pro-inflammatory cytokines. As noted in a previous chapter, several pro-inflammatory cytokines, including TNF-α, IL-1β, and IL-6 exert in many cell types a potent stimulation of ferritin synthesis, with effects virtually restricted to ferritin H subunits.

In some cell types, such as alveolar epithelial lung cells, ferritin synthesis is stimulated by TNF-α much more than by iron.[231,231a] The stimulation of ferritin H gene transcription by TNF-α seems to be dependent on the FER-2 region of the ferritin H gene promoter, located between nucleotides −4776 to −4736 from the start transcription site and containing two DNA sequences, one recognized by the NF-kB transcription factor and the other by another member of the NF-kB/Rel transcription factor family.[232]

Incubation of the cells in the presence of TNF-α elicited an increase of the NF-kB transcription factors to the FER-2 region of the ferritin H gene promoter.[232] NF-kB is a transcription factor that regulates genes involved in the cellular response to stress. NF-kB exists in unstimulated cells in an inactive form in the cytoplasm, complexed with the inhibitory protein IkB. Following activation of the cells with TNF-α, the IkB protein is phosphorylated and translated to the cell nucleus, where it binds to specific DNA sequences present in several gene promoters. It is of interest that the activation of NF-kB requires iron, since iron chelators completely abrogate NF-kB activation elicited by bacterial lipopolysaccharide in human macrophages.[233]

In addition to the mechanisms of transcriptional control of ferritin H gene expression, recent studies suggest that the c-*myc* proto-oncogene may act as a repressor of ferritin H gene expression.[234] In fact, a c-*myc*-responsive element was found in the ferritin H gene promoter between the TATAA box and the start codon[234] (Figure 9.5). The c-*myc* gene encodes the transcription factor c-Myc, which heterodimerizes with a partner protein, termed MAX, to regulate gene expression.[235]

Many genes involved in the control of cell proliferation and differentiation are the targets of c-Myc/MAX heterodimers. The interaction of these heterodimers with the promoters of these genes may either stimulate or repress gene expression.[234] A deregulated increased expression of c-*myc* is a key pathogenetic event in the development of many lymphoid malignancies and is frequently observed in several human tumors.

The mechanism through which the constitutive c-*myc* promotes cell proliferation and transformation was recently clarified. The c-*myc* increases the expression of the transcription factor Id2 which binds and inactivates the active hypophosphorylated forms of retinoblastoma, thus removing the inhibitory control mechanisms on cell cycle progression.[236]

Experiments performed in cell lines induced to the constitutive expression of high levels of c-*myc* showed a repressive effect of the c-Myc/MAX heterodimers on ferritin gene expression, related to a repressive effect on transcription. It was suggested that the c-*myc*-mediated effects of ferritin H gene transcription may play a role in the control of the expression of this gene observed during cell differentiation. It was proposed that the decrease of c-*myc* expression may represent one of the mechanisms responsible for the induction of ferritin H gene expression associated with induction of cell differentiation.[234] It was also suggested that the decrease of ferritin H gene expression induced by c-*myc* is required for the stimulatory effect of this proto-oncogene on cell proliferation and for its oncogenic activity.[234]

A region of the ferritin H gene promoter located at the −272 to −292 promoter region and called G-Fer may act as a repressor.[237] This region contains a G-rich element (a structure of ten Gs) which is able to bind a ubiquitous transcription factor that interacts with G-rich elements and is called inhibitory factor 1. The mutation of the coding sequence of this site elicited an increase in the basal activity of ferritin H promoter activity, thus suggesting an inhibitory role for the G-Fer element and its cognate *trans*-acting factor.[237]

In contrast to the large number of studies of the functional roles of some elements of the ferritin H gene promoter, few studies have been devoted to the characterization of the ferritin L gene promoter. Transient transfection experiments provided evidence that the activity of the L ferritin promoter depends on DNA sequences present within the region corresponding to nucleotides −320 to +60.[237] A negative regulatory element spans nucleotides −268 to −245; this region interacts with transcription factors that bind GC-rich sequences. Footprinting experiments showed that additional *cis*-elements are localized in the region just upstream of the start transcription site and they seem to act as regulatory positive elements. Another positive regulatory element is found at the first exon of the ferritin L gene.[238]

In addition to the transcriptional mechanisms that control ferritin gene expression, post-transcriptional mechanisms also play a relevant role in the control of ferritin gene expression. Particularly interesting are the studies performed on myeloid leukemic cell lines induced to maturation with chemical inducers, such as phorbol esters or retinoic acid. A marked increase in ferritin H mRNA levels was observed.[239] This phenomenon is mainly mediated by post-transcriptional mechanisms dependent upon stabilization of ferritin H mRNA. In control THP-1 cells, the ferritin H mRNA half-life was about 7 hours, while it increased to >24 hours in THP-1 cells induced to monocytic maturation by phorbol myristate acetate, a protein kinase C activator.[240]

The ferritin H mRNA stability was dependent upon a pyrimidine-rich sequence present within the 3′ untranslated region of the ferritin H mRNA. This sequence was recognized by an RNA-binding protein present in cytosolic extracts of THP-1 cells.[241] In THP-1 cells induced to differentiate, the binding of the protein factor to the ferritin H mRNA 3′ untranslated region markedly decreased, thus suggesting that the interaction between the cytosolic protein and the *cis* element serves as a desta-bilizing signal to facilitate the degradation of ferritin H mRNA in THP-1 cells.[241]

It is of interest to note that ferritin was also proposed as a possible modulator of gene expression, due to its presence, although in small amounts, in the cell nucleus. It was proposed that ferritin may bind the proximal promoter region of the human β-globin gene. This issue was recently reinvestigated, providing evidence that: (1) ferritin is present in the nuclei of K-562 human erythroleukemia cells, in addition to its predominant localization in the cytoplasm; (2) gel shift assays have shown that a protein cross-reacting with monoclonal ferritin antibodies, but not with ferritin alone, binds to a region located 150 bp upstream from the β-globin transcription start site.[242]

In addition to the control of mammalian ferritin gene transcription, recent studies provided the first lines of evidence about the transcriptional control of plant ferritin genes. Plant ferritin gene Fer1 was regulated by iron levels through a transcriptional mechanism. Iron loading induces Fer1 gene transcription, a phenomenon inhibited by anti-oxidant agents. This transcriptional induction involves an iron-dependent regulatory sequence (IDRS) present in the promoter of the Fer1 gene that is respon-sible for repression of Fer1 gene transcription under low iron conditions. These observations indicate that different mechanisms developed in plant and animal cells for the control of ferritin gene expression by iron. Low iron conditions repress ferritin synthesis at the translational level in animals, and at the transcriptional level in plants.

9.11 FERRITIN AND ENDOTHELIUM

The endothelium forms a physical barrier that separates blood from tissues. Com-munication between blood and tissues occurs through the delivery of molecules and circulating substances across the endothelial barrier by direct transport either through or between cells. The integrity of endothelium is of critical importance for vessel function. The endothelial cells of arteries and veins synthesize a substance that can cause the vessels to dilate. This substance was identified as nitric oxide and it plays a key role in cardiovascular physiology. In addition to functioning as a potent vasodilator, it also exerts a potent anti-atherogenic effect consisting of inhibition of smooth muscle cell′ proliferation, platelet aggregation, and leukocyte–endothelial cell interactions.

Endothelial cells produce many types of reactive oxidant species, including oxygen peroxide, hydroxyl ions, and peroxynitrite, which act as toxic byproducts of aerobic metabolism.[243] The endothelium is susceptible to damage by oxidants generated from activated macrophages, and this process has been implicated as an important mechanism in a number of pathologies, including adult respiratory distress syndrome, acute tubular necrosis, reperfusion injury, and atherosclerosis.

The cytotoxic action of these oxidant species is greatly potentiated by iron, and several recent studies support the idea that ferritin present in endothelial cells may sequester iron in a form that inhibits its ability to participate in oxidant injury. However, it must be emphasized that reactive oxidant species not only mediate cytotoxic effects on endothelial cells, but also represent mediators of normal physiological processes, such as proliferation, death, and survival of both endothelial and smooth muscle cells of blood vessels.[243]

A series of studies focused on the effect of heme on the synthesis of ferritin by endothelial cells and the consequences of increased synthesis of this protein in these cells. The interest in evaluating the effects of heme on endothelial cells was related to the possible important role of this molecule in mediating endothelial cell damage in pathological conditions associated with intravascular hemolysis, such as disseminated intravascular hemolysis.

Acute exposure of endothelial cells to heme sensitizes these cells to oxidants, but more prolonged exposure to heme protects the endothelium from oxidative damage through the induction of ferritin.[244] During incubation of endothelial cells with heme, the induction of heme oxygenase 1 and ferritin is observed. The cytoprotective effect on endothelial cells is related to ferritin and not to heme oxygenase. In fact, exogenously added apoferritin is taken up by endothelial cells and exerts a cytoprotective effect.[245]

The cytoprotective effect requires the functional integrity of ferritin, as suggested by experiments involving endocytosis into endothelial cells of ferritin mutants lacking ferroxidase activity and iron sequestering activity.[245] It was shown also that heme generated from in vivo intravenous injection of methemoglobin stimulates heme oxygenase 1 and ferritin synthesis in the lung microvascular endothelium.[246] The increase in heme oxygenase activity in endothelium is not required for induction of endothelium ferritin synthesis.[246]

Ferritin endothelial synthesis was sensitive to nitric oxide. Studies on cultured endothelial cells have shown that incubation of these cells with a nitric oxide donor elicited increased heme oxygenase and ferritin synthesis. It was suggested that the increase in ferritin levels was dependent upon the release of iron from heme stores via heme oxygenase I induction since the ferritin increase was inhibited by heme oxygenase inhibitors.[247] On the other hand, agents promoting NO release induce augmentations in ferritin synthesis responsible for the increased resistance to oxidative stress by endothelium.[248]

NO was able to modulate the responses of endothelial cells to heme through nitrosylation of heme which prevents its degradation by heme oxygenase, inhibits heme oxygenase activity, and reduces the rise of ferritin synthesis induced by heme.[249] Interestingly, heme arginate entered endothelial cells similarly to heme, but did not provoke oxidant-mediated endothelial damage; this compound elicited a marked enhancement of heme oxygenase activity, associated with only a weak induction of ferritin synthesis.[250]

The stimulatory effects of hemoglobin and heme on heme oxygenase and ferritin were not restricted only to endothelial cells, but also to smooth muscle cells of blood vessels. Both hemoglobin and heme induced in smooth muscle cells the sequential activation of heme oxygenase and ferritin.[251] The increase of heme oxygenase and ferritin represents a compensatory mechanism to counteract the potentially toxic effects induced by heme uptake into these cells. These observations are relevant for the understanding of the changes in heme oxygenase and ferritin induced by hemoglobin. Hemoglobin may be relevant in diseases in which smooth muscular cells are exposed to these compounds, such as atherosclerosis and vasospasm subarachnoid hemorrhage.[252]

Endothelial ferritin synthesis is increased not only by heme, but also by acetylsalicylic acid, whose stimulatory effect is dependent on a translational mechanism sensitive to iron chelators.[253] It was proposed that this effect of aspirin on endothelial ferritin synthesis may be one of the mechanisms responsible for the protective effect of this drug on endothelial injury in cardiovascular disease.

Several studies focused on the expression of ferritin in atherosclerosis. Atherosclerosis is a pathologic process initiated by injury of the endothelium, followed by the activation and the infiltration of monocytes and T lymphocytes at the level of these lesions; monocytes differentiate to macrophages in the subendothelial tissue. The subsequent step is formation of the plaques, characterized by the accumulation of lipid droplets in foam cells derived from macrophages and smooth muscle cells, the proliferation of smooth muscle cells, and the accumulation of matrix protein in the extracellular spaces, a phenomenon which produces the thickening of the intima of blood vessels.

Ferritin expression, involving both H and L ferritin, was markedly higher in atherosclerotic aortas than in normal counterparts.[254] Studies showed that both H and L ferritin mRNAs were markedly increased in aortas of animals fed high cholesterol diets. The increase in ferritin expression was observed in various cell types in atherosclerotic lesions: endothelial cells, macrophages, and smooth muscle cells.[254] Subsequent studies have clarified that macrophages present in human atheromas exhibit the highest ferritin levels.[255]

Particularly high ferritin levels are observed in apoptotic macrophage-derived foam cells present in atheromas. Ferritin accumulates at the cytoplasm, while low molecular iron chelates accumulate in lysosomal compartments.[255] It was suggested that the lysosomal iron may be exocytosed and contribute to cell-mediated low density lipoprotein (LDL) oxidation.[255] The accumulation of LDL within endothelial cells represents a primary event in the genesis of atherosclerosis.

LDL is endocytosed by monocytes–macrophages, and modified in the vessel walls through various processes including oxidation, lipolysis, proteolysis, and aggregation. These events play roles in the development of the inflammatory response and in the formation of foam cells. A key event is LDL oxidation. Oxidized LDL is responsible for the induction of the inflammatory response promoted through the stimulation of inflammatory cytokines that induce the recruitment of monocytes and lymphocytes, but not neutrophils, to the arterial wall.[256] Furthermore, oxidized LDL inhibits the production of nitric oxide which has several anti-atherogenic properties.

The effect of oxidized LDL on ferritin synthesis in endothelial cells and smooth muscle cells was evaluated, and apparently these lipoprotein species did not affect the rate of synthesis of the iron storage proteins.[254] However, studies performed on monocytic cells provided evidence that oxidized LDL exerts a marked stimulatory effect on L ferritin chain gene expression.[257]

Endogenous peroxisome proliferator-activated receptor gamma (PPARγ) ligands also activate L ferritin expression in a dose-dependent manner.[257] This last observation is interesting, although the link between PPARγ and atherosclerosis remains speculative.[258] On the other hand, there was evidence that ferritin is able to limit the cytotoxic damages induced by oxidized LDL[259] because oxidized LDL inhibits the ability to provoke endothelial cytotoxicity via metal-catalyzed free radical-mediated cytotoxicity.

As outlined in the section on serum ferritin levels, evidence indicates that iron favors the development of atherosclerosis. In this context, the analysis of the mechanisms leading to the increased tendency to develop atherosclerosis in β-thalassemia is interesting. In this disease, the mechanisms favoring the development of atherosclerosis are related to oxidative stress. In fact, β-thalassemia patients show increased synthesis of oxidized LDL, probably due to the oxidative interactions of hemoglobin α-chains (released in the plasma in consequence of hemolysis) with LDL, which act as a trigger of oxidative modifications of LDL.[260]

Despite the direct correlation between serum ferritin levels and the level of oxidized LDL in β-thalassemia patients, all the evidence suggests that the increased tendency to the development of atherosclerosis observed in these patients cannot be related to increased iron stores, but to oxidative stress induced by unbalanced globin chain synthesis. This interpretation is indirectly supported by epidemiologic studies showing that atherosclerosis is not a prominent feature of hereditary hemochromatosis.

Studies performed in experimental models of atherosclerosis in mice (i.e., mice deficient in apo-lipoprotein E) provided evidence that iron restriction elicited a protective effect against the development of atherosclerosis, seemingly making lipoproteins more resistant to oxidative damage. The increase in endothelial ferritin levels was limited.[261]

Finally, recent studies explored the possible protective role of heme oxygenase-1 and ferritin on transplant-induced atherosclerosis.[262,263] These observations open the way for possible therapeutic applications.

9.12 FERRITIN IN TUMOR CELLS

Ferritin levels are increased in several tumor cell types and it was suggested that this phenomenon may play a role in the pathogenesis of some of these tumors. Particularly interesting are the studies of breast carcinoma. Studies on experimental models of immortalization of breast epithelial cells showed that the immortalization of these cells was associated with a marked increase in ferritin levels.[264] In parallel, studies in cancer tissues corresponding to tumors of progressive proliferative capacity and malignancy showed a parallel increase of ferritin levels, with the highest levels observed in infiltrating ductal carcinoma, while ferritin levels were very low in normal breast tissues.[264]

These observations suggest that the increase in ferritin levels might have contributed to the immortalization of breast epithelial cells through two different mechanisms: (1) providing an iron source required to sustain cell proliferation and particularly the expansion of the malignant clone; and (2) providing the iron involved in chemical reactions leading to the generation of oxidative radicals, which produce

DNA damage and mutations.[265] Other studies have shown that progression of invasive breast cancer to the metastatic state is linked to the generation of oxidative radical species.[266]

Additional studies have shown that ferritin may be involved in the mechanism of hydroxyl radical formation. In fact, redox cycling of estrogen metabolites releases Fe^{2+} from ferritin, which in turn is involved in a series of reactions generating hydroxyl radicals. This estrogen-induced hydroxyl radical damage requires ferritin, and may provide an important contribution to tumor induction by estrogens in breast tissue.[267]

The levels of ferritin observed in tumoral breast tissue were compared with levels observed in serum. The result was a constant increase of ferritin content in tumor cells, but not in serum.[268]

Other studies focused on the synthesis of placental isoferritin by breast cancer cells. Placental isoferritin is an acidic form of ferritin and its p43 heavy chain is synthesized by breast cancer cells, but not by their normal counterparts.[269]

Studies of breast cancer biopsies provided evidence that p43 content in the cytosols of breast cancer cells inversely correlates with the degree of proliferation and differentiation of tumor cells.[269] In line with these observations, serum levels of placental ferritin were increased in breast cancer patients with advanced disease and visceral metastases.[270] Evidence indicates that placental isoferritin p43 may exert inhibitory activity on the mitogenesis of cytotoxic CD8+ lymphocytes, and thus mediate anti-tumor activity in breast cancer patients.[271]

Neuroblastoma is a relatively common embryonal malignancy. The progress in the treatment of this tumor was mostly related to the identification of reliable clinical and biologic prognostic criteria. Clinical biological variables associated with poor disease outcome include age older than 1 year at diagnosis, metastatic disease at diagnosis, and unfavorable histopathology. Biologic criteria associated with aggressive disease include amplification of MYCN proto-oncogene and serum ferritin levels.

Initial studies of serum ferritin levels in neuroblastoma patients provided evidence that this parameter was increased in patients with stage IV disease and liver or bone marrow metastases.[272] Subsequent studies confirmed that serum ferritin levels are infrequently increased among stage I and II neuroblastoma patients, while they are frequently increased among stage III and IV patients (37 and 54%, respectively).[273]

According to these studies, ferritin is one of the negative prognostic factors in patients with neuroblastoma. More recent studies confirmed the prognostic value of serum ferritin levels in neuroblastoma patients.[274,275]

One study aimed to prospectively evaluate the prognostic values of several biologic factors in a large number of neuroblastoma patients. The evaluation of MYCN amplification was the most significant prognostic parameter. A patient with non-amplified MYCN had a 3-year disease-free survival corresponding to 93%, while survival rate was only 10% for patients exhibiting MYCN amplification. Ferritin level had prognostic significance in univariate analysis, but not after stratification of patients by MYCN gene copy number.[274]

Tumor tissue seems to be the main source of the increased serum ferritin levels in neuroblastoma patients.[276] Initial studies suggested that exogenous ferritin may stimulate the growth of neuroblastoma cell lines,[277] and that iron chelators may inhibit *in vitro* the growth of neuroblastoma cells.[278] Based on these observations, it was proposed to include iron chelators in standard anti-tumor chemotherapy. However, more recent studies failed to show significant anti-tumor activity of iron chelators in human neuroblastoma xenografts. Iron chelators were unable to produce significant reductions in tumor engraftment, latency, or size.[279]

In spite of these results, *in vitro* studies still suggest a potential role for iron chelators in the treatment of neuroblastoma. These studies have shown that deferoxamine, an iron chelator, induced apoptosis in cultured human neuroblastoma cells preceded by a decrease of M-myc expression due to inhibition of the transcription of this gene. The decrease in M-myc expression does not seem to be required for induction of apoptosis.

Serum ferritin levels have been investigated in patients with leukemia. High serum ferritin levels were observed in patients with acute myeloid leukemias. In chronic myeloid leukemia, normal or slightly increased ferritin levels were detected during the chronic phase, and markedly increased levels were seen during the blastic crisis phase.[280] Response to treatment by acute myeloblastic patients was associated with a complete normalization of serum ferritin levels.

Serum ferritin levels were screened in other myeloproliferative diseases, including chronic granulocytic leukemia, polycythemia vera, and osteomyelosclerosis, and were always within the normal range or slightly decreased in the chronic phases of these diseases.[281] Analysis of serum ferritin levels of patients with acute myeloid leukemia showed that increased ferritin levels were preferentially observed in patients with immature forms of leukemic blasts (M1 and M2 subtypes following the FAB classification), while only moderately increased levels were observed among patients with more mature forms of leukemic blasts (M3, M4 and M5 subtypes).[282]

Several studies focused on serum ferritin levels in patients with different types of lymphomas. The highest ferritin levels were observed in patients with malignant histiocytosis; the particularly high levels of serum ferritin in may be considered pathognomic, and included among the diagnostic criteria for this disease.[283]

Analysis of serum ferritin levels in a large number of lymphoma patients provided clear evidence of an increase of this parameter in a significant proportion of Hodgkin's and non-Hodgkin's lymphomas.[284] Analysis of the ferritin levels in lymphoma patients stratified by stage of disease showed a clear tendency to an increase of ferritin levels associated with the progression of the disease, with the highest values observed in stage III and IV patients.[284] The highest ferritin levels were observed in patients who had advanced Hodgkin's disease with hepatic involvement.[285] A study of basic and acidic isoferritin levels showed that basic ferritins tend to be frankly elevated in patients with systemic disease, while acidic ferritins are increased in almost all patients.[286]

Particularly high levels of ferritin were observed in lymphoma patients with associated hemophagocytic syndrome. This syndrome is observed in lymphomas, leukemias, and other neoplastic conditions and involves the phagocytosis of hematologic elements of various lineages, seemingly by neoplastic cells. Patients with either B or T lymphomas, associated with hemophagocytic syndrome exhibit serum ferritin levels significantly higher than patients with similar pathological conditions not associated with hemophagocytic syndrome.[287]

The lymphomas represent an area of therapeutic applications involving the use of radionucleide-conjugated anti-ferritin monoclonal antibodies. This approach is based on the observation that lymphoma cells possess ferritin receptors on their surfaces and are then able to fix anti-ferritin antibodies. Initial studies showed anti-ferritin antibodies labeled with ^{131}I or ^{111}In for diagnostic purposes or with ^{90}Y for therapeutic purposes.[288] This approach was used for therapeutic purposes on patients with recurrent, end-stage Hodgkin's disease. The result was about 60% response, of which 30% were complete responses.[288] Importantly, complete responders survive longer than partial responders, and partial responders survive longer than patients with progressive disease.[288]

Subsequent studies have shown that *in vivo* injection of anti-ferritin antibodies was able to target more than 95% of tumoral lesions less than 1 cm in size; however, recurrences of the disease were observed in complete responders to the treatment, seemingly due to the incapacity of the anti-ferritin antibody to target tumor lesions larger than 1 cm.[289] A more rational approach was recently adopted. Patients were first injected with [111]I-anti-ferritin antibody to show that this antibody targeted the tumor followed by injection with ^{90}Y-labeled antibody. Using optimal radioactivity dosages (0.5 mCi/kg), 86% of patients responded, and this confirmed that the survival of responding patients was improved.[290]

Ferritin synthesis in colon cancer was the object of several investigations. Initial studies showed a clear increase of ferritin content in colon cancer tissue as compared to normal tissue[291]; within the tumoral tissue, the most intense staining of ferritin was observed in the stromal tissue; epithelial cells stained more weakly.[292] In spite of this high tumor content of ferritin, serum ferritin was lower than in normal controls at diagnosis due to blood loss; in fact, adenomas >1 cm exhibit both serum iron and ferritin levels lower than controls.[293] Interestingly, salindac, a non-steroidal anti-inflammatory drug which was shown to have a protective effect against the incidence and the mortality of colorectal cancer, affects the expression of several genes in colon cancer cells, including ferritin H gene expression.

Serum ferritin levels in ovarian cancer have been explored. Elevated levels were observed in patients; ferritin levels exhibited the tendency to increase with disease progression.[294] However, the low sensitivity of this parameter does not suggest a real utility of serum ferritin determination as a tumor marker in ovarian cancer.[294]

Particularly interesting are the studies of ferritin levels in hepatocarcinoma, since iron overload represents a risk factor for its development and serum ferritin is frequently highly increased in this neoplasia. Hepatocellular carcinoma is for the majority of patients a terminal complication of chronic inflammatory and fibrotic liver disease, the most frequent causes being chronic infection with the hepatitis B and C viruses.[295]

Several studies have suggested that iron overload significantly increases the risk of hepatocellular carcinoma among patients with chronic B and C hepatitis. On the other hand, experimental models of hepatocarcinoma induced by diethylnitrosamine showed that the development of the tumor was associated with a 10-fold over-expression of hepatic ferritin H mRNA.[296] Serum ferritin levels markedly increased (i.e., >400 ng/ml) in more than 80% of patients with hepatocellular carcinoma, while only about 40% of patients with non-neoplastic liver disease showed these values.[297]

It was suggested that the evaluation of acidic ferritin levels coupled with determination of serum ferritin amounts may represent a valuable index for the detection of hepatic cancer.[298]

Serum ferritin levels do not tend to increase in patients with gastric cancer, but there is a relationship between a decrease in ferritin levels and the risk of developing gastric cancer. A combination of low serum ferritin levels and achlorhydria were associated with a 10-fold increased risk of gastric cancer.[299]

The development of lung cancer is frequently associated with high blood levels of tumor markers, including the carcinoembryonic antigen (CEA), carbohydrate antigen 125, carbohydrate antigen 19.9, and ferritin. The carcinoembryonic antigen is the most useful and most commonly studied. The evaluation of ferritin is also of value in the diagnosis and follow-up of lung cancers. Initial studies showed that the evaluation of both CEA and ferritin was a valuable criterion to distinguish patients with lung cancer from those with benign lung diseases and normal controls.[300] Serum ferritin levels are increased in a significant proportion of patients with small lung cancers.[301] Although serum ferritin levels are not of value in staging small lung cancer nor in monitoring its process, they are, however, of prognostic significance in that patients with ferritin levels <600 µg/L survive significantly longer than patients with ferritin values >600 µg/L.[301]

Recent studies have shown the diagnostic and prognostic value of tumor marker evaluation, including ferritin, in pleural effusions of malignant origin. Ferritin values were significantly higher in pleural effusions of malignant origin (adenocarcinoma, small lung cancer, and mesothelioma) as compared to those of benign origin.[302] Similar observations were made of bronchoalveolar lavage fluid.[303]

Another neoplasia for which the evaluations of serum ferritin and other tumor markers, such as lactic dehydrogenase, α-fetoprotein, and CEA are of value is testicular tumor. Immunohistochemical studies have shown that the majority of testicular tumors, particularly seminoma, exhibit marked reactivity with both anti-basic and anti-acidic ferritin antibodies.[304] Serum ferritin levels increased more frequently in seminoma than in other testicular tumors.[305]

Several studies analyzed ferritin levels in renal cell carcinoma. Initial studies showed markedly increased ferritin levels in the cytosols of renal cancer cells as compared to normal counterparts.[306] Studies of serum ferritin levels in renal carcinoma patients showed a high frequency of increased values; serum ferritin levels increased with increasing tumor stage and correlated with tumor volume.[307] The prognostic value of serum ferritin levels in renal cell carcinoma patients was not confirmed in all the studies.[308] However, in patients with tumors at initial stages of development, the increase in serum ferritin levels did not correlate with tumor size and therefore did not seem to be a useful tumor marker for these patients.[309]

Several studies revealed a possible role of iron-binding proteins in the progression of bladder cancer, and suggested the use of agents that interfere with iron uptake to treat metastatic bladder cancer. Experimental models suggest that ferritin has a role in the progression of bladder cancer: (1) rat transitional cell carcinoma cells endowed with high metastatic potential exhibit markedly higher ferritin H chain expression than cells with low metastatic capacity[310]; and (2) metal chelators able to inhibit growth and induce apoptosis of bladder cancer cells display high capacity to inhibit ferritin synthesis.[311] Evaluation of serum ferritin levels in bladder cancer patients did not show significant changes related to the disease; in contrast, increased ferritin levels were observed in the urine of bladder cancer patients.[312]

Some studies explored ferritin synthesis in head and neck tumors, and provided evidence that serum ferritin levels are preferentially increased among patients with cancer of the larynx as compared to patients with malignant neoplasms of maxillo-ethmoid complex.[313,314] Increased serum ferritin levels were also observed in patients with metastatic nasopharyngeal carcinoma.[315] Patients with localized disease and hyperferritinemia have a worse prognosis, related to increased possibility of developing metastatic disease.[315]

9.13 FERRITIN EXPRESSION AND FUNCTION IN SELECTED TISSUES

9.13.1 FERRITIN EXPRESSION AND FUNCTION IN THE EYE

The discovery of the hereditary disease called hereditary hyperferritinemia–cataract syndrome, characterized by elevated serum L ferritin and bilateral cataracts, attracted considerable interest to the study of ferritin synthesis in the eye lens. The disease is due to mutations of the iron regulatory element of ferritin L mRNA (see Chapter 8). A pronounced increase of L ferritin concentration in lenses is observed. The ferritin L content in these patients is >147 µg/g dry weight of lens tissue, as compared to <16 µg/g observed for controls.[316] Histologic examination of the lens opacities showed amorphous, poorly eosinophilic deposits throughout the stroma; electron microscopy analysis showed that these deposits are formed by square crystalline inclusions dispersed within the stroma.[316]

The L ferritin accumulation is particularly pronounced in the lenses, and this phenomenon may be related to two main mechanisms: (1) ferritin L synthesis is increased due to the lack of a repressor mechanism and accumulates within the lenses at high levels since ferritin turnover is particularly low in this tissue; and (2) ferritin L synthesized within the lenses is initially soluble, but with age it may be subjected to post-translational modifications that lead to denaturation and formation of crystalline precipitations.

Initial studies of cultured canine lens epithelial cells provided evidence that ferritin is actively synthesized in these cells and reaches baseline ferritin content ranging from 75 to 160 ng/mg of protein.[317] Lens ferritin synthesis is regulated in other tissues; its synthesis is stimulated by iron loading and inhibited by iron deprivation, and greatly stimulated by ascorbic acid.[317] These observations suggest that ferritin may represent an important mechanism by which the lens epithelium is protected from oxidative damage.[317]

The mechanism by which ascorbic acid stimulates ferritin synthesis in lens epithelial cells was studied. Translation pathways are predominantly involved in this phenomenon.[318] Interestingly, the stimulatory effect of ascorbic acid on ferritin synthesis is counteracted by hydrogen peroxide whose concentration is relatively high in the aqueous humor.[318]

Recent studies provided information about expression of ferritin in human lenses. Gene array analysis showed very high levels of ferritin L mRNA in human lens; these levels were comparable to the levels of crystallins mRNAs and were disproportionately high when compared to the levels of ferritin protein found in normal human lenses.[319]

The distribution of ferritin in normal and cataractous human lenses was recently investigated. In normal lenses ferritin levels were not significantly different in the cortex of the nucleus. However, in cataractous lenses, ferritin was preferentially accumulated in the cortex as an insoluble protein fraction.[320] Additional studies provided evidence of the existence of a chelatable pool of redox-reactive Fe, whose levels are increased in cataractous lenses.[320]

The sensitivity of lenses to oxidative damage and the role of such damage in the development of cataracts led to studies of the effects of anti-oxidants on cultured epithelial lens cells. It was proposed that tempol, a nitroxide, exerted a protective anti-oxidant effect, but this protective effect was transient. Tempol exerted an initial inhibitory effect on ferritin synthesis by lens epithelial cells, followed by a return to pre-treatment values at 20 hours.[321] Despite this biphasic effect, tempol inhibited Fe incorporation into ferritin at all time points.[321] After exerting an initial protective action against oxidative effects, tempol produced toxicity.[321] This observation clearly indicates that Fe sequestration into ferritin in lens epithelial cells plays a key role in the protection of this tissue against oxidative damage.

Ferritin is expressed also in other eye structures. Studies of the development of corneal epithelium during chicken embryogenesis showed that the induction of ferritin synthesis was associated with epithelium stratification and differentiation.[322] Studies using anti-ferritin H monoclonal antibody showed that ferritin in corneal epithelium is localized in the nucleus and not in the cytoplasm.[322] In spite of this unusual localization, corneal ferritin is regulated by iron and cytoplasmic ferritin as observed in other cell types.

It was hypothesized that the nuclear corneal ferritin in these cells may act as a mechanism to prevent oxidative damage to nuclear components.[322] This hypothesis was supported by additional studies in which corneal epithelial cells expressing and not expressing nuclear ferritin were exposed to ultraviolet (UV) radiation which generated oxidant-reactive species. Corneal epithelial cells that express nuclear ferritin are relatively protected from UV-induced DNA strand breaks, while corneal epithelial cells without nuclear ferritin undergo high levels of DNA breakage. The extent of breakage is exacerbated when the cells are exposed to iron.[323] The treatment of corneal epithelial cells with iron chelators markedly inhibits ferritin synthesis and sensitizes these cells to UV-induced damage.[323] Corneal epithelial cells also synthesize lactoferrin which is involved with ferritin in the protection of these cells against oxidative damage. The cooperation of these two molecules in the protection against oxidative damage was observed in a model of reoxygenation injury following hypoxia.[324]

Ferritin is also synthesized in the retina, particularly in the retinal pigmented epithelial cell layer, the choroids, and the inner segments of photoreceptors.[325] It was suggested that ferritin, transferrin, and hemopexin, all of which are synthesized in the retina, participate in protecting the retina against iron and heme-mediated toxicity.[326]

9.13.2 Expression and Function of Ferritin in the Skin

Studies of skin cells, i.e., fibroblasts, adipocytes, and keratinocytes, provided evidence that ferritin is expressed in these cells and plays a key role in the defense against oxidative stress. Exposure to sunlight containing ultraviolet A (UVA) rays generates oxidative stress in skin cells through interaction with chromophores followed by the release of reactive oxygen species such as hydrogen peroxide. These reactive oxygen species induce cellular damage, such as lipid peroxidation, in cell membranes via iron catalyzed oxidative reactions. Mechanisms within these cells limit the iron available for the triggering of these oxidative cytotoxic reactions.

Initial studies of skin fibroblasts showed a four-fold increase in heme oxygenase activity after UVA irradiation, followed by a persistent two-fold increase in ferritin levels.[327] These findings suggest that the effect of UVA irradiation on ferritin levels is mediated through heme oxygenase-dependent release of iron from endogenous heme sources.[327]

Free iron, but not iron bound to ferritin, can induce reactions of oxidative cellular damage, including the breakdown of hydroxyperoxides to thiobarbituric acid-reactive species and the oxidation of sulfhydryl groups of proteins in human skin fibroblasts.[328] Human skin chronically exposed to UV light exhibited the tendency to accumulate iron and contain increased ferritin, as compared to skin areas not exposed to light.[329] Skin fibroblasts develop higher levels of UV-mediated cytotoxicity when grown in iron-rich medium; conversely, iron chelators and radical scavengers such as vitamin E exerted protective effects against the oxidative damage mediated by UVA.[329]

Ferritin is normally present in basal layers of epidermis. Following exposure to UVAI and UVAII radiation, ferritin expression increases in dermal fibroblasts.[330] The kinetics of ferritin rise following UVA irradiation showed an initial increase at 3 hours post-UVA exposure, reached a peak at 6 hours, and remained at peak values up to 48 hours.[330] The stimulatory effect of UVA on ferritin synthesis related minimally to transcriptional mechanisms and to a large extent to translational mechanisms.[331] In addition to the stimulatory effect on ferritin synthesis detectable a few hours after UVA exposure, an intermediate increase in the free iron pool via degradation of the ferritin molecule was observed.[331] This phenomenon was the consequence of lysosomal membranes, subsequent leakage of proteolytic enzymes, and the consequent degradation of ferritin.[331]

Exposure of human fibroblasts to UVA elicited, in addition to ferritin H, the expression of other genes involved in cell cycle control, such as p21/WAF1, or in connective tissue metabolism products, such as metalloproteinase-3.[332]

Recent observations suggest that the protective role exerted by ferritin in human dermal fibroblasts and keratinocytes against oxidative stress is complex and depends on the type of agent triggering the oxidative response.[333] Dermal fibroblasts and keratinocytes stimulated to overexpress ferritin by iron loading are able to markedly inhibit lipid peroxidation, but not the mortality induced by hydroxyperoxide.[333] Ferritin expression in human dermal fibroblasts may be induced by infrared radiation in the absence of induction of oxidative stress or cytotoxic cell effect.[334] According to these observations, a potential role for infrared radiation to promote skin repair was proposed.[334]

Ferritin expression was also explored in adipocytes, cells resident in the derma. Ferritin receptors are expressed on activated lipocytes, and are able to selectively bind ferritin H.[335] On the other hand, studies with iron-loaded ferritin provided evidence that this protein elicited a moderate inhibition of lipocyte proliferation and a marked inhibition of smooth muscle actin expression.[336] This effect was specific for iron-loaded ferritin; it was not displayed by apo-ferritin.[336]

Other studies addressed the issue of ferritin synthesis during adipocyte differentiation by using 3TL-L1 pre-adipocytes induced to terminal adipocyte maturation.[337] These studies showed marked up-modulation of ferritin expression both at mRNA and protein levels. The up-modulation was more evident for ferritin H than for ferritin L.[337] This increase in ferritin expression was paralleled by a concomitant increase in iron regulatory protein 1 expression. This induction of ferritin expression during adipocyte differentiation was associated with increased resistance to lipid peroxidation.[337] This observation suggests that ferritin in adipocytes, as well as in fibroblasts and keratinocytes, is involved in defense against oxidative stress.

9.13.3 FERRITIN EXPRESSION IN THE BRAIN

Iron is required for the metabolism of the different cell types in the brain. Iron is required not only as a constituent of heme and iron–sulfur clusters of many enzymes, but also as a constituent of several enzymes involved in the biosynthetic pathways of myelin and dopamine. Iron is acquired by brain cells through the blood–brain barrier. Iron uptake into the brain is mediated through the binding of serum transferrin to transferrin receptors in capillary brain endothelial cells. The highest density of this receptor was present in the endothelial cells.

The basic distribution of iron and iron-binding proteins in the various cell types within the brain may be briefly summarized as follows:

1. Astrocytes do not contain stainable iron and ferritin, with the exception of astrocytes present in the striatum.
2. Oligodendrocytes possess the majority of stainable iron present in the brain; this iron is acquired via transferrin and ferritin receptors present on their membranes.
3. Resting microglia are rarely biochemically stained for iron, while activated microglia exhibit marked staining for iron.
4. Neurons possess transferrin receptors and Nramp2 involved in the uptake of iron required to sustain their metabolic needs.

Interestingly, neuronal cells express both iron regulatory element⁻ (IRE⁻) and IRE⁺ Nramp2 isoforms. The IRE⁻ Nramp2 isoform is localized in the nuclei of neurons; in contrast, the IRE⁺ Nramp2 isoform is localized in cell membranes and in intracellular vesicles localized in cell bodies, dendrites, and neurites.[338] The IRE⁺ Nramp2 isoform may act in neurons as an iron and metal transporter, while the IRE⁻ Nramp2 isoform may sequester divalent metals in the nucleus.[338]

This section will discuss the localization and possible function of ferritin in various cell types and various areas of the brain. Most of the iron in the brain is present under physiological conditions in the oligodendrocytes. Ferritin is found in virtually all brain regions and is predominantly found in the perikaryal cytoplasm.[339]

Immunostaining of brain sections with anti-ferritin H and L antibodies provided evidence that oligodendrocytes are the only cell types that contain both ferritin H and ferritin L.[340] Immature oligodendrocytes acquire the ability to synthesize ferritin H chain when they become mature and acquire the capacity to synthesize myelin.[341] The induction of ferritin in these cells may be associated with the supply of iron required to sustain oxidative mechanisms and with protection against oxidative injury.

The function of ferritin in oligodendrocytes is related not only to protection against oxidative stress, but also to biochemical mechanisms involved in myelin synthesis. The possible role of ferritin in myelin synthesis is supported by two lines of evidence: (1) iron is required for normal myelination because several iron-dependent enzymes are involved in the synthesis or degradation of myelin lipids[342]; and (2) ferritin mRNA is located in oligodendrocytes at the myelin sheath assembly sites, along with mRNAs that encode myelin basic protein and several myelin-associated oligodendrocytic basic proteins.[343]

Studies of the ferritin-binding capacity of oligodendrocytes were stimulated by the initial observation that ferritin binding was observed mostly in white brain matter,[344,345] where high iron concentrations were reported. Transferrin receptors are preferentially located in the gray matter where iron concentration is low.

These observations suggest that oligodendrocytes bind ferritin and acquire iron through its uptake. This hypothesis was proven by a study of ferritin binding and uptake in cultures of murine oligodendrocyte progenitor cells. These cells were able to bind ferritin in a saturable and competitive manner; they were also able to internalize ferritin.[346]

The inhibition of ferritin internalization into oligodendrocytes activated iron regulatory protein in the cytosols of oligodendrocytes, thus indicating that ferritin is responsible for iron supply to oligodendrocytes.[346] Ferritin receptors are present in both oligodendrocyte progenitors and mature oligodendrocytes, while transferrin receptors are detectable only in immature oligodendrocytes. Oligodendrocytes are also the major cells involved in transferrin production in the brain.

The level of ferritin synthesis in oligodendrocytes is modulated by different stimuli that appear to operate *in vivo* under different physiopathological conditions: iron and hypoxia.[347] Both agents stimulate ferritin expression in oligodendrocytes through a translational mechanism. The stimulatory effect of hypoxia on ferritin translation in oligodendrocytes cannot be enhanced by co-stimulation with iron salts, thus suggesting that both these agents may act through a similar biochemical mechanism that leads to iron regulatory protein inactivation.[347]

Ferritin induction mediated by hypoxia may be enhanced through co-treatment with hydrogen peroxide.[348] The enhancement of oligodendrocyte ferritin synthesis induced by hypoxia and hydrogen peroxide may be inhibited by addition of catalase to the culture medium, thus suggesting the involvement of intracellular radicals in this phenomenon.[348] The great enhancement of ferritin synthesis observed in oligodendrocytes exposed to hypoxic conditions may represent a defense mechanism to protect the cells against oxidative radicals that give rise to lipid peroxidation and subsequent cellular toxicity.

The preferential distribution of ferritin in the oligodendrocytes is supported by studies on fetal and adult human brains.[349,350] In frontal and occipital lobes, ferritin positivity occurs in the subcortical and paraventricular white matter at about 25 weeks of fetal life, concomitant with the induction of the myelination process. Ferritin synthesis in pons and cerebellum is induced after transferrin at fetal stages comparable to those observed for synthesis in frontal and occipital lobes. Ferritin in the developing rat brain is found initially in microglia, then in oligodendrocytes in a temporal and spatial pattern that coincides with the expression of myelin.

The study of ferritin in neurons is particularly important since these cells are often exposed to the damage caused by hypoxia induced by ischemia. Many phenomena observed during brain ischemia and subsequent reperfusion may be dependent on damage to membrane lipids related to lipolysis and particularly to radical-mediated peroxidation of polyunsaturated fatty acids during reperfusion. At the start of reperfusion, a sequence of neuron damages occur, including inhibition of protein synthesis, activation of proteolysis, and activation of endonucleases, with consequent induction of apoptosis. Under conditions of normal blood perfusion and normoxia, specific growth factors sustain anti-oxidant and anti-apoptotic mechanisms in neuronal cells.[351]

The protective effect of ferritin exerted in neuronal cells is mainly due to its capacity to buffer intracellular iron. The buffering into neuronal cells exerts a protective anti-apoptotic effect against oxidative stress. This protective effect may be related to preventing the production of oxidative radicals through Haber–Weiss reactions, or to other mechanisms. A reduction of neuronal free iron prevents the apoptosis elicited by growth factor deprivation[350] and potentiates the expression and activity of transcription factor hypoxia-inducible factor 1 (HIF-1).[352]

The negative effect of iron on neurologic damage in acute cerebral ischemia is supported by clinical observations showing that the severity of initial neurologic deterioration is more pronounced in subjects exhibiting higher iron stores, as evidenced by plasma and CSF ferritin concentrations.[353] These observations clearly indicate that increased body iron stores contribute to the progression of stroke damage by enhancing the cytotoxic mechanisms elicited by stromal ischemia.[351]

Ferritin is present in virtually all neurons of the brain at relatively low levels. H ferritin, but not L ferritin, was observed in brain neurons.[354] H ferritin mRNA was preferentially localized in the dendrites.[355]

Analysis of the sequences of cDNA encoding H ferritin in the brain showed some interesting findings related to the presence of an elongated 3' untranslated region with an additional 279-nucleotide sequence.[356] Ferritin H mRNA of 1 to 2 Kb in size, larger than the known 0.83-Kb ferritin H mRNA was observed in murine oligodendrocytes.[343] In neurons, however, the expression of other proteins involved in iron metabolism, such as transferrin and Nramp2, is markedly higher than that observed for ferritin.[357]

Under physiological conditions, ferritin is present at high concentrations in certain regions of the brain, mainly in subcortical nuclei, such as globus pallidus, substantia nigra, and red nuclei. Ferritin accumulation is related to the presence of oligodendrocytes, mainly at the basal ganglia.[358,359] Additional ferritin accumulation is also observed in the dopaminergic neurons of the substantia nigra, in association with the neuromelanin pigment.[360,361] In some pathological conditions, abnormal accumulation of iron in these and other brain regions produces iron-mediated oxidative damage.

Friedreich's ataxia is an autosomal recessive neurodegenerative disease characterized by progressive ataxia, skeletal abnormalities, and cardiomyopathy. The causative factor responsible for Friedreich's ataxia is the mutation of the frataxin gene encoding a mitochondrial protein expressed in all tissues, particularly in cardiac and skeletal muscles, and dorsal root ganglia.[362]

Reduced expression of a frataxin homologue in yeast is associated with mitochondrial iron accumulation at the expense of cytosolic iron, and a similar phenomenon can be demonstrated in Friedreich's ataxia patients.[363,364] The increased iron deposition in the heart and brain in Friedreich's ataxia suggests that frataxin is involved in intramitochondrial iron metabolism.[365]

The increase in iron accumulation within mitochondria may cause oxidative damage associated with OH production and lipid peroxidation, leading to the inactivation of several proteins of mitochondrial enzymes, including complexes I–III and the Krebs cycle enzyme aconitase, all containing iron–sulfur clusters.[364] Studies on frataxin-deficient cell lines provided evidence that intramitochondrial iron accumulation is a late event occurring through a mechanism independent of the development of Fe–S deficiency in the intramitochondrial enzymes.

A peculiar form of ataxia associated with X-linked sideroblastic anemia also exhibits selective iron accumulation in the mitochondria. This rare anemia is characterized by non-progressive cerebellar ataxia, diminished deep tendon reflexes, loss of coordination, elevated free erythrocyte protoporphyria, and normal parenchymal iron stores associated with increased iron accumulation in mitochondria.[366]

Recent studies have shown that the disease may be related to the mutation of an ABC transporter, ABC7.[366] This gene is required for the maturation of Fe–S clusters present in several cytosolic proteins, including iron regulatory protein 1. However, the linkage of impaired formation of cytosolic Fe–S clusters and reduced heme mitochondrial synthesis associated with increased mitochondrial iron accumulation remains unclear.[366]

Another pathological condition associated with iron and brain ferritin accumulation is Parkinson's disease, caused by the deaths of dopaminergic nerve cells in the substantia nigra, leading to a reduction in striatal dopamine, and producing the key symptoms of bradykinesia, tremor, and rigidity. Autoptic studies showed a consistent increase in iron content in the substantia nigra, and no evidence of increased iron content in other structures of the basal ganglia.[367,368] This finding suggests that iron-related oxidative stress may be an important component of the neurodegenerative processes in such patients. Iron accumulation within the brains of Parkinson's disease patients may be monitored by magnetic resonance imaging (MRI).[369] This technique is an important clinical tool for analysis of brain morphology and detection of metal ions that possess intrinsic magnetic properties.

Non-heme iron in the brain has three molecular forms: low molecular weight complexes, iron bound to medium molecular weight metalloprotein complexes such as transferrin, and high molecular weight complexes such as ferritin and hemosiderin.[369] Ferritin-bound iron is the main storage form present in the brain, predominantly in the extrapyramidal nuclei, where it normally increases with age.[368]

MRI analysis of the brains of Parkinson's disease patients showed increased iron accumulation in the substantia nigra, associated with reduced iron accumulation at the putamen.[370] This peculiar pattern of iron deposition led to an intriguing hypothesis. The putamen is a high-density region of transferrin-binding sites, whose number is significantly increased in Parkinson's disease.[371] It was hypothesized that the increased amounts of iron entering the putamen are transported to the substantia nigra via dopaminergic neurons, probably bound to neuromelanin,[371,372] a dark-colored pigment found in the dopaminergic neurons of the human substantia nigra. Neuromelanin has the ability to bind a variety of metals, including iron, and increased iron-bound neuromelanin is reported in the Parkinsonian substantia nigra.

Any interpretation of the possible role of iron accumulation in the pathogenesis of Parkinson's disease must consider the fact that the iron increase occurs only in the advanced stage of the disease, suggesting that this phenomenon may be a secondary, rather than a primary, initiating event.

The main mechanisms leading to the deaths of neurons of the substantia nigra in Parkinson's disease are related to oxidative stress, resulting in an increase of iron content, reduction of glutathione, and impaired mitochondrial function. The biochemical reactions triggered by oxidative radicals lead to membrane, protein and DNA peroxidation and, as a consequence of these damages, to necrosis and/or apoptosis of dopaminergic neurons. The death of dopaminergic neurons induces an inflammatory response characterized by microglial activation, not accompanied by reactive astrocytosis.[372]

This type of inflammatory response is unique in that the inflammatory reactions occurring in other neurodegenerative diseases, such as multiple sclerosis, Alzheimer's disease, AIDS dementia, and amyotrophic lateral sclerosisis are characterized by both microglia activation and reactive astrocytosis. These conclusions were supported by neuropathological analyses showing the presence of microglial cells in the substantia nigra, markedly labeled by anti-ferritin antibodies.[372] Ferritin immunoreactivity was never observed in dopaminergic neurons or cells with morphologies corresponding to astrocytes.

These findings were confirmed in other neuropathological studies that also showed that the ferritin H/L ratios in the brains of Parkinson's disease patients markedly decreased, a phenomenon seemingly related to the increased L ferritin synthesis in microglia cells.[373,374] It was also proposed that ferritin H synthesis may decrease in the dopaminergic neurons of the substantia nigra in Parkinson's disease, but this hypothesis was not supported by experimental evidence.

At variance with ferritin levels which are restricted in microglial cells in Parkinson's disease lesions, iron staining was observed in various cell types including astrocytes, reactive microglia, and non-pigmented neurons, and also in damaged areas devoid of pigmented neurons.[375] The mechanism of iron-dependent oxidative damage is not well known. It may be related not only to the Fenton pathway involving the reaction of iron with hydrogen peroxide, but also to a different pathway involving the formation of iron–oxygen complexes.[373] Brain tissue is particularly rich in polyunsaturated fatty acids and is very sensitive to lipid peroxidation, whose function is dependent upon the formation of oxidative radicals requiring iron, but apparently not hydrogen peroxide.[376] This observation suggests that iron–oxygen complexes are the primary routes to the initiation of free radical oxidation. The important role of iron in the genesis of oxidative damage in the substantia nigra of Parkinson's disease patients led to the proposal to use iron chelators and radical scavengers in the treatment of these patients.[377,378]

Another illness associated with increased iron accumulation in the basal ganglia is Huntington's disease, caused by an abnormally expanded CAG repeat within the Huntington gene present on chromosome 4. The function of this gene is unknown, but its mutation leads to the loss of medium spiny neurons in the caudate nucleus.[379] Some defects in the oxidative phosporylation mitochondrial respiratory chain are observed, involving severe deficiencies of complexes II and III and aconitase. Increased ferritin levels were observed in the basal ganglia of Huntington's patients, while their ferritin levels in white matter were lower than those observed in normal controls.[380] Interestingly, ferritin iron levels were markedly increased in putamen, caudate, and globus pallidus basal ganglia and were higher than those observed in other neurodegenerative diseases, such as Alzheimer's disease.[381]

Another neurodegenerative disease associated with increased brain iron accumulation is Alzheimer's disease. It is characterized by the deposition of amyloid within the neocortical parenchyma and the cerebrovasculature. Initial studies have shown that ferritin is a component of the neurite plaques of Alzheimer's dementia.[382] Ferritin accumulation was associated with the microglia, whose number was greatly increased.[382] Biochemical studies showed that ferritin was weakly associated with, but was not a component of, the neurofibrillary structures observed in Alzheimer's plaques.[382]

These initial observations suggest that ferritin may have a role in the formation of amyloid or be involved in the removal and processing of the amyloid.[382] These findings were confirmed by studies showing marked ferritin reactivity in the Alzheimer's plaques and some cerebrovascular vessels in microglia cells; minor reactivity was observed also at the pathological plaques.[383]

In contrast, transferrin is distributed around the senile plaques with apparent extracellular localization.[383] Transferrin in white matter was preferentially localized in astrocytes rather than in oligodendrocytes.[383] Reactive microglia cells associated with amyloid plaques express high levels of melanotransferrin.[384] Magnetic resonance imaging studies showed increased ferritin iron accumulation at the caudate and putamen basal ganglia.[381]

Some studies have shown a peculiar similarity between ferritin and the amyloid precursor protein (APP) associated with Alzheimer's disease and processed the β-peptide that accumulates at the amyloid plaques. Two similarities have been observed: (1) the synthesis of ferritin H and APP is stimulated by pro-inflammatory cytokines such as interleukin-1; and (2) the stimulation of APP synthesis by interleukin-1 is mediated through translational mechanisms dependent upon the presence of an mRNA translational enhancer mapping from +55 to +144 nucleotides from the 5' cap site; a similar element was observed for ferritin H and L mRNAs.[385]

Computer alignment studies showed that the sequences in the 5' untranslated region of APP mRNA are similar, but not identical, to those observed in the L ferritin mRNA untranslated region. This slight difference may help to explain why ferritin synthesis is stimulated by interleukin-1 in hepatoma cells, but not in astrocytoma cells.[385]

Increased iron and ferritin content in the brain is observed also in multiple sclerosis, a neurodegenerative disease based on an inflammatory mechanism whose main cellular targets are oligodendrocytes. Progression of the disease is associated with an increase in iron and ferritin accumulation in the brain. It is believed that dying oligodendrocytes are phagocytosed by microglia cells present in the lesions. These macrophagic cells are characterized by their high iron and ferritin content.[386,387] In advanced stages of the disease, the breakdown of the blood–brain barrier allows the leakage of ferritin from plasma into the cerebrospinal fluid, causing the increase in ferritin levels in this fluid in chronic progressive patients.[388]

Microglia cells under physiological conditions are only rarely stainable with iron and anti-ferritin antibodies. However, activated microglial cells are clearly stained for iron and for ferritin.[389] Iron- and ferritin-stained microglia cells are frequently observed in neurodegenerative diseases. Activated microglia cells probably acquire the capacity to uptake iron as a defense mechanism against oxidative stress. Preliminary evidence indicates that microglial cells present in discrete foci of subcortical white matter contain stainable iron and ferritin during early stages of postnatal development and may serve as iron reservoirs for other cell types.[352]

Microglia play a role in the response to cerebral ischemia-induced reperfusion. This condition involves intracellular free radical formation and lipid peroxidation through a series of oxidative reactions that require iron for their cytotoxic effects.[390] Iron chelators were shown to reduce the extent of post-ischemic neurological injury.[391] The cellular source of iron that induces the cytotoxic effects post-ischemia is not well known; ferritin may act as a potentially deleterious iron donor or display an iron-detoxifying function as an iron storage protein.

Increased ferritin levels are observed in the brain tissues after cerebral ischemia, as shown by studies of ferritin immunoreactivity following subarachnoidal hemorrhage[392] or after transient forebrain ischemia.[393,394] Focal cerebral ischemia elicited a significant rise of ferritin H and ferritin L mRNA, starting at 12 hours after the onset of the ischemic insult. The rise of ferritin H and L mRNA was followed by a concomitant rise of ferritin H and L subunits and was limited to the ischemic cerebral areas.[395]

Immunohistochemical analysis of the ischemic tissues showed that the majority of positive cells exhibited morphologies similar to those of oligodendrocytes, while the ferritin-positive cells in the cerebral area surrounding the necrotic region displayed morphological features typical of microglia cells. These findings were confirmed by Panahian et al. who found higher expression of ferritin in both neurons and astroglia 6 hours after the onset of ischemia, with peak expression at 24 hours.[395]

In heme oxygenase-I overexpressing mice, ferritin was expressed primarily at the astroglia and microvascular cells.[396] Post-ischemic neuronal levels of ferritin were significantly higher in animals overexpressing heme oxygenase-I as compared to the levels observed in wild-type animals.

Other studies focused on the effects of experimentally-induced subarachnoid hemorrhage in monkeys on ferritin and heme oxygenase-I expression in cerebral arteries and brain tissues.[397]

Significant increases in ferritin and heme oxygenase-I protein levels, but not of their corresponding mRNAs, were observed in the cerebral arteries, but not in brain tissues.[397] Interestingly, the induction of heme-oxygenase-I synthesis preceded that of ferritin. The differences observed in these two studies may be related to the different experimental settings.

Several recent studies provided evidence that, in addition to ferritin, heme oxygenases are induced following brain ischemic or traumatic damage and play a protective role.[398] Heme oxygenase-I cleaves the heme molecule, produces carbon monoxide and biliverdin, and is essential for iron homeostasis. Several studies have shown that both heme oxygenase-I and heme oxygenase-II exert protective effects against experimental brain ischemia[399,400]; this effect is in part due to the induction of higher ferritin levels in brain tissue.

Astrocytes generally do not contain stainable iron or display reactivity to anti-ferritin and anti-transferrin antibodies. In human and murine brains, the only exceptions are the astrocytes in the striatum.[401,402] Aceruloplasminemia is a pathological condition associated with accumulation of iron and ferritin in the astrocytes. The peculiar iron accumulation observed in this disease is mainly related to the expression in astrocytes of a membrane-bound form of ceruloplasmin.[403] The presence of this glycosylphosphatidylinositol-anchored form of ceruloplasmin is related to the synthesis in astrocytes of an alternative transcript of ceruloplasmin.[403] This GPI-anchored form plays a role similar to the role of the secreted form in oxidizing ferrous iron. The absence of GPI-anchored ceruloplasmin in the central nervous system may lead to the generation of oxidative radicals.

Aceruloplasminemia is an autosomal recessive disorder of iron metabolism; its clinical symptoms include neurologic manifestations, diabetes, and retinal degeneration.[404] Aceruloplasminemia patients exhibit pronounced accumulations of iron at several parenchymal sites, associated with the absence of circulating serum ceruloplasmin caused by inherited mutations in the ceruloplasmin gene.[404]

Studies of animals made deficient in ceruloplasmin synthesis provided evidence that the tendency to accumulate iron in several body compartments depends on a striking impairment in the movement of iron out of cells, and not on increased iron uptake. The increased iron accumulation in the brain occurs in several sites including the retina, basal ganglia, dentate nucleus, and cerebral cortex. The increased iron accumulation in these brain areas leads to progressive neurodegeneration mediated by the generation of iron-free radicals that contribute to the impairment of mitochondrial energy metabolism and induce lipid peroxidation.[405,406]

A peculiar decrease of ferritin levels is observed in another brain disease, Niemann–Pick disease type C (NPC). NPC is a recessive cholesterol storage disease characterized by severe, progressive neurodegeneration caused by mutation of the NPC gene. This gene encodes a protein localized in the endosomes and involved in the control of intracellular trafficking of low density lipoprotein cholesterol.[407]

Initial studies revealed the absence of ferritin L immunoreactivity in spleens and livers of NPC patients.[408,409] More recent studies have shown deficient ferritin H and L immunoreactivity in several tissues of NPC patients.[410] Since this disease affects cholesterol, and not iron metabolism, it was proposed that the molecular defect that blocks intracellular utilization of cholesterol is also responsible for reduced iron utilization and consequent reduced ferritin synthesis.[410]

Few studies have investigated the development of ferritin in different areas of the brain. Brain development is accompanied by changes in ferritin content in humans[411] and rodents.[412] Brain ferritin is high at birth[413] and declines over the first two postnatal weeks.[413]

The distribution of ferritin in the brain also appears to be altered during development. In the early postnatal period, brain ferritin is predominantly localized at microglia. By postnatal day 30, it is preferentially localized in the oligodendrocytes.[354] The mechanisms that control the development of ferritin synthesis in selected areas of the brain are largely unknown. However, a role for thyroid hormones in this process was recently shown.

Ferritin expression was investigated in cerebral tumors. The two most common tumors of the central nervous system are astrocytomas and oligodendroglial tumors. The most common types of astrocytomas include low-grade diffuse astrocytoma, anaplastic astrocytoma, and glioblastoma multiforme. These tumors were traditionally thought to arise from malignant transformations of mature astrocytes and oligodendrocytes. However, recent studies suggest that they arise from a population of glia exhibiting the properties of oligodendrocyte progenitors.[414]

The majority of studies of brain tumors have shown that increased ferritin levels are predominantly associated with astrocytomas. In most astrocytoma cells, ferritin was clearly detectable in the cytoplasm, while in normal cells ferritin is usually present at levels too low to be detectable by cytochemistry.[415]

The levels of ferritin in tumors were inversely related to tumor differentiation, with the highest levels observed in less differentiated tumors.[415] These findings were confirmed in subsequent studies that led to the speculation that astrocytoma cells, due to their reactivity with anti-ferritin and other antibodies reacting with proteins expressed at consistent levels in the macrophage lineage (such as CD11b, CD36, CD68, α-1-antitrypsin, ferritin, and lysozyme), express a macrophage phenotype.[416]

Elevated levels of ferritin were observed in the cerebrospinal fluid (CSF) of patients affected by glioblastoma multiforme. Interestingly, CSF ferritin levels in glioblastoma patients (mean value: 103 ng/ml) were markedly higher than those observed in patients with headache (mean value: 4.3 ng/ml) or meningitis (mean value: 5.4 ng/ml).[417] These observations suggest that the increased levels of CSF ferritin in glioblastoma patients are related to a release of the protein by tumor cells.

Other studies evaluated a possible biological effect of ferritin on astrocytoma cells. Astrocytoma cells, but not normal cells, were sensitive to the stimulatory effect on cell growth exerted by ferritin.[418] This effect is related to ferritin-mediated iron uptake since it is inhibited by gallium and zinc.[418]

The possible mechanisms able to modify ferritin synthesis by astrocytoma and oligodendroglioma cells were investigated. Studies on rat glioma cells showed that insulin and insulin-like growth factor-I significantly stimulated ferritin synthesis through a transcriptional mechanism.[419] Ferritin levels in astrocytoma cells are transiently stimulated by interleukin-1-β; the increase in ferritin levels in turn produced a decrease of the cytoplasmic free iron pool.[420]

The expression of ferritin in astrocytoma cells was confirmed by studies focused on genes whose expression is modulated by iron.[421] One of the genes whose expression is stimulated by iron at mRNA level corresponds to L ferritin. Interestingly, the expression of other genes is also stimulated by iron in astrocytoma cells: two unknown genes expressed by calreticulin and ribosomal protein S9.[421]

9.13.4 FERRITIN EXPRESSION AND FUNCTION IN MACROPHAGES

Reticuloendothelial macrophages play a key role in iron storage and recycling. These cells phagocytose senescent red blood cells and lyse them at the phagolysosomal compartments. Following the lysis of phagocytosed red blood cells, hemoglobin is degraded and iron is liberated from heme through a process involving the heme oxygenase enzyme. Iron is then fixed in ferritin, whose levels of synthesis are very high in macrophagic cells. Iron is then mobilized from ferritin and exported out of the cell through a process that seems to involve ferroportin 1 and ceruloplasmin.

Studies on the expression of ferritin in the macrophagic lineage focused on the changes of the rate of ferritin synthesis during maturation of circulating monocytes to tissue macrophages. Among the mononuclear cells of the peripheral blood, monocytes clearly exhibit the highest ferritin levels.[422] Human monocytes undergoing *in vitro* spontaneous maturation to macrophages exhibit a dramatic upregulation of ferritin expression. Freshly isolated monocytes display a ferritin content of about 10 ng/10^6 cells. This content increases to 350 to 1500 ng/10^6 cells in *in vitro*-differentiated macrophages.[423]

This increase in ferritin expression is paralleled by the concomitant induction of the transferrin receptor on the surfaces of monocytes maturing to macrophages. Analysis of isoferritins in freshly isolated monocytes showed virtually equal proportions of H ferritin and L ferritin. L ferritin was predominant in macrophages and was the only isoferritin whose synthesis increased following iron loading.[424]

H ferritin and L ferritin mRNA levels markedly increased following *in vitro* maturation of monocytes to macrophages.[425] It was proposed that both phenomena based on transcriptional and post-transcriptional mechanisms are involved in up-modulation of ferritin mRNA expression. Parallel experiments analyzing IRP activity showed that during monocyte to macrophage maturation a marked increase in IRP activity was observed[426,427] — a phenomenon that should greatly antagonize ferritin expression through inhibition of translation.

The high rate of expression of ferritin genes in monocytes–macrophages was confirmed by new techniques of serial gene analysis based on cDNA microarrays[428] or on serial analysis.[429] Among 5000 different transcripts, ferritin H and ferritin L mRNAs in freshly isolated monocytes represented the second and fifth most abundantly expressed transcripts.[429] In differentiated macrophages, ferritin H and ferritin L represented the first and third most abundant transcripts, respectively. It is of interest that in differentiated macrophages the ferritin L and ferritin H mRNAs represent 5% of total transcripts.[429]

Several studies investigated the roles of cytokines and other molecules on the control of ferritin synthesis in monocytes–macrophages. The role of interferon-gamma (IFN-γ) on ferritin synthesis in cells of monocytic lineage was investigated in detail. IFN-γ synthesis is induced during several immune and inflammatory processes. Among its different biological effects, IFN-γ is a potent activator of macrophages. Its addition to *in vitro*-grown monocytes elicited a significant reduction of ferritin content, associated with an increase of iron release.[430]

This finding was confirmed in a subsequent study showing that ferritin content was markedly reduced in IFN-γ-treated monocytes–macrophages. In control monocytes, ferritin content was about 361 fg/monocyte, while in IFN-γ treated monocytes, the ferritin content was about 64 fg/monocyte).[431] Ferritin in IFN-γ-treated monocytes was saturated with approximately three times as much ^{59}Fe as ferritin in control monocytes, probably due to reduced iron recycling at the ferritin compartment.[431]

The molecular mechanisms responsible for the modulation of monocyte ferritin synthesis by IFN-γ have been in part elucidated and seem largely related to its capacity to induce the synthesis and the release of nitric oxide (NO) through the stepwise oxidation of L arginine to NO and L citrulline. NO induced by IFN-γ induces activation of iron regulatory protein 1 activity which in turn causes a repression of ferritin synthesis.[432]

Bacterial lipopolysaccharide, which also causes the induction of NO synthesis in cells of the monocyte lineage, induces a decrease of ferritin synthesis in monocytes–macropohages and synergizes with IFN-γ in the induction of this effect.[432] The molecular mechanisms through which NO modulates ferritin synthesis are extensively treated in Chapter 8.

The effects of IFN-γ on monocytes–macrophages are not limited to ferritin. They involve other proteins of iron metabolism, such as transferrin receptor and Nramp2.

The effects of IFN-γ on iron uptake and ferritin expression have some important implications in the bacteriostatic mechanism of this cytokine. The addition of iron-saturated transferrin to monocytes (at a dose twice the level found in plasma) completely abrogated the bacteriostatic activity of IFN-γ.[433]

Recent studies have shown that prostaglandin A_1 may act in monocytes as a modulator of ferritin synthesis through a differential mechanism of ferritin H and ferritin L. Prostaglandin A_1 stimulates ferritin H synthesis through a post-transcriptional mechanism dependent upon inactivation of IRP activity in the absence of any modification of ferritin H mRNA levels, and enhances ferritin L synthesis through a mechanism related to an increase of ferritin L mRNA due to increased transcription or to mRNA stabilization.[434] Prostaglandin A_1 also induces heme oxygenase in human monocytes, but its activity that catalyzes the oxidative cleavage of heme at the α-methene bridge with the consequent formation of iron, carbon monoxide, and biliverdin, does not seem to be required for ferritin induction.[434]

Tumor necrosis factor-α, a pro-inflammatory cytokine, also affects ferritin synthesis in monocytes–macrophages and interferes with iron metabolism in these cells. These effects are mainly related to a decreased iron release from monocytes–macrophages, associated with increased iron storage in ferritin.[435]

These effects, which are induced also by other pro-inflammatory cytokines, seem to be responsible for changes in iron metabolism observed in chronic inflammatory disorders associated with increased iron in the macrophagic body compartments, mainly due to defective iron mobilization. This defective mobilization is mainly observed in chronic diseases and is responsible for the development of anemia associated with these conditions. The induction of hypoferremia elicited by increased iron retention and storage within the macrophage compartment limits iron availability to the erythron and represents a key event in the development of anemia of chronic disease.

In addition to producing defective iron mobilization, TNF-α exerts a stimulatory effect on ferritin synthesis, mainly related to a transcriptional mechanism involving the activity of the transcription factor NF-κB.[232]

Interleukin-1 also stimulates ferritin synthesis in macrophages, mainly through a translational mechanism.[436] Evidence indicates that both pro-inflammatory cytokines and anti-inflammatory cytokines (e.g., interleukin-4, interleukin-10 and interleukin-13) may exert stimulatory effects on iron retention in macrophages, mediated by two mechanisms: (1) inhibition of IFN-γ mediated activation of iron regulatory protein that stimulates ferritin synthesis; and (2) stimulation of transferrin receptor expression.[437] Anti-inflammatory cytokines are also able to promote iron retention by macrophages and, through this mechanism, contribute to the development of anemia associated with chronic disease.[437]

Macrophage ferritin synthesis is stimulated also by other molecules involved in the inflammatory response. The acute-phase protein α-1-antitrypsin induced a significant increase of monocyte ferritin concentration.[438] This effect is mediated independently of iron, through a poorly characterized mechanism that does not involve transcriptional or translational mechanisms, but uses a post-translational mechanism.[438] These findings suggest that α-1-antitrypsin may contribute, with inflammatory cytokines, to the development of anemia of chronic disease.[438]

Under physiologic conditions, macrophages acquire most of their iron through phagocytosis and destruction of senescent red blood cells. This process exposes macrophages to high levels of iron that must be recycled to make it available for erythropoiesis and cellular metabolism. The recognition and destruction of aging red blood cells by macrophages is a regulated process involving the specific recognition of certain membrane molecules exposed on the surfaces of senescent red blood cells. Various receptors (such as CD14) involved in the binding of surface-exposed phosphatidylserine may be implicated in the recognition and uptake of senescent red blood cells.[439,440]

Macrophages acquire iron also through the uptake of hemoglobin released in the circulation as a consequence of intravascular hemolysis. Under physiologic conditions the level of intravascular hemolysis is relatively low, but it can be considerably increased in some pathological conditions, mediated either through autoimmune mechanisms (such as autoimmune hemolytic disease), infections (such as malaria), or inherited conditions (such as sickle cell disease). Hemoglobin released in circulation is captured by the acute-phase protein haptoglobin. The hemoglobin–haptoglobin complex is cleared from the circulation by macrophages through interaction with a scavenger receptor present on their surfaces, the CD163 protein, also known as the acute-phase regulated and signal-inducing macrophage protein.[441]

In vitro studies have shown that after erythrophagocytosis of ^{59}Fe-labeled erythrocytes a complete transfer of iron from hemoglobin to ferritin was observed within 24 hours.[442] Following this transfer of iron from phagocytosed hemoglobin to ferritin, a progressive release of iron from the monocytes in the forms of ferritin and non-protein-bound low molecular weight iron occurred.[442]

In vitro studies using normal monocytes have shown that erythrophagocytosis stimulates H and L ferritin synthesis by these cells.[443] The mechanism underlying this stimulation is complex and seems to involve reactive oxygen species and iron released from phagocytosed red blood cells.[443] These findings suggest that the induction of ferritin observed in monocytes–macrophages following erythrophagocytosis may exert a protective role, in addition to its role in iron recycling.[443]

The enzyme responsible for iron release from hemoglobin within macrophages is heme oxygenase 1 (HO1). It is an inducible enzyme associated with the endoplasmic reticulum, which cleaves the heme ring, giving rise to biliverdin which is rapidly reduced to bilirubin, carbon monoxide, and ferrous iron. This enzyme plays a role in cellular iron metabolism, as suggested by two lines of evidence: (1) mice with targeted genomic deletions of HO1 exhibited increased tissue iron stores, in association with low serum iron levels; and (2) the expression of HO1 induces an increase in iron efflux.[444] The mechanism by which HO1 is linked to iron mobilization also involves a recently identified membrane iron transporter, an iron ATPase.[445] This transporter is localized with HO1 in the microsomal membranes.[445] It is specific for ferrous iron, requires ATP for its activity, and is stimulated by iron.[445] This iron ATPase allows the transport of iron from the cytosol to the lumen of the endoplasmic reticulum.

Based on these observations, it is possible to describe a tentative pattern of macrophage recycling of iron derived from erythrophagocytosis:

1. Recognition of senescent red blood cells through interaction with specific membrane receptors and subsequent phagocytosis.
2. The red blood cells in phagosomes are digested, followed by release and digestion of hemoglobin in macrophage cytosols.
3. The released heme is degraded at the endoplasmic reticulum, generating biliverdin and free iron.
4. The free iron is only transiently in the labile iron pool, and is then transported into the lumen of the endoplamic reticulum.
5. Iron is then fixed by ferritin and either stored or recycled out of the macrophages, again becoming available for iron metabolism.

The coordinated regulation of iron storage within macrophages is relevant to normal iron metabolism and also to appropriate defense against microbial cell invasion. These regulation mechanisms are particularly important since several eukaryotic intracellular pathogens (such as *Candida*, *Cryptococcus*, *Entamoeba*, *Histoplasma*, *Leishmania*, *Naegleria*, *Plasmodium*, *Pneumocystis*, *Schistosoma*, *Toxoplasma*, and *Trypanosoma*) and intracellular prokaryotic invaders (such as *Chlamydia*, *Ehrlichia*, *Francisella*, *Legionella*, *Listeria*, *Mycobacterium*, and *Rickettsia*) capable of developing within phagocytic cells have developed a peculiar strategy to obtain the iron required for their growth. Macrophages respond to invasion by these pathogens by changing their intracellular iron metabolism which limits intracellular pathogen growth.[433] Basically, these defense mechanisms are triggered by the release of interferon-γ by activated T lymphocytes (Th1) which, in turn, leads to down-modulation of transferrin receptors, associated with increased synthesis of nitric oxide and iron transporter Nramp1. The mechanisms used by the pathogens are essentially based on a down-modulation of host defense mechanisms, i.e., enhancement of iron influx through up-modulation of transferrin receptor expression and a decrease of nitric oxide release.

Some pathogens, such as *Mycobacteria*, have developed a unique strategy to obtain within macrophages the iron required to sustain their growth. *Mycobacterium tuberculosis* produces a class of siderophore molecules, called mycobactins, derived from salicylic acid after condensation of a serine residue. These mycobactins act through two different mechanisms. Hydrophilic mycobactin is secreted into the extracellular medium where it competes for iron with iron-binding molecules. Hydrophobic mycobactin operates as an ionophore to shuttle iron across the cell wall. Mycobactins are required for the growth of *Mycobacterium* within macrophages.[446]

Macrophages have also developed a unique mechanism to counteract *Mycobacterium* growth, based on the synthesis of a catabolite of tryptophan metabolism called picolinic acid. This compound is able to form stable chelates with metals and particularly with iron; it synergizes with IFN-γ to inhibit the growth of *Mycobacterium* and stimulates the release of anti-inflammatory cytokines.[447,448]

Other mechanisms of iron metabolism are also involved in the susceptibility to and the defense against *Mycobacterium* infection. Studies on mice strains sensitive to infection with pathogens localized in the reticuloendothelial cells led to the identification of the Nramp1 (natural resistance-associated macrophage protein 1) gene. A mutation in the Nramp1 protein produces high susceptibility of mice to uncontrolled proliferation of many pathogens, including *Mycobacteria*, *Leishmania*, and *Salmonella*. Nramp1 is an integral membrane protein selectively expressed in the lysosomal compartments of macrophages and recruited to the membranes of phagosomes soon after completion of phagocytosis. The Nramp1 contribution to the defense against macrophage infection is the extrusion of divalent cations, including iron, from the phagosomal space. These cations are essential for bacterial growth and their removal from the phagosomal microenvironment impairs bacterial growth.[449]

The role of Nramp1 in macrophage iron homeostasis in not clear. Experiments in Nramp1-transfected RAW macrophagic cell lines led to the conclusion that Nramp1 may transport Fe^{2+} out of the phagosome, as well as out of the cell, via an endocytic pathway.[450] This conclusion was contradicted by the analysis of the phagosomes of Nramp1-transfected cells, which showed increased accumulation of Fe^{2+} in this cell compartment.[451,452] More recent studies provided clear evidence that Nramp1 extrudes Fe^{2+} and other divalent cations from the phagosomal space.[449] Studies of bone marrow-derived macrophages showed that iron loading is a more potent stimulus than IFN-γ or bacterial lipopolysaccharide in inducing Nramp1 expression in these cells.[453]

Macrophages also express a second natural resistance-associated macrophage protein, Nramp2. The expression of this iron transporter is markedly enhanced by bacterial lipopolysaccharide (LPS), while IFN-γ had no effect.[454] As a consequence of increased Nramp2 expression, LPS-treated macrophages exhibited pronounced enhancement of their iron uptake capacity from low molecular mass iron complexes.[454]

A series of studies focused on the capacity to synthesize ferritin by various types of macrophages (i.e., alveolar, peritoneal, and hepatic macrophages). One population studied in detail is alveolar macrophages. These cells are particularly exposed to environmental influences, such as smog or smoking. Extracellular iron is present in the alveolar structures of normal subjects. Under physiological conditions, this extracellular iron is predominantly bound to transferrin.[455] Although iron bound to transferrin is not redox active, after uptake into the cells, it may enhance cellular susceptibility to oxidative injury after intracellular release of iron from transferrin.

The main mechanism of protection against this type of injury to the lung is ferritin present in alveolar macrophages. Alveolar macrophages are able to take up iron in various molecular forms, including transferrin-bound iron, low molecular weight chelates of iron, and iron present in inhaled particles. In addition to the capacity of these cells to synthesize large amounts of ferritin, they also exhibit the ability to actively sequester the different forms of iron in ferritin, thus neutralizing its potential deleterious effects.[456,457] The protective effect of iron sequestration by alveolar macrophages inhibits the generation of extracellular hydroxyl radicals. The protective effect extends beyond the macrophages that incorporate iron; more importantly, it extends to other alveolar cells that are unable to synthesize large amounts of ferritin.[458]

Human alveolar macrophages normally contain substantial amounts of iron and ferritin, and sequestration of iron and ferritin content are markedly enhanced in smokers.[459] Measurement of ferritin content in alveolar macrophages derived from smokers and non-smokers subjects showed that smokers exhibited markedly higher ferritin levels (782 ng/10^6 cells) than non-smokers (133 ng/10^6 cells).[460] Similar observations were made for iron content, which was significantly higher in smokers (27.6 nmol/10^6 cells) than in non-smokers (7.5 nmol/10^6 cells).[460] Subsequent studies showed that alveolar macrophages derived from smokers secrete more ferritin than those of normal controls and the extent of secretion is directly related to the extent of smoking (the level of ferritin secretion was markedly higher in heavy smokers than in light smokers).[461]

Studies of various regions of the lungs of smokers have shown higher ferritin levels in the upper lobes of the lungs.[462] This differential regional concentration of ferritin may have some relevance to explaining the differential sensitivities of the various lung regions to oxidative stress.[462]

Alveolar macrophages mobilize iron from the surfaces of metal oxides such as asbestos and silica, which in turn induces the synthesis of ferritin.[463,464] In additon to the increased ferritin synthesis, macrophages exposed to silica secrete increased amounts of ferritin. However, the increased secretion of ferritin synthesis by alveolar macrophages may have a pro-oxidant and not only a protective effect on the lungs, because iron can be mobilized from extracellular ferritin by various reductants.[465] The stimulation of ferritin synthesis, and particularly its release by alveolar macrophages, was markedly higher for cells isolated from smokers as compared to the values observed in non-smokers.[466] The ferritin released by alveolar macrophages may enhance the cytotoxic effects of asbestos on lung cancer cells.[466] Iron uptake by alveolar macrophages may increase the intracellular iron pool, and intracellular unbound iron can catalyze the generation of highly reactive oxygen species and promote oxidative cell injury.

Alveolar macrophages are particularly exposed to the danger of hypoxia, which promotes intracellular accumulation of reactive oxygen intermediates. The sensitivity of alveolar macrophages to hyperoxia is increased when they are exposed to iron-supplemented media.[467] This effect is related to a pro-oxidant effect of free iron transiently found in cells during iron uptake, and also to an inhibitory effect of hyperoxia on ferritin accumulation by alveolar macrophages in iron-supplemented media.[467]

Several factors, including iron and hyperoxia, regulate the synthesis of pulmonary ferritin. Hyperoxia induces a marked increase in the level of mRNA encoding the ferritin light subunit.[468] This observation suggests a possible link between hyperoxic lung injury, which plays a central role in the pathogenesis of acute and chronic pulmonary diseases, and iron metabolism. The levels of ferritin and lactoferrin contained in alveolar macrophages seem to be major determinants of the sensitivity of lung tissue to hyperoxia damage. In fact, hypotransferrinemic mice, which exhibit elevated expression of both ferritin and lactoferrin in the alveolar macrophages, are tolerant to hyperoxic lung injury.[469]

Hepatic macrophages play a particular role in iron metabolism in that they are contained in a tissue in which two different cell types are involved in iron storage: hepatocytes that function as storage depots for much of the iron that is not needed elsewhere, and hepatic macrophages which are mainly involved in recycling iron derived from the lysis of senescent red blood cells. A significant proportion of iron derived from the phagocytosis and lysis of senescent red blood cells is released by hepatic macrophages as ferritin, which is rapidly bound by hepatic cells[108,109] which possess high numbers of membrane ferritin receptors.

According to these findings, ferritin released by hepatic macrophages may act as an iron donor for hepatocytes. The process of iron transport from hepatic macrophages to hepatocytes through ferritin was explored using mice injected endovenously with iron complexes. The iron complexes are first taken up by hepatic macrophages, and iron is incorporated into ferritin. Iron is later observed within hepatic cells, with concomitant disappearance of hepatic macrophages.[470]

Under some pathological conditions, hepatic macrophages play a key role in hepatic injury. The effector mechanism of these deleterious effects on hepatocytes is the release of cytotoxic and pro-inflammatory cytokines, such as tumor necrosis factor-α, interleukin-1, transforming growth factor-β, and interleukin-6. Tumor necrosis factor-α is particularly cytotoxic for hepatocytes. Its expression is mainly mediated through activation of the NF-κB transcription factor. The induction of NF-κB by bacterial lipopolysaccharide (LPS) in cultured hepatic macrophages may be inhibited by pre-treatment of the cells with an iron chelator.[471] This finding suggests an important role for iron in NF-κB activation. The linkage between iron and NF-κB activation was observed also in pathological conditions, such as alcoholic liver injury. Animal models of liver alcoholic injury demonstrated increased NF-κB activation, nonheme iron and ferritin content, and heme oxygenase activity in hepatic macrophages; treatment with an iron chelator normalized all these parameters.[472] Interestingly, the priming of hepatic macrophages to increase iron content through red blood cell phagocytosis rendered these cells more sensitive to stimulation with LPS for NF-κB activation and TNF-α release.[472]

Peritoneal macrophages are responsive to pro-inflammatory and other stimuli in a manner similar to those displayed by other body macrophage populations. Some studies have shown that peritoneal macrophages are sensitive to hypoxia which acts as a stimulator of ferritin synthesis; interestingly, under hypoxic conditions, the stimulatory effect of iron on ferritin synthesis in peritoneal macrophages was markedly enhanced.[473,474] The stimulation of ferritin synthesis induced by hypoxia is dependent upon inhibition of iron regulatory protein 1 activity and is reversed by O_2^- generating agents, but not by hydrogen peroxide.[473,474]

Particularly interesting are the studies of monocytes–macrophages of patients with hereditary hemochromatosis. In diseases associated with iron overload, iron accumulation in the reticuloendothelial system is different. In secondary iron overload resulting from hemolysis, repeated blood transfusions, or dyserythropoiesis, macrophages exhibit marked increases of iron–ferritin content. In contrast, in primary iron overload resulting from hereditary hemochromatosis, iron accumulation is low in the reticuloendothelial system, while the liver and other parenchymal tissues suffer from iron overload. These findings suggest a defect in iron metabolism in the macrophagic compartments of hereditary hemochromatosis (HH) patients. The reasons for this defective metabolism in reticuloendothelial cells are complex and have been only partially clarified. The reduced iron level in HH macrophages is associated with low intracellular ferritin concentration and augmented iron regulatory protein activity.[475]

In addition, monocytes–macrophages from HH patients also exhibited abnormal responses to IFN-γ and LPS in terms of IRP activity and ferritin synthesis. In monocytes derived from normal controls, a transient stimulation, followed by a marked down-modulation was observed. In contrast, in monocytes derived from HH patients, the transient increase of IRP activity was observed as in normal controls, but not followed by subsequent down-modulation.[476] Furthermore, monocytes from normal controls exhibited significant increases of their ferritin levels 24 hours after IFN-γ/LPS stimulation, a phenomenon not observed in HH monocytes.[476]

The low iron accumulation observed in macrophages of HH patients, in spite of the high levels of parenchymal iron accumulations, may be related to decreased iron uptake, to increased iron extrusion from macrophages, or to both these mechanisms.

Recent studies showed a defect of iron release in HH macrophages. HH monocytes released iron in the form of ferritin with the same relative rate observed for control monocytes; however, HH monocytes released twice as much as iron in the form of non-protein low molecular weight iron complexes as did control monocytes.[477]

More recent observations suggest that the abnormalities in iron metabolism observed in monocytic–macrophagic cells of HH patients may be related to their synthesis of the mutated HFE protein. HFE gene, the gene mutated in HH patients, encodes a class I major histocompatibility complex (MHC)-related protein, called HFE.[478] HFE binds to the transferrin receptor and modulates its iron uptake activity. The mutated HFE protein observed in HH patients fails to interact with the transferrin receptor, with a consequent increase in the level of intestinal iron uptake.[478] The normal HFE protein associates with the transferrin receptor at the cell membrane and in an intracellular compartment of recycling endosomes. The mutated HFE protein was unable to localize at the endosomes, retains the capacity to localize at the basolateral membranes in polarized cells, but is unable to interact with the transferrin receptor.[479]

Transfection experiments in crypt duodenal cells showed that increased HFE expression elicited a decrease in ferritin expression, associated with increased levels of transferrin receptor.[479] Similar studies have been performed on monocytes–macrophages. Peripheral blood monocytes[480] and tissue macrophages[481] express the HFE protein on their surfaces.

Subsequent studies aimed to evaluate a possible role of the HFE protein in the control of iron uptake by monocytes–macrophages. Monocytes derived from HH patients, and thus expressing on their surfaces the mutated HFE protein, were transfected with the normal HFE gene. The transfection resulted in a high level of the normal HFE protein on the surfaces of these cells and an increase of transferrin-mediated iron uptake, with a subsequent increase of ferritin levels.[482] These observations suggest that the HFE protein modulates iron accumulation into macrophages through two different mechanisms, one related to an increase of transferrin-mediated iron uptake, and the other to a decrease of iron extrusion from macrophages.[482] The biologic effects of HFE proteins are different in duodenal crypt cells as compared to monocytes–macrophages.

Ferritin expression has been explored also in lymphocytes. These cells possess ferritin levels distinctly lower than those observed in monocytes. H ferritin was predominant in quiescent T lymphocytes and its concentration was higher in immature thymic precursors.[483] Studies on T lymphocytes activated *in vitro* with mitogenic lectins showed that ferritin levels declined during the initial phases of T lymphocyte proliferation (corresponding to the transition from G0 to G1 phases of the cell cycle). This phenomenon may be related to the utilization of iron for the initial stages of the T cell proliferation program.[484,485]

At later stages of T cell proliferation, T cells express transferrin receptors on their membranes, and actively incorporate iron; ferritin levels increase, reaching values higher than those observed in quiescent lymphocytes. Parallel experiments on membrane ferritin receptor expression showed that among lymphoid cells present in the peripheral blood, B lymphocytes displayed the highest ferritin binding capacity (most cells were able to bind ferritin H), while only about 30% of T lymphocytes (CD4+ and CD8+) displayed ferritin binding capacity.[486]

In vitro experiments showed that ferritins rich in H ferritin subunits display an inhibitory effect on T lymphocyte proliferation.[487] Furthermore, ferritin H and ferritin L do not inhibit B lymphocyte proliferation induced by B lymphoid polyclonal stimulators, but inhibit antibody synthesis induced both by agents acting through T cell-dependent and -independent mechanisms.[488]

The above mentioned studies were performed on the more common T lymphocyte population, i.e., T lymphocytes bearing the T cell receptor (TCR) γ/δ. However, preliminary evidence indicates a unique iron metabolism in T lymphocytes bearing γ/δ TCR. The γ/δ lymphocytes mature extrathymically and accumulate in both epidermal and mucosal epithelia. The evidence of their peculiar iron metabolism is based on different and apparently unrelated observations.

A study of the effects of an anti-transferrin receptor antibody on the differentiation of thymocytes showed that treatment with this antibody completely abrogated the maturation of mature TCR α/β^+ lymphocytes, but did not affect the development of TCR γ/δ^+ lymphocytes.[489] This finding suggests that γ/δ T lymphocytes possess alternative iron-uptake mechanisms.[489] They also possess lactoferrin receptors and their growth is stimulated by lactoferrin addition.[490] While CD8 and TCRα/β knockout mice do not exhibit any detectable alteration in iron metabolism, significant iron overload has been documented in TCRγ/δ knockout mice.[491]

The γ/δ lymphocytes in the intestinal mucosa may play a role in the control of iron metabolism. These lymphocytes interact preferentially with non-conventional class I-like molecules (such as MIC A and B) expressed in the gut. It is tempting to suggest that γ/δ lymphocytes also interact with HFE molecules expressed at high levels in crypt duodenal cells to trigger activation of γ/δ lymphocytes. The activation may generate the production of cytokines, such as keratinocyte growth factor, that can modulate the fate, differentiation, proliferation, and rate of apoptosis of intestinal epithelial cells during their migration/differentiation, starting in the crypt and ending at the tip of the villus.

9.14 UNEXPECTED FUNCTIONS OF FERRITIN

This section summarizes recent studies of ferritin gene knockout or overexpression in various cellular systems, and studies suggesting potential biologic functions of ferritin other than iron storage and recycling.

A series of studies focused on the overexpression of ferritin H chains in various cell lines. Ferritin H chains possess intrinsic ferroxidase activity which appears essential for iron incorporation. Initial studies provided evidence that overexpression of ferritin H chain in mouse erythroleukemia cells produced a decline of the free iron cytoplasmic pool, a phenomenon related to the capacity of the overexpressed ferritin to withdraw iron from this pool.[492]

Analysis of clones of mouse erythroleukemia cells expressing different levels of ferritin H revealed an inverse relationship between the level of ferritin H and the concentration of the free intracellular iron pool.[2] On the other hand, the capacity to withstand oxidative damage triggered by pro-oxidants increased with ferritin H expression.[2] The proliferation of mouse erythroleukemia cells was not affected, even in the case of the clones expressing the highest ferritin H values. Surprisingly, with ferritin H overexpression, the cells concomitantly acquired multidrug resistance properties, due to the induction of the expression of the MDR-1 gene.[2]

Additional experiments were performed in HeLa cells. The overexpression of the ferritin H gene elicited effects similar to those observed in the experiments of ferritin H chain overexpression in mouse erythroleukemia cells, e.g., the induction of a reduced free intracellular iron pool and increased resistance to the cytotoxic effects of oxidative stress.[493] However, at variance with mouse erythroleukemia cells, ferritin H overexpression in HeLa cells induced a significant decrease of cell proliferation, whose extent was directly related to the level of ferritin H expression.[493] Importantly, all these effects were not induced by the overexpression of a ferritin H mutant lacking ferroxidase activity.[493]

In addition to studies of ferritin H overexpression in mammals, studies of overexpression of ferritin gene in plants have been conducted to evaluate its potential as a nutrient supplement or as a defense mechanism against plant pathogens.[12,13]

Two groups reported the development of transgenic plants overexpressing ferritin genes. This aim was achieved in one study by introducing the gene encoding soybean ferritin into rice seeds. The resulting transgenic plants exhibited stable accumulations of ferritin and iron contents as much as three-fold greater than those of their untransformed counterparts.[13] These observations suggest that iron-fortified rice overexpressing ferritin may be used to mitigate iron deficiency in the human diet.

Another group used a more complex approach to produce a rice strain exhibiting the capacity to make β-carotene which is converted after ingestion to vitamin A. Another rice strain expressed ferritin, heat-stable phytase, and metallothionein-like genes.[494] The cross-hybridization of β-carotene and iron-rich rice strains generated hybrids that combined both improvements, grew normally, and may provide a valuable alimentary treatment of vitamin A and iron deficiencies.[495]

These transgenic rice grains appear safe and have potential utility in the diets of populations subject to vitamin A and/or iron deficiency. However, before their introduction into the human food chain, these transgenic plants must be explored for the potential risk of heavy metal loading under various soil conditions.[496]

Ferritin may also have an environmental role in the control of sea water pollution. The use of a ferritin reactor as a means for monitoring the pollution of heavy metal ions in sea water was proposed.[497]

The function of ferritin H was further explored in mice rendered genetically deficient in ferritin H expression by homologous recombination. Mice homozygotes for ferritin H disruption die between 3.5 and 9.5 days of development.[1] This finding suggests that: (1) there is no functional redundancy between the two ferritin chains in that, in the absence of ferritin H chains, L ferritin homopolymers are unable to sustain iron recycling and detoxification; and (2) ferritin H is required for intracellular iron availability during critical periods of development from implantation to gastrulation and early organogenesis.

Recent studies suggest novel functions for ferritin. Studies of iron deficiency suggest a possible relationship between iron metabolism and folate metabolism, as suggested by various lines of evidence: (1) iron deficiency impairs folate utilization in some tissues; (2) iron deficiency determines morphologic alterations in granulocytes as does folate deficiency; and (3) iron deficiency decreases secretion of folate into milk.

The biochemical mechanisms underlying the influence of iron deficiency on folate metabolism were only recently elucidated. Recent studies of the purification and characterization of a folate-catabolizing enzyme provided evidence that this enzyme is similar, and probably identical to heavy ferritin.[498] This finding was confirmed when overexpression of rat ferritin H resulted in increased rates of folate turnover in cultured Chinese hamster ovary cells.

Furthermore, iron chelators inhibit folate metabolism through inhibition of serine hydroxymethyltransferase which catalyzes the conversion of tetrahydrofolate and serine to glycine and methylene tetrahydrofolate.[499]

Other studies suggest a possible role for ferritin in the control of cell proliferation. The relationship between ferritin and cell proliferation goes beyond the effect of ferritin on the size of free iron pool required for the activity of ribonucleotide reductase — the only enzyme involved in the synthesis of deoxyribonucleotides from ribonucleotides. Studies on cell lines made resistant to hydroxyurea (an anti-tumor agent that destabilizes the iron center and inactivates ribonucleotide reductase) and on cell lines overexpressing the R2 subunit of ribonucleotide reductase have shown increases of both ferritin H and ferritin L mRNAs and proteins.[500]

These findings suggest that during the synthesis of ribonucleotide reductase, and particularly of its R2 subunit, ferritin-associated iron is required for the generation of the active R2 subunit, and during R2 subunit turnover, it is important that released iron is sequestered by ferritin.

Recent studies have suggested a possible role of ferritin in the control of nitric oxide (NO) metabolism. NO is produced by many mammalian cells and is responsible for a broad spectrum of signalling functions in physiological and pathophysiological processes. The role of ferritin in NO metabolism is complex in that ferritin synthesis is regulated by NO and ferritin, through the control of the free iron pool, controls the activity of nitric oxide synthase 2. Nitric synthases are isoenzymes located in the cytosols or in the membranes that catalyze the conversion of the L arginine to citrulline and NO.[501] Three types of NO synthase have been identified. NO synthase 2 is expressed in macrophages, microglia, and hepatocytes (all actively involved in the synthesis of elevated levels of ferritin), is highly inducible by cytokines, and requires iron as a cofactor.

Most studies on ferritin and NO metabolism have been carried out in macrophages (for detailed descriptions of these studies, see Chapter 8). NO production in macrophages was stimulated by exposure to bacterial lipopolysaccharide (LPS) and interferon (IFN). Cells exposed to these agents produce the release of endogenous NO which activates iron regulatory protein 1 for binding to iron regulatory elements by a conformational change of the protein involving either NO-mediated disassembly of the 4Fe–4S cluster or NO-mediated sequestration of the free iron.

As a consequence of this activation of IRP1, ferritin translation was inhibited. The reduced synthesis of ferritin causes a decrease of its concentration in the cytosols, with a subsequent increase of free iron and, ultimately, feedback suppression of NOS 2 transcription. Iron depletion induces NOS 2 expression while iron loading suppresses it. The increase of the macrophage free iron pool may be related to the inductive effect of NO on heme oxygenase 1 activity.[501] These cytokines in macrophages also induce the release of thiol-reactive nitrosium ions (NO$^+$) which produce a decrease of IRP2 activity due to S-nitrosylation of crucial cysteine residues.

REFERENCES

1. Ferreira, C. et al. Early embryonic lethality of ferritin-H gene deletion in mice, *J. Biol. Chem.*, 275: 3021–3024, 2000.
2. Epsztejn, S. et al. H-ferritin subunit overexpression in erythroid cells reduces the oxidative stress response and induces multidrug resistance properties, *Blood*, 94: 3593–3603, 1999.
3. Harrison, P.M. and Arosio, P. The ferritins: molecular properties, iron storage function and cellular regulation, *Biochim. Biophys. Acta*, 1275: 161–203, 1996.

4. Briat, J.F. et al. Regulation of plant ferritin synthesis: how and why, *Cell Mol. Life Sci.*, 56: 155–166, 1999.
5. Fobis-Loisy, I. et al. Structure and differential expression in response to iron and abscisic acid of two maize ferritin genes, *Eur. J. Biochem.*, 231: 609–619, 1995.
6. Wardrop, A.J., Wicks, R.E., and Entsch, B. Occurrence and expression of members of the ferritin gene family in cowpeas, *Biochem. J.*, 337: 523–530, 1999.
7. Wei, J. and Theil, E.C. Identification and characterization of the iron regulatory element in the ferritin gene of a plant (soybean), *J. Biol. Chem.*, 275: 17488–17493, 2000.
8. Gaymond, F., Boucherez, J., and Briat, J.F. Characterization of a ferritin mRNA from *Arabidopsis thaliana* accumulated in response to iron through an oxidative pathway independent of abscisic acid, *Biochem. J.*, 318: 67–73, 1996.
9. Proudhon, D. et al. Ferritin gene organization: differences between plants and animals suggest possible kingdom-specific selective constraints, *J. Mol. Evol.*, 42: 325–336, 1996.
10. Lescure, A.M. et al. Ferritin gene transcription is regulated by iron in soybean cell cultures, *Proc. Natl. Acad. Sci. U.S.A.*, 88: 8222–8226, 1991.
11. Lobreaux, S., Hardy, T., and Briat, J.F. Abscisic acid is involved in iron-induced synthesis of maize ferritin, *EMBO J.*, 12: 651–657, 1993.
12. Deak, M. et al. Plants ectopically expressing the iron-binding protein, ferritin, are tolerant to oxidative damage and pathogens, *Nature Biotechnol.*, 17: 192–196, 1999.
13. Goto, F. et al. Iron fortification of rice seed by the soybean ferritin gene, *Nature Biotechnol.*, 17: 282–286, 1999.
14. Yang, X. et al. The iron oxidation and hydrolysis chemistry of *Escherichia coli* bacterioferritin, *Biochemistry*, 39: 4915–4923, 2000.
15. Romao, C.V. et al. A bacterioferritin from the strict anaerobe *Desulfovibrio desulfuricans* ATCC 27774, *Biochemistry*, 39: 6841–6849, 2000.
16. Andrews, S.C. Iron storage in bacteria, *Adv. Microbiol. Physiol.*, 40: 281–351, 1998.
17. Ratnayake, D.B. et al. Ferritin from the obligate anaerobe *Porphyromonas gengivalis*: purification, gene cloning and mutant studies, *Microbiology*, 146: 1119–1127, 2000.
18. Frazier, B.A. et al. Paracrystalline inclusions of a novel ferritin containing nonheme iron, produced by the human gastric pathogen *Helicobacter pylori*: evidence for a third class of ferritins, *J. Bacteriol.*, 175: 966–972, 1993.
19. Evans, D.J. et al. Identification of four new prokaryotic bacterioferritins, from *Helicobacter pylori*, *Anabaena variabilis*, *Bacillus subtilis* and *Treponema pallidum*, by analysis of gene sequences, *Gene*, 153: 123–127, 1995.
20. Bereswill, S. et al. Structural, functional and mutational analysis of the *pfr* gene encoding a ferritin from *Helicobacter pylori*, *Microbiology*, 144: 2505–2516, 1998.
21. Tonello, F. et al. *Helicobacter pylori* neutrophil-activating protein is an iron-binding protein with dodecameric structure, *Mol. Microbiol.*, 34: 238–246, 1999.
22. Bozzi, M. et al. A novel nonheme iron-binding ferritin related to the DNA-binding proteins of the Dps family in *Listeria innocua*, *J. Biol. Chem.*, 272: 3259–3265, 1997.
23. Ilari, A. et al. The dodecameric ferritin from *Listeria innocua* contains a novel inter-subunit iron-binding site, *Nature*, 7: 38–43, 2000.
24. Stefanini, S., et al. Incorporation of iron by unusual dodecameric ferritin from *Listeria innocua*, *Biochem. J.*, 338: 71–75, 1999.
25. Yang, X. et al. Iron oxidation and hyrolysis reactions of a novel ferritin from *Listeria innocua*, *Biochem. J.*, 349: 783–786, 2000.
26. Dietzel, J. et al. Ferritins of *Schistosoma mansoni*: sequence comparison and expression in female and male worms, *Mol. Biochem. Parasitol.*, 50: 145–154, 1992.

27. Schussler, P. et al. Ferritin mRNAs in *Schistosoma manso*ni do not have iron-responsive elements for post-transcriptional regulation, *Eur. J. Biochem.*, 241: 64–69, 1996.

28. Ersfeld, K. and Craig, P.S. Cloning and immunological characterization of *Echinococcus granolosus* ferritin, *Parasitol. Res.*, 81: 382–387, 1995.

29. Nichol, H. and Locke, M. The characterization of ferritin in an insect, *Insect Biochem.*, 19: 587–602, 1989.

30. Dunkov, B.C. et al. Isolation and characterization of mosquito ferritin and cloning of a cDNA that encodes one subunit, *Arch. Insect Biochem. Physiol.*, 29: 293–307, 1995.

31. Pham, D.Q. et al. Structure and location of a ferritin gene of the yellow fever mosquito *Aedes aegypti*, *Eur. J. Biochem.*, 267: 3885–3890, 2000.

32. Charlesworth, A. et al. Isolation and properties of *Drosophila melanogaster* ferritin: molecular cloning of a cDNA that encodes one subunit, and localization of the gene on the third chromosome, *Eur. J. Biochem.*, 247: 470–475, 1997.

33. Dunkov, B.C. and Georgieva, T. Organization of the ferritin genes in *Drosophila melanogaster*, *DNA Cell Biol.*, 18: 937–944, 1999.

34. Adams, M.D. The genome sequence of *Drosophila melanogaster*, *Science*, 287: 2185–2195, 2000.

35. Georgieva, T. et al. Iron availability dramatically alters the distribution of ferritin subunit messages in *Drosophila melanogaster*, *Proc. Natl. Acad. Sci. U.S.A.*, 96: 2716–2721, 1999.

36. Von Darl, M. et al. cDNA cloning and deduced amino acid sequence of two ferritins: soma ferritin and yolk ferritin in the snail *Lymnaea stagnalis* L, *Eur. J. Biochem.*, 222: 353–366, 1994.

37. Van Darl, M. et al. Expression in *Escherichia coli* of a secreted invertebrate ferritin, *Eur. J. Biochem.*, 222: 367–376, 1994.

38. De Graaf, J., Amons, R., and Moller, W. The primary structure of ferritin from *Artemia* cysts, *Eur. J. Biochem.*, 193: 737–750, 1990.

39. Huang, T., Law, J.H., and Soderhall, K. Purification and cDNA cloning of ferritin from the hepatopancreas of the freshwater crayfish *Pacifastacus lenisculus*, *Eur. J. Biochem.*, 236: 450–456, 1996.

40. Andersen, O. et al. Two ferritin subunits of Atlantic salmon (*Salmo salar*): cloning of the liver cDNAs and antibody preparation, *Mol. Mar. Biol. Biotechnol.*, 4: 164–170, 1995.

41. Zinovieva, R.D., Piatigrosky, J., and Tomarev. S,I. O-crystallin, arginine kinase and ferritin from the octopus lens, *Biochim. Biophys. Acta*, 1431: 512–517, 1999.

42. Holland, L.J., Wall, A.A., and Bhattacharya, A. *Xenopus* liver ferritin H subunit: cDNA sequence and mRNA production in the liver following estrogen treatment, *Biochemistry*, 30: 1965–1972, 1991.

43. Trikha, J., Theil, E.C., and Allewell, N.M. High resolution crystal structures of amphibian red-cell ferritin: potential roles for structural plasticity and solvation in function, *J. Mol. Biol.*, 248: 949–967, 1995.

44. Ha, Y. et al. Crystal structure of bullfrog M ferritin at 2.8 Å resolution: analysis of subunit interactions and the binuclear metal center, *J. Biol. Inorg. Chem.*, 4: 243–256, 1999.

45. Lawson, D.M. et al. Solving the structure of human H-ferritin by genetically engineering intermolecular crystal contacts, *Nature*, 349: 541–544, 1991.

46. Hampstead, P.D. et al. Comparison of three-dimensional structures of recombinant human H and horse L ferritin at high resolution, *J. Mol. Biol.*, 268: 424–448, 1997.

47. Yau, S.T., Thomas, B.R., and Vekilov, P.G. Molecular mechanisms of crystallisation and defect formation, *Phys. Rev. Lett.*, 85: 353–356, 2000.

48. Yau, S.T. and Vekilov, P.G. Quasi-planar nucleus structure in apoferritin crystallisation, *Nature,* 406: 494–497, 2000.
49. Santambrogio, P. et al. Evidence that a salt bridge in the light chain contributes the physical stability difference between heavy and light human ferritins, *J. Biol. Chem.,* 267: 14077–14083, 1992.
50. Levi, S. et al. Evidence that residues exposed on the three-fold channels have active roles in the mechanism of ferritin iron incorporation, *Biochem. J.,* 317: 467–473, 1996.
51. Douglas, T. and Ripoli, D.R. Calculated electrostatic gradients in recombinant human H chain ferritin, *Protein Sci.,* 1: 1083–1091, 1996.
52. Treffy, A. et al. Mechanisms of catalysis of Fe (II) oxidation by ferritin H chains, *FEBS Lett.,* 302: 108–122, 1992.
53. Treffy, A. et al. Iron (II) oxidation by H-chain ferritin. Evidence from site-directed mutagenesis that a transient blue species is formed at the dinuclear iron center, *Biochemistry,* 34: 15204–15213, 1995.
54. Treffy, A. et al. Dinuclear center of ferritin: studies of iron binding and oxidation show differences in the two iron sites, *Biochemistry,* 36: 432–441, 1997.
55. Levi, S. et al. The role of the L-chain in ferritin iron incorporation. Studies of homo- and heteropolymers, *J. Mol. Biol.,* 238: 649–654, 1994.
56. Wade, W.J. et al. Influence of site-directed modifications on the formation of iron cores in ferritin, *J. Mol. Biol.,* 221: 1443–1452, 1991.
57. Levi, S. et al. Evidence that H-chains and L-chains have cooperative roles in iron-uptake mechanisms of human ferritin, *Biochem. J.,* 288: 591–596, 1992.
58. Santambrogio, P. et al. Evidence that the specificity of iron incorporation into homopolymers of human ferritin L- and H-chains is conferred by the enucleation and ferroxidase centers, *Biochem. J.,* 314: 139–144, 1996.
59. Perreira, A.S. et al. Rapid and parallel formation of Fe^{3+} multimers, including a trimer, during H-type subunit ferritin mineralization, *Biochemistry,* 36: 7917–7927, 1997.
60. Yang, X, et al. Reaction paths of iron oxidation and hydrolysis in horse spleen ferritin and recombinant human ferritins, *Biochemistry,* 37: 9763–9750, 1998.
61. Pereira, A.S. et al. Direct spectroscopic and kinetic evidence for the involvement of a peroxodoiron intermediate during the ferroxidase reaction in fast ferritin mineralization, *Biochemistry,* 37: 9871–9876, 1998.
62. Moënne-Loccoz, P, et al. The ferroxidase reaction of ferritin reveals a diferric μl, 2-bridging peroxide intermediate in common with other O_2-activating non-heme di-iron proteins, *Biochemistry,* 38: 3290–3299, 1999.
63. Hwang, J. et al. A short Fe–Fe distance in peroxodiferric ferritin: control of Fe substrate versus cofactor decay? *Science,* 287: 122–125, 2000.
64. Bauminger, E.R. Mössbauer spectroscopic investigation of structure function relations in ferritins, *Biochim. Biophys. Acta,* 1118: 48–58, 1991.
65. Chasteen, N.D. and Harrison, P.M. Mineralization in ferritin: an efficient means of iron storage, *J. Struct. Biol.,* 126: 182–194, 1999.
66. Xang, X. and Chasteen, N.D. Ferroxidase activity of ferritin: effect of pH, buffer and Fe (II) and Fe (III) concentrations on Fe (II) autoxidation and ferroxidation, *Biochem. J.,* 338: 615–618, 1999.
67. Cozzi, A. et al. Overexpression on wild-type and mutated human ferritin H-chain in HeLa cells, *J. Biol. Chem.,* 275: 25122–25129, 2000.
68. De Silva, D. et al. *In vitro* loading of apoferritin, *Arch. Biochem. Biophys.,* 293: 409–415, 1992.
69. Juan, S.H., Guo, J.H., and Aust, S.D. Loading of iron into recombinant rat liver ferritin heteropolymers by ceruloplasmin, *Arch. Biochem. Biophys.,* 341: 280–286, 1997.

70. Guo, J.H., Juan, S.H., and Aust, S.D. Mutational analysis of the four alpha-helix bundle iron-loading channel of rat liver ferritin, *Arch. Biochem. Biophys.*, 352: 71–77, 1998.
71. Harris, Z.L. et al. Targeted gene disruption reveals an essential role for ceruloplasmin in cellular iron efflux, *Proc. Natl. Acad. Sci. U.S.A.*, 96: 10812–10817, 1999.
72. Wade, V.J. et al. Structure and composition of ferritin cores from pea seed (*Pisum sativum*) *Biochim. Biophys. Acta*, 1161: 91–96, 1993.
73. Huoang, H, et al. Role of phosphate in Fe²⁺ binding to horse spleen apoferritin, *Biochemistry*, 32: 1681–1687, 1993.
74. Waldo, G.S. et al. Formation of the ferritin iron mineral occurs in plastids, *Plant Physiol.*, 109: 797–802, 1995.
75. Cheng, Y.G. and Chasteen, N.D. Role of phosphate in initial iron deposition in apoferritin, *Biochemistry*, 30: 2947–2953, 1991.
76. Jacobs, D. et al. Fe²⁺ binding to apo- and holo- mammalian ferritin, *Biochemistry*, 28: 9216–9221, 1989.
77. Johnson, J.L. et al. Forming the phosphate layer in reconstituted horse spleen ferritin and the role of phosphate in promoting surface redox reactions, *Biochemistry*, 38: 6706–6713, 1999.
78. Johnson, J.L. et al. Redox reactivity of animal apoferritins and apoheteropolymers assembled from recombinant heavy and light human chain ferritins, *Biochemistry*, 38: 4089–4096, 1999.
79. Takagi, H. et al. Localized unfolding at the junction of three ferritin subunits, *J. Biol. Chem.*, 273: 18685–18688, 1998.
80. Yang, X. and Chasteen, N.D. Molecular diffusion into horse spleen ferritin: a nitroxide radical spin probe study, *Biophys. J.*, 71: 1587–1595, 1996.
81. Yang, X., Arosio, P., and Chasteen, N.D. Molecular diffusion into ferritin: pathways, temperature dependence, incubation time, and concentration effects, *Biophys. J.*, 78: 2049–2059, 2000.
82. Ponka, P. Tissue-specific regulation of heme synthesis: distinct control mechanisms in erythroid cells, *Blood*, 89: 1–25, 1997.
83. Ponka P. Cell biology of heme, *Am. J. Med. Sci.*, 318: 241–256, 1999.
84. Richardson, D.R., Ponka, P., and Vyoral, D. Distribution of iron in reticulocytes after inhibition of heme synthesis with succinylacetone. Examination of cytoplasmic and mitochondrial intermediates involved in iron metabolism, *Blood*, 87: 3477–3488, 1996.
85. Meyron-Holtze, E.G. et al. Binding and uptake of exogenous isoferritins by cultured human erythroid precursor cells, *Br. J. Haematol.*, 86: 635–641, 1994.
86. Gelvan, D. et al. Ferritin uptake by human erythroid precursors is a regulated iron uptake pathway, *Blood*, 88: 3200–3207, 1996.
87. Meyron-Holtz, E.G. et al. Regulation of intracellular iron metabolism in human erythroid precursors by internalized extracellular ferritin, *Blood*, 94: 3205–3211, 1999.
88. Vaisman, B., Fibach, E., and Konijn, A.M. Utilization of intracellular ferritin iron for hemoglobin synthesis in developing human erythroid precursors, *Blood*, 90: 831–838, 1997.
89. Visman B. et al. Ferritin expression in maturing normal human erythroid precursors, *Br. J. Haematol.*, 110: 394–401, 2000.
90. Vyoral, D. and Petrak, J. Iron transport in K562 cells: a kinetic study using native gel electrophoresis and ⁵⁹Fe autoradiography, *Biochim. Biophys. Acta*, 1403: 179–188, 1998.

91. Picard, V. et al. Role of ferritin in the control of the labile iron pool in murine erythroleukemia cells, *J. Biol. Chem.,* 273: 15382–15386, 1998.
92. Konijn, A.M. et al. The cellular labile iron pool and intracellular ferritin in K562 cells, *Blood,* 94: 2128–2134, 1999.
93. Bridges, K.R. Ascorbic acid inhibits lysosomal autophagy of ferritin. *J. Biol. Chem.,* 262: 14773–14778, 1987.
94. Reinheckel, T, et al. Comparative resistance of the 20S and 26S proteasome to oxidative stress, *Biochem. J.,* 335: 637–642, 1998.
95. Mikulits, W. et al. Impaired ferritin mRNA translation in primary erythroid progenitors: shift to iron-dependent regulation by the v-ErbA oncoprotein, *Blood,* 94: 4321–4332, 1999.
96. Van Wyk, C.P., Linder-Horowitz, M., and Munro, H.N. Effect of iron loading of non-heme iron compounds in different liver cell populations, *J. Biol. Chem.,* 296: 1025–1031, 1971.
97. Fleming, R.E. et al. Transferrin receptor 2: continued expression in mouse liver in the face of iron overload and in hereditary hemochromatosis, *Proc. Natl. Acad. Sci. U.S.A.,* 97: 2214–2219, 2000.
98. Mack, U. et al. Characterization of the binding of ferritin to the rat liver ferritin receptors, *J. Biol. Chem.,* 258: 4672–4675, 1983.
99. Mack, U. et al. Characterization of the binding of ferritin to the rat liver ferritin receptor. *Biochim. Biophys. Acta,* 843: 164–170, 1985.
100. Adams, P.C., Powell, L.W., and Halliday, J.W. Isolation of a human hepatic ferritin receptor, *Hepatology,* 8: 719–721, 1988.
101. Sibille, J.C., Kondo, H., and Aisen, P. Uptake of ferritin and iron bound to ferritin by rat hepatocytes: modulation by apotransferrin, iron chelators and chloroquine, *Biochim. Biophys. Acta,* 1010: 204–209, 1989.
102. Moss, D. et al. Functional roles of the ferritin receptors of human liver, hepatoma, lymphoid and erythroid cells, *J. Inorg. Biochem.,* 47: 219–227, 1992.
103. Ramm, G.A., Powell, L.W., and Halliday, J. Effect of colchicines on the clearance of ferritin *in vivo, Am. J. Physiol.,* 258: G707–713, 1990.
104. Ollinger, K. and Roberg, K. Nutrient deprivation of cultured rat hepatocytes increases the desferrioxamine-available iron pool and augments sensitivity to hydrogen peroxide, *J. Biol. Chem.,* 272: 23707–23711, 1997.
105. Radisky, D.C. and Kaplan, J. Iron in cytosolic ferritin can be recycled through lysosomal degradation in human fibroblasts, *Biochem. J.,* 336: 201–205, 1998.
106. Tacchini, L. et al. Induction of ferritin synthesis in ischemic-reperfused rat liver: analysis of the molecular mechanisms, *Gastroenterology,* 113: 946–953, 1997.
107. Ramm, G.A. et al. Identification and characterization of a receptor for tissue ferritin on activated lipocytes, *J. Clin. Invest.,* 94:9–15, 1994.
108. Kondo, H. et al. Iron metabolism in the erythrophagocytosing Kupffer cell, *Hepatology,* 8: 296–301, 1988.
109. Sibille, J.C., Kondott, H., and Aisen, P. Interactions between isolated hepatocytes and Kupffer cells in iron metabolism: a possible role for ferritin as an iron carrier protein, *Hepatology,* 8: 296–301, 1988.
110. Cleton, M.I. et al. Ultrastructural evidence for the presence of ferritin–iron in the biliary system of patients with iron overload, *Hepatology,* 6: 30–35, 1986.
111. Ramm, G.A. et al. Pathways of intracellular trafficking and release of ferritin by the liver *in vivo*: the effect of chloroquine and cytochalasin D, *Hepatology,* 19: 504–513, 1994.

112. Ramm, G.A. et al. Effect of the microtubular inhibitor vinblastine on ferritin clearance and release in the rat, *J. Gastroenterol. Hepatol.*, 11: 1072–1078, 1996.
113. Dickson, D.P.E. et al. Mossbauer spectroscopy, electron microscopy and electron diffraction studies of the iron cores in various human and clinical hemosiderins, *Biochim. Biophys. Acta*, 957: 81–90, 1988.
114. Mann, S. et al. Structural specificity of hemosiderin iron cores in iron-overload diseases, *FEBS Lett.*, 234: 68–72, 1988.
115. Ward, R.J. et al. Further characterization of forms of hemosiderin in iron overloaded tissues, *Eur. J. Biochem.*, 225: 187–194, 1994.
116. St. Pierre, T.G. et al. The form of iron oxide deposits in thalassemic tissues varies between different groups of patients: a comparison between Thai beta-thalassemia/hemoglobin E patients and Australian beta-thalassemia patients, *Biochim. Biophys. Acta*, 1407: 51–60, 1998.
117. Allen, P.D. et al. Low-frequency low-field magnetic susceptibility of ferritin and hemosiderin, *Biochim. Biophys. Acta*, 1500: 186–196, 2000.
118. Andrews, N.C. Disorders of iron metabolism, *New Engl. J. Med.*, 341: 1986–1995, 1999.
119. Olivieri, N.F. The beta-thalassemias, *New Engl. J. Med.*, 341: 99–109, 1999.
120. Angelucci, E. et al. Hepatic iron concentration and total body iron stores in thalassemia major, *New Engl. J. Med.*, 343: 327–331, 2000.
121. Tamura, T. et al. Gender difference in cord serum ferritin concentrations, *Biol. Neonate*, 75: 343–349, 1999.
122. Siimes, A.S. and Siimes, M.A. Changes in the concentration of ferritin in the serum during fetal life in singletons and twins, *Early Hum. Dev.*, 13: 47–52, 1986.
123. Ali, N.A.M. et al. Serum ferritin concentration and bone marrow iron stores: a prospective study, *Can. Med. Assoc. J.*, 118: 945–946, 1978.
124. Jacob, R,A, et al. Utility of serum ferritin as a measure of iron deficiency in normal males undergoing repetitive phlebotomy, *Blood*, 56: 786–791, 1980.
125. Finch, C.A. et al. Effect of blood donation on iron stores as evaluated by serum ferritin, *Blood*, 50:441–447, 1977.
126. Wheby, M.S. Effect of iron therapy on serum ferritin levels in iron-deficiency anemia, *Blood*, 56: 138–140, 1980.
127. Skikne, B.S., Flowers, C.H., and Cook, J.D. Serum transferrin receptor: a quantitative measure of tissue iron deficiency, *Blood*, 75: 1870–1876, 1990.
128. Punnonen, K., Irjala, K., and Rajamaki, A. Serum transferrin receptor and its ratio to serum ferritin in the diagnosis of iron deficiency, *Blood*, 89: 1052–1057, 1997.
129. Suominen, P. et al. Serum transferrin receptor and transferrin receptor–ferritin index identify healthy subjects with subclinical iron deficits, *Blood*, 92: 2934–2939, 1998.
130. Flowers, C.H. and Cook, J.D. Dried plasma spot measurements of ferritin and transferrin receptor as assessing iron status, *Clin. Chem.*, 45: 1826–1832, 1999.
131. Powell, L.W. Primary iron overload, in *Iron Metabolism in Health and Disease*, Brock, J.H. and Halliday, J.W., Eds., London, W.B. Saunders, 227–270, 1994.
132. Nielsen, P. et al. Serum ferritin in iron overload and liver damage: correlation to body iron stores and diagnostic relevance, *J. Lab. Clin. Med.*, 135: 413–418, 2000.
133. Khumolo, H. et al. Serum transferrin receptors are decreased in the presence of iron overload, *Clin. Chem.*, 44: 40–44, 1998.
134. Breuer, W. et al. The assessement of serum nontransferrin-bound iron in chelation therapy and iron supplementation, *Blood*, 95: 2975–2982, 2000.
135. Knisely, A.S. Neonatal hemochromatosis, *Adv. Pediatr.*, 39: 383–403, 1992.

136. Fellman, V. et al. Iron-overload disease in infants involving fetal growth retardation, lactic acidosis, liver hemosiderosis, and aminoaciduria, *Lancet*, 351: 490–493, 1998.

137. Elin, R.J., Wolff, S.M., and Finch, C.A. Effect of induced fever on serum iron and ferritin concentration in man, *Blood*, 49: 147–153, 1977.

138. Mori, S. et al. Dynamic changes in mRNA expression of neutrophils during the course of acute inflammation in rabbits, *Int. Immunol.*, 6: 149–156, 1994.

139. Wesselius, L.J. et al. Alveolar macrophages accumulate tungsten dusts, *J. Lab. Clin. Med.*, 127: 401–409, 1996.

140. Torti, S.V. et al. The molecular cloning and characterization of murine ferritin heavy chain, a tumor necrosis factor-inducible gene, *J. Biol. Chem.*, 263: 12638–12644, 1988.

141. Tsuji, Y. et al. Tumor necrosis factor alpha and interleukin 1 alpha regulate transferrin receptor in human diploid fibroblasts. Relationship to the induction of ferritin heavy chain, *J. Biol. Chem.*, 266: 7257–7261, 1991.

142. Miller, L.L. et al. Iron-independent induction of ferritin H chain by tumor necrosis factor, *Proc. Natl. Acad. Sci. U.S.A.*, 88: 4946–4950, 1991.

143. Hirayama, M.K. et al. Regulation of iron metabolism in HepG2 cells: a possible role for cytokines in the hepatic deposition of iron, *Hepatology*, 18: 874–880, 1993.

144. Fahmy, M. and Young, S.P. Modulation of iron metabolism in monocyte cell line U937 by inflammatory cytokines: changes in transferrin uptake, iron handling and ferritin mRNA, *Biochem. J.*, 296: 175–181, 1993.

145. Kobune, M. et al. Interleukin-6 enhances hepatic transferrin uptake and ferritin expression in rats, *Hepatology*, 19:1468–1475, 1994.

146. Tran, T.N. et al. Secretion of ferritin by rat hepatoma cells and its regulation by inflammatory cytokines and iron, *Blood*, 90: 4979–4986, 1997.

147. Smirnov, I.M. et al. Effects of TNF-alpha and IL-1 beta on iron metabolism by A549 cells and influence on cytotoxicity, *Am. J. Physiol.*, 277: 1257-1263, 1999.

148. Moroz, C. et al. Elevated serum ferritin level in acute myocardial infarction, *Biomed. Pharmacother.*, 51: 126–130, 1997.

149. Bhagat, C.I. et al. Plasma ferritin in acute hepatocellular damage, *Clin. Chem.*, 46: 885–886, 2000.

150. Beesley, R.T. et al. Impact of acute malaria on plasma concentrations of transferrin receptors, *Trans. R. Soc. Trop. Med. Hyg.*, 94: 295–298, 2000.

151. Kumon, Y.K. et al. Ferritin correlates with C-reactive protein and acute phase serum amyloid A in synovial fluid, but not in serum, *Amyloid*, 6: 130–135, 1999.

152. Takahasi, S. et al. Beta-2-microglobulin and ferritin in cerebrospinal fluid for evaluation of patients with meningitis of different etiologies, *Brain Dev.*, 21: 192–199, 1999.

153. Ota, T. et al. Increased serum ferritin levels in adult Still's disease, *Lancet*, I: 562–563, 1987.

154. Pelkonen, P., Swanlijung, K., and Siimes, M.A. Ferritinemia as an indicator of systemic disease activity in children with systemic juvenile rheumatoid arthritis, *Acta Pediatr. Scand.*, 75: 64–68, 1986.

155. Higashi, S., Ota, T., and Eto, S. Biochemical analysis of ferritin subunits in sera from adult Still's disease patients, *Rheumatol. Int.*, 15: 45–50, 1995.

156. Vignes, S. et al. Percentage of glycosylated serum ferritin remains low throughout the course of adult onset Still's disease, *Ann. Rheumatol.*, 59: 347–350, 2000.

157. Sobieska, M. et al. Still's disease in children and adults: a distinct pattern of acute-phase proteins, *Clin. Rheumatol.*, 17: 258–260, 1998.

158. Lin, S.J., Chao, H.C., and Yan, D.C. Different articular outcomes of Still's disease in Chinese children and adults, *Clin. Rheumatol.*, 19: 127–130, 2000.

159. Bell, H. et al. Serum ferritin and transferrin saturation in patients with chronic alcoholic and non-alcoholic liver diseases, *J. Intern. Med.,* 236: 315–322, 1994.
160. Fletcher, L.M., Halliday, J.W., and Powell, L.W. Interrelationships of alcohol and iron in liver disease with particular reference to the iron-binding proteins, ferritin and transferrin, *J. Gastroenterol. Hepatol.,* 14: 202–214, 1999.
161. Bonkovsky, H.L. et al. Iron in liver diseases other than hemochromatosis, *Sem. Liver Dis.,* 16: 65–82, 1996.
162. Bonkovsky, H.L. et al. Non-alcoholic steatohepatitis and iron: increased prevalence of mutations of the HFE gene in non-alcoholic steatohepatitis, *J. Hepatol.,* 31: 421–429, 1999.
163. Hann, H.W. et al. Increased serum ferritin in chronic liver disease: a risk factor for primary hepatocellular carcinoma, *Int. J. Cancer,* 43:376–379, 1989.
164. Bayaraktar, Y. et al. The effect of intereferon and desferrioxamine on serum ferritin and hepatic iron concentrations in chronic hepatitis B, *Hepatogastroenterology,* 45: 2322–2327, 1998.
165. Hérode, C., Cazeneuve, C., and Coué, O. Liver iron accumulation in patients with chronic active hepatitis C: prevalence and role of hemochromatosis gene mutations and relationship with hepatic histological lesions, *J. Hepatol.,* 31: 979–984, 1999.
166. Burt, M.J. and Cooksley, G. The influence of iron on chronic hepatitis C, *J. Gastroenterol. Hepatol.,* 13: 330–333, 1998.
167. Takikawa, T. et al. Correlation between serum levels of alanine aminotransferase and ferritin in male donors with antibody to hepatitis C virus, *J. Gastroenterol.,* 29: 593–597, 1994.
168. Van Thiel, D.H. et al. Responses to alpha-interferon therapy (IFN) are influenced by the iron content of the liver, *J. Hepatol.,* 20: 410–415, 1994.
169. Barbaro, G. et al. Serum ferritin and hepatic glutathione concentrations in chronic hepatitis C patients related to the hepatitis C genotype, *J. Hepatol.,* 30: 774–782, 1999.
170. Weiss, G.F. et al. Association between cellular immune effector function, iron metabolism, and disease activity in patients with chronic hepatitis C virus infection, *J. Infect. Dis.,* 180: 1452–1458, 1999.
171. Sampietro, M., Fiorelli, G., and Fargion, S. Iron overload in porphyria cutanea tarda, *Haematologica,* 84: 248–253, 1999.
172. Bulaj, Z.J. et al. Hemochromatosis genes and other factors contributing to the pathogenesis of porphyria cutanea tarda, *Blood,* 95: 1565–1571, 2000.
173. Scholl, T. and Reilly, T. Anemia, iron and pregnancy outcome, *J. Nutr.,* 130: 4435–4475, 2000.
174. Goldenberg, R.L. et al. Plasma ferritin and pregnancy outcome, *Am. J. Obstet. Gynecol.,* 175: 1356–1359, 1996.
175. Tamura, T. et al. Serum ferritin: a predictor of early spontaneous preterm delivery, *Obstet. Gynecol.,* 87: 360–365, 1996.
176. Scholl, T.O. High third-trimester ferritin concentration: associations with very preterm delivery, infection and maternal nutritional status, *Obstet. Gynecol.,* 92: 161–165, 1998.
177. Scholl, T.O. High third-trimester ferritin concentration: associations with very preterm delivery, infection, and maternal nutritional status, *Obstet. Gynecol.,* 93: 156–157, 1999.
178. Fellman, V. et al. Iron-overload disease in infants involving fetal growth retardation, lactic acidosis, liver hemosiderosis, and aminoaciduria, *Lancet,* 351: 490–493, 1998.
179. Kiechl, S., Aichner, F., and Gestenbrand, F. Body iron stores and presence of carotid atherosclerosis: results from the Bruneck study, *Arterioscler. Thromb.,* 14: 1625–1630, 1994.

180. Salonen, J.T. et al. High stored iron levels are associated with increased risk of myocardial infarction in eastern Finnish men, *Circulation,* 86: 803–811, 1992.
181. Kiechl, S. et al. Body iron stores and the risk of carotid atherosclerosis. Prospective results from the Bruneck study, *Circulation,* 96: 3300–3307, 1997.
182. Sempos, T.C. et al. Body iron stores and the risk of coronary heart disease, *New Engl. J. Med.,* 330: 1119–1124, 1994.
183. Klipstein-Globrush, K. et al. Serum ferritin and the risk of myocardial infarction in the elderly: the Rotterdam study, *Am. J. Clin. Nutr.,* 69: 1231–1236, 1999.
184. Sullivan, J.L. Iron and the genetics of cardiovascular disease, *Circulation,* 100: 1260–1263, 1999.
185. Meyers, D.G. The iron hypothesis: does iron play a role in atherosclerosis? *Transfusion,* 40: 1023–1029, 2000.
186. Ford, E.S. and Cogswell, M.E. Diabetes and serum ferritin concentration among U.S. adults, *Diabetes Care,* 22: 1978–1983, 1999.
187. Carenini, A.P. et al. Serum ferritin in type I diabetes, *Clin. Chim. Acta,* 152: 165–170, 1985.
188. Langlois, M.R. et al. The haptoglobin 2-2 phenotype affects serum markers of iron status in healthy males, *Clin. Chem.,* 46: 1619–1625, 2000.
189. Cragg, S.J., Wagstaff, M., and Worwood, M. Detection of a glycosylated subunit in human serum ferritin, *Biochem. J.,* 199: 565–571, 1981.
190. Linden, M.C. et al. Serum ferritin: does it differ from tissue ferritin? *J. Gastroenterol. Hepatol.,* 11: 1033–1036, 1996.
191. Takakawa, Y.K. et al. The clinical significance of glucosylated ferritin in iron overloads and hematopoietic malignancies, *Rinsho Ketsueki,* 35: 744–750, 1994.
192. Torti, S.V. and Torti, F.M. Kininogen is a ferritin binding protein, *J. Biol. Chem.,* 273: 13630–13635, 1998.
193. Watanabe, K. et al. Characterization of ferritin and ferritin-binding proteins in canine serum, *Biometals,* 13: 57–63, 2000.
194. Roughead, Z.K. and Hunt, J.R. Adaptation in iron absorption: iron supplementation reduces nonheme-iron but not heme-iron absorption from food, *Am. J. Clin. Nutr.,* 72: 982–989, 2000.
195. Arosio, P. et al. Characteristics of ferritins in human milk secretions: similarities to serum and tissue isoferritins, *Clin. Chim. Acta,* 161: 201–208, 1986.
196. Costanzo, F, et al. Cloning and sequencing of a full length cDNA coding for a human apoferritin H chain: evidence for a multigene family, *EMBO J.,* 3: 23–27, 1984.
197. Santoro, C. et al. Cloning of the gene coding for human L apoferritin, *Nucleic Acid Res.,* 14: 2863–2876, 1986.
198. Hentze, M.W. et al. Cloning, characterization, expression, and abnormal chromosomal localization of a human ferritin heavy-chain gene, *Proc. Natl. Acad. Sci. U.S.A.,* 83: 7226–7230, 1986.
199. McGill, J.R. et al. Human ferritin H and L sequences lie on ten different chromosomes, *Hum. Genet.,* 76: 66–72, 1987.
200. Dorner, M.H. et al. Cloning of the cDNA coding for human L ferritin subunit, *Proc. Natl. Acad. Sci. U.S.A.,* 82: 3139–3143, 1985.
201. Santoro, C.F. et al. Cloning of the gene coding for human L-apoferritin, *Nucleic Acid Res.,* 14: 2863–2876, 1986.
202. Mighell, A.J. et al. Vertebrate pseudogenes, *FEBS Lett.,* 468: 109–114, 2000.
203. Gatti, R.A. et al. Human ferritin genes: chromosomal assignments and polymorphisms, *Am. J. Hum. Genet.,* 41: 654–667, 1987.

204. Dugast, I.J. Identification of two human ferritin H genes on the short arm of chromosome 6, *Genomics,* 6: 204–211, 1990.
205. Zheng, H. et al. Conserved mutations in human ferritin H pseudogenes: a second functional sequence or an evolutionary quirk? *Biochim. Biophys. Acta,* 1351: 150–156, 1997.
206. Ursini, M.V. and De Franciscus, V. TSH regulation of ferritin-H chain messenger RNA levels in the rat thyroid, *Biochem. Biophys. Res. Commun.,* 150: 287–295, 1988.
207. Chazenbalk, G.D., Wadsworth, H.L., and Rapaport, B. Transcriptional regulation of ferritin-H messenger RNA levels in FRTL5 rat thyroid cells by thyrotropin, *J. Biol. Chem.,* 265: 666–670, 1990.
208. Miller LL. et al. Iron-independent induction of ferritin-H chain by tumor necrosis factor, *Proc. Natl. Acad. Sci. U.S.A.,* 88: 4946–4950, 1991.
209. Coccia, E.M. et al. Modulation of ferritin-H chain expression in Friend erythroleukemia cells: transcriptional and translational regulation by hemin, *Mol. Cell Biol.,* 12: 3015–3022, 1992.
210. Tsuji, Y. et al. Preferential repression of the H subunit of ferritin by adenovirus E1A in NIH-3T3 mouse fibroblasts, *J. Biol. Chem.,* 268: 7270–7275, 1993.
211. Kwak, E.L., Torti, S.V., and Torti, F.M. Murine heavy ferritin chain: isolation and characterization of functional gene, *Gene,* 94: 255–261, 1990.
212. Chazenbalk, G.D. et al. Thyrotropin and adenosine 3′,5′-monophosphate stimulate the activity of the ferritin-H promoter, *Mol. Endocrinol.,* 4: 1117–1124, 1990.
213. Beaumont, C. et al. Mouse ferritin H subunit gene, *J. Biol. Chem.,* 269: 20281–20288, 1994.
214. Bevilacqua, M.A. et al. Promoter for the human ferritin heavy chain-encoding gene (FERH): structural and functional characterization, *Gene,* 111: 255–260, 1992.
215. Bevilacqua, M.A. et al. Transcriptional regulation of the human H ferritin-encoding gene (FERH) in G418-treated cells: role of the B-box-binding factor, *Gene,* 141: 287–291, 1994.
216. Bevilacqua, M.A. et al. Transcriptional activation of the H-ferritin gene in differentiated Caco-2 cells parallels a change in the activity of the nuclear factor Bbf, *Biochem. J.,* 311: 769–773, 1995.
217. Bevilacqua, M.A. et al. A common mechanism underlying the E1A repression and the cAMP stimulation of H ferritin transcription, *J. Biol. Chem.,* 272: 20736–20741, 1997.
218. Bevilacqua, M.A. et al. Okadaic acid stimulates H ferritin transcription in HeLa cells by increasing the interaction between the p300 co-activator molecule and the transcription factor Bbf, *Biochem. Biophys. Res. Commun.,* 240: 179–182, 1997.
219. Bevilacqua, M.A., et al. P/CAF/p300 complex binds the promoter for the heavy subunit of ferritin and contributes to its tissue-specific expression, *Biochem. J.,* 335: 521–525, 1998.
220. Blobel, G.A. CREB-binding protein and p300: molecular integrators of hematopoietic transcription, *Blood,* 95: 745–755, 2000.
221. Goodman, R.H. and Smolik, S. CBP/p300 in cell growth, transformation and development, *Genes Dev.,* 14: 1553–1577, 2000.
222. Marziali, G. et al. Transcriptional regulation of the ferritin heavy-chain gene: the activity of the CCAAT binding factor NF-Y is modulated in heme-treated Friend leukemia cells and during monocyte-to-macrophage differentiation, *Mol. Cell Biol.,* 17: 1387–1395, 1997.

223. Marziali G. et al. The activity of the CCAAT-box binding factor NF-Y is modulated through the regulated expression in its subunit during monocyte to macrophage differentiation: regulation of tissue-specific genes through a ubiquitous transcription factor, *Blood,* 93: 519–526, 1999.

224. Faniello, M.C. et al. The B-subunit of the CCAAT-binding factor NF-Y binds the central segment of the co-activator p300, *J. Biol. Chem.,* 274: 7623–7626, 1999.

225. Tsuji, Y. et al. FER-1, an enhancer of the ferritin H gene and a target of E1A-mediated transcriptional repression, *Mol. Cell Biol.,* 15: 5152–5164, 1995.

226. Tsuji, Y., Torti, S.V., and Torti, F.M. Activation of the ferritin H enhancer, FER-1, by the cooperative action of members of the AP1 and Sp1 transcription factor families, *J. Biol. Chem.,* 273: 2984–2992, 1998.

227. Tsuji, Y. et al. Transcriptional regulation of the mouse ferritin H gene, *J. Biol. Chem.,* 274: 7501–7507, 1999.

228. Tsuji, Y. et al. Coordinate transcriptional and translational regulation of ferritin in response to oxidative stress, *Mol. Cell Biol.,* 20: 5818–5827, 2000.

229. Wasserman, W.W. and Fahl, W.E. Functional anti-oxidant responsive elements, *Proc. Natl. Acad. Sci. U.S.A.,* 94: 5361–5366, 1997.

230. Orino, K. et al. Adenovirus E1A blocks oxidant-dependent ferritin induction and sensitises cells to pro-oxidant cytotoxicity, *FEBS Lett.,* 461: 334–338, 1999.

231. Smirnov, I.M. et al. Effects of TNF-alpha and IL-1beta on iron metabolism by A459 cells and influence on cytotoxicity, *Am. J. Physiol.,* 277: L257-L263, 1999.

231a. Kwak, E.L. et al. Role for NF-kappa B in the regulation of ferritin H by tumor necrosis factor-alpha, *J. Biol. Chem.,* 270: 15285–15293, 1995.

232. Lin, M. et al. Role of iron in NF-kappa B activation and cytokine gene expression by rat hepatic macrophages, *Am. J. Physiol.,* 272: G1355–G1364, 1997.

233. Wu, K.J., Polack, A., and Dalla Favera, R. Coordinated regulation of iron-controlling genes, H-ferritin and IRP2, by c-Myc, *Science,* 283: 676–679, 1999.

234. Dang, C.V. et al. Function of the c-Myc oncogenic transcription factor, *Exp. Cell. Res.,* 253: 63–77, 1999.

235. Lasorella, A. et al. Id2 is a retinoblastoma protein target and mediates signalling by Myc oncoprotein, *Nature,* 407: 592–598, 2000.

236. Barresi, R. et al. A negative cis-acting G-fer element participates in the regulation of expression of the human H-ferritin-encoding gene (FERH), *Gene,* 140: 195–201, 1994.

237. D'Agostino, P. et al. Negative and positive elements in the promoter region of the human apoferritin L gene, *Biochem. Biophys. Res. Commun.,* 215: 329–337, 1995.

238. Wang, V. et al. Identification of immediate early genes during TPA-induced human myeloblastic leukemia ML-1 cell differentiation, *Gene,* 216: 293–302, 1998.

239. Pang, J.H., Wung, C.J., and Chau, L.Y. Post-transcriptional regulation of H-ferritin gene expression in human monocytic THP-1 cells by protein kinase C, *Biochem. J.,* 319: 185–189, 1996.

240. Ai, L.S. and Chau, L.Y. Post-transcriptional regulation of H-ferritin mRNA, *J. Biol. Chem.,* 274: 30209–30214, 1999.

241. Puntney, D, et al. The identification of ferritin in the nucleus of K562 cells, and investigation of a possible role in the transcriptional regulation of adult beta-globin gene expression, *J. Cell Sci.,* 112: 825–831, 1999.

242. Irani, K. Oxidant signalling in vascular cell growth, death and survival, *Circ. Res.,* 87: 179–183, 2000.

243. Balla, J. et al. Endothelial cell heme oxygenase and ferritin induction by heme proteins: a possible mechanism limiting shock damage, *Trans. Assoc. Am. Phys.,* 105: 1–6, 1992.

244. Balla, G. et al. Ferritin: a cytoprotective anti-oxidant stratagem of endothelium, *J. Biol. Chem.,* 267: 18148–18153, 1992.
245. Balla, G. et al. Endothelial cell heme oxygenase and ferritin induction in rat lung by hemoglobin *in vivo, Am. J. Physiol.,* 268: L321-L327, 1995.
246. Yu, E.L. et al. Effect of nitric oxide on heme metabolism in pulmonary artery endothelial cells, *Am. J. Physiol.,* 271: L512-L518, 1996.
247. Oberle, S. and Schroder, H. Ferritin may mediate SIN-1 induced protection against oxidative stress, *Nitric Oxide,* 1:308–314, 1997.
248. Juckett, M. et al. Heme and the endothelium, *J. Biol. Chem.,* 273: 23388–23397, 1998.
249. Balla, J. et al. Ferriporphyrins and endothelium: a two-edged sword — promotion of oxidation and induction of cytoprotectants, *Blood,* 95: 3442–3450, 2000.
250. Marton, L.S. et al. Effects of hemoglobin on heme oxygenase gene expression and viability of cultured smooth muscle cells, *Am. J. Physiol.,* 279: H2405–H2413, 2000.
251. Ono, S. et al. Heme oxygenase-1 and ferritin are increased in cerebral arteries after subarachnoid hemorrhage in monkeys *J. Cereb. Blood Flow Metab.,* 20: 1066–1076, 2000.
252. Oberle, S. et al. Aspirin increases ferritin synthesis in endothelial cells, *Circ. Res.,* 82: 1016–1020, 1998.
253. Pang, J.H. et al. Increased ferritin gene expression in atherosclerotic lesions, *J. Clin. Invest.,* 97: 2204–2212, 1996.
254. Yum, XM. Apoptotic macrophage-derived foam cells of human atheromas are rich in iron and ferritin, suggesting iron-catalysed reactions to be involved in apoptosis, *Free Radical Res.,* 30: 221–231, 1999.
255. Lusis, A.J. Atherosclerosis, *Nature,* 407: 233–241, 2000.
256. Jang, M.K., Choi, M.S., and Park, Y.B. Regulation of ferritin light chain gene expression by oxidized low-density lipoproteins in human mococytic THP-1 cells, *Biochem. Biophys. Res. Commun.,* 26: 577–583, 1999.
257. Kertsten, S., Desvergne, B., and Wahli, W. Roles of PPARs in health and disease, *Nature,* 405: 421–424, 2000.
258. Juckett, M.B. et al. Ferritin protects endothelial cells from oxidized low density lipoprotein *in vitro, Am. J. Pathol.,* 147: 782–789, 1995.
259. Livrea, M.A. et al. Oxidative modification of low-density lipoproteins and atherogenic risk in beta-thalassemia, *Blood,* 92: 3936–3942, 1998.
260. Lee, T.S. et al. Iron-deficient diet reduces atherosclerotic lesions in ApoE-deficient mice, *Circulation,* 99: 1222–1229, 1999.
261. Hancock, W.W. et al. Antibody-induced transplant arteriosclerosis is prevented by graft expression of anti-oxidant and anti-apoptotic genes, *Nature, Med.,* 4: 1392–1396, 1998.
262. Soares, M.P. et al. Expression of heme oxygenase-I can determine cardiac xenograft survival, *Nature, Med.,* 4: 1073–1077, 1998.
263. Higgy, N.A. et al. Differential expression of human ferritin H gene in immortal human breast epithelial MCF-10F cells, *Mol. Carcinog.,* 20: 332–339, 1997.
264. Russo, J. et al. Developmental, cellular and molecular basis of human breast cancer, *J. Natl. Cancer Inst. Monogr.,* 27: 17–37, 2000.
265. Malins, D.C., Polinar, N.L., and Gunselman, S.J. Progression of human breast cancers to the metastatic state is linked to hydroxyl-radical-induced DNA damage, *Proc. Natl. Acad. Sci. U.S.A.,* 92: 2557–2563, 1996.
266. Willie, S. and Liehr, J.G. Release of iron from ferritin storage by redox cycling of stilbene and steroid estrogen metabolites: a mechanism of induction of free radical damage by estrogen, *Arch. Biochem. Biophys.,* 346: 180–186, 1997.

267. Gimer, G. et al. Cytosol and serum ferritin in breast carcinoma, *Cancer Lett.,* 67: 103–112, 1992.
268. Rosen, H.R. et al. Placental isoferritin associated p43 antigen correlates with features of high differentiation in breast cancer, *Breast Cancer Res. Treat.,* 24: 17–26, 1992.
269. Rosen, H.R. et al. Monoclonal antibody CM-H-9 detects circulating placental isoferritin in the serum of patients with visceral metastases of breast cancer, *Cancer Lett.,* 59: 145–151, 1991.
270. Reinerova, M. et al. Immunosuppressive activity of lymphocyte mitogenesis by breast cancer-associated p43, *Neoplasma,* 43: 363–366, 1996.
271. Hann, H.W. et al. Biologic differences between neuroblastoma stages IV-S and V. Measurement of serum ferritin and E-rosette inhibition in 30 children, *New Engl. J. Med.,* 305: 425–429, 1981.
272. Hann, H.W. et al. Prognostic importance of serum ferritin in patients with stage III and IV neuroblastoma: the Children's Cancer Study Group experience, *Cancer Res.,* 45: 2843–2848, 1985.
273. Schmidt, M.L. et al. Biologic factors determine prognosis in infants with stage IV neuroblastoma: a prospective children's cancer group study, *J. Clin. Oncol.,* 18: 1260–1268, 2000.
274. Maris, J.M. et al. Loss of heterozygosity at 1p36 independently predicts for disease progression but not decreased overall survival probability in neuroblastoma patients: a Children's Cancer Group study, *J. Clin. Oncol.,* 18: 1888–1899, 2000.
275. Iancu, T.C., Shiloh, H., and Kedar, A. Neuroblastomas contain iron-rich ferritin, *Cancer,* 61: 2497–2502, 1988.
276. Blatt, J. and Wharton, V. Stimulation of growth of neuroblastoma cells by ferritin *in vitro, J. Lab. Clin. Med.,* 119: 139–143, 1992.
277. Becton, D.L. and Bryles, P. Deferoxamine inhibition of neuroblastoma viability and proliferation, *Cancer Res.,* 48: 7189–7192, 1988.
278. Salig, R.A. et al. Failure of iron chelators to reduce tumor growth in human neuroblastoma xenografts, *Cancer Res.,* 58:473–478, 1998.
279. Aulbert, E. and Schmidt, C.G. Ferritin: a tumor marker in myeloid leukemia, *Cancer Detect. Prev.,* 8:297–302, 1985.
280. Aulbert, E. and Fromm, H. Ferritin in myeloproliferative diseases, *Med. Klin.,* 85:125–131, 1990.
281. Aulbert, E., Fromm, H., and Hornemann, H. Ferritin in acute leukaemia. Serum ferritin as a non-specific tumor marker for M1 and M2 myeloid leukaemia, *Med. Klin.,* 86: 297–304, 1991.
282. Lukina, E.A. et al. The diagnostic significance of serum ferritin indices in patients with malignant and reactive histiocytosis, *Br. J. Haematol.,* 83: 326–329, 1993.
283. Aulbert, E. and Steffens, O. Serum ferritin: a tumor marker in malignant lymphoma? *Onkologie,* 13: 102–108, 1990.
284. Bozwoda, W.R. et al. Serum ferritin and Hodgkin's disease, *Scand. J. Haematol.,* 35: 505–510, 1985.
285. Cazzola, M. et al. Basic and acidic isoferritins in the serum of patients with Hodgkin's disease, *Eur. J. Cancer Clin. Oncol.,* 19: 339–345, 1983.
286. Miyahara, M. et al. B-cell lymphoma-associated hemophagocytic syndrome: clinicopathological characteristics, *Ann. Hematol.,* 79: 378–388, 2000.
287. Vriesendorp, H.M., Morton, J.D., and Quadri, S.M. Review of five consecutive studies of radiolabeled immunoglobulin therapy in Hodgkin's disease, *Cancer Res.,* 55: 5888–5892, 1995.

288. Vriesendorp, H.M. et al. Recurrence of Hodgkin's disease after indium-111 and yttrium-90 labeled anti-ferritin administration, *Cancer*, 80: 2721–2727, 1997.
289. Vriesendorp, H.M. et al. Fractionated radiolabeled antiferritin therapy for patients with recurrent Hodgkin's disease, *Clin. Cancer Res.*, 9: 3324–3329, 1999.
290. Vaughn, C.B. et al. Ferritin content in human cancerous and noncancerous colonic tissue, *Cancer Invest.*, 5: 7–10, 1987.
291. Campo, E.et al. Ferritin immunohistochemical localization in normal and neoplastic colonic mucosa, *Int. J. Biol. Markers*, 2: 177–183, 1987.
292. Kishida, T. et al. Significance of serum iron and ferritin in patients with colorectal adenomas, *Scand. J. Gastroenterol.*, 32: 233–237, 1997.
293. Pinto, V. Preoperative evaluation of ferritinemia in primary epithelial ovarian cancer, *Tumori*, 83: 927–929, 1997.
294. Schafer, D.F. and Sorrell, M.F. Hepatocellular carcinoma, *Lancet*, 353: 1253–1257, 1999.
295. Wu, G.C. et al. Rat ferritin H: cDNA cloning, differential expression and localization during hepatocarcinogenesis, *Carcinogenesis*, 18: 47–52, 1997.
296. Ola, S.O., Akanji, A.O., and Ayoola, E.A. The diagnostic utility of serum ferritin estimation in Nigerian patients with primary hepatocellular carcinoma, *Nutrition*, 11, Suppl. 5: 532–534, 1995.
297. Jin, J. et al. Radioimmunoassay of human cardiac acidic isoferritin: a new index for hepatic cancer, *Chin. Med. J.*, 111: 150–153, 1998.
298. Akiba, S. et al. Serum ferritin and stomach cancer risk among a Japanese population, *Cancer*, 67: 1707–1712, 1991.
299. Gail, M.H. et al. Multiple markers for lung cancer diagnosis: validation of models for localized lung cancer, *J. Natl. Cancer Inst.*, 80: 97–101, 1988.
300. Cox, R., Gyde, O.H., and Leyland, M.J. Serum ferritin levels in small lung cancer, *Eur. J. Cancer Clin. Oncol.*, 22: 831–833, 1986.
301. Kuralay, F., Tokgoz, Z., and Comlekci, A. Diagnostic uselfuness of tumor marker levels in pleural effusions of malignant and benign origin, *Clin. Chim. Acta*, 300: 43–55, 2000.
302. Fracchia, A. et al. A comparative study on ferritin concentration in serum and bilateral bronchoalveolar lavage fluid of patients with peripheral lung cancer versus control subjects, *Oncology*, 56: 181–188, 1992.
303. Cohen, C., Schulman, G., and Buggeon, L.R. Immunohistochemical ferritin in testicular carcinoma, *Cancer*, 54: 2190–2194, 1985.
304. Takai, K. et al. Significance of serum ferritin level in testicular tumor, *Hinyokika Kijo*, 37: 357–362, 1991.
305. Kirkali, Z. et al. Ferritin: a tumor marker expressed by renal cell carcinoma, *Eur. Urol.*, 28: 131–134, 1995.
306. Partin, A.W. et al. Serum ferritin as a clinical marker for renal cell carcinoma: influence of tumor volume, *Urology*, 45: 211–217, 1995.
307. Citterio, G. et al. Prognostic factors for survival in metastatic renal cell carcinoma: retrospective analysis from 109 consecutive patients, *Eur. Urol.*, 31: 286–291, 1997.
308. D'Addessi, A. et al. Serum ferritin determination: is it useful in the early diagnosis of renal carcinoma? *Arch. Ital. Urol. Nefrol. Androl.*, 69: 283–286, 1997.
309. Vet, J.A. et al. Differential expression of ferritin heavy chain in a rat transitional cell carcinoma progression model, *Biochim. Biophys. Acta*, 1360: 39–44, 1997.
310. Torti, S.V. et al. Tumor cell cytotoxicity of a novel metal chelator, *Blood*, 92: 1384–1389, 1998.

311. Migliari, R. et al. Serum and urine ferritin in patients with transitional cell carcinoma of the bladder, *Arch. Ital. Urol. Nefrol. Androl.,* 63: 141–145, 1991.

312. Gierek, T. et al. The concentration of ferritin in blood serum of patients with neoplasms of the larynx and the maxillary-ethmoidal complex, *Otolaryngol. Pol.,* 49, Suppl. 20: 81–83, 1995.

313. Kimura, Y. et al. Conventional tumor markers are prognostic indicators in patients with head and neck squamous cell carcinoma, *Cancer Lett.,* 155: 163–168, 2000.

314. Ho, S. et al. Strong association between hyperferritinemia and metastatic disease in nasopharyngeal carcinoma, *Eur. J. Cancer B, Oral Oncol.,* 32B: 242–245, 1996.

315. Mumford, A.D. et al. The lens in hereditary hyperferritinemia cataract syndrome contains crystalline deposits of L-ferritin, *Brit. J. Ophthalmol.,* 84: 697–700, 2000.

316. McGahan, M.C. et al. Regulation of ferritin levels in cultured lens epithelial cells, *Exp. Eye Res.,* 59: 551–555, 1994.

317. Goralska, M. et al. Mechanisms by which ascorbic acid increases ferritin levels in cultured lens epithelial cells, *Exp. Eye Res.,* 64: 413–421, 1997.

318. Cheng Q., Gonzales, P., and Zigler, J.S. High levels of ferritin chain mRNA in lens, *Biochem. Biophys. Res. Commun.,* 270: 349–355, 2000.

319. Gorner, B. et al. Distribution of ferritin and redox-reactive transition metals in normal and cataractous human lenses, *Exp. Eye Res.,* 71: 599–607, 2000.

320. Goralska, M., Holley, B., and McGahan, M.C. The effects of Tempol on ferritin synthesis and Fe metabolism in lens epithelilal cells, *Biochim. Biophys. Acta,* 1497: 51–60, 2000.

321. Cai, C.X., Birk, D.E., and Linsenmayer. T.F. Ferritin is a developmentally regulated nuclear protein of avian corneal epithelial cells, *J. Biol. Chem.,* 272: 12831–12839, 1997.

322. Cai, C.X., Birk, D.E., and Linsenmayer, T.F. Nuclear ferritin protects DNA from UV damage in corneal epithelial cells, *Mol. Biol. Cell,* 9: 1037–1051, 1998.

323. Shimmura, S. et al. Reoxygenation injury in a cultured corneal epithelial cell line protected by the uptake of lactoferrin, *Invest. Ophthalmol. Vis. Sci.,* 39: 1346–1351, 1998.

324. Yefimova, M.G. et al. Iron, ferritin, transferrin and transferrin receptor in the adult rat retina, *Invest. Ophthalmol. Vis. Sci.,* 41: 2343–2351, 2000.

325. Hunt, R.C. et al. Hemopexin in the human retina: protection of the retina against heme-mediated toxicity, *J. Cell Physiol.,* 168: 71–80, 1996.

326. Vile, G.F. and Tyrrel, R.M. Oxidative stress resulting from ultraviolet A irradiation of human skin fibroblasts leads to a heme oxygenase-dependent increase in ferritin, *J. Biol. Chem.,* 268: 14678–14681, 1993.

327. Vile, G.F. and Tyrrel, R.M. UVA radiation-induced oxidative damage to lipids and proteins, *Free Radical Biol. Med.,* 18: 721–730, 1995.

328. Morlière, P. et al. Sensitization of skin fibroblasts to UVA by excess iron, *Biochim. Biophys. Acta,* 1334: 283–290, 1997.

329. Applegate, L.A. et al. Evidence that ferritin is UV inducible in human skin: part of a putative defense mechanism, *J. Invest. Dermatol.,* 111: 159–163, 1998.

330. Pourzand, C. et al. Ultraviolet A radiation induces immediate release of iron in human primary skin fibroblasts: the role of ferritin, *Proc. Natl. Acad. Sci. U.S.A.,* 96: 6751–6756, 1999.

331. Wlaschek, M. et al. Isolation and identification of psoralen plus ultraviolet A (PUVA)-induced genes in human dermal fibroblasts by polymerase chain reaction-based subtractive hybridisation, *J. Invest. Dermatol.,* 115: 901–913, 2000.

332. Giordani. A. et al. Contrasting effects of excess ferritin expression on iron-mediated oxidative stress induced by tert-butyl hydroperoxide or ultraviolet-A in human fibroblasts and keratinocytes, *J. Photochem. Photobiol.*, 54: 43–54, 2000.

333. Applegate, L.A. et al. Induction of the putative protective protein ferritin by infrared radiation: implications in skin repair, *Int. J. Mol. Med.*, 5: 247–251, 2000.

334. Ramm, G.A. et al. Ferritin receptors on activated rat lipocytes, *J. Clin. Invest.*, 94: 9–15, 1994.

335. Ramm, G.A. et al. Rat liver ferritin selectively inhibits expression of alpha-smooth muscle actin in cultured rat lipocytes, *Am. J. Physiol.*, 270: G370-G375, 1996.

336. Festa, M. et al. Overexpression of H ferritin and up-regulation of iron regulatory protein genes during differentiation of 3T3-L1 pre-adipocytes, *J. Biol. Chem.*, 275: 36708–36712, 2000.

337. Roth, J.A. et al. Differential localization of divalent metal transporter 1 with and without iron response element in rat PC12 and sympathetic neuronal cells, *J. Neurosci.*, 20: 7595–7601, 2000.

338. Connor, J.R. et al. Cellular distribution of transferrin, ferritin, and iron in normal and aged human brains, *J. Neurosci. Res.*, 27: 595–611, 1990.

339. Connor, J.R. et al. Isoforms of ferritin have a specific cellular distribution in the brain, *J. Neurosci. Res.*, 37: 461–465, 1994.

340. Sanyal, B., Polak, P.E., and Szuchet, S. Differential expression of the heavy-chain ferritin gene in non-adhered and adhered oligodendrocytes, *J. Neurosci. Res.*, 46: 187–197, 1996.

341. Connor, J.R. and Menzies, S.L. Relationship of iron to oligodendrocytes and myelination, *Glia,* 17: 83–93, 1996.

342. Gould, R.M. et al. Messenger RNAs located in myelin sheath assembly sites, *J. Neurochem.*, 75: 1834–1844, 2000.

343. Hulet, S.W. et al. Characterization and distribution of ferritin binding sites in the adult mouse brain, *J. Neurochem.*, 72: 868–874, 1999.

344. Hulet, S.W., Powers, S., and Connor, J.R. Distribution of transferrin and ferritin binding in normal and multiple sclerotic human brains, *J. Neurol. Sci.*, 165: 48–55, 1999.

345. Hulet, S.W. et al. Oligodendrocyte progenitor cells internalise ferritin via clathrin-dependent receptor-mediated endocytosis, *J. Neurosci. Res.*, 61: 52–60, 2000.

346. Qi, Y. and Dawson, G. Hypoxia specifically and reversibly induces the synthesis of ferritin in oligodendrocytes and human oligodendrogliomas, *J. Neurochem.*, 63: 1485–1490, 1994.

347. Qi, Y., Jamindar, T.M., and Dawson, G. Hypoxia alters iron homeostasis and induces ferritin synthesis in oligodendrocytes, *J. Neurochem.*, 64: 2458–2464, 1995.

348. Ozawa, H. et al. Development of ferritin-positive cells in cerebrum of human brain, *Pediatr, Neurol.*, 10: 44–48, 1994.

349. Ozawa, H. and Takashima, S. Immunocytochemical development of transferrin and ferritin immunoreactivity in the human pons and cerebellum, *J. Child. Neurol.*, 13:59–63, 1998.

350. White, B.C. et al. Brain ischemia and reperfusion: molecular mechanisms of neuronal injury, *J. Neurol. Sci.*, 179: 1–33, 2000.

351. Zamm, K. et al. Protection from oxidative stress-induced apoptosis in cortical neuronal cultures by iron chelators is associated with enhanced DNA binding by hypoxia-inducible factor-1 and ATF-1/CREB and increased expression of glycolytic enzymes, p21 and erythropoietin, *J. Neurosci.*, 19: 9821–9830, 1999.

352. Davalos, A. et al. Body iron stores and early neurologic deterioration in acute cerebral infarction, *Neurology,* 54: 1568–1574, 2000.

353. Cheepsunthorn, P., Palmer, C., and Connor, J.R. Cellular distribution of ferritin subunits in postnatal rat brain, *J. Comp. Neurol.,* 400: 73–86, 1998.

354. Ishimoto, T. et al. Dendritic translocation of the rat ferritin H chain mRNA, *Biochem. Biophys. Res. Commun.,* 272: 789–793, 2000.

355. Percy, M.E. et al. Iron metabolism and human ferritin heavy chain cDNA from adult brain with an elongated untranslated region: new findings and insights, *Analyst,* 23: 41–50, 1998.

356. Moos, T., Trinder, T., and Morgan, E.H. Cellular distribution of ferric iron, ferritin, transferrin and divalent metal transporter 1 (DMT1) in substantia nigra and basal ganglia of normal and beta 2-microglobulin-deficient mouse brain, *Cell Mol. Biol.,* 46: 549–561, 2000.

357. Hill, J.M. and Switzer, R.C. The regional distribution and cellular localization of iron in the rat brain, *Neuroscience,* 11: 595–603, 1984.

358. Riederer, P. et al. Transition metals, ferritin, glutathione and ascorbic acid in Parkinsonian brains, *J. Neurochem.,* 52: 515–520, 1989.

359. Zecca, L. et al. Interaction of neuromelanin and iron in substantia nigra and other areas of human brain, *Neuroscience,* 73: 407–415, 1996.

360. Shima, T. et al. Binding of iron to neuromelanin of human substantia nigra and synthetic melanin: an electron paramagnetic resonance spectroscopy study, *Free Radical Biol. Med.,* 23: 110–119, 1997.

361. Koutnikova, H. et al. Studies of human, mouse and yeast homologues indicate a mitochondrial function for frataxin, *Nature Genet.,* 16: 345–351, 1997.

362. Wilson, R.B. and Roof, D.M. Respiratory deficiency due to loss of mitochondrial DNA in yeast lacking the frataxin homologue, *Nature Genet.,* 16: 352–357, 1997.

363. Rotig, A., De Lonlay, P., and Chretien, D. Aconitase and mitochondrial iron–sulphur protein deficiency in Friedreich ataxia, *Nature Genet.,* 17: 215–217, 1997.

364. Gordon, N. Friedreich's ataxia and iron metabolism, *Brain Dev.,* 22: 465–468, 2000.

365. Bekris, S. et al. Human ABC7 transporter: gene structure and mutation causing X-linked sideroblastic anemia with ataxia with disruption of cytosolic iron–sulphur protein maturation, *Blood,* 96: 3256–3264, 2000.

366. Griffiths, P.D. and Crossman, A.R. Distribution of iron in the basal ganglia and neocortex in postmortem tissue in Parkinson's disease and Alzheimer's disease, *Dementia,* 4: 61–65, 1993.

367. Griffiths, P.D. et al. Iron in the basal ganglia in Parkinson's disease. An *in vitro* study using extended x-ray absorption fine structure and cryo-electron microscopy, *Brain,* 122: 667–673, 1999.

368. Vyzmal, J., Urgosik, D., and Bulte. J.W. Differentiation between hemosiderin- and ferritin-bound brain iron using nuclear magnetic resonance and magnetic resonance imaging, *Cell Mol. Biol.,* 46: 835–842, 2000.

369. Graham, J.M. et al. Brain iron deposition in Parkinson's disease imaged using the PRIME magnetic resonance sequence, *Brain,* 123: 2423–2431, 2000.

370. Faucheux, B.A. et al. Autoradiographic localization and density of [^{125}I] ferrotransferrin binding sites in basal ganglia of control subjects, patients with Parkinson's disease and MPTP-lesioned monkeys, *Brain Res.,* 691: 115–124, 1995.

371. Mizza, B. et al. The absence of reactive astrocytosis is indicative of a unique inflammatory process in Parkinson disease, *Neuroscience,* 95: 425–432, 2000.

372. Connor, J.R. et al. A quantitative analysis of isoferritins in selected regions of aged, Parkinsonian, and Alzheimer's diseased brains, *J. Neurochem.,* 65: 717–724, 1995.

373. Dexter, D.T. et al. Alterations in the levels of iron, ferritin and other trace metals in Parkinson's disease and other neurodegenerative diseases affecting the basal ganglia, *Brain,* 114: 1953–1975, 1991.

374. Jellinger, K. et al. Brain iron and ferritin in Parkinson's and Alzheimer's diseases, *J. Neural Transm. Park. Dis. Dement. Sect.,* 2: 327–340, 1990.

375. Schafer, F.Q., Qian, S.Y., and Buettner, G.R. Iron and free radical oxidations in cell membranes, *Cell Mol. Biol,* 46: 657–662, 2000.

376. Kidd, P.M. Parkinson's disease as multifactorial oxidative neurodegeneration: implications for integrative management, *Altern. Med. Rev.,* 5: 502–509, 2000.

377. Yodim, M.B., Grunblatt, E., and Mandel, S. The potential role of iron in NF-kappa B activation and nigrostriatal dopaminergic neurodegeneration. Prospects for neuroprotection in Parkinson's disease with iron chelators, *Ann. N.Y. Acad. Sci.,* 890: 7–25, 1999.

378. Leonard, J.V. and Shapira, H.V. Mitochondrial respiratory chain disorders II: neurogenerative disorders and nuclear gene defects, *Lancet,* 355: 389–394, 2000.

379. Bartzokis, G. et al. Increased basal ganglia iron levels in Huntington disease, *Arch. Neurol.,* 56: 569–574, 1999.

380. Bartzokis, G. and Tishler, T.A. MRI evaluation of basal ganglia ferritin iron and neurotoxicity in Alzheimer's and Huntington's disease, *Cell Mol. Biol.,* 46: 821–833, 2000.

381. Grundke-Iqbal, I. et al. Ferritin is a component of the neuritic (senile) plaque in Alzheimer dementia, *Acta Neuropathol.,* 81: 105–110, 1990.

382. Connor, J.R. et al. A histochemical study of iron, transferrin, and ferritin in Alzheimer's diseased brains, *J. Neurosci. Res.,* 31: 75–83, 1992.

383. Jeffries, W. et al. Reactive microglia specifically associated with amyloid plaques in Alzheimer's disease brain tissue express melanotransferrin, *Brain Res.,* 712: 122–126, 1996.

384. Rogers, J.T. et al. Translation of the Alzheimer amyloid precursor protein mRNA is up-regulated by interleukin-1 through 5′ untranslated region sequences, *J. Biol. Chem.,* 274: 6421–6431, 1999.

385. Craelius, W. et al. Iron deposits surrounding multiple sclerosis plaques, *Arch. Pathol. Lab. Med.,* 106: 397–399, 1982.

386. Le Vine, S.M. Iron deposits in multiple sclerosis and Alzheimer's disease brains, *Brain Res.,* 760: 298–303, 1997.

387. Le Vine, S.M. et al. Ferritin, transferrin and iron concentrations in the cerebrospinal fluid of multiple sclerosis patients, *Brain Res.,* 821: 511–515, 1999.

388. Malecki, E.A. et al. Existing and emerging mechanisms for transport of iron and manganese to the brain, *J. Neurosci. Res.,* 36: 113–122, 1999.

389. Chan, P.H. Role of oxidants in ischemic brain damages, *Stroke,* 27: 1124–1129, 1996.

390. Rosenthal, R.E. et al. Prevention of post-ischemic brain lipid conjugated diene production and neurological injury by hydroxyethyl starch-conjugated deferoxamine, *Free Radical Biol. Med.,* 12: 29–33, 1992.

391. Ozawa, H. et al. Immunohistochemical study of ferritin-positive cells in the cerebellar cortex within subarachnoidal hemorrhage in neonates, *Brain Res.,* 651:345–348, 1994.

392. Ishimaru, H. et al. Activation of iron handling system within the gerbil hippocampus after cerebral ischemia, *Brain Res.,* 726: 23–30, 1996.

393. Kondo, Y. et al. Regional differences in late-onset iron deposition, ferritin, transferrin, astrocyte proliferation and microglia activation after transient forebrain ischemia in rat brain, *J. Cereb. Blood Flow Metab.,* 15: 216–226, 1995.

394. Chi, S.I. et al. Differential regulation of H- and L-ferritin messenger RNA subunits, ferritin protein and iron following focal cerebral ischemia–reperfusion, *Neuroscience,* 100: 475–484, 2000.

395. Panahian, N. et al. Overexpression of heme-oxygenase I is neuroprotective in a model of permanent middle cerebral artery occlusion in transgenic mice, *J. Neurochem.,* 72: 1187–1203, 1999.

396. Ono, S. et al. Heme-oxygenase-1 and ferritin are increased in cerebral arteries after subarachnoid hemorrhage in monkeys, *J. Cereb. Blood Flow Metab.,* 20: 1066–1076, 2000.

397. Berschorner, R. et al. Long-term expression of heme oxygenase-1 (HO-1, HSP-32) following cerebral infarctions and traumatic brain injury in human, *Acta Neuropathol.,* 100: 377–284, 2000.

398. Chen, K., Gunter, K., and Maines, M.D. Neurons overexpressing heme oxygenase-1 resist oxidative stress-mediated cell death, *J. Neurochem.,* 75: 304–313, 2000.

399. Dorè, S. et al. Heme oxygenase-2 acts to prevent neuronal death in brain cultures and following transient cerebral ischemia, *Neuroscience,* 99: 587–592, 2000.

400. Connor, J.R. Iron regulation in the brain at the cell and molecular level, *Adv. Exp. Med. Biol.,* 356: 229–238, 1994.

401. Dickinson, T.K. and Connor, J.R. Cellular distribution of iron, transferrin and ferritin in the hypotransferrinemic (Hp) mouse brain, *J. Comp. Neurol.,* 335: 67–80, 1995.

402. Patel, B.N., Dunn, R.J, and David, S. Alternative RNA splicing generates a glycosyl-phosphatidylinositol-anchored form of ceruloplasmin in mammalian brain, *J. Biol. Chem.,* 275: 4305–4310, 2000.

403. Gitlin, J.D. Aceruloplasminemia, *Pediatr. Res.,* 44: 271–276, 1998.

404. Kohno, S. et al. Defective electron transfer in complexes I and IV in patients with aceruloplasminemia, *J. Neurol. Sci.,* 182: 57–60, 2000.

405. Yoshida, K. et al. Increased lipid peroxidation in the brain of aceruloplasminemia patients, *J. Neurol. Sci.,* 175: 91–95, 2000.

406. Ioannu, Y.A. The structure and function of the Niemann–Pick C1 protein, *Mol. Genet. Metab.,* 71: 175–181, 2000.

407. Christomanou, H. and Harzer, K. Deficient ferritin immunoreactivity in visceral organs from patients with Niemann–Pick disease type C, *Biochem. Mol. Med.,* 55: 105–115, 1995.

408. Christomanou, H. and Harzer, K. Outcherlony double immunodiffusion method demonstrates absence of ferritin immunoreactivity in visceral organs from nine patients with Niemann–Pick disease type C, *Biochem. Mol. Med.,* 58: 176–183, 1996.

409. Christomanou, H. et al. Deficient ferritin immunoreactivity in tissues from Niemann–Pick type C disease: extension of findings to fetal tissues, H and L ferritin isoforms, but also one case of the rare Nieman–Pick C2 complementation group, *Mol. Genet. Metab.,* 70: 196–202, 2000.

410. Ozawa, H. et al. Development of ferritin-containing cells in the pons and cerebellum of the human brain, *Brain Dev.,* 16: 92–95, 1994.

411. Connor, J.R. Iron acquisition and expression of iron regulatory proteins in the developing brain: manipulation by ethanol exposure, iron deprivation and cellular dysfunction, *Dev. Neurosci.,* 16: 233–247, 1994.

412. Roskams, A.J. and Connor, J.R. Iron, transferrin, and ferritin in the rat brain during development and aging, *J. Neurochem.,* 63: 709–716, 1994.

413. Shoshan, Y. et al. Expression of oligodendrocyte progenitor cell antigens by gliomas: implications for the histogenesis of brain tumors, *Proc. Natl. Acad. Sci. U.S.A.,* 96: 10361–10366, 1999.

414. Liu, Y.F., Yang, W.L., and Liu. J.A. Ferritin in astrocytomas, *Chin. Med. J.,* 104: 326–329, 1998.

415. Leenstra, S. et al. Human malignant astrocytes express macrophage phenotype, *J. Neuroimmunol.,* 56: 17–25, 1995.

416. Sato, Y. et al. Cerebrospinal fluid ferritin in glioblastoma: evidence for tumor synthesis, *J. Neuroncol.,* 40: 47–50, 1998.

417. Beljanski, M. and Crochet, S. Differential effects of ferritin, calcium, zinc, and gallic acid on *in vitro* proliferation of human glioblastoma cells and normal astrocytes, *J. Lab. Clin. Med.,* 123: 547–555, 1994.

418. Yokomori, N. et al. Transcriptional regulation of ferritin messenger ribonucleic acid levels by insulin in cultured rat glioma cells, *Endocrinology,* 128: 1474–1480, 1991.

419. Piero, D.J. et al., Interleukin-1 beta increases binding of the iron regulatory protein and the synthesis of ferritin by increasing the labile iron pool, *Biochim. Biophys. Acta,* 1497: 279–288, 2000.

420. Ye, Z, and Connor, J.R. Identification of iron responsive genes by screening of cDNA libraries from suppression subtractive hybridization with anti-sense probes from three iron conditions, *Nucleic Acid Res.,* 28: 1802–1807, 2000.

421. Summers, M., White, G., and Jacobs, A. Ferritin synthesis in lymphocytes, poly morphs and monocytes, *Br. J. Haematol.,* 30: 425–434, 1975.

422. Andreesen, R. Expression of transferrin receptors and intracellular ferritin during terminal differentiation of human monocytes, *Blut,* 49: 195–202, 1984.

423. Worrall, M. and Worwood, M. Immunological properties of ferritin during *in vitro* maturation of human monocytes, *Eur. J. Haematol.,* 47: 223–228, 1991.

424. Testa, U. et al. Iron up-modulates the expression of transferrin receptors during monocyte–macrophage maturation, *J. Biol. Chem.,* 264: 13181–13187, 1989.

425. Testa U. et al. Differential regulation of iron regulatory element-binding proteins in cell extracts of activated lymphocytes versus monocytes–macrophages, *J. Biol. Chem.,* 266: 13925–13930, 1991.

426. Sposi, N.M. et al. Mechanisms of diffrential transferrin receptor expression in normal hematopoiesis, *Eur. J. Biochem.,* 267: 1–14, 2000.

427. Heller, R.A. et al. Discovery and analysis of inflammatory disease-related genes using cDNA microarrays, *Proc. Natl. Acad. Sci. U.S.A.,* 94: 2150–2155, 1997.

428. Hashimoto, S. et al. Serial analysis of gene expression in human monocytes and macrophages, *Blood,* 94: 837–844, 1999.

429. Teatle, R. and Honeysett, J.M. Gamma interferon modulates human monocyte/macrophage transferrin receptor expression, *Blood,* 71: 1590–1595, 1988.

430. Byrd, T.F. and Horwitz, M.A. Regulation of transferrin receptor expression and ferritin content in human mononuclear phagocytes. Coordinate upregulation by iron transferrin and downregulation by interferon gamma, *J. Clin. Invest.,* 91: 969–976, 1993.

431. Drapier, J.C. et al. Biosynthesis of nitric oxide activates iron regulatory factor in macrophages, *EMBO J.,* 12: 3643–3649, 1993.

432. Weinberg, E.D. Modulation of intramacrophage iron metabolism during microbial cell invasion, *Microbes Infect.,* 2: 85–89, 2000.

433. Elia, G. et al. Induction of ferritin and heat shock proteins by prostaglandin A1 in human monocytes, *Eur. J. Biochem.,* 264: 736–745, 1999.

434. Lee, G.R. The anemia of chronic disease, *Semin. Hematol.,* 20: 61–80, 1983.

435. Rogers, J.T. et al. Translational control during the acute phase response. Ferritin synthesis in response to interleukin 1, *J. Biol. Chem.,* 265: 14572–14578, 1990.

436. Weiss, G., Bogdan, C., and Hentze, M.W. Pathways for the regulation of macrophage iron metabolism by the anti-inflammatory cytokines IL-4 and IL-13, *J. Immunol.,* 158: 420–425, 1997.

437. Graziadei, I. et al. Modulation of iron metabolism in monocytic THP-1 and cultured human monocytes by the acute-phase protein alpha 1-antitrypsin, *Exp. Hematol.,* 26: 1053–1060, 1998.

438. Devitt, A. Human CD14 mediates recognition and phagocytosis of apoptotic cells, *Nature,* 392: 505–509, 1998.

439. Fadok, V.A. A receptor for phosphatidylserine-specific clearance of apoptotic cells, *Nature,* 405: 85–90, 2000.

440. Kristiansen, M. et al. Identification of the haemoglobin scavenger receptor, *Nature,* 409: 198–201, 2001.

441. Moura, E. et al. Iron release from human monocytes after erythrophagocytosis *in vitro*: an investigation in normal subjects and hereditary hemochromatosis patients, *Blood,* 92: 2511–2519, 1998.

442. Bornman, L. et al. Differential regulation and expression of stress proteins and ferritin in human monocytes, *J. Cell Physiol.,* 178: 1–8, 1999.

443. Ferris, C.D. et al. Haem oxygenase-I prevents cell death by regulating cellular iron, *Nature Cell. Biol.,* 1: 152–157, 1999.

444. Baranano, D.E. et al. A mammalian iron ATPase induced by iron, *J. Biol. Chem.,* 275: 15166–15173, 2000.

445. De Voss, J. et al. The salicylate-derived mycobactin siderophores of *Mycobacterium tuberculosis* are essential for growth in macrophages, *Proc. Natl. Acad. Sci. U.S.A.,* 97: 1252–1257, 2000.

446. Bosco, M.C. et al. The tryptophan catabolite picolinic acid selectively induces the chemokines macrophage inflammatory protein-1-alpha and 1-beta in macrophages, *J. Immunol.,* 164: 3283–3291, 2000.

447. Pais, T.F. and Appelberg, R. Macrophage control of mycobacterial growth induced by picolinic acid is dependent on host cell apoptosis, *J. Immunol.,* 164: 389–397, 2000.

448. Jabado, N.S. et al. Natural resistance to intracellular infections: natural resistance-associated macrophage protein 1 (NRAMP1) functions as a pH-dependent manganese transporter at the phagosomal membrane, *J. Exp. Med.,* 192: 1237–1247, 2000.

449. Atkinson, P.G.P. and Barton, C.H. High level expression of Nramp G169 in RAW 264.7 cell transfectants: analysis of intracellular iron transport, *Immunology,* 16: 656–662, 1999.

450. Zwilling, B.S. et al. Role of iron in Nramp1-mediated inhibition of mycobacterial growth, *Infect. Immun.,* 67: 1386–1392, 1999.

451. Kuhn, D.E. et al. Differential iron transport into phagosomes isolated from the RAW 264.7 macrophage cell lines transfected with Nramp1 Gly 169 or Nramp1 Asp 169, *J. Leukoc. Biol.,* 66: 113–119, 1999.

452. Baker, S.T., Barton, C.H., and Biggs, F.E. A negative autoregulatory link between Nramp1 function and expression, *J. Leukoc. Biol.,* 67: 501–507, 2000.

453. Wardrop, S.L. and Richardson, D.R. Interferon-gamma and lipopolysaccharide regulate the expression of Nramp2 and increase the uptake from low relative molecular mass complexes by macrophages, *Eur. J. Biochem.,* 267: 6586–6593, 2000.

454. Gutteridge, J.M.C. et al. Pro-oxidant iron is present in human pulmonary epithelial lining fluid: implications for oxidative stress in the lung, *Biochem. Biophys. Res. Commun.,* 220: 1024–1027, 1996.

455. Olakanmi, O., Stokes, J.B., and Britigan, B.E. Acquisition of iron bound to low molecular weight chelates by human monocyte-derived macrophages, *J. Immunol.*, 153: 2691–2703, 1994.

456. Wesselius, L.J. et al. Alveolar macrophages accumulate iron and ferritin after *in vivo* exposure to iron or tungsten dusts, *J. Lab. Clin. Med.*, 127: 401–409, 1996.

457. Olakanmi, O.S. et al. Iron sequestration by macrophages decreases the potential for extra-cellular hydroxyl radical formation, *J. Clin. Invest.*, 91: 889–899, 1993.

458. Wesselius, L.J., Flowers, C.H., and Skikne, B.S. Alveolar macrophage content of isoferritins and transferrin, *Am. Rev. Resp. Dis.*, 145: 311–316, 1992.

459. McGowan, S.E. and Henley, S.A. Iron and ferritin contents and distribution in human alveolar macrophages, *J. Lab. Clin. Med.*, 111: 611–617, 1988.

460. Wesselius, L.J., Nelson, M.E., and Skikne, B.S. Increased release of ferritin and iron by iron-loaded alveolar macrophages in cigarette smokers, *Am. J. Respir. Crit. Care Med.*, 150: 690–695, 1994.

461. Nelson. M.E. et al. Regional variation in iron-binding proteins within the lungs of smokers, *Am. J. Crit. Care Med.*, 153: 1353–1358, 1996.

462. Ghio, A.J. et al. Phagocyte-generated superoxide reduces Fe^{3+} to displace it from the surface of asbestos, *Arch. Biochem. Biophys.*, 315: 219–225, 1994.

463. Ghio, A.J. et al. Ferritin expression after *in vitro* exposures of human alveolar macrophages to silica is iron dependent, *Am. J. Respir. Cell. Mol. Biol.*, 17: 533–540, 1994.

464. Reiff, D.W. Ferritin as a source of iron for oxidative damage, *Free Radical Biol. Med.*, 12: 417–427, 1992.

465. Plautz, M.W., Bailey, K., and Wesselius, L.J. Influence of cigarette smoking in crocidolite-induced ferritin release by human alveolar macrophages, *J. Lab. Clin. Med.*, 136: 449–456, 2000.

466. Wesselius, L.J., et al. Iron uptake promotes hyperoxic injury to alveolar macrophages, *Am. J. Respir. Crit. Care Med.*, 159: 100–106, 1999.

467. Ryan, T.P. et al. Pulmonary ferritin: differential effects of hyperoxic lung injury on subunit mRNA levels, *Free Radical Biol. Med.*, 22: 901–908, 1997.

468. Yang, F. et al. Resistance of hypotransferrinemic mice to hyperoxia induced lung injury, *Lung Cell. Mol. Physiol.*, 21: L1214-L1223, 1999.

469. Ono, T. and Seno, S. Transport of ferritin from Kupffer cells to liver parenchymal cells. Morphological and immunocytochemical observations, *Int. J. Hematol.*, 54: 93–102, 1991.

470. Lin, M.R. et al. Role of iron in NF-κB activation and cytokine gene expression by rat hepatic macrophages, *Am. J. Physiol.*, 272: G1355-G1364, 1997.

471. Tsukamoto, H. et al. Iron primes hepatic macrophages for NF-κB activation in alcoholic liver injury, *Am. J. Physiol.*, 277: G1240-G1250, 1999.

472. Sato, H., Yamaguchi, M., and Sannai, S. Regulation of ferritin synthesis in macrophages by oxygen and a sulphydryl-reactive agent, *Biochem. Biophys. Res. Commun.*, 201: 38–44, 1994.

473. Kuriyama-Matsumura, K. et al. Regulation of ferritin synthesis and iron regulatory protein 1 by oxygen in mouse peritoneal macrophages, *Biochem. Biophys. Res. Commun.*, 249: 241–246, 1998.

474. Cairo, G. et al. Inappropriately high iron regulatory protein activity in monocytes of patients with genetic hemochromatosis, *Blood*, 89: 2546–2563, 1997.

475. Recalcati, S., et al. Response of monocyte iron regulatory protein activity to inflammation: abnormal behaviour in genetic hemochromatosis, *Blood,* 91: 2565–2572, 1998.

476. Moura, E. et al. Iron release from human monocytes after erythrophagocytosis *in vitro*: an investigation in normal subjects and hereditary hemochromatosis patients, *Blood,* 92: 2511–2519, 1998.

477. Ehrlich, R. and Lemonnier, F.A. HFE. A novel nonclassical class I molecule that is involved in iron metabolism, *Immunity,* 13: 585–588, 2000.

478. Remalingman, T.S. et al. Binding of the transferrin receptor is required for endocytosis of HFE and regulation of iron homeostasis, *Nature Cell. Biol.,* 2: 953–957, 2000.

479. Parkkila, S. et al. Cell surface expression of HFE protein in epithelial cells, macrophages, and monocytes, *Haematologica,* 85: 340–345, 2000.

480. Bastin, J.M. et al. Kupffer cell staining by an HFE-specific monoclonal antibody: implications for hereditary hemochromatosis, *Br. J. Haematol.,* 103: 931–941, 1998.

481. Montosi, G. et al. Wild-type HFE protein normalizes transferrin iron accumulation in macrophages from subjects with hereditary hemochromatosis, *Blood,* 96: 1125–1129, 2000.

482. Verroni, P. et al. Ferritin in malignant lymphoid cells, *Br. J. Haematol.,* 62: 105–110, 1986.

483. Pelosi, E. et al. Expression of transferrin receptors in phytohemagglutinin-stimulated human T-lymphocytes. Evidence for a three-step model, *J. Biol. Chem.,* 261: 3036–3042, 1986.

484. Pelosi, E. et al. Mechanisms underlying T-lymphocyte activation: mitogen initiates and IL-2 amplifies the expression of transferrin receptors via intracellular iron levels, *Immunology,* 64: 273–279, 1988.

485. Fargion, S. et al. Specific binding sites for H-ferritin on human lymphocytes: modulation during cellular proliferation and potential implications in cell growth control, *Blood,* 78: 1056–1061, 1991.

486. Matzner, Y. et al. Differential effect of isolated placental isoferritins on *in vitro* T-lymphocyte function, *Br. J. Haematol.,* 59: 443–448, 1985.

487. Morikawa, K., Oseko, F., and Morikawa, S. H- and L-rich ferritins suppress antibody production, but not proliferation of human B lymphocytes *in vitro, Blood,* 83: 737–743, 1994.

488. Brekelemans, P. et al. Inhibition of proliferation and differentiation during early T cell development by anti-transferrin receptor antibody, *Eur. J. Immunol.,* 24: 2896–2902, 1994.

489. Mincheva-Nilsson, L., Hammarstrom, S., and Hammarstrom, M.L. Activated human gamma/delta T lymphocytes express functional lactoferrin receptors, *Scand. J. Immunol.,* 46: 609–618, 1997.

490. Ten Elshof, A.E. et al. Gamma/delta intraepithelial lymphocytes drive tumor necrosis factor-alpha responsiveness to intestinal iron challenge: relevance to hemochromatosis, *Immunol. Rev.,* 167: 223–232, 1999.

491. Picard, V. et al. Overexpression of the ferritin H subunit in cultured erythroid cells changes the intracellular iron distribution, *Blood,* 87: 2057–2066, 1996.

492. Cozzi, A. et al. Overexpression of wild-type and mutated human ferritin H-chain in HeLa cells, *J. Biol. Chem.,* 275: 25122–25129, 2000.

493. Ye, X. et al. Engineering of the pro-vitamin A (beta-carotene) biosynthetic pathway into (carotenoid-free) rice endosperm, *Science,* 287: 303–305, 2000.

494. Gura, T. New genes boost rice nutrients, *Science,* 285:994–995, 1999.

495. Briat, J.F. Plant ferritin and human iron deficiency, *Nature Biotechnol.,* 17:621, 1999.

496. Huang, H.Q., Lin, Q.M., and Lou, Z.B. Construction of a ferritin reactor: an efficient means for trapping various heavy metal ions in flowing seawater, *J. Prot. Chem.,* 19: 441–447, 2000.

497. Suh, J.R. et al. Purification and properties of a folate-catabolizing enzyme, *J. Biol. Chem.*, 275: 35646–35655, 2000.

498. Oppenheim, E.W. et al. Mimosine is a cell-specific antagonist of folate metabolism, *J. Biol. Chem.*, 275: 19268–19274, 2000.

499. Fan, H., Villegas, C., and Wright, J.A. A link between ferritin gene expression and ribonucleotide reductase R2 protein, as demonstrated by retroviral vector mediated stable expression of R2 cDNA, *FEBS Lett.*, 382: 145–148, 1996.

500. Bogdan, C. Nitric oxide and the regulation of gene expression, *Trends Cell Biol.*, 11: 66–75, 2001.

501. Bouton, C. and Demple, B. Nitric oxide-inducible expression of heme oxygenase-1 in human cells, *J. Biol. Chem.*, 275: 32688–32693, 2000.

Index

A

ABC7 transporter, 498
Abscisic acid, 451
Absorption
  iron, 6–14
    during pregnancy, 3
    heme, 26–28
    human milk and lactoferrin synthesis,
      119–120
    physiologic regulation, 14–19
    steps, 1
  lactoferrin spectra, 75, 76, 77
Aceruloplasminemia, 392–394, 302
Acetylsalicylic acid, 486
Achlorhydria, 5
Acidification, 270
Acinar cells, 287
Aconitase
  iron regulatory protein
    comparison, 416–417, 418, 419, 428
    evolution, 426
    oxidative stress, 433, 434
    nitric oxide target, 429
Actin, 266, 267, 270
Active transport, 6
Acute inflammatory proteins, 283
Acute phase response, 182–183
Acute promyeleocytic leukemia (APL), 118, see
  also Leukemia cell line
Adapter proteins (AP), transferrin receptor
  expression and regulation, 303
  endocytosis, 263, 264, 265, 266
Adenocarcinoma, 121–122, 288
Adenomas, 61, 216
Adenosine diphosphate ribosylation factors
  (ARFs), 269
Adenosine-rich elements (AREs), 424–425
Adenosine triphosphatase (ATPase), 263–264
Adenosine triphosphate (ATP), 173
Adipocytes, 495
Advanced glycation end (AGE) products, 101–102
Aedes aegypti, 149, 455
Affinity chromatography, 177, 199–200
AFR, *see* Ascorbate free-radical reductase
AGE, *see* Advanced glycation end

Aggregation, 97
Alcoholism, 178, 470
Algae, 149–150, *see also* Individual entries
Alveolar macrophages, 509, *see also*
      Macrophages; Monocytes–Macrophages
Alzheimer's disease, 394, 500
Amino acid sequence
  ceruloplasmin, 392
  ferritins, 455
  iron regulatory proteins, 415–416, 420, 421
  lactoferrin, 74, 75, 78–79
  Nramp genes, 387, 388
  transferrin
    family proteins, 146–147, 148, 151–152
    receptor, 205, 251, 254, 258
    soluble receptor, 371
    transferrin-binding protein, 199
δ-Aminolevulinic acid, 436
Amphibians, 457
Amphiphysins, 265, 266
Amyloid, 500
Amyloid precursor protein (APP), 500
Ancestor gene, 202, 203, 428
Anemia
  chronic and iron absorption, 19
  diagnosis and serum ferritin, 468
  soluble transferrin receptor, 374, 375, 376, 378
Anemia of chronic disease, 376, *see also* Anemia
Animal models, 150, 178, 219
Anions, iron-binding, *see also* Bicarbonate
  transferrin, 168–169, 172
  lactoferrin, 77, 84
α-1-Antitrypsin, 506
Antibacterial activity, 72, 81, 92
Antibody response, 208
Anti-ferritin antibodies, 490, *see also* Ferritin
Antigens, 306, *see also* T lymphocytes; Transferrin
      receptor
Anti-IgD antibodies, 316, *see also* T lymphocytes;
      Transferrin receptor
Antimicrobial activity, 72, 73, 319
Antioxidants, 493
Anti-transferrin monoclonal antibodies, 49–50
Anti-transferrin receptor antibody, 220, 221, 330
Anti-tumor activity, 114
Anti-tumor drugs, 55–56, 286

AP1 binding element, 86, 87, *see also* Lactoferrin
AP-1 site, 480
AP-2 site, 116
APL, *see* Acute promyelocytic leukemia
Apoferritin, 15, 457–458, 461, *see also* Ferritin
Apolactoferrin, 83, 99, 169, *see also* Lactoferrin
Apoptosis, 56
Apotransferrin, 11, 159, 165, 271, *see also* Iron, binding; Transferrin
APP, *see* Amyloid precursor protein
APs, *see* Adapter proteins
Aqueous solutions, 143
Arabidopsis thaliana, 451
AREs, *see* Adenosine-rich elements
ARFs, *see* Adenosine diphosphate ribosylation factors
Arginine, iron binding
    lactoferrin, 76, 80, 84–85
    transferrin family proteins, 152, 174
Artemia salina, 456
Arteries, 484, *see also* Ferritin
Arteriosclerotic disease, 472
Ascorbate, 212, 465
Ascorbate free-radical reductase (AFR), 212
Ascorbic acid, 20–21, 492
Asialotransferrin, 177–178
Asialotransferrin receptor, 277, 278
Asparagine, 176, 260, 261
Aspartic acid
    iron binding/release by transferrin, 165, 166, 167, 168, 174
    lactoferrin gene mutagenesis, 82
Aspergillus nidulans, 88
Assimilation, 195
Astrocytes
    ceruloplasmin, 394
    ferritin expression, 501, 502
    transferrin synthesis, 186, 187
Astrocytomas, 503–504
ATF-1 transcription factor, 304
Atheromas, 486
Atherosclerosis, 486–487
Atlantic salmon, 457
ATP, *see* Adenosine triphosphate
ATPase, *see* Adenosine triphosphatase
Autologous hematopoietic stem cell
    transplantation (HSCT), 112, 113, 376,
    see also Lactoferrin
Azurophil granules, 115, 117

B

B box binding factors, 478, 479
B lymphocytes, 316, *see also* T lymphocytes
Bacillus subtilis, 426

Bacteria, 102, 452, *see also* Individual entries
Bacterioferritins, 452–453
Bacteriostatic activity, 98–103
Bacteroferrins, 461
Baculovirus expression system, 49
Balance mechanism, 16
Bap 31, 268
Basal ganglia, 497
Basal membranes, 293
Base pairing, 411, 413
Basic fibroblast growth factor (b-FGF), 110
Basolateral membrane, 6, 335
Basolateral transferring receptors, 14
b-FGF, *see* Basic fibroblast growth factor
b-*fos*, 480
BFU, *see* Burst-forming unit
BFU–E, *see* Burst-forming unit–Erythroid
Bicarbonate, iron binding, *see also* Anions, iron-binding
    lactoferrin, 75–76, 81, 85
    transferrin, 159, 162, 163, 171
Bile, 466
Bi-lobed transferrin, 153, *see also* Transferrin
Bioavailability, iron, 19–26
    breast feeding, 94
Biochemistry
    aceruloplasminemia, 393
    lactoferrin, 74–88
    transferrin receptor, 251, 252
Biomagnetometry, 468
Biosynthetic studies, 394, *see also* Ceruloplasmin
2,2′-Bipyridyl-6-carbothioamide, 54, 55
Blaberus discoidalis, 146
Bladder, 288–289
    cancer, 289, 491–492
Blast cells, 325
Blood, 6, 325, 326
Blood–brain barrier
    ceruloplasmin, 394
    drug delivery, 220
    ferritin expression in brain, 495
    multiple sclerosis, 501
    transferrin receptor expression/distribution, 284–286
    transferrin synthesis, 185
Blood–cerebral fluid barrier, 185
Bone marrow
    lactoferrin, 111, 115, 116, 117
    transferrin receptor, 289–290, 324, 325, 326, 327
Bordetella spp., 104, 203
Bovine lactoferrin, 85, *see also* Lactoferrin
Brain
    ferritin expression, 495–504
    iron homeostasis and ceruloplasmin, 393–394

nonheme iron bioavailability, 22
  transferrin, 185
  transferrin gene, 191, 193
  transferrin receptor, 284
  tumors, 51, 286
Brain stem, 284
Breast, 96–97
  cancer, 97, 121, 288, 487
Breast feeding, 94
Broken-wing conformation, 178
Bronchial epithelial cells, 282
Brush border membrane, 6, 7, 19, 93–94
Bulge structure, 410, 411, 413
Burst-forming unit (BFU), 110
Burst-forming unit–Erythroid (BFU–E), 324, 325,
    326–327, 330

C

C lobes
  transferrin, 162, 164, 165, 166, 171, 174
  transferrin–transferrin receptor interaction, 175
C/EBP transcription factors, 189, 190, 191
C282Y mutation, 397, 398, 399
Caco-2 cells, 7, 8, 11, 271
Calcein probe, 209
Calcium carbonate, 24
Calcium
  lactoferrin, 93
  nonheme iron bioavailability, 24–25
  transferrin receptor, 279–281, 306
Calmodulin, 280
Calreticulin, 13
cAMP, see Cyclic adenosine monophosphate
cAMP responsive elements, 302
Cancer, 59–61, 111, see also Individual entries
Candida albicans, 103, 106
Cap structure, 425, see also Iron regulatory
    elements
Capillaries, 284, 285
Carbohydrate, 178–179
Carbohydrate-deficient glycoprotein syndrome
    (CDGS), 178–179
Carbonate, 81, 164
Carcinoembryonic antigen (CEA), 491
β-Carotene, 22
Carotid artery disease, 472
Cartilage, 292
CASTing method, 113
CAT assay, 303
Cathepsin G, 115
CCAAT box, 116, 478, 481
CCAAT displacement protein (CDP), 116, 117
CD+ lymphocytes, 187
CD14 antigen, 109

CD34+ hemopoietic progenitors, 438
CD4+ T lymphocytes, 307, 315, see also T
    lymphocytes, activated
CD-56 antigen, 114
Cdc2 cyclin, 54, 58
CDGS, see Carbohydrate-deficient glycoprotein
    syndrome
Cdk, see Cyclin-dependent kinases
Cdk2 gene, 48
cDNA, see Complementary deoxyribonucleic acid
CDP, see CCAAT displacement protein
CEA, see Carcinoembryonic antigen
Cell cycle
  arrest and iron chelators, 54
  iron and control, 45–48
  ribonucleotide reductase, 44
  transferrin, 213
  transferrin receptor, 295
Cell growth, inhibition
  agents that interfere with intracellular iron
      incorporation, 51–53
  monoclonal antibodies to transferrin receptor,
      48–51
Cell proliferation
  iron role in control, 514–515
  transferrin
    control, 308
    growth factor, 208, 214
    receptor relationship, 295, 296
Cellubrevin, 267, 268
Central nervous system (CNS), 185, 285–286, 393,
    395
Central region I-binding protein (CRI-BP), 191
Cerebral ischemia, 497, 501–502
Cerebral tumors, 503
Cerebrospinal fluid, 177, 503
Ceruloplasmin
  anchored form and brain damage, 502
  ferroxidase activity in ferritin, 461
  iron
    transport mechanisms, 390–395
    uptake, 335
c-fos proteins, 303, 310, 311, 480
CFU, see Colony-forming units
CFU–E, see Colony-forming unit–erythroid
CFU–GM, see Colony-forming
    units–granulocytes–macrophage
Check points, 45, 46
Chelators, iron
  cell proliferation inhibition, 53–59
  cyclin activation, 47
  folate metabolism inhibition, 514
  iron regulatory protein activity, 430, 433
  low molecular weight, 319, 383, 385
  neuroblastoma treatment, 489

oxidase activity, 392
regulation of transferrin synthesis, 182
release from transferrin, 173–174, 175
ribonucleotide reductase, 43
transferrin receptor, 297, 299, 300, 312
synergism, 49–50
Chemoprevention, 114
Chicken ovalbumin upstream promoter
transcription factor (COUP-TF)
lactoferrin, 86, 87, 116, 120
transferrin gene expression, 190, 193
Chicken transferrin receptors, 253, 256, *see also*
Transferrin receptors
Chimeric molecules, 221, 386
Chinese hamster ovary (CHO) cells
melanotransferrin, 400
transferrin receptor endocytosis, 268, 269, 270,
274
Chlorate, 90
Chloroquine, 318
CHO, *see* Chinese hamster ovary cells
Cholesterol, 502
Choroid plexus, 185–186, 191, 395
Chromosome 3, 180, 181, 254
Chromosome 4, 500
Chromosome 11, 474
Chromosome pumping technique, 254
Chronic overload, 15
Chylomicrons remnant receptor, 95
Circulation, 374, *see also* Soluble transferrin
receptor
c-*jun*, 303, 480
Clathrin-coated pits, 262, 263–264,
Clearance, lactoferrin, 96
Cloning
iron regulatory protein genes, 416, 421
lactoferrin, 88
transferrin receptor 2, 255, 256, 257
c-*myc*
ferritin gene expression, 482, 483
iron regulatory protein activity, 438
transferrin, 213, 296
CNS, *see* Central nervous system
Coalbumin, *see* Ovotransferrin
Coatomer proteins (COPs), 268
Coenzyme Q, 210–211
Colchicine, 465
Collagenase, 115
Colon cancer, 60, 61, 490
Colony-forming units (CFU), myelopoiesis and
lactoferrin, 110
Colony-forming unit–erythroid (CFU–E), 324,
325, 327, 330–331
Colony-forming units–granulocytes–macrophage
(CFU–GM), 326

Colony-stimulating factors (CSFs), 110, 111
Colorectal cancer, 60–61, 490
Colostrum, 72
Competitive binding assays, 105
Complementary deoxyribonucleic acid (cDNA),
85, 87–88, *see also* Lactoferrin
Complex carbohydrates, 23–24
Computer alignment studies, 500
Computer modeling, 194
Con A, *see* Concanavalin
Concanavalin (Con A), 311
Conformational change
iron regulatory protein, 417
periplasmic proteins, 169
transferrin, 163, 165–166
Congenital atransferrinemia, 218–219
Consensus sequence, 410, 411, *see also* Iron
regulatory element
Continuous cell lines, 295, *see also* Individual
entries
Contraception, 2–3
Copper, 166, 386, 391, 392
COPs, *see* Coatomer proteins
Corneal epithelium, 493
Cortex, 284, *see also* Brain
COS-1 cells, 388
COUP-TF, *see* Chicken ovalbumin upstream
promoter transcription factor
Cow milk, *see* Milk, cow
Coxiella burnetii, 319
CREB, *see* Cyclic-AMP response binding
protein
CRI-BP, *see* Central region I-binding protein
Cross-over, 153
Cross-talk, 338
Crypt cells
ferritin, 512, 513
HFE gene expression, 333, 398
iron
absorption, 6, 8–9
uptake, 336, 337
Cryptococcus neoformans, 318
Crystallography
aconitase, 417
lactoferrin, 75, 80, 83
ribonucleotide reductase, 42
siderophores pathway, 195
transferrin protein family, 165
CSFs, *see* Colony-stimulating factors
Cyclic adenosine monophosphate (cAMP), 477
Cyclic-AMP response binding protein (CREB), 87,
191, 193
Cyclins, 46, 47–48
Cyclin-dependent kinases (Cdk), 46, 47
Cytochrome oxidase, 284

Cytokines
  ferritin, 477, 482, 504, 510
  lactoferrin, 107–109, 112
  serum ferritin levels, 469
  transferrin receptor, 283, 295, 306, 320, 323
  transferrin synthesis, 182–183, 184
Cytoplasmic iron pools, 55
Cytoprotection, 485
Cytotoxicity, 52
Cytotrophoblasts, 214–215, 294

D

D lactoferrin, 86, *see also* Lactoferrin
DCT1, *see* Divalent cation/metal ion transporter
    gene
Deferipone, 55, 56, 58
Deferoxamine
  cell proliferation inhibition, 53–54, 55, 56–59
  iron and cyclin activation, 47–48
  transferrin receptor, 297, 312
Deficiency
  congenital atransferrinemia, 219
  iron
    ferritin overexpression in rice, 514
    gastrectomy and achlorhydria, 5
    physiologic regulation of absorption, 16,
        17–18
    plants, 144
    serum ferritin levels, 468
    soluble transferrin receptor, 374, 375, 376
    transferrin gene, 192, 194
    transferrin receptor, 298
  lactoferrin, 112, 117
Degradation, ferritin, 465–466
Deletion analysis, 273
Density shift technique, 278
Deoxynojirimycin, 260–261
Deoxyribonucleic acid (DNA)
  lactoferrin interaction, 113
  synthesis
    iron, 215
    ribonucleotide reductase, 40–44
    transferrin receptor, 296, 297
Deoxyribonucleotides, 43, 44
Desferrioxamine, 6, 14, 55, 393
Desulfovibrio desulfuricans, 453
Development
  brain, 186, 503
  fetal and transferrin synthesis, 181
  lactoferrin expression, 122–123
  plants and ferritin evolution, 451
Diabetes, 472
Diatoms, 145
Diet, 2, 514

Dietary iron, 5, 436, *see also* Iron
Diferric transferrin, 11, 159, 161, *see also*
    Transferrin
Differential scanning calorimetry (DSC)
    technique, 171
Differentiation, 217
Digestion, 3–6
Digestive gut, 93
Dileucine motifs, 264, 273–274
Di-lysine trigger, 85, 167, 168, 169
Dimethyl sulfoxide (DMSO), 327–330
Diphosphate anions, 172, *see also* Anions
Diphtheria toxin, 220
Direct atomic force microscopy, 457
Direct binding assays, 105
Disease, 157–158
Distribution
  transferrin, 151, 186
  transferrin alleles, 157
  transferrin receptor, 281–294
Disulfide bridges, 152, 161, 251, 253
Divalent cation/metal ion transporter gene
    (DCT1), 334–335, 386–387, 388
Divalent ions, 93
d-*jun*, 480
DMSO, *see* Dimethyl sulfoxide
DNA, *see* Deoxyribonucleic acid
Domains
  iron regulatory protein, 417
  melanotransferrin, 399–400
  soluble transferrin receptor, 371
  transferrin, 162, 171–172, 191
  transferrin receptor, 254, 255–256, 257
Dopaminergic neurons, 499
Doxorubicin, 50, 434, *see also* Transferrin
Drosophila melanogaster, 149, 427, 428, 456
Drugs, 44, 219–222
DSC, *see* Differential scanning calorimetry
    technique
Dunalialla salina, 149–150
Duodenum, 11, 14, 334
Dynamin, 263–264, 265, 266

E

E1A protein, 477, 478, 480
E-box response element, 189, 192, 193
Echinococcus granulosus, 455
EFA6 proteins, 269, *see also* Guanosine nucleotide
    exchange factors
EGF, *see* Epidermal growth factor
eIF4F factor, 425
Elastase, 115
Electophoretic mobility, 74
Electromobility shift assays, 426

Electron spin echo envelope modulation
      (ESEEM), 77
Electron spin resonance spectroscopy, 42
Embryogenesis, 122–123, 181, 182
Endocrine glands, 291
Endocytosis
   receptor-mediated and iron absorption, 9
   transferrin receptor, 251
      basic features, 262–272
      structural requirement for internalization,
         272–275
Endometrial adenocarcinomas, 121–122
Endometrium, 120
Endosomal compartments, 373, 386
Endosomes, 262, 263, 267, 268, see also
      Endocytosis
Endothelial cells, 218, 484–487
Enhancers, nonheme iron bioavailability, 20–21
Enterocytes
   heme iron absorption, 26–27
   iron
      absorption, 9, 10, 11, 12, 14, 16
      import and export, 336, 337
   lactoferrin binding, 93
   Nramp2 localization, 389
Epidermal growth factor (EGF), 274, 295, 437
Epidermis, 286
Epididymis, 122
Epithelial cells, 96–97, 281, 394
Eps 15 protein, 265, 266
Epsins, 265, 266
ERE, see Estrogen response element
Erythroblastosis, 464
Erythroblasts, 375
Erythroid cells, see also Erythropoiesis
   ferritin, 463–464
   iron
      –heme metabolism peculiarities, 331–332
      physiologic regulation of absorption, 15
      uptake role, 338, 339, 389
   transferrin, 214, 215
   transferrin receptor
      erythropoietic differentiation/maturation,
         324, 325, 326, 327, 330
      regulation, 304–305
Erythroleukemia cells, see also Friend
      erthyroleukemia cells
   ferritin, 513
   ferritin gene, 478, 484
   iron regulatory protein activity, 434
   soluble transferrin receptor release, 375
   transferrin-independent transport of iron, 384
Erythrophagocytosis, 506–507
Erythropoiesis

rate
   iron uptake, 2
   physiologic regulation of iron absorption, 15,
      16–17, 18–19
   soluble transferrin receptor, 376, 377
   transferrin receptor, 324–332
Erythropoietin, 378–379, 438
ESAG, see Expression site-associated genes
Escherichia coli
   aconitases and iron regulatory protein evolution,
      426
   lactoferrin, 99, 104
   ribonucleotide reductase, 41–42
   transferrin-binding protein, 197, 199
ESEEM, see Electron spin echo envelope
      modulation
Esophagus, 397–398
Estradiol, 120, 121
Estrogen, 120, 121, 123, 182
Estrogen receptor, 120
Estrogen receptor protein, 97
Estrogen response element (ERE), 86, 87, 120
Ethnic groups, 121, 397
Ets family, 305
Eukarytic bacteria, 507–508, see also Individual
      entries
Evolution
   ferritins, 450–457
   iron regulatory protein, 426–428
   transferrin family proteins, 147, 153, 202, 203
Exfoliation, 2
Exochelins, 207
Exocytosis, 373
Exon–intron structure, 256–257
Exosomes, 373
Expression site-associated genes (ESAG), 205,
      206
Externalization, 270
Eye, 394, 492–493

F

Familial hypotransferrinemia, 219
Fatty acids, 213
FbpA protein, see Periplasmic proteins
Feedback, 110–111
Fenton pathway, 383, 499
FER-1 region, 480, 481, 482
Fer1/2 gene, 451, 456
FerH alleles, 450, 452, see also Ferritin
Ferric-chelate reductase, 144
Ferric ion-binding protein (hFBP), 202
Ferricyanide, 210, 212, 308, 383–384
Ferrihydrite nucleation, 459

Ferrireductase, 145, 383, 384
Ferritin
  endothelium, 484–487
  evolution, 450–457
  expression and function
    brain, 495–504
    eye, 492–493
    macrophages, 504–513
    skin, 494–495
  ferroxidase activity, 459–461
  genes, 474–477
  hemosiderin, 466–467
  iron
    absorption role, 9, 11, 15
    -dependent regulation of transferrin
      receptors, 298, 299
    -dependent regulation of synthesis, 408–409
    phosphate role in oxidation and core
      formation, 461–462
    release, 462–466
  iron regulatory element structure, 413, 414, 415
  iron regulatory protein activity, 425–426, 428,
    434, 437
  serum, 467–473
  three-dimensional structure, 457–459
  transcriptional control of gene expression,
    477–484
  transferrin receptor expression
    activated T lymphocytes, 311, 312
    distribution, 283, 285
    erythropoietic differentiation and maturation,
      330, 332
    monocytes–macrophages, 322, 323
    tumor cells, 487–492
    unexpected functions, 513–515
Ferritin H, 496, 497, 505, 513
Ferritin H gene
  structure, 474, 475
  transcription control of expression, 482, 483,
    484
  transcriptional activity, 477
Ferritin L, 496, 505
Ferritin L gene
  expression and mutations, 414
  structure, 474, 476
  transcription control of expression, 482, 483
Ferritin-like iron responsive element, 415
Ferroportin 1 gene, 335
Ferrous ions, hydrated, 143, see also Iron
Ferrous sulfate, 433
Ferroxidase, 459–461, 513
Fet3/Fet4 protein, 390–391
Fetus, 181
feu genes, 426

Fever, 469
FhuA, 195
Fiber, dietary, 23–24
Fibroblasts, 432, 494
Fisher rat thyroid cell (FRT), 291
5' Flanking region, 301, 302
Flavin mono-oxygenase (FMO), 12, 13
Flow cytometry, 102, 109
FMO, see Flavin mono-oxygenase
Folate metabolism, 514
Follicle cells, 184, 291
Footprinting experiments, 411, 419, 422–423
Franscisella tularensis, 318
Frataxin, 338
Frataxin gene, 498
Free radical scavengers, see Scavengers
Free radicals, see also Hydroxyl radicals; Oxygen
    radicals
  iron, 39, 250
  lactoferrin, 72
  ribonucleotide reductase, 41
  transferrin alleles, 158
  transferrin receptor expression, 317–318
Freeze quenching, 460
Friedreich's ataxia, 498
Friend erythroleukemia cells (FLCs), see also
    Erythroleukemia cells
  biological effects of FerH overexpression, 450
  ferritin as iron donor, 464
  transferrin receptor expression/regulation, 305,
    327–330
Frps protein, 197
FRT, see Fisher rat thyroid cell
Ftr1 transporter, 390, 391, 386
Full-length human cDNA clone, 301

G

G protein, 270
Galactosemia, 179
Gallium, 51–52
Gallium nitrate, 52–53
Gastrectomy, 5
Gastric acidity, 21
Gastric cancer, 491
Gastric secretions, 1
Gastroferrin, 5
Gastrointestinal system, 3, 102–103
GATA-1 sites, 116
G-CSF, see Granulocytic colony-stimulating factor
GEFs, see Guanosine nucleotide exchange factors
Gel retardation assay, 426
Gel shift assays, 304
Gelatinase, 115–116, 117

Gender, 1, 60, 375, 467–468
Gene array analysis, 493
Gene, *see also* Individual entries
  ferritin
    expression, 474–484
    plant, 451
    structure, 455–456
  lactoferrin, 85, 113–114
  stability of mRNA, 408
  transferrin as transporter in drug therapy, 221,
    222
  transferrin
    cellular studies of synthesis, 182–187
    expression, 148, 154–155
    overview, 179–182
    regulatory sequences involved in regulation,
      187–195
    variants, 155–158
  transferrin receptor, 256–257, 301–305
Genetic markers, 157
Genomic libraries, 85
Geotria australis, 146
Germ cells, 182, 183, 184
Gestation, 3
G-Fer, 483
Glial cells, 186, 191, 286, 394
Glioblastoma, 286
Glioblastoma multiforme, 220, 503
β-Globin gene, 484
GLP, *see* Glycosylphosphatidylinositol
Glutamate carboxypeptidase II, 258
Glutamic acid, 167–168
Glutaredoxin, 42
Glutathione, 42, 431
Glycans, 77, 80–81, 177–178
Glycosylation
  lactoferrin, 81, 93
  transferrin family proteins, 152–153, 176–179,
    182
  transferrin receptor, 258–262, 372
Glycosylphosphatidylinositol (GPI), 395, 400
Golgi complex
  intracellular pools of transferrin receptors,
    277–279
  transferrin receptor endocytosis, 264, 268–269
Gonadotropins, 184–185
gp330, 95
GPI, *see* Glycosylphosphatidylinositol
GPI-anchored ceruloplasmin, 502
Graft, survival, 316
Gram-positive/negative bacteria, 98, 101, 195, *see
  also* Individual entries
Granules, primary/secondary, 115, 117
Granulocytic colony-stimulating factor (G-CSF),
    107–108, 112

Granulopoiesis, 72, 111
Granulosa cells, 184, 185
Gray matter, 285, 496
Growth, 208–218, *see also* Transferrin
Growth factor, 295
GTPase, *see* Guanosine triphosphatases
GTP-binding protein, 264, 265, 266, 267, 269
Guanosine nucleotide exchange factors (GEFs), 269
Guanosine triphosphatases (GTPase), 263, 264,
    266, 267, 269
Guar gum, 24

H

H68.4 epitope, 49
Haber–Weiss reaction, 99–100, 319, 320, 497
Haemophilus influenzae, 101, 103, 200–202
Haemophilus spp., 195–196, 199, 208
Hallervorden–Spatz syndrome, 395
HBED, *see* N,N-bis-2-hydroxybenzyl-
    ethylenediamide-N,N-diacetic acid
Head tumors, 492
HeLa cells, 333, 384, 386, 513
Helicobacter pylori, 106, 453
Hematopoiesis, 73
Heme, *see also* Non-heme iron
  bacterioferritins, 452
  iron
    absorption, 26–28
    derivation and digestion role, 5, 6
    metabolism and transferrin receptor,
      331–332
    uptake and synthesis in erythroid cells, 463
  iron regulatory protein activity regulation, 436
  meat as enhancer of nonheme iron
    bioavailability, 21
  oxidant injury in endothelial cells, 485
  transcription control of ferritin gene expression,
    477
Heme–hemopexin complex, 298, 309
Heme oxidase, 280
Heme oxygenase
  absorption of heme iron, 27–28
  brain, 501–502
  endothelium and nitric oxide, 485–486
  iron-dependent regulation of transferrin
    receptors, 298
  iron regulatory protein activity, 436
Heme oxygenase 1 (HO1), 507
Hemin, 297–298, 299
Hemochromatosis, 377
Hemoglobin
  iron
    acquisition and ferritin expression, 506
    requirements and replacement, 2

iron regulatory protein activity, 436
synthesis, 181, 338
Hemoglobin depletion assay, 24
Hemolytic anemia, 376
Hemophagocytic syndrome, 489
Hemopoietic growth factors (HGFs), 324
Hemosiderin, 466–467
Hemosiderosis, 393
Hepatic cells, 394, 464–466
Hepatic macrophages, 510, *see also* Macrophages; Monocytes–macrophages
Hepatic tissue
gene expression of transferrin receptor 2, 258
transferrin receptor expression and distribution, 281, 282, 283–284
Hepatitis B/C, 471, 490
Hepatocarcinogenesis, 61
Hepatocarcinoma cells, 259, 490
Hepatocellular carcinoma, 58, 179
Hepatocyte nuclear factor (HNF), 190, 191
Hepatocytes
ceruloplasmin synthesis, 391, 395
iron import and export, 337
non-receptor-mediated mechanism of iron transport, 384–385
transferrin synthesis, 182
Hepatoma cell line
iron chelation and cell proliferation inhibition, 56, 58
oxidative stress and iron regulatory protein activity, 434
transferrin-independent transport of iron, 384
transferrin receptor, 270, 278
Hereditary hemochromatosis (HH)
ferritin expression in macrophages, 511–512
HFE gene, 334, 395–398
iron relation, 9, 61, 465
serum ferritin levels, 470
Hereditary hyperferritinemia–cataract syndrome (HHCS), 413, 414, 415, 492
Hereditary hypotransferrinemia, 145
Hereditary microcytic anemia, 388–389
Heregulin differentiation factors, 184
hFBP, *see* Ferric ion-binding protein
HFE gene
ferritin expression in macrophages, 511–512
iron
absorption, 9–10
alternative uptake systems, 395–399
uptake control, 327, 328, 329
transferrin receptors, 294, 332–334
H-ferritin, 213
HGFs, *see* Hemopoietic growth factors
HH, *see* Hereditary hemochromatosis

HHCS, *see* Hereditary hyperferritinemia-cataract syndrome
HIF, *see* Hypoxia-inducible factor
Histidine, 82, 84, 152, 165
Histoplasma capsulatum, 207, 318
HIV, *see* Human immunodeficiency virus
H-kininogen, 473
HLA-A2 gene, 395–396
HLA-H protein, 395, 396, 397, 398, 399
HNF, *see* Hepatocyte nuclear factor
Hodgkin's disease, 489–490, *see also* Non-Hodgkin's disease
Homologies
Fet3 to ceruloplasmin, 391
melanotransferrin, 400
transferrin family proteins, 152, 154, 158
transferrin receptor, 256
Hormone-dependent cells, 290
Host defense, 71–72, 92, 101
HPO, *see* 3-Hydroxypyridin-4-one
HSCT, Autologous hematopoietic stem cell transplantation
HTLV-1 virus, 314
Human immunodeficiency virus (HIV), 315
Human milk, *see* Milk, human
Human populations, 156, 157
Human transferring gene family cluster, 85
Huntington's disease, 394, 500
Hybridization analysis, 388
Hydrochloric acid, 5
Hydrogen peroxide
ferritin expression in brain, 496
ferroxidase reaction, 460
lactoferrin, 73
regulation of iron regulatory protein activity, 432, 433
Hydrophilic/hydrophobic regions, 458
3-Hydroxypyridin-4-one (HPO), 54, 55
Hydroxybenzhydroxamic acid, 44
N,N-bis-2-Hydroxybenzyl-ethylenediamide-N,N-diacetic acid (HBED), 55
Hydroxyl radicals, *see also* Free radicals; Oxygen radicals
ferritin, 488
iron uptake through Fenton chemistry, 383
lactoferrin, 73
bacteriostatic activity, 100–101
transferrin receptor, 317–318
transferrin-independent transport of iron, 385
Hydroxyperoxides, 494
Hydroxyurea, 41
Hyperoxia, 510
Hypochromic anemia, 335
Hypoferremia, 505–506
Hypoferremic response, 92

Hypoproliferation, 378
Hypoxia
   ferritin, 496, 510, 511
   IREG-1 gene, 336
   iron regulatory protein regulation, 434
   physiologic regulation of iron absorption, 18, 19
   transferrin, 182, 183
   transferrin gene, 194
   transferrin receptors, 300
Hypoxia-inducible factor (HIF)
   ferritin expression in brain, 497
   transferrin gene expression, 194
   transferrin receptors, 300–301, 305
   transferrin synthesis, 183

I

IDRS, *see* Iron-dependent regulatory sequence
IEC-6 cell line, 7
IFN-γ, *see* Interferon-γ
IGF, *see* Insulin-like growth factor
IL-1, *see* Interleukin-1
IL-2 receptor (IL-2R), 307, 309, 310, 311, 312
IL-3, *see* Interleukin-3
IL-6, *see* Interleukin-6
IL-13, *see* Interleukin-13
IL-2R, *see* IL-2 receptor
Immortalization, 487
Immunoassay, 467
Immunoblotting studies, 177
Immunocytochemical studies, 106
Immunoenzymatic quantification, 374
Immunoglobulins, 473
Immunohistochemical studies, 9, 122, 185, 186
Immunohistological studies, 8–9, 98
Immunoliposomes, 50
Immunoperoxidase studies, 114, 119
Immunoprecipitation studies, 116
Immunotoxin, 50–51, 220
Infants, 3, 21, 94, 468
Infectious diseases, 107–109
Inflammation
   iron handling in monocytes–macrophages, 317
   lactoferrin, 73
   nitric oxide regulation of iron regulatory protein
      activity, 429
   Parkinson's disease and ferritin expression, 499
   serum ferritin levels, 469, 470
Inflammatory agents, 321
Inflammatory disorders, 283, 505
Inhibitors, nonheme iron bioavailability, 22–26
Insects, 147, 148, 455
Insulin-like growth factor (IGF), 292, 295, 503
Integrins, 12, 13, 14
Interferon-γ (IFN-γ)

ferritin expression in macrophages, 504–505
iron regulatory protein activity, 430, 431
transferrin receptor expression, 296, 320, 323
Interleukin-1 (IL-1), 73, 107, 111, 506
Interleukin-2 (IL-2)
   iron regulatory protein activity, 438
   transferrin receptor expression, 306, 307, 309,
      310, 311, 312
Interleukin-3 (IL-3), 110
Interleukin-4 (IL-4), 323
Interleukin-6 (IL-6), 107, 108, 111
Interleukin-13 (IL-13), 323
Internal homology, *see* Homologies
Internalization
   hepatic cells and ferritin, 465
   transferrin receptor
      calcium ion, 279–280
      continual and intracellular pools, 277
      structural requirement, 272–275
      transferrin–transferrin receptor, 264, 265, 266
Internalization motif, 264, 273
Internalization signal, 373
Interstitial fluid, 286
Intestine, 17
Intracellular iron pool, 299, 430, 432
Intracellular pools, transferrin receptors,
   277–279
Intragenic duplication, 153
Intrinsic tag technique, 24
Invertebrates, 146
IREG-1, 335–336
IREs, *see* Iron regulatory elements
IROMPs, *see* Iron-repressed outer membrane
   proteins
Iron
   absorption
      background, 6–14
      heme, 26–28
      HFE gene, 333
      physiologic regulation, 14–19
   agents that interfere with intracellular
      incorporation, 51–53
   alternative systems of uptake
      cell proliferation, 295
      HFE gene, 395–399
      mechanisms and role of ceruloplasmin,
         390–395
      melanotransferrin, 399–401
      proton-coupled metal ion transporter,
         386–390
      transferrin-dependent, 383–386
   anemia and soluble transferrin receptor, 375
   binding
      hFBP, 202–203
      lactoferrin, 73, 75–77, 82

Tf B/C/D transferrin genetic variations, 156
transferrin, 152, 159, 160, 162–172
bioavailability, 19–26
breast feeding and lactoferrin, 94
cell growth in tissue culture, 215
cell proliferation
cancer, 59–61
cell cycle control, 45–48
inhibition, 48–59
ribonucleotide reductase, 40–44
deficiency, *see* Deficiency, iron
digestion, 3–6
ferritin, 451, 495, 496
Fet3 synthesis, 391
germ cells, 183
iron regulatory protein, 417, 421, 430, 434,
436–437
overload, *see* Overload, iron
oxidation and phosphate, 461–462
Parkinson's disease, 499
proliferation of activated T lymphocytes, 307,
308
release
ferritin, 462–467
transferrin, 172–174, 175
requirements, 2–3
saturation, 111, 469
sequestering mechanisms, 201, 206–207,
319–320
status, 89, 182, 373
storage, 72, 457–458
transferrin, 181, 196, 211, 212
transferrin gene, 192, 194
transferrin receptor
affinity, 258
-dependent regulation, 297–301
distribution, 284, 285–286
heme metabolism, 330–332
intracellular pools, 279
monocytes–macrophages, 317, 318–319
pathways in Gram-negative bacteria, 195,
196
regulation and cell cycle requirements, 297
synthesis and mRNA stability, 408, 409
transferrin receptor, 250, 271
transferrin-binding proteins in *Haemophilus*
spp., 201–202
various cell types, 334–339
Iron-binding pocket, 167, 168
Iron-binding proteins, 11
Iron-binding sites, *see* Iron, binding
Iron-core, 461–462
Iron-dependent regulatory sequence (IDRS), 484
Iron chelators, *see* Chelators, iron
Iron regulatory elements (IREs)

iron regulatory protein, 422, 423, 425
–Nramp2 isoform and ferritin expression in
brain, 495
secondary and tertiary structures, 410–415
transferrin receptor, 257–258
Iron regulatory protein (IRP)
characterization, 415–422
cloning genes, 416
coordination of *in vivo* iron metabolism,
436–437
evolution, 426–429
ferritin, 425–426, 505, 511
heme, 436
iron regulatory element selective binding, 413
nitric oxide, 429–432
oxidative stress, 432–435
phosphorylation role and control of activity, 435
tissue-dependent regulation of activity, 437–439
transferrin receptor
activated T lymphocytes, 312
erythropoietic differentiation and maturation,
331
HFE gene interaction 334
internalization, 280
iron-dependent regulation, 300
monocytes–macrophages, 322–323
mRNA stability, 422–425
Iron regulatory protein, 213, 282–283, 286
Iron-repressed outer membrane proteins
(IROMPs), 104
Iron salts, 312, 321–322
Iron–sulfur cluster, iron regulatory protein, 416,
417, 418, 420
evolution, 426
oxidative stress, 432, 433, 434
peroxynitrate regulation, 432
IRP, *see* Iron regulatory proteins
Ischemic heart disease, 472
Isoelectric point, 74
Isothermal titration calorimetry, 171

J

J774 cells, 430–431
Jurkat T cell line, 306

K

Keratinocytes, 287, 494
Kidneys, 287, 378
Kinetics, 7, 75, 77, 453
Knock-out mice, 121, 334
KRDS tetrapeptide, 97
Kupffer cells, 281
Kyte–Doolittle analysis, 385

L

Labile iron pool, 209
Lactation, 72, 120
Lactoferricin, 102–103, 114
Lactoferrin
    background, 71–74
    biochemistry and molecular biology, 74–88
    expression
        development, 122–123
        eye, 493
        neutrophils and mammary glands, 114–122
    functions, 98–114
    N lobes comparison with periplasmic proteins,
        169
    receptors, 88–98
    bacterial, 103–109
    transferrin comparison of conformational
        change, 165
Lamprey eel, 428
LBP, *see* LPS-binding protein
LDL, *see* Low-density lipoprotein
LDL-related receptor (LRP), 95–96
Lectin, 310, 312
Legionella pneumophila, 99, 318, 321, 426
Leishmania spp., 204, 206, 319, 427
Lenses, 492, 493
Leukemia, 56–57, 377, 489
Leukemic cell line
    lactoferrin, 113–114, 118
    lactoferrin receptors, 88–89
    retinoic acid, 116–117
    transferrin receptor, 49, 275, 276
Lf$_N$, *see* Recombinant N terminal lobe
Ligands, 5
    binding, 251, 252, 253
Limnaea stagnalis, 457
Lipid peroxidation, 494, 498
Lipocytes, 466
Lipopolysaccharide (LPS)
    ferritin expression in macrophages, 505, 510
    iron regulatory protein activity, 431
    lactoferricin, 101, 102
    lactoferrin, 107, 108, 111
    transferrin receptor, 321
Listeria spp., 318, 454–455
Liver
    iron metabolism, 337, 338
    iron regulatory protein activity, 432
    lactoferrin receptors, 94–96
    serum ferritin levels, 470
    transferrin, 181, 182
    transferrin gene, 188–191, 192
    transferrin receptor, 282–283
        development, 325, 327

Low-density lipoprotein (LDL), 272, 273, 274, 486
    receptor, 258
    receptor gene family, 95
LPS, *see* Lipopolysaccharide
LPS-binding protein (LBP), 109
LRP, *see* LDL-related receptor
Lung, 182, 281–282, 509
    cancer, 158, 491, 509–510
Lymnae stagnalis, 456
Lymphoid cell line, 187, 280, 512
Lymphomas, 52, 489–490
Lymphoproliferative disorders, 377
Lysine, 167, 168, 169, 174
Lysis, 102
Lysozyme, 101, 115

M

Macrophages, *see also* Monocytes–macrophages
    ferritin
        as iron donor, 466
        expression and function, 504–513
    iron regulatory protein activity, 430
    iron storage and release role, 338, 339
    Nramp1 stimulation, 388
    transferrin receptor, 282
Madin–Darby kidney (MDCK) cells, 274, 275
MafB protein, 305
Magnetic resonance imaging (MRI), 498, 499
Maize, 451
Major histocompatibility complex (MHC)
        molecules, 332
Malaria, 206–207
Mammary glands, 119–122
Mammary tissue, 288
Manduca sexta, 146, 147
Manganese, 25–26
Marker, 178
MAX protein, 482
M-CSF, 320
MDCK, *see* Madin–Darby kidney cells
Meat, 21, 28
Medulloblastoma, 220
Megaloblastic anemia, 377
Melanoma, 59, 308, 399, 400
Melanotransferrin, 166, 399–401
Membrane receptors, 88, 90, 92
Membrane trafficking, 276
Menadione, 281, 433, 434
Meningitis, 470
Menstruation, 2
Mesencephalon, 97, 98
Messenger ribonucleic acid (mRNA), 408,
        422–425
Metal contamination, 144

Metal ion transporter, proton-coupled, 386–390
Metal-binding site
  lactoferrin, 77, 82
  transferrin, 159, 160, 162, 163, 166
Metals, 13, 150
Metamyelocytes, 116
Methotrexate, 56
MHC, see Major histocompatibility complex
    molecules
Microalgae, 150
Microbiology, transferrin, 195–208
Microcytic anemia, 12, see also Anemia
Microglia cells, 501
β-2 Microglobulin
  HFE gene association, 395, 396, 397, 398
  iron absorption, 12, 13
Microvilli, 293
  membranes, 6, 7, 8–9
Milk
  cow, 24, 25, 119
  human, 24, 72, 74, 473
Mimosine, 48
Minerals, 23, 25–26
Missense mutation, 389
Mitochondria, 338, 463
Mitochondrial aconitase, 418, see also
    Aconitase
Mobilferrin, 11, 12–13, 14
Molecular biology
  lactoferrin, 74–88
  transferrin receptor, 254–258
Molecular modeling studies, 178
Monoclonal antibodies, 48–51, 219
Monocytes, 512, see also
    Monocytes–macrophages
Monocytes–macrophages
  ferritin expression, 504
  iron regulatory protein activity, 438
  iron transport, 385, 393–394
  lactoferrin receptors, 91–92
  myelopoiesis and lactoferrin, 110, 111
  serum ferritin levels, 472–473
  transferrin receptor, 317–324
Mono-lobed transferrin, 153, see also Transferrin
Moraxella spp., 105, 208
Mosquito, see Aedes aegypti
Mouse, 181–182
MRI, see Magnetic resonance imaging
mRNA, see Messenger ribonucleic acid
Mucins, 5
Mucosa cells, 16
Mucosal block theory, 15
Multiple sclerosis, 501
Murine, 182
Muscle cells, 215, 217–218, 292

Mutagenesis, 82, see also Site-directed
    mutagenesis
Mutation
  ceruloplasmin gene and aceruloplasminemia,
    392–393
  iron regulatory protein, 419, 423
  iron regulatory element, 411, 413, 414
  Nramp2 gene, 389
  transferrin and iron binding, 167–168, 172
MYCN amplification, 488
Mycobacterium tuberculosis, 207, 318, 319, 508
Mycobactins, 207, 508
Myelin synthesis, 187, 496, 497
Myeloblast, 115, 116
Myelocytes, 115, 116, 118
Myelodysplasia, 376
Myeloid cells, 116, 117
Myeloid leukemia, 112, 118, see also Leukemia
    cell line
Myeloperoxidase, 100, 115, 117
Myelopoiesis, 109–113
Myeloproliferative disease, 489
Myoblasts, 292
Myocardial infarction, 470, 472
Myogenesis, 218

N

N lobe
  transferrin, 162, 164, 165, 166, 171, 172, 174
  transferrin–transferrin receptor interaction,
    175
Na$^+$/H$^+$ antiport, 210
NADH-diferric transferrin reductase, 210
NADH ferricyanide oxidoreductase, 210, 308
NADH oxidoreductase, 210
Natural killer (NK) cells, 114, 323, 482, 510
Neck tumors, 492
Neisseria spp.
  iron uptake, 196
  lactoferrin, 104, 105–106
  transferrin receptor, 197, 198, 199, 208
Neonatal hemosiderosis, 472
Neonates, 8–9, see also Infants
Neoplasias, 467
Neuraminidase, 372
Neuroblastoma cells
  iron chelation and cell proliferation inhibition,
    56, 57–58
  serum ferritin levels, 488
  transferrin gene, 191
Neurodegenerative disorders
  aceruloplasminemia, 393, 394
  ferritin expression, 499–501
Neuronal cells

ferritin expression, 495
lactoferrin receptors, 97–98
transferrin, 186
transferrin gene, 191, 192
transferrin receptor, 284, 286
Neutropenia, 107
Neutrophils, lactoferrin
  degranulation, 72
  expression, 114–118
  regeneration, 111–112
  release and infectious disease, 107
Neutrophil-specific granule deficiency, 117–118
NFY factor, 479, 480, 481
Niemann–Pick disease type C (NPC), 502
Nilaparvata lugens, 456
Nitric acid synthase (NOS), 322–323, 429
Nitric oxide (NO)
  endothelial cell response to heme, 484, 485
  ferritin, 505, 515
  iron regulatory protein activity, 429–432
  transferrin receptor, 322–323
NK, *see* Natural killer
NMR, *see* Nuclear magnetic resonance
NO, *see* Nitric oxide
Nocodazole, 279
Nodulation process, 451, 452
Non-alcoholic steatohepatitis, 470–471
Non-heme iron, *see also* Heme
  absorption and brush border membrane, 6–7
  bioavailability
    enhancers, 20–21
    gastric digestion, 3–5
    inhibitors, 22–26
  -containing ferritins and structural
    characteristics, 452, 453–454
  forms in brain, 498
  physiologic regulation of iron absorption, 15–16
Non-Hodgkin's disease, 52–53, see also Hodgkin's
  disease; Lymphomas
Non-receptor-mediated mechanism, 384
Non-transferrin bound iron (NTBI), 332
NOS, *see* Nitric oxide synthases
NPC, *see* Niemann-Pick disease type C
Nramp family, 387
Nramp1, 319, 387–388, 508
Nramp2
  ferritin expression in brain, 495
  gene expression, 11–12, 16
  iron uptake in enterocytes/hepatocytes, 336,
    337, 338, 339
  lipopolysaccharide enhancement in
    macrophages, 508–509
  transferrin receptor endocytosis, 271
NTBI, *see* Non-transferrin bound iron
Nuclear magnetic resonance (NMR), 413

O

Obligatory iron loss, 2
Ocean water, 145
Octopus, 457
Okadaic acid, 281, 478
Oligodendrocytes
  ferritin expression in brain, 496
  transferrin, 185, 186, 187
  transferrin gene, 191
Oligosaccharide, 80, 258, 259–260
One open–one closed structure, 82, 83
Ontogenesis, 123, 374
Ovary, 291
  cancer, 490
Overload, iron
  ceruloplasmin, 393
  congenital atransferrinemia, 218
  consequences, 40
  serum ferritin levels, 468–469
  transferrin gene, 194
  transferrin receptor, 283
Ovotransferrin, 145, 148, 166, *see also* Transferrin
Oxalamate, 420
Oxidant injury, 484–485
Oxidase, 392
Oxidation
  iron, 250, 459, 460
  transferrin receptor internalization, 279–281
Oxidative stress
  ferritin, 494, 496, 498, 509, 513
  ferritin gene, 482
  iron regulatory protein activity, 432–435
  transferrin receptor, 280
Oxygen ligands, 164
Oxygen radicals, 59, *see also* Free radicals;
  Hydroxyl radicals
Oxygen, 194

P

P/CAF protein, 478
p300/CBP co-activator, 480
p300/pCAF proteins, 479, 480
p43, 488
p45, 96
p53, 56
p97, 3, 153, 181
Pacifasacus leniusculus, 150, 456
Palmitoylation, 274
Pancreas, 281, 287
Parabactin, 55
Paraferritin, 14, *see also* Ferritin
Parasites, 427, 455
Parenchymal cells, 94–96, 282

Parkinson's disease
  ceruloplasmin relation, 394
  ferritin expression in brain, 498–500
  lactoferrin receptors, 97, 98
Parvovirus initiation factor (PIF), 304
Pasteurella haemolytica, 203
pcD-TR1 gene elements, 301
PCR, *see* Polymerase chain reaction
PCT, *see* Porphyria cutanea tarda
PDF, *see* Platelet-derived factor
Pea, 451, 452
Peptide mapping, 85
Perhexiline, 280
Perinatal hemochromatosis, 469
Periplasmic proteins, 83–84, 169, 202
Peritoneal macrophages, 511, *see also*
  Macrophages
Peritubular cells, 184
Peroxisome proliferator-activated receptor gamma
  (PPARγ), 487
Peroxynitrate, 431–432
PGF, *see* Platelet-derived growth factor
pH
  absorption of heme iron, 26
  ferroxidase activity in ferritin, 461
  iron
    absorption, 7, 14
    digestion role, 4, 5
    lactoferrin binding, 73, 82, 83, 84, 85
    transferrin binding, 162, 165, 168, 172
    release from transferrin, 174
  lactoferrin, 75, 92
  transferrin receptor
    endocytosis, 262, 270
    HFE gene interaction, 333, 398–399
PHA, *see* Phytohemagglutinin
Phagocytic cells, 99, 100
Phagosomes, 207
Phaseolus vulgaris, 452
Phorbol ester
  iron regulatory protein activity, 435
  lactoferrin, 115
  -responsive elements, 302
  transferrin receptor, 276, 304
Phorone, 432
Phosphorylation, 435
Phosphate, 461–462
Phosphatidylinositol 4,5-diphosphate (PIP$_2$), 306
Phosphatidylinositol 3'-kinase, 270
Phosphatidylinositol pathway, 316
Phospholipase C, 306
Phosphorus-containing compounds, 25
Phosphorylation
  transferrin receptor, 275–277, 304
  tyrosine and activated T lymphocytes, 306

Phylogenetic analysis, 147
Physiologic regulation, iron, 14–19
Phytates, 20, 21, 22, 23
Phytohemagglutinin (PHA), T lymphocyte
    activation
  lactoferrin binding, 89
  transferrin receptor expression, 309, 310, 311,
      312, 313
Phytoplankton, 145
Picolinic acid, 312, 508, *see also* Chelators, iron
PIF, *see* Parvovirus initiation virus
PIH, *see* Pyridoxal isonicotinoyl hydrazone
PIP$_2$, *see* Phosphatidylinositol 4,5-diphosphate
Pituitary gland, 281, 292, 437
Pituitary tumor cells, 215, 291
PKC, *see* Protein kinase C
Placenta, 259, 292–294, 398
Placental syncytiotrophoblasts, 281
Plants
  aconitases and iron regulatory protein, 427
  ferritins, 450–451
  ferritin gene, 484
  ion acquisition, 144
  transferrin, 150
  transgenic, 514
Plasma membrane reductase, 212
Plasmodium falciparum, 321
Plastids, 451, 452
Platelet-derived factor (PDF), 295
Platelet-derived growth factor (PGF), 276
Platelets, 97
PmodS, 184
Polyadenylation signals, 421
Polymerase chain reaction (PCR), 256, 416
Polymorphisms, 156, 257, 472–473
  inherited, 155–156
Polyphenols, 20, 21, 23
Polyvalent cationic metals, 385
Porphyria cutanea tarda (PCT), 399, 471
Positive/silencer region, 191, 192, 193
Post-menopausal women, 1, 24
Post-translational modifications, 251, 252, 258
PPARγ, *see* Peroxisome proliferator-activated
    receptor gamma
Pregnancy
  iron requirements, 3
  physiologic regulation of iron absorption, 18
  serum ferritin levels, 471–472
  soluble transferrin receptor, 375
  transferrin glycosylation, 178
Premature birth, 158
Prepubertal period, 184
Prevertebrates, 154
Prevotella spp., 204
Primordial gene, 148, 153

Primordium, 182
Progenitor cells, 110, 111
Progesterone, 121, 122
Prolactin, 119, 120
Promoter elements, 86, 87, 188
Promoter region
    ferritin evolution, 451
    ferritin gene expression, 477–481
    transferrin gene expression, 191, 192, 193
    transferrin receptor, 301–302
Promyelocytes, 115, 116
Prostaglandin A_1, 505
Prostate, 122
    cancer, 59, 216, 289, 290
Prostate-specific membrane antigen, 258
Protease cleavage mechanism, 372
Protease inhibitors, 372
Proteases, 5, 373
Protective effect, 431
Protein kinase C (PKC)
    iron regulatory protein activity, 435, 437
    transferrin, 212–213
    transferrin receptor, 276, 306, 313
Proteins, 160–161, 274, see also Individual
    entries
Proteolysis, 372–373
Proteosome inhibitors, 422, 431
Proton pumps, 270
Protoporphyrin IX, 14, 299
Pseudogenes, 474
PU.1 boxes, 116
Purification, lactoferrin, 74
PYBP protein, 189
Pyramidal cells, 185
Pyridoxal isonicotinoyl hydrazone (PIH), 55, 215
Pyura spp., 146, 154

Q

Quinones, 433
quoxD gene, 426

R

R1/R2 proteins, 41–43
Rab proteins, 267–268
Rabaptin, 267
Rabbits, 161, 163
Rabphillin, 267, 268
Radiolabeling, 259
Rana catasbesiana, 154
RARE, see Retinoic acid receptor element
Rats, 182
RAW 264.7 cells, 430, 431
Rb, see Retinoblastoma protein

Recombinant human lactoferrin, 88, see also
    Lactoferrin
Recombinant N terminal lobe (Lf_N), 84
Recombinant ribonucleotide reductase, 42, see
    also Ribonucleotide reductase
Recombinant transferrin receptor, 49, see also
    Transferrin receptor
Recycling, iron, 2
Red blood cells, 281, 506, 510
Redox activity, 462
Reducing agents, 418
Reduction, 42, 250
Renal cell carcinoma, 491
Renal failure, 378
Reporter genes, 113
Reproductive hazards, 158
Restriction point, 45, 46
Reticulocyte index, 378
Reticulocytes, 173, 389
Reticuloendothelial cells, 72, 466
Reticuloendothelial macrophages, 504, see also
    Macrophage; Monocytes–macrophages
Retina, 493
Retinoblastoma protein (Rb), 46, 47
Retinoic acid, 116, 118, 287
Retinoic acid receptor element (RARE), 87, 116,
    118
Retinoids, 123
Rheumatoid arthritis, 158
RHL-1 receptor, 96
Ribonuclease (RNase), 74, 121
Ribonucleotide reductase
    DNA synthesis, 40–44, 297
    inhibition by gallium, 52
    iron chelators, 54
    R2 subunit and ferritin-associated iron, 515
    transferrin, 146, 213
    transferrin receptor, 307
Rice, 514
RNase, see Ribonuclease

S

S phase, 49
Saccharomyces cerevisae, 262, 390, 419–420
Salicylaldehyde isonicotinoyl hydrazone (SIH),
    215
Salindac, 490
Salmonella enterica, 319
Salt bridges, 165
Sarcophaga perigrina, 148–149
Saxiphilin, 154
Scavengers, 98, 309, 401, 433
Schistosoma mansoni, 204–205, 455
Schwann cells, 186

Second messengers, 306
Secondary/tertiary structures, iron regulatory
    element, 410–415
Semen, 122
Seminal plasma, 122
Seminal vesicles, 122
Seminiferous tubules, 183, 291, 290
Sepharose beads, 384
Serine, 435
Serotransferrin variants, 177
Sertoli cells
    transferrin, 183, 184
    transferrin gene, 188–191, 193
    transferrin receptor, 290
Serum ferritin, see also Ferritin
    characterization, 467–473
    physiologic regulation of iron absorption, 18
    soluble transferrin receptor, 374–375
Serum iron concentration, 60
Serum lactoferrin, 72, see also Lactoferrin
SFT, see Stimulator of iron transport
Shigella flexneri, 103
Sideroblastic anemia, 498, see also Anemia
Siderophilin, see Transferrin
Siderophores, 98, 195
    microbial, 54, 55
Signal peptides, 455
Signal sequence, 272, 273
SIH, see Salicylaldehyde isonicotinoyl hydrazone
SILEX procedure, 413
Silica, 509
Site-directed mutagenesis
    lactoferrin gene, 82, 84
    ribonucleotide reductase, 42
    transferrin, 166, 172
    transferrin receptor, 259–260, 261
Skeletal regeneration, 292
Skin, 494–495
Skin fibroblasts, 384
Skin tumors, 287
Small intestine, 6, 93–94
Small cell lung cancer, 59, 217, 282
Smf1/Smf2 transport divalent cations, 390
Smokers, 509
Soluble transferrin receptor, see also Transferrin
    receptor
    circulation in human serum, 374
    clinical relevance, 374–379
    generation, 371–373
Soy products, 22
Soybeans, 451
Sp1 site, 478, 481
Sp1 transcription factor, 302, 303
Sp3 site, 480
Spectroscopic analysis, 159

Sperm, 122
Spermatogenesis, 183, 290
Spi/Ssa secretion system, 319
Spinal cord, 284
Spleen, 394
Staphlococcus spp., 203, 204
Stem cells, 109–110, 325
Stem loops, 422–423
Stem–loop structure, 410
Steric effects, 77
Steroid hormones, 182
Still's disease, 470
Stimulator of iron transport (SFT), 385–386, 387
Stomach, 397–398
    cancer, 60
Storage, 151, 338, 449, 450
Stores, iron, 15, 468
Stress, oxidative, see Oxidative stress
Stroke, 497
Structural studies, 410–413
Subarachnoidal hemorrhage, 501
Subcortical nuclei, 497
Substantia nigra, 497–498
Subunits, ferritin, 450, 451
Superoxide, 73, 433
SV-40 promoter, 301
Swainsonine, 260–261
Synaptobrevin, 275
Syncytiotrophoblast cells, 293–294, 398
Synergism, 52, 172, 173
Synthesis
    ceruloplasmin in brain, 394
    lactoferrin
        mammary glands, 119–122
        neutrophils, 115–118
    transferrin, 151, 181, 182–187

T

T cell receptor (TCR), 306, 512
T lymphocytes
    activated
        lactoferrin receptors, 89–91
        transferrin receptor, 305–317
    ferritin expression, 512
    iron regulatory protein activity, 437–438
    mitogen-activated, 49
    transferrin as growth factor, 213, 214
T-ALL blasts, 314, 315
Tamoxifen, 270
Tannins, 23
TATA box, 301
TBP1, see Transferrin-binding protein 1
TBP2, see Transferrin-binding protein 2
TCR, see T cell receptor

Tea, 23
Tempol, 493
Testis, 183, 290
   cancer, 291, 491
Tf B/C/D alleles, 155–157
TfR, *see* Transferrin receptor
β-Thalassemia, 19, 376–377, 487
Thiol groups, 308
Thioredoxin, 42, 431
Thiyl radical, 43
THP-1 cells, 483–484
Three dimensional structure
   ferritin, 457–459
   hFBP, 202
   iron regulatory element, 411
   lacteroferricin, 102
   lactoferrin, 75, 80, 81, 84, 85
      human, murine, and bovine homology,
         90, 91
   transferrin, 161–162
   transferrin receptor, 175–176, 254, 255
Three-fold channel, 458
Thymic cells, 316, 512
Thymocytes, 513
Thyroid cells, 216
Thyroid gland, 291
Thyroid-stimulating hormone (TSH), 437, 477
Thyrotropin (TRH), 437
TIBC, *see* Total iron-binding capacity
Tinca tinca, 146
Tissue-dependent regulation, 437–439
TNF, *see* Tumor necrosis factor
Tobacco, 427
TonB, 195
Total iron-binding capacity (TIBC), 60
Toxicology studies, 220
TRA box, 303, 304
TRAC complex, 304
Trans-acting element, 423
Transcription
   ferritin gene, 477–484
   inhibitors, 423
   studies, 44
   transferrin, 183
Transcription factors, 189, 190, 191
   binding motifs, 116
Transfection experiments, 413, 424, 425
Transferrin
   characterization, 143–151
   comparative analysis of family proteins,
      151–155
   congenital atransferrinemia, 218–219
   gene expression
      cellular studies of synthesis regulation,
        182–187

      overview, 179–182
      regulatory sequences, 187–195
   genetic variants, 155–158
   growth factor, 208–218
   iron absorption, 10, 11
   lactoferrin, 72–73, 90
   microbiology, 195–208
   monoclonal antibodies and cell growth
      inhibition, 50
   saturation, 60
   serum
      biochemical properties and basic structure,
        158–174
      glycosylation, 176–179
      structural basis of interaction with transferrin
        receptors, 174–176
   transporter of drugs, 219–222
Transferrin-binding protein 1 (TBP1)
   structure and function
      Gram-negative bacteria, 197–198, 199, 201
      Gram-positive bacteria, 203, 204
   vaccines, 208
Transferrin-binding protein 2 (TBP2), 197, 199,
      201, 208
Transferrin-independent transport, 383–386
Transferrin-like genes, 153
Transferrin receptor
   Ca^{2+} effect and oxidation upon internalization,
      279–281
   endocytosis
      basic features, 262–272
      structural requirement for efficient
        internalization, 272–275
   expression and regulation
      erythropoietic differentiation and maturation,
        324–332
      monocytes–macrophages, 317–324
      T lymphocytes, 305–317
   gene mapping of human chromosome 3, 181
   general features, 251–254
   glycosylation, 258–262
   HFE interaction, 332–334
   intracellular pools, 277–279
   iron absorption, 7, 8–10
   iron uptake mechanism, 334–339
   molecular biology, 254–258
   monoclonal antibodies to and inhibition of cell
      growth, 48–51
   oxidative stress, 434
   phosphorylation, 275–277
   rearrangement and cell proliferation, 209
   regulation, 294–305
   structure and function, 197–198
   tissue distribution and expression, 281–294
   –transferrin interaction, 174–176

Transferrin receptor 2, 257
Transferrin receptor (TfR) gene, 315
Transferrin receptor-like iron responsive element, 415
Transferrin serum receptor, 18
Transgenic plants, 514
Translation inhibitors, 423
Translational mechanisms, 194
Transport, iron, 72, 145
TRB box, 303, 304
TRH, *see* Thyrotropin
Tricarboxylate acid pathway, 420
Trophoblast cells, 294
Trypanosoma spp., 204, 205, 206, 321, 427
TSH, *see* Thyroid-stimulating hormone
Tuberculosis, 319
Tumor aneuploidity, 289
Tumor markers, 491
Tumor necrosis factor (TNF)
    ferritin expression in macrophages, 505, 506
    lactoferrin interaction, 107, 108, 111
    serum ferritin levels, 470
    transcription control of ferritin gene expression, 477, 482
    transferrin synthesis, 185
Tumors
    ferritin, 487–492
    monoclonal antibodies and inhibition of cell growth, 49
    transferrin as growth factor, 216, 217
    transferrin receptor expression and distribution, 283
    transferrin-conjugated diphtheria toxin, 220
Tunicamycin, 260–261
Tyr–Gln–Arg–Phe sequence, 273
Tyrosine
    -based motif, 264, 273–274
    iron binding
        lactoferrin, 84
        transferrin family proteins, 152, 166–167, 168
    lactoferrin gene mutagenesis, 82
    phosphorylation, 280, 304
Tyrosine kinase, 306
Tyrosyl radical, 42, 43

U

Ubiquitination, 422
Ultraviolet absorption spectra, 160, 172, 173
Ultraviolet radiation, 493, 494
Unexpected functions, ferritin, 513–515
Uptake mechanisms, iron, 144
Urinary tract, 109
Urochordates, 153
Urogenital system, 288–289
Uterus, 120, 121

V

Vaccines, 200, 207–208
Vasodilators, 484
Veins, 484
VHL, *see* Von Hippel–Lindau tumor suppressor gene product
Viral hepatitis, 471
Viral infections, 109
Vitamin A, 22, 514
Vitamin B12, 115
Von Hippel–Lindau (VHL) tumor suppressor gene product, 300

W

White matter, 496, 497, *see also* Brain
Wortmannin, 270

X

Xanthine oxidase, 432
Xenopus spp., 146, 385
X-ray crystallography, 148, 161
X-ray diffraction, 164, 457
X-ray solution scattering, 83, 162, 166, 167

Y

YTRF sequence, 273

Z

Zinc, 390

Milton Keynes UK
Ingram Content Group UK Ltd.
UKHW021931071024
449327UK00022B/1764